现代兽医基础研究经典著作

COLOR ATLAS OF ANIMAL HISTOEMBRYOLOGY

动物组织学与胚胎学

彩色图谱

彭克美　主编

中国农业出版社

北　京

图书在版编目（CIP）数据

动物组织学与胚胎学彩色图谱/彭克美主编. —北京：中国农业出版社，2021.5
（现代兽医基础研究经典著作）
国家出版基金项目
ISBN 978-7-109-28192-9

Ⅰ.①动… Ⅱ.①彭… Ⅲ.①动物组织学-图谱②动物胚胎学-图谱 Ⅳ.①Q954-64

中国版本图书馆CIP数据核字（2021）第076163号

DONGWU ZUZHIXUE YU PEITAIXUE CAISE TUPU

中国农业出版社出版
地址：北京市朝阳区麦子店街18号楼
邮编：100125
责任编辑：肖 邦
版式设计：王 晨　责任校对：刘丽香　沙凯霖
责任印制：王 宏
印刷：北京通州皇家印刷厂
版次：2021年5月第1版
印次：2021年5月北京第1次印刷
发行：新华书店北京发行所
开本：880mm×1230mm 1/16
印张：58.25
字数：1805千字
定价：680.00元

版权所有·侵权必究
凡购买本社图书，如有印装质量问题，我社负责调换。
服务电话：010-59195115　010-59194918

编 审 人 员

主　编
　　彭克美（华中农业大学）

编　者（按汉语拼音排序）
　　安铁洙（东北林业大学）
　　曹贵方（内蒙古农业大学）
　　曹维维（华中农业大学）
　　陈　敏（信阳农林学院）
　　陈秋生（南京农业大学）
　　陈文钦（湖北生物科技职业学院）
　　程福生（武汉回盛生物科技股份有限公司）
　　程佳月（华中农业大学）
　　戴博杰（美国俄克拉荷马大学）
　　丁文格（武汉回盛生物科技股份有限公司）
　　杜安娜（中国科学院武汉病毒研究所）
　　范瑞文（山西农业大学）
　　冯悦平（华中农业大学）
　　葛晓红（华中农业大学）
　　哈西卜·哈利克（华中农业大学）
　　何文波（华中农业大学）
　　赫晓燕（山西农业大学）
　　胡　满（河北农业大学）
　　黄海波（华中农业大学）
　　黄　立（信阳农林学院）
　　黄丽波（山东农业大学）
　　金春艳（华中农业大学）
　　靳二辉（安徽科技学院）

荆海霞（青海大学）

柯妍妍（厦门医学院）

李德雪（中国人民解放军军事医学科学院）

李福宝（安徽农业大学）

李升和（安徽科技学院）

李艳和（华中农业大学）

李　勇（江西农业大学）

李玉谷（华南农业大学）

刘华珍（华中农业大学）

刘婷婷（华中农业大学）

刘晓丽（华中农业大学）

刘志伟（华中农业大学）

刘忠虎（河南农业大学）

陆　军（武汉回盛生物科技股份有限公司）

卢　顺（华中农业大学）

逯志强（华中农业大学）

栾维民（吉林农业大学）

罗厚强（温州科技职业学院）

逢莎莎（华中农业大学）

彭克美（华中农业大学）

彭　森（天津大学）

卿素珠（西北农林科技大学）

丘伟巍（华中农业大学）

宋　卉（华中农业大学）

宋学雄（青岛农业大学）

孙鹏鹏（华中农业大学）

孙艳芳（信阳师范学院）

唐　娟（华中农业大学）

唐　丽（四川农业大学）

王家乡（长江大学）

王　静（华中农业大学）

王　蕾（华中农业大学）

王树迎（山东农业大学）

王　为（华中农业大学）

王　巍（华中农业大学）

王　岩（临沂大学）
王　英（宿州学院）
王　云（华中农业大学）
王玉珍（华中农业大学）
王政富（佛山科学技术学院）
王子旭（中国农业大学）
位　兰（河南科技大学）
温　乐（华中农业大学）
肖　珂（华中农业大学）
杨　隽（黑龙江八一农垦大学）
杨克礼（华中农业大学）
杨　倩（南京农业大学）
杨　智（华中农业大学）
殷　俊（扬州大学）
尹　伟（天津大学）
岳占碰（吉林大学）
张登荣（河北工程大学）
张高英（华中农业大学）
张乐萃（青岛农业大学）
张　玲（信阳农林学院）
张　媛（华南农业大学）
张玉丹（华中农业大学）
赵晓玲（西藏大学）
郑昕婷（华中农业大学）
钟菊明（美国奥本大学）
朱黛云（华中农业大学）

主　审

陈焕春（华中农业大学）
陈耀星（中国农业大学）

Editorial Board

Editor

    PENG Kemei (Huazhong Agricultural University)

Associate Editors

    AN Tiezhu (Northeast Forestry University)

    CAO Guifang (Inner Mongolia Agricultural University)

    CAO Weiwei (Huazhong Agricultural University)

    CHEN Min (Xinyang Agriculture and Forestry University

    CHEN Qiusheng (Nanjing Agricultural University)

    CHEN Wenqin (Hubei Vocational College of Bio-Technology)

    CHENG Fusheng (Wuhan HVSEN Biotechnology Co., Ltd)

    CHENG Jiayue (Huazhong Agricultural University)

    DAI Bojie (University of Oklahoma, USA)

    DING Wenge (Wuhan HVSEN Biotechnology Co., Ltd)

    DU Anna (Wuhan Institute of Virology, Chinese Academy of Science)

    FAN Ruiwen (Shanxi Agricultural University)

    FENG Yueping (Huazhong Agricultural University)

    GE Xiaohong (Huazhong Agricultural University)

    HASEEB Hkaliq (Huazhong Agricultural University)

    HE Wenbo (Huazhong Agricultural University)

    HE Xiaoyan (Shanxi Agricultural University)

    HU Man (Hebei Agricultural University)

    HUANG Haibo (Huazhong Agricultural University)

    HUANG Li (Xinyang Agriculture and Forestry University)

    HUANG Libo (Shandong Agricultural University)

    JIN Chunyan (Huazhong Agricultural University)

    JIN Erhui (Anhui Science and Technology University)

    JING Haixia (Qinghai University)

KE Yanyan (Xiamen Medical College)

LI Dexue (Academy of Military Medical Science China)

LI Fubao (Anhui Agricultural University)

LI Shenghe (Anhui Science and Technology University)

LI Yanhe (Huazhong Agricultural University)

LI Yong (Jiangxi Agricultural University)

LI Yugu (South China Agricultural University)

LIU Huazhen (Huazhong Agricultural University)

LIU Tingting (Huazhong Agricultural University)

LIU Xiaoli (Huazhong Agricultural University)

LIU Zhiwei (Huazhong Agricultural University)

LIU Zhonghu (Henan Agricultural University)

LU Jun (Wuhan HVSEN Biotechnology Co., Ltd)

LU Shun (Huazhong Agricultural University)

LU Zhiqiang (Huazhong Agricultural University)

LUAN Weimin (Jilin Agricultural University)

LUO Houqiang (Wenzhou Vocational College of Science and Technology)

PANG Shasha (Huazhong Agricultural University)

PENG Sen (Tianjin University)

QING Suzhu (Northwest A & F University)

QIU Weiwei (Huazhong Agricultural University)

SONG Hui (Huazhong Agricultural University)

SONG Xuexiong (Qingdao Agricultural University)

SUN Pengpeng (Huazhong Agricultural University)

SUN Yanfang (Xinyang Normal University)

TANG Juan (Huazhong Agricultural University)

TANG Li (Sichuan Agricultural University)

WANG Jiaxiang (Yangtze University)

WANG Jing (Huazhong Agricultural University)

WANG Lei (Huazhong Agricultural University)

WANG Shuying (Shandong Agricultural University)

WANG Wei (Huazhong Agricultural University)

WANG Wei (Huazhong Agricultural University)

WANG Yan (Linyi University)

WANG Ying (Suzhou University)

WANG Yun (Huazhong Agricultural University)

WANG Yuzhen (Huazhong Agricultural University)

WANG Zhengfu (Foshan University of Science and Technology)

WANG Zixu (China Agricultural University)

WEI Lan (Henan Institute of Science and Technology)

WEN Le (Huazhong Agricultural University)

XIAO Ke (Huazhong Agricultural University)

YANG Jun (Heilongjiang Bayi Agricultural University)

YANG Keli (Huazhong Agricultural University)

YANG Qian (Nanjing Agricultural University)

YANG Zhi (Huazhong Agricultural University)

YIN Jun (Yangzhou University)

YIN Wei (Tianjin University)

YUE Zhanpeng (Jilin University)

ZHANG Dengrong (Hebei University of Engineering)

ZHANG Gaoying (Huazhong Agricultural University)

ZHANG Lecui (Qingdao Agricultural University)

ZHANG Ling (Xinyang Agriculture and Forestry University)

ZHANG Yuan (South China Agricultural University)

ZHANG Yudan (Huazhong Agricultural University)

ZHAO Xiaoling (Tibet University)

ZHENG XinTing (Huazhong Agricultural University)

ZHONG Juming (Auburn University, USA)

ZHU Daiyun (Huazhong Agricultural University)

Reviewers

CHEN Huanchun (Huazhong Agricultural University)

CHEN Yaoxing (China Agricultural University)

序

进入21世纪以来，我国的畜牧兽医事业取得了飞速的发展，已经成为农业经济中的支柱产业，是整个国民经济中的重要组成部分。作为科学之本、技术之源，基础研究是科技发展的内在动力，深刻影响着一个国家基础性的创新能力。随着"同一个世界，同一个医学，同一个健康"和"健康动物—健康食品—健康人类"新观念的提出，畜牧兽医事业正在不断地扩大和延伸，在国民经济和社会稳定中发挥着越来越重要的作用。恩格斯曾经说过，"科学的发生和发展一开始就是由生产决定的"。社会生产发展的需要推动了自然科学的发展。科学来自生产实践；同时，科学又必须经受实践的检验。

实践证明，组织学与胚胎学在医学（包括动物医学和人类医学）和畜牧业等领域都有着广阔的前景。例如，要想研究动物生长、生殖、消化与吸收等生理机能，首先要了解执行这些生理功能的器官系统的组织结构。因为当机体形态结构发生变化时，生理功能也会随之发生变化；同样，当生理环境发生改变时，形态结构也会随之发生改变。形态结构与生理功能是相互作用的。组织结构与病理学的关系极为密切，只有在熟悉了正常的组织结构以后，才能认识患病机体的各种病理变化。又如在养殖业中，要想开发利用某个特别优良的动物品种，首先要了解它的繁殖习性、性成熟期以及胚胎发育等规律，然后才能提出合理的措施。高产奶牛、优质赛马等的人工繁殖，更需要有扎实的胚胎学知识基础。而应用胚胎发育的规律，进行人工繁殖之前，首先必须掌握不同胚胎发育时期的形态结构。胚胎的结构还与遗传育种关系密切。精子和卵子中的染色体是主要的遗传物质基础，而精子和卵子成熟后的受精正是胚胎发育的开始。因此，采用人工杂交、性别控制、胚胎移植和克隆动物等技术改变动物的遗传特性，从而培育更加优质的动物品种，这是胚胎学和遗传学的共同任务。

20世纪70年代和90年代，动物组织学相关彩色图谱分别出版了2个版本，在内容和形式上各有特色。但目前尚无系统、全面和科学实用的动物组织学与胚胎学彩色图谱。当前，新理论、新知识、新技术和新方法不断充实生命科学，又促进了组织学与胚胎学的发展。各个大专院校动物科学、动物医学专业和生命科学专业的师生们、广大临床兽医和科技工作者，以及养殖企业与专业户，普遍感到显微镜下观察与识别各种动物细胞、组织和器官的细微形态、结构特征是一件相当困难的事情。因此，特别希望能有一本得心应手的动物组织学与胚胎学彩色图谱。

综上所述，出版一部图像清晰、色彩明快、使用方便的动物组织胚胎彩色图谱，对于掌握组织学与胚胎学知识，提升动物医学、临床兽医、动物科学、广大养殖工作者和科技人员的业务水平，促进畜牧兽医事业的可持续发展具有十分重要的意义。

欣闻彭克美教授主编出版《动物组织学与胚胎学彩色图谱》这一专著并诚邀我作序。该图谱收集了1 600余幅彩色图像，这些珍贵的图片都是从作者几十年来亲自在显微镜下拍摄积累的数十万张显微图片中精选出来的，编排合理、内容丰富、色彩鲜明、层次分明、结构清晰、印刷精美，堪称图谱中的精品。我深信这一专著的出版发行，必将受到广大读者的热烈欢迎，对组织学与胚胎学的教学、科研和畜牧兽医事业的发展发挥重要作用。

中国工程院院士、华中农业大学教授　陈焕春

2020年8月28日

Preface

Since the beginning of the 21st century, our country's animal husbandry and veterinary medicine have achieved rapid development and become a pillar industry in the agricultural economy, which is a very important part in the entire national economy. As the foundation of science and the source of technology, basic research is the internal driving force for the development of science and technology and profoundly affects a country's fundamental innovation capabilities. With the new concepts of "one health" and "healthy animals-healthy food-healthy people", animal husbandry and veterinary medicine are continuously expanding and extending, playing an increasingly important role in the national economy and social stability. Friedrich Engels once said that from the very beginning, the origin and development of the sciences have been determined by production. It is the needs to develop social production that promote the development of natural sciences. The source of science comes from production practice; meanwhile, the authenticity of science must be tested by practice.

Practice has proved that histology and embryology have broad application prospects in medicine (including animal medicine and human medicine), agriculture and animal husbandry. For example, to investigate the physiological functions of animals such as growth, reproduction, digestion and absorption, we must first understand the histological structure of the organ systems that perform these physiological functions. When the morphological and histological structure of the organ changes, the physiological function will change accordingly; similarly, when the physiological environment changes, the morphological and histological structure will change. In other words, morphological and histological structure of the organ is highly correlated with its function. The relationship between histological structure and pathological changes is very interactive. Only after familiarizing with the normal tissue structure can we understand the various pathological changes of the diseased body. In the case of animal breeding, to develop and utilize a particularly excellent animal species, we must first understand its reproductive habits, sexual maturity, and embryonic development, and then set up reasonable breeding strategies. For instance, artificial breeding of high-yielding cows and high-quality racehorses requires a solid foundation of embryology. Moreover, before using the laws of embryonic development for artificial breeding, we must first know the morphological and histological structure of different embryonic development stages. The structure of the embryo is also closely related to genetic breeding. The chromosomes in sperm and egg are the main bases of genetic materials, and the fertilization of sperm and egg after they mature is the beginning of embryonic development. Therefore, it is the shared task of embryology and genetics to change the heredity of animals by using techniques such as artificial hybridization, sex control, embryo transfer, and animal cloning, so as to breed better animal breeds.

In the 1970s and 1990s, two versions of the color atlas related to animal histology were published, each with its own characteristics in content and form. However, there is no systematic, comprehensive and scientifically practical "Color Atlas of Animal Histoembryology ". At present, new theories, new knowledge, new technologies and new methods continue to facilitate life sciences, and promote the development of histology and embryology. However, it is quite difficult for teachers and students engaged in animal science, veterinary medicine and life sciences in universities and colleges to observe and identify the histology of various animal cells, tissues and organs. It is the same for clinical veterinarians and workers, as well as breeding companies and professional households. Therefore, I particularly hope to have a very useful "Color Atlas of Animal Histoembryology ".

In summary, publication of a color atlas of animal histology and embryology with clear images, bright colors, and easy access will be very useful for mastering knowledge of histology and embryology. More importantly, its publication will improve the professional level of scientific and technical personnel in animal medicine, clinical veterinary medicine, animal science, and breeding. Together, its publication is of great significance to promote the sustainable development of animal husbandry and veterinary medicine.

I am delighted to know that Professor Kemei Peng wrote, and edited the book " Color Atlas of Animal Histoembryology " and kindly invited me to write a prologue. It is my pleasure to do it. The atlas has collected more than 1 600 color images. These precious pictures are selected from hundreds of thousands of microscopic pictures that the author has personally taken under the microscope for decades. The pictures are well-organized, rich in content, bright in color, and well-defined. In addition, the printing is exquisite, so it is the most extraordinary atlas. I truly believe that the publication and distribution of this atlas will be warmly welcomed by readers, and will play an important role in the teaching and researching of histology and embryology and the development of animal husbandry and veterinary industry.

Academician of Chinese Academy of Engineering,

Professor of Huazhong Agricultural University

Huanchun Chen

2020.8.28

前 言

组织学与胚胎学是生物科学与医学（包括动物医学和人类医学）相互交叉融合的科学，是基础医学中的骨干学科，也是生命科学中发展迅猛、成就卓著的前沿学科之一。21世纪是生命科学和医学飞速发展的新时期，胚胎移植、干细胞理论和干细胞工程、生殖医学和生殖工程、哺乳动物克隆技术、组织工程等新理论和新技术的出现，发育生物学的兴起等，都与组织学与胚胎学的发展密切相关，这些新理论和新技术的出现又促进了组织学与胚胎学更加深入和快速地发展。

1978年和1994年，我国先后出版了两部动物组织学相关的彩色图谱，其内容和形式上各有特色。当前，新理论、新知识、新技术和新方法不断地充实生命科学，又给组织胚胎学科提出了新的要求，但目前尚无系统、全面和科学实用的动物组织学与胚胎学彩色图谱。广大动物科学和动物医学专业师生、临床兽医工作者和生物科技工作者，以及成千上万的养殖人员，普遍感到在观察与鉴别各种动物器官的组织结构和细胞的特征时困难重重，特别渴望能有一部高清晰的，真实反映动物器官、组织、细胞形态结构的彩色图谱或专用工具书，为他们的学习和工作提供实用的帮助。为此，我们参阅了大量国内外同类书籍，编著了这部《动物组织学与胚胎学彩色图谱》。

本书共分为22章，内容包括绪论、细胞的形态结构与细胞分裂、上皮组织、固有结缔组织、软骨组织和骨组织、血液与淋巴、肌组织、神经组织、神经系统、循环系统、免疫系统、内分泌系统、消化管、消化腺、呼吸系统、泌尿系统、雌性生殖系统、雄性生殖系统、被皮系统、感觉器官、禽类的主要组织结构特征、畜禽早期胚胎发育；涉及牛、马、猪、羊、犬、猫、兔、鸡、鸭、鸽、鹅11种动物；采用了HE、镀银、美蓝、硫堇、卡红、荧光染色等多种技术。作者40余年的积累，收集了数十万张图片，最终从中精选出来光镜和电镜图片1 643幅。本书不仅图片影像清晰、色彩靓丽、设计精美，每一章的开始还有各种组织和器官的发生及其组织学结构概述以及动物种间差异的精炼文字描述，全面系统地涵盖了动物细胞、基本组织、器官组织与胚胎发育的形态结构。它可以作为高等院校广大动物科学和动物医学专业师生、临床兽医工作者和生物科技工作者，以及广大养殖人员的工具书。为了便于读者阅读和使用，并与国际接轨，本书对组织学与胚胎学的专业名词、术语采用了中、英文双语注释的方式，查阅方便、科学实用。

为这样一门具有深厚科学内涵而且充满发展活力的学科编著出版一部图谱具有极大的挑战性。组织学与胚胎学是以形态学为主的学科，全书采用图像达1 600余幅，每章中的模式图也全部精心设计、绘制成了精美的彩图。众所周知，图片的面积越大和放大倍数越高，对组织切片和显微摄影等各方面的质量要求就越高。本书的编著工作于2010年8月正式启动，历经十年磨一剑的艰辛。在从选购动物、取材固定、制作组织切片标本、光学显微镜和电子显微镜检验、显微摄影、图像采集、精选照片，到电脑修图、手工绘图、拉线、注字及中英文双语注释的无数个环节中，每一步都凝聚了大量心血和汗水。

特别要指出的是，为了凸显科学实用和扩展读者的知识面，既做到与当前兽医临床、畜牧生产实际和生命科学热点相结合，又能紧跟科学前沿进展，本书在编写和排版技巧上进行了改进，一改以往传统的做法，直接在彩图上注字或拉线注字，并在图下配有英汉对照的简要说明。这样就大大提高了读者查阅和掌握图内信息的速度，同时也有利于英语读者查阅及与国际接轨。

本书从启动、编著到出版，得到了华中农业大学的大力支持。各位编者为此书的完成付出了诸多辛苦和努力，在本书即将面世之际，谨向支持和关心本书编著和出版工作的所有单位和同仁致以最诚挚的谢意！

武汉回盛生物科技有限公司对本书的出版给予了大力支持，特此感谢！

由衷地感谢在百忙之中担任本书主审和作序的兽医学家、中国工程院院士、中国畜牧兽医学会名誉理事长、华中农业大学陈焕春教授的大力支持！衷心感谢中国畜牧兽医学会动物解剖及组织胚胎学分会理事长、中国农业大学陈耀星教授担任本书主审！

在本书的编著过程中，笔者参阅了大量中外文书籍。本书中有几幅模式图和示意图是根据所附参考文献中的有关插图而改绘或仿绘的。在此，我们特别对原书作者和出版者致以衷心的感谢！

由于编者的水平能力所限，书中难免有不当之处，敬请广大读者批评和指正。

2020年8月18日于武汉

Foreword

Histology and embryology is the interdisciplinary science of biological science and medicine (including veterinary medicine and human medicine), and it is the backbone discipline of basic medicine. It is one of the frontier disciplines of life science with rapid development and outstanding achievements. The 21st century witnessed the rapid development of the life science and medical science. The emergence of the new theories and new technologies in embryo transfer, stem cell and stem cell engineering, reproductive medicine, reproductive engineering, mammalian cloning technology and tissue engineering, are closely associated with the development of histology and embryology. The rise of developmental biology and so on is also closely associated with it. In a way, the new theories and technologies also push forward the development of histology and embryology.

Two Color Atlas of Animal Histology books were published in China in 1978 and 1994, each with its own characteristics in content and form. At present, new theories, new knowledge, new techniques and new methods are constantly enriching the life science and putting forward new requirements for the subject of tissue and embryo. However, there is no systematic, comprehensive and scientifically practical color atlas about Animal Histology and Embryology. Teachers and students engaged in animal science and animal medicine, clinical veterinarians, biotechnology researchers, and farm breeding personnel have trouble in observing and identifying the tissue structures of different animal organs and their cell characteristics, and are in urgent need of a well-illustrated color atlas about animal organs, tissues and cell morphological structures. Thus we complied this Color Atlas of Animal Histoembryology on the basis of a large number of relevant domestic and foreign publications.

This Color Atlas has altogether 22 chapters, including Introduction, Cell Morphology and Cell Division, Epithelial Tissue, Connective Tissue Proper, Cartilage and Bone Tissue, Blood and Lymph, Muscle Tissue, Nerve Tissue, Nervous System, Circulatory System, Immune System, Endocrine System, Digestive Tract, Digestive Glands, Respiratory System, Urinary System, Female Reproductive System, Male Reproductive System, Integumentary System, Sensory Organs, The Main Structural Features of Fowl Tissues, and Early Embryonic Development of Livestock and Fowl. Eleven species of animals are mentioned in the book, covering cattle, horses, pigs, sheep/goats, dogs, cats, rabbits, chicken, ducks, pigeons and geese. It also employs various staining technologies like HE, silver stain, methylene blue, thionine, carmine, fluorescent stain and so on. 1 643 images taken by light microscope and the electron microscope technique were selected from tens of thousands of pictures that have been collected after 40 years of hard work. They are clear, rich in color and exquisitely designed.

At the beginning of each chapter, there is a refined text description about the occurrence and histological structures of various tissues and the differences among animal species, providing a comprehensive knowledge about the morphological structure of animal cells, basic tissues, organ tissues and embryonic development. This color atlas can be used as a reference book by teachers and students in the field of animal science and veterinary medicine, clinical veterinarians and biotechnology workers, as well as the vast number of breeding staff in colleges and universities. In order to facilitate readers and meet with the international standards, professional terms and terminologies appear in both Chinese and English in the book. We hope the color atlas is easy to use and able to provide practical help to people in need.

It is a great challenge to write and publish a monograph atlas for histology and embryology science, which is mainly a subject of morphology. There are more than 1 600 images in the book, and the pattern diagrams in each chapter are all elaborately designed and of delicate colors. It is well known that the larger the image area is and the higher the magnification is, the more demanding tissue sections and microphotography become. The compilation of this atlas was officially launched in August 2010 and took a decade of painstaking work. From the selection of animals, collection and fixation, tissue section, specimen preparation, optical microscope and electron microscope examination, microphotography, image collection, photo selection, to computer editing, manual drawing, tagging, character annotation and Chinese and English annotations, each step has accumulated a lot of hard work and sweat.

In particular, it should be noted that the book, different from traditional approaches, writes the notes directly on the colored images or by way of tagging and provides under each image both English and Chinese descriptions, so that readers, from home and abroad, are able to grasp the information at a glance.

This atlas has received great support from Huazhong Agricultural University all the way from its startup, to compilation and to publication. As the book is about to be published, I would like to extend my most sincere thanks to all the units and colleagues who have supported and cared about the compilation and publication of this book.

Wuhan HVSEN Biotechnology Co., Ltd has given strong support for the publication of this book, thank you!

I also want to express my sincere gratitude to Professor Huanchun Chen from Huazhong Agricultural University who is an academician of the Chinese Academy of Engineering and the honorary president of Chinese Association of Animal Science and Veterinary Medicine, for writing the prologue for this book. And special thanks go to Professor Yaoxing Chen from China Agricultural University, the president of Anatomy and Histology Embryology Branch of Chinese Association of Animal Science and Veterinary Medicine, for being the deputy chief reviewer of this book!

While compiling this book, the authors referred to a large number of Chinese and foreign books. Several schematic diagrams in this book were simulated according to the illustrations in these references. Therefore, we are especially grateful to the author and publishers!

Due to my limited abilities, the book inevitably has some deficiencies and is subject to readers' suggestions and correction.

Kemei Peng
Wuhan, China
2020.8.18

目 录

序
Preface
前言
Foreword

第一章 绪 论
Introduction

Outline ··· 1
一、组织切片标本概述 ··· 2
二、研究组织胚胎的技术与设备 ·· 3
三、观察切片标本的注意事项 ··· 5
四、组织胚胎学研究技术与设备图谱 ··· 5

第二章 细胞形态结构与细胞分裂
Cell Morphology and Cell Division

Outline ·· 21
一、细胞的基本概念 ·· 22
二、细胞的起源 ··· 22
三、细胞的组织结构概述 ··· 22
四、细胞分裂 ·· 23
五、细胞形态结构与细胞分裂图谱 ··· 23

第三章　上皮组织
Epithelial Tissue

Outline ... 47
一、上皮组织的发生 ... 48
二、上皮组织的组织学结构概述 ... 48
三、上皮组织图谱 ... 49

第四章　固有结缔组织
Connective Tissue Proper

Outline ... 70
一、结缔组织的发生 ... 71
二、结缔组织的组织学结构概述 ... 71
三、结缔组织图谱 ... 71

第五章　软骨组织和骨组织
Cartilage and Bone Tissue

Outline ... 89
一、软骨和骨的发生 ... 90
二、软骨和骨的组织学结构概述 ... 90
三、软骨组织和骨组织图谱 ... 91

第六章　血液与淋巴
Blood and Lymph

Outline ... 129
一、血细胞的发生 ... 130
二、血细胞的组织学结构概述 ... 130
三、骨髓 ... 131
四、淋巴 ... 131
五、血细胞与骨髓图谱 ... 131

第七章 肌组织
Muscle Tissue

Outline	150
一、肌组织的发生	151
二、肌组织的组织学结构概述	151
三、肌组织图谱	152

第八章 神经组织
Nervous Tissue

Outline	176
一、神经组织的发生	177
二、神经组织的组织学结构概述	177
三、神经组织图谱	178

第九章 神经系统
Nervous System

Outline	211
一、神经系统的发生	211
二、神经系统的组织学结构概述	212
三、神经系统图谱	213

第十章 循环系统
Circulatory System

Outline	270
一、循环系统的发生	271
二、循环系统的组织学结构概述	272
三、循环系统图谱	273

第十一章　免疫系统
Immune System

Outline	300
一、免疫器官的发生	301
二、免疫器官的组织学结构概述	302
三、免疫系统图谱	303

第十二章　内分泌系统
Endocrine System

Outline	353
一、内分泌系统的发生	354
二、内分泌系统的组织学结构概述	354
三、内分泌系统图谱	355

第十三章　消化管
Digestive Tract

Outline	404
一、消化管的发生	405
二、消化管的组织结构概述	406
三、消化管图谱	408

第十四章　消化腺
Digestive Glands

Outline	484
一、消化腺的发生	485
二、消化腺的组织学结构概述	486
三、消化腺图谱	487

第十五章　呼吸系统
Respiratory System

Outline	548

一、呼吸系统的发生 ... 549
二、呼吸器官的组织学结构概述 ... 549
三、呼吸系统图谱 ... 550

第十六章 泌尿系统
Urinary System

Outline ... 571

一、泌尿系统的发生 ... 572
二、泌尿器官的组织学结构概述 ... 572
三、泌尿系统图谱 ... 573

第十七章 雌性生殖系统
Female Reproductive System

Outline ... 609

一、生殖系统的发生 ... 610
二、雌性生殖器官的组织学结构概述 ... 610
三、雌性生殖系统图谱 ... 611

第十八章 雄性生殖系统
Male Reproductive System

Outline ... 657

一、雄性生殖器官的发生 ... 658
二、雄性生殖器官的组织学结构概述 ... 658
三、雄性生殖系统图谱 ... 659

第十九章 被皮系统
Integumentary System

Outline ... 703

一、被皮系统的发生 ... 704
二、被皮系统的组织学结构概述 ... 704
三、被皮系统图谱 ... 705

第二十章 感觉器官
Sensory Organs

Outline ··· 736

一、眼和耳的发生 ·· 737

二、眼和耳的组织学结构概述 ··· 737

三、感觉器官图谱 ·· 738

第二十一章 禽类的主要组织结构特征
Main Structural Features of Fowl Tissues

Outline ··· 757

一、禽类的主要组织结构特点概述 ·· 759

二、禽类的主要组织结构特征图谱 ·· 763

第二十二章 畜禽早期胚胎发育
Early Embryonic Development of Livestock and Fowl

Outline ··· 852

一、家畜的早期胚胎发育 ··· 853

二、家禽的早期胚胎发育 ··· 855

三、早期胚胎发育图谱 ·· 856

主要参考文献 ··· 908

第一章
绪 论
Introduction

> **Outline**
>
> The study of the microscopic structure of animal tissues and organs and of embryogenesis and development is a branch of morphology. It is not only the basis for understanding many basic veterinary disciplines, such as physiology, biochemistry, pathology and reproduction, revealing the life activities of the normal body and the process of material metabolism and other basic scientific issues, but also the basis for exploring the whole life science. The techniques used to study animal tissues and embryos include tissue sectioning, light microscopy, electron microscopy, histochemistry and cytochemistry, autography, tissue culture, morphometry, in situ hybridization and embryo transfer, etc. To master the characteristics of morphological science, it is more important to be able to recognize a large number of microscopic structures of various tissues and organs under light microscope and ultrastructural images under electron microscope in addition to the basic theoretical knowledge. Clear microscopic and ultrastructural images of light microscope and electron microscope were obtained from high quality tissue sections and ultrathin section specimens.
>
> How to use this Color Atlas of Animal Histoembryology? The correct studying methods are very important. Histoembryology is a science which studies the microstructure and ultra-microstructure of animals using microscope and electron microscope.
>
> 1. Plane and three-dimensional object. Tissue slices or pictures are the partial two-dimensional plane image. However, cells, tissues and organs are all three-dimensional solid structures. Different cross sections in the same organ can appear different morphous. Tridimensional and holistic conception is needed in study. Use the imagination to aggregately analyze the image seen, and master the relationship of plane and three-dimensional, part and integrity.
>
> 2. Structure and function. Some cells, tissues or organs have definite morphous and structure. The specific structure is the foundation to exercise function and activity. For example, there is abundant rough

endoplasmic reticulum and developed Golgi complex in cytoplasm of plasmocyte which have the function of synthesis and secretion. The muscle fiber in intestinal villi can stretch or shorten the intestinal villi and promote the nutritive material conveying. Therefore, structural morphous and physiological functions are closely and intensely related.

3. Static and dynamic state. Living cells are under dynamic change. The structure is changing with the cell differentiation, metabolism and function exercising. Cells propagate, decease, renovate unceasingly and the change in the process of embryonic development is more predominant and complicated. However, the structure expressed by slice or picture is only the resting image at a moment. Therefore, we cannot learn this subject well unless we are good at understanding static state in dynamic changing.

4. Theory and experiment. While learning theories of this subject, great importance must be attached to the training of practical manipulation skill. Various light microscopic samples, electron microscopic images, paraffin sections, embryo models and etc should be observed seriously. Try to aggregately analyze what you have learned and combine the sensible recognition and theoretical knowledge together to deepen comprehension, so as to build a solid foundation for further studies.

研究动物机体的组织、器官的微细结构和胚胎发生、发育的学科隶属于形态学科。它既是认识基础兽医诸多相关学科，如生理学、生物化学、病理学和繁殖学，揭示正常机体的生命活动和物质代谢的过程等基本科学问题的基础，也是探索整个生命科学的基础。研究动物组织胚胎的技术手段包括组织切片制作技术、光镜技术、电镜技术、组织化学和细胞化学技术、放射自显影技术、组织培养技术、形态测量技术、原位杂交技术和胚胎移植技术等。掌握形态科学的特点，除了掌握基本理论知识外，更重要的是必须能够识别各种组织和器官的光学显微镜下的微细结构图像和电子显微镜下的超微结构图像。清晰的光镜微细结构图像和电镜超微结构图像来源于优质的组织切片和超薄切片标本。

一、组织切片标本概述

为了正确、清晰地显示动物器官、组织和细胞的微细结构，必须制备适合显微镜下观察的切片样本。取新鲜组织，先用固定剂固定，使组织中的蛋白质迅速凝固，保持其原有的组织结构；然后采用石蜡、火棉胶或树胶等包埋，使用专门的切片机切成薄片。

因为研究目的的不同或器官、组织材料的差异，切片标本的制作方法也有区别。若需保存细胞内酶的活性或快速制片，可选用冰冻切片法，将组织在低温条件下快速冷冻，直接制成冰冻切片；血液、骨髓等液体组织可直接涂于玻片上制成涂片；无定形的疏松结缔组织和肠系膜等软组织，可铺于载玻片上制成铺片；骨等坚硬的组织可制成磨片。

动物细胞的直径一般为10μm左右，为了更好地显示其微细结构，以免细胞重叠影响观察效果，切片的厚度一般在3～6μm。为使组织保持一定的硬度以利切片，必须在切片之前在组织内渗入某种支持物。根据所用支持物的不同，可分为石蜡切片、火棉胶切片、冰冻切片、振动切片、半薄切片及超薄切片等。其中最常用的为石蜡切片和冰冻切片。

未经染色的切片标本是无色透明的。因其细胞各部分结构的折光率很低而难以分辨，需要通过染色才能使微细结构变得清晰。染色方法有多种多样，其中最常用的染色方法是苏木素-伊红染色法（hematoxylin-eosin staining），简称HE染色。这种染色方法可将细胞核等嗜碱性成分染成蓝紫色，而细胞质等嗜酸性成分染成红色，形成明显不同的色泽。苏木精为碱性染料，能被其染色的特性称嗜碱性（basophilia）；伊红为酸性染料，能被其染色的特性称嗜酸性（acidophilia）。对碱性和酸性染料的亲和力均不强者，称中性（neutrophilia）。组织内有些结构经硝酸银染色后，可使硝酸银还原成棕黑色的银微粒附着在组织结构上，这种特性称亲银性（argentaffin）；有的结构本身不能使硝酸银还原，需加还原剂才能使其还原，这种特性称嗜银性（argyrophilia）。有的细胞或组织用某些碱性染料染色时，其染色结果与染料的原有颜色不同，这种颜色的变异性称异染性（metachromasia），如用甲苯胺蓝染肥大细胞时，胞质内的颗粒被染成紫红色而不是蓝色。

组织学标本的制作程序十分复杂，以石蜡切片HE染色为例，归纳起来需要经过取材、固定、冲洗、脱水、透明、浸蜡、包埋、切片、贴片、烘片、复水、染色、脱水、透明、封片等一系列重要的步骤。

二、研究组织胚胎的技术与设备

研究组织胚胎以微细结构的形态描述为基本内容，主要利用显微镜进行观察研究。光学显微镜（light microscope，LM，简称光镜）下所见的结构称光镜结构，其分辨率约为0.2μm，可将物体放大约1 500倍。电子显微镜（electron microscope, EM，简称电镜）下所见的结构称超微结构（ultrastructure），其分辨率可达0.2 nm，可将物体放大上百万倍。在光镜和电镜下常用的计量单位和换算关系如下：

1 nm（纳米，nanometre）= 10^{-3} μm（微米，micrometre）= 10^{-6} mm（毫米，millimetre）

随着生物科学技术的飞速发展，现代组织胚胎学的研究手段在不断更新，涉及面也越来越宽。这里简要介绍几种常用的研究技术设备。

（一）光学显微镜

光学显微镜是观察组织切片标本的基本技术设备。显微镜有多种型号，但基本构造大致相同，包括机械部分和光学部分。

1. 显微镜的结构

（1）机械部分

①镜座和镜柱　支持和稳定整个镜体的主要部件，通常由铸铁制成。

②镜臂　连在镜柱上端的转折部分。

③载物台　放置切片标本的平台，其中央有圆孔，可透光。载物台上面有片夹，下面有前后左右移动的推进器。

④镜筒　是成像光柱的通道，镜筒上端装目镜，下端装物镜转换器。物镜转换器上有4个物镜孔，分别连接不同放大倍数的物镜。物镜的螺纹和口径按国际标准统一设计，因此物镜可以互换。

⑤调焦旋钮　分粗调和微调，以调节焦距使物像清晰。粗调的幅度大，每转一周可使载物台升降约10 mm；微调的幅度小，每转一周，载物台仅升降0.1～0.2 mm。

（2）光学部分

①物镜　装于物镜转换器上，常有4倍、10倍、40倍及100倍(油镜)等数种。物镜的外壁上除标有Achromatic外，还有焦距和数值孔径(NA)等数值。NA值越大，分辨力越高。物镜的作用是分辨标本的细节，形成有效的初级图像。因此，物镜的质量决定图像的优劣。

②目镜　常用的有10倍和15倍两种。目镜装在镜筒上端，外壁标有放大倍数和表示透镜光学校正程度的符号，如P或Plan表示平场，即视野弯曲已被校正。

图像的放大倍数=目镜放大倍数×物镜放大倍数

③聚光器　位于载物台下方，可聚光并通过载物台的中央孔透过标本。聚光器也有调节旋钮，可使其在一定范围内升降，从而调节光线进入物镜的聚散程度。聚光器上常配有光圈，可开大或缩小以调节进入聚光器的光束。聚光器下还配有滤光片框，并可向内外移动，便于更换滤光片。

④基座　内有变压器，将220 V交流电变为6～12 V的直流电光源。亮度钮可连续调节光源强度。

2. 显微镜的使用及注意事项

（1）转移和安放　搬运显微镜时，一手握镜臂，另一手托基座，保持镜体直立。安放时，显微镜靠近身体胸前略偏左，以便右手记录。

（2）调节照明　转动物镜转换器，使低倍镜对准聚光器，两眼睁开，注视目镜。打开光圈，先将亮度钮调节至最小，然后打开电源开关，适当调节亮度。上升聚光器，使光线进入物镜。要求视野全部照明，亮度均匀。根据光源的强弱、标本的情况和物镜的倍数，灵活运用聚光器和光圈。如观察染色较浅的标本时，要降低聚光器并缩小光圈，以增加标本的明暗对比；用高倍镜和油镜观察时，要升高聚光器并开大光圈，使视野明亮。

（3）放置标本　将标本的盖玻片向上，放置于载物台上，用片夹固定。应注意盖玻片过厚时，高倍镜无法对被检标本聚焦；放反的标本也无法聚焦。粗心大意时常常看不清物像，甚至压碎切片或损坏镜头。

（4）调焦观察　将标本移至物镜下方，一边从目镜中观察，一边转动粗调旋钮，直至找到观察目标并将物像调至清晰。低倍镜观察视野范围较大，有利于了解标本的整体情况。要想观察细节，可将局部移至视野中央，再换中、高倍镜观察。如果低倍镜的焦距已经调好，换中、高倍镜后只需微调少许即可。

（5）油镜的使用　用油镜观察标本，必须先在低倍镜、高倍镜看清要观察的物像，并将其移至视野中央。移开高倍镜头，在标本的观察部位滴一小滴香柏油，转换油镜，并使镜头浸入油内紧贴玻片，缓慢调节微调旋钮，直至物像清晰。用油镜观察时，必须升高聚光器并开大光圈，使视野明亮。观察完毕，移开镜头，用蘸有乙醚-乙醇（1∶1）混合液的擦镜纸清洁镜头和标本。

（6）保养和收藏　显微镜是精密仪器，使用后须保养。不得随意拆卸显微镜的部件。显微镜光学部分有污垢时，要用擦镜纸或绸布轻轻擦拭，勿用其他物品擦拭，以免损坏镜面。显微镜的机械部分可用纱布擦拭。使用完毕，应降下载物台，将物镜转成"八"字形垂于镜筒下。收藏显微镜应避免潮湿和灰尘，禁止与化学试剂或药品接触，保持干燥，以防显微镜的光学部件长霉和金属部件生锈。

3. 几种特殊显微镜

（1）荧光显微镜　荧光显微镜(fluorescence microscope, FM)用来观察标本中的自发荧光物质或以荧光素染色的细胞和结构。组织中的自发荧光物质如神经元和心肌细胞内的脂褐素呈棕黄色荧光，肝贮脂细胞和视网膜色素上皮细胞内的维生素A呈绿色荧光，某些神经内分泌细胞和神经纤维内的单胺类物质在甲醛作用下呈不同颜色的荧光，组织内含有的奎宁、四环素等药物也呈现一定的荧光。细胞内的某些成分可与荧光素结合而显荧光，如溴化乙锭与吖啶橙可与DNA结合，进行细胞内DNA含量测定。荧光显微镜更广泛用于免疫细胞化学研究，即以荧光素标记抗体，以检测相应抗原的存在与分布。

（2）相差显微镜和倒置显微镜　相差显微镜（phase contrast microscope，PCM）用于观察组织培养中活细胞的形态结构。活细胞无色透明，一般光镜下不易分辨细胞轮廓及其结构。相差显微镜能将活细胞不同厚度及细胞内各种结构对光的不同折射作用，转换为光密度差异即明暗差而得到辨认。组织培养研究常用倒置相差显微镜（inverted phase microscope，IPM），它的光源和聚光器在载物台的上方，物镜在载物台的下方，便于观察贴附在培养皿底壁上的活细胞。

（3）共聚焦激光扫描显微镜　共聚焦激光扫描显微镜（confocal laser scanning microscope, CLSM）以激光为光源，采用共聚焦成像系统和电子光学系统，经过微机图像分析系统对组织或细胞进行二维和三维分析处理。CLSM可用于细胞内各种荧光标记物的微量分析，细胞的受体移动和膜电位变化的观测，酶活性和物质转运的测定，DNA精确分析等，因此，CLSM可更准确、快速地对细胞内的微细结构进行定性和定量测定。

（二）电子显微镜

电子显微镜（electron microscope, EM）以电子发射器（电子枪）代替光源，以电子束代替光线，以电磁透镜代替光学透镜，最后将放大的物像投射到荧光屏上以便观察。常用的电镜有透射电镜和扫描电镜。

1. 透射电镜　透射电镜（transmission electron microscope, TEM）用于观察细胞内部的超微结构。一般采用戊二醛或锇酸作固定剂，合成树脂包埋，在超薄切片机上制成超薄切片（厚度为50～100 nm），经铅或铀等重金属盐电子染色，然后置电镜下观察。标本在荧光屏上呈黑白反差的结构图像。被重金属盐深染的结构特性称电子密度高，被浅染的结构特性称电子密度低，这种染色称正染色（positive staning）；若被染结构着色浅，其周围部分着色深，则称负染色（negative staning）。

2. 扫描电镜　扫描电镜（scanning electron microscope, SEM）用于观察组织、细胞表面的超微立体结构。扫描电镜标本不需制成超薄切片，组织经固定、脱水、干燥后，在其表面喷涂金属膜即可上镜观察。扫描电镜的视场大、景深长、图像的立体感强，但分辨率较低。常用于观察细胞表面的突起、微绒毛、纤毛等。

三、观察切片标本的注意事项

（一）平面与立体的关系

组织切片标本或图片只是器官或细胞某个局部的二维平面图像，而细胞、组织和器官都是三维的立体结构。同一器官的不同切面可呈现出不同的形态结构。因此，在阅读和使用时，应将平面图像与立体结构相结合，发挥想象力，建立起立体和整体概念；综合分析所观察到的图像，掌握平面与立体、局部与整体的关系。

（二）结构与功能的关系

各种细胞、组织或器官都具有一定的形态结构，特定的结构是行使其功能活动的基础，如浆细胞的胞质内含有丰富的粗面内质网和发达的高尔基复合体，具有合成与分泌功能；肠绒毛内含有肌纤维可使其舒缩，促进营养物质的转运。因此，形态结构与生理功能是紧密相关的。

（三）静态与动态的关系

活细胞处于动态变化中，在细胞分化、新陈代谢和行使功能的过程中，其结构也随之而发生变化；细胞还不断进行增殖、死亡和更新；胚胎发育过程中的变化更为显著和复杂，而切片或图片所表现的结构只是某个瞬间的静息图像。因此，应善于从动态变化来理解静态时相，才能真正掌握细胞动态。

（四）理论与实际的关系

阅读和使用本图谱的同时，应认真观察各种器官、组织的低、中、高倍的光镜图像和电镜图像，把感性认识与理论知识有机结合起来，进行综合分析，加深对生物机体微观结构的理解。

（五）切片中的人为现象

（1）裂纹裂开　在切片制作过程中，乙醇等脱水剂的浓度梯度相差太大，或脱水速度过快，所以组织内出现了不应有的皱裂现象。

（2）破损　在切片制作过程时破坏了组织；或封片时错动了盖玻片，因而导致组织内出现了破损现象。

（3）重叠　在切片时刀片与组织块之间的角度过大，出现了严重卷曲；或在展片过程中，水温过低而没有展平，因而导致蜡带-组织薄片出现了折叠现象。

（4）杂质　在切片制作过程中，流水冲洗不够，多余的染料没有冲洗干净，导致切片中有残留的染料；或封片时异物落入切片中，致使组织内出现了杂质。

（5）气泡　在封片时中性树胶内的气体没有排尽，或配制封片树胶时的浓度过低即太稀，而二甲苯挥发极快，致使组织内进入了大小不等的气泡。

（6）刀痕　切片时所使用的刀片不快或有小口，或切片过程中碰到了微小的硬物例如砂砾，导致组织中出现了大小不等的划痕甚至裂口。

（7）显微摄影中的问题　拍摄组织切片显微照片时，曝光不匀，或曝光不足，或曝光过度等，都会影响显微照片的质量。

四、组织胚胎学研究技术与设备图谱

1.**切片机**　图1-1～图1-7。

2.**组织切片的制作程序**　图1-8。

3.**显微镜**　图1-9～图1-17。

4.**切片中的人为现象**　图1-18～图1-30。

图1-1 轮式切片机
Fig.1-1 wheel type microtome

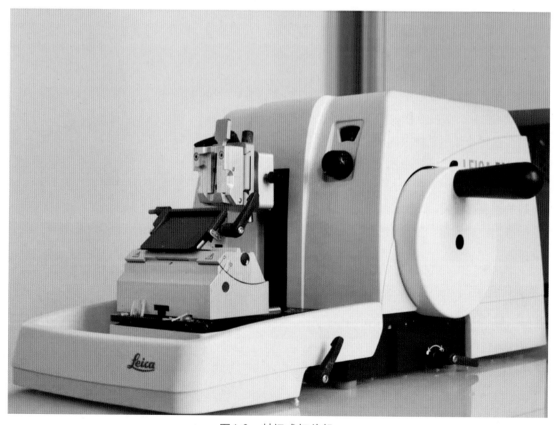

图1-2 封闭式切片机
Fig.1-2 closed type microtome

第一章 绪 论 Introduction

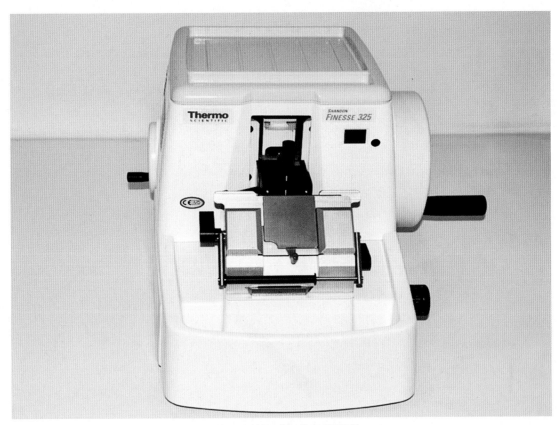

图1-3 封闭式切片机正面观

Fig.1-3 anterior view of closed type microtome

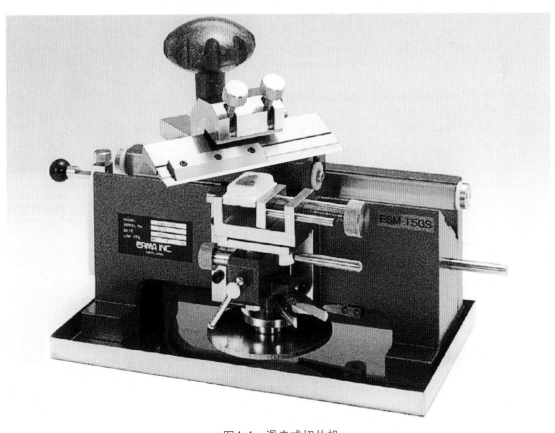

图1-4 滑走式切片机

Fig.1-4 sliding microtome

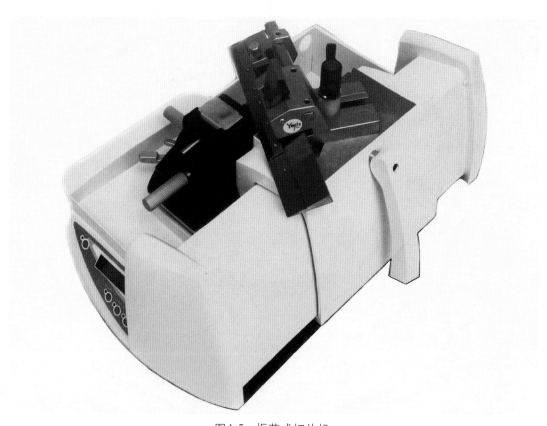

图1-5　振荡式切片机

Fig.1-5　oscillating microtome

图1-6　冰冻切片机

Fig.1-6　freezing microtome

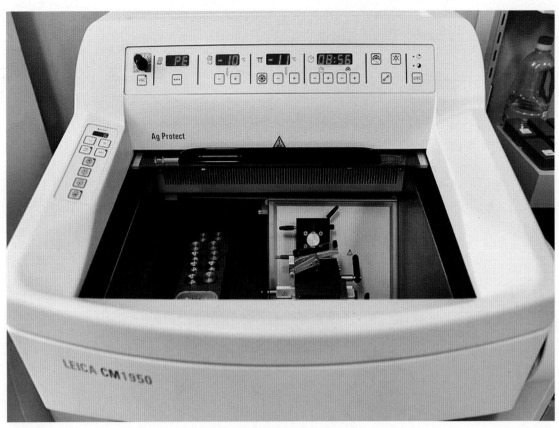

图1-7 冰冻切片机内部构造
Fig.1-7 internal construction of freezing microtome

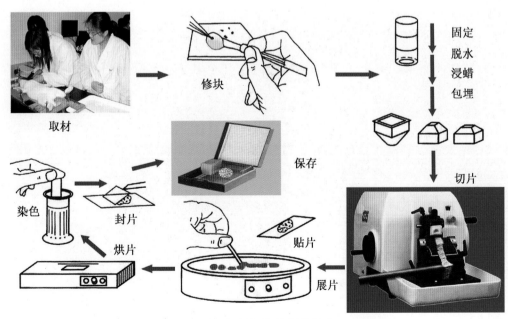

图1-8 组织切片的制作程序
Fig.1-8 the procedure of the sectioning

取材—sampling 修块—repair 固定—fixing 脱水—dehydration 浸蜡—waxing 包埋—embedding
切片—sectioning 展片—unfolding 贴片—paster 烘片—drying 染色—staining 封片—mounting 保存—preserving

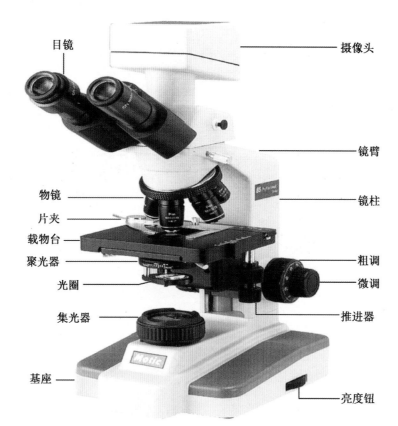

图 1-9 光学显微镜

Fig.1-9 light microscope

目镜—eyepiece 物镜—objective 片夹—clamp 载物台—stage 聚光器—condenser 光圈—diaphragm 集光器—collecting mirror 基座—base 摄像头—camera 镜臂—arm 镜柱—pillar 粗调—coarse tuning 微调—fine tuning 推进器—propel 亮度钮—luminance tuning

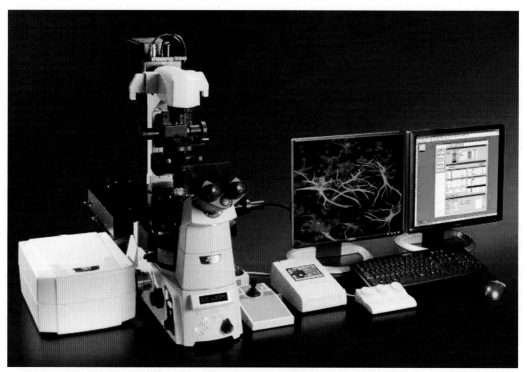

图 1-10 荧光显微镜

Fig.1-10 fluorescence microscope

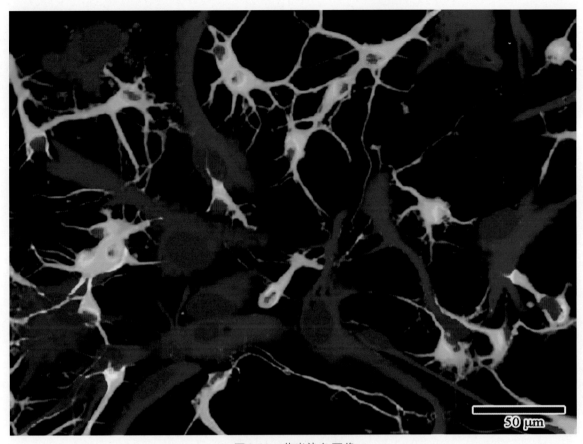

图1-11 荧光染色图像

Fig.1-11 fluorescence staining image

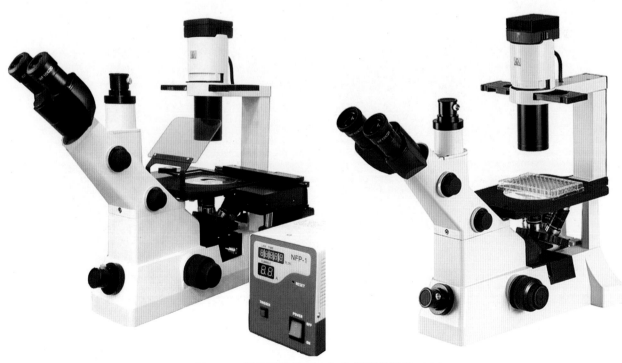

图1-12 相差显微镜（左）和倒置显微镜（右）

Fig.1-12 phase contrast microscope（left）and inverted microscope（right）

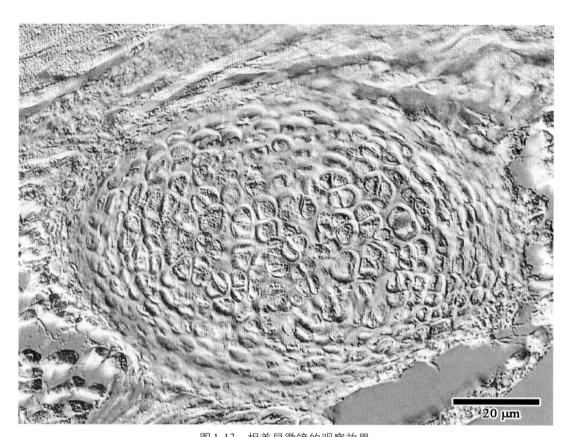

图1-13　相差显微镜的观察效果

Fig.1-13　the effect of phase contrast microscope

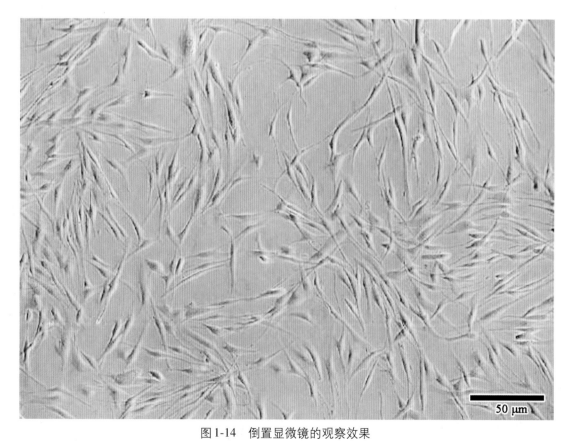

图1-14　倒置显微镜的观察效果

Fig.1-14　the effect of inverted microscope

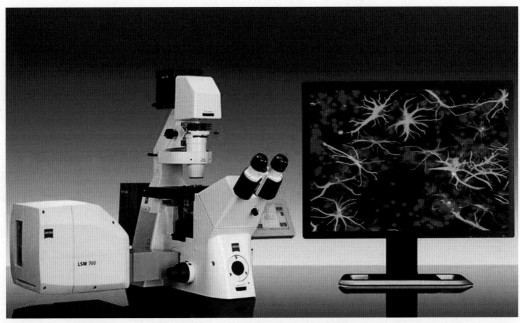

图1-15 共聚焦激光扫描显微镜

Fig.1-15 confocal laser scanning microscope

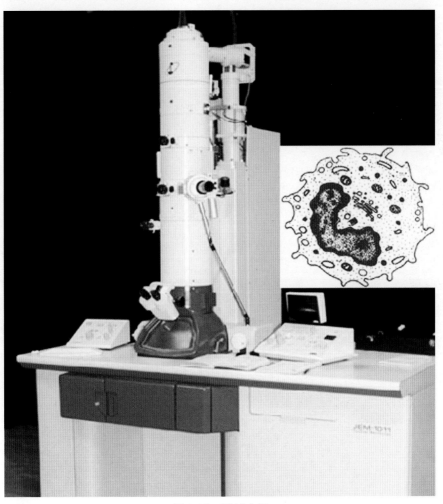

图1-16 透射电镜

Fig.1-16 transmission electron microscope

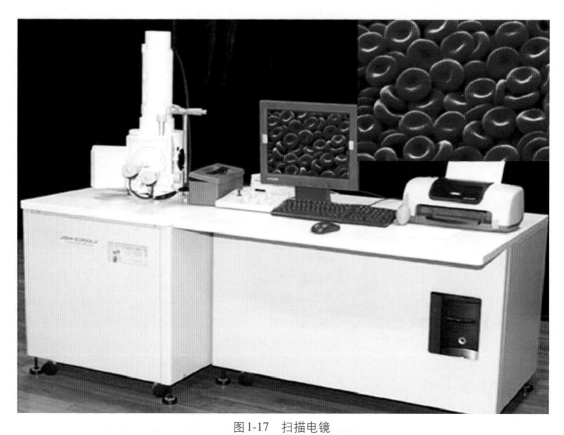

图1-17　扫描电镜

Fig.1-17　scanning electron microscope

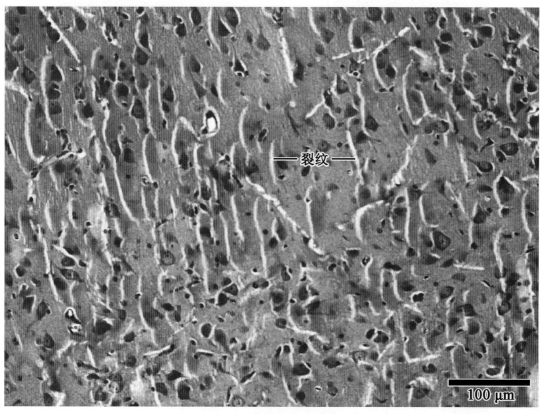

图1-18　皱裂现象（HE）

Fig.1-18　craze crack（HE）

裂纹—crack

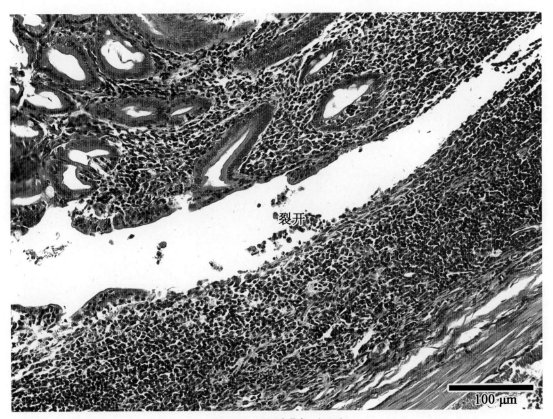

图1-19 开裂现象（HE）

Fig.1-19 cracking (HE)

裂开—fissuration

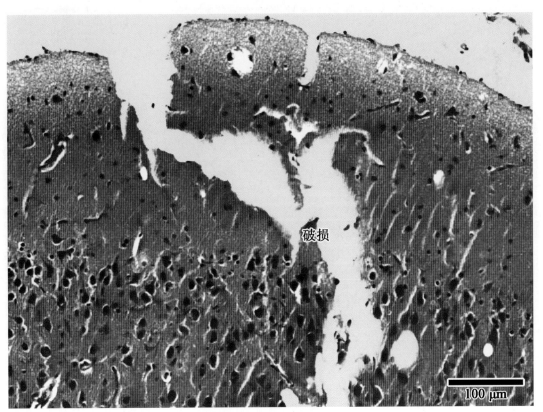

图1-20 破损现象（HE）

Fig.1-20 breakage (HE)

破损—damage

图1-21 折叠现象（HE）

Fig.1-21 overlapping（HE）

折叠—fold up

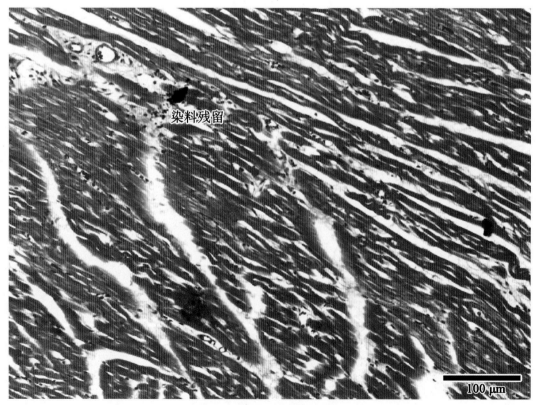

图1-22 染料残留现象（HE）

Fig.1-22 dye residues（HE）

染料残留—dye residues

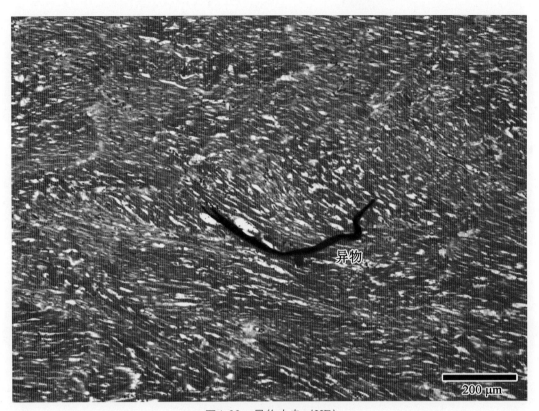

图1-23 异物夹杂（HE）

Fig.1-23　foreign matter（HE）

异物—foreign matter

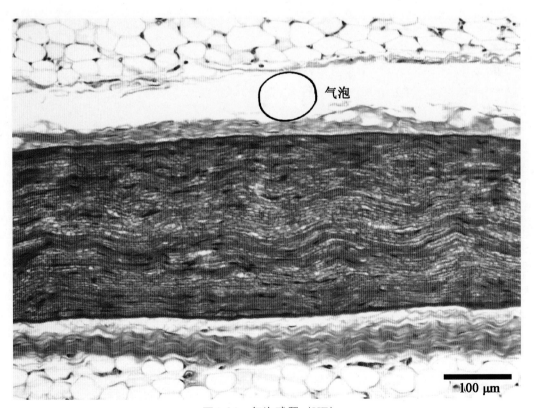

图1-24 气泡残留（HE）

Fig.1-24　air bubble residues（HE）

气泡—air bubble

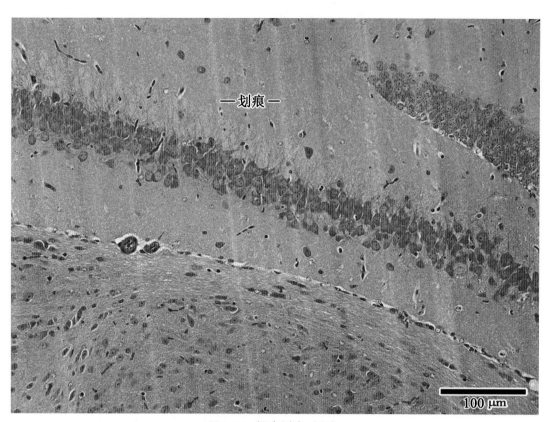

图1-25 轻度划痕（HE）

Fig.1-25 slight scratch（HE）

划痕—scratch

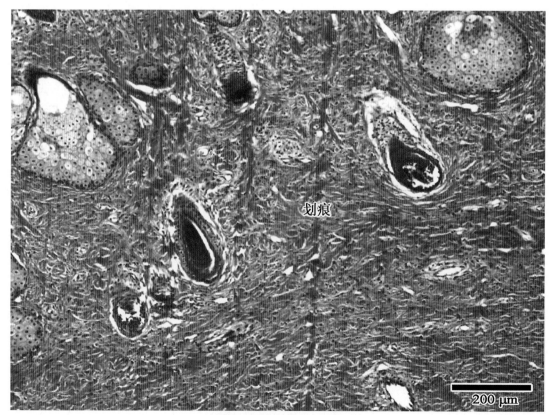

图1-26 划痕现象（HE）

Fig.1-26 scratch（HE）

划痕—scratch

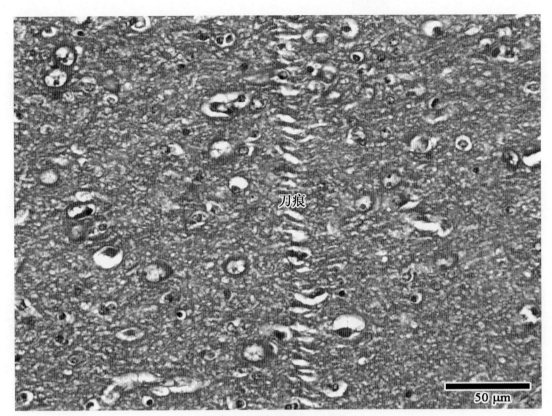

图1-27 刀痕现象（HE）

Fig.1-27 relief lines（HE）

刀痕—tool marks

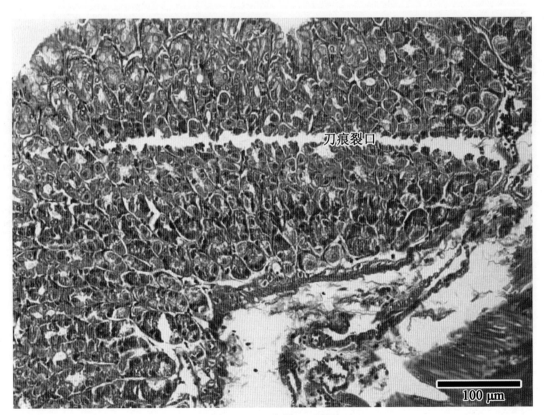

图1-28 刀痕裂口（HE）

Fig.1-28 knife gap（HE）

刀痕裂口—tool marks

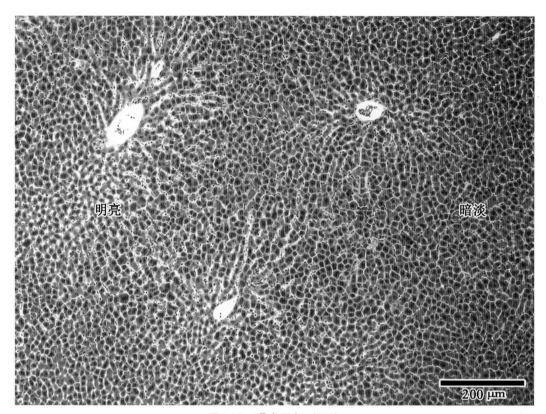

图1-29　曝光不匀（HE）

Fig.1-29　uneven exposure（HE）

明亮—light　暗淡—dark

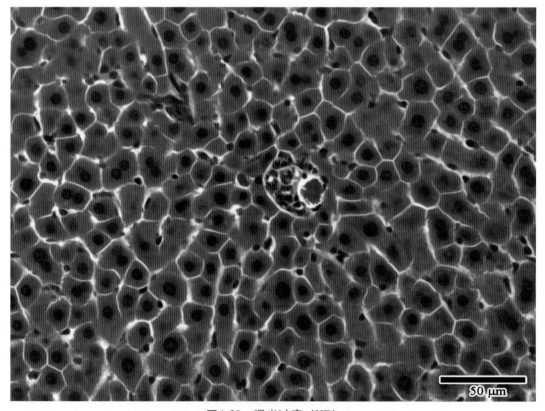

图1-30　曝光过度（HE）

Fig.1-30　overexposure（HE）

第二章
细胞形态结构与细胞分裂
Cell Morphology and Cell Division

Outline

The cell, as the basic unit of an organism, is composed of 3 basic parts: cell membrane, cytoplasm and nucleus. All the study on lives is derived from the investigation of cells. Cells are bounded by a cell membrane which is not resolved in thin section viewed with a light microscope, but in electron micrographs of high magnification, the membrane appears as two electron-dense layers separated by an electron-lucent intermediate zone. They consist of a bimolecular layer of mixed phospholipid with their hydrophilic portions at the outer and inner surface of the membrane and their hydrophobic chains projecting toward the middle of the bilayer. Cholesterol and varying amounts of proteins, glycoproteins and glycolipids are intercalated in the phospholipids bilayer. The above described "mosaic" is called the fluid mosaic model.

The cytoplasm is composed of several kinds of cell organelles that carry out different functions that are essential to cell metabolism. Rough endoplasmic reticulum has ribosomes attached to the outer surface of its membrane; and smooth endoplasmic reticulum, which lacks adherent ribosomes. The rough endoplasmic reticulum is most abundant in glandular cells that secrete proteins. In the liver, smooth endoplasmic reticulum plays an important role in the synthesis of the lipid component of very-low-density lipoproteins. It is also the principal site of detoxification and metabolism of lipid-soluble exogenous drugs. Striated muscle contains a specialized form of smooth endoplasmic reticulum which forms networks around all of the myofibrils of the myocytes. Its principal function is the sequestration of calcium ions that control muscle contraction. Proteins synthesized in the rough endoplasmic reticulum are transported to the Golgi complex for further processing, concentrating, and packaging in secretory granules for discharge from the cell. Mitochondria are present in all eukaryotic cells. These organelles transform, with high efficiency, the chemical energy of the metabolites present in cytoplasm into available energy.

Nucleus is the most important organelle of the cell, which is centrally (not really) situated and usually round or ellipsoidal. Nucleus contains nuclear envelope, nucleolus, chromatin and nuclear matrix. Chromatin is composed mainly of coiled strands of DNA bound to histones proteins; which composed of nucleosomes. Studies have identified the thin filament connecting the nucleosomes as a double-stranded DNA molecule and have shown that the core of the nucleosomes is an octamer of two tetramers. DNA, the genetic material of the nucleus, resides in the chromosomes.

一、细胞的基本概念

细胞（cell）是生命活动的基本单位。一切有机体都由细胞构成，细胞是有机体结构的基本单位；细胞具有独立有序的自控代谢体系，细胞是代谢与功能的基本单位；有机体的生长和发育以细胞的增殖与分化为基础，细胞是有机体生长发育的基本单位；细胞具有遗传的全能性，它是遗传的基本单位。

二、细胞的起源

地球上生命的进化历史包含了从原始的前细胞进化出原核细胞进而到真核细胞的复杂过程。但是，细胞的起源依然是现代生命科学中未解的最大谜团之一。原核细胞的成型标志着原始地球上生命的正式诞生，而真核细胞的诞生带来了地球生命的空前繁荣。通常认为，原始细胞的起源，是一个由多种原始生物大分子协同驱动的动力学系统有序的自组织过程，该系统的各主要阶段都受内部的动力学稳定和对外环境的适应等因素的选择。

三、细胞的组织结构概述

构成动物体的细胞种类繁多，大小、形态、结构和功能各异，但具有共同的特征：都由细胞膜、细胞质及细胞核构成。

（一）细胞膜（cell membrane）

细胞膜是包围在细胞质外面的薄层生物膜。用高倍电镜观察，细胞膜结构分3层：内外两层电子密度高，中间层电子密度低，通常将具有这样三层结构的膜称为单位膜。除细胞膜外，细胞内某些细胞器的膜也是这种膜。

细胞膜的分子结构即流体镶嵌模型，由液态的脂质双分子层中镶嵌着可移动的球形蛋白质构成。每个脂质分子有一个头部和两个尾部。头部具有亲水性，分别朝向膜的内、外表面；而尾部具有疏水性，伸入膜的中央。蛋白质分子镶嵌在脂质分子之间，称为嵌入蛋白。

（二）细胞质（cytoplasm）

执行细胞生理功能和化学反应，填充在细胞膜与细胞核之间，由基质、细胞器和内含物组成。

基质呈均匀、透明而无定形的胶状，内含蛋白质、糖类、脂类、水和无机盐等。各种细胞器、内含物和细胞核悬浮于基质中。细胞器是具有一定形态结构和执行一定功能的结构，包括线粒体、核糖体、内质网、高尔基复合体、溶酶体、过氧化物酶体、中心体、微丝、微管和中间丝等。

1. **线粒体**（mitochondria） 是双层膜形成的囊状细胞器，结构包括外膜、内膜、膜间隙和内室，位于除成熟红细胞以外的所有细胞。主要功能是进行氧化磷酸化，为细胞生命活动提供能量。

2. **核糖体**（ribosome） 几乎所有细胞内都有，由核糖体RNA与蛋白质构成，是合成蛋白质的场所。

3. **内质网**（endoplasmic reticulum） 根据其表面是否附着有核糖体，可分为粗面内质网（rough endoplasmic reticulum）和滑面内质网（smooth endoplasmic reticulum）。前者的主要功能是合成和运输蛋白质，后者是脂质合成的重要场所。

4. **高尔基复合体**（Golgi complex） 位于细胞核附近，主要功能与细胞的分泌、溶酶体的形成及糖类的合成有关。

5.溶酶体（lysosome） 由单层膜包裹的内含多种酸性水解酶的囊泡状细胞器。主要功能是进行细胞内消化作用，消化分解进入细胞的异物和微生物或细胞自身失去功能的细胞器，有细胞内消化器之称。

6.过氧化物酶体或微体（peroxisome or microbody） 由单层膜围绕的内含氧化酶类的细胞器。与细胞内物质的氧化以及过氧化氢的形成有关。

7.中心体（centrosome） 位于细胞的中央或细胞核附近，其功能与细胞分裂有关，此外还参与纤毛和鞭毛的形成。

8.微管、微丝和中间丝（microtube，microfilament and intermediate filament） 参与组成细胞骨架结构。

（三）细胞核（nucleus）

细胞核是细胞的重要组成部分，遗传信息的贮存场所，控制细胞的遗传和代谢活动。在动物体内除成熟的红细胞没有核外，所有细胞都有细胞核。多数细胞有1个核，但也有2个和多个核的（如肝细胞和骨骼肌细胞）。细胞核主要由核膜、核质、核仁和染色质组成。核膜是细胞核与细胞质之间的界膜，上有许多散在的核孔，是细胞核与细胞质之间进行物质交换的通道。核质是无结构、透明的胶状物质，又称核液，成分与细胞质的基质很相似，含多种酶和无机盐。核仁有1～2个，也有3～5个的，它是rRNA合成、加工和核糖体亚单位的装配场所。染色质是指细胞核内能被碱性染料着色的物质，当细胞进入有丝分裂期时，每条染色质丝均高度螺旋化，变粗变短，成为一条条的染色体。

四、细胞分裂

在细胞的生命活动中，细胞从上一次分裂结束到下一次分裂结束的历程称为一个细胞周期。细胞周期经过G_1、S、G_2各期后，DNA合成加倍；进入M期，在一段短暂的时间内，染色质凝缩成染色体，细胞分裂成两个子细胞。细胞分裂形式包括有丝分裂、减数分裂和无丝分裂。

有丝分裂（mitosis）在真核生物中最常见，细胞核的变化最大，需经过前、中、后、末四期，是一个连续变化的过程。

减数分裂（meiosis）比较特殊，只存在于生殖细胞的成熟过程中，其特点是分裂后子细胞的染色体数目比亲代细胞减少了一半。

无丝分裂（amitosis）是最简单的分裂方式，在原核细胞中常见，真核生物的间质组织、肌组织和乳腺组织中较常见。细胞体积增大，细胞核形成哑铃状，中部缩细断裂，胞质缢缩，最后形成两个子细胞。

五、细胞形态结构与细胞分裂图谱

1.细胞的形态与结构　图2-1～图2-8。

2.细胞膜　图2-9～图2-10。

3.细胞质　图2-11～图2-27。

4.细胞核　图2-28～图2-32。

5.细胞分裂　图2-33～图2-45。

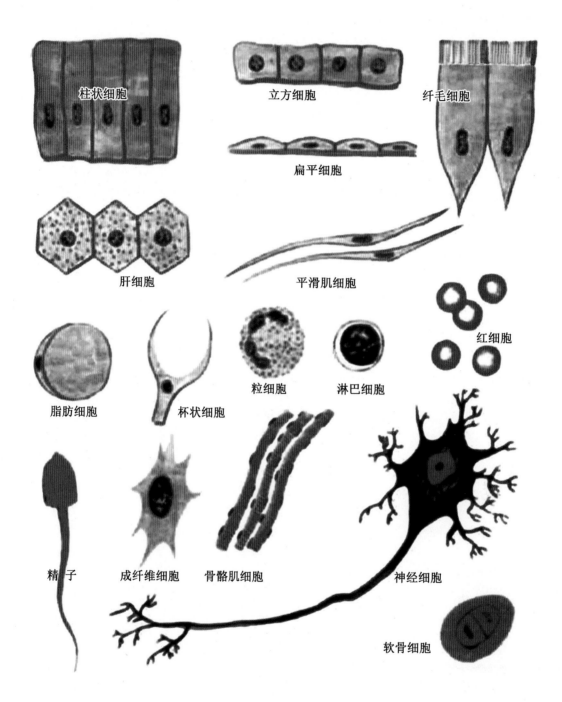

图2-1 细胞的形态

Fig.2-1 cell morphology

柱状细胞—columnar cell　立方细胞—cuboid cell　扁平细胞—flat cell　纤毛细胞—ciliated cell
肝细胞—liver cell　平滑肌细胞—smooth muscle cell　脂肪细胞—fat cell　杯状细胞—goblet cell
粒细胞—granulocyte　淋巴细胞—lymphocyte　红细胞—red blood cell　精子—sperm　成纤维细胞—fibroblast
骨骼肌细胞—skeletal muscle cell　神经细胞—nerve cell　软骨细胞—chondrocyte

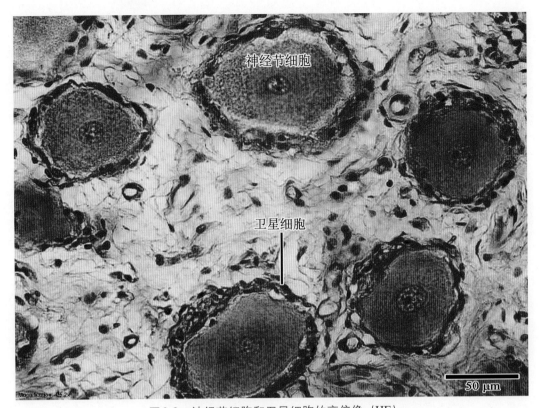

图2-2 神经节细胞和卫星细胞的高倍像（HE）

Fig.2-2 high magnification of ganglion cell and satellite cells（HE）

神经节细胞—ganglion cell 卫星细胞—satellite cell

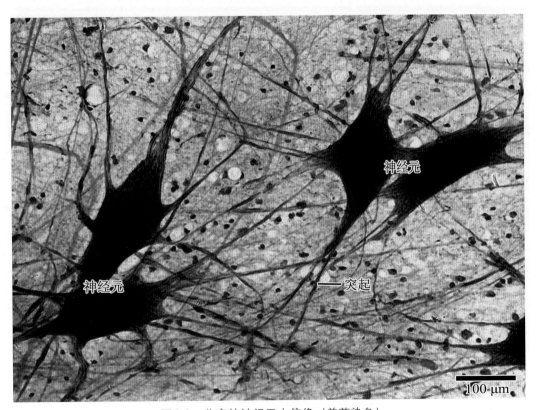

图2-3 分离的神经元中倍像（美蓝染色）

Fig.2-3 mid magnification of isolated neurons（methylene blue stain）

神经元—neuron 突起—neural process

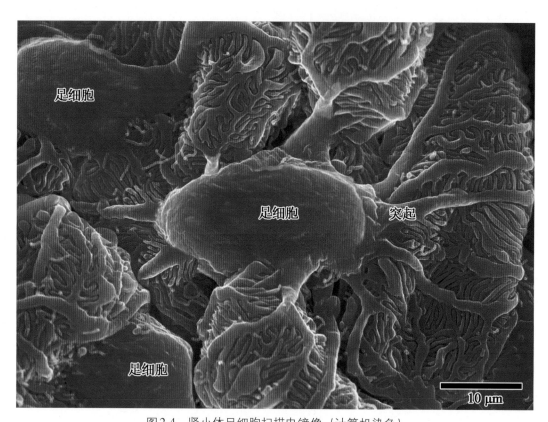

图2-4　肾小体足细胞扫描电镜像（计算机染色）

Fig.2-4　scanning electron microimage of renal corpuscular podocyte（computer stain）

足细胞—podocyte　突起—protuberance

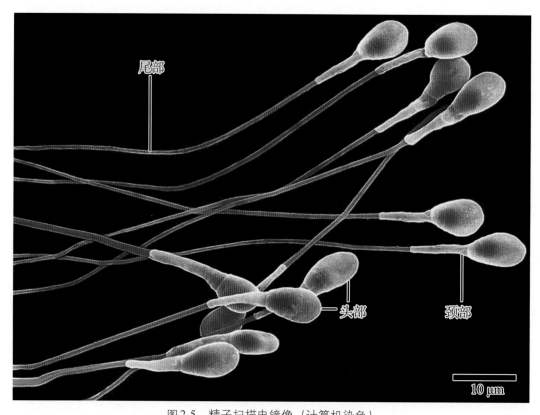

图2-5　精子扫描电镜像（计算机染色）

Fig.2-5　scanning electron microimage of sperm（computer stain）

尾部—tail　头部—head　颈部—neck

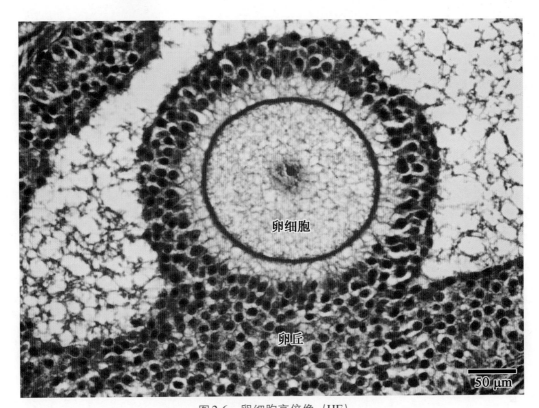

图2-6 卵细胞高倍像（HE）

Fig.2-6 high magnification of oocyte（HE）

卵细胞—oocyte 卵丘—germ hillock

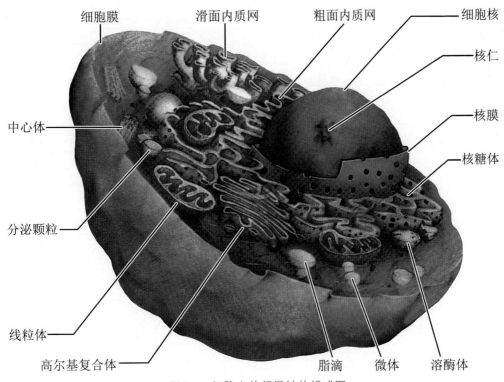

图2-7 细胞立体超微结构模式图

Fig. 2-7 stereo ultrastructural pattern of cell

细胞膜—membrane 中心体—centrosome 分泌颗粒—secretory particles 线粒体—mitochondria 高尔基复合体—Golgi complex
滑面内质网—smooth endoplasmic reticulum 粗面内质网—rough endoplasmic reticulum 细胞核—nucleus 核仁—nucleoli
核膜—nuclear membrane 核糖体—ribosome 脂滴—lipid droplet 微体—microbodies 溶酶体—lysosome

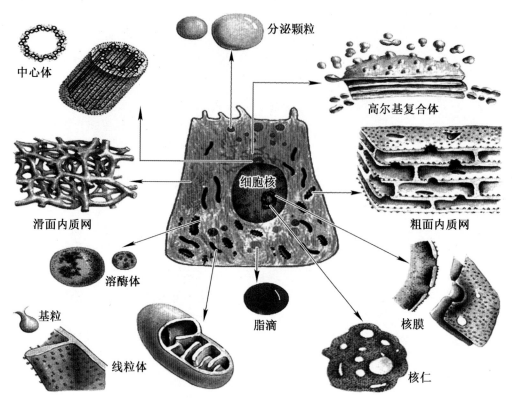

图2-8 细胞超微结构模式图

Fig. 2-8 cell ultrastructural pattern

中心体—centrosome 滑面内质网—smooth endoplasmic reticulum 溶酶体—lysosome 线粒体—mitochondria 基粒—granum
分泌颗粒—secretory particles 高尔基复合体—Golgi complex 粗面内质网—rough endoplasmic reticulum
细胞核—nucleus 核膜—nuclear membrane 核仁—nucleoli 脂滴—lipid droplet

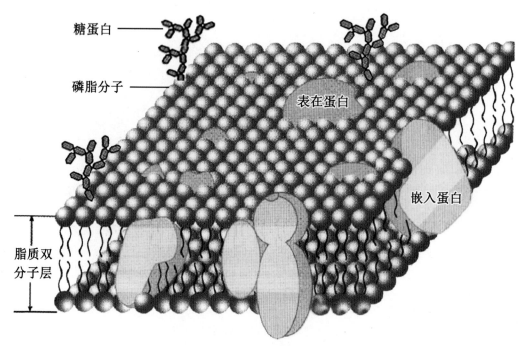

图2-9 细胞膜的超微结构模式图1

Fig. 2-9 ultrastructural pattern of cell membrane 1

糖蛋白—glycoprotein 磷脂分子—phospholipid molecule 脂质双分子层—lipid bilayer 表在蛋白—surface protein
嵌入蛋白—embedded protein

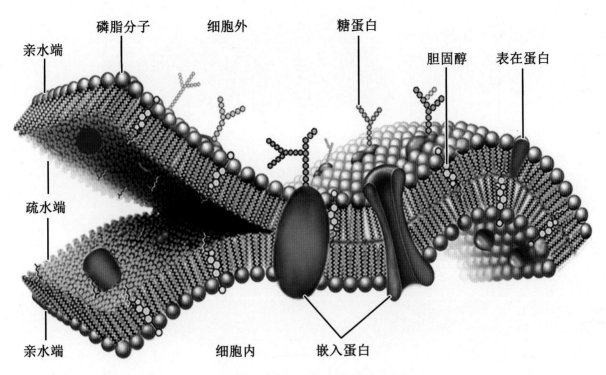

图2-10 细胞膜的超微结构模式图

Fig. 2-10 ultrastructural pattern of cell membrane 2

细胞外—extracellular 细胞内—intracellular 磷脂分子—phospholipid molecule 亲水端—hydrophilic end
疏水端—hydrophobic end 糖蛋白—glycoprotein 胆固醇—cholesterin 表在蛋白—surface protein
嵌入蛋白—embedded protein

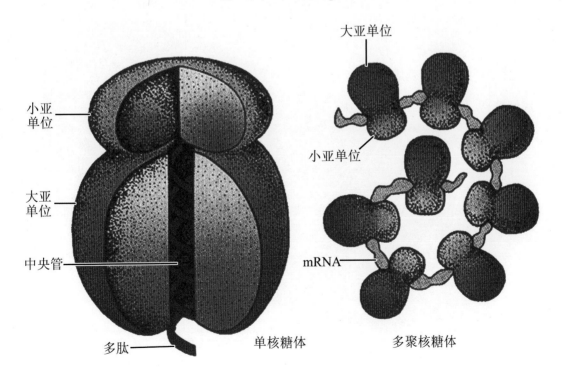

图2-11 核糖体超微结构模式图

Fig. 2-11 ultrastructural pattern of ribosome

单核糖体—single ribosome 多聚核糖体—polyribosome 小亚单位—small subunit 大亚单位—large subunit
中央管—central canal 多肽—polypeptide mRNA—信使RNA

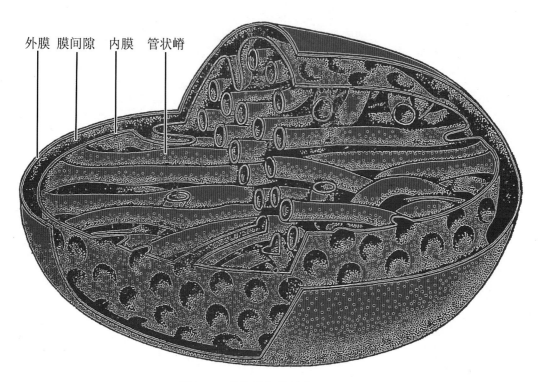

图 2-12　线粒体超微结构模式图 1

Fig.2-12　ultrastructural pattern of mitochondria 1

外膜—outer membrane　膜间隙—intermembrane space　内膜—inner membrane　管状嵴—tubular cristae

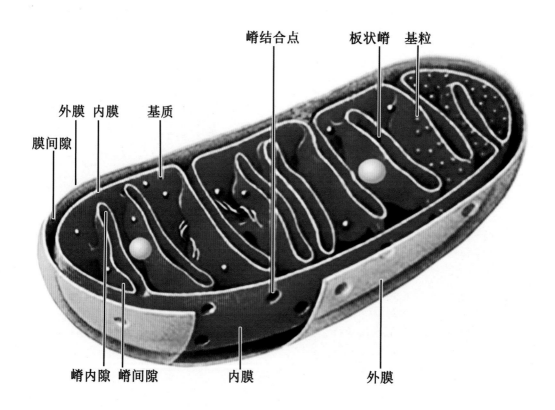

图 2-13　线粒体超微结构模式图 2

Fig.2-13　ultrastructural pattern of mitochondria 2

膜间隙—intermembrane space　外膜—outer membrane　内膜—inner membrane　基质—matrix　嵴结合点—cristae point
板状嵴—plate cristae　基粒—granum　嵴内隙—intracristal space　嵴间隙—intercristal space

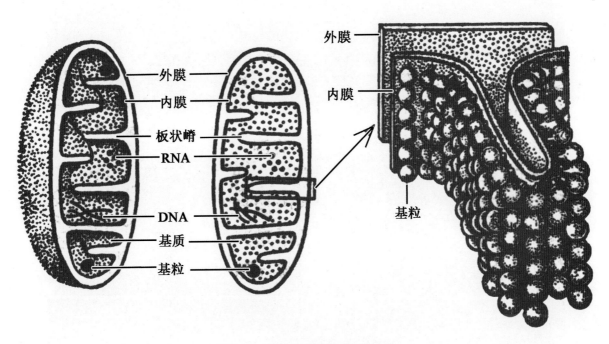

图2-14　线粒体超微结构模式图3

Fig.2-14　ultrastructural pattern of mitochondria 3

外膜—outer membrane　　内膜—inner membrane　　板状嵴—plate cristae
RNA—核糖核酸　　DNA—脱氧核糖核酸　　基质—matrix　　基粒—granum

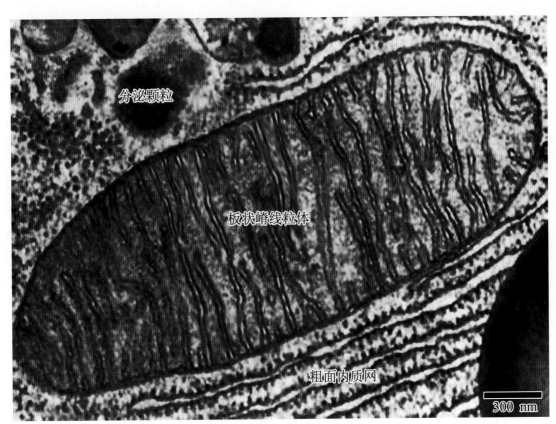

图2-15　线粒体透射电镜像1

Fig.2-15　transmission electron microimage of mitochondria 1

分泌颗粒—secretory granule　　板状嵴线粒体—mitochondria with plate cristae　　粗面内质网—rough endoplasmic reticulum

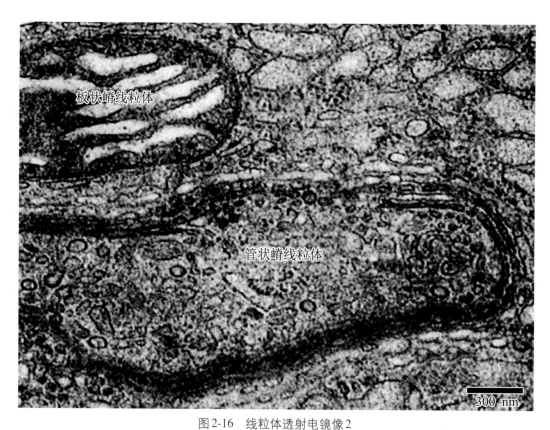

图2-16 线粒体透射电镜像2

Fig.2-16 transmission electron microimage of mitochondria 2

板状嵴线粒体—mitochondria with plate cristae 管状嵴线粒体—mitochondria with tubular cristae

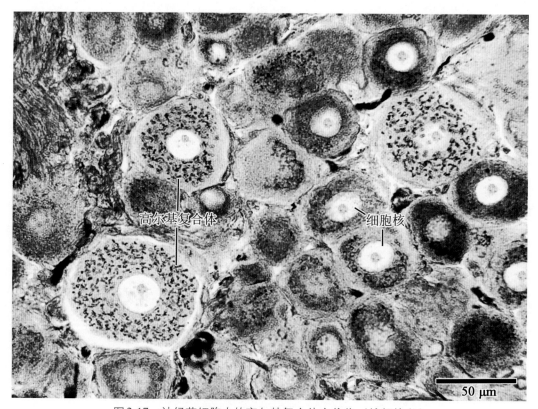

图2-17 神经节细胞内的高尔基复合体高倍像（镀银染色）

Fig.2-17 high magnification of Golgi complex in ganglion cells（silver stain）

高尔基复合体—Golgi complex 细胞核—nucleus

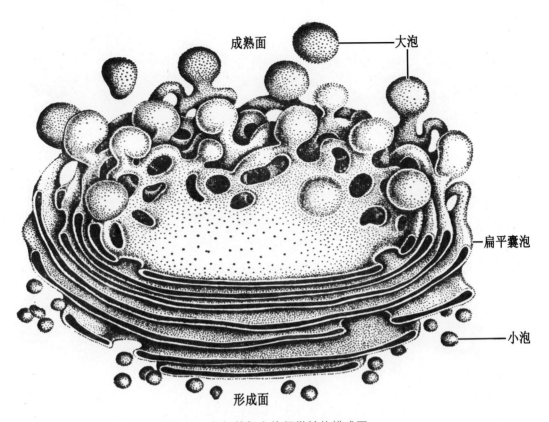

图2-18 高尔基复合体超微结构模式图

Fig.2-18 ultrastructural pattern of Golgi complex

成熟面—mature surface　形成面—forming surface　大泡—big vesicle　扁平囊泡—flat vesicle　小泡—small vesicle

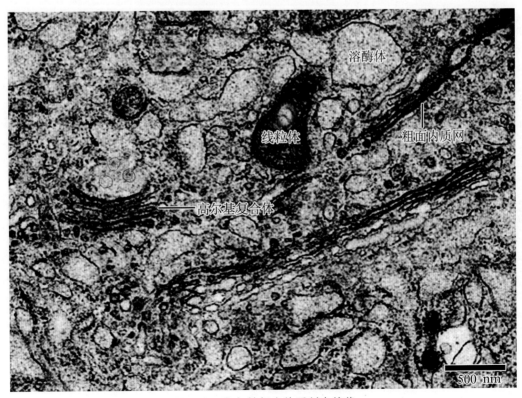

图2-19 高尔基复合体透射电镜像

Fig.2-19 transmission electron microimage of Golgi complex

高尔基复合体—Golgi complex　线粒体—mitochondria　溶酶体—lysosome　粗面内质网—rough endoplasmic reticulum

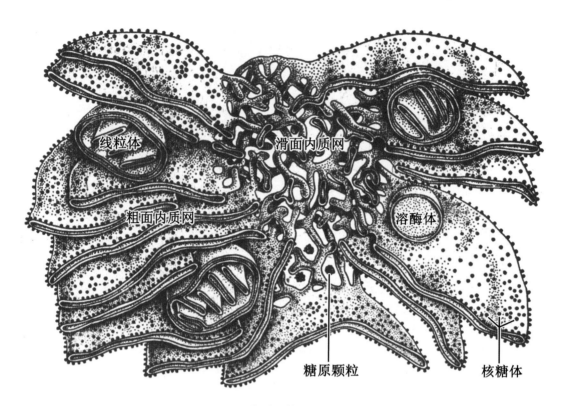

图2-20　内质网超微结构模式图

Fig.2-20　ultrastructural pattern of endoplasmic reticulum

线粒体—mitochondria　滑面内质网—smooth endoplasmic reticulum　粗面内质网—rough endoplasmic reticulum

溶酶体—lysosome　糖原颗粒—glycogen particle　核糖体— ribosome

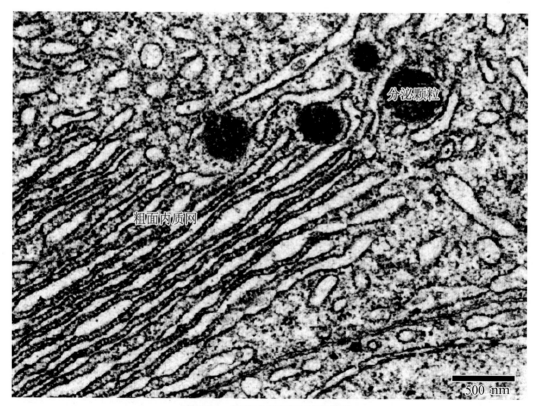

图2-21　粗面内质网透射电镜像

Fig.2-21　transmission electron microimage of rough endoplasmic reticulum

粗面内质网—rough endoplasmic reticulum　分泌颗粒—secretory granule

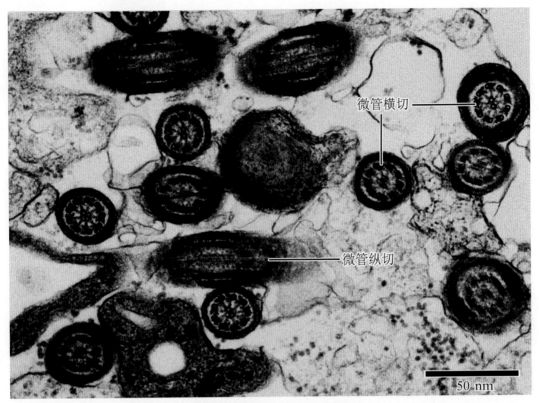

图2-22 微管的透射电镜像1

Fig.2-22 transmission electron microimage of microtubule 1

微管横切—microtubules transection 微管纵切—microtubule longitudinal section

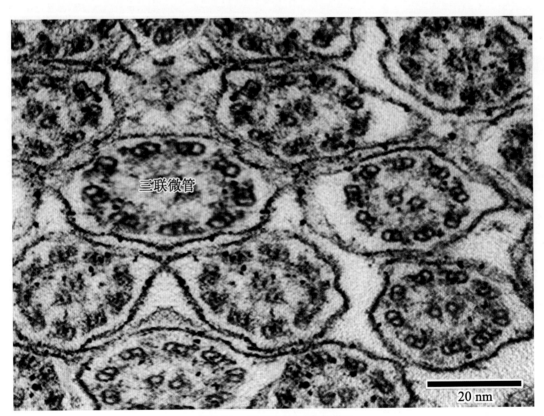

图2-23 微管的透射电镜像2

Fig.2-23 transmission electron microimage of microtubule 2

三联微管—triplomicrotubule

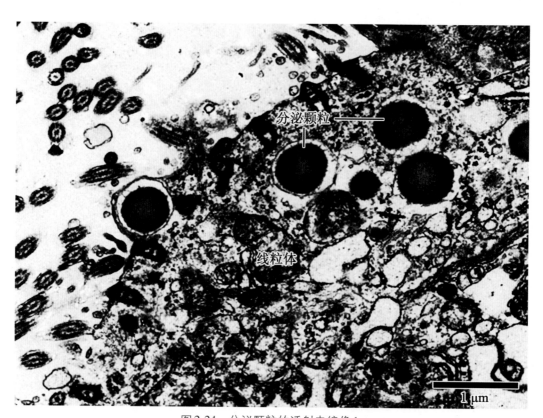

图 2-24　分泌颗粒的透射电镜像 1

Fig.2-24　transmission electron microimage of secretory granule 1

分泌颗粒—secretory granule　线粒体—mitochondria

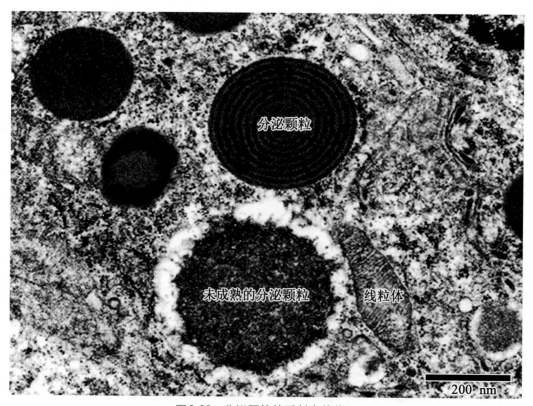

图 2-25　分泌颗粒的透射电镜像 2

Fig.2-25　transmission electron microimage of secretory granule 2

分泌颗粒—secretory granule　未成熟的分泌颗粒—immature secretory granule　线粒体—mitochondria

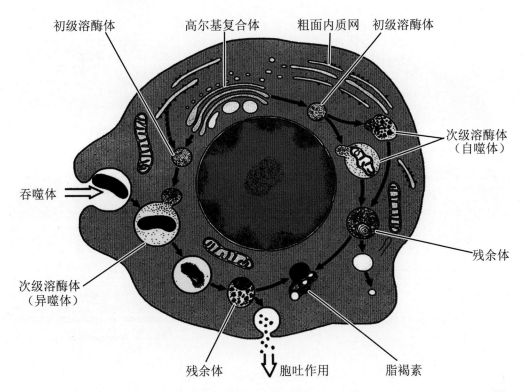

图 2-26　溶酶体的形成和转化示意图

Fig.2-26　formation and transformation diagram of lysosome

高尔基复合体—Golgi complex　粗面内质网—rough endoplasmic reticulum　初级溶酶体—primary lysosome
吞噬体—phagosome　次级溶酶体（异噬体）—secondary lysosome（heterophagosome）
次级溶酶体（自噬体）—secondary lysosome（autophagosome）
残余体—residual body　脂褐素—lipofuscin　胞吐作用—exocytosis

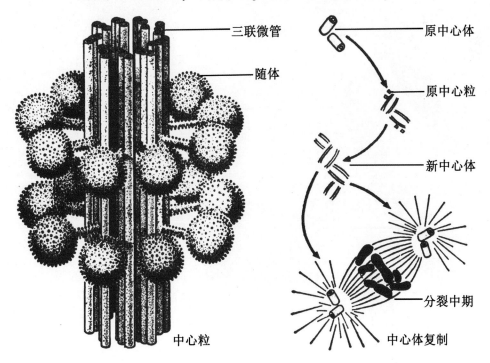

图 2-27　中心体的超微结构和复制示意图

Fig.2-27　ultrastructure and replication diagram of centrosome

中心粒—centriole　中心体复制—centrosome replication　三联微管—triplex microtubules　随体—trabant
原中心体—protocentrosome　原中心粒—procentrioles　新中心体—neocentrosome　分裂中期—metaphase

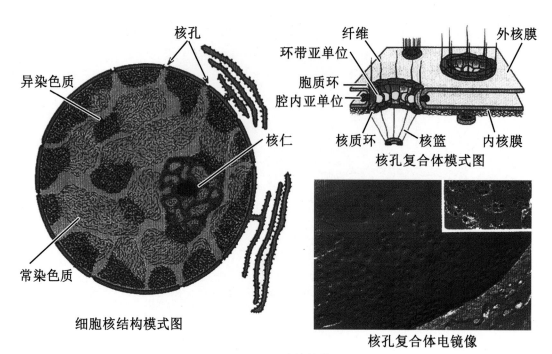

图 2-28　细胞核结构

Fig.2-28　structure of nucleus

细胞核结构模式图—ideograph of nuclear structure　核孔复合体模式图—ideograph of nuclear pore complex
核孔复合体电镜像—electron microimage of nuclear pore complex　异染色质—heterochromatin　常染色质—euchromatin
核孔—nuclear pore　核仁—nucleolus　纤维—fiber　环带亚单位—ring subunit　胞质环—cytoplasmic ring
腔内亚单位—luminal subunit　核质环—nuclear cytoplasmic ring　核篮—nuclear basket
外核膜—outer nuclear membrane　内核膜—inner nuclear membrane

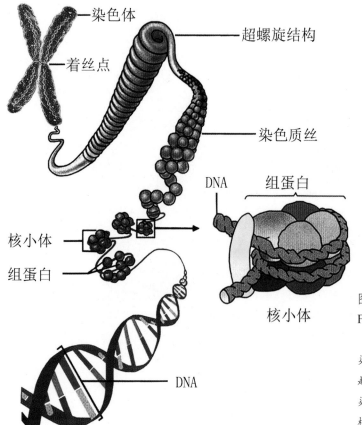

图 2-29　染色体、染色质和核小体的结构

Fig.2-29　structure of chromosomes, chromatin and nucleosome

染色体—chromosome　着丝点—centromere
超螺旋结构—superhelical structure
染色质丝—chromatic fibrils
核小体—nucleosome　组蛋白—histone
DNA—脱氧核糖核酸

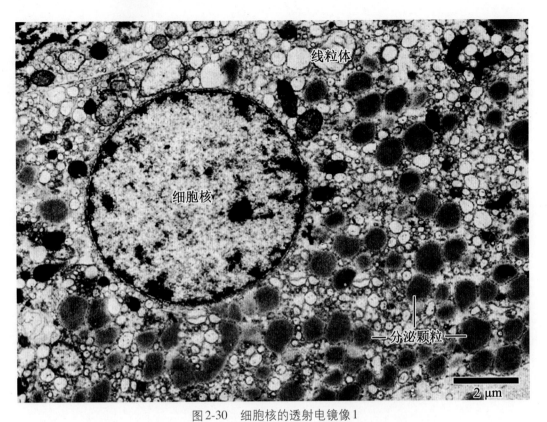

图 2-30 细胞核的透射电镜像 1

Fig.2-30 transmission electron microimage of nucleus 1

细胞核—nucleus　线粒体—mitochondria　分泌颗粒—secretory granule

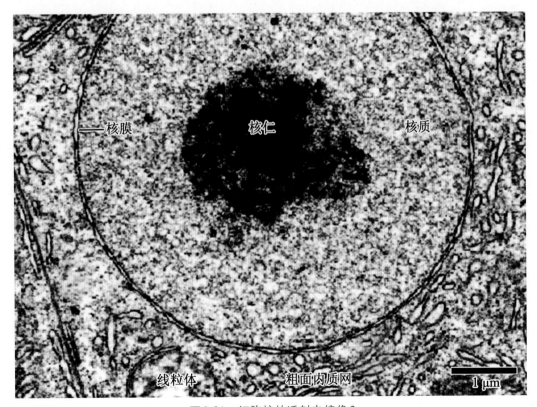

图 2-31 细胞核的透射电镜像 2

Fig.2-31 transmission electron microimage of the nucleus 2

核膜—nuclear membrane　核仁—nucleolus　核质—nucleoplasm　线粒体—mitochondria

粗面内质网—rough endoplasmic reticulum

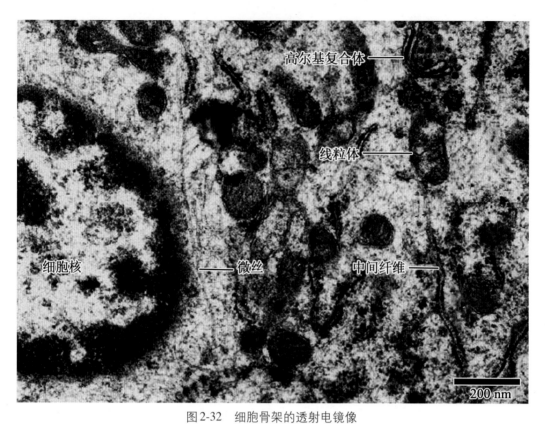

图2-32 细胞骨架的透射电镜像

Fig.2-32 transmission electron microimage of cytoskeleton

细胞核—nucleus 高尔基复合体—Golgi complex 线粒体—mitochondria

微丝—microfilament 中间纤维—intermediate filament

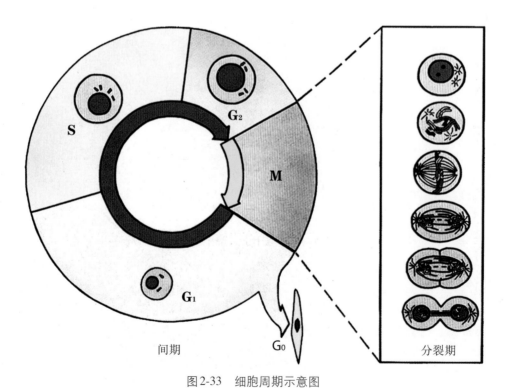

图2-33 细胞周期示意图

Fig.2-33 diagram of cell cycle

G_0—细胞休眠期 cell dormant period G_1—DNA合成前期 DNA presynthetic phase S—DNA合成期 DNA synthetic phase

G_2—DNA合成后期 DNA post-synthetic phase M—细胞分裂期 phase of cell division 分裂期—division stage

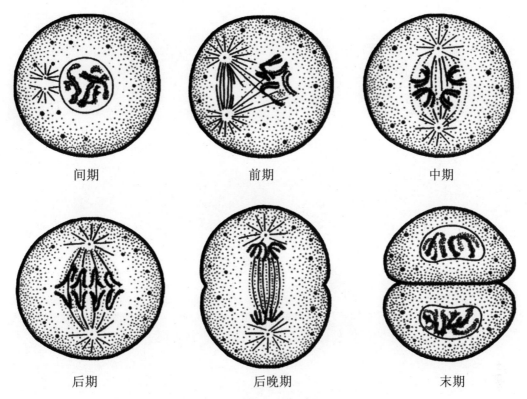

图2-34 细胞有丝分裂模式图

Fig.2-34 diagram of cell mitosis

间期—interphase 前期—prophase 中期—metaphase 后期—anaphase 后晚期—post-anaphase 末期—telophase

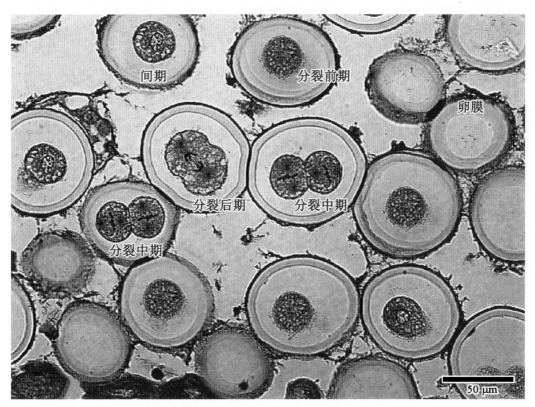

图2-35 细胞有丝分裂高倍像1（马蛔虫，铁苏木精+洋红染色）

Fig.2-35 high magnification of cell mitosis 1 （*Ascaris equi*, iron hematoxylin + carmine stain）

间期—interphase 分裂前期—prophase 分裂中期—metaphase 分裂后期—anaphase 卵膜—oolemma

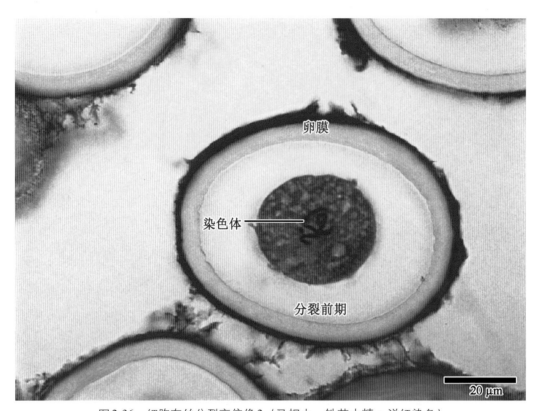

图 2-36　细胞有丝分裂高倍像 2（马蛔虫，铁苏木精 + 洋红染色）
Fig.2-36　high magnification of cell mitosis 2（*Ascaris equi*, iron hematoxylin + carmine stain）
卵膜—oolemma　染色体—chromosome　分裂前期—prophase

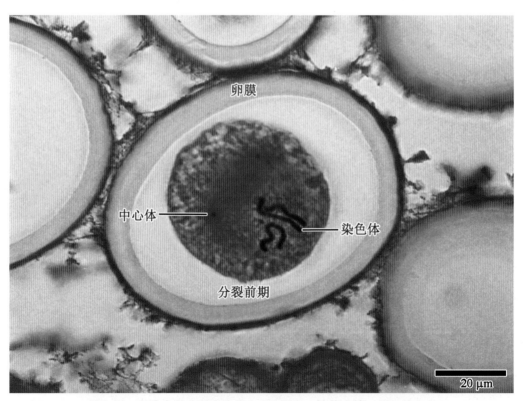

图 2-37　细胞有丝分裂高倍像 3（马蛔虫，铁苏木精 + 洋红染色）
Fig.2-37　high magnification of cell mitosis 3（*Ascaris equi*, iron hematoxylin + carmine stain）
卵膜—oolemma　中心体—centrosome　染色体—chromosome　分裂前期—prophase

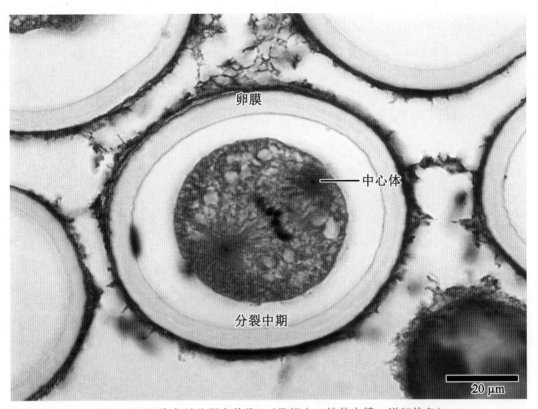

图2-38 细胞有丝分裂高倍像4（马蛔虫，铁苏木精＋洋红染色）

Fig.2-38 high magnification of cell mitosis 4（*Ascaris equi*, iron hematoxylin + carmine stain）

卵膜—oolemma 中心体—centrosome 分裂中期—metaphase

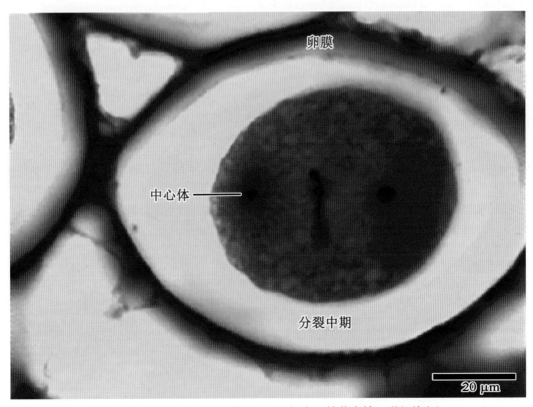

图2-39 细胞有丝分裂高倍像5（马蛔虫，铁苏木精＋洋红染色）

Fig.2-39 high magnification of cell mitosis 5（*Ascaris equi*, iron hematoxylin + carmine stain）

卵膜—oolemma 中心体—centrosome 分裂中期—metaphase

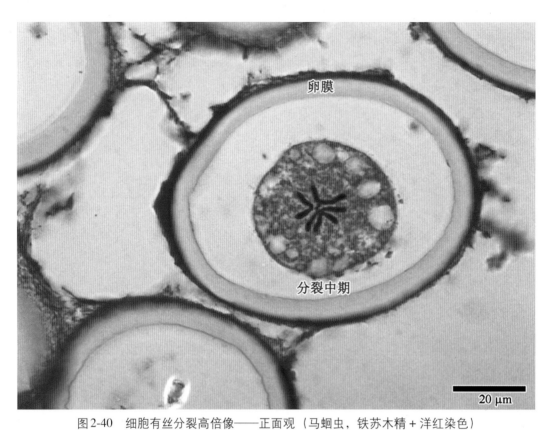

图2-40 细胞有丝分裂高倍像——正面观（马蛔虫，铁苏木精+洋红染色）
Fig.2-40 high magnification of cell mitosis—anterior view（*Ascaris equi*, iron hematoxylin + carmine stain）
卵膜—oolemma 分裂中期—metaphase

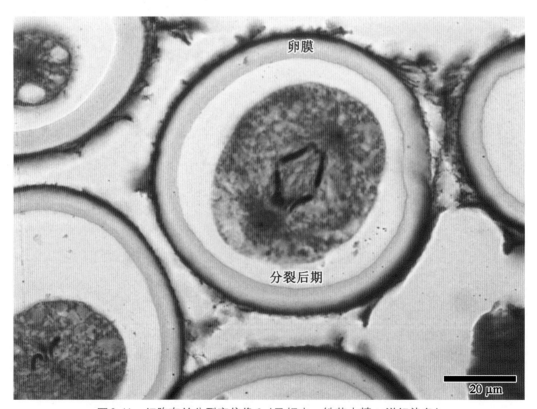

图2-41 细胞有丝分裂高倍像6（马蛔虫，铁苏木精+洋红染色）
Fig.2-41 high magnification of cell mitosis 6（*Ascaris equi*, iron hematoxylin + carmine stain）
卵膜—oolemma 分裂后期—anaphase

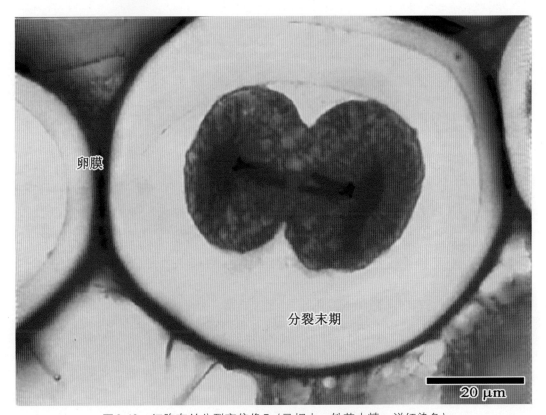

图2-42 细胞有丝分裂高倍像7（马蛔虫，铁苏木精+洋红染色）
Fig.2-42 high magnification of cell mitosis 7（*Ascaris equi*, iron hematoxylin + carmine stain）
卵膜—oolemma 分裂末期—telophase

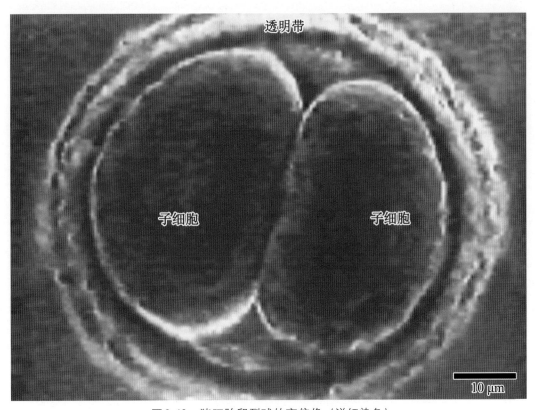

图2-43 猪胚胎卵裂球的高倍像（洋红染色）
Fig.2-43 high magnification of pig embryo cleavage（carmine stain）
透明带—zona pellucida 子细胞—daughter cells

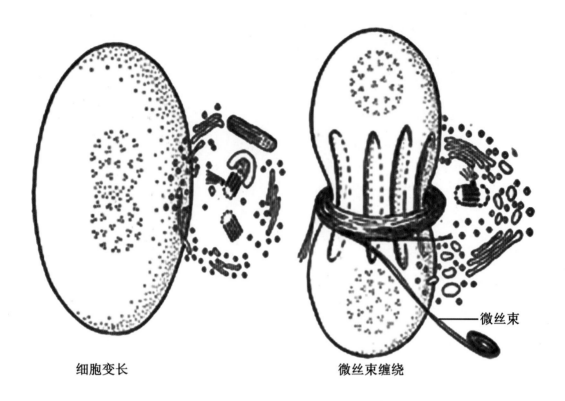

细胞变长　　　　　　　　　　微丝束缠绕

图2-44　细胞无丝分裂模式图1

Fig.2-44　diagram of cell amitosis 1

细胞变长—cell gets longer　微丝束缠绕—microfilament bundles convolved　微丝束—microfilament bundle

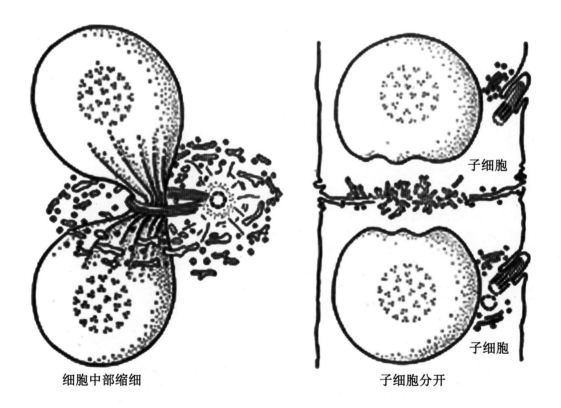

细胞中部缩细　　　　　　　　子细胞分开

图2-45　细胞无丝分裂模式图2

Fig.2-45　diagram of cell amitosis 2

细胞中部缩细—cell shrinks in middle　子细胞分开—daughter cells separation　子细胞—daughter cells

第三章 上皮组织
Epithelial Tissue

Outline

Epithelial tissues include closely aggregated polyhedral cells with very little extracellular substance. Polarity is an important feature of epithelia: they have a free surface, and a basal surface that rests on the basal membrane. Blood vessels do not normally penetrate an epithelium, so all nutrients have to pass out of the capillaries in the underlying lamina propria. According to their structure and function, epithelia are divided into two main groups, namely, covering epithelia and glandular epithelia. Covering epithelia are tissues in which the cells are organized in layers that cover the external surface or line the cavities of the body. They can be further classified into six groups in terms of the number of cell layers and the morphologic features of the cells in the surface layer. The six groups include simple squamous epithelium, simple cuboidal epithelium, simple columnar epithelium, pseudostratified columnar ciliated epithelium, stratified squamous epithelium and transitional epithelium. Epithelial tissues perform the principle functions of the covering and lining of the surface, absorption, and secretion. Glands are usually divided into exocrine and endocrine. A gland of the exocrine type releases its secretion to a duct system and thus to a body surface. An endocrine gland releases its secretion directly or indirectly into the blood or into the lymph. Epithelium contains specialized structures on the cell surface, lateral surface and basal surface in order to make a perfect adaptation to various roles.

上皮组织(epithelial tissue)由密集排列的细胞和少量细胞间质共同组成。根据上皮组织的形态结构和生理功能，可分为被覆上皮(covering epithelium)、腺上皮(glandular epithelium)、感觉上皮(sensory epithelium)、肌上皮(myoepithelium)和生殖上皮（germinal epithelium）等。被覆上皮被覆于动物体的外表面或衬于体内各种管、腔及囊的内表面。腺上皮是以分泌功能为主的上皮。感觉上皮是一种特殊分化的上皮，具有特殊感觉机能。肌上皮是某些器官中特化为具有收缩能力的上皮。生殖上皮是位于睾丸生精小管和卵巢表面的上皮。

一、上皮组织的发生

在胚胎发生和胚体卷曲的过程中，上皮组织由三个不同的胚层分化而来。胚胎早期的上皮均为结构简单的单层上皮，此后上皮层次和细胞形态随着胚体的发育进一步分化，与其功能特点相适应，或保持单层、或增生为复层上皮，并产生了不同的结构和功能特征。一种上皮可能含有来自不同胚层的细胞，细胞的结构和功能也有很大差异，甚至截然不同。如表皮除由外胚层来源的上皮细胞构成其主体外，还有中胚层起源的具有免疫功能的朗格汉斯细胞。被覆上皮细胞可进一步增生和分化，形成腺上皮，有些部位的上皮则分化出特殊的功能。

二、上皮组织的组织学结构概述

（一）被覆上皮的分类

（1）单层扁平上皮（simple squamous epithelium） 全部是扁平细胞的单层上皮。

（2）单层立方上皮（simple cuboidal epithelium） 全部是立方细胞的单层上皮。

（3）单层柱状上皮（simple columnar epithelium） 全部是柱状细胞的单层上皮。

（4）假复层纤毛柱状上皮（pseudostratified ciliated columnar epithelium） 形状不同、高度各异的细胞构成的单层上皮。

（5）复层扁平上皮（stratified squamous epithelium） 为扁平细胞的复层上皮。复层扁平上皮为未角化、角化不全或角化型。

（6）变移上皮（transitional epithelium） 以游离面大而圆的细胞为特点的复层上皮。该细胞随生理状态而发生形态变化，使泌尿道的不同部位在充盈时维持上皮的完整。

（7）复层柱状上皮（stratified columnar epithelium） 表层为柱状细胞的、由两层或多层细胞构成的上皮。

（二）上皮组织的特化结构

1. 上皮细胞游离面的特化结构

（1）微绒毛（microvillus） 细胞膜和细胞质共同突向腔面形成的细小指状突起。如小肠上皮细胞游离面的微绒毛排列整齐，密集而细长，称为纹状缘。肾小管上皮细胞游离面微绒毛排列不规则，称为刷状缘。微绒毛可扩大细胞的表面积，有利于吸收和分泌。

（2）纤毛（cilium） 上皮细胞游离面的细胞膜和细胞质伸向腔面的、能摆动的小突起。纤毛比微绒毛粗而且长，结构复杂，光镜下能见到。纤毛可摆动，许多纤毛同步摆动，可推送上皮表面的尘埃和细菌等异物。

2. 上皮细胞的侧面

（1）紧密连接（tight junction） 在柱状上皮细胞近端的侧面。电镜下观察，相邻细胞的细胞膜外层呈网格状融合，融合处无细胞间隙，未融合处有一定间隙，外观呈带状，环绕细胞的顶端。可阻止大分子物质通过，防止组织液的流失。

（2）中间连接（intermediate junction） 在紧密连接下方，呈连续的环腰带状。可加强细胞间连接和维持细胞形状。

（3）桥粒（desmosome） 在中间连接的下方，在上皮细胞间呈不连续的斑点状。是上皮细胞间较牢固的连接方式，多见于易受机械性摩擦的部位，如皮肤的表皮。

（4）缝隙连接（gap junction） 在桥粒深部。连接处呈圆的平板状，相邻的细胞膜呈间断的融合，在未融合处，细胞间有2nm左右的间隙。缝隙连接可供细胞相互交换某些小分子物质和离子，以传递化学信息，便于传递电冲动。常位于吸收上皮细胞或分泌上皮细胞之间，以及肌细胞、神经细胞及骨细胞之间。

（5）镶嵌连接（interdigitation） 上皮细胞侧面基部的细胞膜凹凸不平，与相邻细胞膜相互对插，密切镶嵌，可增强细胞间连接和扩大细胞间接触面积。

3. 基底面的特化结构

（1）基膜（basement membrane） 光镜观察，基膜是由上皮组织形成的基板和结缔组织形成的网板组成，网板可缺如。

（2）质膜内褶（plasma membrane infolding） 上皮细胞基底面的细胞膜向胞质内折入形成的内褶。内褶周围有许多纵向的线粒体，提供物质转运时所需能量。质膜内褶的主要作用是扩大细胞基底面的表面积，以利于水和电解质的物质交换。

（3）半桥粒（hemidesmosome） 位于上皮细胞基底面与基膜接触处，只在上皮细胞的细胞膜内侧形成一增厚的斑，其微细结构是桥粒结构的一半。其作用是加强上皮细胞与基膜的连接。

（三）腺

1. 外分泌腺（exocrine gland） 其分泌物经导管排放到上皮表面，分为单细胞腺的杯状细胞，以及多细胞腺。多细胞腺按导管的分支形式分类：导管无分支的称为单腺；有分支的称复腺。按分泌单位的形状分为管状、泡状或管泡状腺。其他分类方式还有：

（1）分泌物类型 浆液腺如腮腺、胰腺，黏液腺如腭腺，以及混合腺如舌下腺、颌下腺。混合腺含浆液、黏液性腺泡和浆半月。

（2）分泌方式 局部分泌，仅分泌物释放，如腮腺；顶浆分泌，分泌物与细胞顶端的细胞质一起释放，如乳腺；全浆分泌，整个细胞形成分泌物，如皮脂腺、睾丸和卵巢。

腺被结缔组织进一步分为叶或小叶，导管分布于叶间、叶内、小叶间和小叶内如纹状管、闰管。肌上皮细胞分布于腺实质的基膜，其长突起围绕着腺泡，通过收缩协助分泌物进入导管。

2. 内分泌腺（endocrine gland） 内分泌腺是无管腺，其分泌物释放于血液中。详见内分泌系统。

三、上皮组织图谱

1. **被覆上皮的分类** 图3-1～图3-28。
2. **上皮组织的特化结构** 图3-29～图3-34。
3. **腺** 图3-35～图3-39。

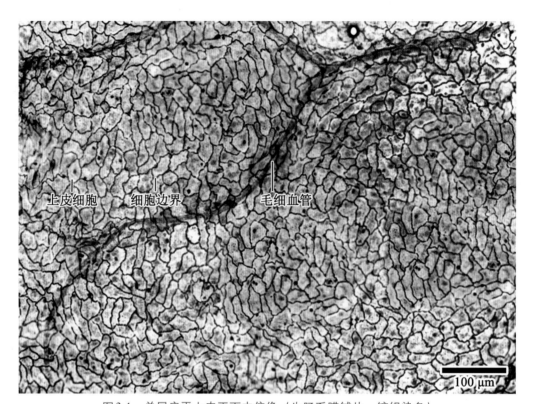

图3-1 单层扁平上皮正面中倍像（牛肠系膜铺片，镀银染色）

Fig.3-1 mid magnification of simple squamous epithelium, surface view (cow mesentery stretched preparation, silver stain)

上皮细胞—epithelium cell 细胞边界—cell border 毛细血管—capillary

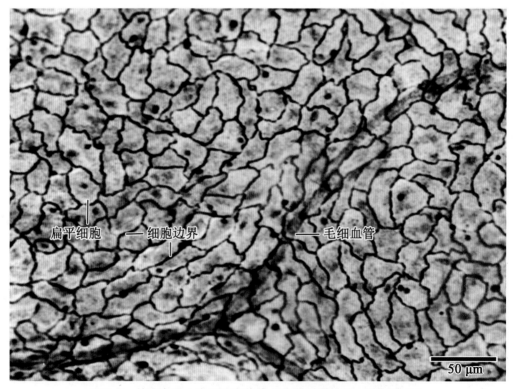

图3-2 单层扁平上皮正面高倍像（牛肠系膜铺片，镀银染色）

Fig.3-2 high magnification of simple squamous epithelium, surface view (cow mesentery stretched preparation, silver stain)

扁平细胞—flat cell 细胞边界—cell borders 毛细血管—capillary

第三章 上皮组织 Epithelial Tissue

图3-3 单层扁平上皮正面高倍像（马肠系膜铺片，镀银染色）

Fig.3-3 high magnification of simple squamous epithelium, surface view (horse mesentery stretched preparation, silver stain)

扁平细胞—flat cell 细胞边界—cell borders

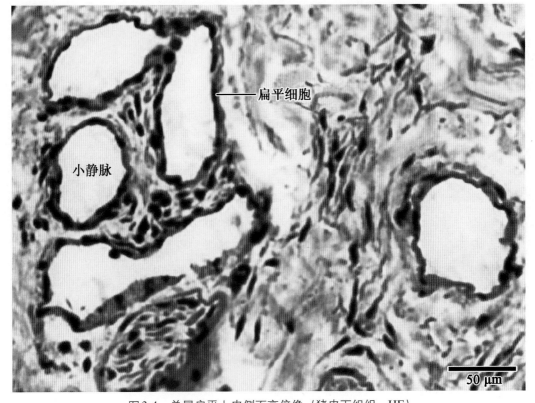

图3-4 单层扁平上皮侧面高倍像（猪皮下组织，HE）

Fig.3-4 high magnification of lateral simple squamous epithelium（pig hypodermis, HE）

扁平细胞—flat cell 小静脉—veinule

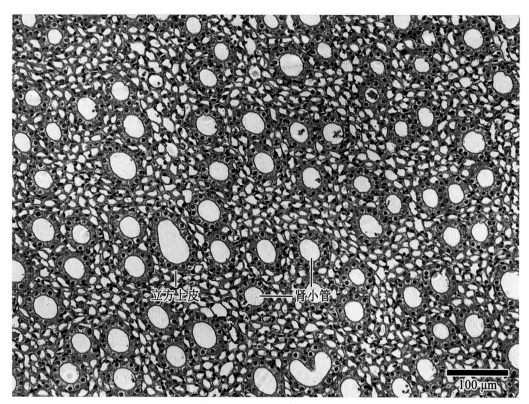

图3-5 单层立方上皮中倍像（兔肾髓质，HE）
Fig.3-5 mid magnification of simple cuboidal epithelium (rabbit renal medulla, HE)
立方上皮—cuboidal epithelium 肾小管—renal tubule

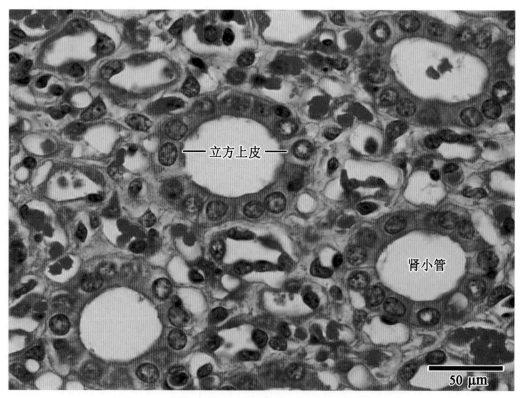

图3-6 单层立方上皮高倍像1（牛肾髓质，HE）
Fig.3-6 high magnification of simple cuboidal epithelium 1 (ox renal medulla, HE)
立方上皮—cuboidal epithelium 肾小管—renal tubule

第三章 上皮组织 Epithelial Tissue

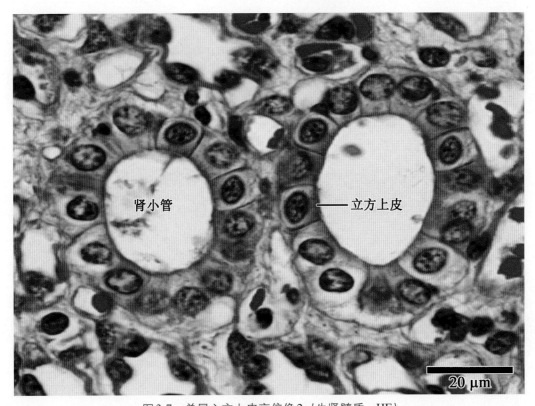

图3-7 单层立方上皮高倍像2（牛肾髓质，HE）

Fig.3-7 high magnification of simple cuboidal epithelium 2（ox renal medulla, HE）

立方上皮—cuboidal epithelium 肾小管—renal tubule

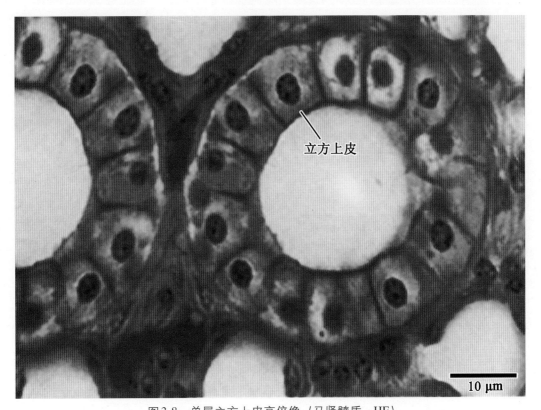

图3-8 单层立方上皮高倍像（马肾髓质，HE）

Fig.3-8 high magnification of simple cuboidal epithelium（horse renal medulla, HE）

立方上皮—cuboidal epithelium

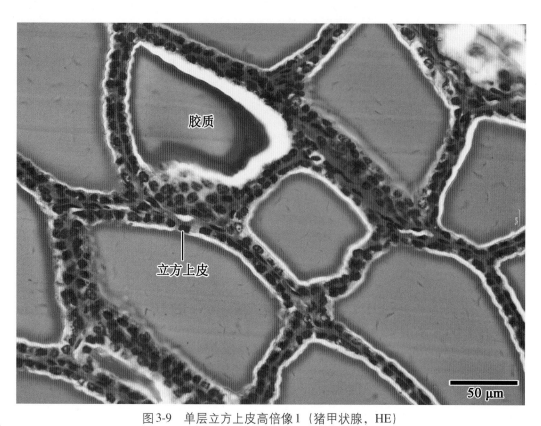

图3-9 单层立方上皮高倍像1（猪甲状腺，HE）
Fig.3-9 high magnification of simple cuboidal epithelium 1（pig thyroid gland, HE）
胶质—colloid　立方上皮—cuboidal epithelium

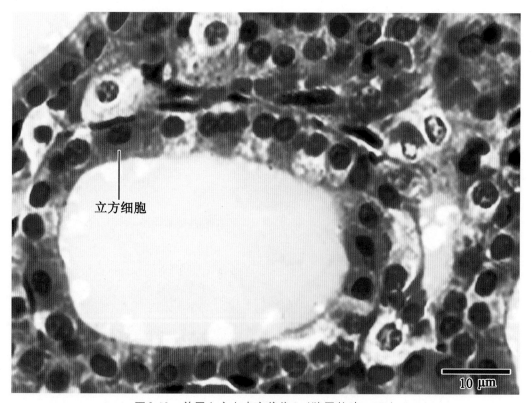

图3-10 单层立方上皮高倍像2（猪甲状腺，HE）
Fig.3-10 high magnification of simple cuboidal epithelium 2（pig thyroid gland, HE）
立方细胞—cuboid cell

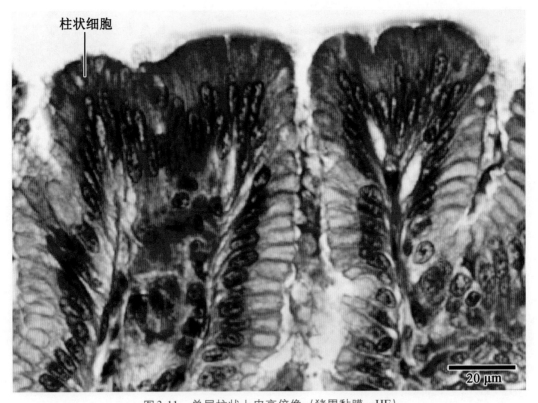

图3-11 单层柱状上皮高倍像（猪胃黏膜，HE）

Fig.3-11 high magnification of simple columnar epithelium（pig stomach mucosa, HE）

柱状细胞—columnar cell

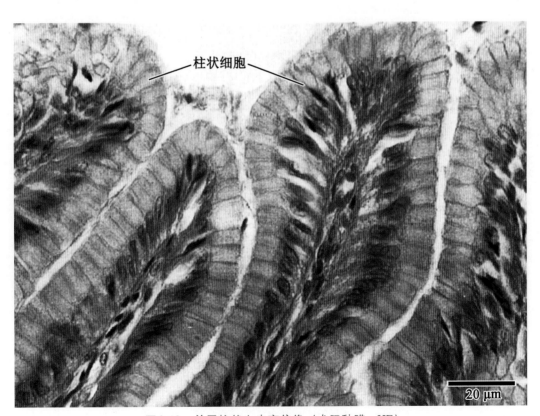

图3-12 单层柱状上皮高倍像（犬肠黏膜，HE）

Fig.3-12 high magnification of simple columnar epithelium（dog intestine mucosa, HE）

柱状细胞—columnar cell

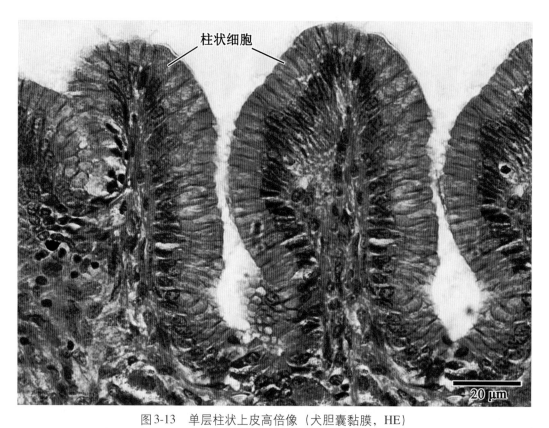

图3-13 单层柱状上皮高倍像（犬胆囊黏膜，HE）
Fig.3-13 high magnification of simple columnar epithelium (dog gallbladder mucosa, HE)

柱状细胞—columnar cell

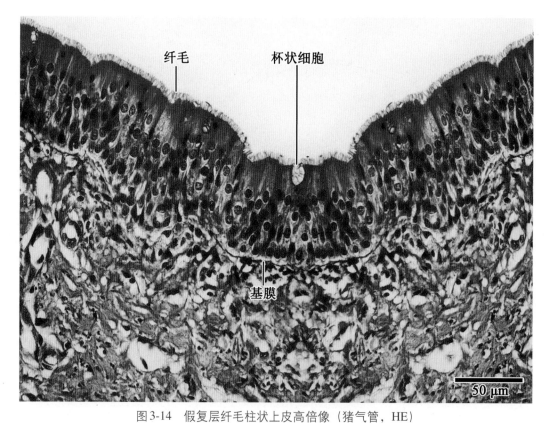

图3-14 假复层纤毛柱状上皮高倍像（猪气管，HE）
Fig.3-14 high magnification of pseudostratified ciliated columnar epithelium (pig trachea, HE)

纤毛—cilium　杯状细胞—goblet cell　基膜—basement membrane

第三章 上皮组织 Epithelial Tissue

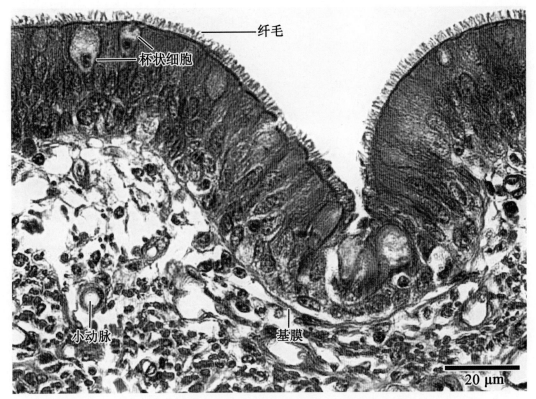

图 3-15　假复层纤毛柱状上皮高倍像 1（犬气管，HE）
Fig.3-15　high magnification of pseudostratified ciliated columnar epithelium 1　(dog trachea, HE)
纤毛—cilium　杯状细胞—goblet cell　小动脉—arteriole　基膜—basement membrane

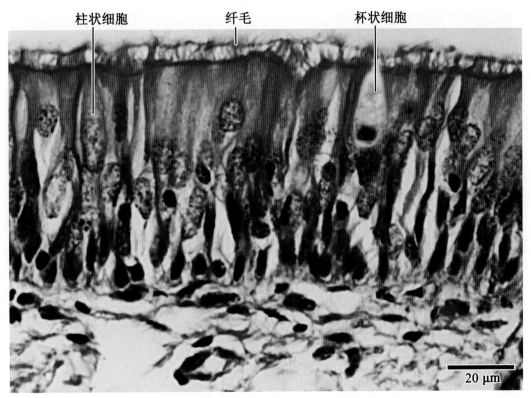

图 3-16　假复层纤毛柱状上皮高倍像 2（犬气管，HE）
Fig.3-16　high magnification of pseudostratified ciliated columnar epithelium 2 (dog trachea, HE)
柱状细胞—columnar cell　纤毛—cilium　杯状细胞—goblet cell

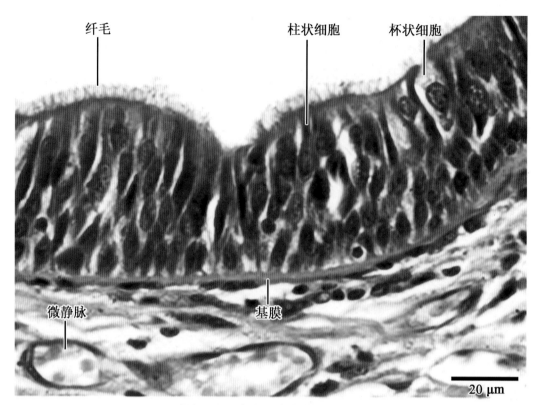

图3-17　假复层纤毛柱状上皮高倍像（牛气管，HE）

Fig.3-17　high magnification of pseudostratified ciliated columnar epithelium (cow trachea, HE)

纤毛—cilium　柱状细胞—columnar cell　杯状细胞—goblet cell　基膜—basement membrane　微静脉—venule

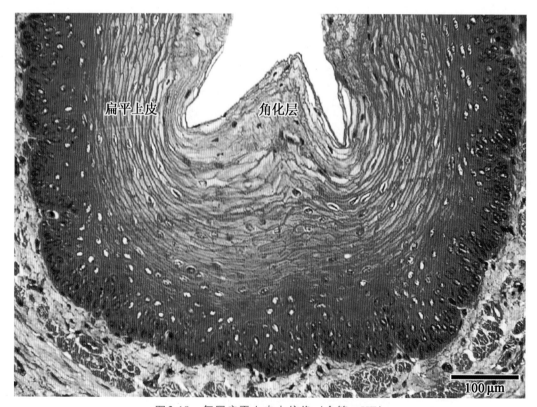

图3-18　复层扁平上皮中倍像（食管，HE）

Fig.3-18　high magnification of stratified squamous epithelium (esophagus, HE)

扁平上皮—squamous epithelium　角化层—cuticular layer

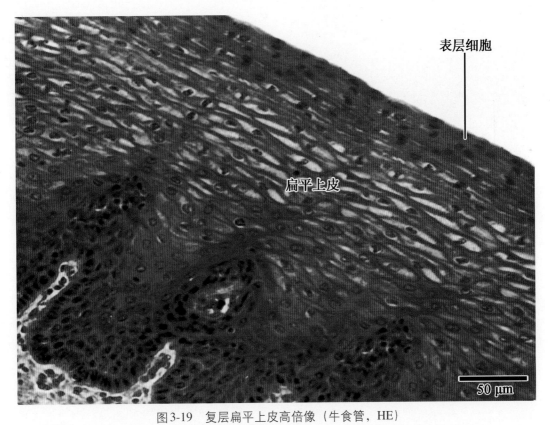

图3-19　复层扁平上皮高倍像（牛食管，HE）

Fig.3-19　high magnification of stratified squamous epithelium (ox esophagus, HE)

扁平上皮—squamous epithelium　表层细胞—surface cell

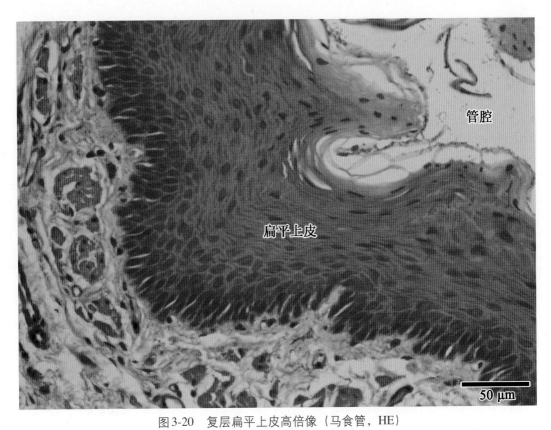

图3-20　复层扁平上皮高倍像（马食管，HE）

Fig.3-20　high magnification of stratified squamous epithelium (horse esophagus, HE)

扁平上皮—squamous epithelium　管腔—lumen

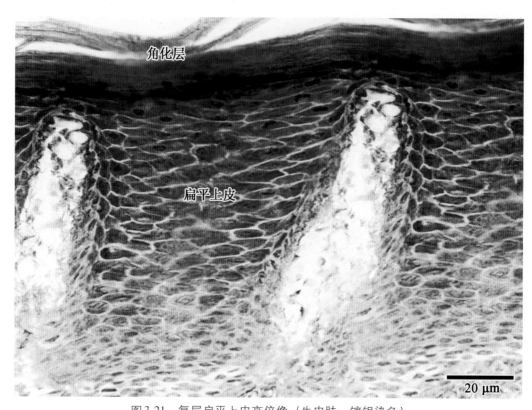

图 3-21 复层扁平上皮高倍像（牛皮肤，镀银染色）

Fig.3-21 high magnification of stratified squamous epithelium (cow skin, silver stain)

角化层—cuticular layer　扁平上皮—squamous epithelium

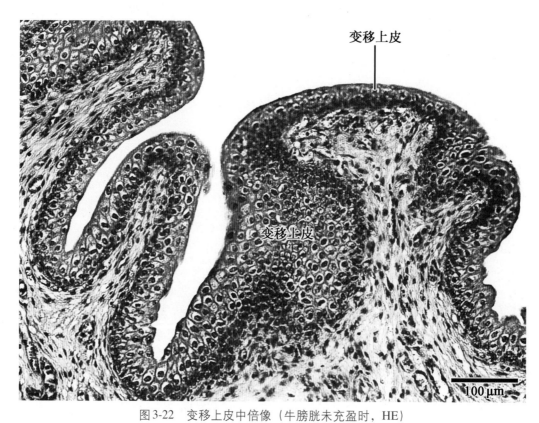

图 3-22 变移上皮中倍像（牛膀胱未充盈时，HE）

Fig.3-22 mid magnification of transitional epithelium (cow empty bladder, HE)

变移上皮—transitional epithelium

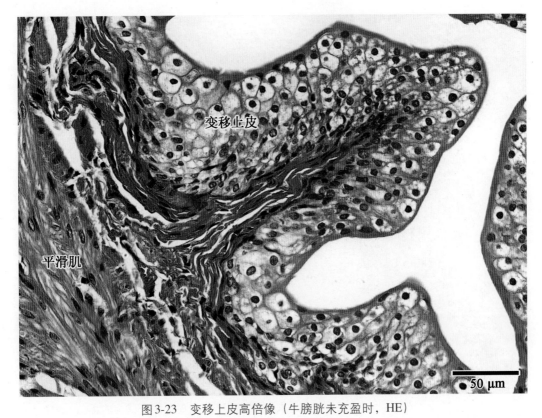

图3-23 变移上皮高倍像（牛膀胱未充盈时，HE）
Fig.3-23 high magnification of transitional epithelium (cow empty bladder, HE)
变移上皮—transitional epithelium 平滑肌—smooth muscle

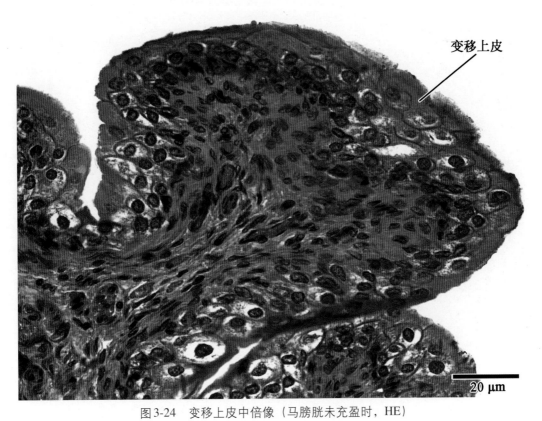

图3-24 变移上皮中倍像（马膀胱未充盈时，HE）
Fig.3-24 high magnification of transitional epithelium (horse empty bladder, HE)
变移上皮—transitional epithelium

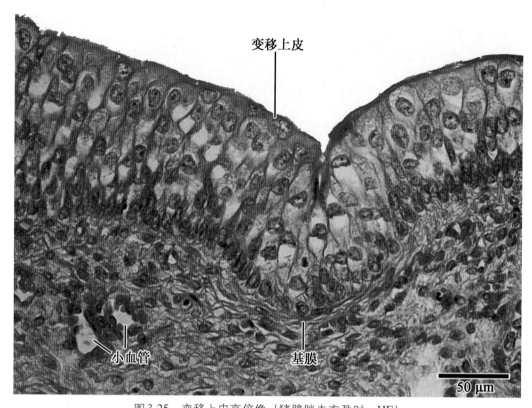

图3-25 变移上皮高倍像（猪膀胱未充盈时，HE）

Fig.3-25 high magnification of transitional epithelium (pig empty bladder, HE)

变移上皮—transitional epithelium 小血管—small blood vessel 基膜—basement membrane

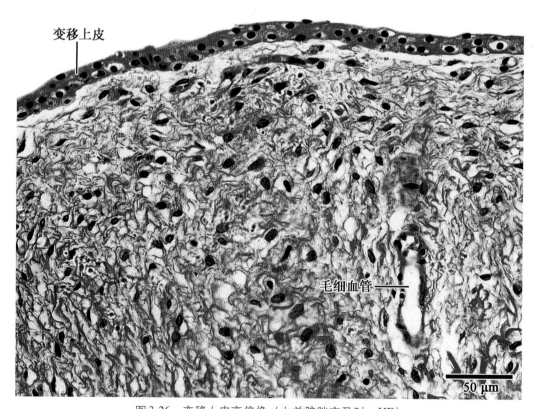

图3-26 变移上皮高倍像（山羊膀胱充盈时，HE）

Fig.3-26 high magnification of transitional epithelium (goat full bladder, HE)

变移上皮—transitional epithelium 毛细血管—capillary

第三章 上皮组织 Epithelial Tissue

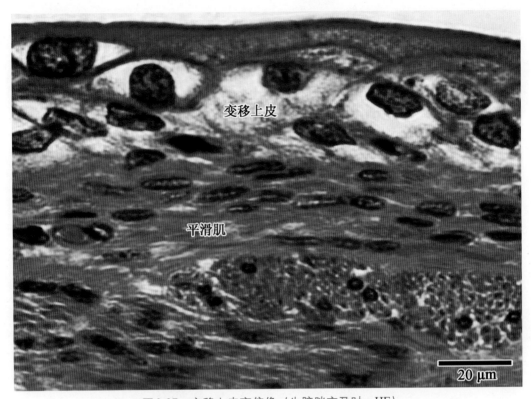

图3-27 变移上皮高倍像（牛膀胱充盈时，HE）
Fig.3-27 high magnification of transitional epithelium (cow full bladder, HE)
变移上皮—transitional epithelium 平滑肌—smooth muscle

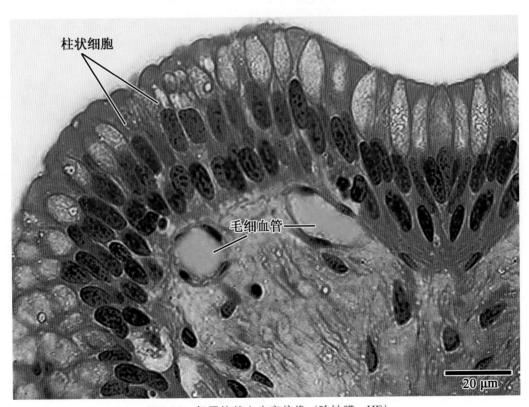

图3-28 复层柱状上皮高倍像（睑结膜，HE）
Fig.3-28 high magnification of stratified columnar epithelium (palpebral conjunctiva, HE)
柱状细胞—columnar cell 毛细血管—capillary

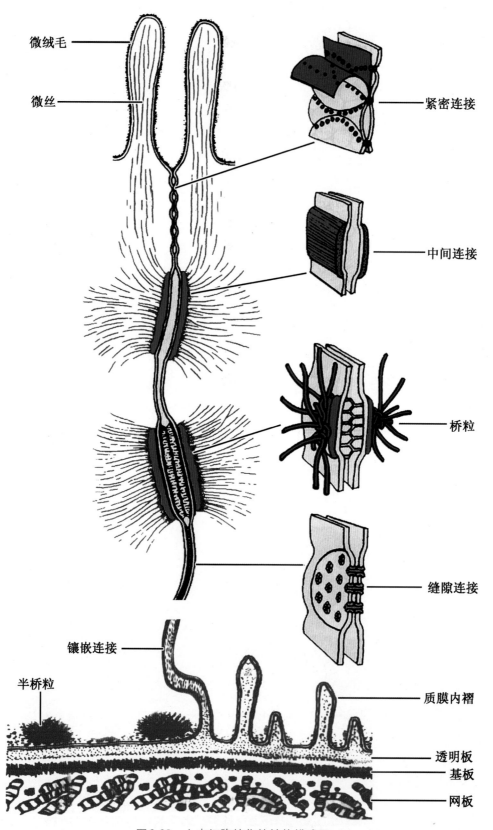

图3-29 上皮细胞特化的结构模式图

Fig.3-29 structural pattern of epithelial cell specialization

微绒毛—microvilli 微丝—microfilaments 紧密连接—tight junction 中间连接—intermediate junction 桥粒—desmosome 缝隙连接—gap junction 镶嵌连接—mosaic junction 半桥粒—hemidesmosome 质膜内褶—plasma membrane infolding 透明板—lamina lucida 基板—basal lamina 网板—reticular lamina

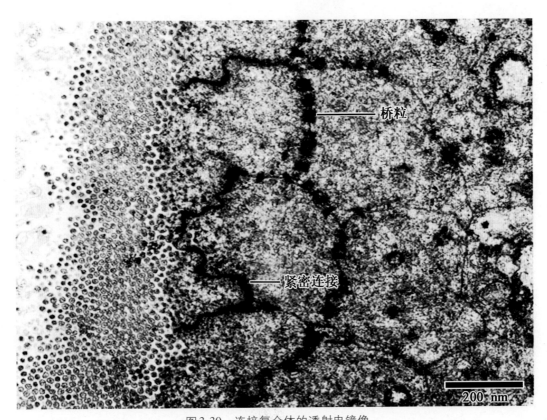

图3-30 连接复合体的透射电镜像

Fig.3-30 transmission electrical image of junctional complex

桥粒—desmosome 紧密连接—tight junction

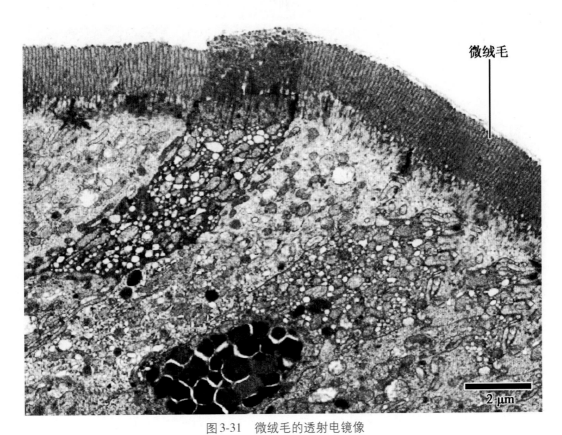

图3-31 微绒毛的透射电镜像

Fig.3-31 transmission electrical image of microvilli

微绒毛—microvilli

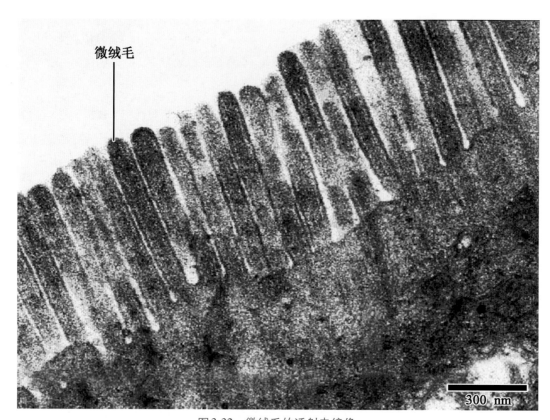

图 3-32 微绒毛的透射电镜像

Fig.3-32 transmission electrical image of microvilli

微绒毛—microvilli

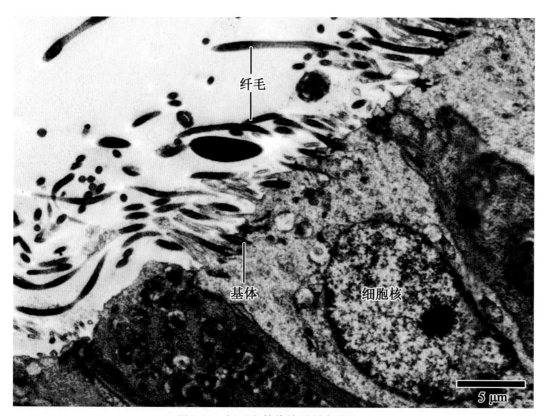

图 3-33 纤毛和基体的透射电镜像 1

Fig.3-33 transmission electrical image of cilia and basal body 1

纤毛—cilium 基体—basal body 细胞核—nucleus

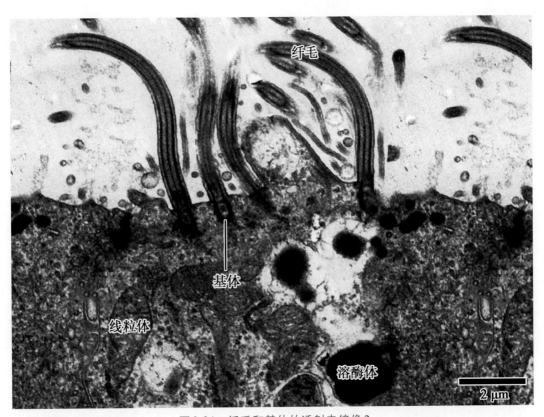

图3-34 纤毛和基体的透射电镜像2

Fig.3-34 transmission electrical image of cilia and basal body 2

纤毛—cilium 基体—basal body 线粒体—mitochondria 溶酶体—lysosome

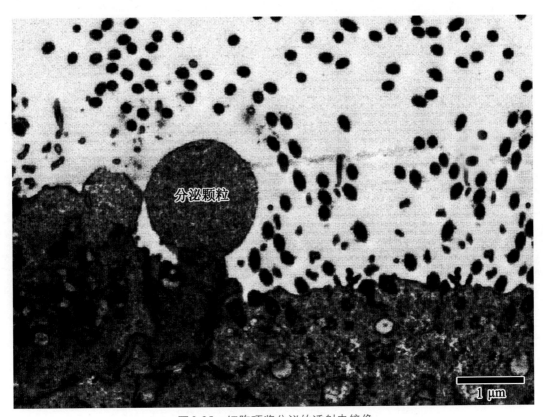

图3-35 细胞顶浆分泌的透射电镜像

Fig.3-35 transmission electrical image of apocrine secretion

分泌颗粒—secretory granule

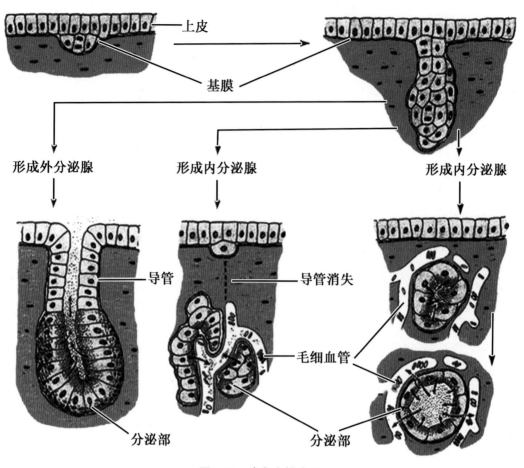

图 3-36　腺发生模式图

Fig.3-36　ideograph of glandular genesis

上皮—epithelium　基膜—basement membrane　形成外分泌腺—forming exocrine gland
形成内分泌腺—forming endocrine gland　导管—duct　导管消失—duct disappear　分泌部—secretory part
毛细血管—capillary

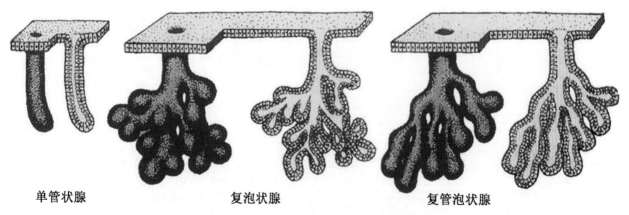

图 3-37　外分泌腺的形态和分类模式图

Fig.3-37　morphology and classification ideograph of exocrine glands

单管状腺—simple tubular gland　复泡状腺—multi-acinar gland　复管泡状腺—multi-tubuloacinar gland

第三章　上皮组织　Epithelial Tissue

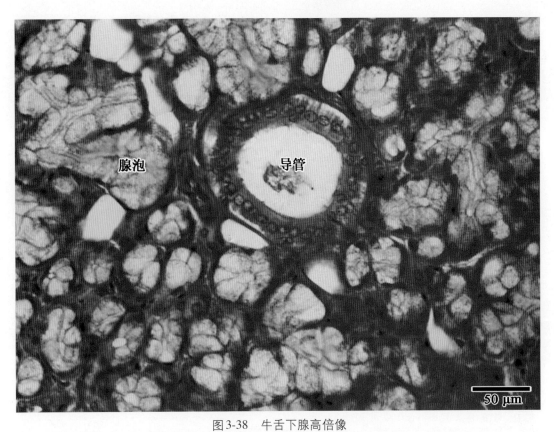

图3-38　牛舌下腺高倍像

Fig.3-38　high magnification of cow sublingual gland

腺泡—glandular acinar　导管—duct

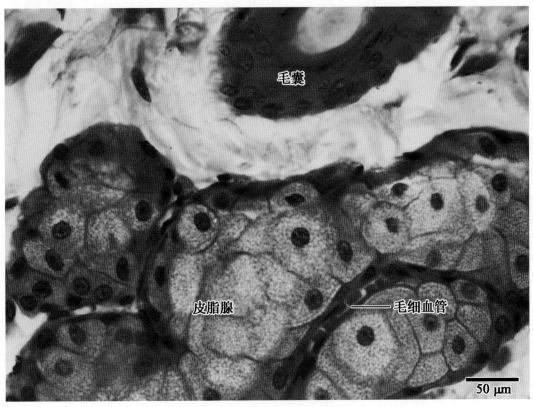

图3-39　绵羊皮脂腺高倍像

Fig.3-39　high magnification of sheep sebaceous gland

毛囊—hair follicle　皮脂腺—sebaceous gland　毛细血管—capillary

第四章
固有结缔组织
Connective Tissue Proper

Outline

The connective tissues include connective tissue proper, cartilage tissue, bone tissue and blood. Connective tissue that is usually referred to is actually connective tissue proper which can be classified into loose connective tissue, dense connective tissue, adipose tissue and reticular tissue. The connective tissue provides and maintains body form. It contains small number of cells and large amount of extracellular matrix. So, the major constituent of connective tissue is extracellular matrix that is composed of fibers, ground substance, and tissue fluid. All of the connective tissues originate from embryonal mesenchyme, and have functions of connecting, supporting, protecting, defending and transporting, etc.

Loose connective tissue, also called areolar tissue, is characterized by different cell types, loosely arranged thin fibers and abundance of ground substances. There are seven types of cells in the loose connective tissue: fibroblast, microphage, mast cell, plasma cell, adipocyte, undifferentiated mesenchymal cell and leukocyte. Meanwhile, there are three kinds of fibers in the loose connective tissue: collagenous fiber, elastic fiber and reticular fiber. Dense connective tissue has much more fibers and can be subdivided into regular and irregular types according to whether the fibers have an ordered or disordered arrangement. Adipose tissue is a specialized form of connective tissue consisting of adipocytes. There are two types of adipose tissue: white and brown one. Reticular Tissue consists of reticular cells and reticular fibers. It provides a special architectural framework for the hematopoietic and lymphoid organs.

结缔组织（connective tissue）在体内分布最广，由细胞和细胞间质构成，具有支持、连接、填充、营养、保护、储水、修复和防御等作用。其细胞的类型和数量随结缔组织的类型不同而异。细胞间质由细胞产生，包括纤维、基质和基质中的组织液。基质有液态、固态和凝胶态3种状态。根据细胞和细胞间质的不同，广义的结缔组织包括固有结缔组织、软骨组织、骨组织、血液和淋巴。本章展示固有结缔组织。

第四章 固有结缔组织 Connective Tissue Proper

一、结缔组织的发生

所有的结缔组织都是由胚胎时期的间充质演变而来。间充质(mesenchyme)是胚胎时期填充在外胚层和内胚层之间，散在的中胚层组织，由间充质细胞及液体状的基质组成，无纤维成分。间充质细胞(mesenchymal cell)呈星形，有许多胞质突起；胞质弱嗜碱性，核较大，卵圆形，核仁明显；相邻细胞的突起彼此连接成网。间充质细胞的分化程度低，不但能分化为多种结缔组织细胞，还能分化为内皮细胞和平滑肌细胞等。

二、结缔组织的组织学结构概述

（一）疏松结缔组织（loose connective tissue）

1. **细胞** 成纤维细胞最多，呈梭形，形态与巨噬细胞很相似。巨噬细胞椭圆形的细胞核比成纤维细胞的核小，染色深，胞质中有粗大的吞噬颗粒。肥大细胞位于血管附近，胞质中有许多细小颗粒，细胞核大且圆，位于细胞中央。偶见有脂肪细胞呈圆形，空泡周围环绕薄层细胞质，细胞核被脂滴挤到边缘，使细胞呈环形。在有些部位，如肠壁的固有层，常有浆细胞和白细胞。浆细胞呈圆形，细胞核为圆形，染色质排列呈车轮状。还有淋巴细胞、中性粒细胞和少量嗜酸性粒细胞。

2. **细胞间质** 粗大的胶原纤维束呈长带状；细长的弹性纤维有分支、染色深；网状纤维在HE染色的切片中看不到，镀银等特殊染色可显示。

（二）不规则致密结缔组织（dense irregular connective tissue）

1. **细胞** 主要有成纤维细胞、巨噬细胞等。

2. **细胞间质** 无序排列的波浪状、粗大胶原纤维束，并有少量弹性纤维和网状纤维。

（三）规则致密结缔组织（dense regular connective tissue）

1. **细胞** 少量平行排列的扁平成纤维细胞，夹在致密的纤维束之间。

2. **细胞间质** 胶原蛋白密集而有规律地排列成平行的纤维束。

（四）网状组织（reticular tissue）

1. **细胞** 网状细胞仅见于网状结缔组织，细胞呈星形，常遮盖由其产生的网状纤维；大细胞核呈椭圆形，染色浅，光镜不易看到细胞质。空隙中有淋巴细胞、巨噬细胞等。

2. **细胞间质** 主要为网状纤维，经镀银染色为黑褐色并有分支的微细纤维。

（五）脂肪组织（adipose tissue）

1. **细胞** 脂肪组织由脂肪细胞紧密地堆积而成，使细胞正常的球形发生改变。脂肪细胞群被疏松结缔组织鞘分隔成小叶，间隔中有肥大细胞、血管内皮细胞及神经与血管。

2. **细胞间质** 脂肪细胞被网状纤维包裹，网状纤维固定在结缔组织间隔的胶原纤维上。

三、结缔组织图谱

1. **结缔组织的发生** 图4-1～图4-3。
2. **疏松结缔组织** 图4-4～图4-21。
3. **规则致密结缔组织** 图4-22～图4-24。
4. **不规则致密结缔组织** 图4-25～图4-27。
5. **网状组织** 图4-28～图4-29。
6. **脂肪组织** 图4-30～图4-34。

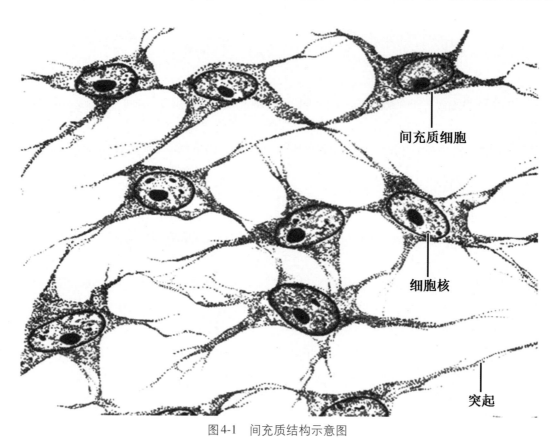

图4-1 间充质结构示意图

Fig.4-1 structural diagram of mesenchyme

间充质细胞—mesenchymal cell 细胞核—nucleus 突起—protuberance

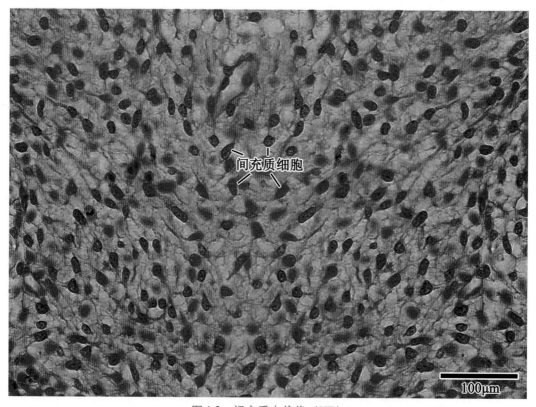

图4-2 间充质中倍像（HE）

Fig.4-2 mid magnification of mesenchyme（HE）

间充质细胞—mesenchymal cell

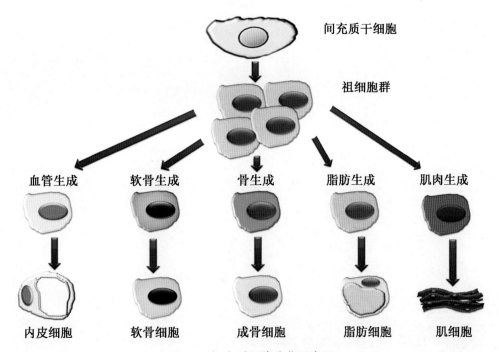

图4-3 间充质细胞分化示意图

Fig.4-3 differentiation diagram of mesenchymal cell

间充质干细胞—mesenchymal stem cell 祖细胞群—progenitor cell population 血管生成—angiogenesis 内皮细胞—endothelial cell 软骨生成—chondrogenesis 软骨细胞—chondrocyte 骨生成—osteogenesis 成骨细胞—osteoblast 脂肪生成—adipogenesis 脂肪细胞—adipocyte 肌肉生成—muscle generation 肌细胞—muscle cell

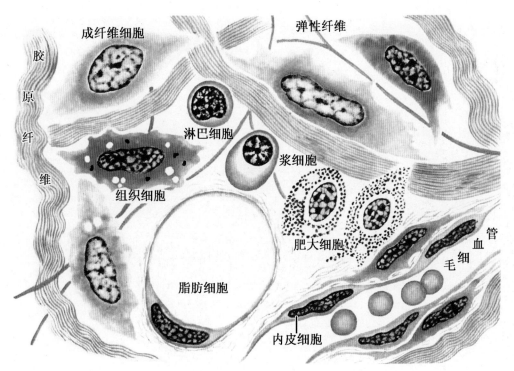

图4-4 疏松结缔组织结构示意图

Fig.4-4 structural diagram of loose connective tissue

胶原纤维—collagen fiber 弹性纤维—elastic fiber 成纤维细胞—fibroblast 淋巴细胞—lymphocyte 组织细胞—histocyte 浆细胞—plasma cell 肥大细胞—mast cell 脂肪细胞—fat cell 内皮细胞—endothelium 毛细血管—capillaries

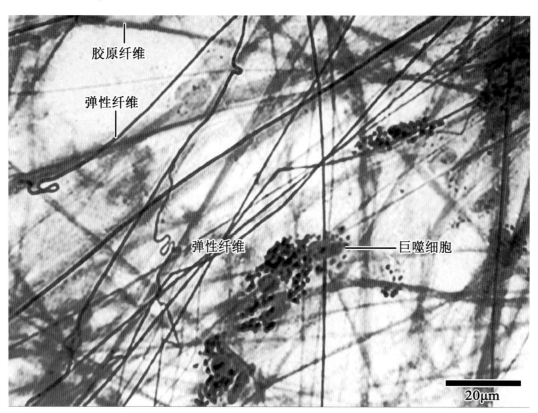

图4-5 疏松结缔组织铺片高倍像1（HE）

Fig.4-5 high magnification of loose connective tissue stretched preparation 1 (HE)

胶原纤维—collagen fiber 弹性纤维—elastic fiber 巨噬细胞—macrophage

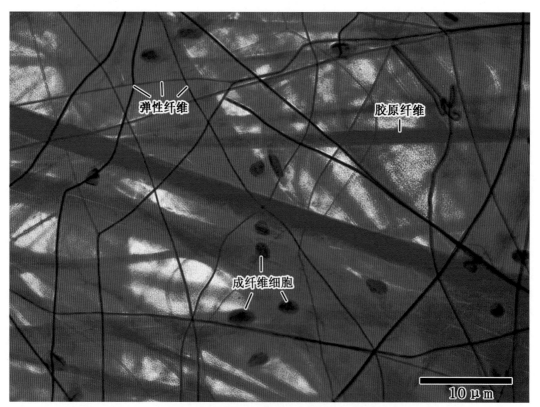

图4-6 疏松结缔组织铺片高倍像2（HE）

Fig.4-6 high magnification of loose connective tissue stretched preparation 2 (HE)

胶原纤维—collagen fiber 弹性纤维—elastic fiber 成纤维细胞—fibroblast

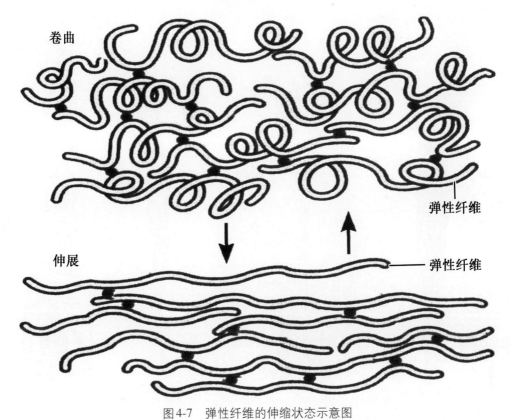

图4-7 弹性纤维的伸缩状态示意图

Fig.4-7 crispation and stretch state diagram of elastic fibers

卷曲—crispation 伸展—extension 弹性纤维—elastic fiber

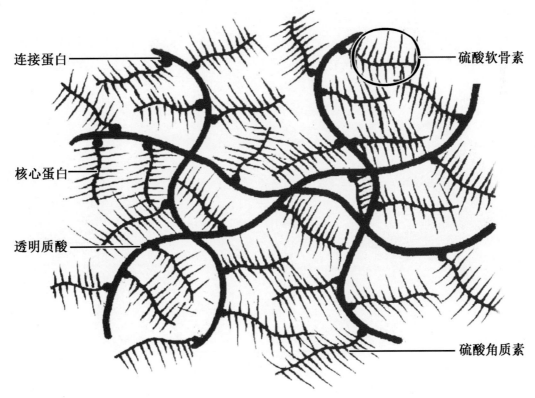

图4-8 基质分子筛结构示意图

Fig.4-8 structural diagram of matrix molecular sieve

连接蛋白—ligandin 核心蛋白—core protein 透明质酸—hyaluronic acid

硫酸软骨素—chondroitin sulfate 硫酸角质素—keratan sulfate

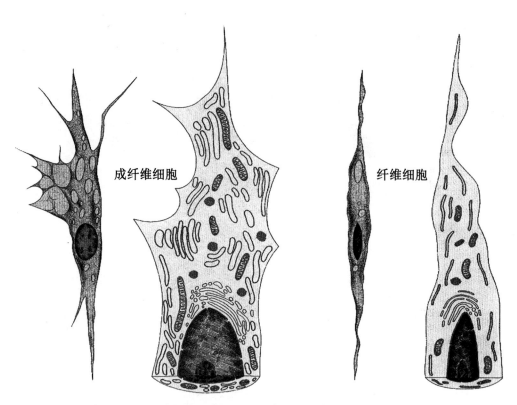

图 4-9　成纤维细胞和纤维细胞光镜、电镜结构示意图

Fig.4-9　diagram of fibroblast and fiber cell under light and electron microscope

成纤维细胞—fibroblast　纤维细胞—fiber cell

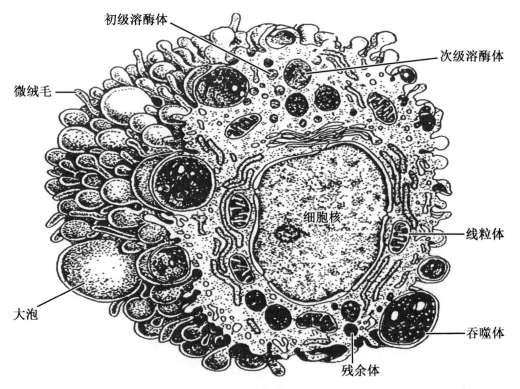

图 4-10　巨噬细胞超微结构示意图

Fig.4-10　ultrastructure diagram of macrophage

初级溶酶体—primary lysosome　次级溶酶体—secondary lysosome　线粒体—mitochondria

吞噬体—phagosome　残余体—remnant　细胞核—nucleus　大泡—bullae　微绒毛—microvilli

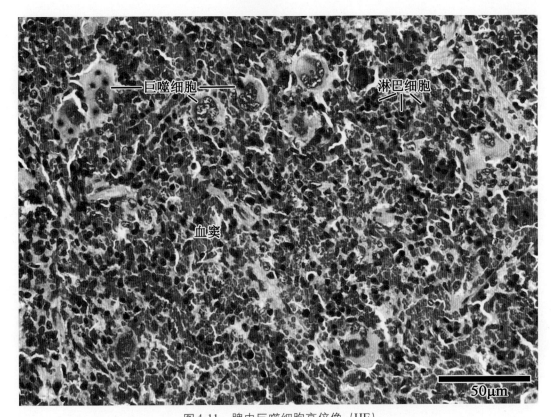

图4-11 脾中巨噬细胞高倍像（HE）

Fig.4-11 high magnification of macrophage in spleen（HE）

巨噬细胞—macrophage 淋巴细胞—lymphocyte 血窦—blood sinusoid

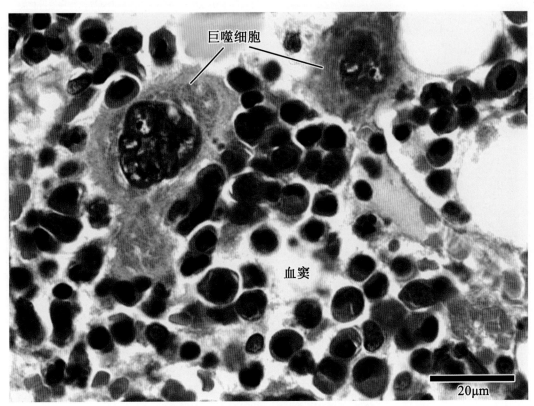

图4-12 骨髓中巨噬细胞高倍像（HE）

Fig.4-12 high magnification of macrophage in bone marrow（HE）

巨噬细胞—macrophage 血窦—blood sinusoid

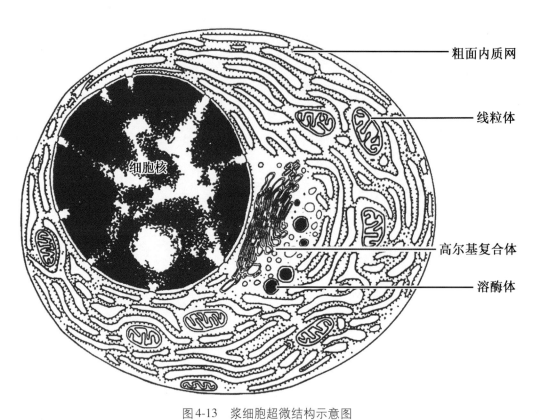

图4-13 浆细胞超微结构示意图

Fig.4-13 ultrastructure diagram of plasmacyte

细胞核—nucleus 粗面内质网—rough endoplasmic reticulum 线粒体—mitochondria

高尔基复合体—Golgi complex 溶酶体—lysosome

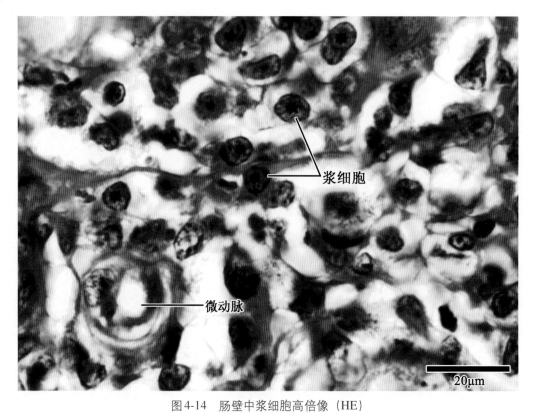

图4-14 肠壁中浆细胞高倍像（HE）

Fig.4-14 high magnification of plasmacyte in intestinal wall（HE）

浆细胞—plasmocyte 微动脉—arteriole

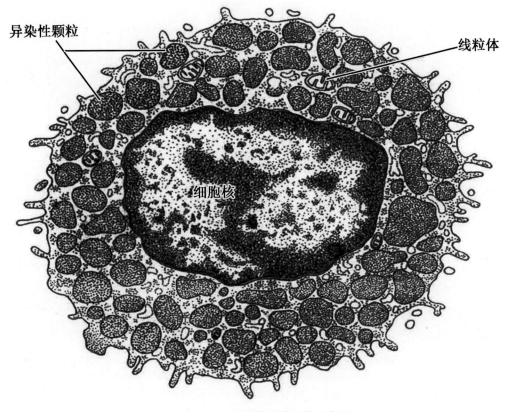

图4-15 肥大细胞超微结构示意图

Fig.4-15 ultrastructure diagram of mastocyte

异染性颗粒—metachromatic granule 线粒体—mitochondria 细胞核—nucleus

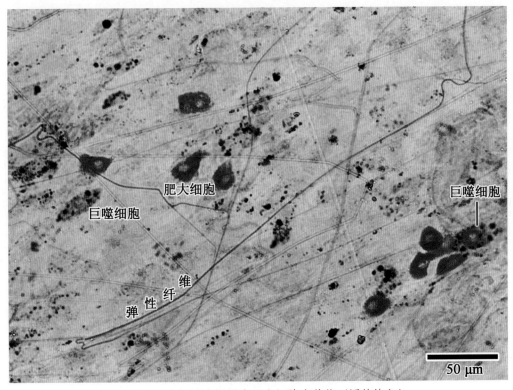

图4-16 疏松结缔组织中肥大细胞高倍像（活体染色）

Fig.4-16 high magnification of mast cell in loose connective tissue（vital stain）

肥大细胞—mast cell 巨噬细胞—macrophage 弹性纤维—elastic fiber

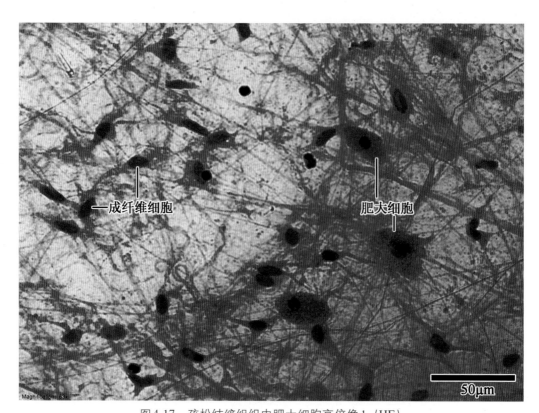

图4-17 疏松结缔组织中肥大细胞高倍像1（HE）

Fig.4-17 high magnification of mast cell in loose connective tissue 1 （HE）

成纤维细胞—fibroblast 肥大细胞—mast cell

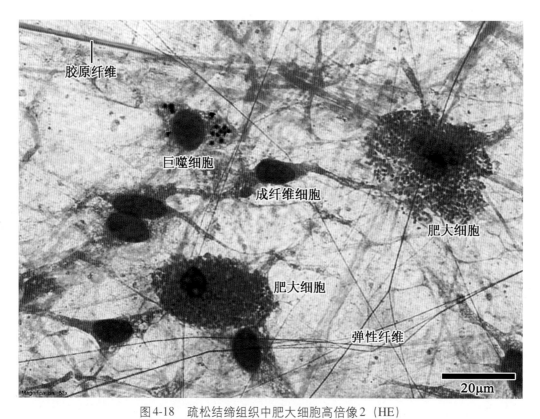

图4-18 疏松结缔组织中肥大细胞高倍像2（HE）

Fig.4-18 high magnification of mast cell in loose connective tissue 2 （HE）

胶原纤维—collagen fiber 巨噬细胞—macrophage 成纤维细胞—fibroblast

肥大细胞—mast cell 弹性纤维—elastic fiber

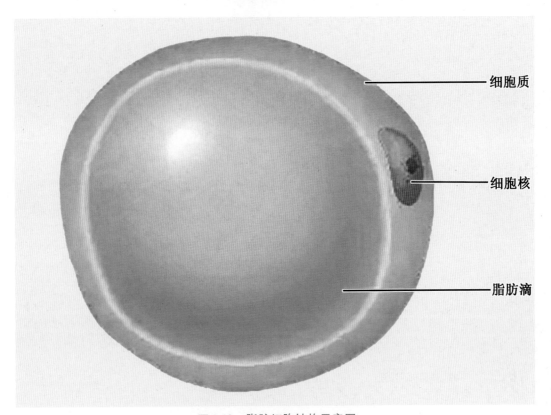

图4-19 脂肪细胞结构示意图

Fig.4-19 structural diagram of adipocyte

细胞质—cytoplasm 细胞核—nucleus 脂肪滴—fat droplet

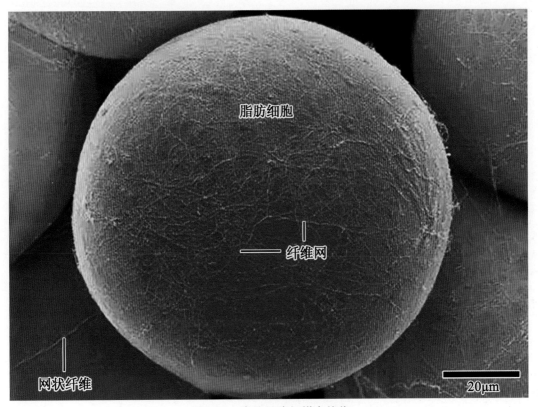

图4-20 脂肪细胞扫描电镜像

Fig.4-20 scanning electrical image of adipocyte

脂肪细胞—adipocyte 纤维网—fibroreticulate 网状纤维—reticular fiber

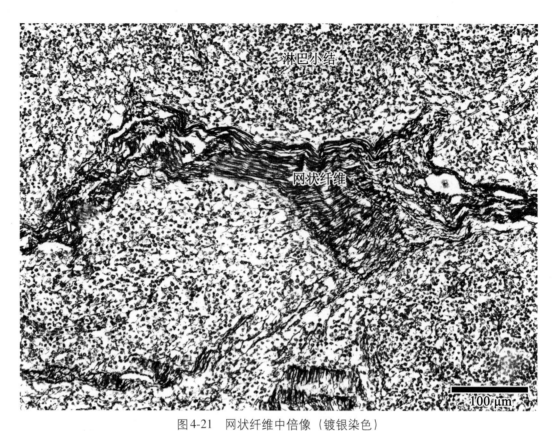

图 4-21　网状纤维中倍像（镀银染色）
Fig.4-21　mid magnification of reticular fibers（silver stain）
淋巴小结—lymphoid nodule　网状纤维—reticular fiber

图 4-22　规则致密结缔组织中倍像（牛跟腱，HE）
Fig.4-22　mid magnification of dense regular connective tissue（ox tendon, HE）
胶原纤维束—collagenous fiber bundle　腱细胞—tendon cell

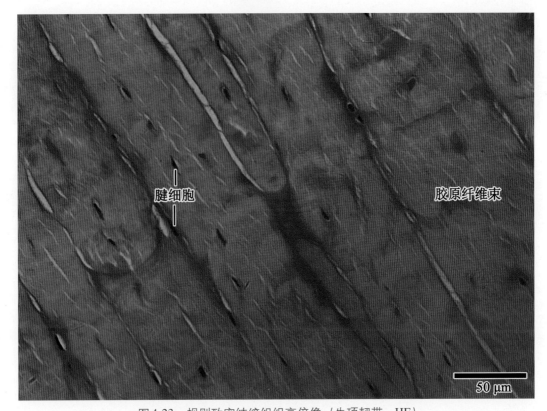

图4-23 规则致密结缔组织高倍像（牛项韧带，HE）

Fig.4-23 high magnification of dense regular connective tissue (ox nuchal ligament, HE)

腱细胞—tendon cell　胶原纤维束—collagenous fiber bundle

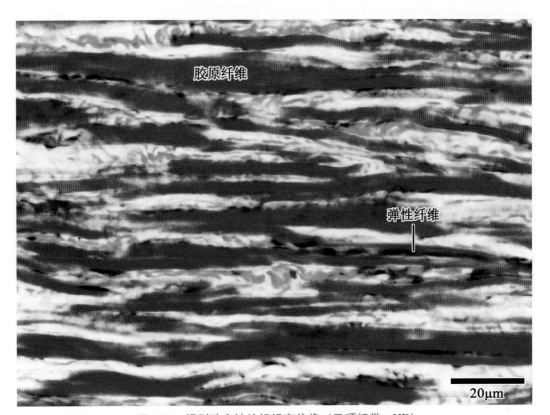

图4-24 规则致密结缔组织高倍像（马项韧带，HE）

Fig.4-24 high magnification of dense regular connective tissue (horse nuchal ligament, HE)

胶原纤维—collagenous fiber　弹性纤维—elastic fiber

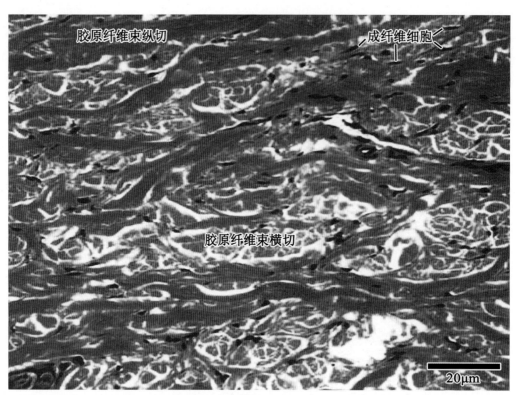

图4-25 不规则致密结缔组织高倍像（马真皮，HE）
Fig.4-25 high magnification of dense irregular connective tissue （horse dermis, HE）
胶原纤维束纵切—collagenous fiber bundle longitudinal section　成纤维细胞—fibroblast
胶原纤维束横切—collagenous fiber bundle transverse section

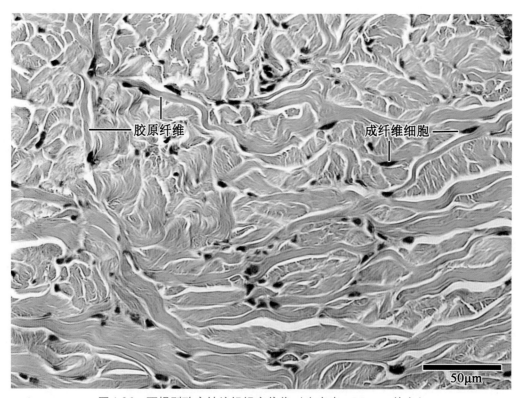

图4-26 不规则致密结缔组织高倍像（牛真皮，Masson染色）
Fig.4-26 high magnification of dense irregular connective tissue （ox dermis, Masson stain）
胶原纤维—collagenous fiber　成纤维细胞—fibroblast

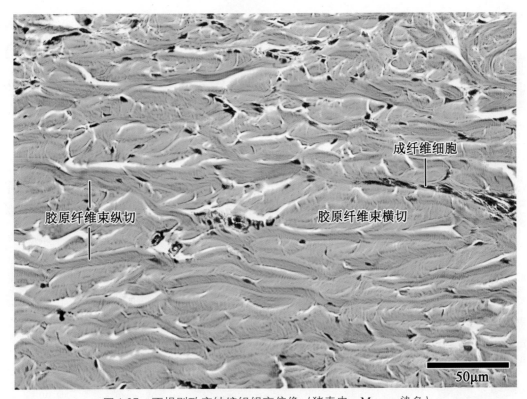

图 4-27 不规则致密结缔组织高倍像（猪真皮，Masson 染色）

Fig.4-27 high magnification of dense irregular connective tissue (pig dermis, Masson stain)

胶原纤维束纵切—collagenous fiber bundle longitudinal section　成纤维细胞—fibroblast

胶原纤维束横切—collagenous fiber bundle transverse section

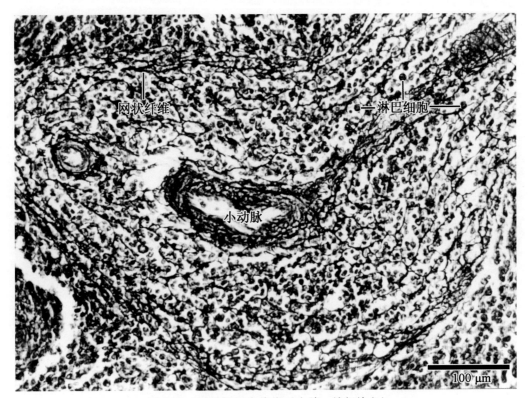

图 4-28 网状组织中倍像（牛脾，镀银染色）

Fig.4-28 mid magnification of reticular tissue（ox spleen, silver stain）

网状纤维—reticular fiber　淋巴细胞—lymphatic cell　小动脉—arteriole

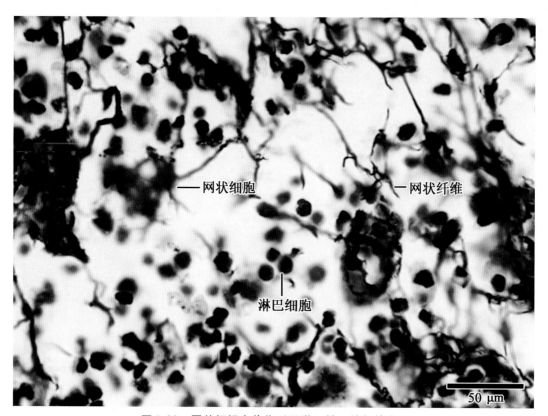

图4-29 网状组织高倍像（马淋巴结，镀银染色）

Fig.4-29 high magnification of reticular tissue（horse lymphnode, silver stain）

网状细胞—reticular cell　网状纤维—reticular fiber　淋巴细胞—lymphatic cell

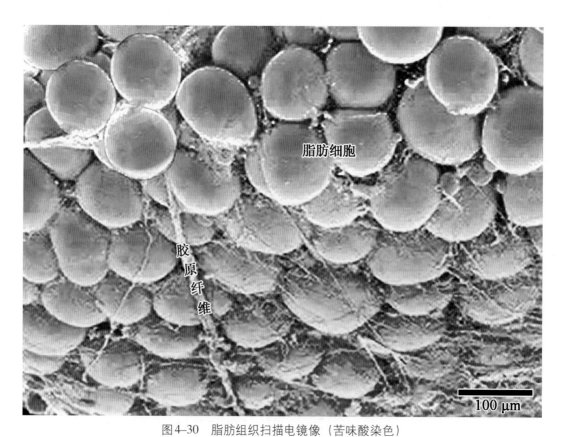

图4-30 脂肪组织扫描电镜像（苦味酸染色）

Fig.4-30 scanning electric image of adipose tissue (picric acid stain)

脂肪细胞—adipocyte　胶原纤维—collagenous fiber

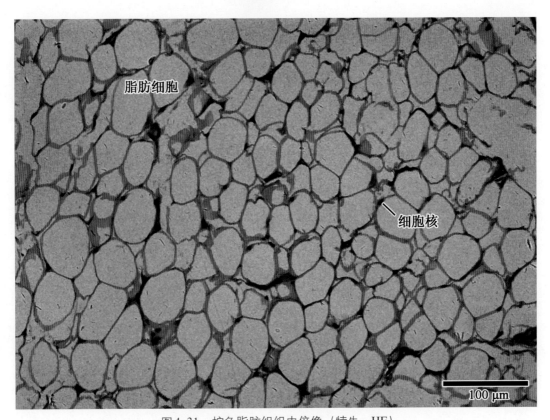

图4-31 棕色脂肪组织中倍像（犊牛，HE）

Fig.4-31 mid magnification of brown adipose tissue（calf, HE）

脂肪细胞—adipose cell 细胞核—nucleus

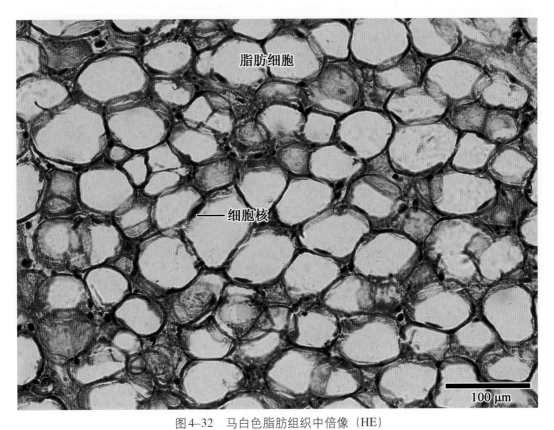

图4-32 马白色脂肪组织中倍像（HE）

Fig.4-32 mid magnification of horse white adipose tissue（HE）

脂肪细胞—adipose cell 细胞核—nucleus

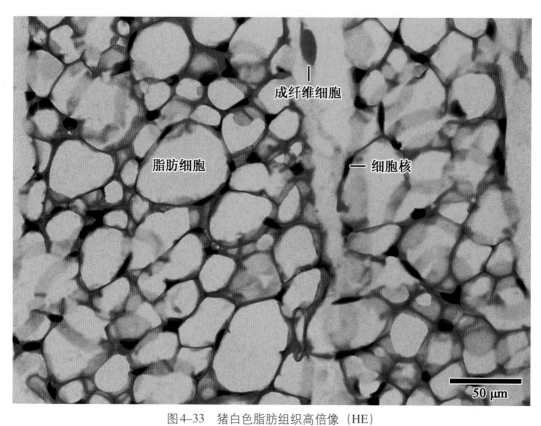

图4-33 猪白色脂肪组织高倍像（HE）

Fig.4-33 high magnification of pig white adipose tissue（HE）

脂肪细胞—adipose cell　成纤维细胞—fibroblast　细胞核—nucleus

图4-34 山羊白色脂肪组织高倍像（HE）

Fig.4-34 high magnification of goat white adipose tissue（HE）

脂肪细胞—adipose cell　细胞核—nucleus

第五章
软骨组织和骨组织
Cartilage and Bone Tissue

> **Outline**
>
> The cartilage is composed of cartilage tissue and perichondrium. It is one of the connective tissues. It contains chondrocytes and an extensive extracellular matrix composed of fibers and ground substance. Chondrocytes synthesize and secrete the matrix, which contains proteoglycans. There are 3 types of cartilage: Hyaline cartilage, elastic cartilage and fibrous cartilage that differ from one another mainly by the type of fibers embedded in the matrix. Cartilage is devoid of blood vessels, nourished by diffusion from the blood vessels in the perichondrium. Perichondrium is capable of forming new cartilage.
>
> The bone consists of osseous tissue, bone marrow, endosteum and periosteum. Osseous tissue is a specialized connective tissue which composed of intercellular material, the bone matrix, and 4 different cell types: osteogenic cells, osteoblasts, osteocytes and osteoclasts. The osteogenic cells act as stem cells of bone tissue, and may divide and differentiate into osteoblasts. Osteoblasts are associated with bone formation. Osteocytes are found in cavities within the matrix. Osteoclasts are involved in the resorption and remodeling of osseous tissue. Inorganic matter which forms hydroxyapatite crystals represents about 65% of the dry weight of the bone. The organic matter consists of collagen fibers and amorphous matrix.
>
> During the embryo stage, development of the bones is formed by two ways: one is intramembranous ossification, such as the skull, that occurs within a membrane of condensed mesenchymal tissue. This process begins in areas occupied by mesenchymal, the packing tissue of the embryo. Another is endochondral ossification, such as the limb bones, takes place within a cartilaginous model. The model is gently destroyed and replaced by bone formed by incoming cells from surrounding periosteal connective tissues. In this way, the reconstruction and repair of the bone continue the lifetime.

一、软骨和骨的发生

（一）软骨的发生

软骨（cartilage）由胚胎时期的中胚层间充质分化而来。例如，牛胚龄至第5周时，在将发生软骨的部位，多突起的间充质细胞其突起萎缩、细胞变圆，增殖并聚集成团，称前软骨组织或软骨形成中心。前软骨组织中央的间充质细胞继续增殖和分化，体积增大，称成软骨细胞(chondroblast)。成软骨细胞可合成和分泌软骨基质和纤维。随着软骨基质的不断增多，相邻成软骨细胞之间的距离变大，相互分隔开，包埋在软骨基质内。成软骨细胞进一步分化为成熟的软骨细胞(chondrocyte)。成软骨组织周围的间充质分化为软骨膜。早期的软骨细胞大多单独存在于软骨基质内，细胞仍保留分裂增殖和产生纤维和基质的潜能。

（二）骨的发生

骨（bone）的发生有两种方式。

第一种为膜内成骨(intramembranous ossification)，是间充质细胞增殖分化、形成含骨原细胞的结缔组织膜。骨原细胞不断增殖，部分细胞分化为成骨细胞；成骨细胞产生骨间质的有机成分即纤维和基质，形成类骨质(osteoid)。类骨质继而钙化成骨基质，成骨细胞被包埋在骨基质内，成为骨细胞。部分颅骨的扁骨和面骨，以及锁骨的一部分以膜内成骨方式形成。最早形成骨组织的部位称骨化中心(ossification center)。新形成的骨组织为骨松质，由许多针状或片状的骨小梁组成，骨小梁之间为骨髓。骨松质周围的间充质分化为骨膜，进而在骨松质周围继续造骨，形成骨密质，使骨不断增厚变大。

第二种为软骨内成骨(endochondral ossification)，是由间充质先分化形成软骨，在胚胎发育中软骨又先后退化并被分解吸收；在此基础上，再由骨原细胞增殖分化为成骨细胞而造骨。四肢骨的长骨、躯干骨和颅底骨等主要以软骨内成骨方式形成。在长骨发生的部位，由间充质先形成透明软骨，其外形与将形成的长骨外形相似，称软骨雏形(cartilage model)，随胚胎的生长发育，软骨逐渐长大增粗，并出现软骨退化，逐渐被骨质取代的过程。

二、软骨和骨的组织学结构概述

（一）软骨

1. 透明软骨（hyaline cartilage）

（1）软骨膜（perichondrium） 分两层，外侧为纤维层，内侧为软骨发生层。前者含胶原纤维和成纤维细胞，后者含软骨形成细胞和成软骨细胞。

（2）基质（matrix） 基质呈嗜碱性，围绕在陷窝周围的基质颜色较深，其他部位的颜色较浅。胶原纤维被基质掩盖，在镜下看不到。

（3）细胞 在软骨陷窝中有一个或多个软骨细胞，称同源细胞群，其周围环绕着深蓝色的软骨囊。

2. 弹性软骨（elastic cartilage）

（1）软骨膜 与透明软骨的软骨膜相同。

（2）基质 除了胶原纤维，还有大量深染的弹性纤维。

（3）细胞 与透明软骨的相同，有软骨细胞、成软骨细胞和软骨形成细胞。

3. 纤维软骨（fibrous cartilage）

（1）软骨膜 一般无软骨膜。

（2）基质 基质很少，许多粗大的胶原纤维束位于平行排列成行的软骨细胞之间。

（3）细胞 软骨细胞比透明软骨和弹性软骨的细胞小，夹在粗大的胶原纤维束之间，平行纵向地排列成行。

（二）骨

1. 脱钙的密质骨

（1）骨外膜（periosteum） 有两层，外侧的纤维层和内侧的成骨层。前者含有胶原纤维和成纤维细胞，后者含有骨原细胞和成骨细胞。骨外膜由穿通纤维固定在骨上。

（2）骨板系统（bone plate system）　包括外环骨板、内环骨板、骨单位(哈弗斯系统)和间骨板。

（3）骨内膜（endosteum）　为一层薄膜，衬在骨髓腔内侧，容纳黄骨髓。

（4）细胞　骨细胞占据的小腔隙称骨陷窝。成骨细胞和骨原细胞可见于骨外膜、骨内膜和中央管内膜的成骨层。破骨细胞沿着骨的吸收面，位于吸收腔隙内。类骨质，即未钙化的骨基质，分布于骨的细胞和钙化组织之间。

2.未脱钙的密质骨

（1）骨板系统　清晰地展示骨组织的薄板层结构：排列成外环骨板、内环骨板、骨单位和间骨板。

（2）骨单位（osteon）　由同心圆骨板构成圆柱状结构。干制标本上骨陷窝是空的，但生长期的骨陷窝内含骨细胞。骨小管以骨陷窝为中心，呈放射状排列。生长期的中央管内，含有血管、成骨细胞和骨原细胞。黏合线标志着每个骨单位的边界。相邻的中央管由穿通管相互连接。

3.脱钙的松质骨

（1）骨板系统　包括骨针和骨小梁。

（2）细胞　骨细胞占据骨陷窝。成骨细胞衬附全部骨小梁和骨针。偶见多核的大破骨细胞占据吸收腔隙。未钙化的骨基质，即类骨质，分布于骨细胞和钙化组织之间。小梁内和小梁之间的腔隙内填充骨髓。

三、软骨组织和骨组织图谱

1.**软骨**　图5-1～图5-30。

2.**骨**　图5-31～图5-62。

3.**骨的发生**　图5-63～图5-73。

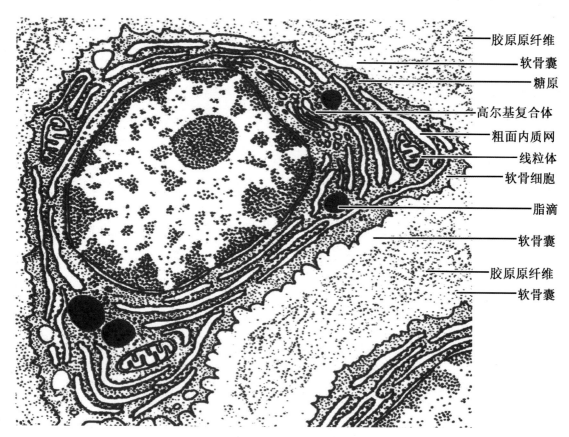

图5-1 软骨超微结构模式图

Fig.5-1 ultrastructure diagram of cartilage

胶原原纤维—collagen fibril　软骨囊—cartilage capsule　糖原—glycogen　高尔基复合体—Golgi complex　粗面内质网—rough endoplasmic reticulum　线粒体—mitochondria　软骨细胞—cartilage cell　脂滴—lipid droplet

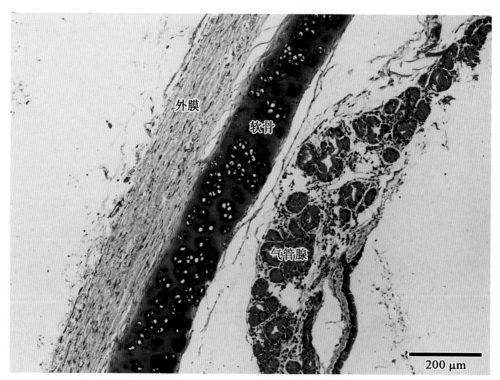

图5-2 气管透明软骨低倍像（美蓝染色）

Fig.5-2 low magnification of trachea hyaline cartilage (methylene blue stain)

外膜—outer membrane　软骨—cartilage　气管腺—tracheal gland

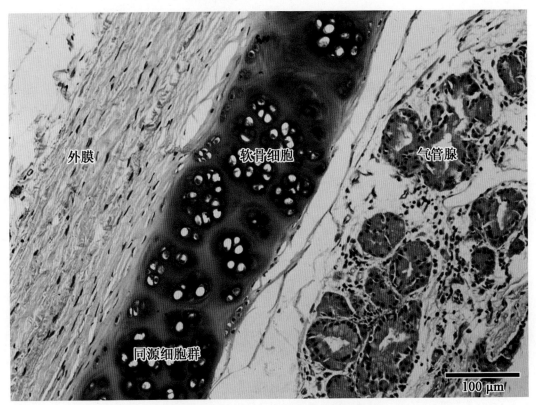

图5-3 气管透明软骨中倍像（美蓝染色）
Fig.5-3 mid magnification of trachea hyaline cartilage (methylene blue stain)
外膜—outer membrane 软骨细胞—cartilage cell 气管腺—tracheal gland 同源细胞群—isogenous cell group

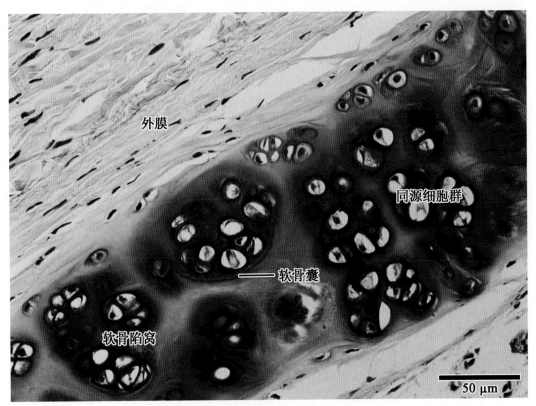

图5-4 气管透明软骨高倍像（美蓝染色）
Fig.5-4 high magnification of trachea hyaline cartilage (methylene blue stain)
外膜—outer membrane 同源细胞群—isogenous cell group 软骨囊—cartilage capsule 软骨陷窝—cartilage lacuna

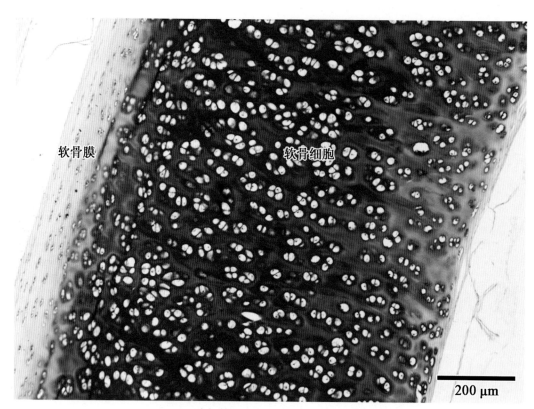

图5-5　猪气管透明软骨低倍像（美蓝染色）
Fig.5-5　low magnification of pig trachea hyaline cartilage (methylene blue stain)
软骨膜— perichondrium　软骨细胞—cartilage cell

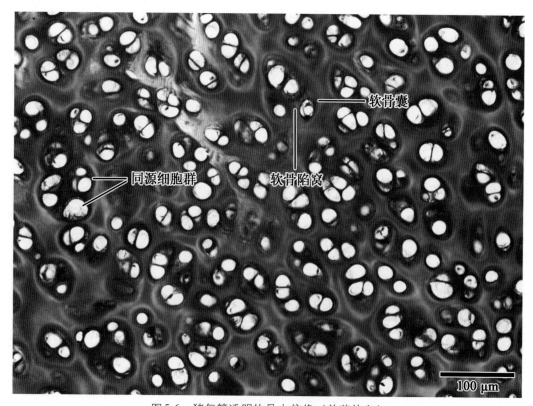

图5-6　猪气管透明软骨中倍像（美蓝染色）
Fig.5-6　mid magnification of pig trachea hyaline cartilage (methylene blue stain)
同源细胞群—isogenous cell group　软骨囊—cartilage capsule　软骨陷窝—cartilage lacuna

第五章 软骨组织和骨组织　Cartilage and Bone Tissue

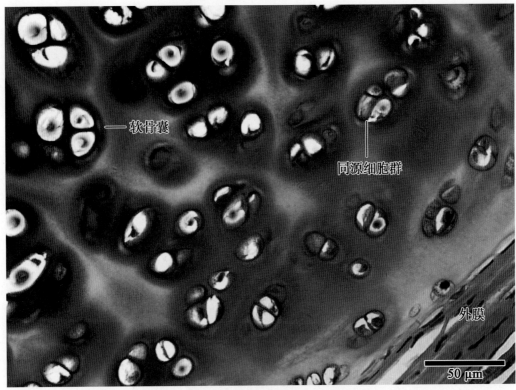

图5-7　猪气管透明软骨高倍像（美蓝染色）

Fig.5-7　high magnification of pig trachea hyaline cartilage (methylene blue stain)

软骨囊—cartilage capsule　同源细胞群—isogenous cell group　外膜—outer membrane

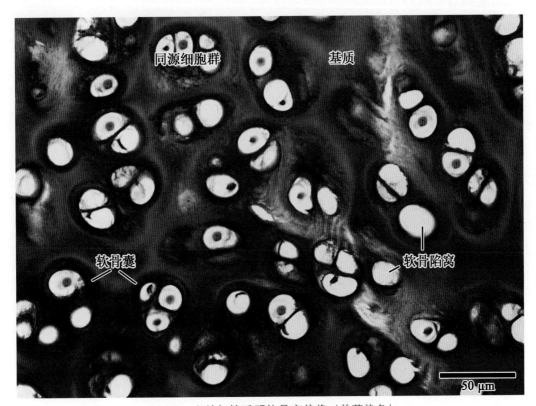

图5-8　山羊气管透明软骨高倍像（美蓝染色）

Fig.5-8　high magnification of goat trachea hyaline cartilage (methylene blue stain)

同源细胞群—isogenous cell group　基质—matrix　软骨囊—cartilage capsule　软骨陷窝—cartilage lacuna

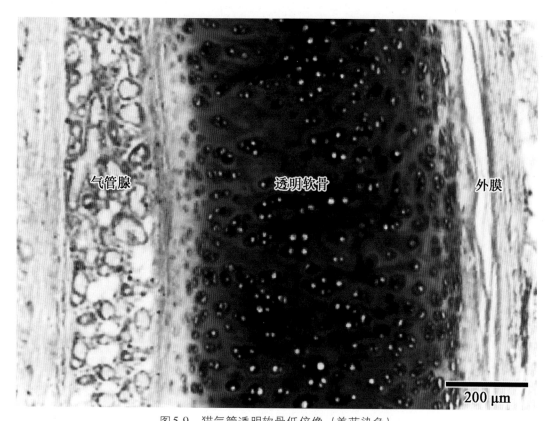

图5-9 猫气管透明软骨低倍像（美蓝染色）
Fig.5-9 low magnification of cat trachea hyaline cartilage (methylene blue stain)
气管腺—tracheal gland 透明软骨—hyaline cartilage 外膜—outer membrane

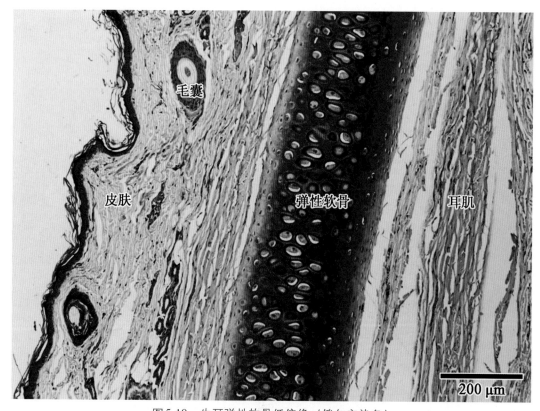

图5-10 牛耳弹性软骨低倍像（俄尔辛染色）
Fig.5-10 low magnification of ox ear elastic cartilage (Orsin stain)
皮肤—skin 毛囊—hair follicle 弹性软骨—elastic cartilage 耳肌—ear muscle

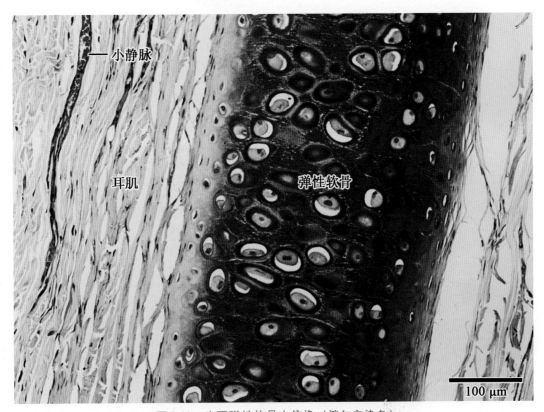

图5-11 牛耳弹性软骨中倍像（俄尔辛染色）
Fig.5-11 mid magnification of ox ear elastic cartilage (Orsin stain)
小静脉—venule 耳肌—ear muscle 弹性软骨—elastic cartilage

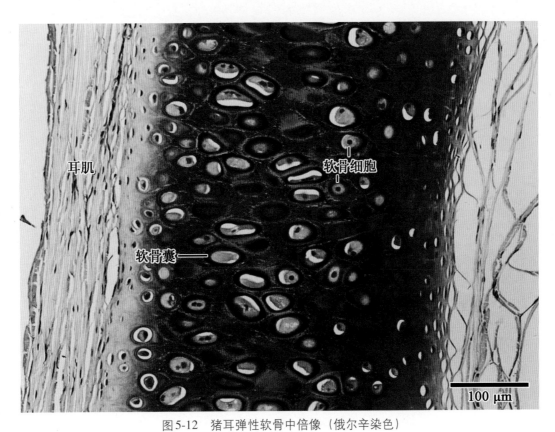

图5-12 猪耳弹性软骨中倍像（俄尔辛染色）
Fig.5-12 mid magnification of pig ear elastic cartilage (Orsin stain)
耳肌—ear muscle 软骨囊—cartilage capsule 软骨细胞—cartilage cell

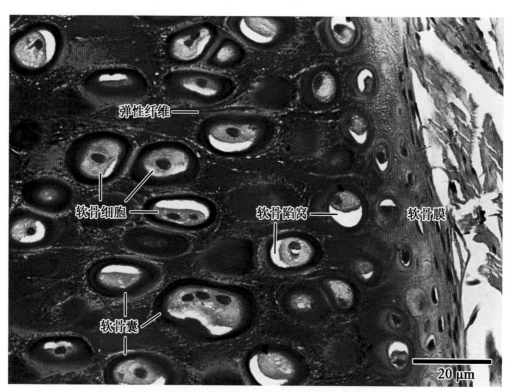

图5-13　马耳弹性软骨高倍像（俄尔辛染色）

Fig.5-13　high magnification of horse ear elastic cartilage (Orsin stain)

弹性纤维—elastic fiber　软骨细胞—cartilage cell　软骨陷窝—cartilage lacuna　软骨囊—cartilage capsule

软骨膜—perichondrium

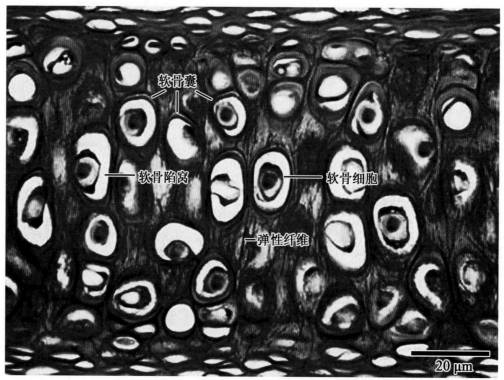

图5-14　猪耳弹性软骨高倍像（俄尔辛染色）

Fig.5-14　high magnification of pig ear elastic cartilage (Orsin stain)

软骨囊—cartilage capsule　软骨陷窝—cartilage lacuna　软骨细胞—cartilage cell　弹性纤维—elastic fiber

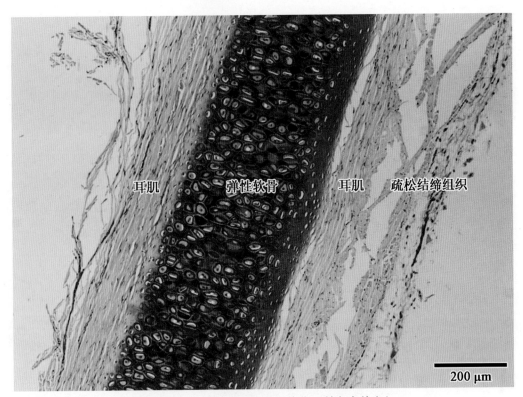

图5-15 山羊耳弹性软骨低倍像（俄尔辛染色）
Fig.5-15 low magnification of goat ear elastic cartilage (Orsin stain)
耳肌—ear muscle　弹性软骨—elastic cartilage　疏松结缔组织—loose connective tissue

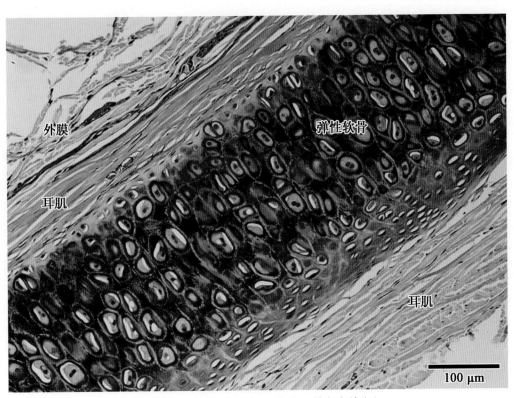

图5-16 犬耳弹性软骨中倍像（俄尔辛染色）
Fig.5-16 mid magnification of dog ear elastic cartilage (Orsin stain)
外膜—outer membrane　耳肌—ear muscle　弹性软骨—elastic cartilage

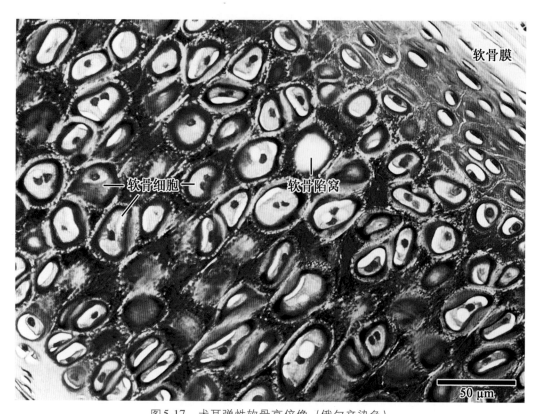

图5-17 犬耳弹性软骨高倍像（俄尔辛染色）

Fig.5-17 high magnification of dog ear elastic cartilage (Orsin stain)

软骨膜—perichondrium 软骨细胞—cartilage cell 软骨陷窝—cartilage lacuna

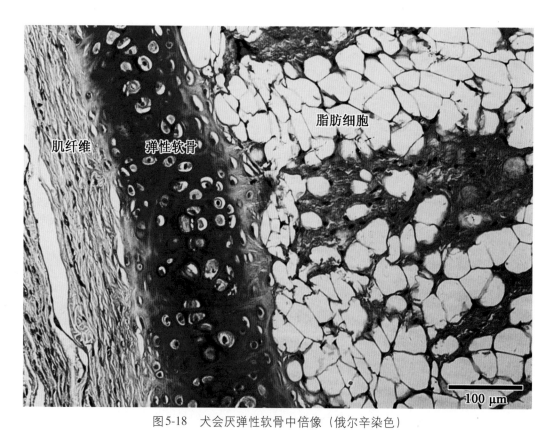

图5-18 犬会厌弹性软骨中倍像（俄尔辛染色）

Fig.5-18 mid magnification of dog epiglottis elastic cartilage (Orsin stain)

肌纤维—muscle fiber 弹性软骨—elastic cartilage 脂肪细胞—fat cell

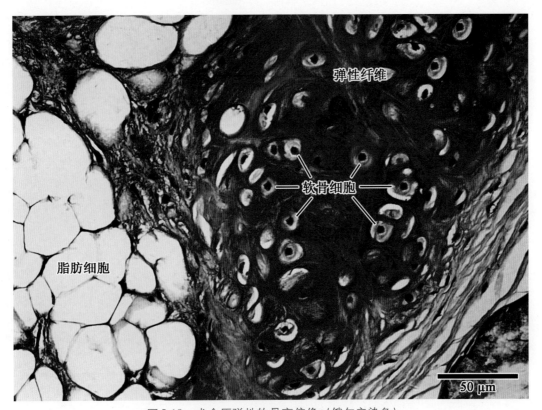

图5-19 犬会厌弹性软骨高倍像（俄尔辛染色）
Fig.5-19 high magnification of dog epiglottis elastic cartilage (Orsin stain)
脂肪细胞—fat cell 弹性纤维—elastic fiber 软骨细胞—cartilage cell

图5-20 牛椎间盘纤维软骨低倍像1（HE）
Fig.5-20 low magnification of cow intervertebral disc fibrous cartilage 1（HE）
胶原纤维束—collagen fiber bundle 软骨细胞—cartilage cell

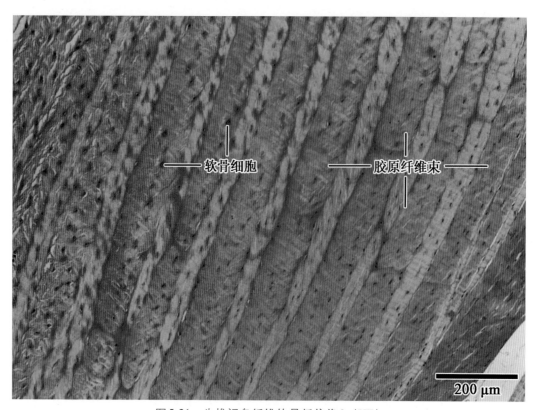

图5-21　牛椎间盘纤维软骨低倍像2（HE）

Fig.5-21　low magnification of cow intervertebral disc fibrous cartilage 2（HE）

软骨细胞—cartilage cell　胶原纤维束—collagen fiber bundle

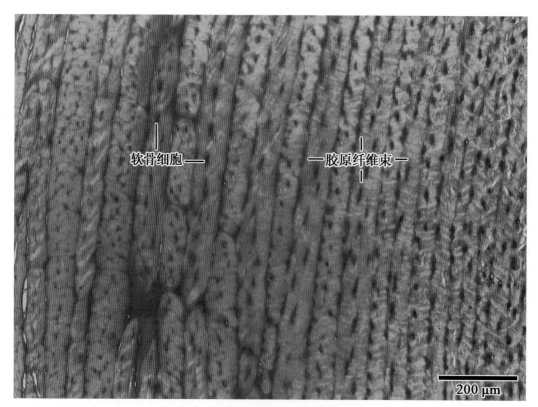

图5-22　马椎间盘纤维软骨低倍像（HE）

Fig.5-22　low magnification of horse intervertebral disc fibrous cartilage（HE）

软骨细胞—cartilage cell　胶原纤维束—collagen fiber bundle

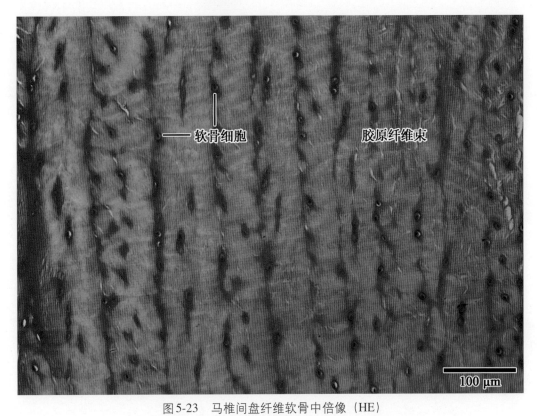

图5-23 马椎间盘纤维软骨中倍像（HE）

Fig.5-23　mid magnification of horse intervertebral disc fibrous cartilage（HE）

软骨细胞—cartilage cell　胶原纤维束—collagen fiber bundle

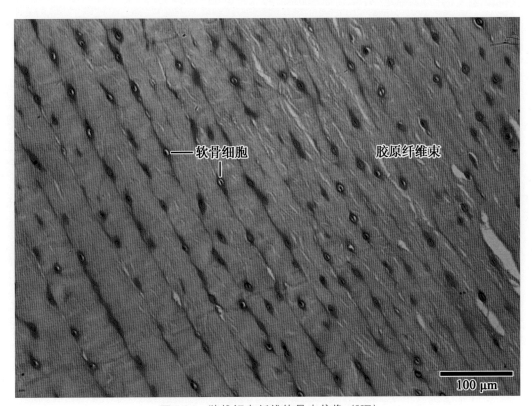

图5-24 猪椎间盘纤维软骨中倍像（HE）

Fig.5-24　mid magnification of pig intervertebral disc fibrous cartilage（HE）

软骨细胞—cartilage cell　胶原纤维束—collagen fiber bundle

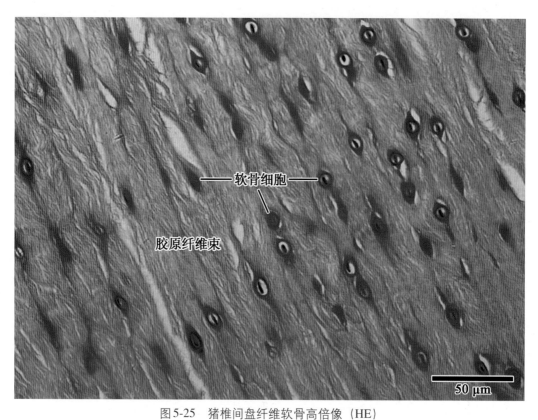

图 5-25　猪椎间盘纤维软骨高倍像（HE）

Fig.5-25　high magnification of pig intervertebral disc fibrous cartilage（HE）

软骨细胞—cartilage cell　胶原纤维束—collagen fiber bundle

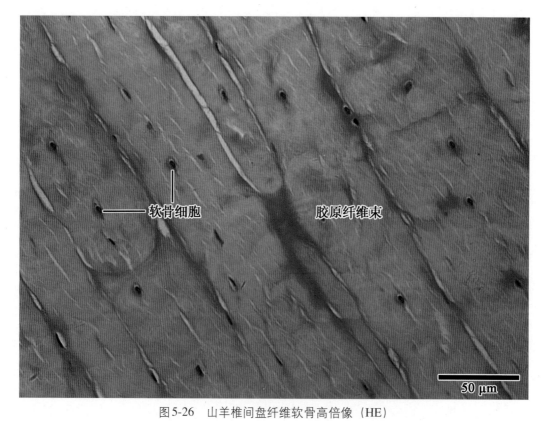

图 5-26　山羊椎间盘纤维软骨高倍像（HE）

Fig.5-26　high magnification of goat intervertebral disc fibrous cartilage（HE）

软骨细胞—cartilage cell　胶原纤维束—collagen fiber bundle

第五章 软骨组织和骨组织 Cartilage and Bone Tissue

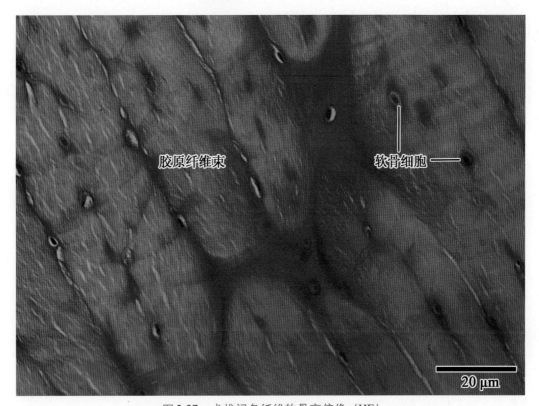

图 5-27 犬椎间盘纤维软骨高倍像（HE）
Fig.5-27 high magnification of dog intervertebral disc fibrous cartilage（HE）
胶原纤维束—collagen fiber bundle 软骨细胞—cartilage cell

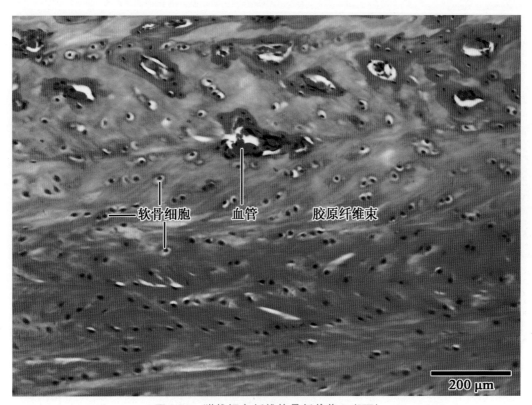

图 5-28 猫椎间盘纤维软骨低倍像1（HE）
Fig.5-28 low magnification of cat intervertebral disc fibrous cartilage 1（HE）
软骨细胞—cartilage cell 血管—blood vessel 胶原纤维束—collagen fiber bundle

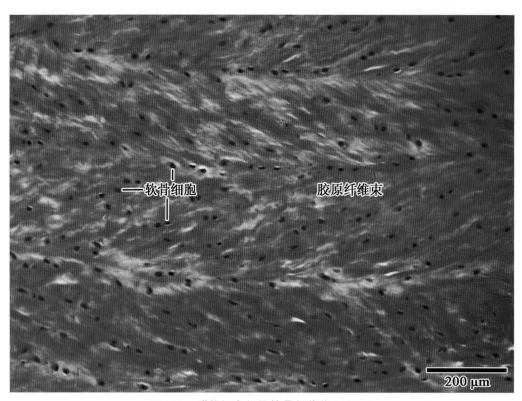

图5-29 猫椎间盘纤维软骨低倍像2（HE）

Fig.5-29 low magnification of cat intervertebral disc fibrous cartilage 2 (HE)

软骨细胞—cartilage cell　胶原纤维束—collagen fiber bundle

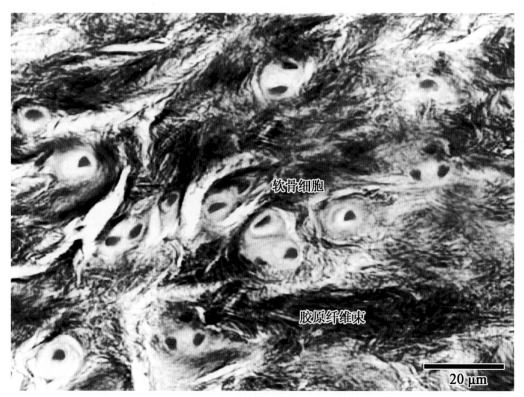

图5-30 兔椎间盘纤维软骨高倍像（里昂蓝染色）

Fig.5-30 high magnification of rabbit intervertebral disc fibrous cartilage (Lyon blue stain)

软骨细胞—cartilage cell　胶原纤维束—collagen fiber bundle

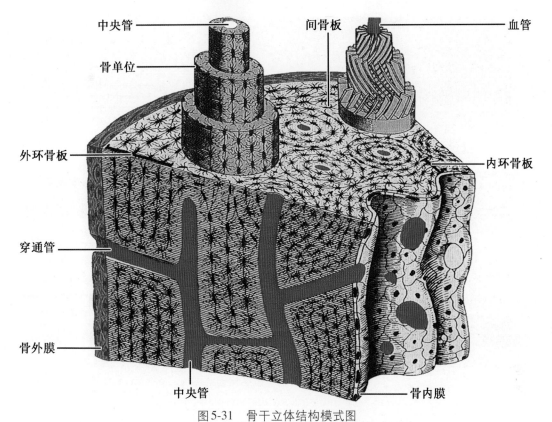

图5-31 骨干立体结构模式图

Fig.5-31 stereo-structure diagram of diaphysis

中央管—central canal 骨单位—osteon 外环骨板—outer circumferential lamella 穿通管—perforating canal
骨外膜—periosteum 间骨板—interstitial lamella 血管—blood vessel 内环骨板—inner circumferential lamella
骨内膜—endosteum

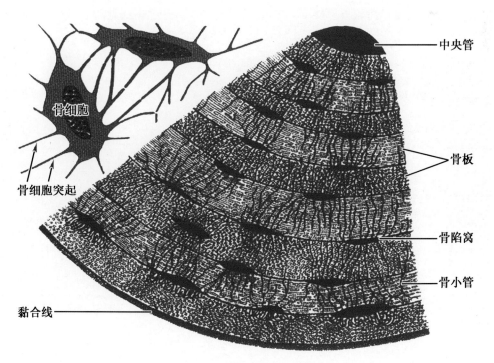

图5-32 骨细胞和骨板结构模式图

Fig.5-32 structural diagram of bone cells and lamina

骨细胞—bone cell 骨细胞突起—protuberance 黏合线—cement line 中央管—central canal
骨板—bone lamella 骨陷窝—bone lacuna 骨小管—bone canaliculus

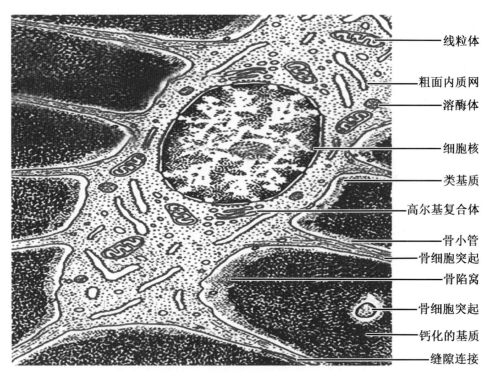

图5-33　骨细胞超微结构模式图

Fig.5-33　ultrastructure diagram of osteocyte

线粒体—mitochondria　粗面内质网—rough endoplasmic reticulum　溶酶体—lysosome　细胞核—cell nucleus
类基质—matrix　高尔基复合体—Golgi complex　骨小管—bone canaliculus　骨细胞突起—osteocyte process
骨陷窝—bone lacuna　钙化的基质—calcified matrix　缝隙连接—gap junction

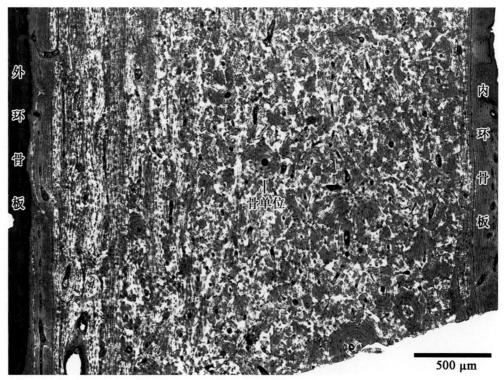

图5-34　牛股骨横切低倍像1（硫堇染色）

Fig.5-34　low magnification of cow femoral transection 1 （thionine stain）

外环骨板—outer circumferential lamella　骨单位—osteon　内环骨板—inner circumferential lamella

第五章 软骨组织和骨组织　Cartilage and Bone Tissue

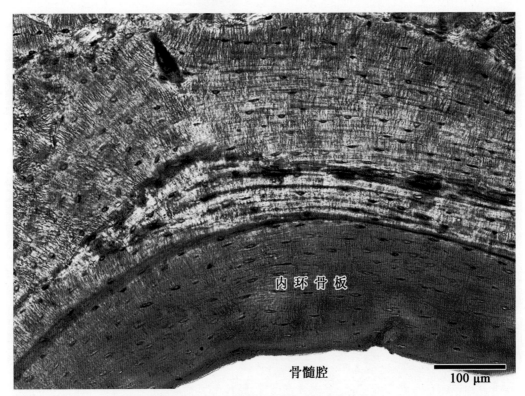

图5-35　牛股骨横切中倍像1（硫堇染色）

Fig.5-35　mid magnification of cow femoral transection 1（thionine stain）

内环骨板—inner circumferential lamella　骨髓腔—marrow cavity

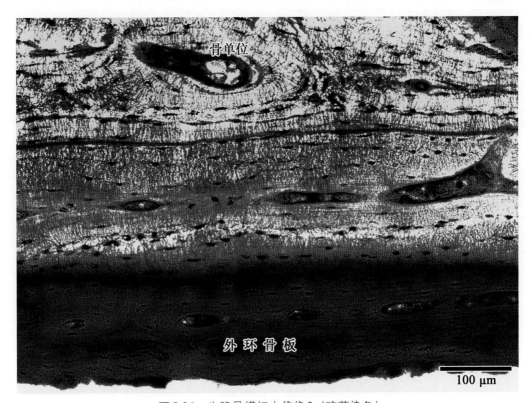

图5-36　牛股骨横切中倍像2（硫堇染色）

Fig.5-36　mid magnification of cow femoral transection 2（thionine stain）

骨单位—osteon　外环骨板—outer circumferential lamella

图 5-37 牛股骨纵切中倍像（硫堇染色）
Fig.5-37 mid magnification of cow femoral longitudinal section（thionine stain）
骨单位—osteon 中央管—central canal

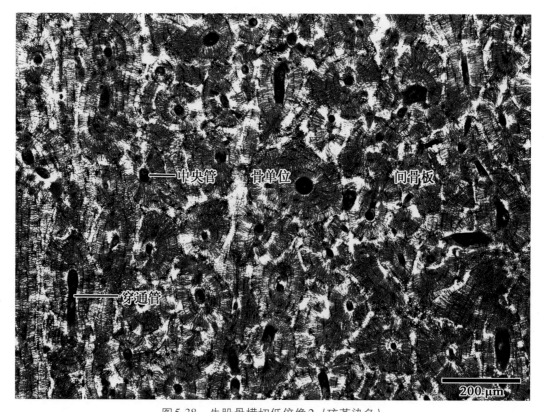

图 5-38 牛股骨横切低倍像 2（硫堇染色）
Fig.5-38 low magnification of cow femoral transection 2（thionine stain）
中央管—central canal 骨单位—osteon 间骨板—interstitial lamella 穿通管—perforating canal

第五章 软骨组织和骨组织　Cartilage and Bone Tissue

图5-39　牛股骨横切中倍像3（硫堇染色）

Fig.5-39　mid magnification of cow femoral transection 3（thionine stain）

骨单位—osteon　间骨板—interstitial lamella　穿通管—perforating canal　中央管—central canal

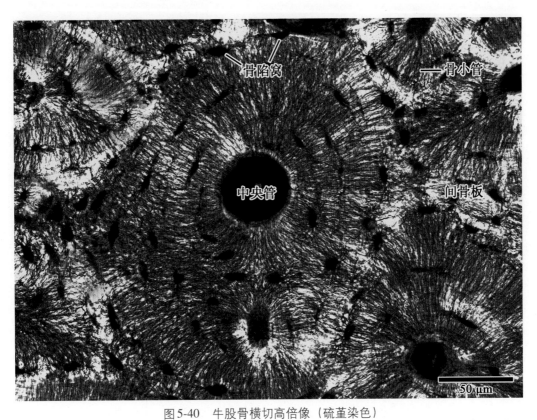

图5-40　牛股骨横切高倍像（硫堇染色）

Fig.5-40　high magnification of cow femoral transection（thionine stain）

骨陷窝—bone lacuna　骨小管—bone canaliculus　中央管—central canal　间骨板—interstitial lamella

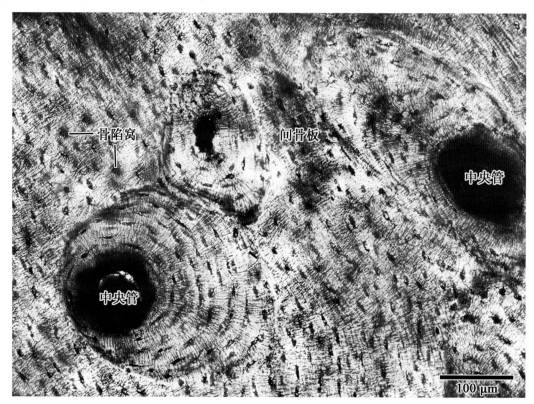

图5-41 马股骨横切中倍像1（复红染色）
Fig.5-41 mid magnification of horse femoral transection 1 (fuchsin stain)
骨陷窝—bone lacuna 间骨板—interstitial lamella 中央管—central canal

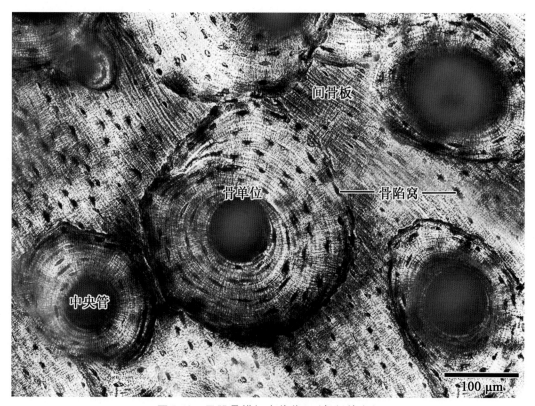

图5-42 马股骨横切中倍像2（复红染色）
Fig.5-42 mid magnification of horse femoral transection 2 (fuchsin stain)
间骨板—interstitial lamella 骨单位—osteon 骨陷窝—bone lacuna 中央管—central canal

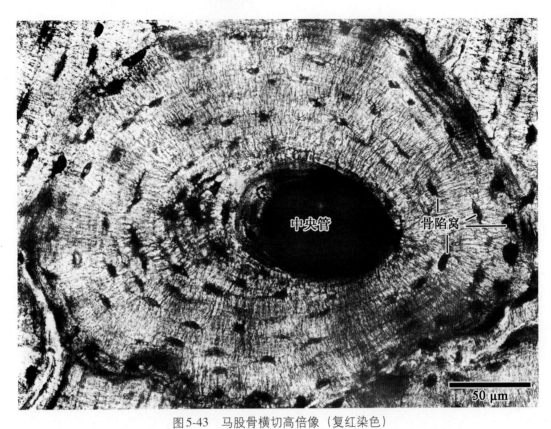

图 5-43 马股骨横切高倍像（复红染色）

Fig.5-43 high magnification of horse femoral transection (fuchsin stain)

中央管—central canal 骨陷窝—bone lacuna

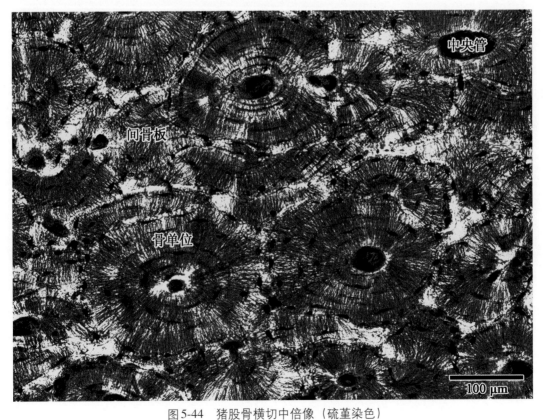

图 5-44 猪股骨横切中倍像（硫堇染色）

Fig.5-44 mid magnification of pig femoral transection (thionine stain)

中央管—central canal 间骨板—interstitial lamella 骨单位—osteon

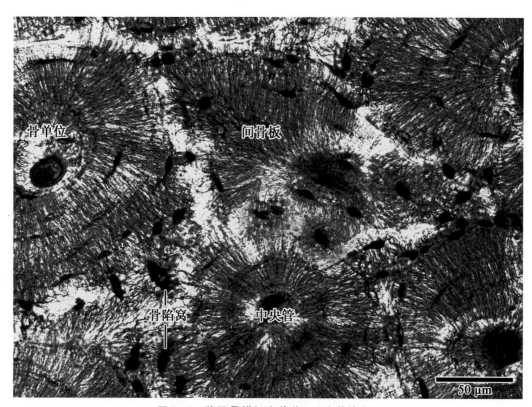

图5-45 猪股骨横切高倍像1（硫堇染色）

Fig.5-45 high magnification of pig femoral transection 1 (thionine stain)

骨单位—osteon 间骨板—interstitial lamella 骨陷窝—bone lacuna 中央管—central canal

图5-46 猪股骨横切高倍像2（硫堇染色）

Fig.5-46 high magnification of pig femoral transection 2 (thionine stain)

骨小管—bone canaliculus 中央管—central canal 骨陷窝—bone lacuna 间骨板—interstitial lamella

第五章 软骨组织和骨组织　Cartilage and Bone Tissue

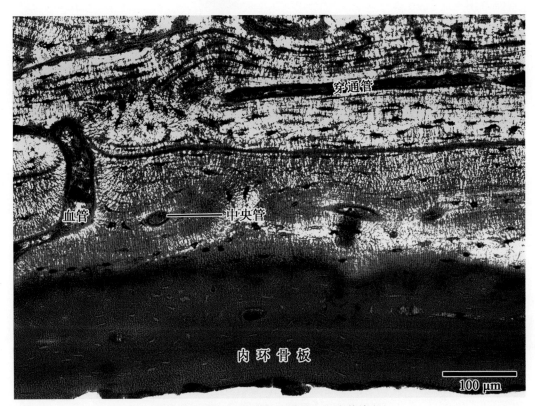

图5-47　绵羊胫骨横切中倍像1（硫堇染色）

Fig.5-47　mid magnification of sheep tibia transection 1 （thionine stain）

血管—blood vessel　穿通管—perforating canal　中央管—central canal　内环骨板—inner circumferential lamella

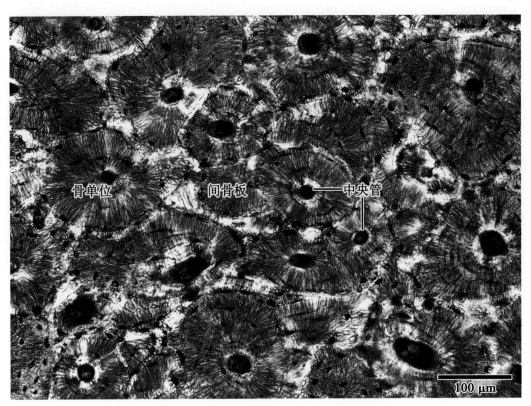

图5-48　绵羊胫骨横切中倍像2（硫堇染色）

Fig.5-48　mid magnification of sheep tibia transection 2 （thionine stain）

骨单位—osteon　间骨板—interstitial lamella　中央管—central canal

115

图5-49 绵羊胫骨横切高倍像（硫堇染色）

Fig.5-49 high magnification of sheep tibia transection (thionine stain)

骨单位—osteon 中央管—central canal 骨小管—bone canaliculus 骨陷窝—bone lacuna

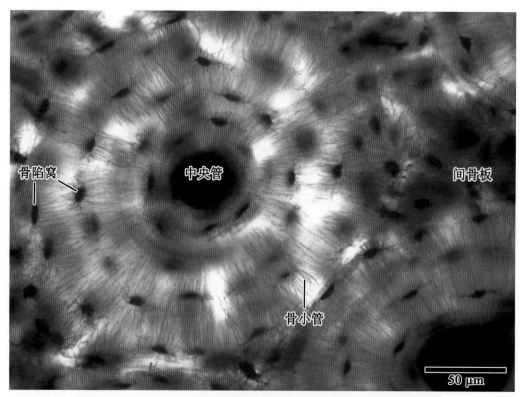

图5-50 绵羊胫骨横切高倍像（碱性品红染色）

Fig.5-50 high magnification of sheep tibia transection (basic fuchsin stain)

骨陷窝—bone lacuna 中央管—central canal 间骨板—interstitial lamella 骨小管—bone canaliculus

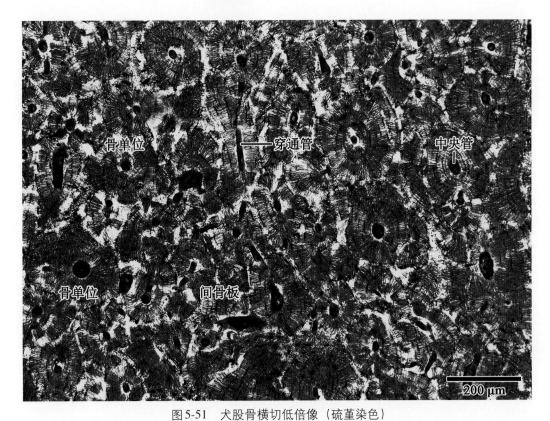

图5-51 犬股骨横切低倍像（硫堇染色）

Fig.5-51 low magnification of dog femoral transection (thionine stain)

骨单位—osteon 穿通管—perforating canal 中央管—central canal 间骨板—interstitial lamella

图5-52 犬股骨横切中倍像（硫堇染色）

Fig.5-52 mid magnification of dog femoral transection (thionine stain)

骨单位—osteon 中央管—central canal 间骨板—interstitial lamella 骨陷窝—bone lacuna

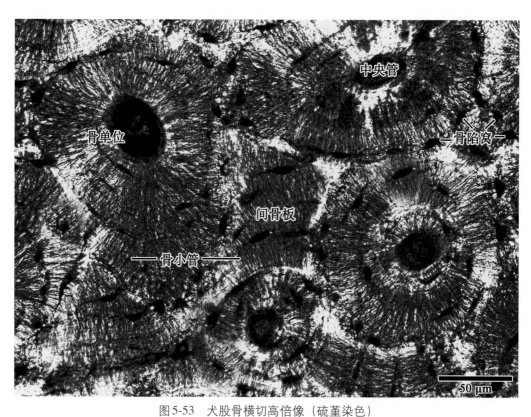

图5-53 犬股骨横切高倍像（硫堇染色）

Fig.5-53 high magnification of dog femoral transection (thionine stain)

中央管—central canal 骨单位—osteon 骨陷窝—bone lacuna 间骨板—interstitial lamella

骨小管—bone canaliculus

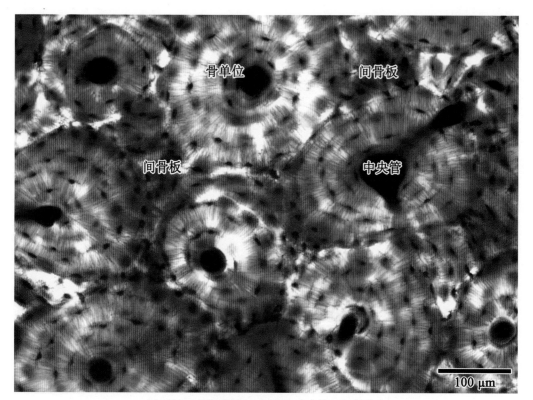

图5-54 犬胫骨横切中倍像（碱性品红染色）

Fig.5-54 mid magnification of dog tibia transection (basic fuchsin stain)

骨单位—osteon 间骨板—interstitial lamella 中央管—central canal

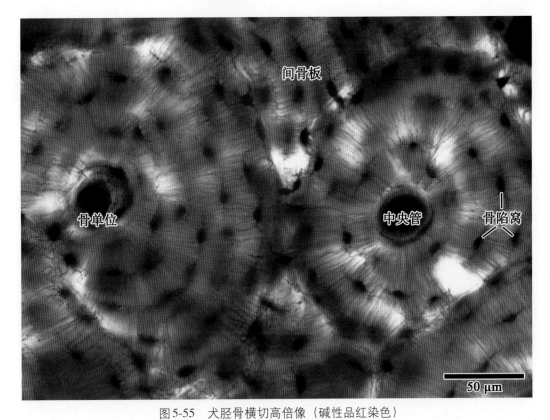

图5-55 犬胫骨横切高倍像（碱性品红染色）

Fig.5-55 high magnification of dog tibia transection (basic fuchsin stain)

骨单位—osteon 间骨板—interstitial lamella 中央管—central canal 骨陷窝—bone lacuna

图5-56 猫股骨横切中倍像1（硫堇染色）

Fig.5-56 mid magnification of cat femoral transection 1 (thionine stain)

骨单位—osteon 中央管—central canal 黏合线—cement line 内环骨板—inner circumferential lamella

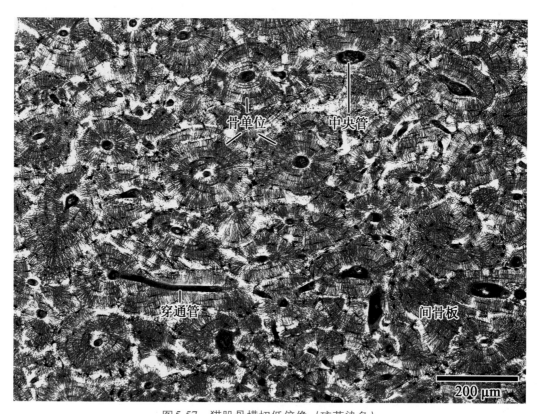

图5-57 猫股骨横切低倍像（硫堇染色）

Fig.5-57 low magnification of cat femoral transection (thionine stain)

骨单位—osteon 中央管—central canal 穿通管—perforating canal 间骨板—interstitial lamella

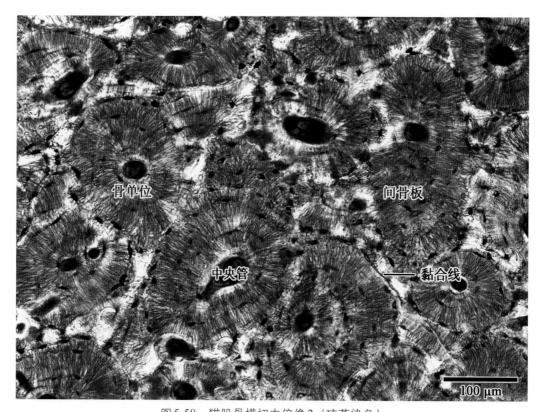

图5-58 猫股骨横切中倍像2（硫堇染色）

Fig.5-58 mid magnification of cat femoral transection 2 (thionine stain)

骨单位—osteon 间骨板—interstitial lamella 中央管—central canal 黏合线—cement line

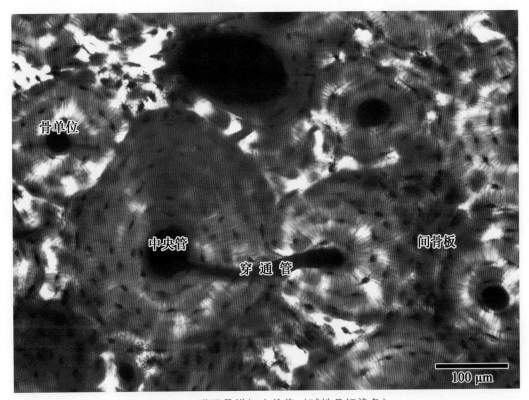

图 5-59 猫胫骨横切中倍像（碱性品红染色）

Fig.5-59 mid magnification of cat tibia transection (basic fuchsin stain)

骨单位—osteon 中央管—central canal 穿通管—perforating canal 间骨板—interstitial lamella

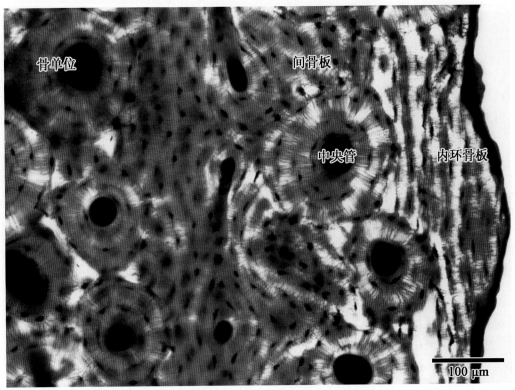

图 5-60 兔股骨横切中倍像（碱性品红染色）

Fig.5-60 mid magnification of rabbit femoral transection (basic fuchsin stain)

骨单位—osteon 间骨板—interstitial lamella 中央管—central canal 内环骨板—inner circumferential lamella

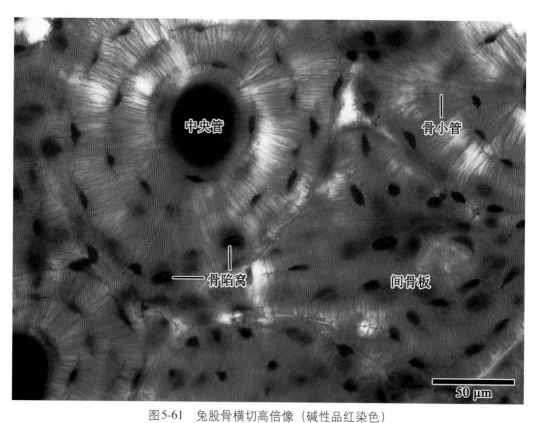

图5-61 兔股骨横切高倍像（碱性品红染色）

Fig.5-61 high magnification of rabbit femoral transection (basic fuchsin stain)

中央管—central canal 骨小管—bone canaliculus 骨陷窝—bone lacuna 间骨板—interstitial lamella

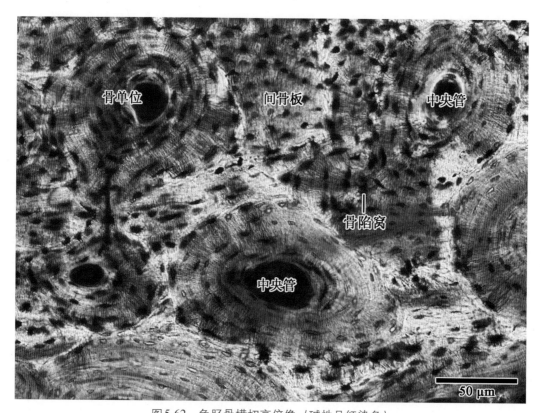

图5-62 兔胫骨横切高倍像（碱性品红染色）

Fig.5-62 high magnification of rabbit tibia transection (basic fuchsin stain)

骨单位—osteon 间骨板—interstitial lamella 中央管—central canal 骨陷窝—bone lacuna

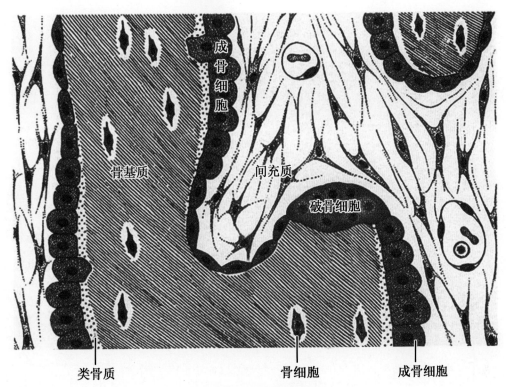

图5-63 膜内成骨模式图

Fig.5-63 diagram of intramembranous ossification

类骨质—osteoid　骨基质—bone matrix　成骨细胞—osteoblast　间充质—mesenchyme

破骨细胞—osteoclast　骨细胞—bone cell

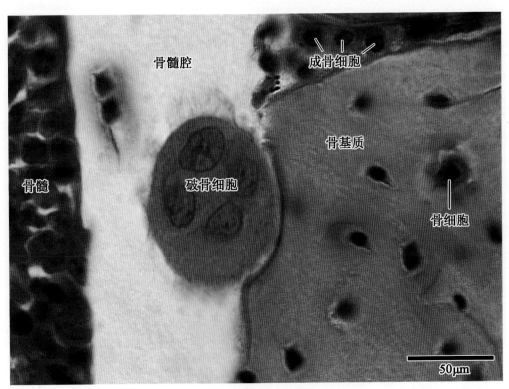

图5-64 骨发育高倍像1

Fig.5-64 high magnification of bone development 1

骨髓— bone marrow　骨髓腔—marrow cavity　成骨细胞—osteoblast　破骨细胞—osteoclast

骨基质—bone matrix　骨细胞—bone cell

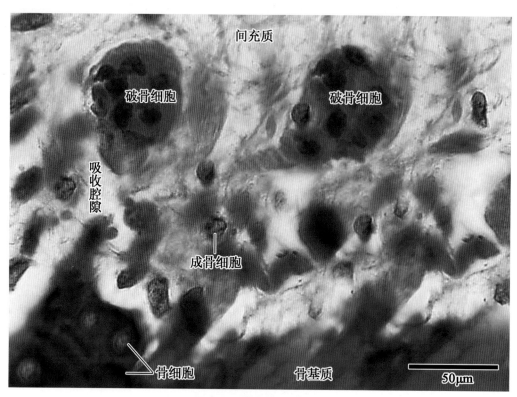

图 5-65 骨发育高倍像 2

Fig.5-65 high magnification of bone development 2

间充质—mesenchyme 破骨细胞—osteoclast 吸收腔隙—absorption lacuna

成骨细胞—osteoblast 骨细胞—bone cell 骨基质—bone matrix

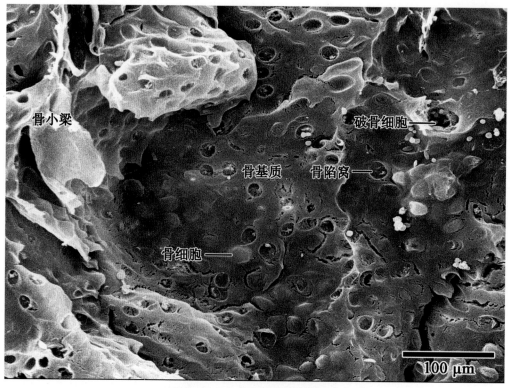

图 5-66 骨发育扫描电镜像 1

Fig.5-66 scanning electric image of bone development 1

骨小梁—bone trabecula 破骨细胞—osteoclast 骨基质—bone matrix 骨陷窝—bone lacuna 骨细胞—bone cell

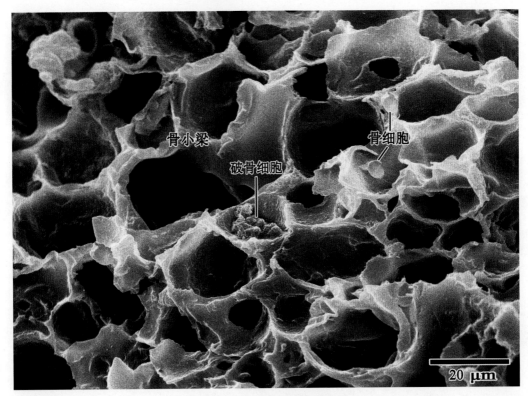

图5-67 骨发育扫描电镜像2

Fig.5-67 scanning electric image of bone development 2

骨小梁—bone trabecula 破骨细胞—osteoclast 骨细胞—bone cell

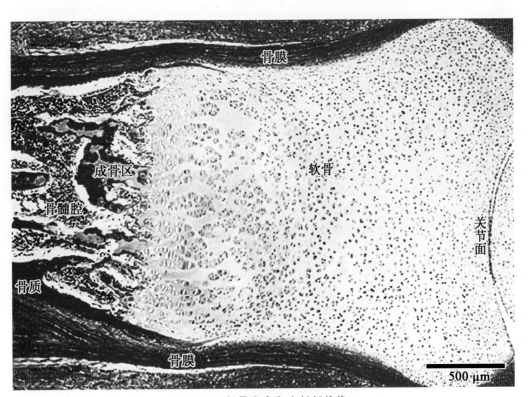

图5-68 长骨发育和生长低倍像

Fig.5-68 low magnification of long bone development and growth

骨膜—periosteum 成骨区—ossification zone 软骨—cartilage 骨髓腔—marrow cavity 骨质—sclerotin
关节面—articular surface

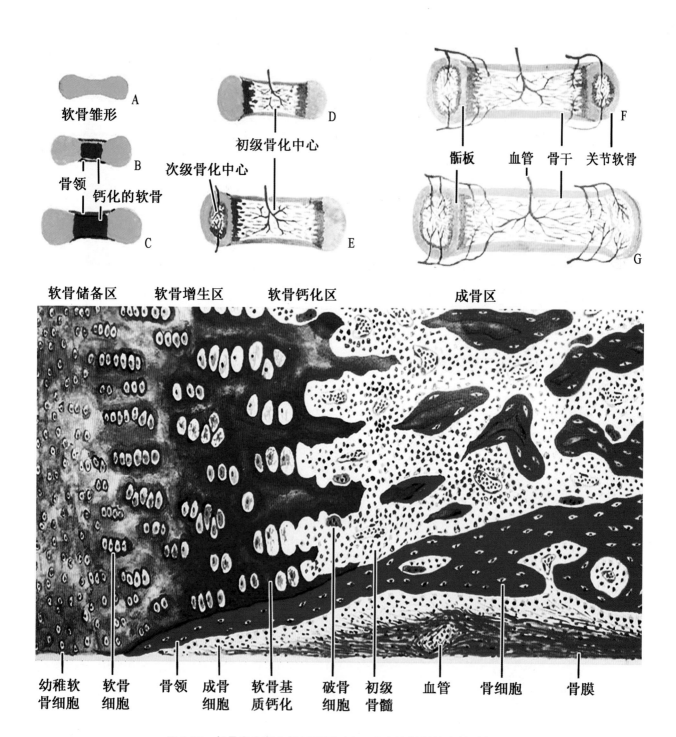

图5-69 长骨发生和生长示意图（A～G表示长骨的生长过程）

Fig.5-69 diagram of long bone development and growth (A～G indicating long bone growth process)

软骨雏形—cartilage　骨领—bone collar　钙化的软骨—calcified cartilage　初级骨化中心—primary ossification center　次级骨化中心—secondary ossification center　骺板—epiphyseal plate　血管—blood vessel　骨干—backbone　关节软骨—articular cartilage　软骨储备区—cartilage reserve zone　软骨增生区—cartilage multiplication zone　软骨钙化区—cartilage calcified zone　成骨区—ossification zone　幼稚软骨细胞—infantile cartilage cell　软骨细胞—cartilage cell　成骨细胞—osteoblast　软骨基质钙化—cartilage matrix calcification　破骨细胞—osteoclast　初级骨髓—primary bone marrow　骨细胞—osteocyte　骨膜—periosteum

第五章 软骨组织和骨组织　Cartilage and Bone Tissue

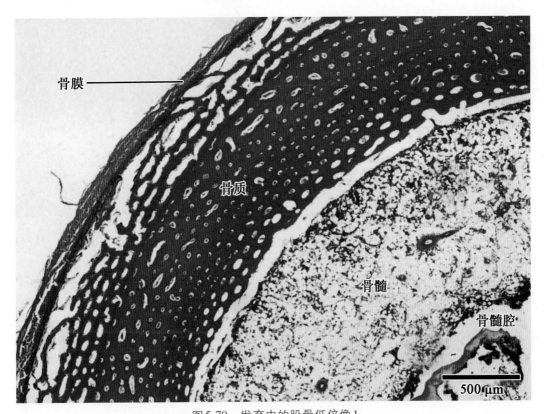

图5-70　发育中的股骨低倍像1

Fig.5-70　low magnification of developing thighbone 1

骨膜— periosteum　骨质—sclerotin　骨髓— bone marrow　骨髓腔—marrow cavity

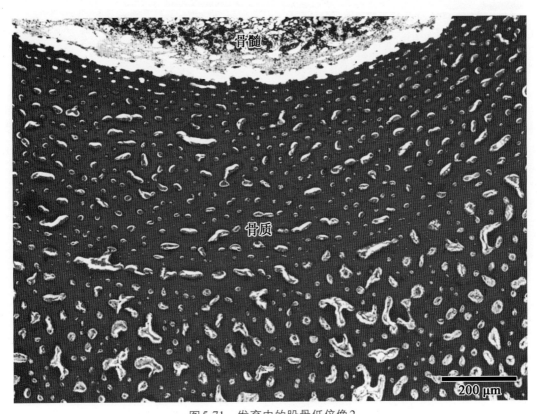

图5-71　发育中的股骨低倍像2

Fig.5-71　low magnification of developing thighbone 2

骨髓— bone marrow　骨质—sclerotin

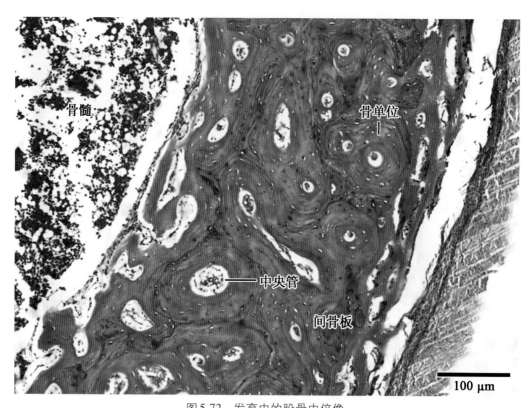

图5-72 发育中的股骨中倍像

Fig.5-72 mid magnification of developing thighbone

骨髓— bone marrow 骨单位—osteon 中央管—central canal 间骨板—interstitial lamella

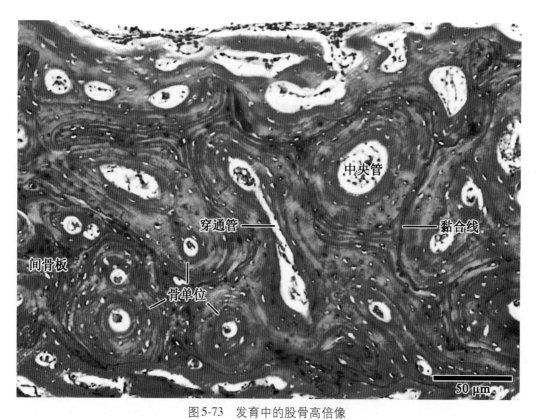

图5-73 发育中的股骨高倍像

Fig.5-73 high magnification of developing thighbone

间骨板—interstitial lamella 骨单位—osteon 穿通管—perforating canal

中央管—central canal 黏合线—cement line

第六章
血液与淋巴
Blood and Lymph

Outline

Blood and lymph are fluid connective tissues. They consist of cells and liquid intercellular substances. Blood is composed of blood cells and blood plasma. All of the blood cells are suspended in the fluids. Blood plasma, pH 7.3-7.4, contains plasma protein (albumin, globulin, fibrinogen), lipoprotein, enzyme, hormone, vitamin, inorganic salt and products of metabolism. When the fibrinogen has been removed from the blood plasma by clotting, the residual liquid is called serum.

Blood cells are divided into erythrocytes or red blood cells, leukocytes or white blood cells, and platelets. Mature erythrocytes of mammalian are biconcave disks without nuclei, but avian erythrocytes are oval with nuclei. The functions of the erythrocyte are transporting oxygen and carbon dioxide. Leukocytes are classified into two groups according to the type of granules in their cytoplasm: granulocytes and agranulocytes. Granulocytes have nuclei with two or more lobes and specific granules, and include neutrophils, eosinophils and basophils. Agranulocytes contain no specific granules. This group includes lymphocytes and monocytes. Leukocytes involve in the cellular and humoral defense of the body against foreign materials. Thrombocytes are nonnucleated, disk-like cell fragments, which originate from the fragmentation of giant polyploidy megakaryocytes that reside in the bone marrow. Thrombocytes promote blood clotting and help repair gaps in the walls of blood vessels, preventing loss of blood. The roles of platelets include primary aggregation, secondary aggregation, blood coagulation, clot retraction and clot removal. Mature blood cells have a relatively short life span, and consequently the population must be continuously replaced with the progeny of stem cells produced in the hematopoietic organs. In the earliest stages of embryogenesis, blood cells arise from the yolk sac mesoderm. Sometime later, the liver and the spleen serve as temporary hematopoietic tissue. Erythrocytes, granular leukocytes, monocytes and platelets are derived from pluripotential hematopoietic stem cells located in bone marrow. The proliferating stem form daughter cells with reduced potentiality.

> Lymph is made up of lymphocytes and tissue fluid. All of the lymphocytes are suspended in the tissue fluids. Lymphocytes are round in the microscope. They consist of small (6-8μm), medium-sized (9-12μm), and large (13-20 μm) in diameter. Lymphocyte has a round nucleus with indentation, chromatin appears as spot-liked. The cytoplasm is basophilic: bright blue in color. Lymphocytes can be classified into two groups: thymus dependent lymphocytes, TLC, which involve in cellular immune reaction and regulate immune response, bone marrow dependent lymphocytes, BLC, which become into plasma cell, involve in humoral immune response.

血液（blood）是循环流动的液态结缔组织，由红细胞、白细胞、血小板和血浆组成。血浆是流动的细胞外成分。血液环流全身，具有运送营养、氧气、二氧化碳、激素、细胞、废物和其他物质，维持体温相对恒定的重要功能。

一、血细胞的发生

血细胞最初起源于胚外卵黄囊血岛，以后造血器官迁移到肝、脾和骨髓；哺乳动物出生后，骨髓是主要的造血器官。血细胞的生成是一个较长的细胞增殖、分化、成熟和释放的过程。全部血细胞由全能造血干细胞 (pluripotential hemopoietic stem cell, PHSC) 分化而来。这些细胞经历有丝分裂，产生两种多能造血干细胞 (multipotential hemopoietic stem cell, MHSC) ——脾集落生成单位和淋巴细胞集落生成单位。大多数造血干细胞位于短骨和扁骨的红骨髓。未成年时长骨的骨髓是红色的，而成年后脂肪沉积，变成黄白色，称黄骨髓。在不同的造血生长因子刺激下，干细胞经历细胞分裂来维持循环中红细胞、白细胞和血小板的数量。

（一）红细胞的发生

红细胞由脾集落生成单位发育而来。红细胞生成素水平增高时，脾集落生成单位产生红细胞爆裂型集落生成单位的细胞，而当红细胞生成素水平降低时，遂产生红细胞集落生成单位。虽然有几代红细胞集落生成单位的产物，但最后从组织学上可辨认的是原红细胞。这些细胞产生早幼红细胞，随后经历细胞分裂形成中幼红细胞，后者进行有丝分裂，形成晚幼红细胞。这阶段的细胞不再分裂，排出细胞核分化为网织红细胞，再发育为成熟红细胞。

（二）粒细胞的发生

粒细胞系的发育开始于多潜能的脾集落生成单位。该系列中第1个从组织学上可分辨的细胞是原粒细胞，经有丝分裂形成早幼粒细胞，后者同样经历细胞分裂再形成中幼粒细胞。中幼粒细胞是该系列中最早具有特殊颗粒的细胞。所以，可分辨出中性中幼粒细胞、嗜酸性中幼粒细胞和嗜碱性中幼粒细胞。该系列下一个阶段的细胞是晚幼粒细胞，它们不再分裂，而分化为幼稚型的杆状核细胞，后者再演变为成熟粒细胞进入血流。

二、血细胞的组织学结构概述

（一）红细胞（erythrocyte）

哺乳动物成熟的红细胞呈粉红色、双面凹陷的圆盘状，直径为7～8μm，胞质内充满血红蛋白，无细胞核。

（二）有粒白细胞（granular leukocyte）

1.**中性粒细胞**（neutrophil granulocyte） 在白细胞中数量最多，直径为9～12μm，胞质浅粉红色，含有许多嗜天青颗粒和较小的特殊颗粒。特殊颗粒不易着色，这些细胞由此得名。深蓝色的细胞核，粗糙且多呈分叶状，多数为2～3叶，中间有细丝相连接。

2.**嗜酸性粒细胞**（eosinophil） 细胞直径为10～14μm，有许多有折光性的较大的橘红色球形特殊颗粒，胞质中也有嗜天青颗粒。棕色细胞核多为双叶，叶间以细丝连接。

3.嗜碱性粒细胞（basophil） 在白细胞中数目最少，直径为8～10μm。胞质中常充满粗大深染的嗜碱性特殊颗粒。这些特殊颗粒常遮盖嗜天青颗粒。细胞核呈S形，染成蓝色。

（三）无粒白细胞（agranular leukocyte）

1.淋巴细胞（lymphocyte） 组织学上淋巴细胞按体积分为小、中和大三种。大多数淋巴细胞是小型的，直径8～10μm，具有一个蓝色致密的非中心位的细胞核，占据细胞的大部分，核周仅有一薄层浅蓝色的胞质。胞质中嗜天青颗粒即溶酶体很明显。

2.单核细胞（monocyte） 单核细胞的体积最大，直径12～15μm。大量灰蓝色胞质中含有许多嗜天青颗粒。细胞核呈肾形，染色质呈空隙清晰的粗网状。

（四）血小板（blood platelet）

血小板呈小圆形、直径2～4μm的细胞碎片。没有细胞核，常聚在一起，有一个称颗粒区的深蓝色中央位的颗粒状部位和一个称作透明区的浅蓝色周边位的明亮部位。

三、骨髓

骨髓（bone marrow）是主要的造血器官，由网状组织形成支架，填充在骨髓腔内。网状组织中的网状细胞与网状纤维一起交织成网，形成血细胞生成的特定微环境。网架结构的网眼中存在着各种造血干细胞，干细胞在特定的微环境中可逐渐分化为各种细胞系的细胞。骨髓中有丰富的有孔毛细血管，形成形状不规则的迂回腔隙，分化成熟的红细胞、白细胞和血小板等成分，穿越血管内皮进入毛细血管腔内。

四、淋巴

淋巴（lymph）是流动在淋巴管内的液体，由淋巴细胞和淋巴浆组成。小部分组织液渗入毛细淋巴管内形成淋巴浆，无色透明，与血浆成分相似，在流经淋巴结后，其中的细菌等异物被清除，加入淋巴细胞和抗体。在机体不同部位和不同生理状况下，淋巴的成分和数量不同，如小肠淋巴管中的淋巴含许多脂滴，呈乳白色，又称乳糜；毛细淋巴管中的淋巴只有淋巴浆而无淋巴细胞，流经淋巴结后加入淋巴细胞。淋巴在淋巴管内向心流动，最终注入静脉，协助体液回流，是血液循环的辅助部分，在维持全身组织液动态平衡和防御中发挥重要作用。

五、血细胞与骨髓图谱

1. **血细胞** 图6-1～图6-26。
2. **白细胞** 图6-27～图6-31。
3. **骨髓** 图6-32。

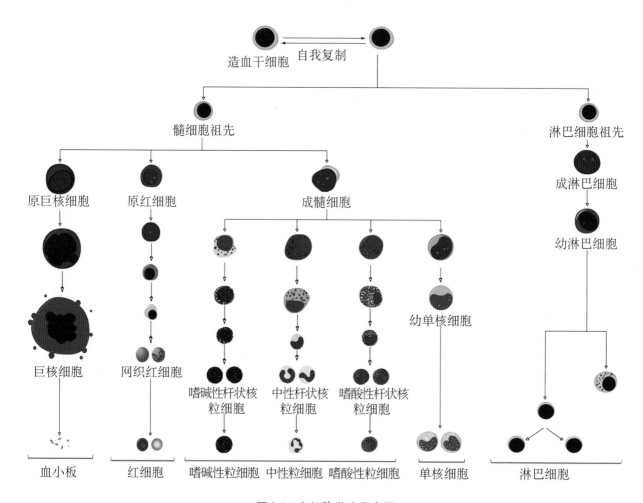

图6-1 血细胞发生示意图

Fig.6-1 hematopoiesis diagram

造血干细胞—haematopoietic stem cell　自我复制—self reproduction　髓细胞祖先—myeloid progenitor
原巨核细胞—megakaryoblast　巨核细胞—megakaryocyte　血小板（血栓细胞）—blood platelet（thrombocyte）
原红细胞—proerythroblast　网织红细胞—reticulocyte　红细胞—erythrocyte
嗜碱性杆状核粒细胞—basophilic granulocyte band form　嗜碱性粒细胞—basophil　成髓细胞—myeloblast
中性杆状核粒细胞—neutrophilic granulocyte band form　中性粒细胞—neutrophil
嗜酸性杆状核粒细胞—eosinophilic granulocyte band form　嗜酸性粒细胞—eosinophil　幼单核细胞—promonocyte
单核细胞—monocyte　淋巴细胞祖先—common lymphoid progenitor　成淋巴细胞—lymphoblast
幼淋巴细胞—prolymphocyte　淋巴细胞—lymphocyte

第六章 血液与淋巴 Blood and Lymph

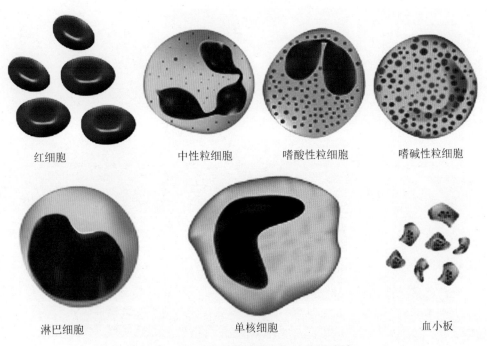

图6-2 哺乳动物血细胞光镜结构模式图

Fig.6-2 mammalia blood cells light microscope diagram

红细胞—erythrocyte 中性粒细胞—neutrophil granulocyte 嗜酸性粒细胞—eosinophil 嗜碱性粒细胞—basophil
淋巴细胞—lymphocyte 单核细胞—monocyte 血小板—blood platelet

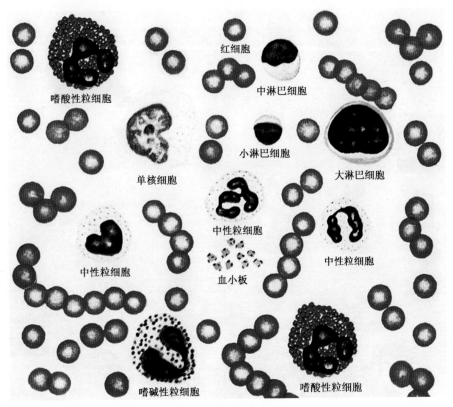

图6-3 牛血涂片模式图

Fig.6-3 cow blood smear diagram

嗜酸性粒细胞—eosinophil 红细胞—erythrocyte 中淋巴细胞—mid lymphocyte 单核细胞—monocyte
小淋巴细胞—small lymphocyte 大淋巴细胞—big lymphocyte 中性粒细胞—neutrophil 血小板—blood platelet
嗜碱性粒细胞—basophil

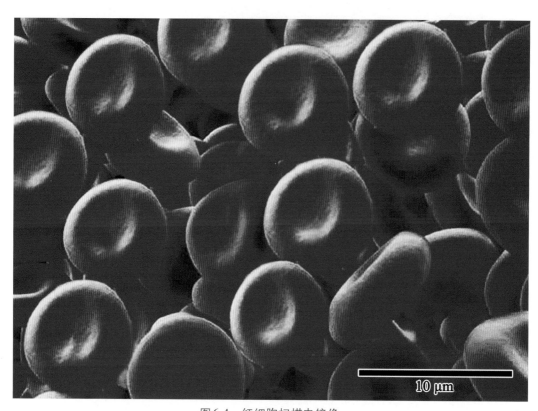

图6-4 红细胞扫描电镜像

Fig.6-4 erythrocyte scanning electron image

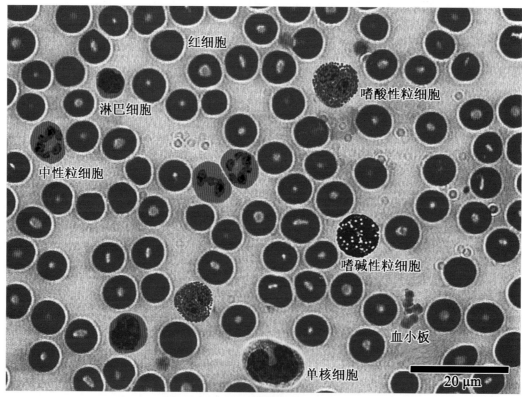

图6-5 牛血涂片高倍像（Giemsa染色）

Fig.6-5 cow blood smear high magnification（Giemsa stain）

红细胞—erythrocyte 淋巴细胞—lymphocyte 中性粒细胞—neutrophil 嗜酸性粒细胞—eosinophil
嗜碱性粒细胞—basophil 血小板—blood platelet 单核细胞—monocyte

第六章 血液与淋巴　Blood and Lymph

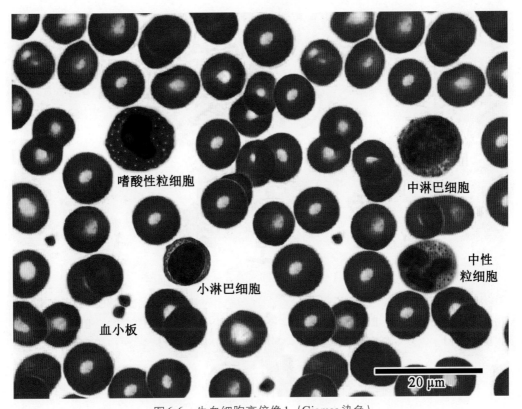

图6-6　牛血细胞高倍像1（Giemsa染色）

Fig.6-6　cow blood cells high magnification 1（Giemsa stain）

嗜酸性粒细胞—eosinophil　小淋巴细胞—small lymphocyte　血小板—blood platelet

中淋巴细胞—mid lymphocyte　中性粒细胞—neutrophil

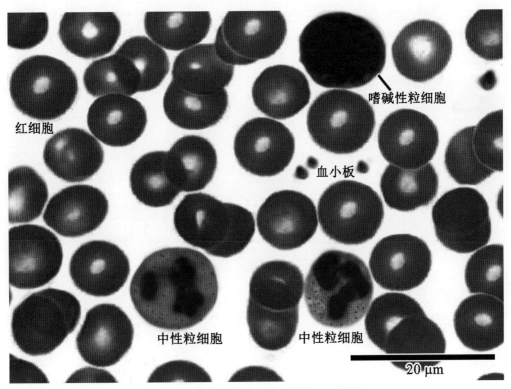

图6-7　牛血细胞高倍像2（Giemsa染色）

Fig.6-7　cow blood cells high magnification 2（Giemsa stain）

红细胞—erythrocyte　嗜碱性粒细胞—basophil　血小板—blood platelet　中性粒细胞—neutrophil

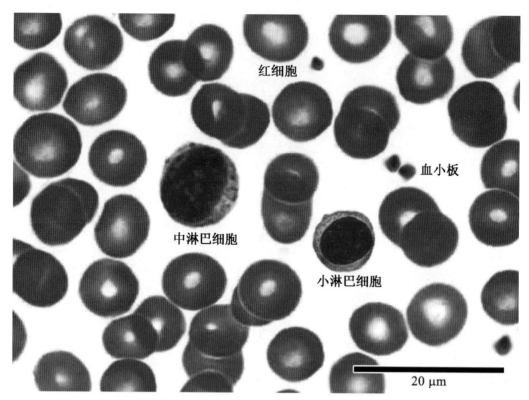

图6-8　牛血细胞高倍像3（Giemsa染色）

Fig.6-8　cow blood cells high magnification 3（Giemsa stain）

红细胞—erythrocyte　中淋巴细胞—mid lymphocyte　小淋巴细胞—small lymphocyte　血小板—bolood platelet

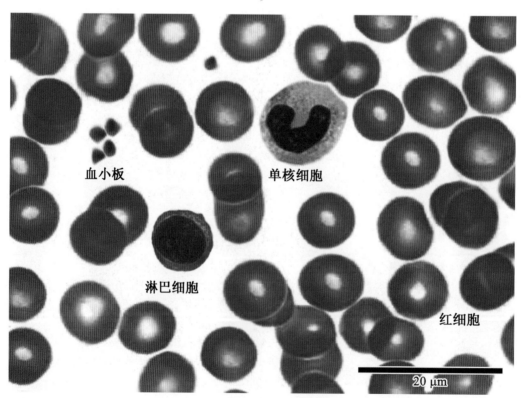

图6-9　牛血细胞高倍像4（Giemsa染色）

Fig.6-9　cow blood cells high magnification 4（Giemsa stain）

血小板—blood platelet　单核细胞—monocyte　淋巴细胞—lymphocyte　红细胞—erythrocyte

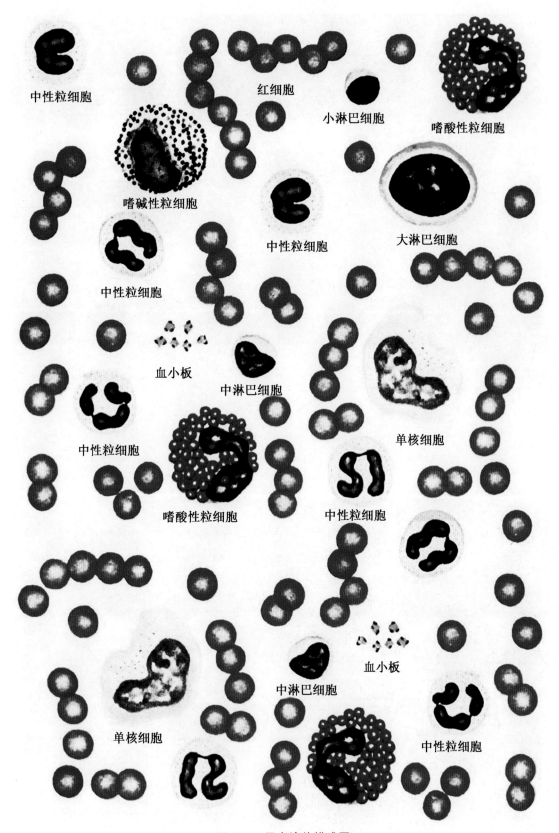

图6-10 马血涂片模式图

Fig.6-10 horse blood smear diagram

中性粒细胞—neutrophil 红细胞—erythrocyte 小淋巴细胞—small lymphocyte 嗜酸性粒细胞—eosinophil
嗜碱性粒细胞—basophil 大淋巴细胞—big lymphocyte 血小板—blood platelet
中淋巴细胞—mid lymphocyte 单核细胞—monocyte

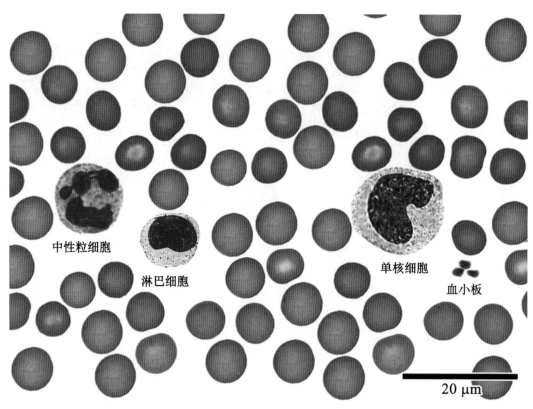

图6-11 马血细胞高倍像1（Wright染色）

Fig.6-11 horse blood cells high magnification 1（Wright stain）

中性粒细胞—neutrophil 淋巴细胞—lymphocyte 单核细胞—monocyte 血小板—blood platelet

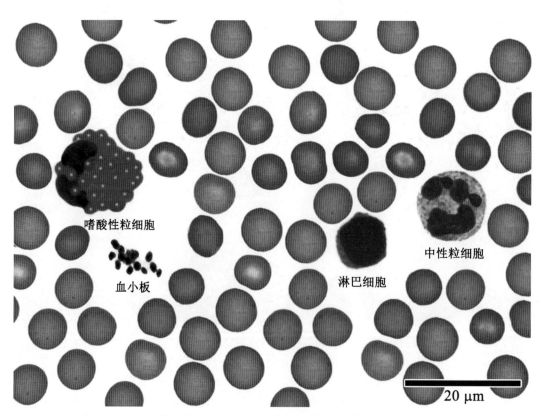

图6-12 马血细胞高倍像2（Wright染色）

Fig.6-12 horse blood cells high magnification 2（Wright stain）

嗜酸性粒细胞—eosinophil 血小板—blood platelet 淋巴细胞—lymphocyte 中性粒细胞—neutrophil

第六章 血液与淋巴　Blood and Lymph

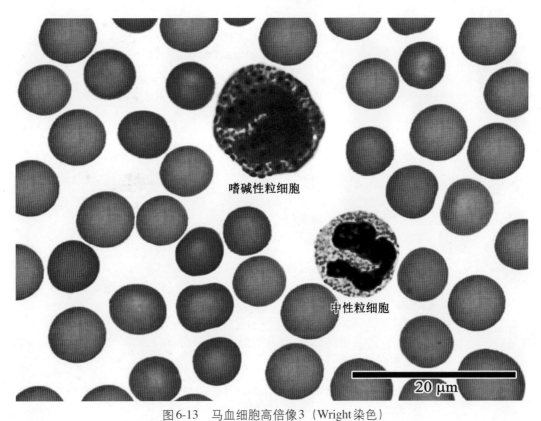

图6-13　马血细胞高倍像3（Wright染色）

Fig.6-13　horse blood cells high magnification 3（Wright stain）

嗜碱性粒细胞—basophil　中性粒细胞—neutrophil

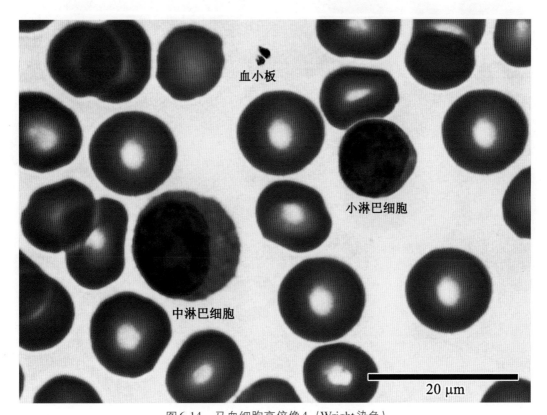

图6-14　马血细胞高倍像4（Wright染色）

Fig.6-14　horse blood cells high magnification 4（Wright stain）

血小板—blood platelet　小淋巴细胞—small lymphocyte　中淋巴细胞—mid lymphocyte

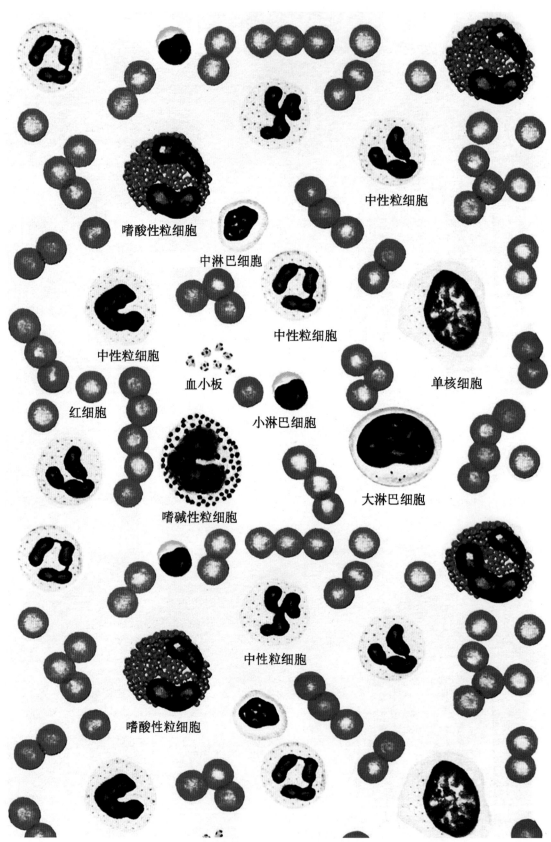

图6-15 猪血涂片模式图

Fig.6-15 pig blood smear diagram

嗜酸性粒细胞—eosinophil 中淋巴细胞—mid lymphocyte 中性粒细胞—neutrophil 血小板—blood platelet

单核细胞—monocyte 红细胞—erythrocyte 嗜碱性粒细胞—basophil 小淋巴细胞—small lymphocyte

大淋巴细胞—big lymphocyte

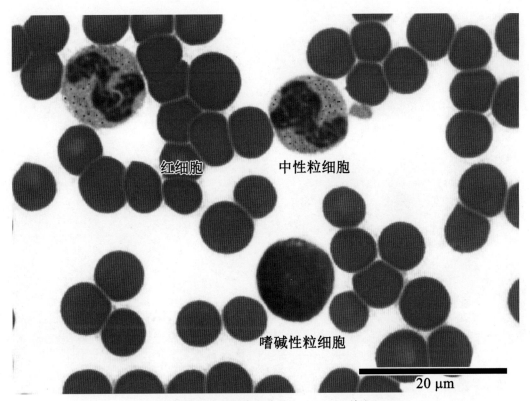

图6-16 猪血细胞高倍像1（Giemsa染色）

Fig.6-16 pig blood cells high magnification 1（Giemsa stain）

红细胞—erythrocyte 中性粒细胞—neutrophil 嗜碱性粒细胞—basophil

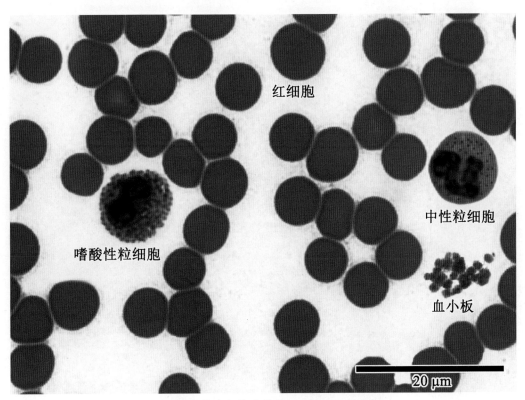

图6-17 猪血细胞高倍像2（Giemsa染色）

Fig.6-17 pig blood cells high magnification 2（Giemsa stain）

红细胞—erythrocyte 嗜酸性粒细胞—eosinophil 中性粒细胞—neutrophil 血小板—blood platelet

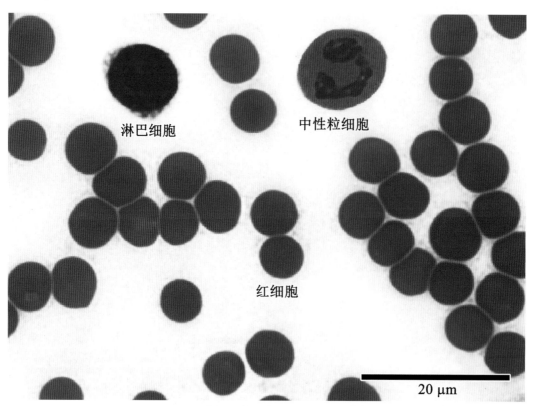

图6-18 猪血细胞高倍像3（Giemsa染色）

Fig.6-18 pig blood cells high magnification 3（Giemsa stain）

淋巴细胞—lymphocyte 中性粒细胞—neutrophil 红细胞—erythrocyte

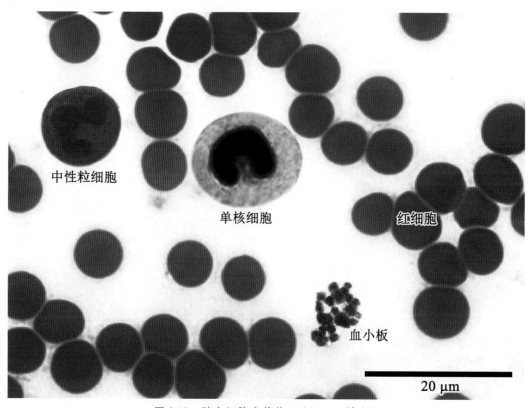

图6-19 猪血细胞高倍像4（Giemsa染色）

Fig.6-19 pig blood cells high magnification 4（Giemsa stain）

中性粒细胞—neutrophil 单核细胞—monocyte 红细胞—erythrocyte 血小板—blood platelet

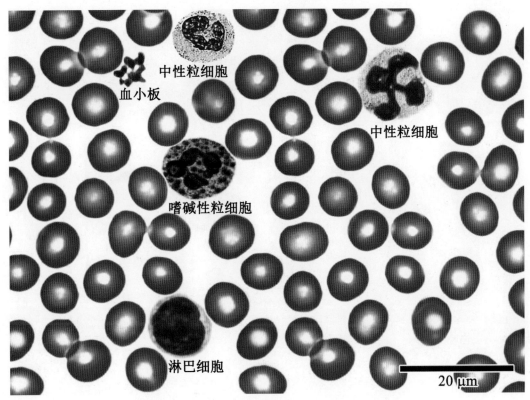

图6-20 绵羊血涂片高倍像1（Giemsa染色）

Fig.6-20 sheep blood cells high magnification 1（Giemsa stain）

血小板—blood platelet 中性粒细胞—neutrophil 嗜碱性粒细胞—basophil 淋巴细胞—lymphocyte

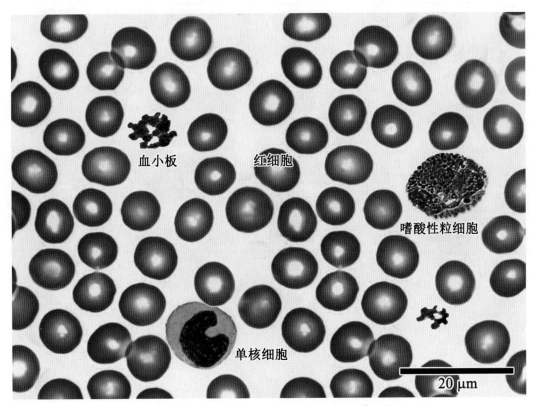

图6-21 绵羊血涂片高倍像2（Giemsa染色）

Fig.6-21 sheep blood cells high magnification 2（Giemsa stain）

血小板—blood platelet 红细胞—erythrocyte 嗜酸性粒细胞—eosinophil 单核细胞—monocyte

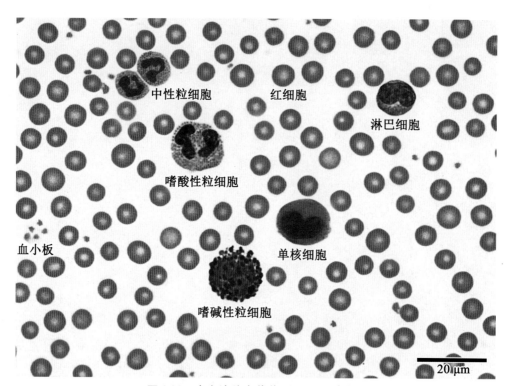

图6-22 犬血涂片高倍像（Giemsa染色）

Fig.6-22 dog blood smear diagram（Giemsa stain）

中性粒细胞—neutrophil 红细胞—erythrocyte 淋巴细胞—lymphocyte 嗜酸性粒细胞—eosinophil
血小板—blood platelet 嗜碱性粒细胞—basophil 单核细胞—monocyte

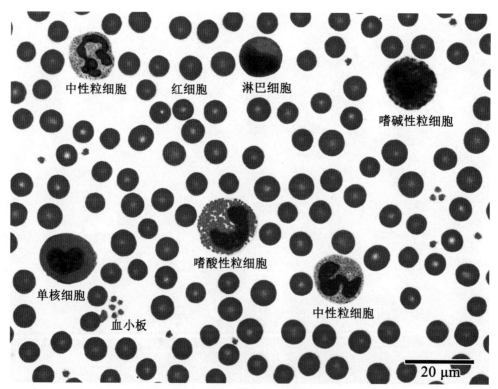

图6-23 猫血涂片高倍像（Giemsa染色）

Fig.6-23 cat blood smear diagram（Giemsa stain）

中性粒细胞—neutrophil 红细胞—erythrocyte 淋巴细胞—lymphocyte 嗜碱性粒细胞—basophil
单核细胞—monocyte 嗜酸性粒细胞—eosinophil 血小板—blood platelet

第六章 血液与淋巴 Blood and Lymph

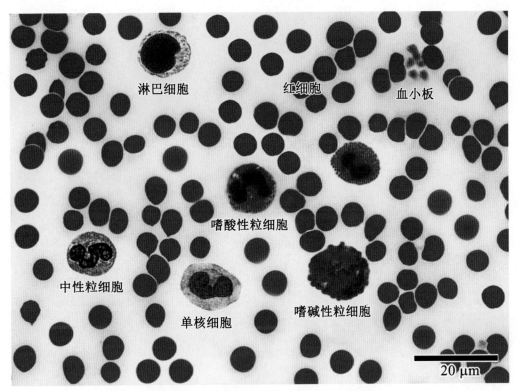

图6-24 兔血涂片高倍像（Giemsa 染色）

Fig.6-24 rabbit blood smear diagram（Giemsa stain）

淋巴细胞—lymphocyte 红细胞—erythrocyte 血小板—blood platelet 嗜酸性粒细胞—eosinophil
中性粒细胞—neutrophil 单核细胞—monocyte 嗜碱性粒细胞—basophil

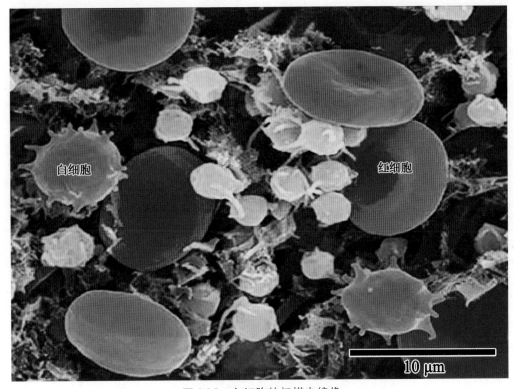

图6-25 血细胞的扫描电镜像

Fig.6-25 scanning electron microscope image of blood cells

白细胞—white cell 红细胞—erythrocyte

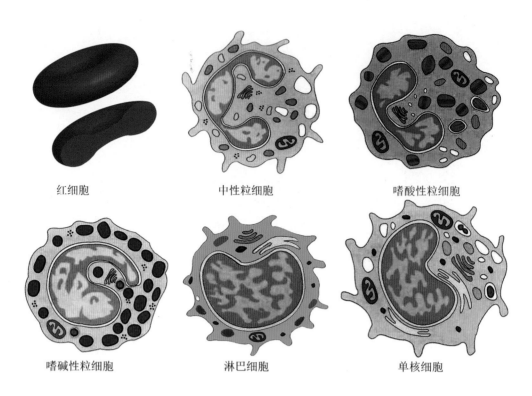

图6-26 血细胞超微结构模式图

Fig.6-26 blood cells ultrastructure diagram

红细胞—erythrocyte 中性粒细胞—neutrophil 嗜酸性粒细胞—eosinophil
嗜碱性粒细胞—basophil 淋巴细胞—lymphocyte 单核细胞—monocyte

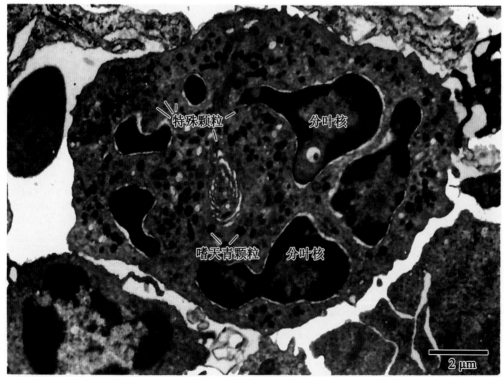

图6-27 中性粒细胞的超微结构

Fig.6-27 ultrastructure of neutrophil

特殊颗粒—specific granule 嗜天青颗粒—azurophilic granule 分叶核—lobulated nuclear

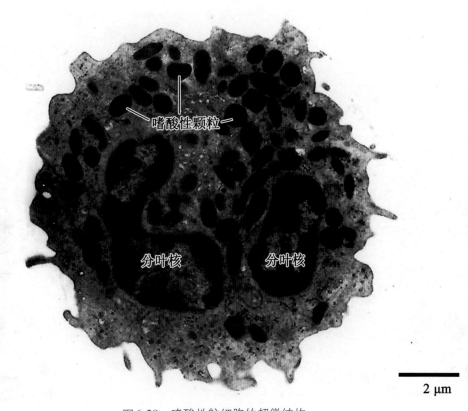

图6-28 嗜酸性粒细胞的超微结构
Fig.6-28 ultrastructure of eosinophil
嗜酸性颗粒—eosinophilic granule 分叶核—lobulated nuclear

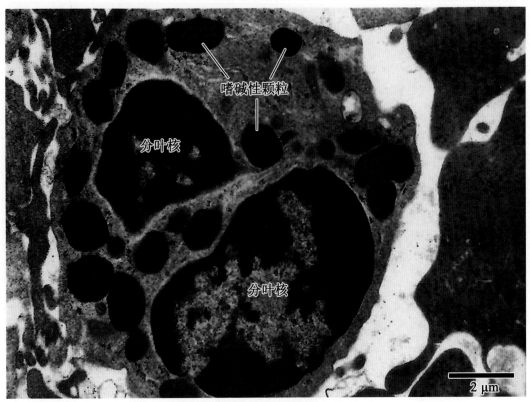

图6-29 嗜碱性粒细胞的超微结构
Fig.6-29 ultrastructure of basophil
嗜碱性颗粒—basophilic granule 分叶核—lobulated nuclear

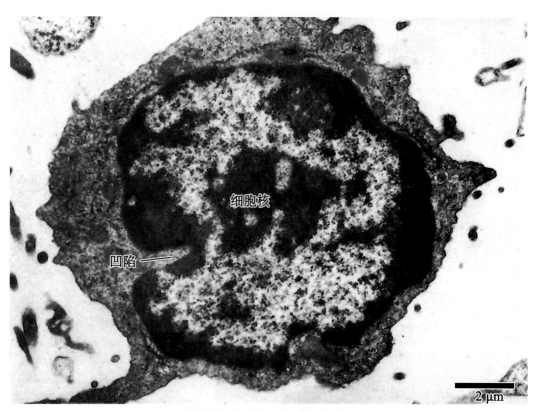

图6-30 淋巴细胞的超微结构

Fig.6-30 ultrastructure of lymphocyte

凹陷—pit 细胞核—nucleus

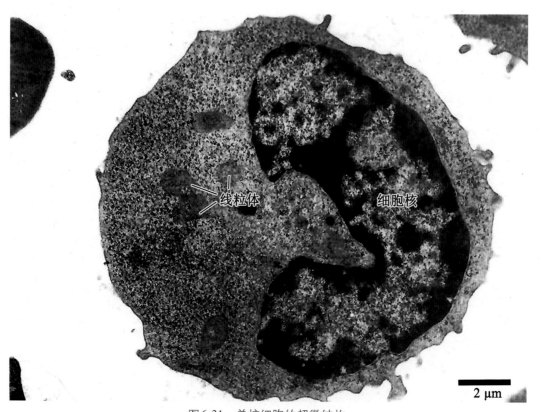

图6-31 单核细胞的超微结构

Fig.6-31 ultrastructure of monocyte

线粒体—mitochondria 细胞核—nucleus

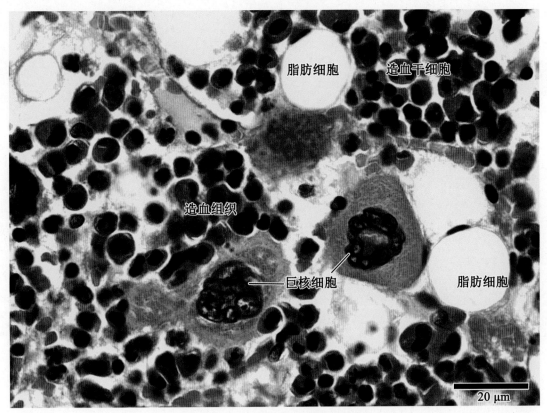

图6-32 骨髓高倍像（Giemsa染色）

Fig.6-32 high magnification of bone marrow (Giemsa stain)

脂肪细胞—fat cell　造血干细胞—hematopoietic stem cell　造血组织—hematopoietic tissue　巨核细胞—megakaryocyte

第七章 肌组织
Muscle Tissue

Outline

Muscle tissue is unique because it can contract and perform mechanical work. Muscle cells are elongated with the long axis in the direction of contractions, so they are usually referred to as muscle fibers. According to their structural and functional characteristics, muscle tissue can be divided into three kinds: skeletal muscle, cardiac muscle and smooth muscle. Muscle fibers are surrounded and supported by connective tissues that also supply their blood and nerve innervation.

Skeletal and cardiac muscles are often called cross-striated muscles in that the intracellular contractile proteins form an alternating series of transverse bands along the cell when observed with a light microscope. Skeletal muscle consists of bundles of long and cylindrical fibers which contains a lot of peripherally placed nuclei. Under voluntary nervous control, skeletal muscle is also called voluntary muscle.

Cardiac muscle is composed of elongated and branched individual cells in which one or two nuclei are centrally located and surrounded by a pale-staining cytoplasmic region. It is also with cross-striation like that of skeletal muscle. There is a special step-like structure between muscle fibers which is called intercalated disc. The intercalated discs are points of end-to-end contact between contiguous muscle fibers. Meanwhile, cardiac muscle is involuntary and innervated by the autonomic nervous system.

Smooth muscle contains spindle-shaped cells, in which the contractile proteins are not arranged in the same orderly manner so that there is no transverse striation. A single nucleus is located at the middle part. Smooth muscle is also involuntary and innervated by the autonomic nervous system.

  肌组织（muscle tissue）主要由肌细胞组成，肌细胞间有少量结缔组织和血管及神经。肌细胞因细而长，又称肌纤维。根据结构和功能特点，肌组织分为骨骼肌（skeletal muscle）、心肌（cardiac muscle）和平滑肌（smooth muscle）。它们都起源于胚胎时期的间充质。首先由间充质细胞分化为成肌细胞(myoblast)，成肌细胞再进一步发育为成熟的肌细胞。

一、肌组织的发生

（一）骨骼肌的发生

骨骼肌起源于生肌节、体壁中胚层或腮弓内的间充质细胞。间充质细胞分化为成肌细胞，后者是具单个核的梭形或有突起的细胞，胞质因含多量核糖体而呈嗜碱性，核大、椭圆形，核仁明显。成肌细胞能快速地进行分裂。有些分裂后的细胞失去分裂能力，排列成束，并互相融合成长柱状多核细胞，称肌管 (myotube)。肌管细胞内开始出现肌原纤维，核糖体则随之减少，胞质由嗜碱性渐变为嗜酸性，细胞的形态也逐渐变长。在一个肌管细胞内有数个乃至十多个细胞核，排列在肌管的中央。肌管周围的成肌细胞可继续附加、融合在肌管上。随着肌原纤维的增多，位于肌管中央的细胞核向周边移动，肌管逐渐发育成为骨骼肌细胞。在肌管时期，附着在肌管表面的单个核细胞分化为肌卫星细胞 (muscle satellite cell)。肌卫星细胞为扁平有突起的细胞，位于肌细胞膜与基膜之间，是储留在骨骼肌组织中的生肌干细胞，与骨骼肌的再生有关。

（二）心肌的发生

心肌细胞起源于原始心脏的心肌外套层间充质细胞。心肌外套层来自侧中胚层的脏壁中胚层。当左右两条心内皮管逐渐靠拢并在中线融合成一条单管时，围绕心内皮管的脏壁中胚层则形成心肌外套层，其外层将分化成心外膜，内层分化为心肌层。心肌外套层内层的间充质细胞首先分化为成肌细胞，细胞分裂增殖后，胞质内出现肌丝成为心肌细胞。血管长入心肌时，密集的心肌细胞被血管分开，显出单个核的有分支的细胞在心肌细胞间以桥粒、中间连接和缝隙连接相连，典型的闰盘到出生后才发育明显并增多。

（三）平滑肌的发生

平滑肌由胚胎时期的间充质细胞分化而来。间充质细胞首先聚集到消化管、呼吸管及血管等管道上皮的周围，细胞分裂增殖，且逐渐变为长梭形的成肌细胞，胞质中出现具有收缩功能的蛋白质。在成肌细胞发育成平滑肌细胞的过程中，具有收缩功能的蛋白质组装成粗肌丝和细肌丝，另有蛋白质组装成中间丝、密体和密斑，构成细胞骨架网。

二、肌组织的组织学结构概述

（一）骨骼肌 (skeletal muscle)

1. 纵切面 肌束膜的结缔组织成分中有神经、血管、胶原纤维和成纤维细胞等。肌内膜由细的网状纤维和基膜组成，光镜下二者都不明显。骨骼肌纤维的直径相近，平行排列成长柱状。细胞核很多，位于细胞周边。卫星细胞的细胞核明显。在高倍镜和油镜下，横纹，即明带、暗带和Z线都很清晰。

2. 横切面 结缔组织中含有成纤维细胞、毛细血管、其他的小血管和神经。肌纤维呈不规则的多角形，断面大小一致。肌原纤维呈细小点状，常形成明显的小区。在肌纤维的边缘，常有1～2个细胞核。肌束紧密地聚集在一起，肌内膜很薄，清晰地划分出肌纤维的轮廓。

（二）心肌 (cardiac muscle)

1. 纵切面 结缔组织成分清晰可见，这是因为其细胞核明显小于心肌纤维的细胞核。结缔组织富含毛细血管。肌内膜不明显。心肌纤维长而有分支并相互吻合。椭圆形细胞核比较大，位于细胞的中央，略呈泡状。细胞内有明带和暗带，但不如骨骼肌的明显；闰盘是相邻心肌纤维的端部边界标记，使用碘酸钠-苏木精特殊染色，闰盘清晰可见。心内膜下层可见到浦肯野纤维。

2. 横切面 结缔组织将心肌纤维分隔得界限明显，这些细胞的细胞核比心肌细胞的核小得多。心肌纤维的横切面不规则，大小不等。1～2个大细胞核位于细胞中央。肌原纤维聚集成放射状排列。有时可见浦肯野纤维。

（三）平滑肌（smooth muscle）

1. 纵切面 肌纤维之间的结缔组织很少，由细的网状纤维构成。一片平滑肌纤维被含有血管和神经的疏松结缔组织分隔。平滑肌纤维呈密集存在、交错排列的梭形结构。细胞核为椭圆形，位于中央；平滑肌纤维收缩时，细胞核呈现一种特有的螺旋形。

2. 横切面 结缔组织很少，主要是网状纤维，可在细胞间隙中观察到。平滑肌块和肌束彼此由疏松结缔组织分隔，其中有神经与血管。因平滑肌纤维是密集交错的梭形结构，因此横切面产生圆形、直径不同、外表均匀的断面。1个细胞核位于横断面的最宽处，故横切面上的细胞核很少。

三、肌组织图谱

1. **骨骼肌** 图 7-1 ～图 7-23。
2. **心肌** 图 7-24 ～图 7-38。
3. **平滑肌** 图 7-39 ～图 7-46。

第七章 肌组织 Muscle Tissue

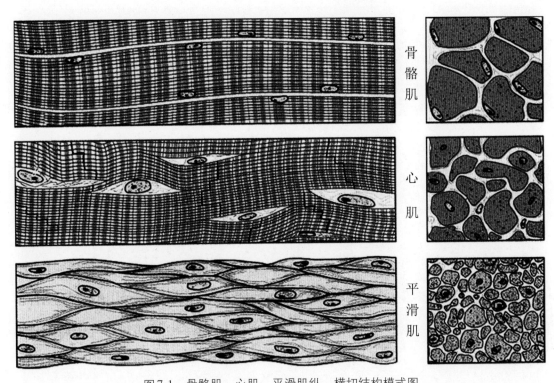

图 7-1 骨骼肌、心肌、平滑肌纵、横切结构模式图
Fig.7-1 structure pattern diagram of skeletal, cardiac and smooth muscles
骨骼肌—skeletal muscle 心肌—cardiac muscle 平滑肌—smooth muscle

图 7-2 骨骼肌结构模式图
Fig.7-2 structure pattern diagram of skeletal muscle
肌细胞—muscle fiber 肌束—muscle bundle 肌束膜—perimysium 肌外膜—epimysium
肌原纤维—myofibril 肌细胞核—muscle nucleus 肌内膜—endomysium 肌膜—sarcolemma

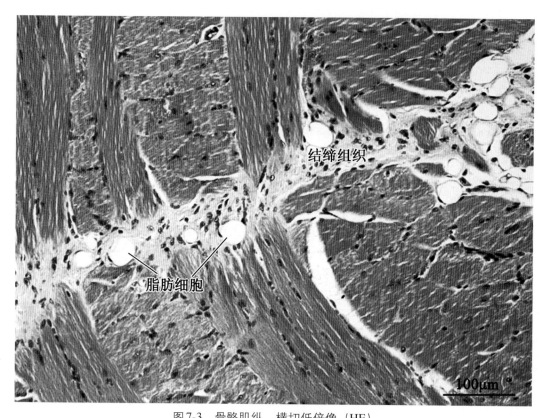

图7-3 骨骼肌纵、横切低倍像（HE）

Fig.7-3 low amplification of skeletal muscle longitudinal and cross section（HE）

结缔组织—connective tissue 脂肪细胞—fat cell

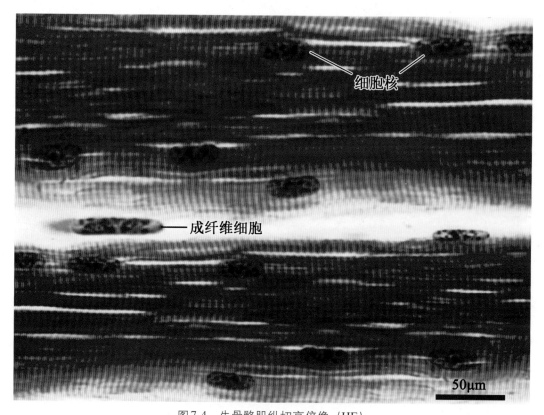

图7-4 牛骨骼肌纵切高倍像（HE）

Fig.7-4 high amplification of ox skeletal muscle longitudinal section（HE）

细胞核—nucleus 成纤维细胞—fibroblast

第七章 肌组织 Muscle Tissue

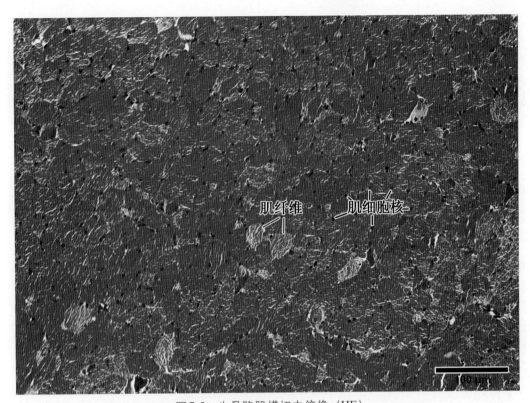

图7-5 牛骨骼肌横切中倍像（HE）
Fig.7-5 mid amplification of ox skeletal muscle cross section（HE）
肌纤维—muscle fiber　肌细胞核—muscle nucleus

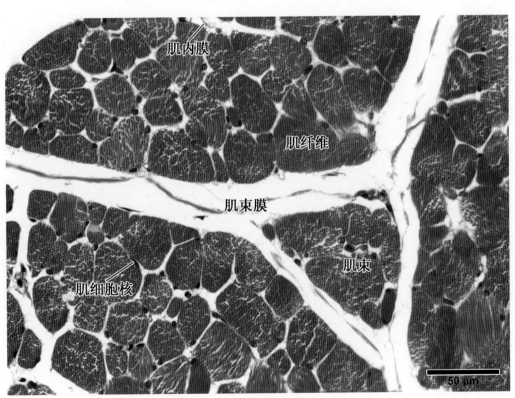

图7-6 牛骨骼肌横切高倍像（HE）
Fig.7-6 high amplification of ox skeletal muscle cross section（HE）
肌内膜—endomysium　肌纤维—muscle fiber　肌束膜—perimysium
肌细胞核—muscle nucleus　肌束—muscular bundle

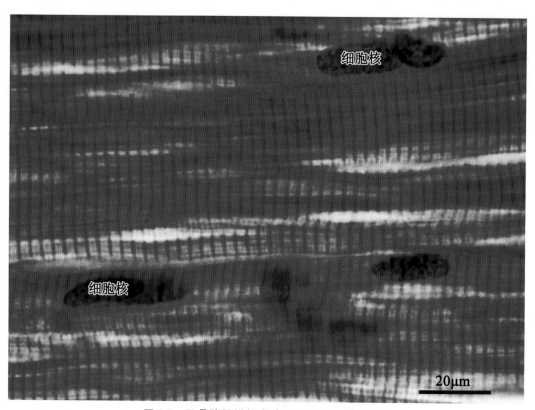

图7-7　马骨骼肌纵切高倍像（铁苏木精染色）
Fig.7-7　high amplification of horse skeletal muscle longitudinal section（iron hematoxylin stain）
细胞核—nucleus

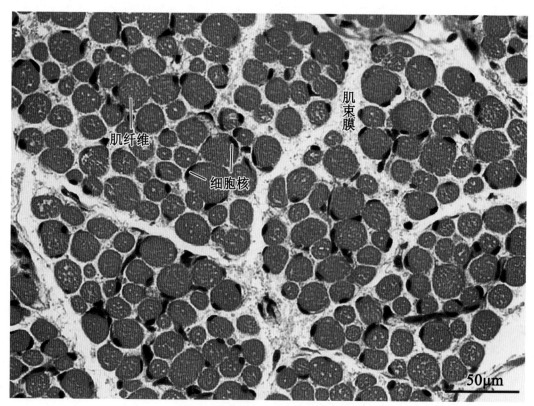

图7-8　马骨骼肌横切中倍像（HE）
Fig.7-8　mid amplification of horse skeletal muscle cross section（HE）
肌纤维—muscle fiber　细胞核—nucleus　肌束膜—perimysium

第七章 肌组织 Muscle Tissue

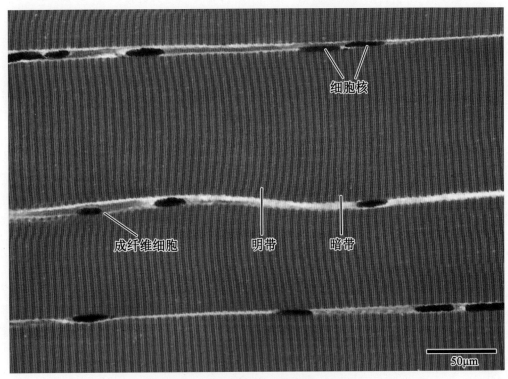

图7-9 猪骨骼肌纵切高倍像（HE）
Fig.7-9 high amplification of pig skeletal muscle longitudinal section（HE）
细胞核—nucleus　成纤维细胞—fibroblast　明带—light band　暗带—dark band

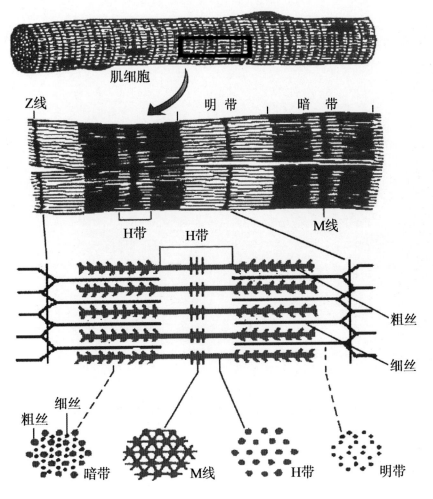

图7-10 骨骼肌纤维结构模式图
Fig.7-10 structure pattern diagram of skeletal muscle fiber
肌细胞—muscle fiber　Z线—Z line
明带—light band　暗带—dark band
H带—H band　M线—M line
粗丝—thick filament
细丝—thin filament

157

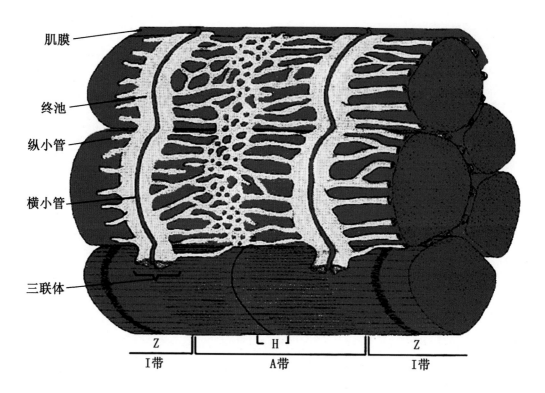

图7-11 骨骼肌纤维超微结构模式图

Fig.7-11 ultrastructure pattern diagram of skeletal muscle fiber

肌膜—sarcolemma 终池—terminal cisterna 纵小管—longitudinal tubule 横小管—transverse tubule
三联体—triplet Z—Z line I带—light band A带—dark band H—H band

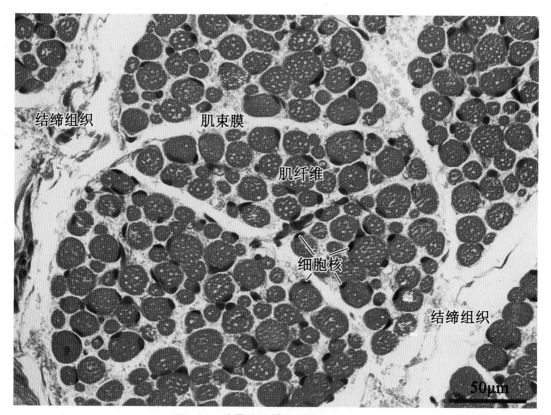

图7-12 猪骨骼肌横切高倍像1（HE）

Fig.7-12 high amplification of pig skeletal muscle cross section 1（HE）

结缔组织—connective tissue 肌束膜—perimysium 肌纤维—muscle fiber 细胞核—nucleus

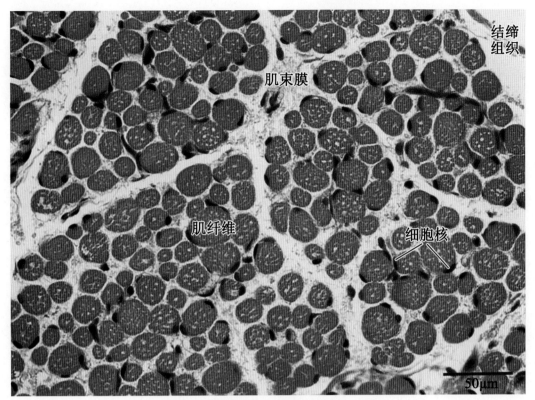

图7-13 猪骨骼肌横切高倍像2（HE）

Fig.7-13 high amplification of pig skeletal muscle cross section 2（HE）

肌束膜—perimysium 结缔组织—connective tissue 肌纤维—muscle fiber 细胞核—nucleus

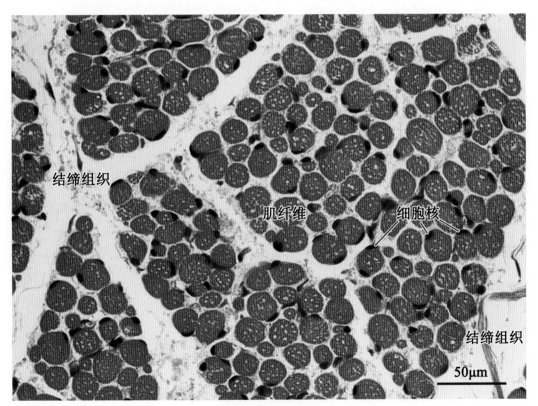

图7-14 猪骨骼肌横切高倍像3（HE）

Fig.7-14 high amplification of pig skeletal muscle cross section 3（HE）

结缔组织—connective tissue 肌纤维—muscle fiber 细胞核—nucleus

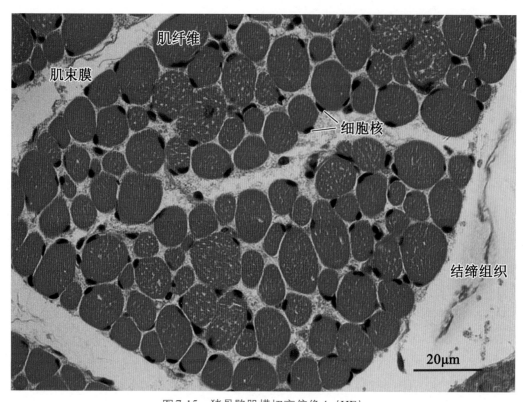

图7-15　猪骨骼肌横切高倍像4（HE）

Fig.7-15　high amplification of pig skeletal muscle cross section 4（HE）

肌束膜—perimysium　肌纤维—muscle fiber　细胞核—nucleus　结缔组织—connective tissue

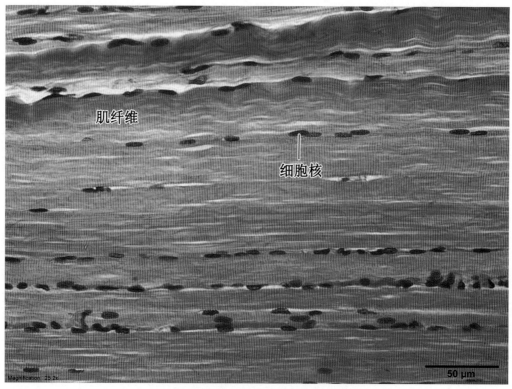

图7-16　绵羊骨骼肌纵切中倍像（铁苏木精染色）

Fig.7-16　mid amplification of sheep skeletal muscle longitudinal section（iron hematoxylin stain）

肌纤维—muscle fiber　细胞核—nucleus

第七章 肌组织 Muscle Tissue

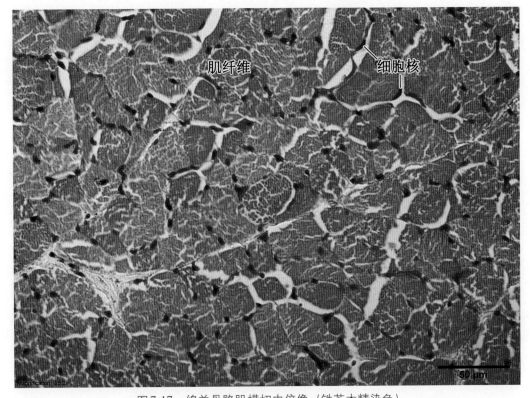

图7-17 绵羊骨骼肌横切中倍像（铁苏木精染色）
Fig.7-17 mid amplification of sheep skeletal muscle cross section（iron hematoxylin stain）
肌纤维—muscle fiber 细胞核—nucleus

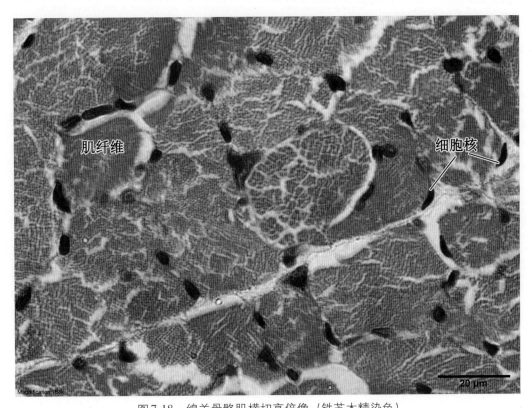

图7-18 绵羊骨骼肌横切高倍像（铁苏木精染色）
Fig.7-18 high amplification of sheep skeletal muscle cross section（iron hematoxylin stain）
肌纤维—muscle fiber 细胞核—nucleus

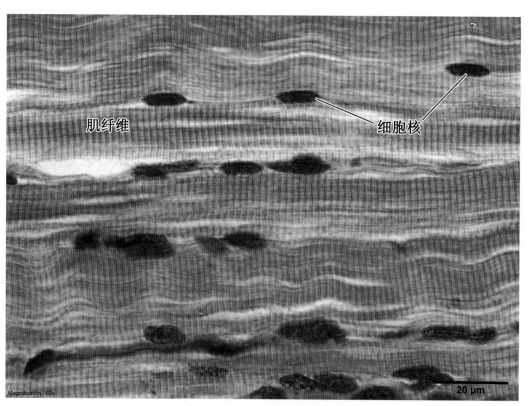

图7-19 犬骨骼肌纵切高倍像（铁苏木精染色）

Fig.7-19 high amplification of dog skeletal muscle longitudinal section（iron hematoxylin stain）

肌纤维—muscle fiber　细胞核—nucleus

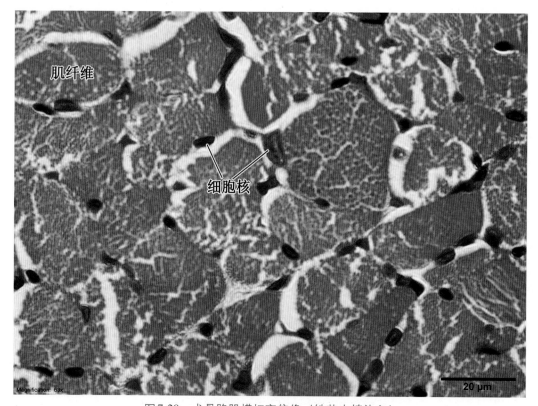

图7-20 犬骨骼肌横切高倍像（铁苏木精染色）

Fig.7-20 high amplification of dog skeletal muscle cross section（iron hematoxylin stain）

肌纤维—muscle fiber　细胞核—nucleus

第七章 肌组织 Muscle Tissue

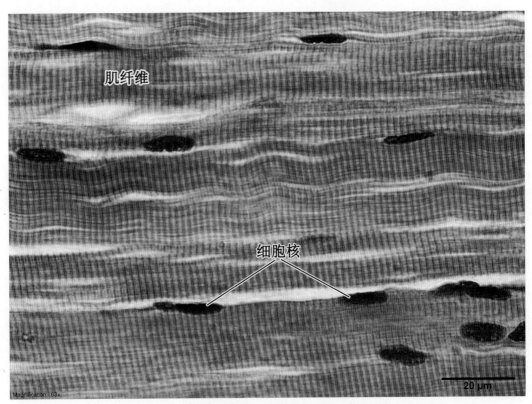

图 7-21 猫骨骼肌纵切高倍像（铁苏木精染色）

Fig.7-21 high amplification of cat skeletal muscle longitudinal section（iron hematoxylin stain）

肌纤维—muscle fiber 细胞核—nucleus

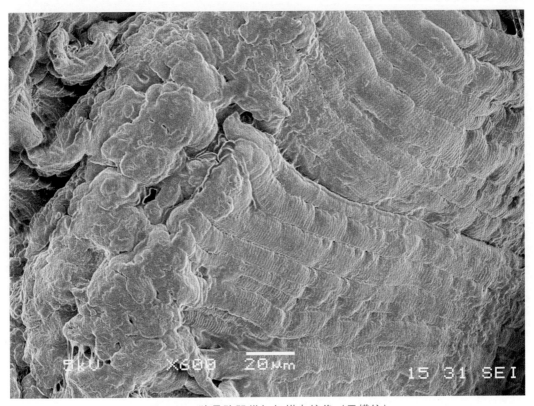

图 7-22 猪骨骼肌纵切扫描电镜像（示横纹）

Fig.7-22 scanning electron micrograph of pig skeletal muscle longitudinal section（show the band）

图7-23 猪骨骼肌纵切透射电镜像

Fig.7-23 transmission electron micrograph of pig skeletal muscle longitudinal section

暗带—dark band 明带—light band M线—M line Z线—Z line 肌节—sarcomere

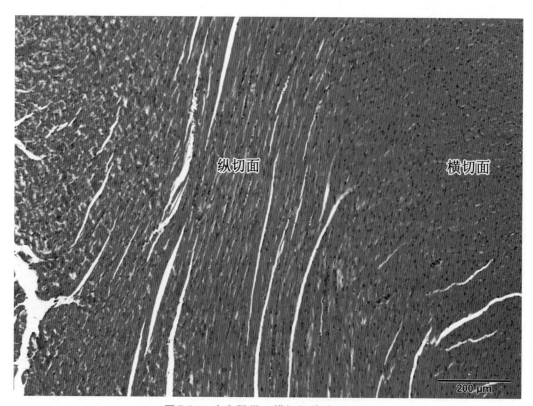

图7-24 牛心肌纵－横切低倍像（HE）

Fig.7-24 low amplification of ox cardiac muscle longitudinal-cross section（HE）

纵切面—longitudinal section view 横切面—cross section view

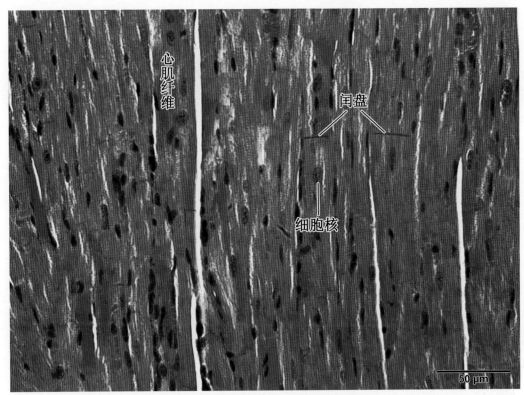

图7-25 牛心肌纵切中倍像（HE）

Fig.7-25 mid amplification of ox cardiac muscle longitudinal section（HE）

心肌纤维—muscle fiber 闰盘—intercalated disc 细胞核—nucleus

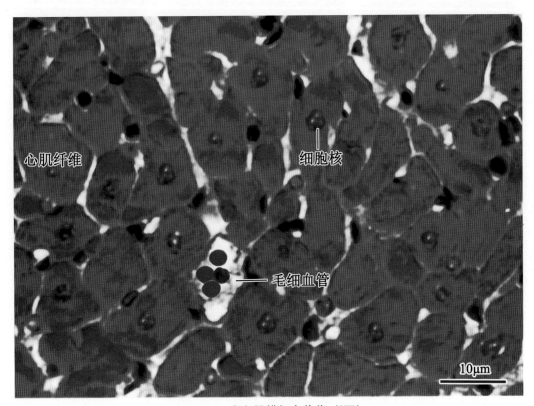

图7-26 牛心肌横切高倍像（HE）

Fig.7-26 high amplification of ox cardiac muscle cross section（HE）

心肌纤维—muscle fiber 细胞核—nucleus 毛细血管—capillary

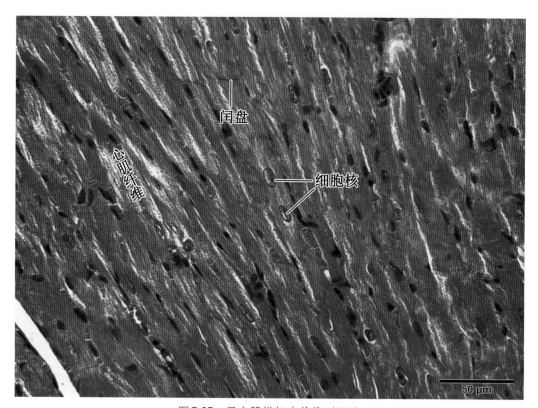

图7-27 马心肌纵切中倍像（HE）
Fig.7-27 mid amplification of horse cardiac muscle longitudinal section（HE）
心肌纤维—muscle fiber　闰盘— intercalated disc　细胞核—nucleus

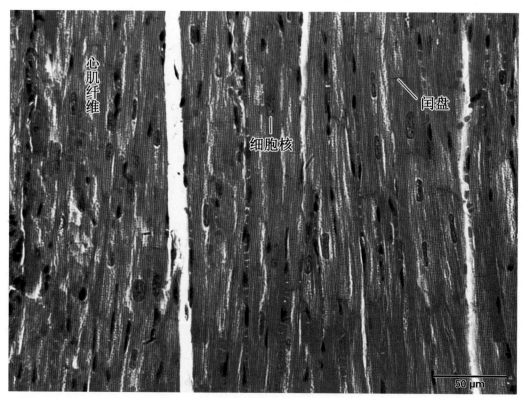

图7-28 猪心肌纵切中倍像（HE）
Fig.7-28 mid amplification of pig cardiac muscle longitudinal section（HE）
心肌纤维—muscle fiber　细胞核—nucleus　闰盘— intercalated disc

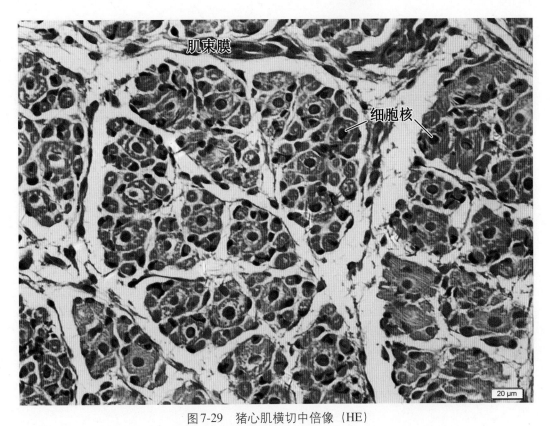

图7-29 猪心肌横切中倍像（HE）

Fig.7-29 mid amplification of pig cardiac muscle cross section（HE）

肌束膜—perimysium 细胞核—nucleus

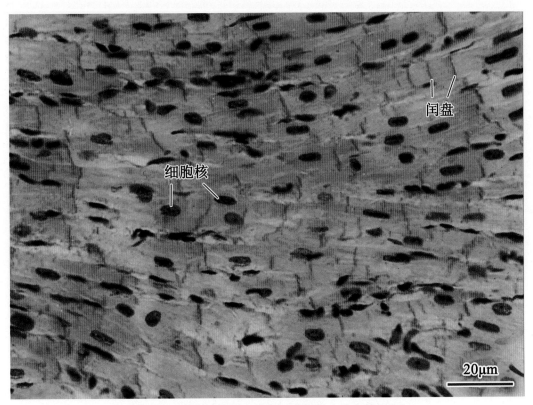

图7-30 牛心肌纵切中倍像（碘酸钠－苏木精染色）

Fig.7-30 mid amplification of ox cardiac muscle longitudinal section（sodium iodate-hematoxylin stain）

细胞核—nucleus 闰盘—intercalated disc

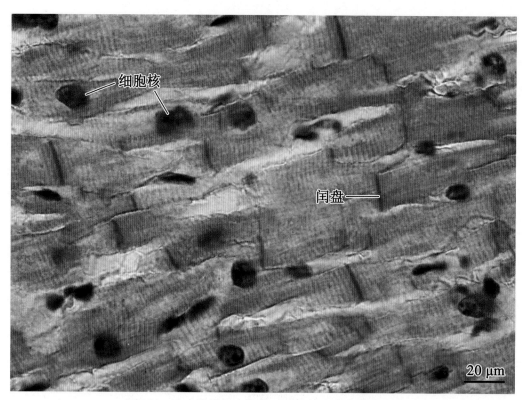

图7-31 牛心肌纵切高倍像（碘酸钠－苏木精染色）

Fig.7-31 high amplification of ox cardiac muscle longitudinal section（sodium iodate-hematoxylin stain）

细胞核—nucleus 闰盘—intercalated disc

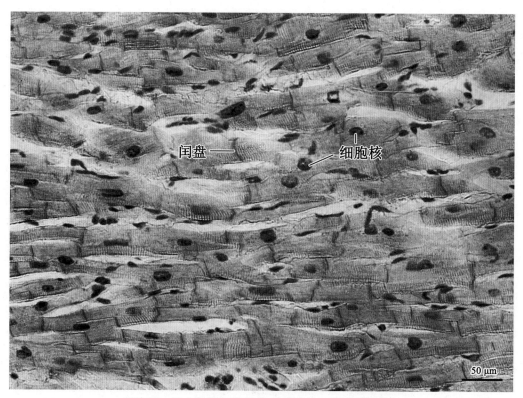

图7-32 马心肌纵切高倍像（碘酸钠－苏木精染色）

Fig.7-32 high amplification of horse cardiac muscle longitudinal section（sodium iodate-hematoxylin stain）

闰盘—intercalated disc 细胞核—nucleus

第七章 肌组织 Muscle Tissue

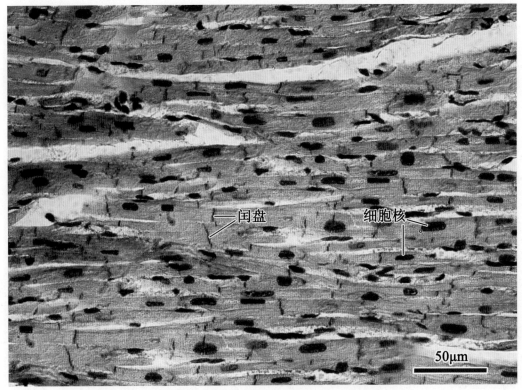

图 7-33 猪心肌纵切中倍像（碘酸钠-苏木精染色）
Fig.7-33 mid amplification of pig cardiac muscle longitudinal section (sodium iodate-hematoxylin stain)
闰盘— intercalated disc　细胞核—nucleus

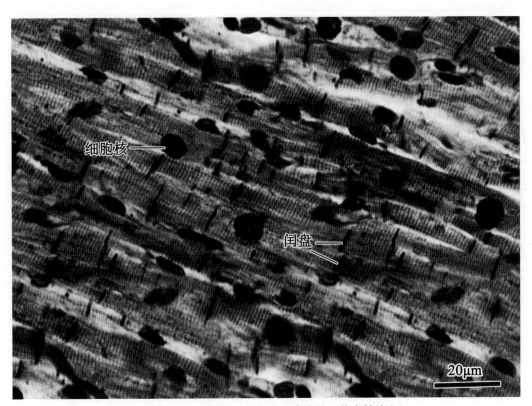

图 7-34 猪心肌纵切高倍像（碘酸钠-苏木精染色）
Fig.7-34 high amplification of pig cardiac muscle longitudinal section (sodium iodate-hematoxylin stain)
细胞核—nucleus　闰盘— intercalated disc

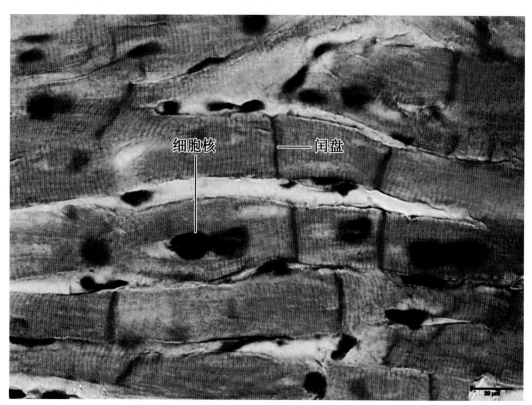

图7-35 绵羊心肌纵切高倍像（碘酸钠-苏木精染色）

Fig.7-35 high amplification of sheep cardiac muscle longitudinal section（sodium iodate-hematoxylin stain）

细胞核—nucleus 闰盘—intercalated disc

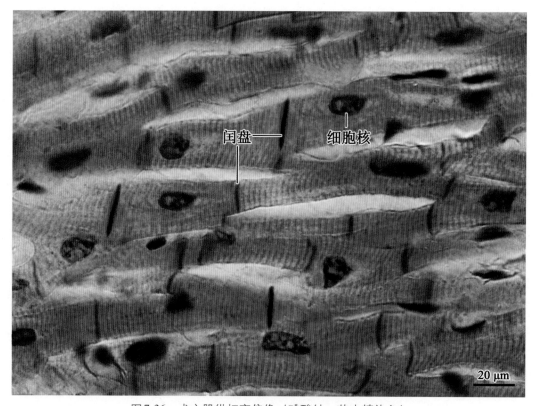

图7-36 犬心肌纵切高倍像（碘酸钠-苏木精染色）

Fig.7-36 high amplification of dog cardiac muscle longitudinal section（sodium iodate-hematoxylin stain）

闰盘—intercalated disc 细胞核—nucleus

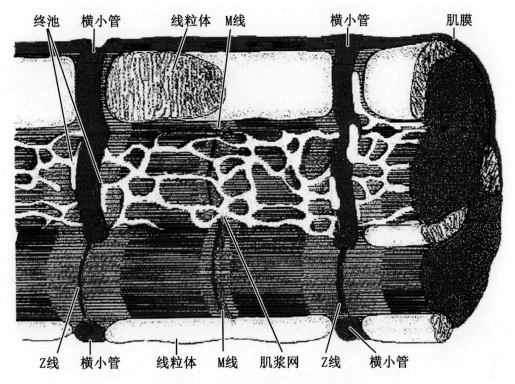

图7-37 心肌纤维超微结构模式图

Fig.7-37 ultrastructure pattern diagram of cardiac muscle fiber

终池—terminal cisterna　横小管—transverse tubule　线粒体—mitochondria　M线—M line　肌膜—sarcolemma
Z线—Z line　肌浆网—sarcoplasmic reticulum

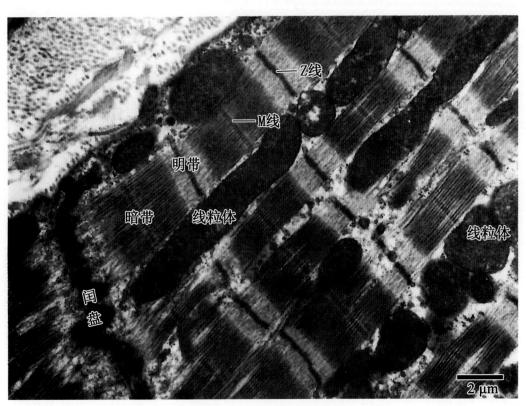

图7-38 心肌纤维的超微结构

Fig.7-38 ultrastructure pattern of cardiac muscle fiber

闰盘—intercalated disc　暗带—dark band　明带—light band　M线—M line　Z线—Z line　线粒体—mitochondria

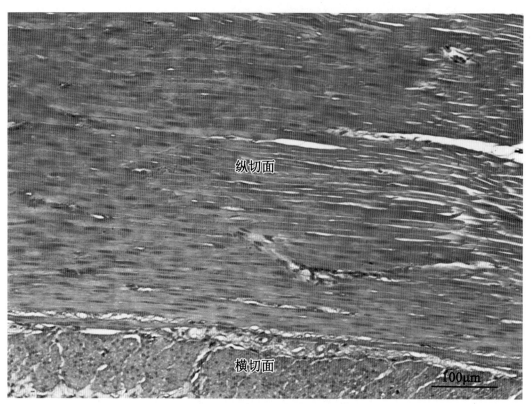

图7-39 牛平滑肌纵-横切中倍像（HE）
Fig.7-39 mid amplification of ox smooth muscle longitudinal-cross section（HE）
纵切面—longitudinal section view 横切面—cross section view

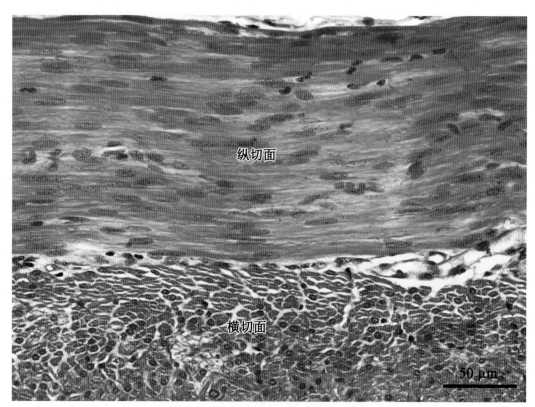

图7-40 马平滑肌纵-横切中倍像（HE）
Fig.7-40 mid amplification of horse smooth muscle longitudinal-cross section（HE）
纵切面—longitudinal section view 横切面—cross section view

第七章 肌组织 Muscle Tissue

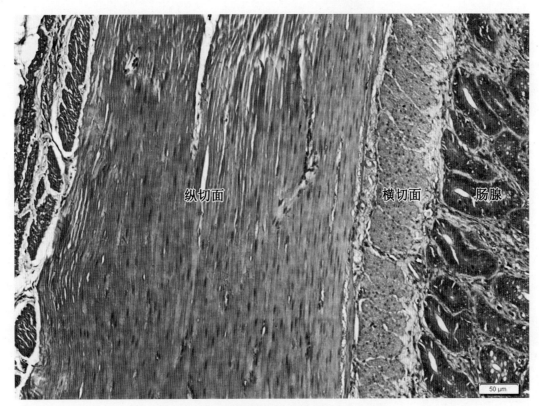

图7-41 山羊平滑肌纵－横切中倍像（HE）
Fig.7-41 mid amplification of goat smooth muscle longitudinal-cross section（HE）
纵切面—longitudinal section view 横切面—cross section view 肠腺—intestinal gland

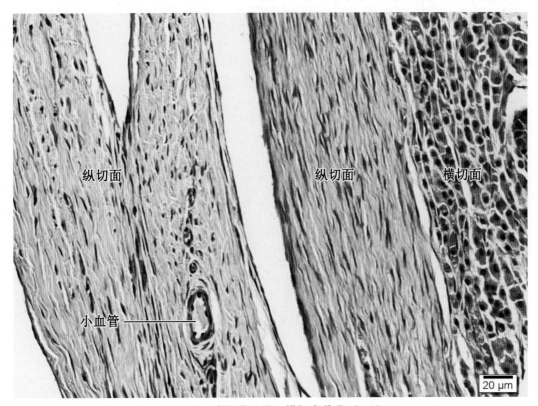

图7-42 猪平滑肌纵－横切高倍像（HE）
Fig.7-42 high amplification of pig smooth muscle longitudinal-cross section（HE）
纵切面—longitudinal section view 横切面—cross section view 小血管—small blood vessel

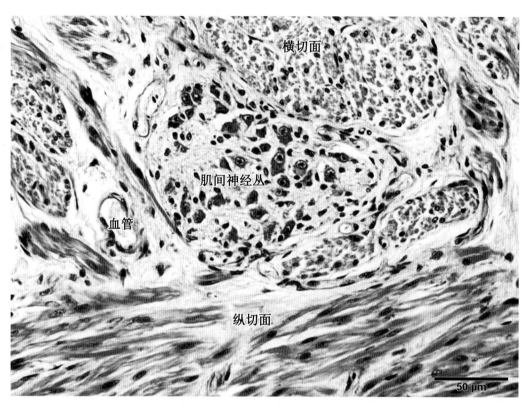

图7-43 犬平滑肌纵－横切高倍像（HE）

Fig.7-43 high amplification of dog smooth muscle longitudinal-cross section（HE）

血管—blood vessel 横切面—cross section view 纵切面—longitudinal section view

肌间神经丛—myenteric nerve plexus

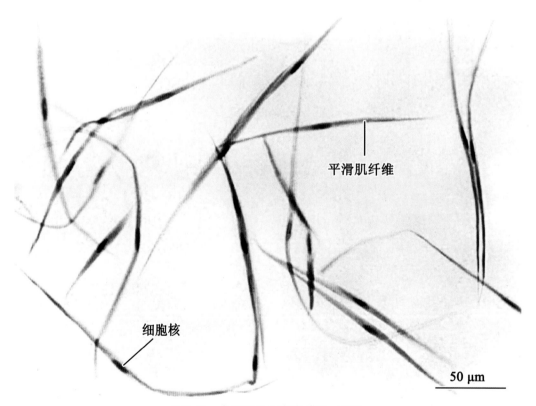

图7-44 犬平滑肌分离高倍像（HE）

Fig.7-44 high amplification of separated dog smooth muscle section（HE）

平滑肌纤维—smooth muscle fiber 细胞核—nucleus

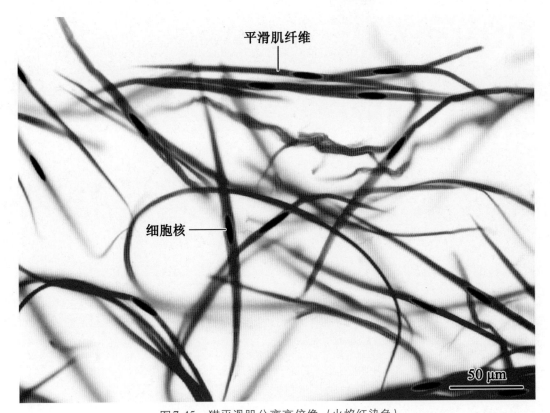

图7-45 猫平滑肌分离高倍像（火焰红染色）
Fig.7-45 high amplification of cat separated smooth muscle section（flame red stain）
平滑肌纤维—smooth muscle fiber 细胞核—nucleus

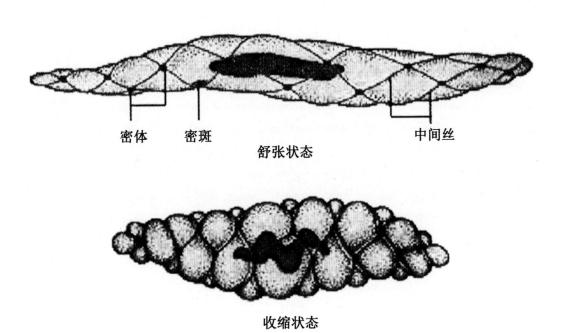

图7-46 平滑肌纤维超微结构模式图
Fig.7-46 ultrastructure pattern diagram of smooth muscle fiber
舒张状态—relaxation state 密体—dense body 密斑—dense patch
中间丝—intermediate filament 收缩状态—contraction state

第八章
神经组织
Nervous Tissue

Outline

As an integrated communication network, nerve tissue distributes throughout the body. It is structurally and functionally composed of nerve cells or neurons, and several types of glial cells or neuroglia. Neurons can receive stimulation, conduct impulse, and regulate the activities of different organs in the body. Neuroglia can not conduct impulse, but they have the roles of support, nutrition, protection and insulation to the neurons.

Most neurons include 2 parts: the cell body or perikaryon, and the dendrites and axon. Perikaryon consists of nucleus, Nissl body, nerofilaments, mitochondria, Golgi apparatus, lipofuscin and so on. On the basis of the numbers of their processes, most neurons can be classified into three kinds: pseudounipolar neurons, bipolar neurons, and multipolar neurons. In the light of their functional roles, neurons can also be divided into motor neurons and sensory neurons. According to their containing neurotransmitters, the neurons can also be classified into cholinergic neurons, aminergic neurons, amino acidergic neurons and peptidergic neurons.

Synapses are the sites where neurons or neurons and other effector cells contact with one another. Most of them are chemical synapses. Under an electron microscope, the plasma membranes of chemical synapses at the pre-and postsynaptic regions are reinforced and appear thicker than membranes adjacent to the synapse. The thin intercellular space is referred to as the synaptic cleft. The presynaptic terminal always contains synaptic vesicles and numerous mitochondria. The vesicles contain neurotransmitters.

Nerve fibers are composed of axons enveloped by neuroglia. The sheath cell is the Schwann cell in peripheral nerve fibers, but the oligodendrocyte in central nerve fibers. Thicker fibers enveloped by myelin sheath are known as myelinated nerves. Axons of small diameter are usually unmyelinated nerve fibers: they consist of sensory nerve endings and motor nerve endings.

Neuroglia, which in the central nervous system, consists of several varieties: Astrocytes, oligodendrocytes, microglia, and ependymal cell. Peripheral nervous system has Schwann cell and satellite cell within it. Peripheral nervous system comprises nerve fibers and small aggregates of nerve cells called nerve ganglia.

第八章 神经组织 Nervous Tissue

神经组织（nervous tissue）主要由神经元（神经细胞）和神经胶质细胞组成。神经元是神经系统的结构和功能单位，具有接受刺激、整合信息和传导冲动的功能，其通过突触形成神经通路和网络。神经胶质细胞遍布于神经元胞体和突起之间，对神经元起支持、营养、保护、绝缘和引导等作用。

一、神经组织的发生

神经组织除小胶质细胞外，其余神经元和神经胶质细胞均来自胚胎早期室管膜上皮（外胚层）产生的神经上皮细胞（neuroepithelium），神经上皮早期为单层柱状上皮，然后变为假复层柱状上皮。神经上皮部分细胞分化为成神经细胞后，又分化为成神经胶质细胞。成神经细胞起初为圆形，称无极成神经细胞，然后发生两个突起，称双极成神经细胞；双极成神经细胞先成为单极成神经细胞，接着发生为多极成神经细胞，进一步生长分化为多极神经细胞。

神经胶质细胞的发生晚于神经细胞。在神经胶质细胞的发生过程中，先由成胶质细胞分化为各类胶质细胞的前体细胞，即成星形胶质细胞（astroblast）和成少突胶质细胞（oligodendroblast）。前者分化为原浆性和纤维性星形胶质细胞，后者分化为少突胶质细胞。小胶质细胞发生较晚，多数人认为来源于血液的单核细胞，属于单核巨噬细胞系统。神经胶质细胞始终保持分裂增殖能力。

二、神经组织的组织学结构概述

（一）神经元

神经元（neuron）按形态分为多极神经元、双极神经元和假单极神经元；按功能分为感觉神经元（传入神经元）和运动神经元（传出神经元）。神经元的基本结构分为胞体和突起两部分。

1. 胞体 是神经元的主体，其细胞质称神经浆。胞体中有2种特殊的结构。

（1）尼氏体（Nissl bodies） 光镜下呈形态不同的嗜碱性颗粒或小块，电镜下由排列紧密的粗面内质网和游离核糖体组成。

（2）神经原纤维（neurofibril） 细丝状，由微丝或微管交叉排列成网，并伸入树突和轴突内。构成神经元的细胞骨架。

2. 突起

（1）树突（dendrite） 神经元从胞体发出的树枝状突起，内含尼氏体。多数神经元都有几个树突，主干反复分支以扩大表面积。

（2）轴突（axon） 神经元从胞体发出的细长的突起。只有一个，可发出侧支，末端有多个分支形成轴突终末。轴突内无尼氏体。轴突可将胞体发出的冲动传给另一神经元或效应器。

（二）突触

突触（synapse） 是神经元之间的接触部位，或神经元与效应细胞之间的一种特化的细胞连接，是传递信息的结构。突触由突触前膜、突触间隙和突触后膜组成。

（三）神经纤维

神经纤维（nerve fiber） 由神经元的轴突和外面包围的神经胶质细胞构成。中枢内神经纤维包围的是少突胶质细胞；外周神经纤维包围的是施万细胞（神经膜细胞）（Schwann cell）。根据有无神经胶质细胞形成髓鞘，将神经纤维分为2种：

1. 有髓神经纤维（myelinated nerve fiber） 由轴突外包髓鞘构成。髓鞘由神经胶质细胞绕轴突呈同心圆状反复缠绕形成。一个神经膜细胞组成一段髓鞘，包绕轴突构成一个节间段。相邻两个节间段之间缩窄的部分称郎飞结。

2. 无髓神经纤维（unmyelinated nerve fiber） 轴突外无髓鞘包裹，无郎飞结。

（四）神经末梢

神经末梢（nerve ending） 是周围神经纤维的终末部分，终止于全身各组织器官内形成的特殊结构。按功能分为感觉神经末梢和运动神经末梢。

1.感觉神经末梢（sensory nerve ending）　是感觉神经元（假单极神经元）周围突的终末部分，接受刺激，将神经冲动经感觉神经纤维传至中枢。感觉神经末梢分为4种。

（1）游离神经末梢（free nerve ending）　由较细的有髓或无髓神经纤维的终末反复分支，以裸露的末梢分布在上皮细胞之间。感受冷、热、轻触、痛等感觉。

（2）触觉小体（tactile corpuscle）　呈卵圆形，外包结缔组织囊，内有许多横列的扁平细胞。有髓神经纤维进入小体时失去髓鞘，轴突分成细支盘绕在扁平细胞间。感受触觉。

（3）环层小体（Pacinian corpuscle）　卵圆形或球形，小体被有数十层呈同心圆排列的扁平细胞，小体中央有一条内棍。有髓神经纤维进入小体失去髓鞘，裸露的轴突穿行于小体内棍中。感受压觉和振动觉。

（4）肌梭（muscle spindle）　分布在骨骼肌内的梭形小体，外有结缔组织被囊，内含若干条细小的骨骼肌纤维。感觉神经纤维进入肌梭时失去髓鞘，其轴突细支呈环状包绕梭内肌纤维的两端。肌梭是本体感受器，在调节骨骼肌的活动中起重要作用。

2.运动神经末梢（motor nerve ending）　是运动神经元轴突的终末部分，包括运动终板和内脏运动神经末梢。

（1）运动终板（motor end-plate）　运动神经元的轴突在肌组织的终末结构，终止在骨骼肌上，支配骨骼肌收缩。

（2）内脏运动神经末梢（visceral motor nerve ending）　分布于内脏及血管平滑肌、心肌和腺上皮细胞等处。神经纤维细，无髓鞘，其轴突终末分支呈串珠样，支配肌纤维收缩和腺体分泌。

（五）神经胶质细胞

神经胶质细胞（neuroglial cell）　数量比神经元多十倍。细胞有突起，但不分轴突和树突，无尼氏体和神经原纤维，无传导功能，起支持、保护、营养和绝缘作用。

1.星形胶质细胞（astrocyte）　体积大、数量多，细胞呈星形，核大，圆形或卵圆形。又分为纤维性星形胶质细胞和原浆性星形胶质细胞。前者的突起细长，分支少，胞质内含大量胶质丝，多分布在白质；后者的突起短粗，分支多，胞质内胶质丝少，多分布在灰质。

2.少突胶质细胞（oligodendrocyte）　胞体比星形胶质细胞小，核圆形，染色较深。胞体突起末端包裹神经元的轴突形成髓鞘。

3.小胶质细胞（microglia）　体积小，胞体细长或椭圆，核小，扁平或三角形，染色深。来源于血液中的单核细胞，具有吞噬能力。

4.室管膜细胞（ependymal cell）　衬于脑室及脊髓中央管腔面的一层立方或柱状细胞，游离端有纤毛。具有分泌脑脊液、支持和再生作用。

5.施万细胞（Schwann cell）　形成周围神经纤维的髓鞘，在神经纤维的再生中起诱导作用。

6.卫星细胞（satellite cell）　也称被囊细胞，是神经节内包裹神经元胞体的一种扁平或立方形细胞，细胞核圆形或卵圆形，染色较深，具有营养和保护作用。

三、神经组织图谱

1.神经元　图8-1～图8-18。

2.突触　图8-19～图8-21。

3.神经纤维　图8-22～图8-34。

4.神经末梢　图8-35～图8-48。

5.神经胶质细胞　图8-49～图8-64。

第八章 神经组织 Nervous Tissue

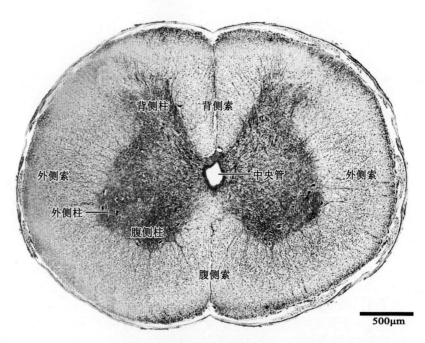

图8-1 牛腰段脊髓横切低倍像（镀银染色）
Fig.8-1 low magnification of transverse section of cow lumbar spinal cord (silver stain)
背侧柱—dorsolateral column　背侧索—dorsolateral funiculus
外侧索—lateral funiculus　中央管—central canal
外侧柱—lateral column　腹侧柱—ventral column
腹侧索—ventral funiculus

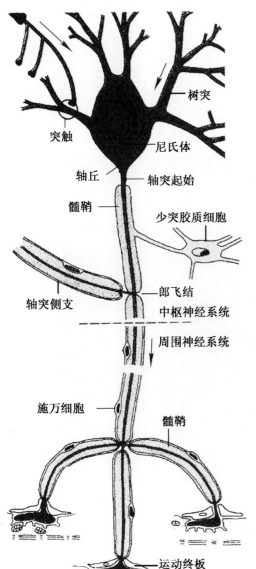

图8-2 神经元结构模式图
Fig.8-2 diagram of neuron structure
树突—dendrite　突触—synapse　尼氏体—Nissl body
轴丘—axon hillock　轴突起始—axon beginning
髓鞘—myelin sheath　少突胶质细胞—oligodendroglia cell
轴突侧支—axon branch　郎飞结—Ranvier node
中枢神经系统—central nervous system
周围神经系统—peripheral nervous system
施万细胞—Schwann cell　运动终板—motor end-plate

179

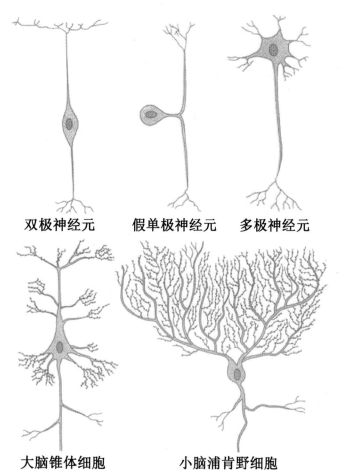

图8-3 神经元的类型模式图

Fig.8-3 diagram of neuron classification

双极神经元—bipolar neuron

假单极神经元—pseudo-unipolar neuron

多极神经元—multipolar neuron

大脑锥体细胞—cerebrum pyramidal cell

小脑浦肯野细胞—cerebellar Purkinje cell

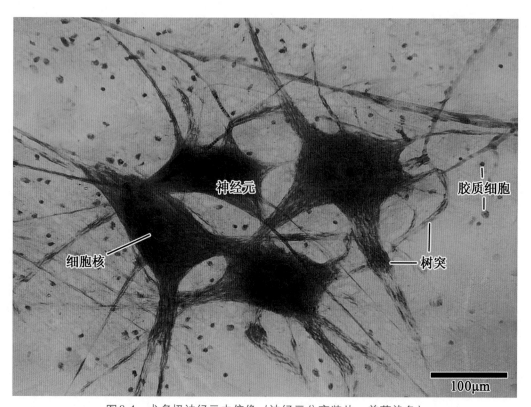

图8-4 犬多极神经元中倍像（神经元分离装片，美蓝染色）

Fig.8-4 mid magnification of dog multipolar neuron (separated neurons, methylene blue stain)

神经元—neuron 细胞核—nucleus 胶质细胞—neuroglia 树突—dendrite

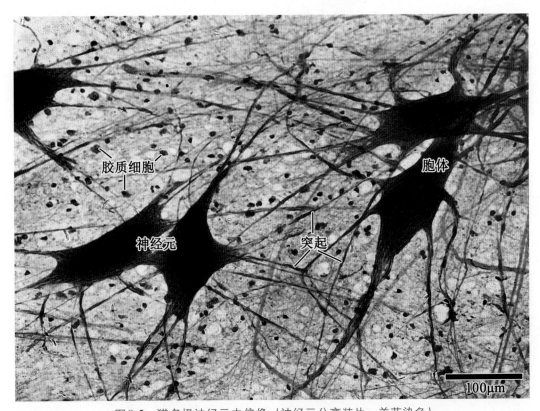

图8-5 猫多极神经元中倍像（神经元分离装片，美蓝染色）

Fig.8-5 mid magnification of cat multipolar neuron（separated neurons，methylene blue stain）

胶质细胞—neuroglia 神经元—neuron 突起—neurite 胞体—cell body

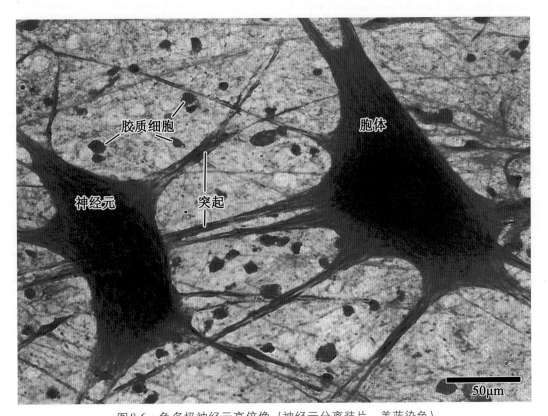

图8-6 兔多极神经元高倍像（神经元分离装片，美蓝染色）

Fig.8-6 high magnification of rabbit multipolar neuron（separated neurons，methylene blue stain）

胶质细胞—neuroglia 神经元—neuron 突起—neurite 胞体—cell body

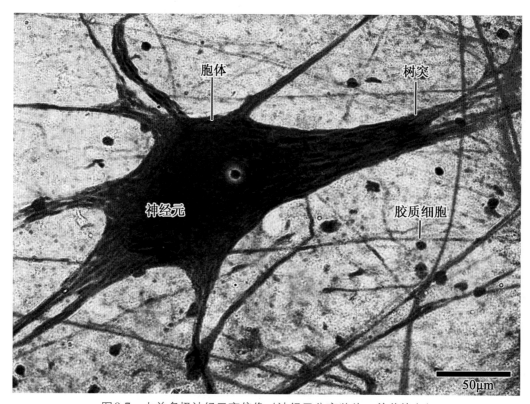

图8-7 山羊多极神经元高倍像（神经元分离装片，美蓝染色）

Fig.8-7 high magnification of goat multipolar neuron (separated neurons, methylene blue stain)

神经元—neuron 胞体—cell body 树突—dendrite 胶质细胞—neuroglia

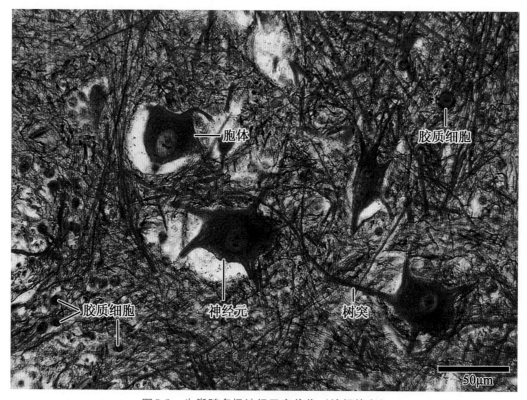

图8-8 牛脊髓多极神经元高倍像（镀银染色）

Fig.8-8 high magnification of cow multipolar neuron in spinal cord (silver stain)

胞体—cell body 神经元—neuron 树突—dendrite 胶质细胞—neuroglia

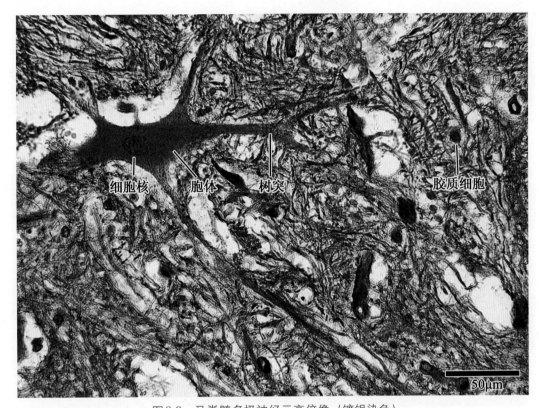

图8-9 马脊髓多极神经元高倍像（镀银染色）

Fig.8-9 high magnification of horse multipolar neuron in spinal cord（silver stain）

细胞核—nucleus 胞体—cell body 树突—dendrite 胶质细胞—neuroglia

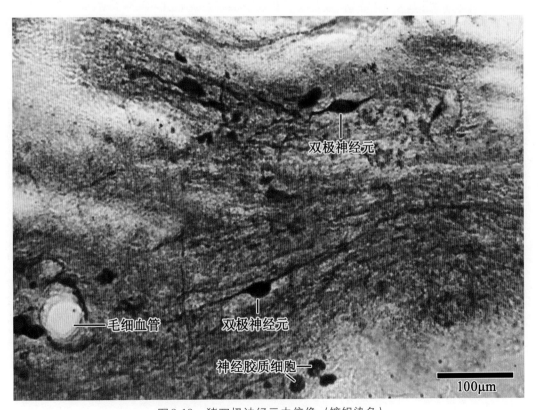

图8-10 猪双极神经元中倍像（镀银染色）

Fig.8-10 mid magnification of pig bipolar neuron（silver stain）

双极神经元—bipolar neuron 毛细血管—capillary 神经胶质细胞—neuroglia cell

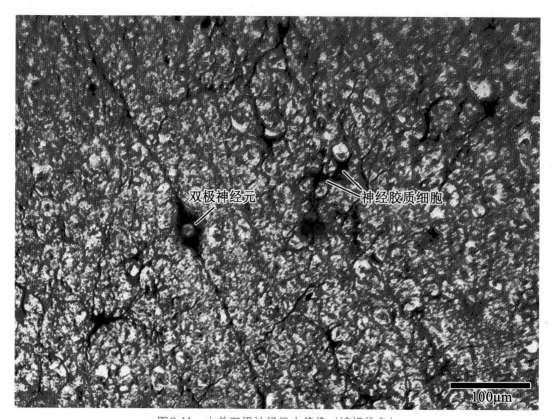

图8-11 山羊双极神经元中倍像（镀银染色）

Fig.8-11 mid magnification of goat bipolar neuron（silver stain）

双极神经元—bipolar neuron 神经胶质细胞—neuroglia cell

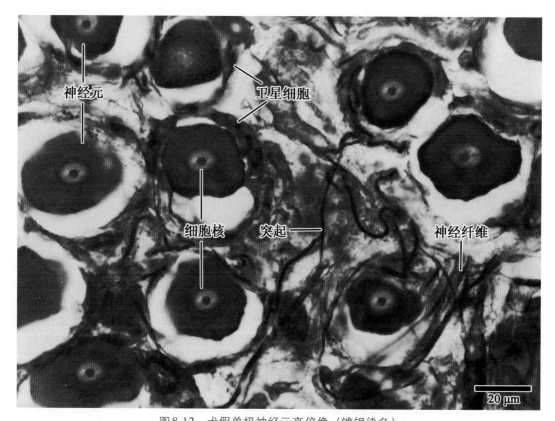

图8-12 犬假单极神经元高倍像（镀银染色）

Fig.8-12 high magnification of dog pseudounipolar neuron（silver stain）

神经元—neuron 卫星细胞—satellite cell 细胞核—nucleus 突起—neurite 神经纤维—nerve fiber

第八章 神经组织 Nervous Tissue

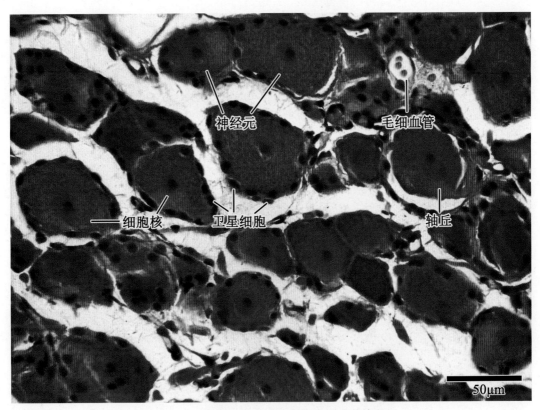

图 8-13 猫假单极神经元高倍像（硫堇染色）

Fig.8-13　high magnification of cat pseudo-unipolar neuron（thionine stain）

神经元—neuron　细胞核—nucleus　卫星细胞—satellite cell　轴丘—axon hillock　毛细血管—capillary

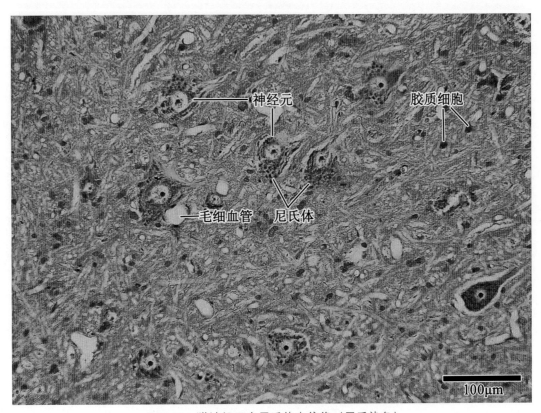

图 8-14 猫神经元内尼氏体中倍像（尼氏染色）

Fig.8-14　mid magnification of nissl body in cat neuron（Nissl stain）

神经元—neuron　尼氏体—Nissl body　毛细血管—capillary　胶质细胞—neuroglia cell

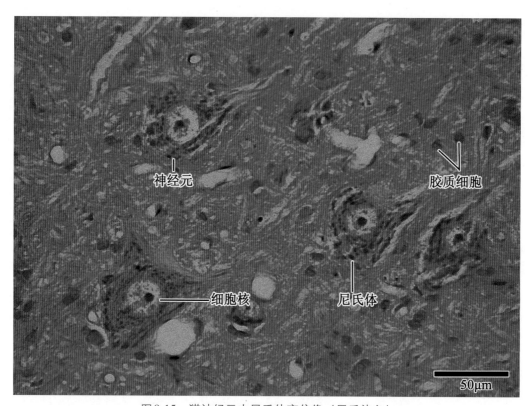

图8-15 猫神经元内尼氏体高倍像（尼氏染色）

Fig.8-15 high magnification of nissl body in cat neuron（Nissl stain）

神经元—neuron 细胞核—nucleus 尼氏体—Nissl body 胶质细胞—neuroglia cell

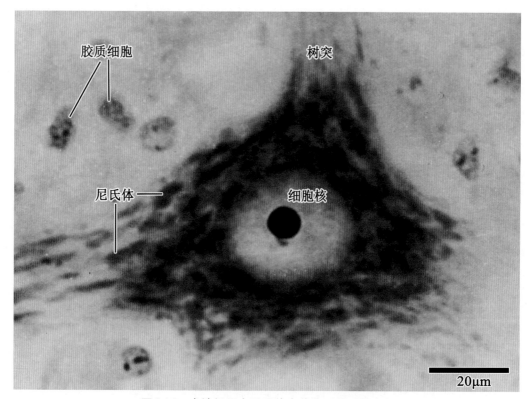

图8-16 犬神经元内尼氏体高倍像（尼氏染色）

Fig.8-16 high magnification of nissl body in dog neuron（Nissl stain）

胶质细胞—neuroglia cell 尼氏体—Nissl body 树突—dendrite 细胞核—nucleus

第八章 神经组织 Nervous Tissue

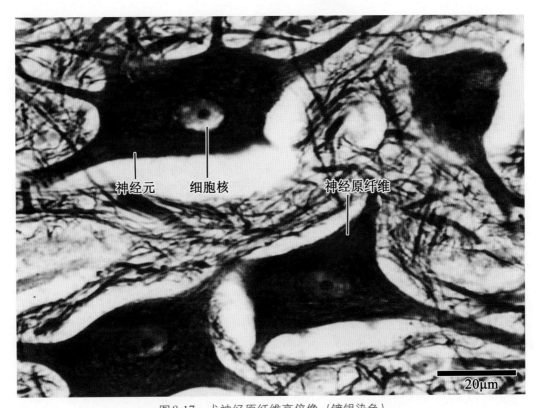

图8-17 犬神经原纤维高倍像（镀银染色）
Fig.8-17 high magnification of neurofibril in dog neuron（silver stain）
神经元—neuron 细胞核—nucleus 神经原纤维—neurofibril

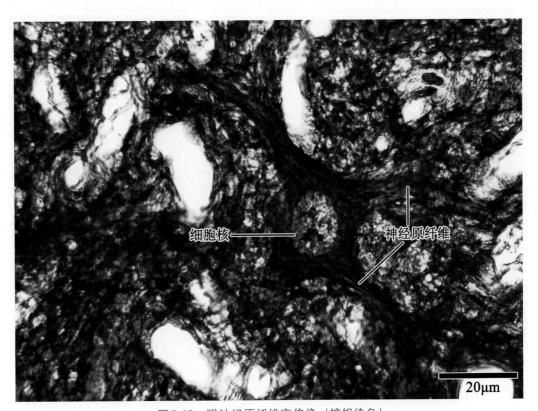

图8-18 猫神经原纤维高倍像（镀银染色）
Fig.8-18 high magnification of neurofibril in cat neuron（silver stain）
细胞核—nucleus 神经原纤维—neurofibril

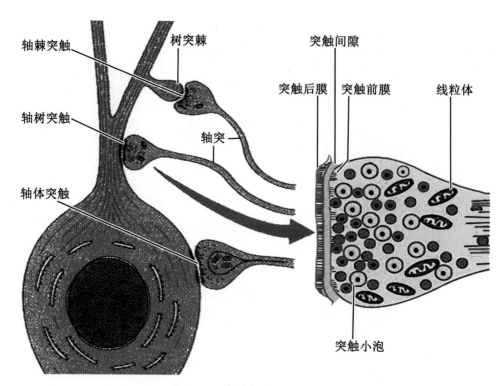

图8-19　突触电镜结构模式图

Fig.8-19　diagram of synaptic electron microscopy

轴体突触—axosomatic synapse　轴树突触—axodendritic synapse　轴棘突触—axospinous synapse
树突棘—dendritic spine　轴突—synapse　突触后膜—postsynaptic membrane　突触间隙—synaptic cleft
突触前膜—presynaptic membrane　突触小泡—synaptic vesicle　线粒体—mitochondria

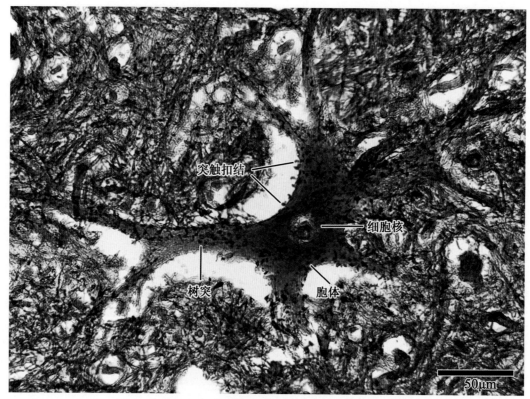

图8-20　牛脊髓多极神经元上的突触扣结1（镀银染色）

Fig.8-20　synaptic boutons on cow spinal cord neuron 1（silver stain）

突触扣结—synaptic bouton　细胞核—nucleus　树突—dendrite　胞体—cell body

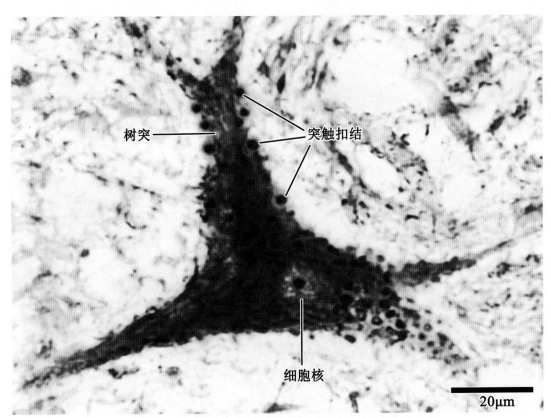

图8-21 牛脊髓多极神经元上的突触扣结2（镀银染色）
Fig.8-21 synaptic boutons on cow spinal cord neuron 2 (silver stain)
树突—dendrite 突触扣结—synaptic bouton 细胞核—nucleus

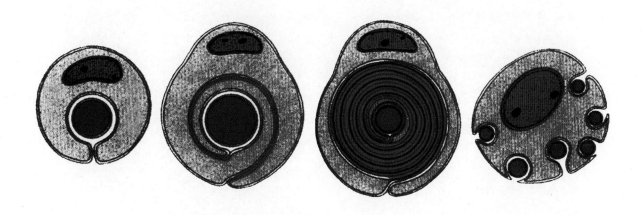

有髓神经纤维髓鞘形成过程　　　　　　　　　无髓神经纤维

图8-22 神经纤维及其髓鞘形成模式图
Fig.8-22 diagraph of nerve fiber and medullary sheath formation
有髓神经纤维髓鞘形成过程—myelin formation of myelinated nerve fiber 无髓神经纤维—unmyelinated nerve fiber

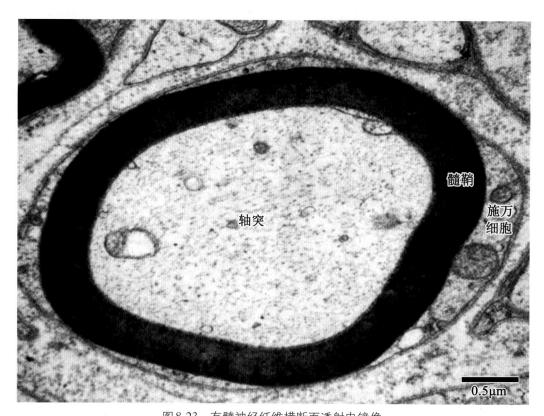

图8-23　有髓神经纤维横断面透射电镜像

Fig.8-23　electron image of myelinated nerve fiber transverse section

轴突—axon　髓鞘—myelin sheath　施万细胞—Schwann cell

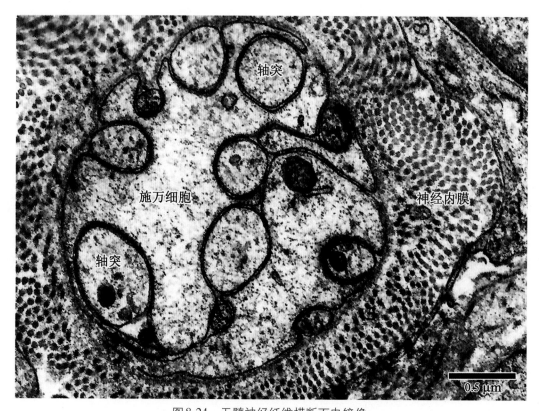

图8-24　无髓神经纤维横断面电镜像

Fig.8-24　electron image of unmyelinated nerve fiber transverse section

轴突—axon　施万细胞—Schwann cell　神经内膜—endoneurium

第八章 神经组织 Nervous Tissue

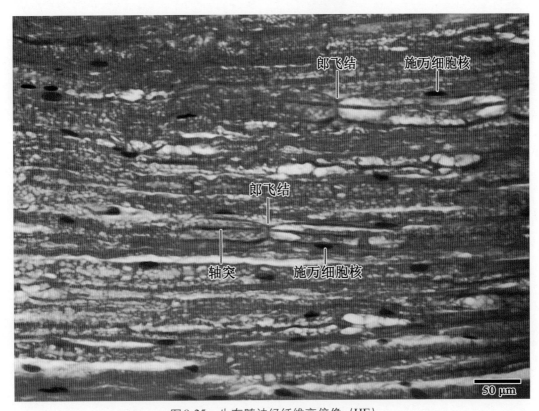

图8-25 牛有髓神经纤维高倍像（HE）
Fig.8-25 high magnification of cow myelinated nerve fiber（HE）
郎飞结—Ranvier node 施万细胞核—Schwann cell nucleus 轴突—axon

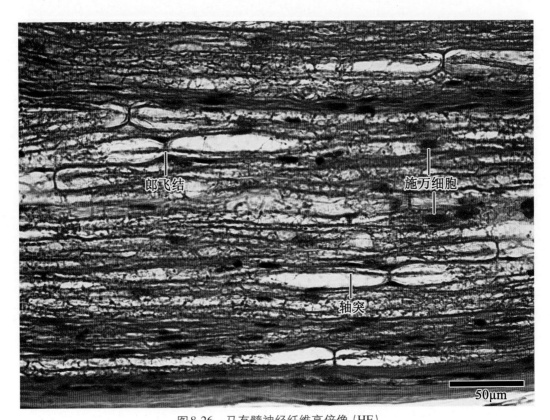

图8-26 马有髓神经纤维高倍像（HE）
Fig.8-26 high magnification of horse myelinated nerve fiber（HE）
郎飞结—Ranvier node 施万细胞—Schwann cell 轴突—axon

191

图 8-27 猪有髓神经纤维高倍像（HE）

Fig.8-27 high magnification of pig myelinated nerve fiber（HE）

郎飞结—Ranvier node 施万细胞—Schwann cell 轴突—axon

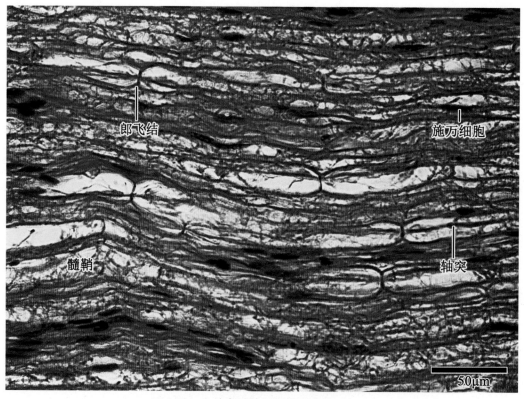

图 8-28 山羊有髓神经纤维高倍像（HE）

Fig.8-28 high magnification of goat myelinated nerve fiber（HE）

郎飞结—Ranvier node 髓鞘—myelin sheath 施万细胞—Schwann cell 轴突—axon

图 8-29 犬有髓神经纤维高倍像（HE）

Fig.8-29　high magnification of dog myelinated nerve fiber（HE）

郎飞结—Ranvier node　施万细胞—Schwann cell　轴突—axon

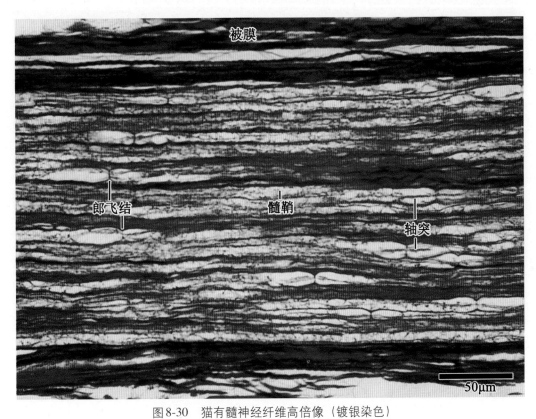

图 8-30 猫有髓神经纤维高倍像（镀银染色）

Fig.8-30　high magnification of cat myelinated nerve fiber（silver stain）

被膜—capsule　郎飞结—Ranvier node　髓鞘—myelin sheath　轴突—axon

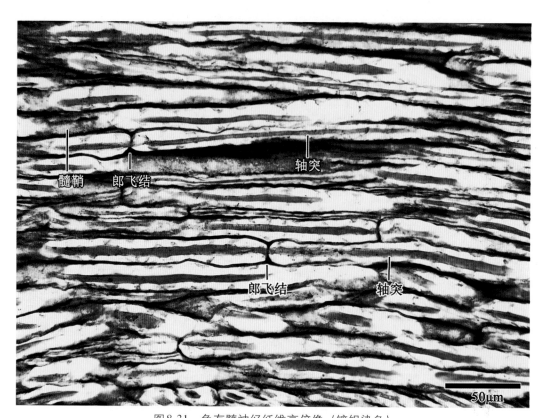

图8-31 兔有髓神经纤维高倍像（镀银染色）
Fig.8-31 high magnification of rabbit myelinated nerve fiber（silver stain）
髓鞘—myelin sheath 郎飞结—Ranvier node 轴突—axon

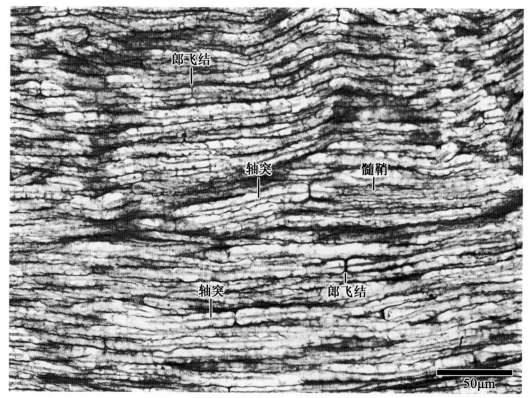

图8-32 犬有髓神经纤维高倍像（镀银染色）
Fig.8-32 high magnification of dog myelinated nerve fiber（silver stain）
郎飞结—Ranvier node 轴突—axon 髓鞘—myelin sheath

第八章 神经组织 Nervous Tissue

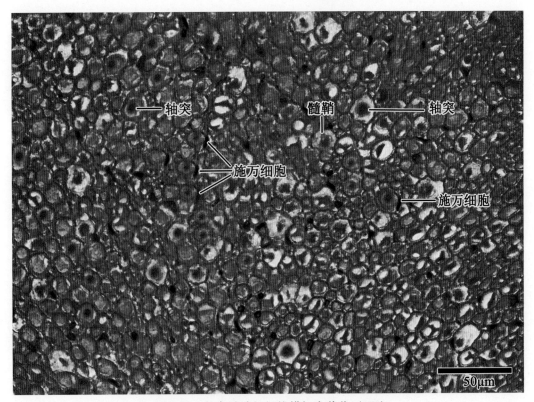

图 8-33 兔有髓神经纤维横切高倍像（HE）
Fig.8-33 high magnification of rabbit myelinated nerve fiber transection（HE）
轴突—axon 髓鞘—myelin sheath 施万细胞—Schwann cell

图 8-34 兔无髓神经纤维高倍像（镀银染色）
Fig.8-34 high magnification of rabbit unmyelinated nerve fiber（silver stain）
被膜—capsule 轴突—axon

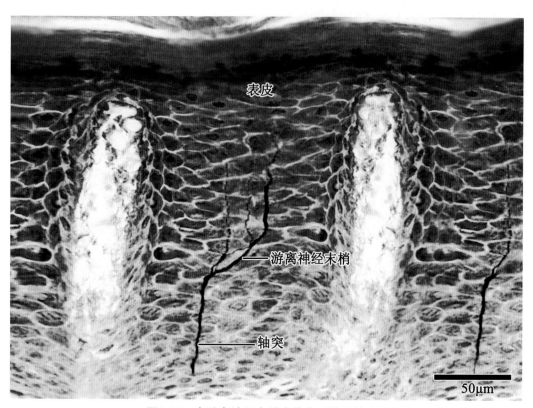

图 8-35 犬游离神经末梢高倍像（镀银染色）

Fig.8-35 high magnification of dog free nerve ending (silver stain)

表皮—epidermis 游离神经末梢—free nerve ending 轴突—axon

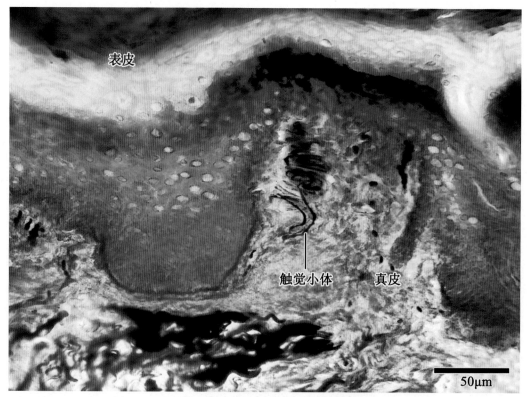

图 8-36 猫触觉小体高倍像（镀银染色）

Fig.8-36 high magnification of cat tactile corpuscle (silver stain)

表皮—epidermis 触觉小体—tactile corpuscle 真皮—dermis

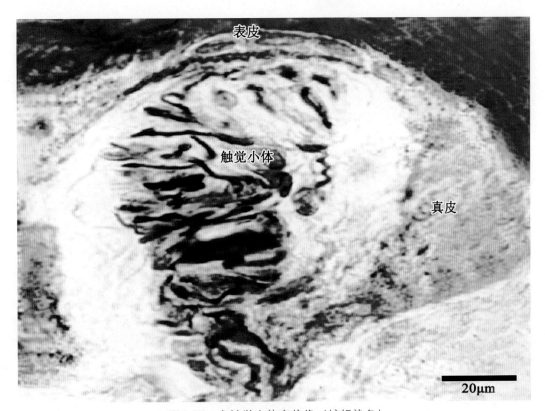

图8-37 犬触觉小体高倍像（镀银染色）
Fig.8-37 high magnification of dog tactile corpuscle（silver stain）
表皮—epidermis 触觉小体—tactile corpuscle 真皮—dermis

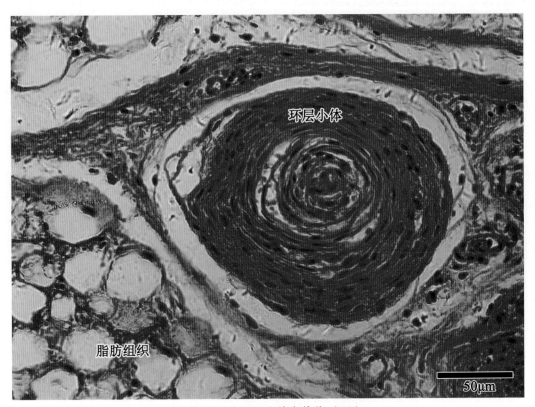

图8-38 牛环层小体高倍像（HE）
Fig.8-38 high magnification of cow Pacinian corpuscle（HE）
环层小体—Pacinian corpuscle 脂肪组织—adipose tissue

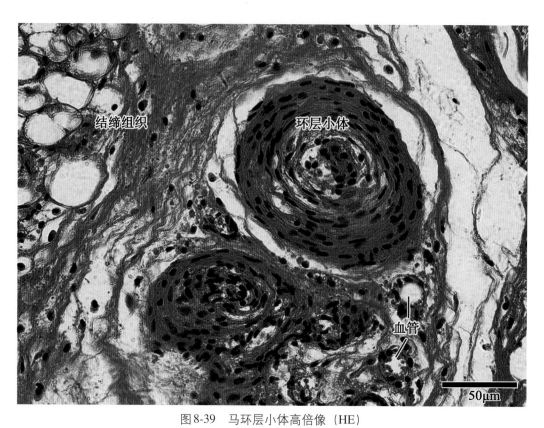

图8-39 马环层小体高倍像（HE）

Fig.8-39 high magnification of horse Pacinian corpuscle（HE）

结缔组织—connective tissue 环层小体—Pacinian corpuscle 血管—blood vessel

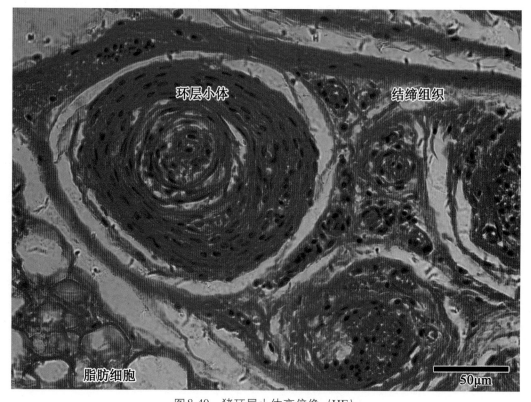

图8-40 猪环层小体高倍像（HE）

Fig.8-40 high magnification of pig Pacinian corpuscle（HE）

环层小体—Pacinian corpuscle 结缔组织—connective tissue 脂肪细胞—adipocyte

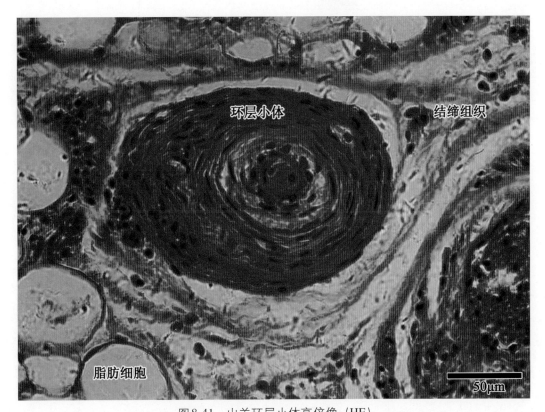

图8-41　山羊环层小体高倍像（HE）

Fig.8-41　high magnification of goat Pacinian corpuscle（HE）

环层小体—Pacinian corpuscle　结缔组织—connective tissue　脂肪细胞—adipocyte

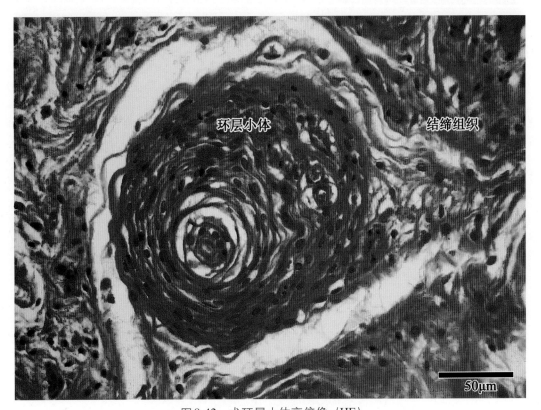

图8-42　犬环层小体高倍像（HE）

Fig.8-42　high magnification of dog Pacinian corpuscle（HE）

环层小体—Pacinian corpuscle　结缔组织—connective tissue

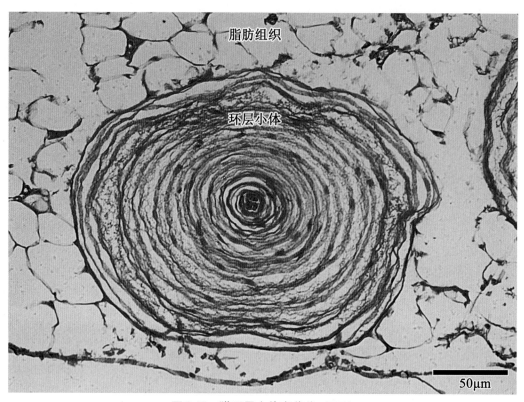

图 8-43 猫环层小体高倍像（HE）

Fig.8-43 high magnification of cat Pacinian corpuscle（HE）

脂肪组织—adipose tissue　环层小体—Pacinian corpuscle

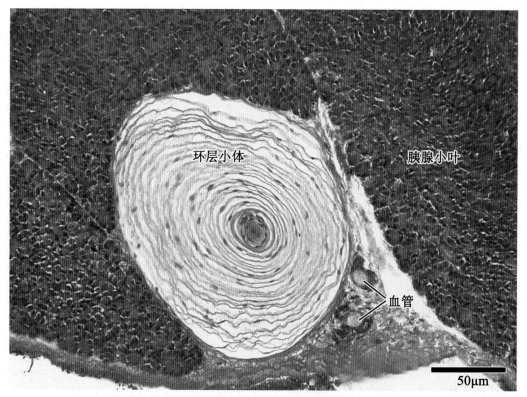

图 8-44 兔环层小体高倍像（HE）

Fig.8-44 high magnification of rabbit Pacinian corpuscle（HE）

环层小体—Pacinian corpuscle　胰腺小叶—pancreatic lobule　血管—blood vessel

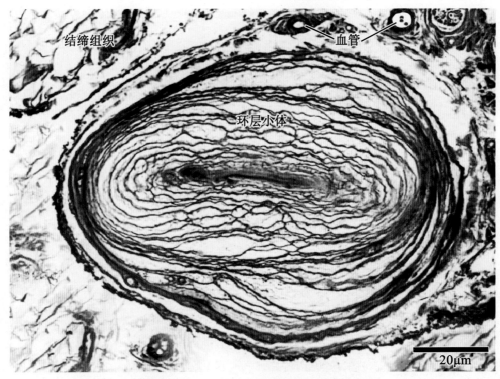

图8-45 犬环层小体纵切高倍像（镀银染色）

Fig.8-45 high magnification of dog Pacinian corpuscle longitudinal section（silver stain）

结缔组织—connective tissue 环层小体—Pacinian corpuscle 血管—blood vessel

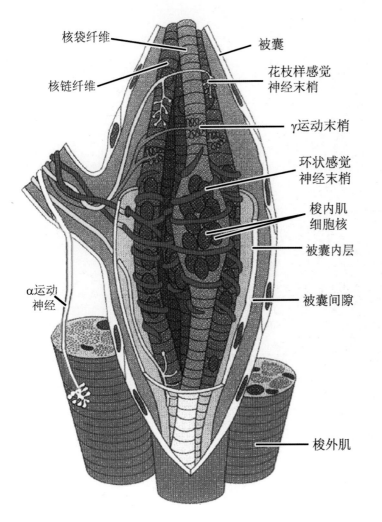

图8-46 肌梭结构模式图

Fig.8-46 diagram of muscle spindle structure

核袋纤维—nuclear bag fiber

核链纤维—nuclear chain fiber

α运动神经—α-motor nerve 被囊—capsule

花枝样感觉神经末梢—branch sensory nerve endings

γ运动神经—r-motor nerve

环状感觉神经末梢—ring sensory nerve endings

梭内肌细胞核—intra spindle muscle nucleus

被囊内层—capsule inside 被囊间隙—capsule clearance

梭外肌—extra spindle muscle

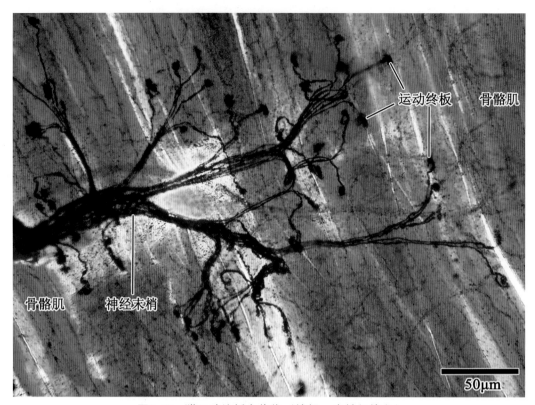

图8-47　猫运动终板高倍像（镀银＋中性红染色）

Fig.8-47　high magnification of cat motor end-plate（silver + neutral red stain）

骨骼肌—skeletal muscle　神经末梢—nerve endings　运动终板—motor end-plate

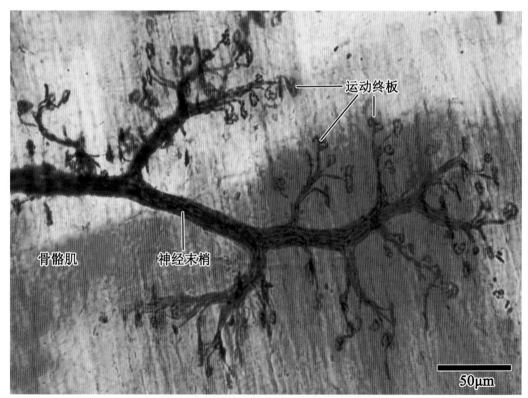

图8-48　猪运动终板高倍像（镀银＋中性红染色）

Fig.8-48　high magnification of pig motor end-plate（silver + neutral red stain）

骨骼肌—skeletal muscle　神经末梢—nerve endings　运动终板—motor end-plate

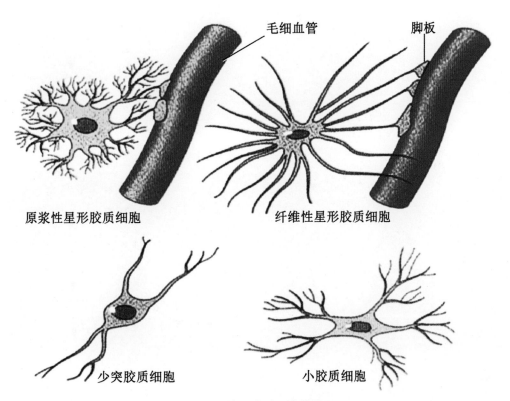

图 8-49　神经胶质细胞模式图

Fig.8-49　neuroglia diagram

毛细血管—capillary　脚板—foot plate　原浆性星形胶质细胞—protoplasmic astrocyte
纤维性星形胶质细胞—fibrous astrocyte　少突胶质细胞—oligodendrocyte　小胶质细胞—microglia cell

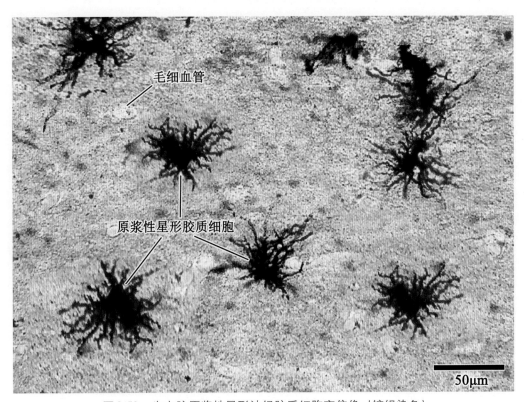

图 8-50　牛大脑原浆性星形神经胶质细胞高倍像（镀银染色）

Fig.8-50　high magnification of cow cerebrum protoplasmic astrocyte（silver stain）

毛细血管—capillary　原浆性星形胶质细胞—protoplasmic astrocyte

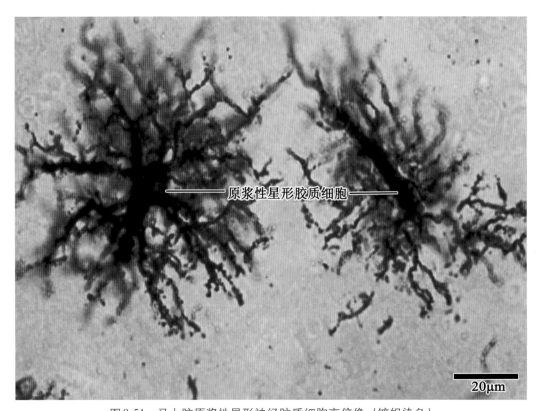

图 8-51 马大脑原浆性星形神经胶质细胞高倍像（镀银染色）
Fig.8-51 high magnification of horse cerebrum protoplasmic astrocyte（silver stain）
原浆性星形胶质细胞—protoplasmic astrocyte

图 8-52 猪大脑原浆性星形神经胶质细胞中倍像（镀银染色）
Fig.8-52 mid magnification of pig cerebrum protoplasmic astrocyte（silver stain）
原浆性星形胶质细胞—protoplasmic astrocyte　毛细血管—capillary

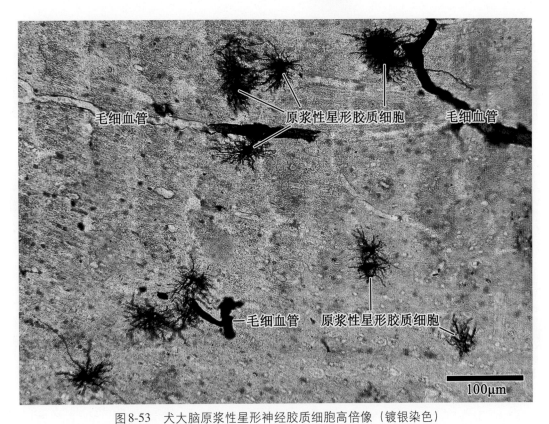

图 8-53 犬大脑原浆性星形神经胶质细胞高倍像（镀银染色）
Fig.8-53 high magnification of dog cerebrum protoplasmic astrocyte（silver stain）

毛细血管—capillary 原浆性星形胶质细胞—protoplasmic astrocyte

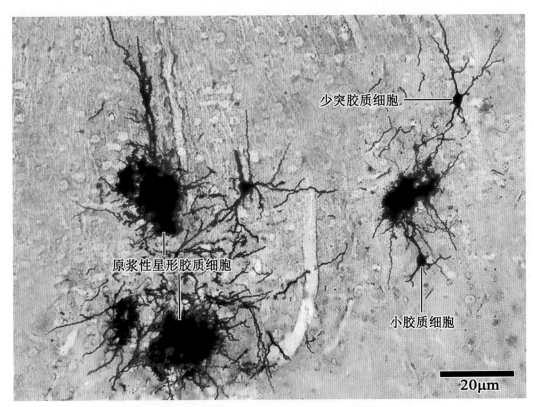

图 8-54 猫大脑神经胶质细胞高倍像（镀银染色）
Fig.8-54 high magnification of cat cerebrum neuroglial cell（silver stain）

原浆性星形胶质细胞—protoplasmic astrocyte 少突胶质细胞—oligodendroglia cell 小胶质细胞—microglia

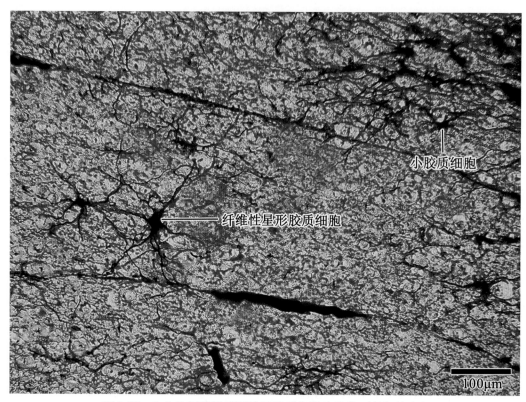

图 8-55 牛大脑纤维性星形神经胶质细胞中倍像（镀银染色）

Fig.8-55 mid magnification of cow cerebrum fibrous astrocyte（silver stain）

纤维性星形胶质细胞—fibrous astrocyte　小胶质细胞—microglia

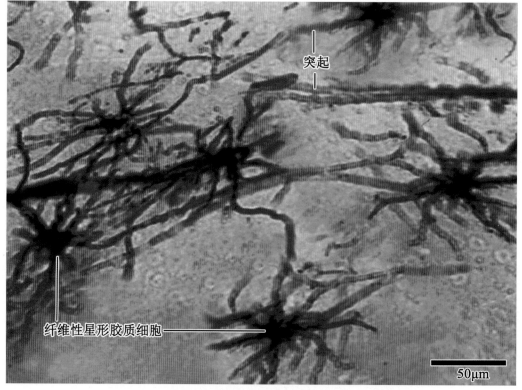

图 8-56 马大脑纤维性星形神经胶质细胞高倍像（镀银染色）

Fig.8-56 high magnification of horse cerebrum fibrous astrocyte（silver stain）

纤维性星形胶质细胞—fibrous astrocyte　突起—neurite

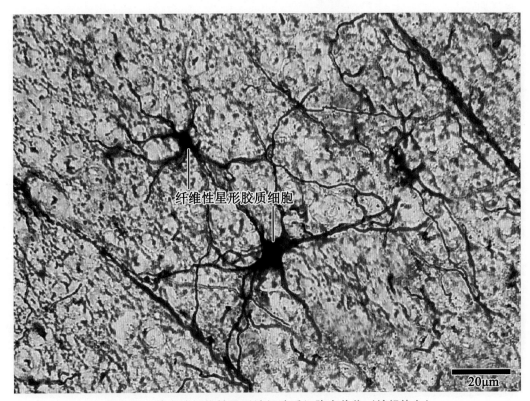

图 8-57 猪大脑纤维性星形神经胶质细胞高倍像（镀银染色）
Fig.8-57 high magnification of pig cerebrum fibrous astrocyte（silver stain）
纤维性星形胶质细胞—fibrous astrocyte

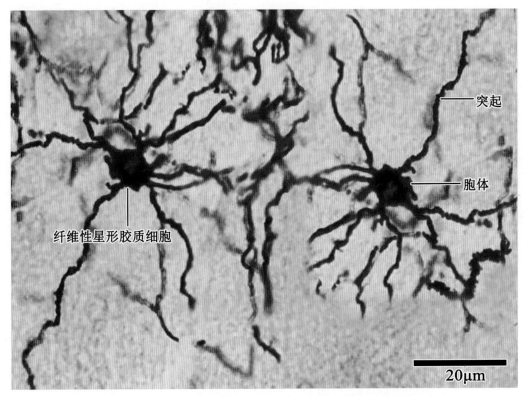

图 8-58 山羊大脑纤维性星形神经胶质细胞高倍像（镀银染色）
Fig.8-58 high magnification of goat cerebrum fibrous astrocyte（silver stain）
纤维性星形胶质细胞—fibrous astrocyte　突起—neurite　胞体—cell body

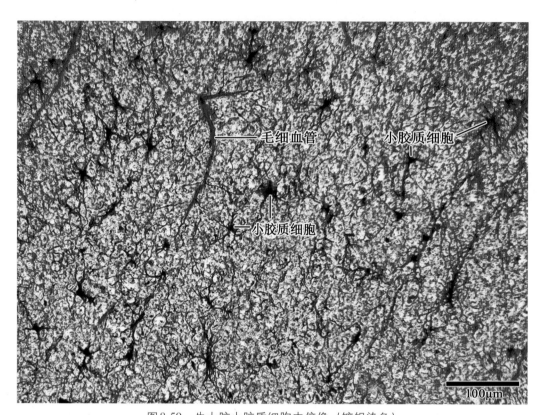

图8-59　牛大脑小胶质细胞中倍像（镀银染色）

Fig.8-59　mid magnification of cow cerebrum microglia（silver stain）

毛细血管—capillary　小胶质细胞—microglia

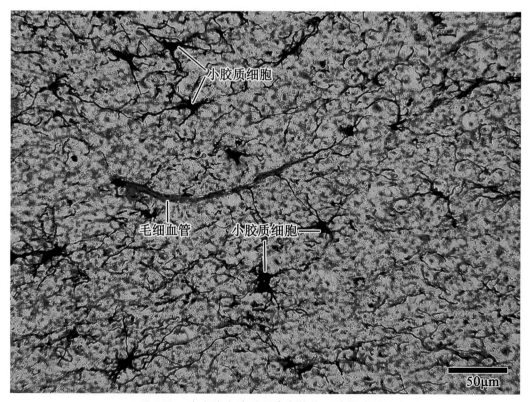

图8-60　牛大脑小胶质细胞高倍像（镀银染色）

Fig.8-60　high magnification of cow cerebrum microglia（silver stain）

小胶质细胞—microglia　毛细血管—capillary

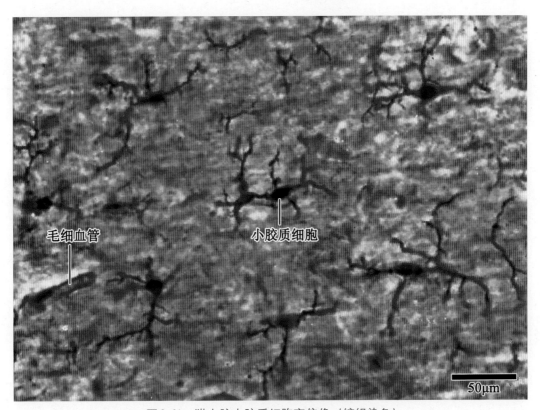

图 8-61　猫大脑小胶质细胞高倍像（镀银染色）

Fig.8-61　high magnification of cat cerebrum microglia (silver stain)

毛细血管—capillary　小胶质细胞—microglia

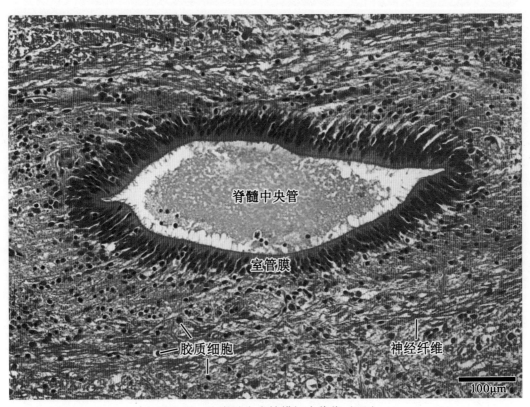

图 8-62　牛脊髓中央管横切中倍像（HE）

Fig.8-62　mid magnification of transverse section of cow spinal cord central canal (HE)

脊髓中央管—spinal cord central canal　室管膜—ependyma　胶质细胞—glial cell　神经纤维—nerve fiber

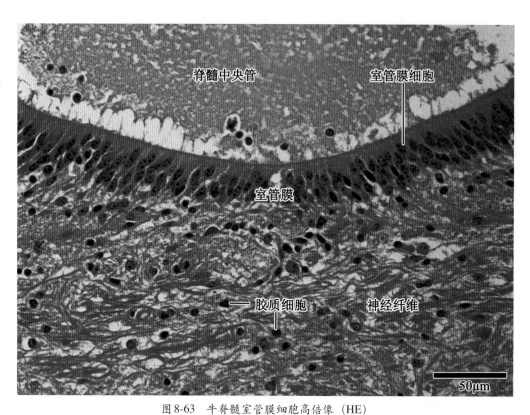

图 8-63 牛脊髓室管膜细胞高倍像（HE）

Fig.8-63 high magnification of cow spinal cord ependymocyte（HE）

脊髓中央管—spinal cord central canal 室管膜—ependyma 室管膜细胞—ependymocyte

胶质细胞—glial cell 神经纤维—nerve fiber

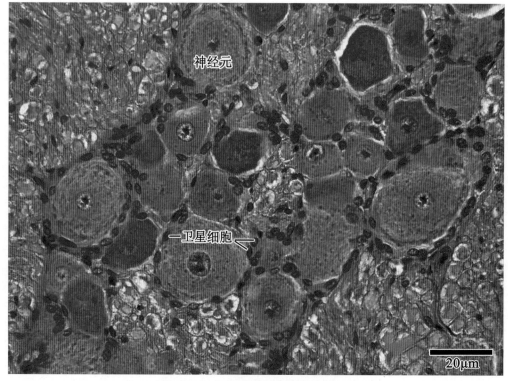

图 8-64 犬卫星细胞高倍像（HE）

Fig.8-64 high magnification of dog satellite cells（HE）

神经元—neuron 卫星细胞—satellite cell

第九章
神经系统
Nervous System

Outline

Nervous system is a complex group of organs which are divided into central and peripheral nervous system. Both of which are derived from ectoderm. Central nervous system (CNS) includes the brain and the spinal cord, consists of gray matter and white matter. Peripheral nervous system (PNS) consists of ganglion, nerve and nerve ending. The spinal cord which forms the caudal end of the CNS, its white matter locates peripherally and gray matter locates centrally assuming the shape of a butterfly. In the middle of this butterfly is an opening space, which is called as the central canal. The cerebellum, which has a cortex of gray matter and a central area of white matter, composed of stellate cells, basket cells, Purkinje cells, granular cells and Golgi cells. It can be divided into three layers: the outer molecular layer, the central layer of Purkinje cell and the inner granular layer. The cerebrum also has a cortex of gray matter and a central area of white matter in which nuclei of gray matter are found. The cerebral cortex can be divided into six layers: the molecular layer, the external granular layer, the external pyramidal layer, the internal granular layer, the internal pyramidal layer, and the polymorphic layer.

There are three membranes (meninges) around the spinal cord and brain. They consist of dura mater, arachnoid and pia mater.

The blood-brain barrier (BBB) is a protectively functional structure that prevents the passage of some substances, such as antibiotics and chemical and bacterial toxic matters, from the blood to nerve tissue. It is composed of the endothelial cells of continuous capillaries, basal lamina and neuroglial membrane.

神经系统（nervous system）分为中枢神经系统（central nervous system）和周围神经系统（peripheral nervous system）。中枢神经系统有脑和脊髓，周围神经系统包括脑神经、脊神经、植物性神经及其神经节。

一、神经系统的发生

神经系统起源于外胚层。在胚胎约一周龄时，胚盘中轴部的外胚层细胞增厚形成神经板（neural plate）。随后，神经板沿其长轴凹陷形成神经沟（neural groove），沟两侧的隆起称神经褶（neural fold）。随着进一步发育，神经褶开始在神经沟的多个部位发生闭合，最后神经沟完全闭合形成神经管。神经管是中枢神经系统的原基。在神经管关闭的过程中，一些细胞迁移至神经管的背外侧，形成两条纵行的细胞索，即神经嵴（neural crest）。神经嵴主要分化为周围神经系统的结构。

（一）脊髓的发生

在神经管头段演变为脑时，神经管尾段即脊髓部保持着较细的直管状。早期的神经管脊髓部，横断面的管腔呈菱形。随着神经管的管壁增厚，管腔逐渐变小。以后，由于神经管背侧部左、右侧壁的合并，该部管腔逐渐消失；腹侧部的管腔则变圆并演化为脊髓中央管，尾端的管腔形成终室。

（二）脑的发生

脑泡的发生和演变约在胚胎第3周，神经沟在融合成神经管时，其头端发育成较宽的两叶片状，即未来的前脑。前脑区后缘至第1对体节的一段神经沟，将形成中脑和菱脑头部，脑原基尾段部分将形成菱脑的尾部。约在胚胎第4周，前神经孔和后神经孔先后关闭。胚胎第4周末，神经管头段逐渐演变形成3个膨大区域，称脑泡(brian vesicle)，由前向后分别为前脑泡、中脑泡和菱脑泡。中脑泡与菱脑泡之间的缩窄区域称脑峡。胚胎第5周，前脑泡的头端向两侧膨大，形成左右两个端脑泡，以后演变为大脑两半球，而位于两端脑泡之间的前脑泡尾部则扩大形成间脑。中脑泡演变为中脑，菱脑泡演变为头侧的后脑和尾侧的末脑。后脑经脑峡与头端的中脑相连，末脑则与脊髓连续。最终，后脑演变为脑桥和小脑，末脑演变为延髓。脑的内腔成为脑室和中脑导水管。至此，神经管的脑部由前向后依次为左右成对的端(大)脑、间脑、中脑、后脑(脑桥)和末脑(延髓)5部分。

二、神经系统的组织学结构概述

中枢神经系统内，神经元胞体聚集成灰质，神经元突起则构成白质。脊髓的灰质位于内部，白质在灰质的周围。大脑和小脑的灰质位于脑的表层，又称皮质（cortex）。白质中分布的一些灰质团块，称为神经核（nucleus）。

（一）脊髓（spinal cord）

1. 灰质（gray matter） 由神经元、神经胶质细胞及毛细血管等构成。神经元胞体主要分布在灰质的两翼，大多数为多极神经元。其大小和形态各不相同，均为运动神经元和中间神经元。脊髓灰质的横切面呈蝴蝶形，分为腹侧柱、背侧柱与外侧柱：腹侧柱内有运动神经元的胞体，支配骨骼肌；外侧柱内有内脏运动神经元，为植物性神经节前神经元的胞体；背侧柱内含有各种类型的中间神经元的胞体，这些中间神经元接受脊神经节内的感觉神经元的冲动，传导至运动神经元或下一个中间神经元。

2. 白质（white matter） 含由来自脊髓灰质和脑的神经纤维束构成的三种传导索（束）：即背侧索、腹侧索和外侧索。上行束分布于背侧索、腹侧索和外侧索，可将感觉冲动传向脑的反射中枢；下行束分布于腹侧索和外侧索，将脑反射中枢的神经冲动传至运动神经元。

（二）小脑（cerebellum）

1. 小脑皮质 由多极神经元、神经纤维、神经胶质细胞和血管构成，由外向内明显分为三层：分子层、浦肯野细胞层和颗粒层。

（1）分子层（molecular layer） 神经元较少，主要由无髓神经纤维组成。

（2）浦肯野细胞层（Purkinje cell layer） 是一层排列规则的浦肯野细胞，胞体大，呈梨形，顶端发出2～3个粗大的树突伸向分子层，并在水平面上发出许多较短的分支，胞体底端发出1个长轴突伸入髓质。

（3）颗粒层（granular layer） 由密集排列的颗粒细胞（granular cell）和少量高尔基细胞（Golgi cell）组成。颗粒细胞呈球形，胞体小，核大，细胞质少。高尔基细胞少，胞体大，树突很多，分支发达，轴突较短，只位于颗粒层，与颗粒细胞的轴突和苔藓纤维的末端膨大形成小脑小球。

2. 小脑髓质 由三种有髓神经纤维构成。

（1）浦肯野细胞轴突 是小脑皮质唯一的传出纤维，止于小脑髓质齿状核。

（2）攀登纤维（climbing fiber） 小脑皮质的传入纤维，主要来自延髓下橄榄核。

（3）苔藓纤维（mossy fiber） 小脑皮质的传入纤维，来自脊髓背核和前庭核。

（三）大脑（cerebrum）

1. 大脑皮质 是神经系统的高级中枢，由大量多极神经元、神经胶质细胞和神经纤维构成不同的功能区，并与中枢神经的其他部分和外周神经发生联系。大脑皮质的神经元按形态可分为锥体细胞、颗粒细胞和梭形细胞。锥体细胞（pyramidal cell）数量较多，分为大、中、小三种。胞体呈锥形或三角形。颗粒细胞（granular cell）数量最多，散在分布于皮质。胞体较小，呈颗粒状。细胞的形态多样，有星形细胞、篮状细胞和水平细胞，以星形细胞最多。梭形细胞（spindle cell）数量较少，主要分布于皮质深层。胞体呈梭形，其长轴与皮质表面垂直。大脑皮质的神经元胞体成层排列，从浅到深可分为6层。

（1）分子层（molecular layer） 位于皮质的最表层。以平行的神经纤维为主，神经元较少，主要是小型的星形细胞和水平细胞。

（2）外颗粒层（external granular layer） 位于分子层的内侧，神经元较多，主要是大量的星形细胞和少量的小锥体细胞。

（3）外锥体细胞层（external pyramidal layer） 神经元很多，主要是中、小型锥体细胞和星形细胞。

（4）内颗粒层（internal granular layer） 由密集的星形细胞组成，胞体很小。

（5）内锥体细胞层（internal pyramidal layer） 神经元较少，主要是大、中型锥体细胞和小星形细胞。

（6）多形细胞层（polymorphic layer） 位于皮质最内层，胞体形态不规则，主要有梭形细胞，还有一些星形细胞和小锥体细胞。

2. 大脑白质 在大脑皮质的深层，主要是大量的有髓神经纤维和与之相关的神经胶质细胞。这些纤维分3种。

（1）联络纤维 连接同侧大脑半球各脑回的纤维。

（2）联合纤维 连接两侧大脑半球皮质的纤维，构成胼胝体的纤维。

（3）投射纤维 连接大脑皮质与脑干、脊髓等低级中枢的纤维。

（四）脉络丛（choroid plexus）

脉络丛由来自脑软膜、蛛网膜的小血管丛构成，表面被覆一层立方形的室管膜细胞，与脑脊液的形成有关。

（五）脑脊膜（meninges）

是包在脑和脊髓外面的结缔组织膜。脑膜和脊髓膜互相延续，结构相似，可保护、固定、营养脑和脊髓。由外向内分为硬膜、蛛网膜和软膜。

（六）神经节（ganglion）

包括脑脊神经节和植物性神经节。脑脊神经节属感觉神经节，包括位于脊神经背根上的脊神经节和脑感觉神经上的神经节。植物性神经节属运动神经节，包括交感神经节和副交感神经节，细胞是多极运动神经元。神经节外包有结缔组织被膜，并伸入神经节内，将神经纤维分隔成束，神经节细胞被神经纤维束分成小群。

（七）神经（nerve）

外周神经由成束的轴突和树突组成。轴突和树突束的外面包有几层扁平上皮细胞，形成神经束膜。每条轴突和树突外面都被施万细胞缠绕，形成神经内膜，具有绝缘和保护作用。

三、神经系统图谱

1. **脊髓** 图 9-1 ～图 9-25。

2. **小脑** 图 9-26 ～图 9-54。

3. **大脑** 图 9-55 ～图 9-85。

4. **脉络丛** 图 9-86 ～图 9-89。

5. **神经节和神经** 图 9-90 ～图 9-112。

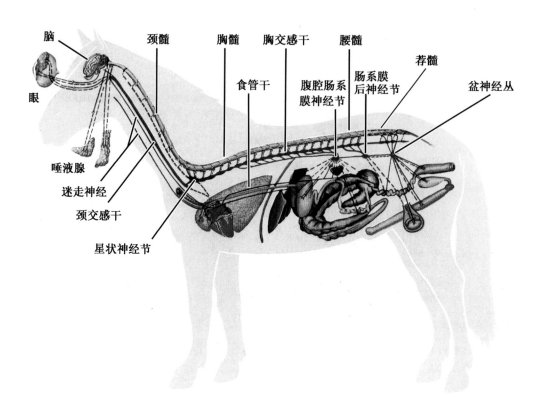

图 9-1 动物主要器官及主要神经分布示意图
Fig.9-1 diagram of animal main organs and nerves distribution

脑—brain　颈髓—cervical cord　胸髓—thoracic cord　胸交感干—thoracic sympathetic trunk　腰髓—lumbar cord　荐髓—sacral cord　食管干—esophageal trunk　腹腔肠系膜神经节—abdominal mesenteric ganglion　肠系膜后神经节—posterior mesenteric ganglion　盆神经丛—pots plexus　眼—eye　唾液腺—salivary gland　迷走神经—vagus　颈交感干—cervical sympathetic trunk　星状神经节—stellate ganglion

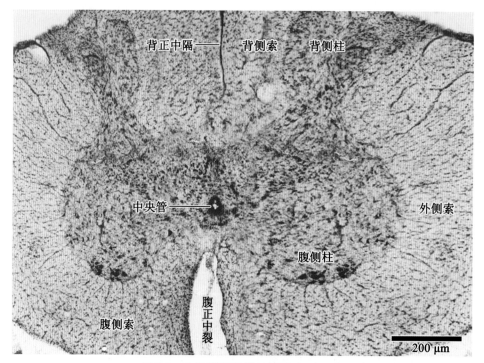

图 9-2 牛颈段脊髓横切低倍像（HE）
Fig.9-2 low magnification of transverse section of cow cervical spinal cord（HE）

背正中隔—dorsal septum　背侧索—dorsolateral funiculus　背侧柱—dorsolateral column　中央管—central canal　外侧索—lateral funiculus　腹侧柱—ventral column　腹侧索—ventral funiculus　腹正中裂—ventral fissure

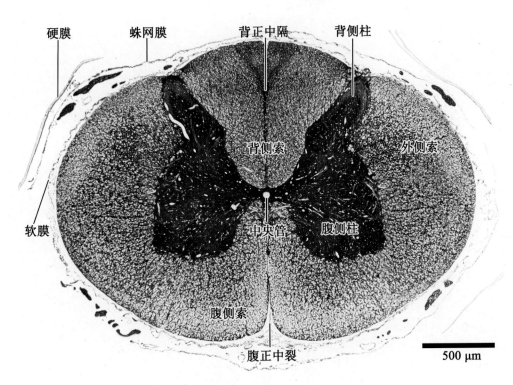

图9-3 马颈段脊髓横切低倍像（镀银染色）
Fig.9-3 low magnification of transverse section of horse cervical spinal cord（silver stain）
硬膜—dura mater 蛛网膜—arachnoid mater 软膜—pia mater 背正中隔—dorsal septum
背侧柱—dorsolateral column 背侧索—dorsolateral funiculus 外侧索—lateral funiculus 中央管—central canal
腹侧柱—ventral column 腹侧索—ventral funiculus 腹正中裂—ventral fissure

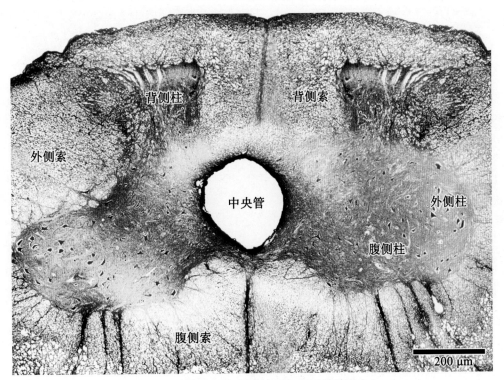

图9-4 猪脊髓颈膨大横切低倍像（镀银染色）
Fig.9-4 low magnification of transverse section of pig cervical enlargement（silver stain）
背侧柱—dorsolateral column 背侧索—dorsolateral funiculus 外侧索—lateral funiculus 中央管—central canal
外侧柱—lateral column 腹侧柱—ventral column 腹侧索—ventral funiculus

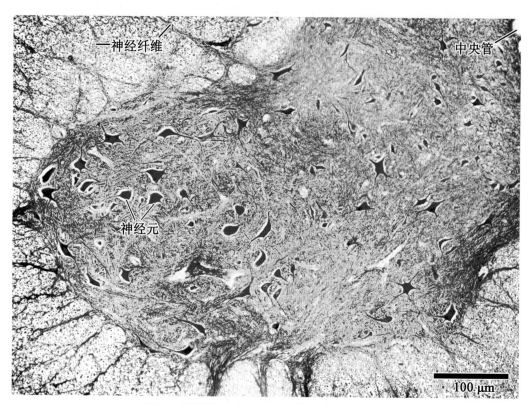

图9-5 猪脊髓腹侧柱横切中倍像（镀银染色）

Fig.9-5 mid magnification of transverse section of pig spinal ventral column （silver stain）

神经纤维—nerve fiber 中央管—central canal 神经元—neuron

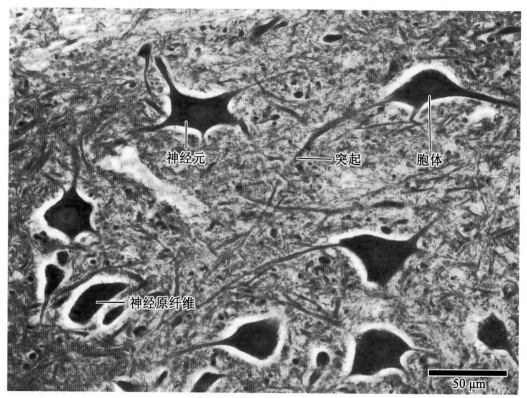

图9-6 猪脊髓腹侧柱横切高倍像（镀银染色）

Fig.9-6 high magnification of transverse section of pig spinal ventral column （silver stain）

神经元—neuron 突起—dendrite 胞体—cell body 神经原纤维—neurofibril

第九章 神经系统 Nervous System

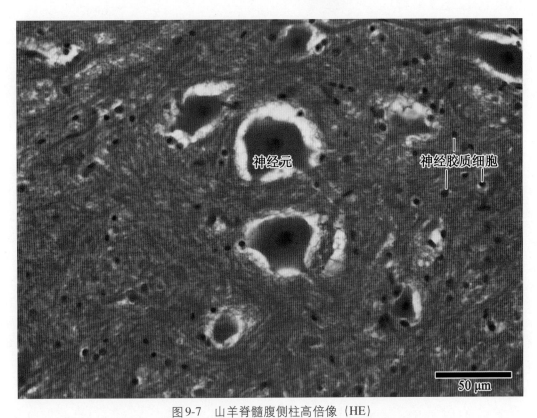

图 9-7 山羊脊髓腹侧柱高倍像（HE）

Fig.9-7 high magnification of goat spinal ventral column（HE）

神经元—neuron 神经胶质细胞—neurogliocyte

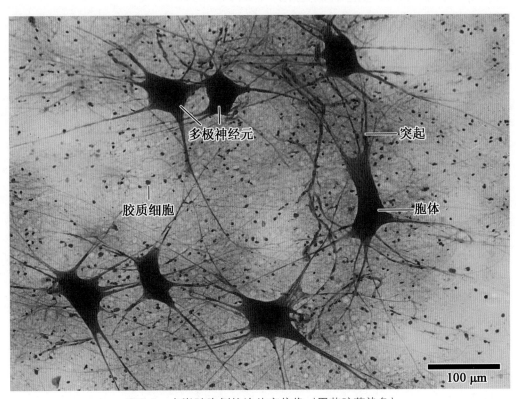

图 9-8 犬脊髓腹侧柱涂片高倍像（甲苯胺蓝染色）

Fig.9-8 high magnification of dog spinal ventral column smear（toluidine blue stain）

多极神经元—multipolar neuron 胶质细胞—glial cell 突起—dendrite 胞体—cell body

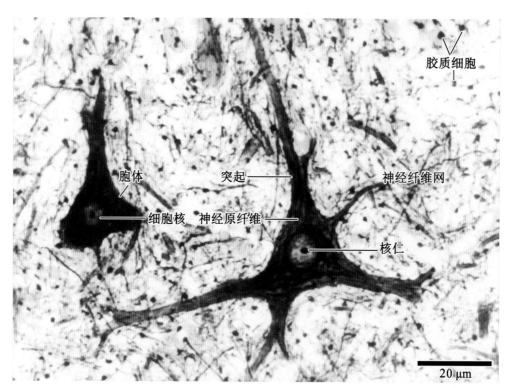

图9-9 猫脊髓多极神经元高倍像（镀银染色）

Fig.9-9 high magnification of multipolar neurons in cat spinal cord (silver stain)

胞体—cell body 细胞核—nucleus 突起—dendrite 神经原纤维—neurofibril 胶质细胞—glial cell
神经纤维网—neuropil 核仁—nucleolus

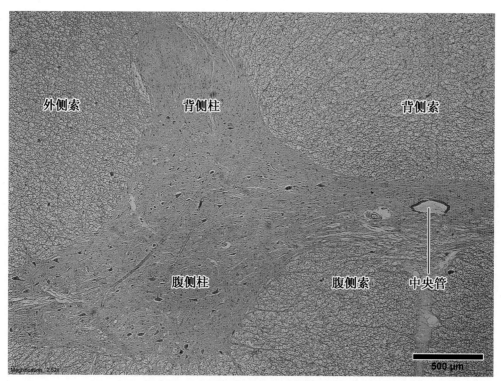

图9-10 牛腰段脊髓横切低倍像1（HE）

Fig.9-10 low magnification of transverse section of cow cervical spinal cord 1（HE）

外侧索—lateral funiculus 背侧柱—dorsolateral column 背侧索—dorsolateral funiculus
腹侧柱—ventral column 腹侧索—ventral funiculus 中央管—central canal

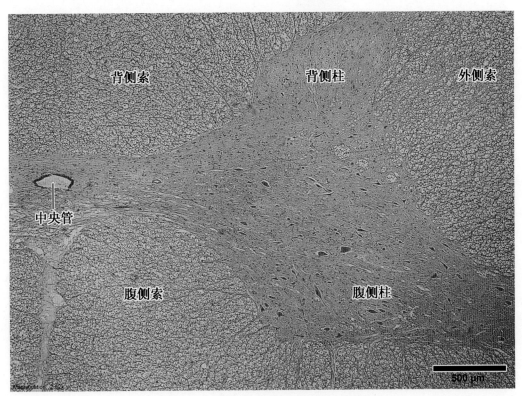

图 9-11 牛腰段脊髓横切低倍像 2（HE）

Fig.9-11 low magnification of transverse section of cow lumbar spinal cord 2（HE）

背侧索—dorsolateral funiculus 背侧柱—dorsolateral column 外侧索—lateral funiculus

中央管—central canal 腹侧索—ventral funiculus 腹侧柱—ventral column

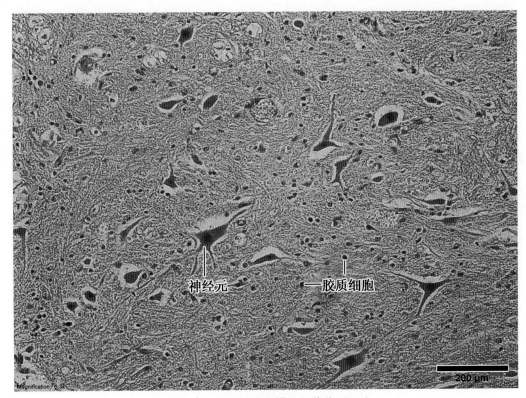

图 9-12 马腰段脊髓横切低倍像（HE）

Fig.9-12 low magnification of transverse section of horse lumbar spinal cord（HE）

神经元—neuron 胶质细胞—neuroglia cell

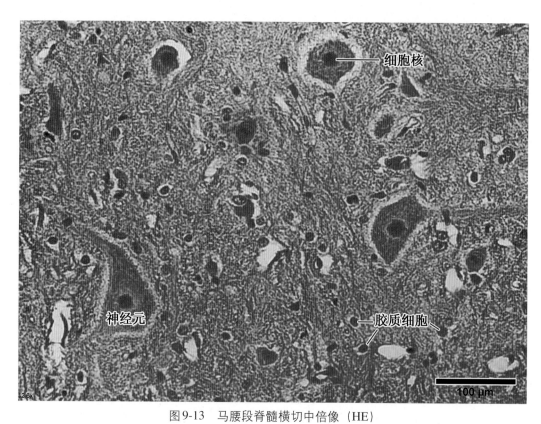

图9-13 马腰段脊髓横切中倍像（HE）

Fig.9-13 mid magnification of transverse section of horse lumbar spinal cord （HE）

神经元—neuron 细胞核—nucleus 胶质细胞—neuroglia cell

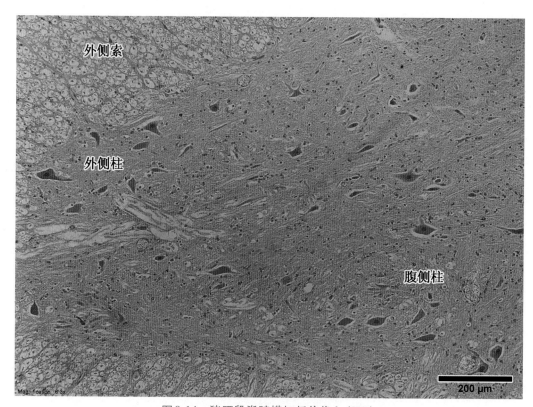

图9-14 猪腰段脊髓横切低倍像1（HE）

Fig.9-14 low magnification of transverse section of pig lumbar spinal cord 1 （HE）

外侧索—lateral funiculus 外侧柱—lateral column 腹侧柱—ventral column

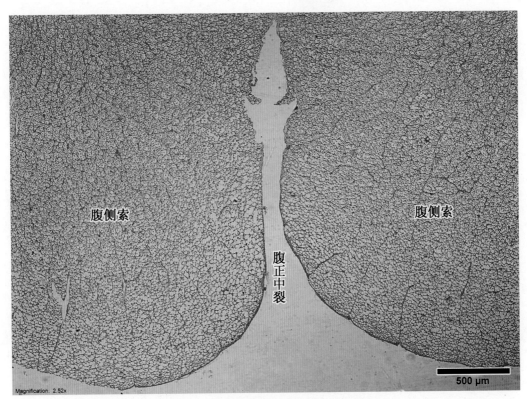

图9-15 猪腰段脊髓横切低倍像2（HE）

Fig.9-15 low magnification of transverse section of pig lumbar spinal cord 2 （HE）

腹侧索—ventral funiculus 腹正中裂—ventral fissure

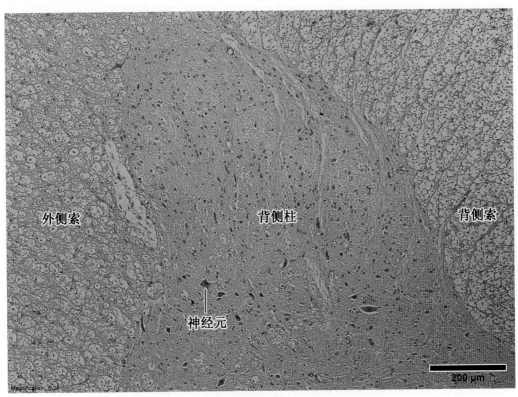

图9-16 绵羊腰段脊髓横切低倍像（HE）

Fig.9-16 low magnification of transverse section of sheep lumbar spinal cord （HE）

外侧索—lateral funiculus 背侧柱—dorsolateral column 背侧索—dorsolateral funiculus 神经元—neuron

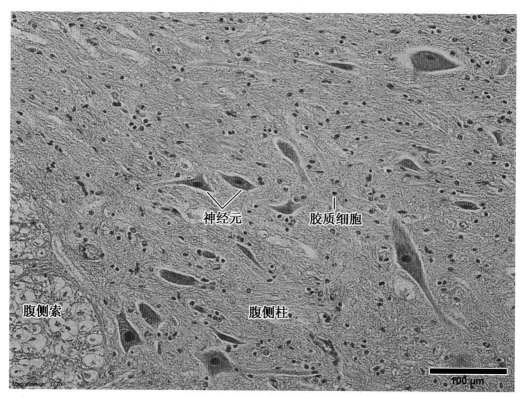

图9-17 绵羊腰段脊髓腹侧柱横切中倍像（HE）

Fig.9-17 mid magnification of transverse section of sheep lumbar spinal cord（HE）

腹侧索—ventral funiculus 神经元—neuron 胶质细胞—neuroglia cell 腹侧柱—ventral column

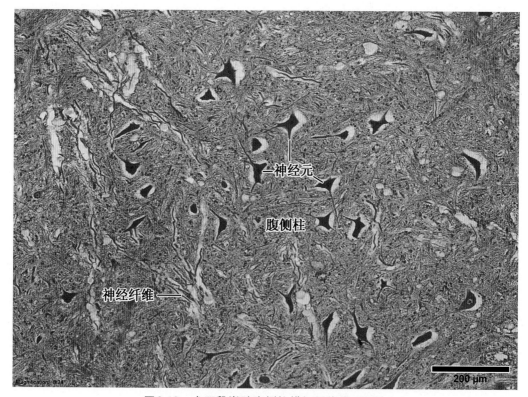

图9-18 犬腰段脊髓腹侧柱横切低倍像（HE）

Fig.9-18 low magnification of transverse section of dog lumbar spinal cord（HE）

神经元—neuron 腹侧柱—ventral column 神经纤维—nerve fiber

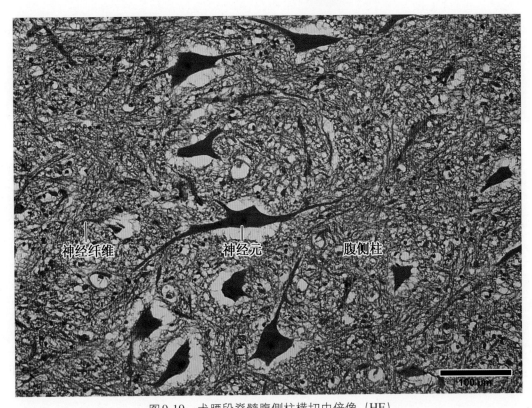

图 9-19 犬腰段脊髓腹侧柱横切中倍像（HE）

Fig.9-19 mid magnification of transverse section of dog lumbar spinal cord（HE）

神经元—neuron 神经纤维—nerve fiber 腹侧柱—ventral column

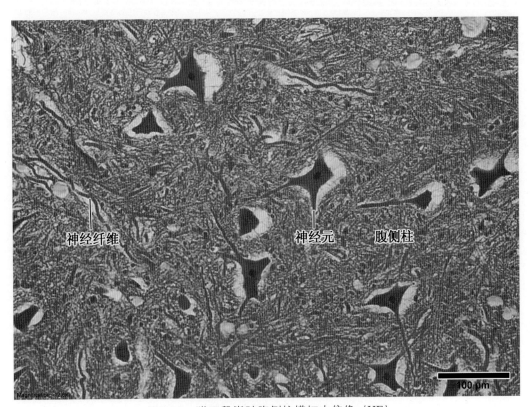

图 9-20 猫腰段脊髓腹侧柱横切中倍像（HE）

Fig.9-20 mid magnification of transverse section of cat lumbar spinal cord（HE）

神经纤维—nerve fiber 神经元—neuron 腹侧柱—ventral column

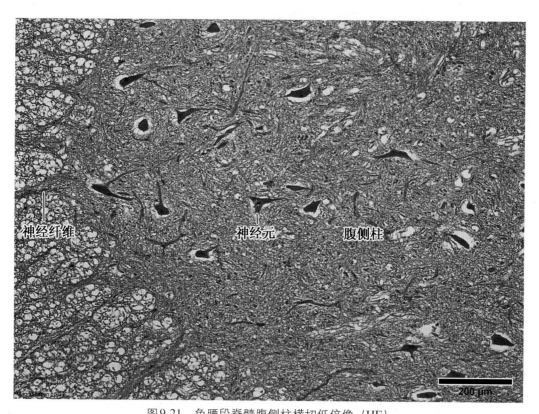

图9-21　兔腰段脊髓腹侧柱横切低倍像（HE）
Fig.9-21　low magnification of transverse section of rabbit lumbar spinal cord（HE）
神经纤维—nerve fiber　神经元—neuron　腹侧柱—ventral column

图9-22　牛脊髓中央管横切中倍像（HE）
Fig.9-22　mid magnification of transverse section of cow spinal cord central canal（HE）
中央管—central canal　室管膜—ependyma

第九章 神经系统 Nervous System

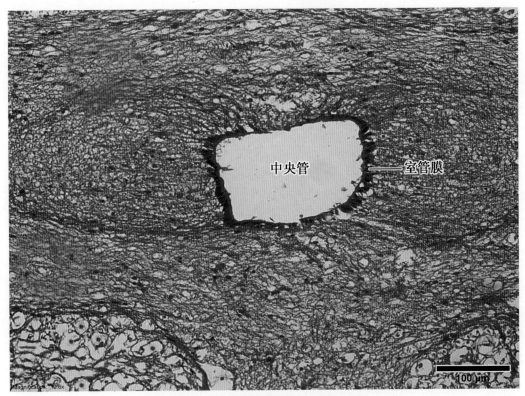

图9-23 马脊髓中央管横切低倍像（HE）

Fig.9-23 low magnification of transverse section of horse spinal cord central canal（HE）

中央管—central canal 室管膜—ependyma

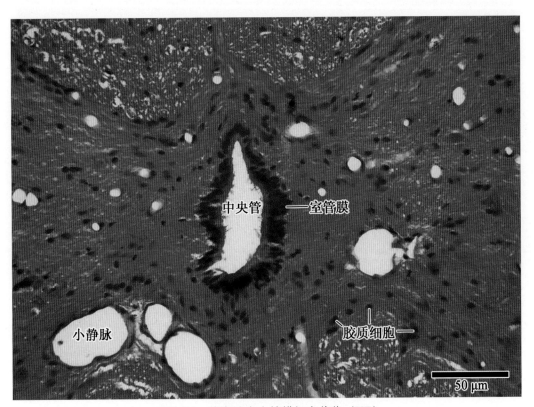

图9-24 猪脊髓中央管横切高倍像（HE）

Fig.9-24 high magnification of transverse section of pig spinal cord central canal（HE）

中央管—central canal 室管膜—ependyma 小静脉—venule 胶质细胞—neuroglia cell

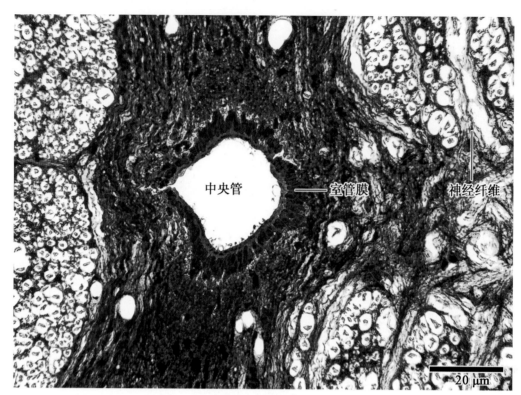

图 9-25　山羊脊髓中央管高倍像（镀银染色）

Fig.9-25　high magnification of goat spinal cord central canal(silver stain)

中央管—central canal　室管膜—ependyma　神经纤维—nerve fiber

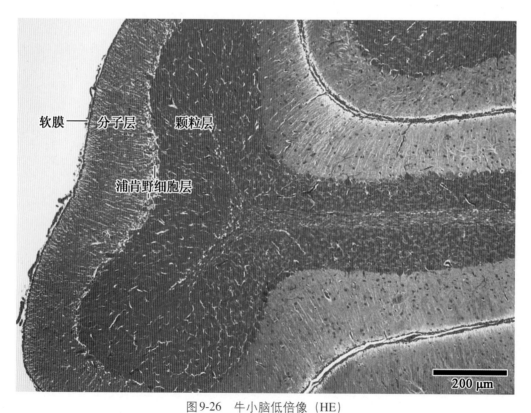

图 9-26　牛小脑低倍像（HE）

Fig.9-26　low magnification of cow cerebellum（HE）

软膜—pia mater　分子层—molecular layer　颗粒层—granular layer　浦肯野细胞层—Purkinje cell layer

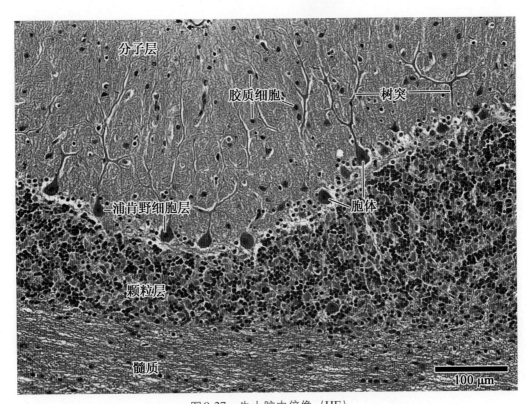

图9-27 牛小脑中倍像（HE）

Fig.9-27 mid magnification of cow cerebellum（HE）

分子层—molecular layer 浦肯野细胞层—Purkinje cell layer 颗粒层—granular layer
髓质—medulla 胶质细胞—glial cell 树突—dendrite 胞体—cell body

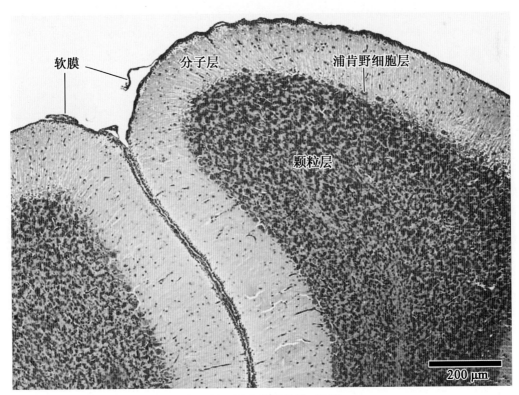

图9-28 马小脑低倍像（HE）

Fig.9-28 low magnification of horse cerebellum（HE）

软膜—pia mater 分子层—molecular layer 浦肯野细胞层—Purkinje cell layer 颗粒层—granular layer

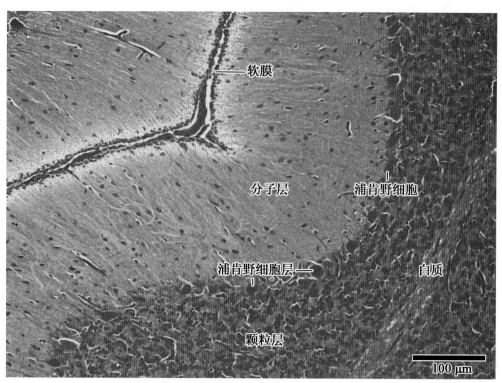

图9-29 马小脑中倍像（HE）

Fig.9-29 mid magnification of horse cerebellum（HE）

软膜—pia mater 分子层—molecular layer 浦肯野细胞—Purkinje cell 浦肯野细胞层—Purkinje cell layer
颗粒层—granular layer 白质—white matter

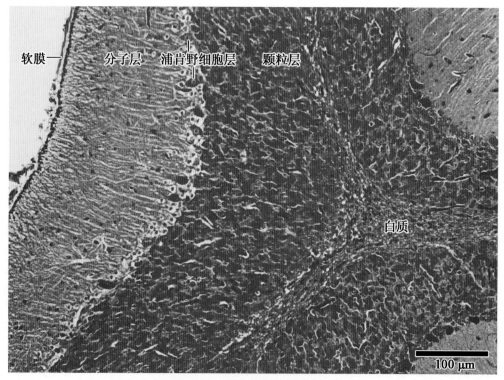

图9-30 猪小脑中倍像（HE）

Fig.9-30 mid magnification of pig cerebellum（HE）

软膜—pia mater 分子层—molecular layer 浦肯野细胞层—Purkinje cell layer
颗粒层—granular layer 白质—white matter

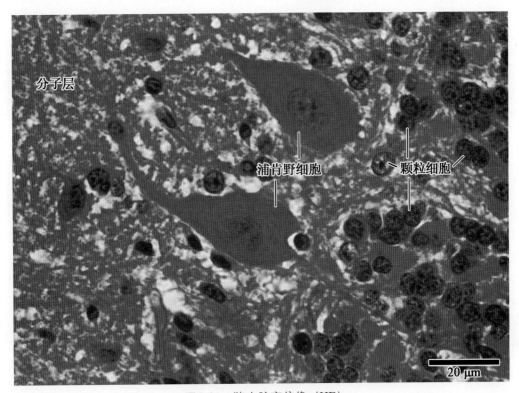

图9-31 猪小脑高倍像（HE）

Fig.9-31 high magnification of pig cerebellum（HE）

分子层—molecular layer 浦肯野细胞—Purkinje cell 颗粒细胞—granular cell

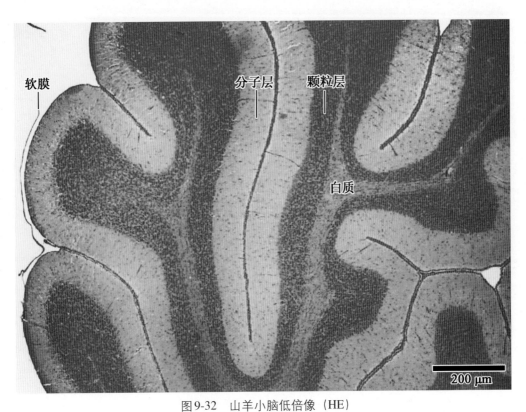

图9-32 山羊小脑低倍像（HE）

Fig.9-32 low magnification of goat cerebellum（HE）

软膜—pia mater 分子层—molecular layer 颗粒层—granular layer 白质—white matter

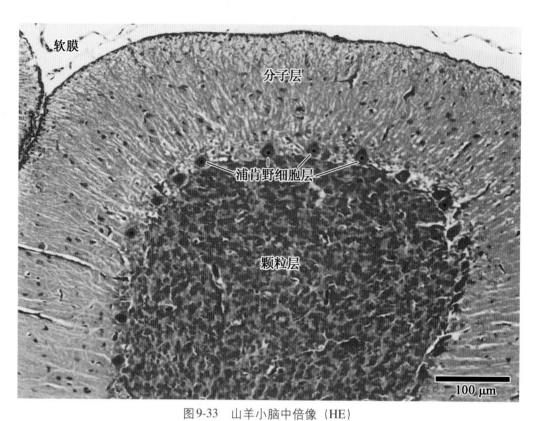

图 9-33　山羊小脑中倍像（HE）

Fig.9-33　mid magnification of goat cerebellum（HE）

软膜—pia mater　分子层—molecular layer　浦肯野细胞层—Purkinje cell layer　颗粒层—granular layer

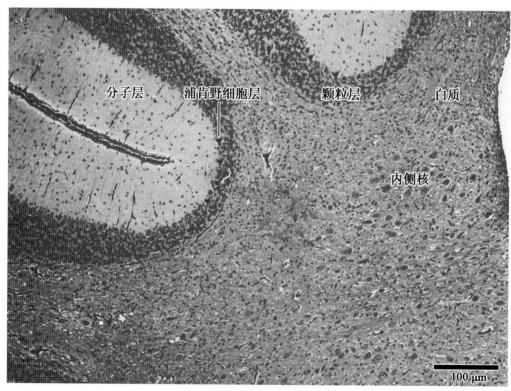

图 9-34　犬小脑中倍像 1（HE）

Fig.9-34　mid magnification of dog cerebellum 1（HE）

分子层—molecular layer　浦肯野细胞层—Purkinje cell layer　颗粒层—granular layer
白质—white matter　内侧核—medial nucleus

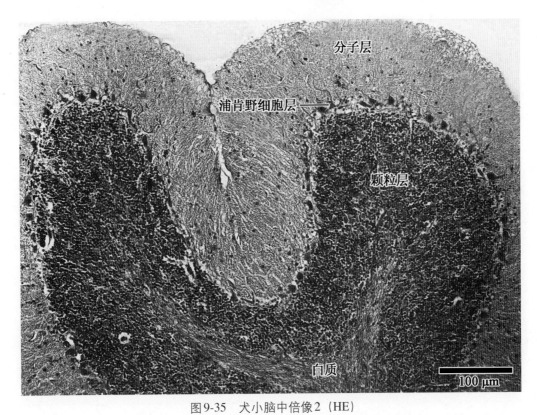

图9-35 犬小脑中倍像2（HE）

Fig.9-35 mid magnification of dog cerebellum 2（HE）

分子层—molecular layer　浦肯野细胞层—Purkinje cell layer　颗粒层—granular layer　白质—white matter

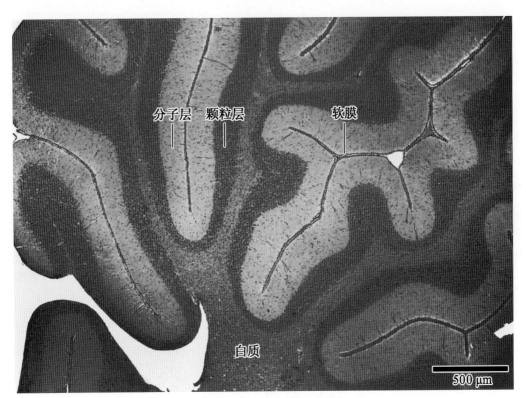

图9-36 猫小脑低倍像（HE）

Fig.9-36 low magnification of cat cerebellum（HE）

分子层—molecular layer　颗粒层—granular layer　软膜—pia mater　白质—white matter

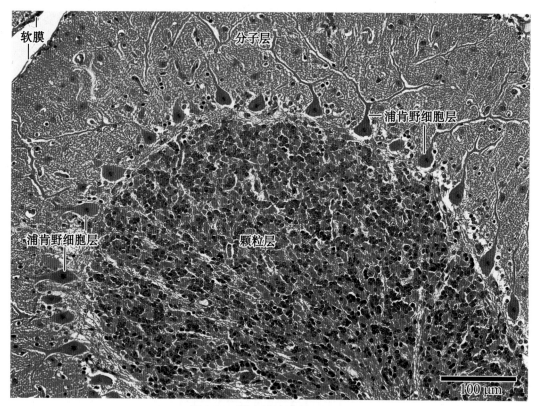

图9-37 猫小脑中倍像（HE）

Fig.9-37 mid magnification of cat cerebellum（HE）

软膜—pia mater　分子层—molecular layer　浦肯野细胞层—Purkinje cell layer　颗粒层—granular layer

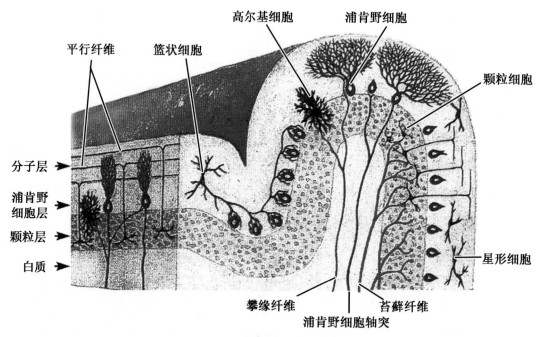

图9-38 小脑皮质神经元和纤维分布示意图

Fig.9-38 distribution diagram of cerebellum neurons and nerve fibers

分子层—molecular layer　浦肯野细胞层—Purkinje cell layer　颗粒层—granular layer　白质—white matter
平行纤维—parallel fiber　篮状细胞—basket cell　高尔基细胞—Golgi cell　浦肯野细胞—Purkinje cell
颗粒细胞—granular cell　星形细胞—stellate cell　攀缘纤维—climbing fiber　浦肯野细胞轴突—Purkinje cell axon
苔藓纤维—mossy fiber

第九章 神经系统 Nervous System

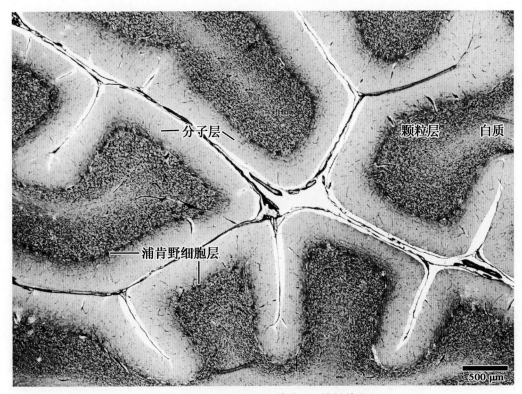

图 9-39 牛小脑镀银低倍像 1（镀银染色）

Fig.9-39 low magnification of cow cerebellum 1 (silver stain)

分子层—molecular layer 浦肯野细胞层—Purkinje cell layer 颗粒层—granular layer 白质—white matter

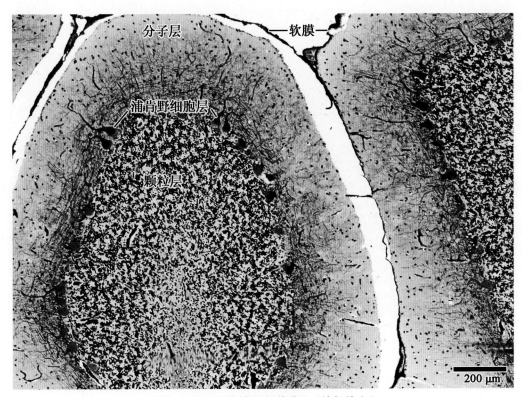

图 9-40 牛小脑镀银低倍像 2（镀银染色）

Fig.9-40 low magnification of cow cerebellum 2 (silver stain)

软膜—pia mater 分子层—molecular layer 浦肯野细胞层—Purkinje cell layer 颗粒层—granular layer

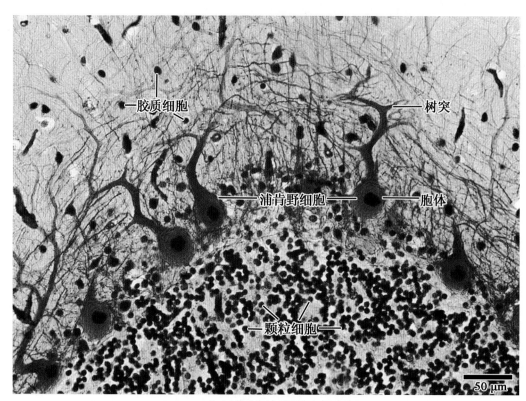

图 9-41　牛小脑镀银高倍像（镀银染色）

Fig.9-41　high magnification of cow cerebellum(silver stain)

胶质细胞—glial cell　浦肯野细胞—Purkinje cell　树突—dendrite　胞体—cell body　颗粒细胞—granular cell

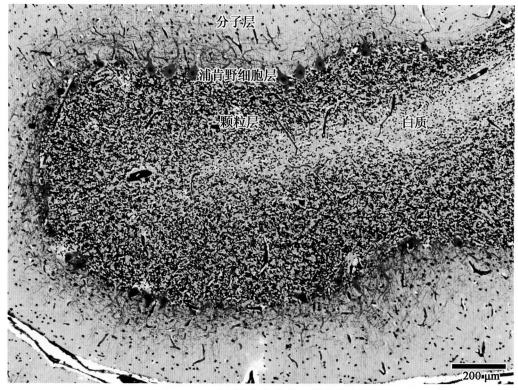

图 9-42　马小脑镀银低倍像（镀银染色）

Fig.9-42　low magnification of horse cerebellum(silver stain)

分子层—molecular layer　浦肯野细胞层—Purkinje cell layer　颗粒层—granular layer　白质—white matter

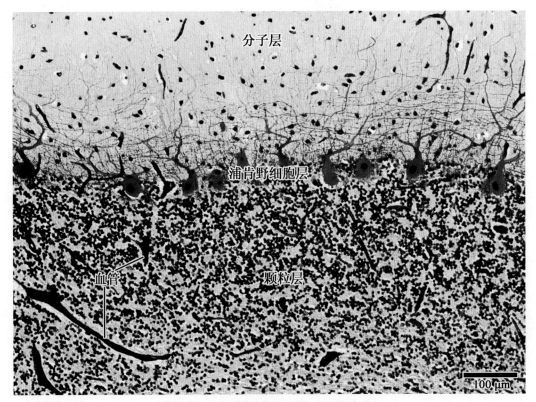

图9-43 马小脑镀银中倍像（镀银染色）

Fig.9-43 mid magnification of horse cerebellum(silver stain)

分子层—molecular layer 浦肯野细胞层—Purkinje cell layer 颗粒层—granular layer 血管—blood vessel

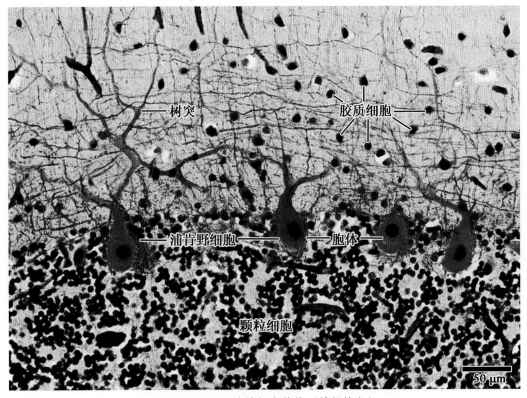

图9-44 马小脑镀银高倍像（镀银染色）

Fig.9-44 high magnification of horse cerebellum(silver stain)

树突—dendrite 胶质细胞—glial cell 浦肯野细胞—Purkinje cell 胞体—cell body 颗粒细胞—granular cell

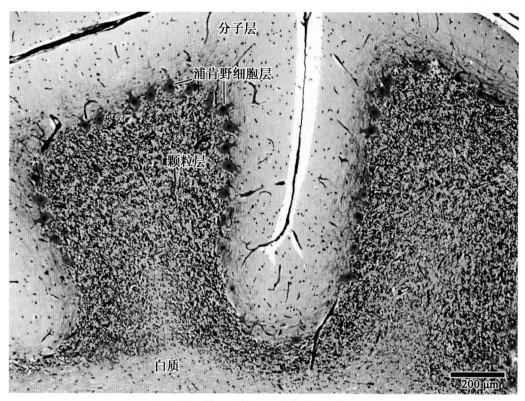

图 9-45 猪小脑镀银低倍像（镀银染色）

Fig.9-45 low magnification of pig cerebellum(silver stain)

分子层—molecular layer 浦肯野细胞层—Purkinje cell layer 颗粒层—granular layer 白质—white matter

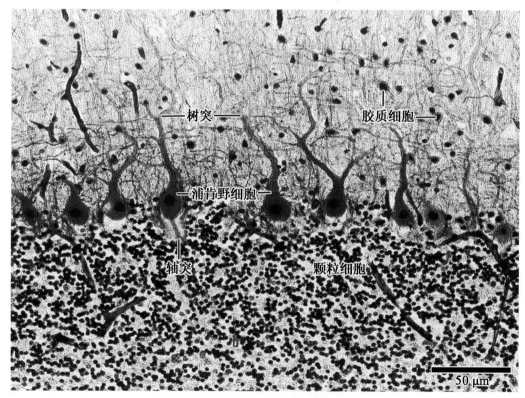

图 9-46 猪小脑镀银高倍像（镀银染色）

Fig.9-46 high magnification of pig cerebellum(silver stain)

树突—dendrite 浦肯野细胞—Purkinje cell 轴突—axon 胶质细胞—glial cell 颗粒细胞—granular cell

第九章 神经系统 Nervous System

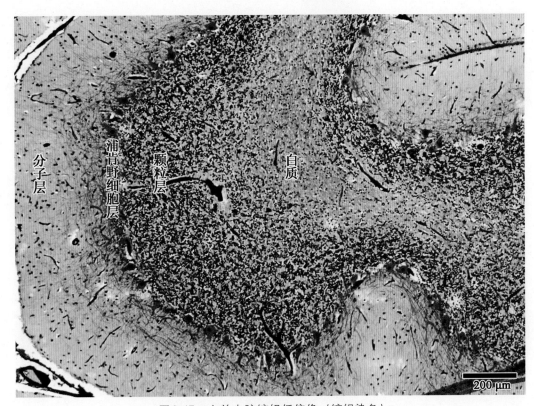

图9-47 山羊小脑镀银低倍像（镀银染色）
Fig.9-47 low magnification of goat cerebellum(silver stain)
分子层—molecular layer 浦肯野细胞层—Purkinje cell layer 颗粒层—granular layer 白质—white matter

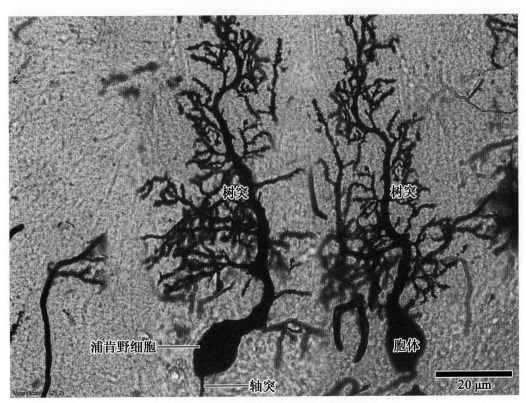

图9-48 山羊小脑镀银高倍像（镀银染色）
Fig.9-48 high magnification of goat cerebellum(silver stain)
树突—dendrite 浦肯野细胞—Purkinje cell 轴突—axon 胞体—cell body

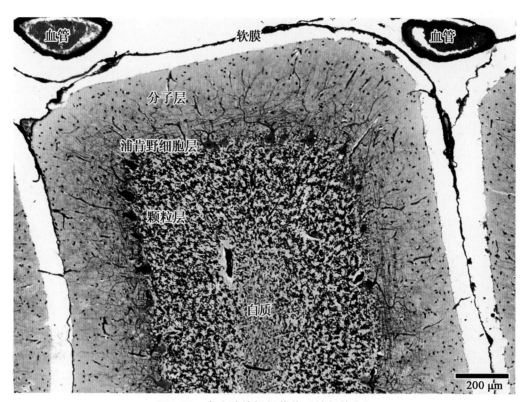

图 9-49 犬小脑镀银低倍像（镀银染色）

Fig.9-49 low magnification of dog cerebellum(silver stain)

血管—blood vessel 软膜—pia mater 分子层—molecular layer 浦肯野细胞层—Purkinje cell layer
颗粒层—granular layer 白质—white matter

图 9-50 犬浦肯野细胞高倍像（镀银染色）

Fig.9-50 high magnification of dog Purkinje cell(silver stain)

分子层—molecular layer 树突—dendrite 胞体—cell body 轴突—axon 颗粒层—granular layer

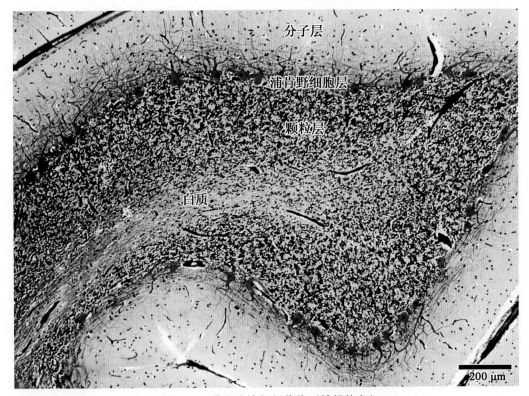

图 9-51　猫小脑镀银低倍像（镀银染色）

Fig.9-51　low magnification of cat cerebellum(silver stain)

分子层—molecular layer　浦肯野细胞层—Purkinje cell layer　颗粒层—granular layer　白质—white matter

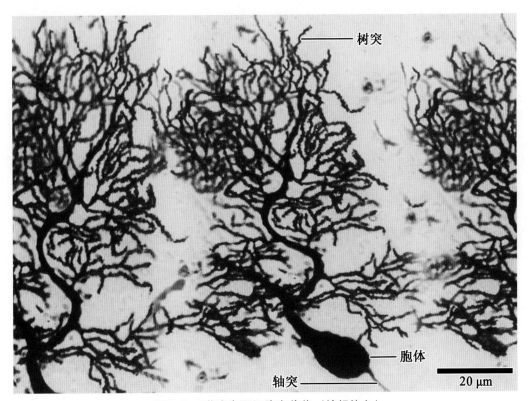

图 9-52　猫浦肯野细胞高倍像（镀银染色）

Fig.9-52　high magnification of cat Purkinje cell(silver stain)

树突—dendrite　胞体—cell body　轴突—axon

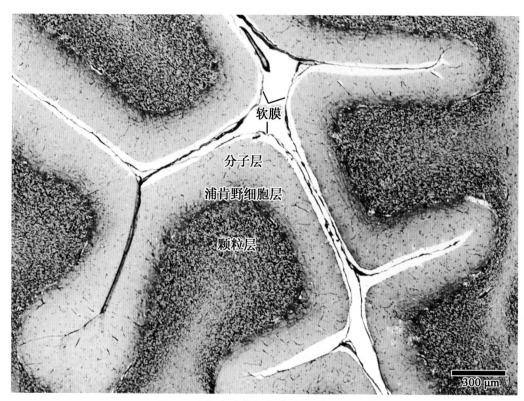

图 9-53　兔小脑镀银低倍像（镀银染色）

Fig.9-53　low magnification of rabbit cerebellum(silver stain)

软膜—pia mater　分子层—molecular layer　浦肯野细胞层—Purkinje cell layer　颗粒层—granular layer

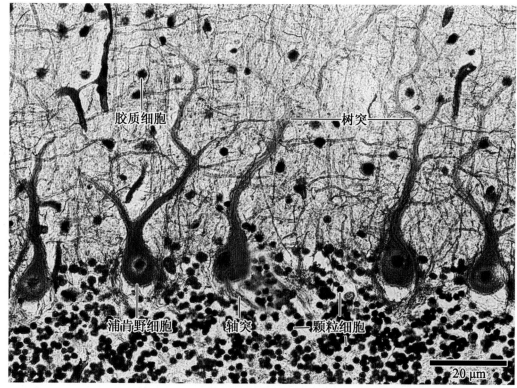

图 9-54　兔小脑镀银高倍像（镀银染色）

Fig.9-54　high magnification of rabbit cerebellum(silver stain)

胶质细胞—glial cell　浦肯野细胞—Purkinje cell　树突—dendrite　轴突—axon　颗粒细胞—granular cell

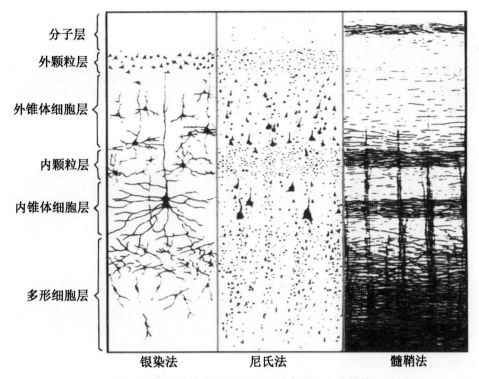

图9-55 三种染色方法显示大脑皮质的分层结构

Fig.9-55 three stain methods show the layered structure of cerebral cortex

银染法—silver stain 尼氏法—Nissl stain 髓鞘法—myelin stain 分子层—molecular layer
外颗粒层—external granular layer 外锥体细胞层—external pyramidal layer 内颗粒层—internal granular layer
内锥体细胞层—internal pyramidal layer 多形细胞层—polymorphic layer

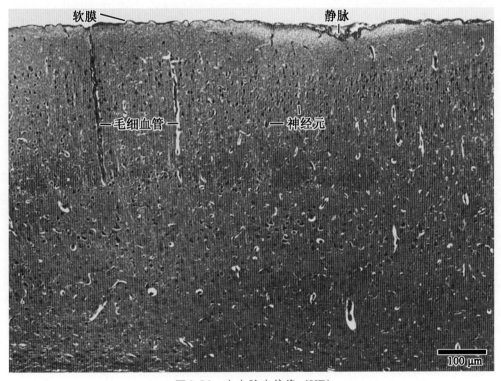

图9-56 牛大脑中倍像（HE）

Fig.9-56 mid magnification of cow cerebrum（HE）

软膜—pia mater 静脉—vein 毛细血管—capillary 神经元—neuron

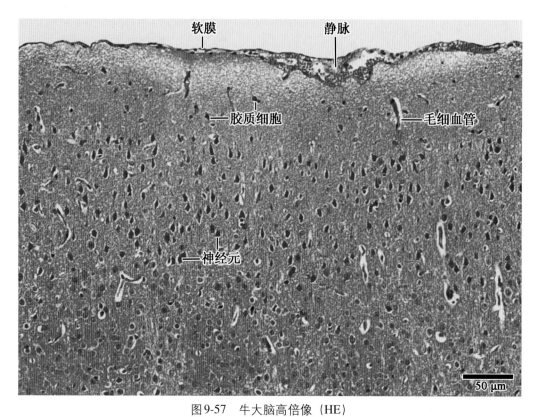

图9-57 牛大脑高倍像（HE）

Fig.9-57 high magnification of cow cerebrum（HE）

软膜—pia mater　静脉—vein　胶质细胞—glial cell　毛细血管—capillary　神经元—neuron

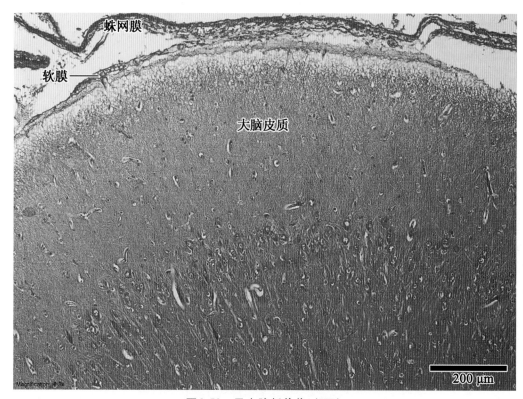

图9-58 马大脑低倍像（HE）

Fig.9-58 low magnification of horse cerebrum（HE）

蛛网膜—arachnoid　软膜—pia mater　大脑皮质—cerebrum cortex

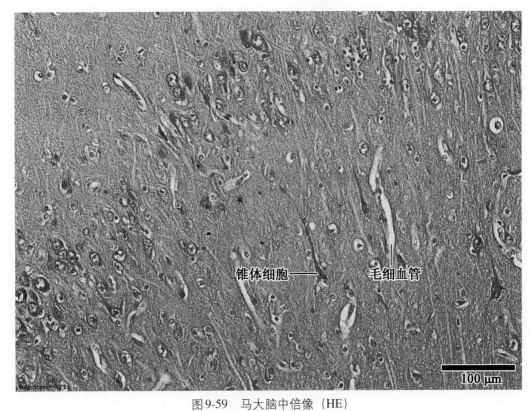

图9-59 马大脑中倍像（HE）

Fig.9-59 mid magnification of horse cerebrum（HE）

锥体细胞—pyramidal cell 毛细血管—capillary

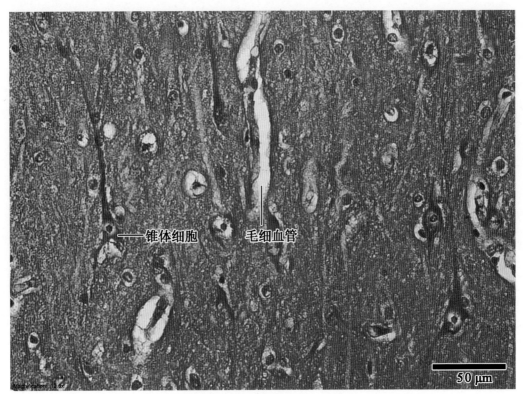

图9-60 马大脑高倍像1（HE）

Fig.9-60 high magnification of horse cerebrum 1（HE）

锥体细胞—pyramidal cell 毛细血管—capillary

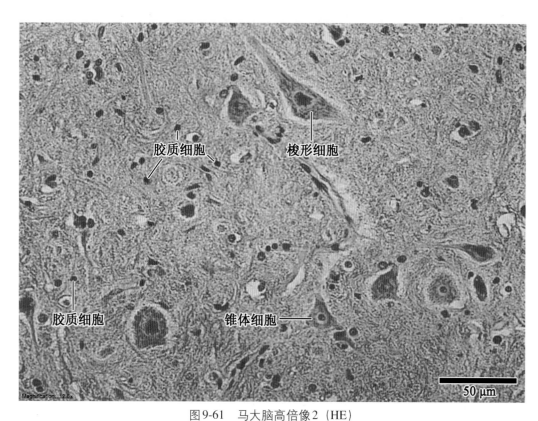

图 9-61　马大脑高倍像 2（HE）

Fig.9-61　high magnification of horse cerebrum 2（HE）

胶质细胞—glial cell　梭形细胞—spindle cell　锥体细胞—pyramidal cell

图 9-62　猪大脑中倍像（HE）

Fig.9-62　mid magnification of pig cerebrum（HE）

软膜—pia mater　静脉—vein　毛细血管—capillary　神经元—neuron

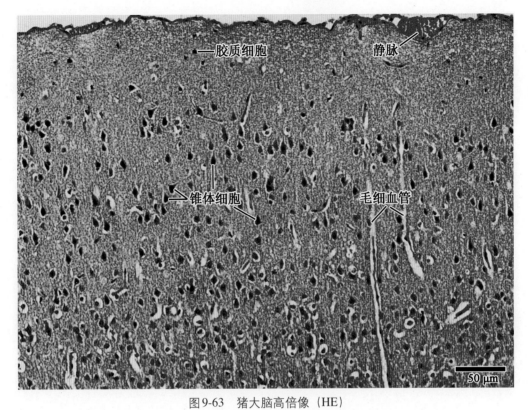

图9-63 猪大脑高倍像（HE）

Fig.9-63 high magnification of pig cerebrum（HE）

胶质细胞—glial cell 静脉—vein 锥体细胞—pyramidal cell 毛细血管—capillary

图9-64 山羊大脑中倍像（HE）

Fig.9-64 mid magnification of goat cerebrum（HE）

软膜—pia mater 血管—blood vessel 神经元—neuron

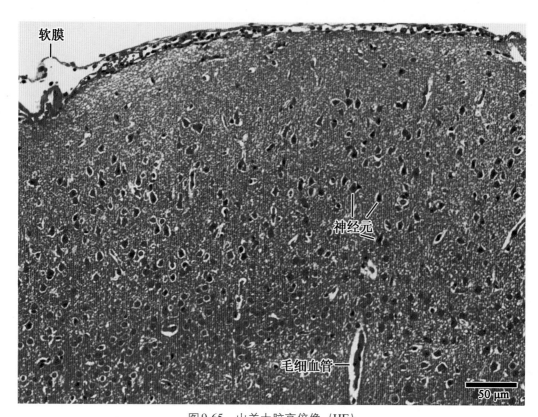

图9-65 山羊大脑高倍像（HE）
Fig.9-65 high magnification of goat cerebrum（HE）
软膜—pia mater 神经元—neuron 毛细血管—capillary

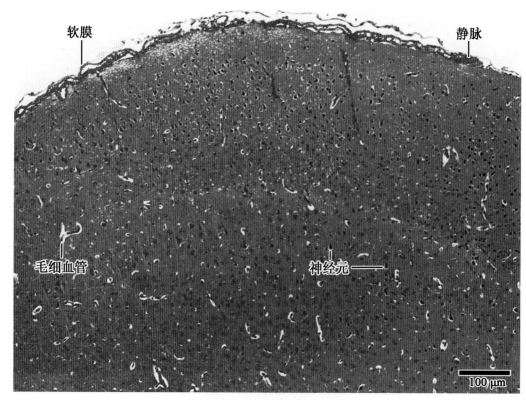

图9-66 犬大脑中倍像（HE）
Fig.9-66 mid magnification of dog cerebrum（HE）
软膜—pia mater 静脉—vein 毛细血管—capillary 神经元—neuron

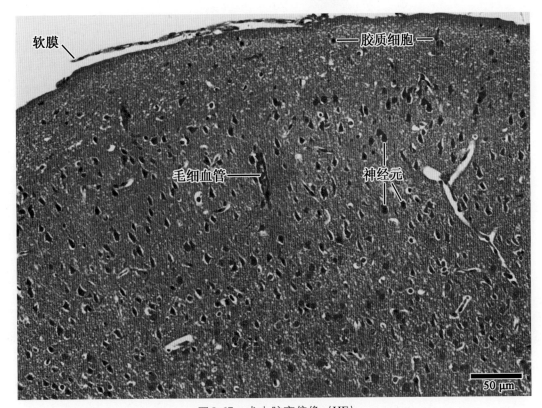

图9-67 犬大脑高倍像（HE）

Fig.9-67 high magnification of dog cerebrum（HE）

软膜—pia mater 胶质细胞—glial cell 毛细血管—capillary 神经元—neuron

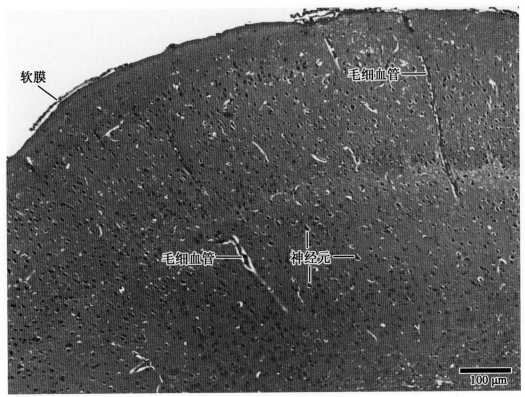

图9-68 猫大脑中倍像（HE）

Fig.9-68 mid magnification of cat cerebrum（HE）

软膜—pia mater 毛细血管—capillary 神经元—neuron

图9-69 猫大脑高倍像（HE）

Fig.9-69 high magnification of cat cerebrum（HE）

软膜—pia mater　胶质细胞—glial cell　毛细血管—capillary　神经元—neuron

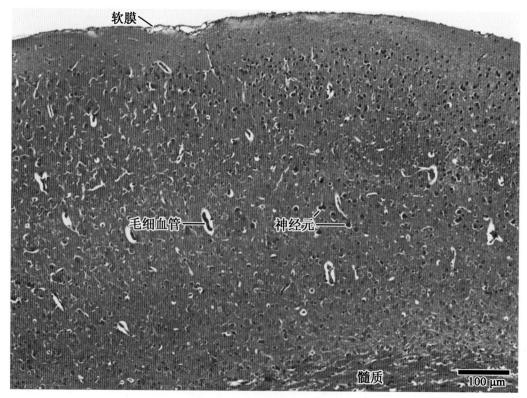

图9-70 兔大脑中倍像（HE）

Fig.9-70 mid magnification of rabbit cerebrum（HE）

软膜—pia mater　毛细血管—capillary　神经元—neuron　髓质—medulla

图9-71 兔大脑高倍像（HE）

Fig.9-71 high magnification of rabbit cerebrum（HE）

软膜—pia mater 毛细血管—capillary 神经元—neuron

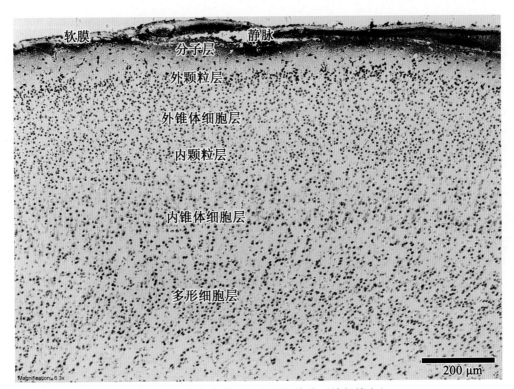

图9-72 牛大脑皮质的分层低倍像（镀银染色）

Fig.9-72 low magnification of cow cerebral cortex layers (silver stain)

软膜—pia mater 静脉—vein 分子层—molecular layer 外颗粒层—external granular layer
外锥体细胞层—external pyramidal layer 内颗粒层—internal granular layer 内锥体细胞层—internal pyramidal layer
多形细胞层—polymorphic layer

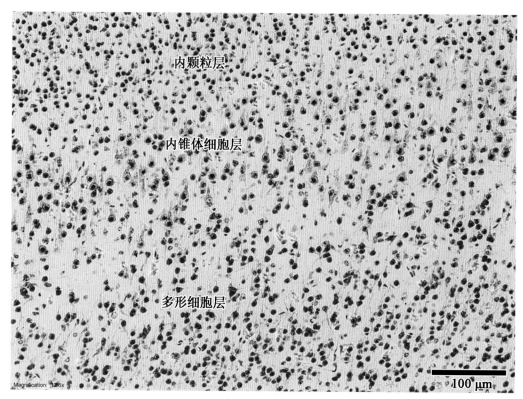

图9-73 牛大脑皮质的分层中倍像（镀银染色）

Fig.9-73 mid magnification of cow cerebral cortex layers(silver stain)

内颗粒层—internal granular layer　内锥体细胞层—internal pyramidal layer　多形细胞层—polymorphic layer

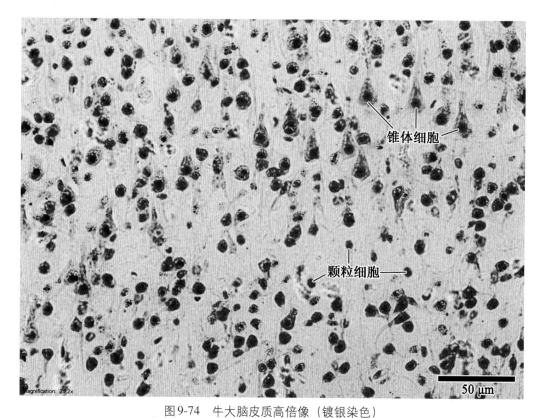

图9-74 牛大脑皮质高倍像（镀银染色）

Fig.9-74 high magnification of cow cerebral cortex(silver stain)

锥体细胞—pyramidal cell　颗粒细胞—granular cell

第九章 神经系统 Nervous System

图9-75 马大脑皮质高倍像（镀银+中性红染色）

Fig.9-75 high magnification of horse cerebral cortex(silver + neutral red stain)

锥体细胞—pyramidal cell 树突—dendrite

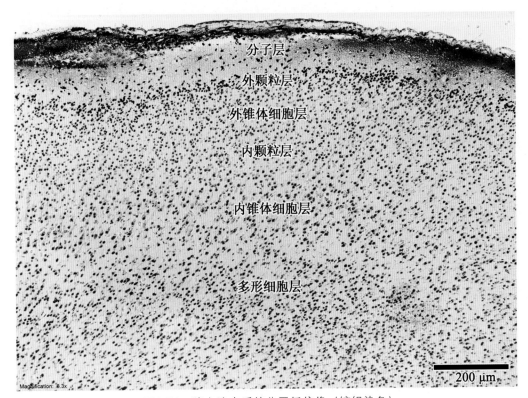

图9-76 猪大脑皮质的分层低倍像（镀银染色）

Fig.9-76 low magnification of pig cerebral cortex layers (silver stain)

分子层—molecular layer 外颗粒层—external granular layer 外锥体细胞层—external pyramidal layer
内颗粒层—internal granular layer 内锥体细胞层—internal pyramidal layer 多形细胞层—polymorphic layer

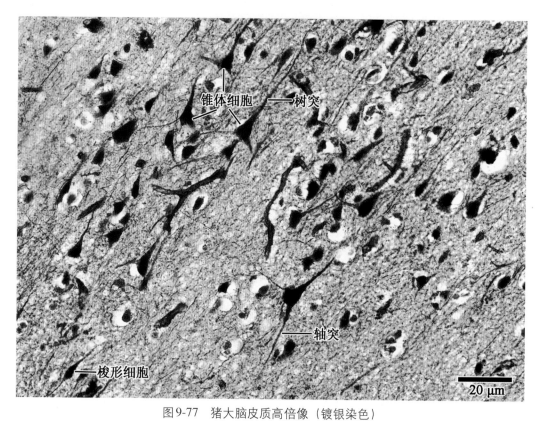

图9-77 猪大脑皮质高倍像（镀银染色）
Fig.9-77 high magnification of pig cerebral cortex(silver stain)
锥体细胞—pyramidal cell 树突—dendrite 轴突—axon 梭形细胞—spindle cell

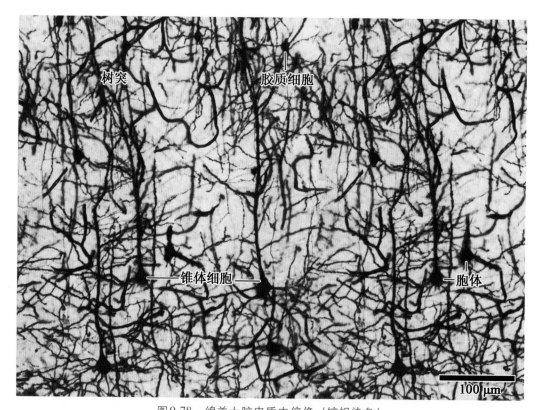

图9-78 绵羊大脑皮质中倍像（镀银染色）
Fig.9-78 mid magnification of sheep cerebral cortex(silver stain)
树突—dendrite 胶质细胞—glial cell 锥体细胞—pyramidal cell 胞体—cell body

第九章 神经系统 Nervous System

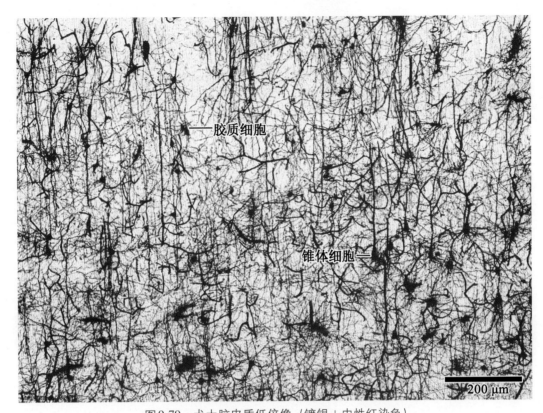

图9-79 犬大脑皮质低倍像（镀银+中性红染色）
Fig.9-79 low magnification of dog cerebral cortex(silver + neutral red stain)
胶质细胞—glial cell 锥体细胞—pyramidal cell

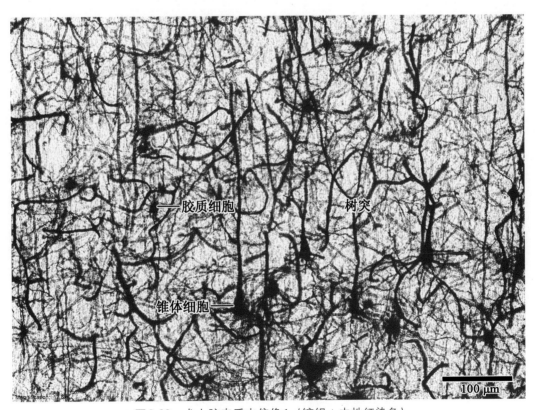

图9-80 犬大脑皮质中倍像1（镀银+中性红染色）
Fig.9-80 mid magnification of dog cerebral cortex 1 (silver + neutral red stain)
胶质细胞—glial cell 锥体细胞—pyramidal cell 树突—dendrite

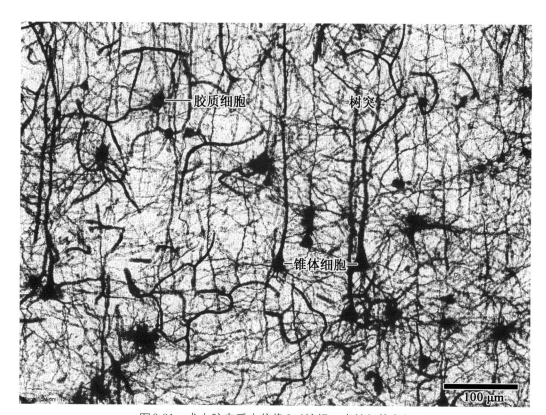

图9-81 犬大脑皮质中倍像2（镀银＋中性红染色）

Fig.9-81 mid magnification of dog cerebral cortex 2 (silver + neutral red stain)

胶质细胞—glial cell 树突—dendrite 锥体细胞—pyramidal cell

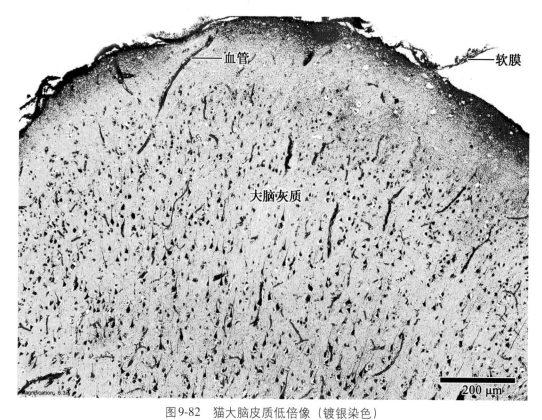

图9-82 猫大脑皮质低倍像（镀银染色）

Fig.9-82 low magnification of cat cerebral cortex(silver stain)

血管—blood vessel 软膜—pia mater 大脑灰质—cerebral gray matter

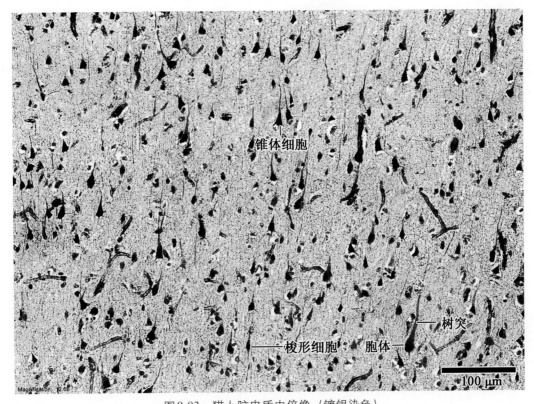

图 9-83　猫大脑皮质中倍像（镀银染色）
Fig.9-83　mid magnification of cat cerebral cortex(silver stain)
锥体细胞—pyramidal cell　梭形细胞—spindle cell　树突—dendrite　胞体—cell body

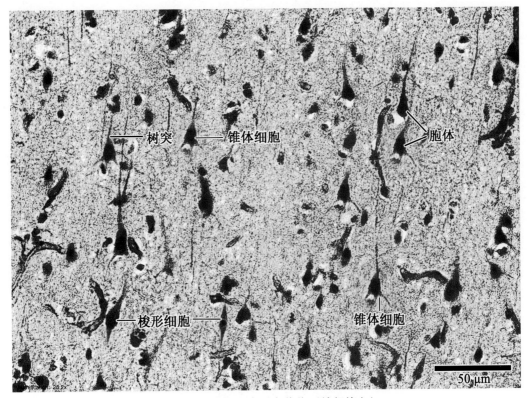

图 9-84　兔大脑皮质高倍像（镀银染色）
Fig.9-84　high magnification of rabbit cerebral cortex(silver stain)
锥体细胞—pyramidal cell　树突—dendrite　胞体—cell body　梭形细胞—spindle cell

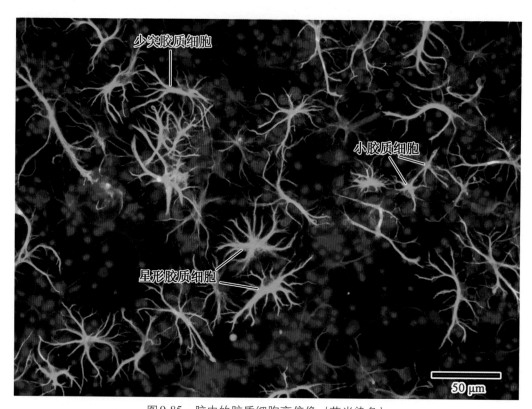

图 9-85 脑中的胶质细胞高倍像（荧光染色）

Fig.9-85　high magnification of cerebral glial cells (fluorescent stain)

少突胶质细胞—oligodendrocyte　小胶质细胞—microglia　星形胶质细胞—astrocyte

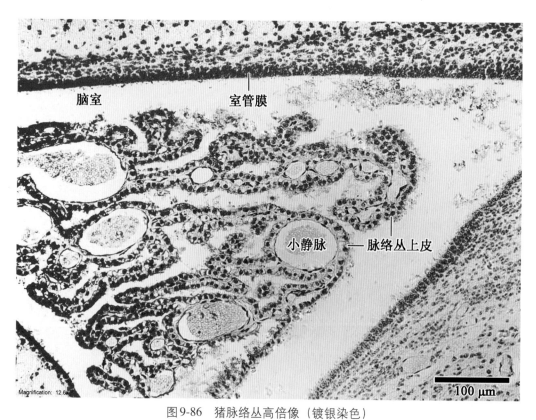

图 9-86 猪脉络丛高倍像（镀银染色）

Fig.9-86　high magnification of pig cerebral choroid plexus (silver stain)

脑室—ventricle　室管膜—ependyma　小静脉—venule　脉络丛上皮—choroid epithelium

第九章 神经系统 Nervous System

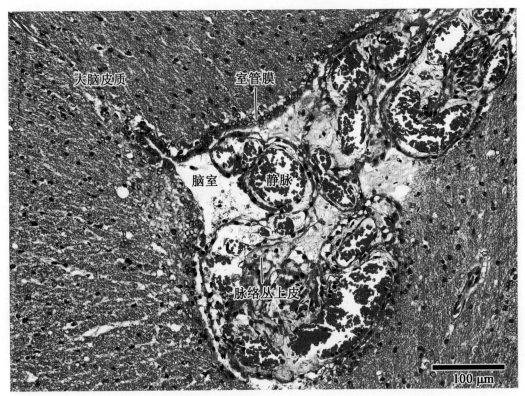

图9-87 犬脉络丛中倍像（HE）

Fig.9-87 mid magnification of dog cerebral choroid plexus (HE)

大脑皮质—cerebral cortex 室管膜—ependyma 脑室—ventricle 静脉—vein 脉络丛上皮—choroid epithelium

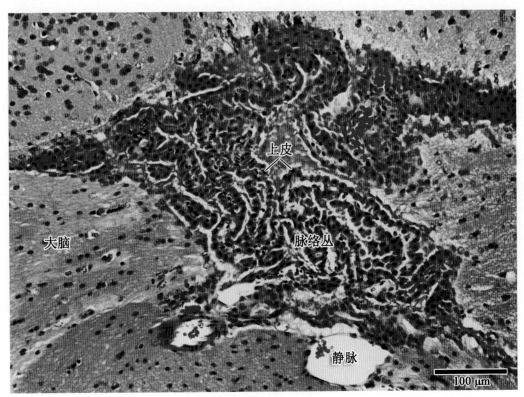

图9-88 猫脉络丛中倍像（HE）

Fig.9-88 mid magnification of cat cerebral choroid plexus (HE)

大脑—cerebrum 上皮—epithelium 脉络丛—choroid plexus 静脉—venule

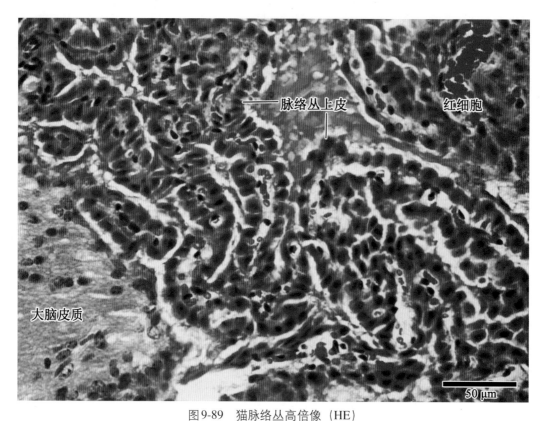

图 9-89 猫脉络丛高倍像（HE）

Fig.9-89 high magnification of cat cerebral choroid plexus (HE)

大脑皮质—cerebrum cortex 脉络丛上皮—choroid plexus epithelium 红细胞—erythrocyte

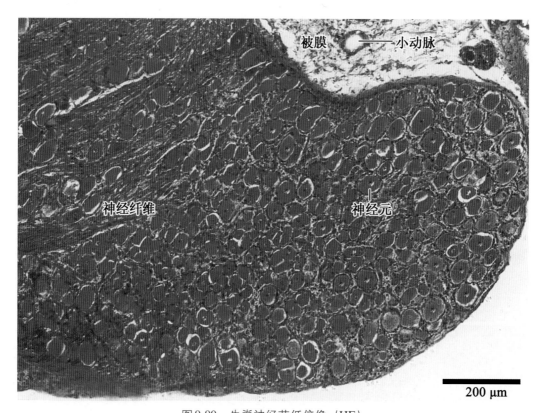

图 9-90 牛脊神经节低倍像（HE）

Fig.9-90 low magnification of cow spinal ganglion(HE)

被膜—capsule 小动脉—arteriole 神经纤维—nerve fiber 神经元—neuron

第九章 神经系统 Nervous System

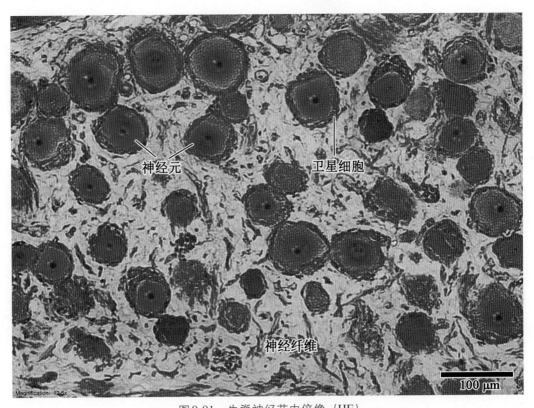

图9-91　牛脊神经节中倍像（HE）

Fig.9-91　mid magnification of cow spinal ganglion(HE)

神经元—neuron　卫星细胞—satellite cell　神经纤维—nerve fiber

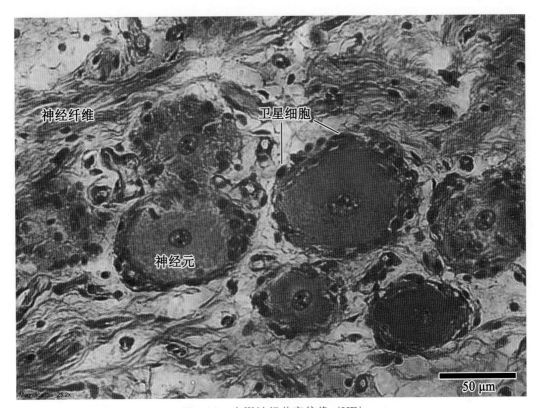

图9-92　牛脊神经节高倍像（HE）

Fig.9-92　high magnification of cow spinal ganglion(HE)

神经纤维—nerve fiber　卫星细胞—satellite cell　神经元—neuron

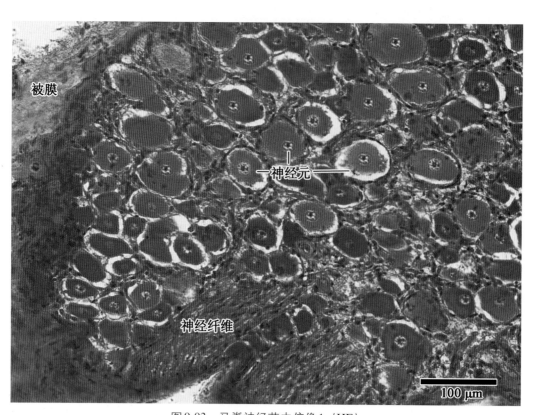

图 9-93　马脊神经节中倍像 1（HE）

Fig.9-93　mid magnification of horse spinal ganglion 1　(HE)

被膜—capsule　神经元—neuron　神经纤维—nerve fiber

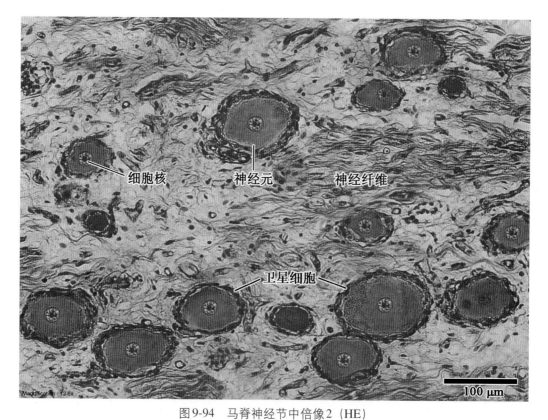

图 9-94　马脊神经节中倍像 2（HE）

Fig.9-94　mid magnification of horse spinal ganglion 2　(HE)

细胞核—nucleus　神经元—neuron　神经纤维—nerve fiber　卫星细胞—satellite cell

第九章 神经系统 Nervous System

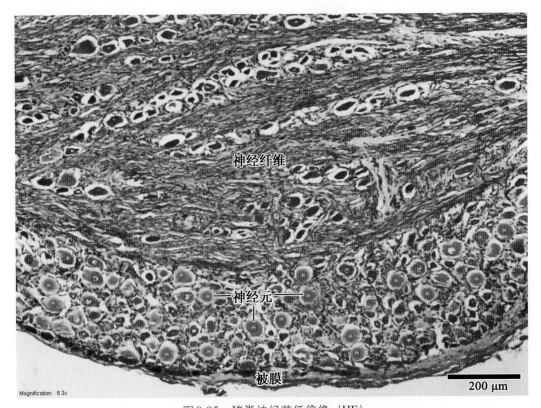

图 9-95　猪脊神经节低倍像（HE）

Fig.9-95　low magnification of pig spinal ganglion(HE)

神经纤维—nerve fiber　神经元—neuron　被膜—capsule

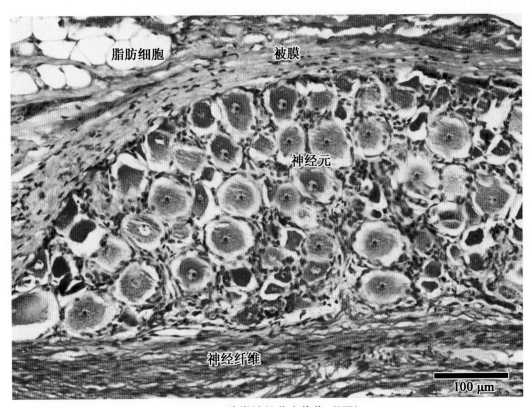

图 9-96　猪脊神经节中倍像（HE）

Fig.9-96　mid magnification of pig spinal ganglion(HE)

脂肪细胞—fat cell　被膜—capsule　神经元—neuron　神经纤维—nerve fiber

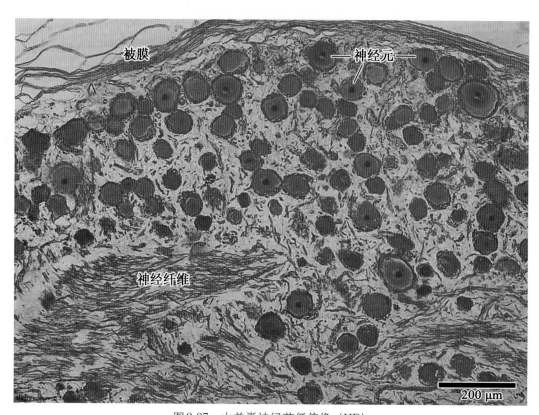

图 9-97　山羊脊神经节低倍像（HE）

Fig.9-97　low magnification of goat spinal ganglion(HE)

被膜—capsule　神经元—neuron　神经纤维—nerve fiber

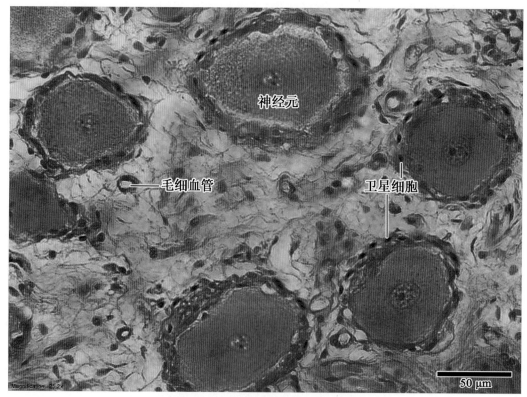

图 9-98　山羊脊神经节高倍像（HE）

Fig.9-98　high magnification of goat spinal ganglion(HE)

神经元—neuron　毛细血管—capillary　卫星细胞—satellite cell

第九章 神经系统 Nervous System

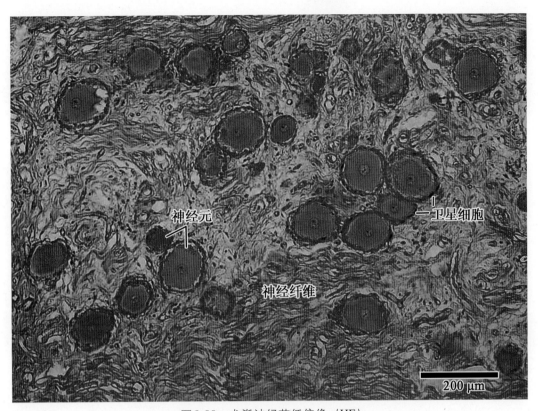

图9-99 犬脊神经节低倍像（HE）
Fig.9-99 low magnification of dog spinal ganglion(HE)
神经元—neuron 神经纤维—nerve fiber 卫星细胞—satellite cell

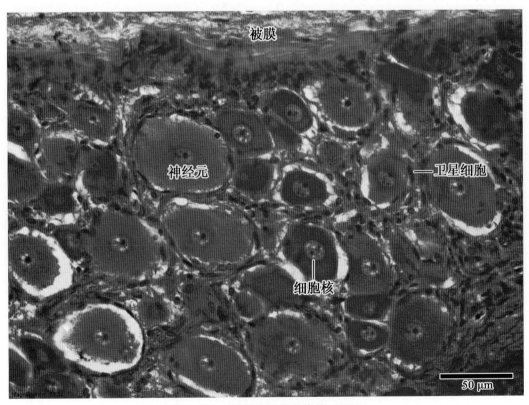

图9-100 猫脊神经节高倍像（HE）
Fig.9-100 high magnification of cat spinal ganglion(HE)
被膜—capsule 神经元—neuron 细胞核—nucleus 卫星细胞—satellite cell

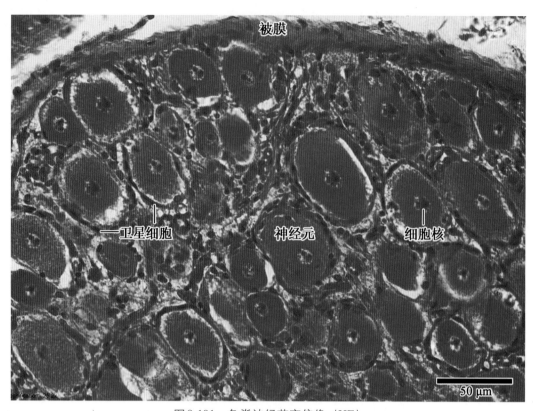

图 9-101　兔脊神经节高倍像（HE）

Fig.9-101　high magnification of rabbit spinal ganglion(HE)

被膜—capsule　卫星细胞—satellite cell　神经元—neuron　细胞核—nucleus

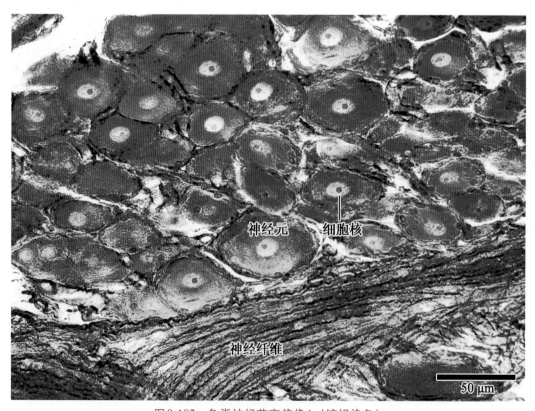

图 9-102　兔脊神经节高倍像 1（镀银染色）

Fig.9-102　high magnification of rabbit spinal ganglion 1　(silver stain)

神经元—neuron　细胞核—nucleus　神经纤维—nerve fiber

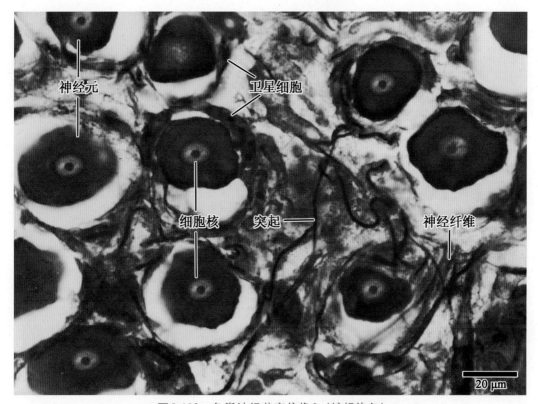

图9-103 兔脊神经节高倍像2（镀银染色）

Fig.9-103 high magnification of rabbit spinal ganglion 2 (silver stain)

神经元—neuron 卫星细胞—satellite cell 细胞核—nucleus 神经纤维—nerve fiber 突起—neurite

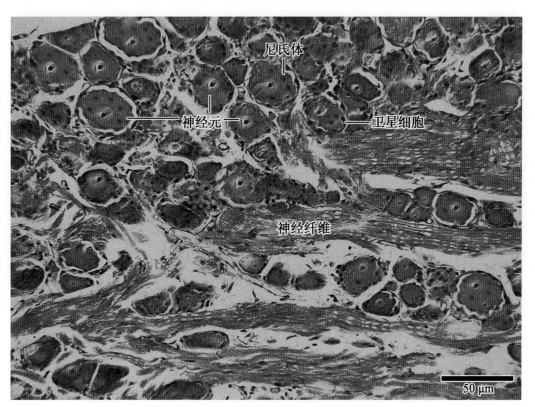

图9-104 猫交感神经节高倍像（尼氏染色）

Fig.9-104 high magnification of cat spinal ganglion (Nissl stain)

尼氏体—Nissl body 神经元—neuron 卫星细胞—satellite cell 神经纤维—nerve fiber

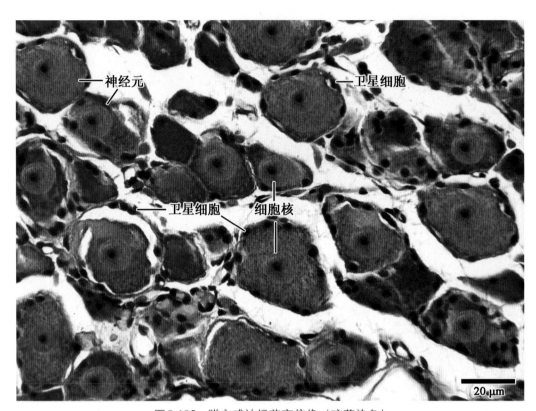

图9-105　猫交感神经节高倍像（硫堇染色）

Fig.9-105　high magnification of cat spinal ganglion (thionine stain)

神经元—neuron　卫星细胞—satellite cell　细胞核—nucleus

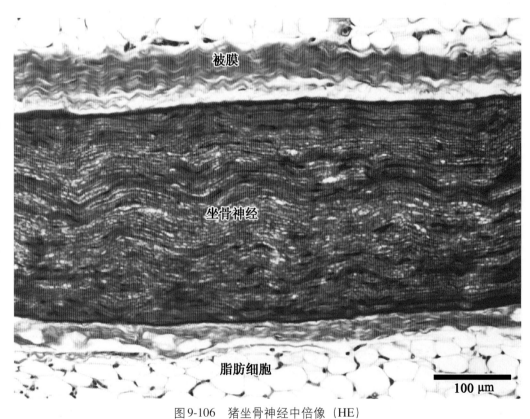

图9-106　猪坐骨神经中倍像（HE）

Fig.9-106　mid magnification of pig ischiadic nerve (HE)

被膜—capsule　坐骨神经—ischiadic nerve　脂肪细胞—fat cell

第九章 神经系统 Nervous System

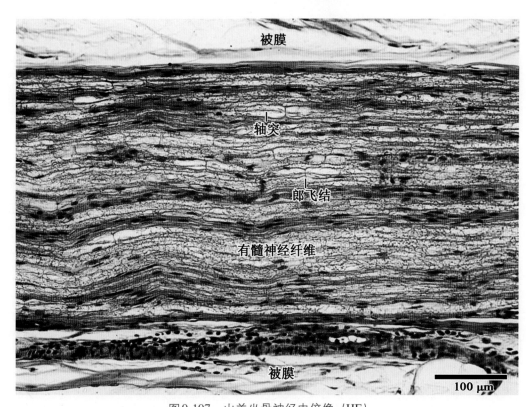

图9-107 山羊坐骨神经中倍像（HE）
Fig.9-107 mid magnification of goat ischiadic nerve (HE)
被膜—capsule 轴突—axon 郎飞结—Ranvier node 有髓神经纤维—myelinated nerve fiber

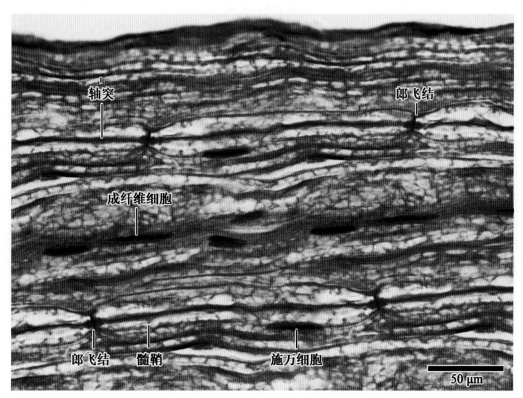

图9-108 犬坐骨神经高倍像（HE）
Fig.9-108 high magnification of dog ischiadic nerve (HE)
轴突—axon 郎飞结—Ranvier node 成纤维细胞—fibroblast 髓鞘—myelin sheath 施万细胞—Schwann cell

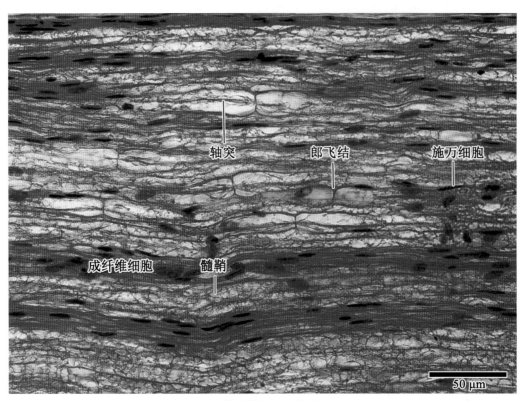

图 9-109　猫坐骨神经高倍像（HE）

Fig.9-109　high magnification of cat ischiadic nerve (HE)

轴突—axon　郎飞结—Ranvier node　施万细胞—Schwann cell　成纤维细胞—fibroblast　髓鞘—myelin sheath

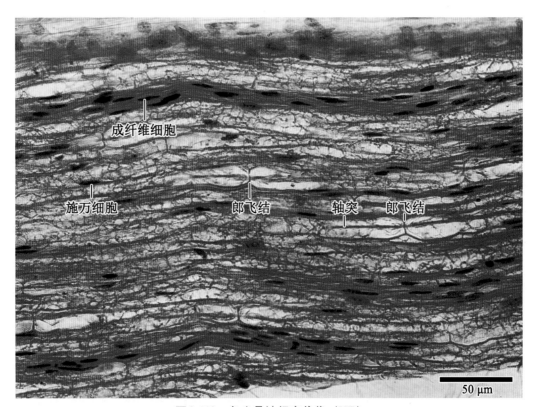

图 9-110　兔坐骨神经高倍像（HE）

Fig.9-110　high magnification of rabbit ischiadic nerve (HE)

成纤维细胞—fibroblast　施万细胞—Schwann cell　郎飞结—Ranvier node　轴突—axon

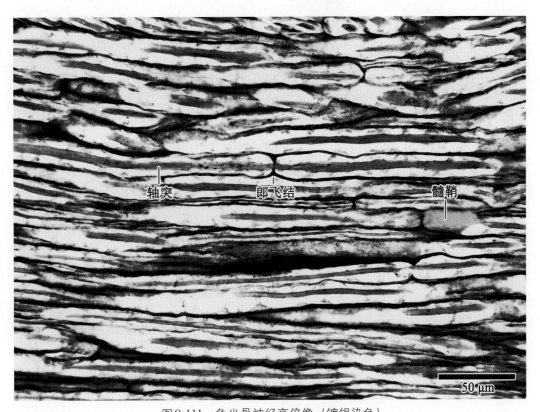

图9-111 兔坐骨神经高倍像（镀银染色）
Fig.9-111 high magnification of rabbit ischiadic nerve(silver stain)
轴突—axon 郎飞结—Ranvier node 髓鞘—myelin sheath

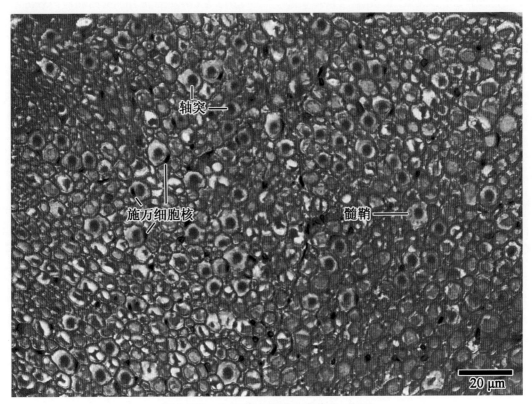

图9-112 兔坐骨神经横切高倍像（HE）
Fig.9-112 high magnification of rabbit ischiadic nerve transection(HE)
轴突—axon 施万细胞核—Schwann cell nucleus 髓鞘—myelin sheath

第十章
循环系统
Circulatory System

> **Outline**
>
> Circulatory system consists of cardiovascular and lymphatic vascular systems. Cardiovascular system includes heart, arteries, veins and capillaries. Lymphatic vascular system consists of lymphatic capillaries, lymphatic vessels and lymphatic ducts. Their walls of the different organs in the circulatory system, except the capillaries and lymphatic capillaries, are divided into three layers.
>
> In the heart, the three layers are termed endocardium, myocardium, and epicardium respectively. Endocardium of heart consists of three layers: endothelium, subendothelial layer and subendocardial layer. In the subendocardial layer, there are some special cells called Purkinje fibers, which are the branches of the impulse-conducting system of heart. This impulse-conducting system consists of sinoatrial node, atrioventricular node and atrioventricular bundle and its branches. Myocardium is the thickest of the layers of heart and consists of cardiac muscle cells. Epicardium is a serosa enclosing the heart.
>
> In the blood vessel walls, these three layers are termed tunica intima, tunica media and tunica adventitia. Arteries can be classified according to their sizes into four groups: large, medium-sized, small arteries and arterioles. The structure of the medium-size arteries is most representative. Tunica intima consists of the endothelium, the subendothelial layer and the internal elastic lamina. Tunica media are mostly composed of concentrically arranged smooth muscle cells. Therefore they are termed muscular arteries. Tunica media has a thinner external elastic lamina, which separates it from tunica adventitia. Whereas large arteries contain a lot of elastic fibers and a series of concentrically arranged elastic lamina in the tunica media. So, they are termed elastic arteries.
>
> The characteristics of the veins are: a large diameter and thinner walls than their accompanying arteries; the lumen of veins is usually irregular, but that of arteries is rounded; the boundaries between the three tunics of veins are not as clear as there in the arteries. The reason is that the internal and the external elastic lamina are often absent in veins; the veins, especially large ones, have a well-developed tunica

第十章 循环系统 Circulatory System

intima, but tunica media is much thinner, with few layers of smooth muscle cells and abundant connective tissue; adventitia layer is the thickest, and frequently contains longitudinal bundles of smooth muscle in large veins. Valves are presented in most of the veins.

Capillaries have the simplest structure with a layer of endothelial cells, a basal lamina and pericytes. Capillaries can be divided according to their structure into three types: continuous, fenestrated and sinusoidal capillaries.

The structure of lymphatic capillaries is similar to that of capillaries. Lymphatic vessels and ducts are similar to corresponding veins, but they have thinner walls and lack a clear-cut separation between three tunics. They have more numerous internal valves, as well.

循环系统（circulatory system）包括心血管系统和淋巴管系统两部分：心血管系统由心脏、动脉、毛细血管和静脉组成，淋巴管系统由毛细淋巴管、淋巴管和淋巴导管组成。

一、循环系统的发生

心血管系统和淋巴系统由胚胎时期中胚层的间充质发育而来。沿着主要血管发生的位置，间充质细胞集中成团索状，分化形成血岛（blood island）。血岛内间充质细胞由不规则的多突起形变为圆形，血岛周围细胞变扁平，成为血管内皮，而中间的细胞变成原始血细胞。众多的血岛联系起来，形成血管网。

心脏（heart）由一对位于咽下部的血管原基演化而来。在前肠门两侧，由胚胎脏壁中胚层分离出一些细胞，形成2条管道，称心内膜管。随着前肠的延长和前肠门的后移，这2条心内膜管相互靠近，并于前肠腹侧融合成1条心内膜管。靠近心内膜管处的胚胎脏壁中胚层加厚，并将心内膜管包围起来，形成心肌外套层。将来，心内膜管分化为心内膜；心肌外套层分化为心肌膜和心外膜。心肌外套层外面的胚胎脏壁中胚层分化为心包。

（一）血管的发生

胚胎中胚层内首先出现许多血岛（blood island），血岛周边的细胞变扁，分化成内皮细胞围成内皮管，即原始血管。血岛中央的游离细胞分化成为原始血细胞，即造血干细胞。内皮管不断向外出芽延伸，与相邻血岛的内皮管互相融合通连，逐渐形成一个丛状分布的内皮管网。随着胚胎不断发育，一个弥散的内皮管网逐渐形成，分布于胚体内外的间充质中。此后，其中有的内皮管因相互融合及血液汇流而增粗，有的则因血流减少而萎缩或消失。原始血管系统便逐渐形成。

（二）心脏的发生

心脏发生于生心区。生心区是指胚盘前缘脊索前板（口咽膜）前面的中胚层，此区前方的中胚层为原始横膈。

1. 原始心脏的形成 首先生心区的中胚层内出现围心腔，围心腔腹侧出现生心板，板中央变空形成一对心管。当胚体出现头褶和侧褶时，一对心管向中线靠拢，融合为一条。此时，在心管的背侧出现了心背系膜，将心管悬连于心包腔的背侧壁。心背系膜的中部退化，形成心包横窦。心背系膜仅在心管的头、尾端存留。当心管融合和陷入心包腔时，其周围的间充质密集，形成一层厚的心肌外套层，将来分化成为心肌膜和心外膜。内皮和心肌外套层之间的组织为疏松的胶样结缔组织，将来参与组成心内膜。

2. 心脏外形的建立 心管的头端与动脉连接，尾端与静脉相连，两端连接固定在心包上。心管首先出现三个膨大，由头端向尾端依次称心球、心室和心房。以后在心房的尾端又出现一个膨大，称静脉窦（venous sinus）。心房和静脉窦早期位于原始横膈内。在心管发生过程中，心球和心室形成U形弯曲，然后心房和静脉窦相继离开原始横膈，静脉窦位于心房的背面尾侧，以窦房孔与心房通连。此时的心脏外形呈S形弯曲，心房扩大，房室沟加深，房室之间形成狭窄的房空管。心球则可分为三段：远侧段细长，为动脉干；中段较膨大，为心动脉球；近侧段被心室吸收，成为原始右心室。原来的心室成为原始左心室，左、右心室之间的表面出现室间沟。至此，心脏已初具外形，但内部仍未完全分隔。随着胚胎的发育，心脏各部的分隔同时进行。

3. 房室管的分隔 房室管背、腹侧壁的心内膜下组织增生，分别形成背、腹心内膜垫；二者融合后，便将房室管分隔成左、右房室孔。围绕房室孔的间充质局部增生并向腔内隆起，逐渐形成房室瓣，右侧为三尖瓣，左侧为二尖瓣。

二、循环系统的组织学结构概述

（一）血管（blood vessel）

根据结构和功能不同，血管分为动脉、静脉和毛细血管。动脉从心脏发出后，反复分支，管径逐渐变细。除毛细血管外，血管壁由内向外依次分为内膜、中膜和外膜。

1. 动脉（artery） 分为大动脉、中动脉、小动脉和微动脉，它们之间没有明显的分界。

（1）大动脉（弹性动脉） 包括主动脉、肺动脉、颈总动脉和锁骨下动脉。内膜衬有多边形的内皮细胞，内皮下层为弹性纤维组织，内含纵行的平滑肌纤维。内弹性膜不很明显。中膜特征是有多层螺旋状或同心圆状排列的弹性纤维。弹性纤维层之间是环行的平滑肌纤维、胶原纤维和网状纤维。外膜为薄层结缔组织，含弹性纤维和少量纵行的平滑肌纤维。还可见到自养血管。

（2）中动脉（肌性动脉） 包括除大动脉以外的有名称的动脉。内膜衬有多边形扁平的内皮细胞，血管收缩时细胞凸入管腔。内皮下层含有细小的胶原纤维和极少的纵行平滑肌纤维。内弹性膜明显。中膜特征为有多层环形排列的平滑肌。在平滑肌纤维之间是弹性纤维、胶原纤维和网状纤维。外弹性膜明显。外膜为很厚的胶原和弹性组织，还有纵行平滑肌纤维，并有营养血管。

（3）小动脉 直径小于100μm的动脉。内膜衬有内皮和结缔组织。较大的小动脉可见内弹性膜，较小的小动脉则没有。中膜螺旋状排列的平滑肌纤维可达三层。较大的小动脉可见外弹性膜，较小的小动脉缺如。外膜是胶原和弹性结缔组织，厚度与中膜相近。

2. 毛细血管（capillary） 为直径小于10μm的薄壁血管。毛细血管分布最广、分支最多、管径最小、管壁最薄，由内皮和基膜构成。偶尔切片中可见到内皮细胞核和红细胞。管壁常塌陷，外围有周细胞，光镜下不明显。分为连续毛细血管、有孔毛细血管和血窦。

3. 静脉（vein） 将血液运回心脏，常与动脉伴行。由毛细血管移行而来，多为小静脉、中静脉和大静脉。

（1）小静脉 比相应的小动脉管腔大，管壁薄。内膜的内皮下方为很薄的结缔组织。内皮下层的厚度随管径增大而增加，通常与周细胞相关。较细的小静脉无中膜，而较大的小静脉的中膜可见1～2层平滑肌。外膜为胶原性结缔组织，内含成纤维细胞和弹性纤维。

（2）中静脉 内膜衬内皮和少量结缔组织。内壁有明显的瓣膜。中膜比相应的动脉薄得多，但有多层平滑肌。偶尔肌纤维纵行排列而非环行排列。可见胶原纤维束，其间有散在的少量弹性纤维。外膜很厚，由胶原纤维和弹性纤维组成。偶见纵行的平滑肌纤维，营养血管甚至穿入中膜。

（3）大静脉 内膜与中静脉的相同，但内皮下层结缔组织较厚。有的大静脉有明显的瓣膜。中膜不明显，在胶原纤维和弹性纤维之间，有散在的平滑肌纤维。外膜最厚，在很厚的胶原纤维和弹性纤维之间，常有纵行的平滑肌纤维束和营养血管。

（二）心脏（heart）

心脏是壁厚、中空的肌性器官，是心血管系统的动力装置。其节律性收缩和舒张推动血液在血管中不断地循环流动，使体内各组织器官得到充足的血液供应。心脏分为左、右心房和左、右心室4个腔，心壁的组织结构分3层，由内向外依次为心内膜、心肌膜和心外膜。

心内膜由内皮、内皮下层和心内膜下层构成，心内膜下层的结缔组织中分布着具有传导功能的浦肯野纤维（Purkinje fiber）。

心肌膜最厚，分心房肌和心室肌，主要由心肌纤维构成，可分内纵、中环和外斜三层，心肌纤维之间具有闰盘（intercalated disc）结构，实际上是心肌细胞之间的特殊连接。

心外膜是心包浆膜的脏层，外面被覆间皮，间皮下是薄层结缔组织。

（三）淋巴管（lymph vessel）

毛细淋巴管、淋巴管和淋巴导管共同组成了淋巴管系统。淋巴管常因塌陷而在镜下看不到，当充满淋巴时，可见到清晰的、衬有内皮细胞的管道，与小静脉相似，管腔内有淋巴细胞，但无红细胞，偶尔可见到瓣膜。

毛细淋巴管的管腔大而不规则，管壁薄，仅由内皮和极薄的结缔组织构成，无周细胞。内皮细胞间有较宽的间隙，无基膜，故通透性大，大分子物质易进入其中。

淋巴管的结构与静脉相似，但管径大而壁薄，管壁由内皮、少量平滑肌和结缔组织构成，瓣膜较多，可防止淋巴倒流。

淋巴导管的结构与大静脉相似，但管壁薄，三层膜分界不明显。

三、循环系统图谱

1. **心脏**　图10-1～图10-20。
2. **血管**　图10-21～图10-50。
3. **淋巴管**　图10-51～图10-52。

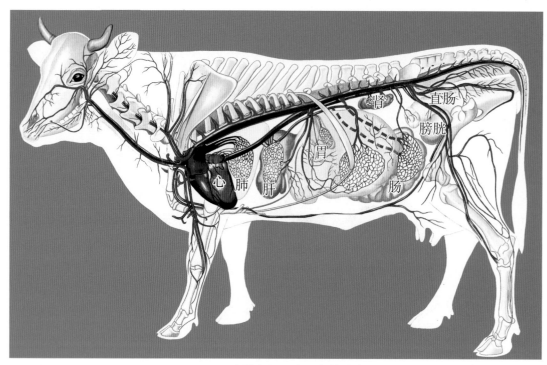

图 10-1 牛的主要器官及血管分布

Fig.10-1 The main organs and distribution of blood vessels in cattle

心—heart 肺—lung 肝—liver 胃—rumen 肠—intestine 肾—kidney 膀胱—bladder 直肠—rectum

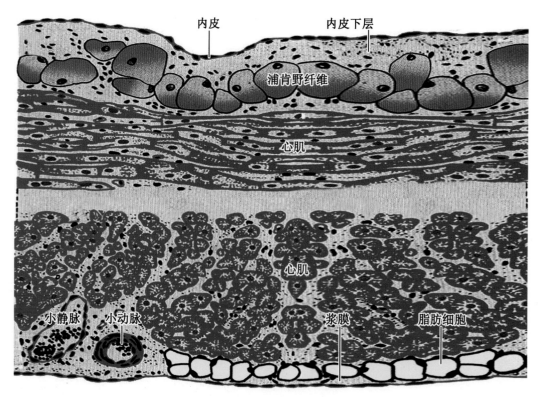

图 10-2 心壁结构模式图

Fig.10-2 ideograph of heart wall

内皮—endothelium 内皮下层—subendothelial layer 浦肯野纤维—Purkinje fiber 心肌—myocardium
小静脉—small vein 小动脉—small artery 浆膜—serosa 脂肪细胞—fat cell

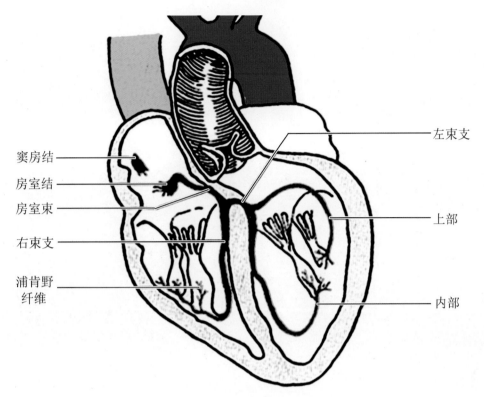

图 10-3　心传导系统模式图

Fig.10-3　ideograph of heart conduction system

窦房结—atrionector　房室结—atrioventricular node　房室束—atrioventricular bundle　右束支—right bundle
浦肯野纤维—Purkinje fiber　左束支—left bundle　上部—up bundle　内部—inner bundle

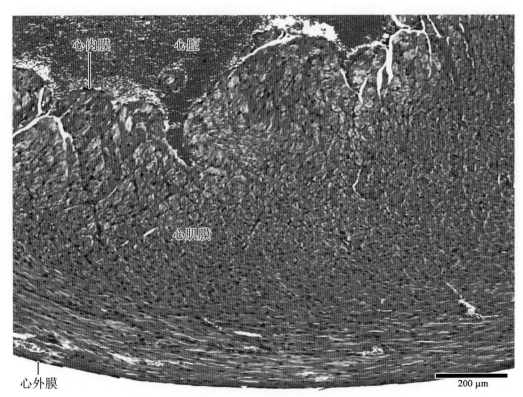

图 10-4　牛心房壁低倍像（HE）

Fig.10-4　low magnification of cattle atrial wall（HE）

心内膜—endocardium　心腔—heart chamber　心肌膜—myocardium　心外膜—epicardium

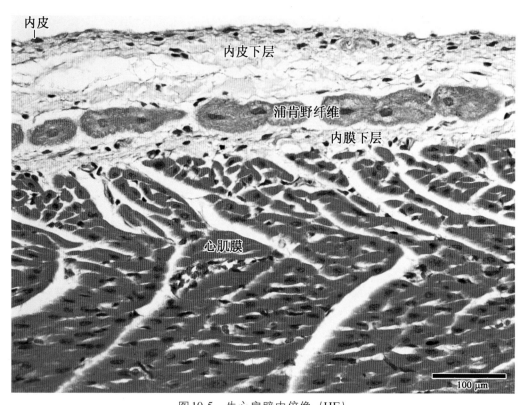

图 10-5　牛心房壁中倍像（HE）

Fig.10-5　mid magnification of cattle atrial wall（HE）

内皮—endothelium　内皮下层—subendothelial layer　浦肯野纤维—Purkinje fiber

内膜下层—subendocardial layer　心肌膜—myocardium

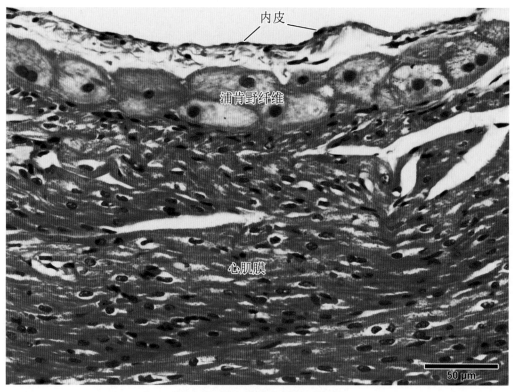

图 10-6　猪心房壁高倍像（HE）

Fig.10-6　high magnification of pig atrial wall（HE）

内皮—endothelium　浦肯野纤维—Purkinje fiber　心肌膜—myocardium

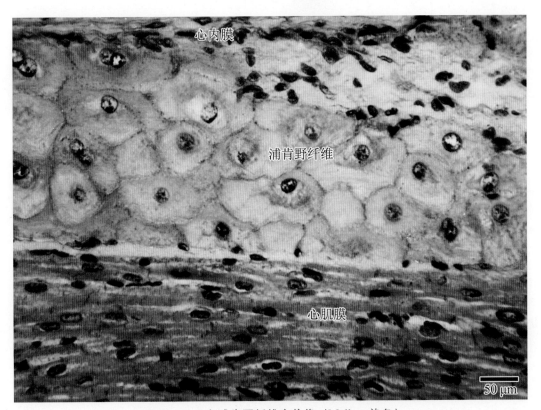

图10-7 马心浦肯野纤维高倍像（Mallory染色）

Fig.10-7 high magnification of horse Purkinje fiber（Mallory stain）

心内膜—endocardium 浦肯野纤维—Purkinje fiber 心肌膜—myocardium

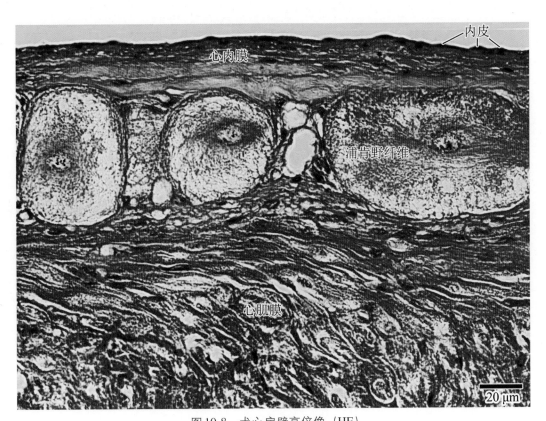

图10-8 犬心房壁高倍像（HE）

Fig.10-8 high magnification of dog atrial wall（HE）

心内膜—endocardium 内皮—endothelium 浦肯野纤维—Purkinje fiber 心肌膜—myocardium

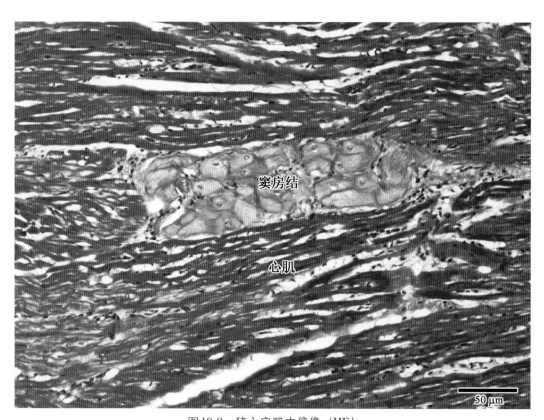

图10-9 猪心房壁中倍像（HE）

Fig.10-9 mid magnification of pig atrial wall（HE）

窦房结—atrionector 心肌—cardiac muscle

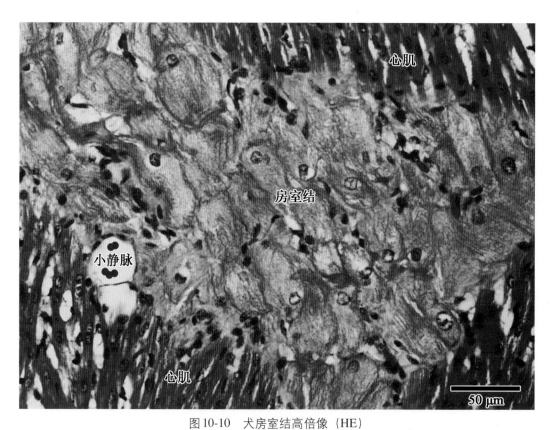

图10-10 犬房室结高倍像（HE）

Fig.10-10 high magnification of dog atrioventricular node（HE）

心肌—cardiac muscle 房室结—atrioventricular node 小静脉—venule

第十章 循环系统 Circulatory System

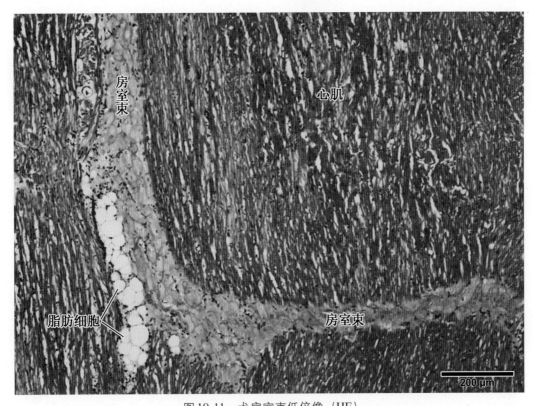

图 10-11 犬房室束低倍像（HE）

Fig.10-11 low magnification of dog atrioventricular bundle（HE）

房室束—atrioventricular bundle 心肌—cardiac muscle 脂肪细胞—fat cell

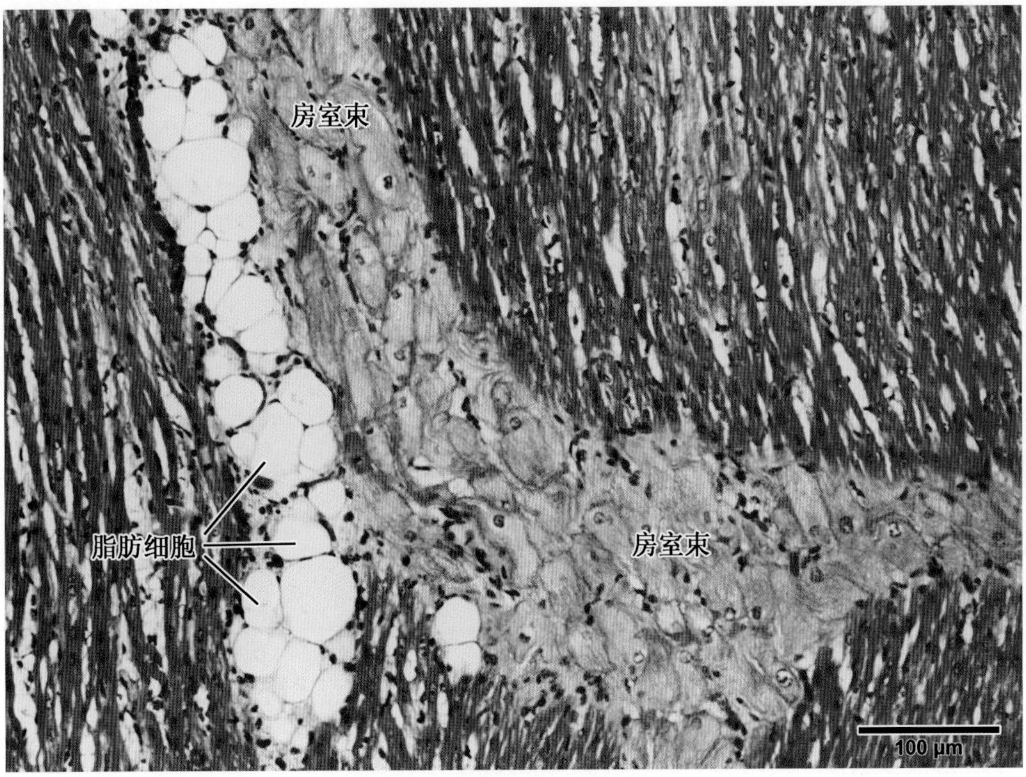

图 10-12 犬房室束中倍像（HE）

Fig.10-12 mid magnification of dog atrioventricular bundle（HE）

房室束—atrioventricular bundle 脂肪细胞—fat cell

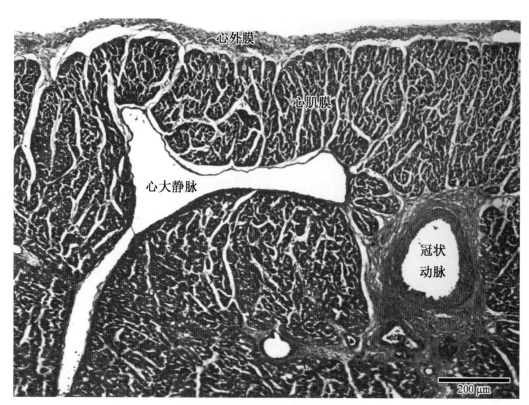

图10-13 猪心壁低倍像（HE）

Fig.10-13　low magnification of pig heart wall（HE）

心外膜—epicardium　心肌膜—myocardium　心大静脉—venae cordis magna　冠状动脉—coronary artery

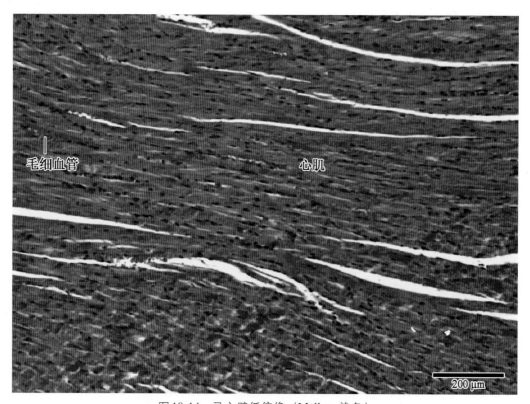

图10-14 马心壁低倍像（Mallory染色）

Fig.10-14　low magnification of horse heart wall（Mallory stain）

毛细血管—capillary　心肌—myocardium

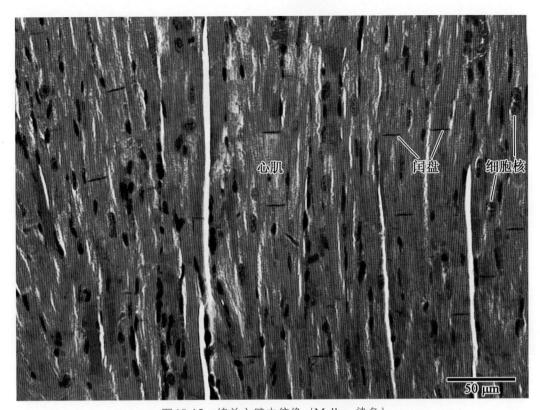

图 10-15　绵羊心壁中倍像（Mallory 染色）

Fig.10-15　mid magnification of sheep heart wall（Mallory stain）

心肌—myocardium　闰盘—intercalated disc　细胞核—nucleus

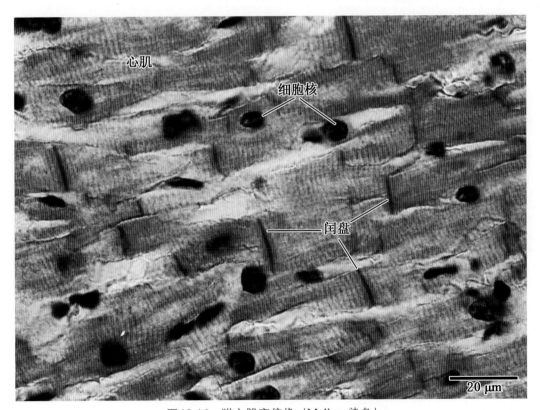

图 10-16　猫心肌高倍像（Mallory 染色）

Fig.10-16　high magnification of cat cardiac muscle（Mallory stain）

心肌—myocardium　细胞核—nucleus　闰盘—intercalated disc

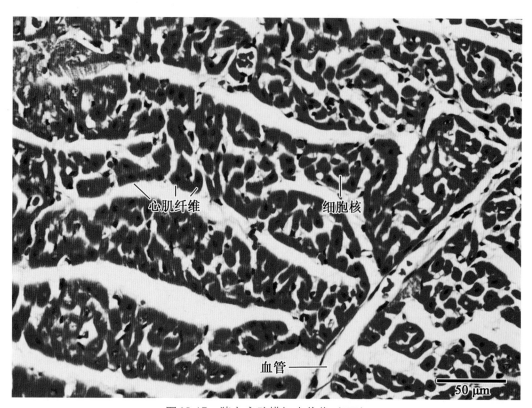

图10-17 猪心房壁横切中倍像（HE）

Fig.10-17 mid magnification of transection of pig atrium wall（HE）

心肌纤维—myocardial fiber　细胞核—nucleus　血管—blood vessel

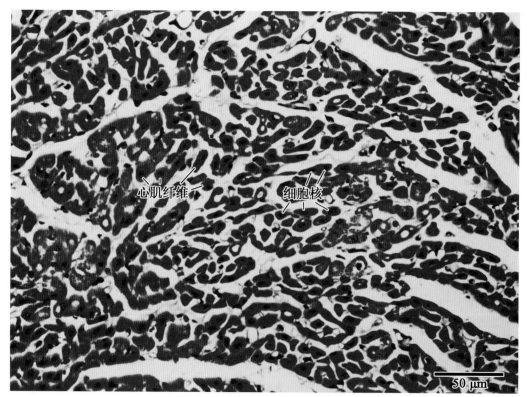

图10-18 犬心房壁横切中倍像（HE）

Fig.10-18 mid magnification of transection of dog atrium wall（HE）

心肌纤维—myocardial fiber　细胞核—nucleus

第十章 循环系统　Circulatory System

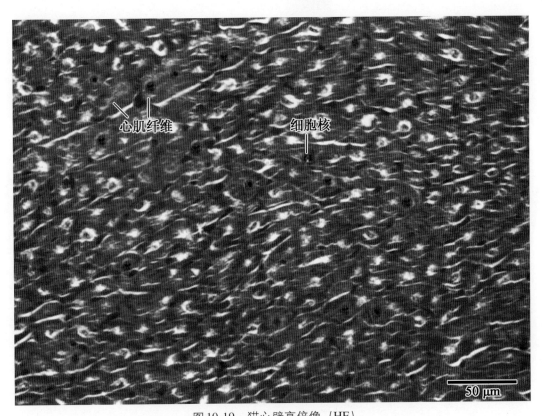

图 10-19　猫心壁高倍像（HE）

Fig.10-19　high magnification of cat cardiac wall（HE）

心肌纤维—myocardial fiber　　细胞核—nucleus

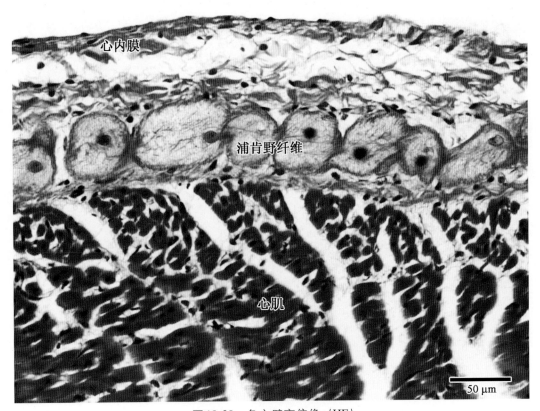

图 10-20　兔心壁高倍像（HE）

Fig.10-20　high magnification of rabbit cardiac wall（HE）

心内膜—endocardium　　浦肯野纤维—Purkinje fiber　　心肌—myocardium

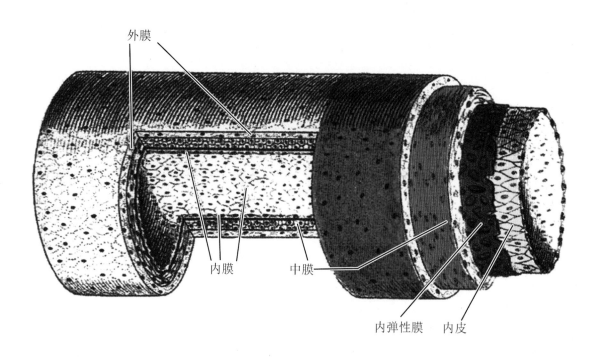

图 10-21 血管壁结构模式图

Fig.10-21 ideograph of vascular wall

外膜—outer membrane 内膜—internal membrane 中膜—mid membrane

内弹性膜—internal elastic membrane 内皮—endothelium

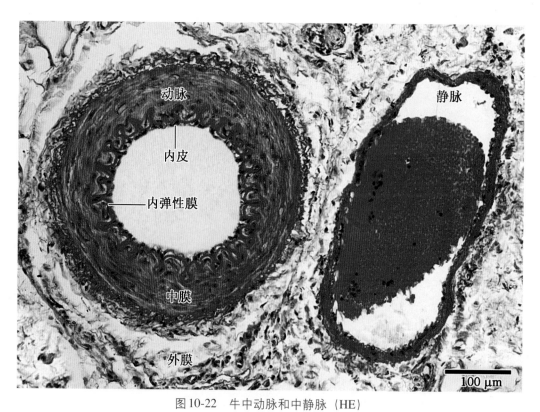

图 10-22 牛中动脉和中静脉（HE）

Fig.10-22 medium artery and medium vein in cow（HE）

动脉—artery 内皮—endothelium 内弹性膜—internal elastic membrane 中膜—mid membrane

外膜—outer membrane 静脉—vein

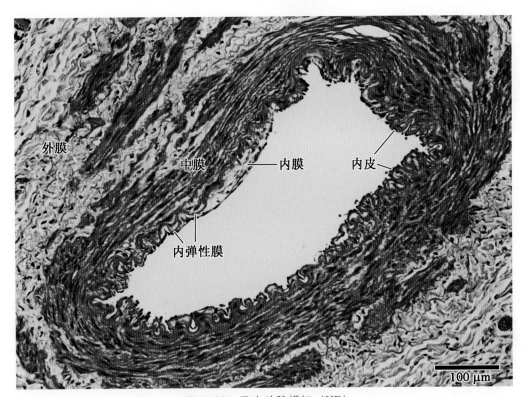

图10-23 马中动脉横切（HE）

Fig.10-23 transection of medium artery in horse（HE）

外膜—outer membrane 中膜—mid membrane 内膜—inner membrane
内弹性膜—internal elastic membrane 内皮—endothelium

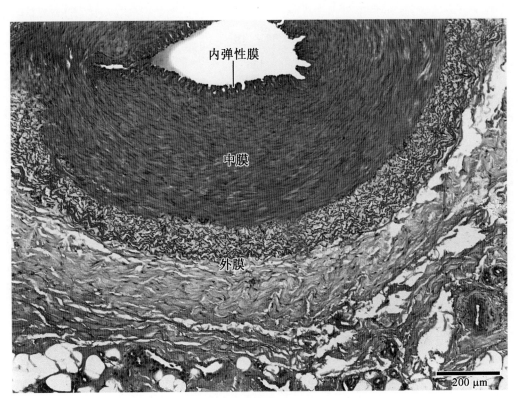

图10-24 猪中动脉横切低倍像（HE）

Fig.10-24 low magnification of medium artery transection in pig（HE）

内弹性膜—internal elastic membrane 中膜—mid membrane 外膜—outer membrane

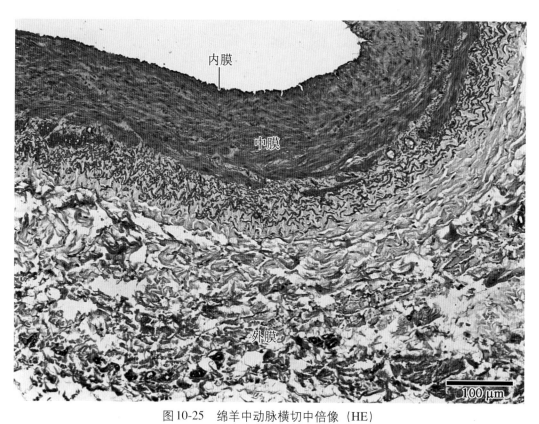

图10-25 绵羊中动脉横切中倍像（HE）

Fig.10-25 mid magnification of medium artery transection in sheep（HE）

内膜—inner membrane 中膜—mid membrane 外膜—outer membrane

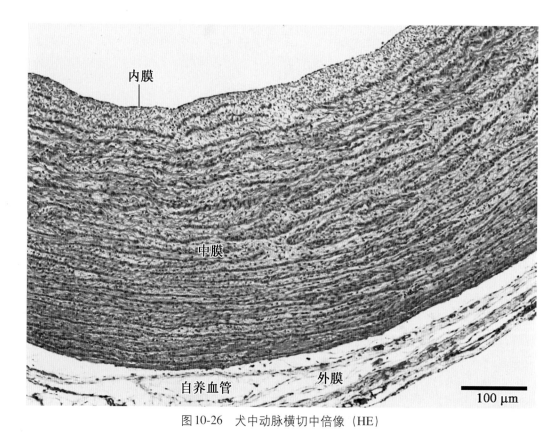

图10-26 犬中动脉横切中倍像（HE）

Fig.10-26 mid magnification of medium artery transection in dog（HE）

内膜—inner membrane 中膜—mid membrane 外膜—outer membrane 自养血管—autotrophy vessel

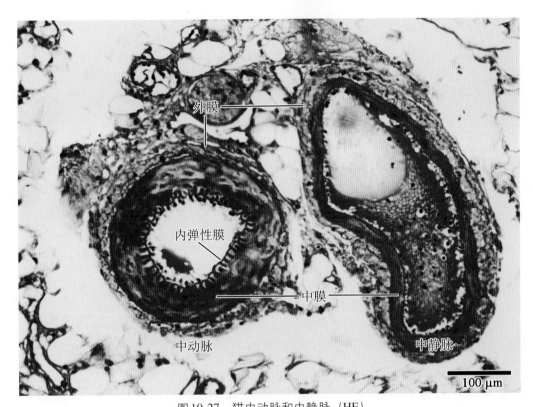

图10-27 猫中动脉和中静脉（HE）

Fig.10-27 medium artery and medium vein in cat（HE）

外膜—outer membrane　内弹性膜—internal elastic membrane　中膜—mid membrane

中动脉—medium artery　中静脉—medium vein

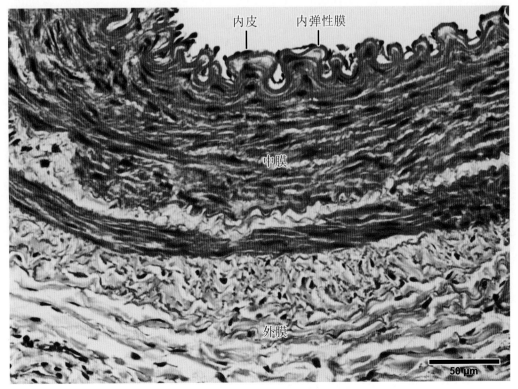

图10-28 猫中动脉横切高倍像（HE）

Fig.10-28 high magnification of medium artery transection in cat（HE）

内皮—endothelium　内弹性膜—internal elastic membrane　中膜—mid membrane　外膜—outer membrane

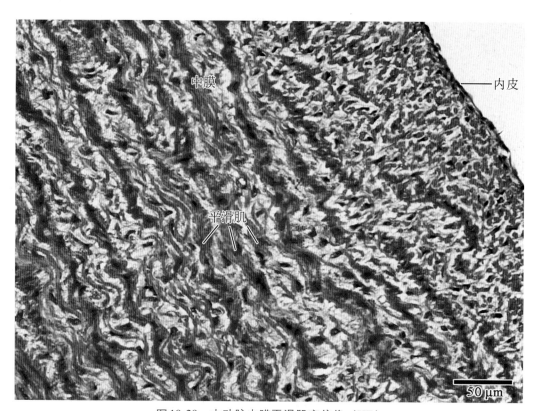

图 10-29 中动脉中膜平滑肌高倍像（HE）

Fig.10-29　high magnification of medium artery smooth muscle（HE）

中膜—mid membrane　平滑肌—smooth muscle　内皮—endothelium

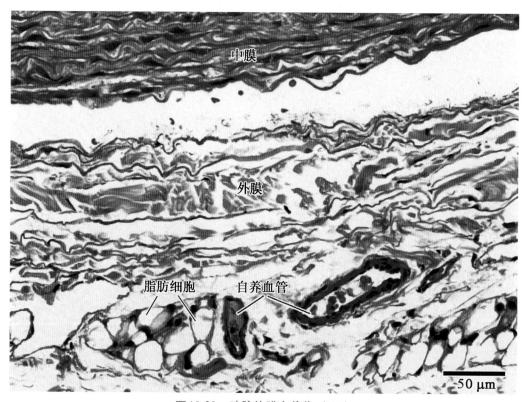

图 10-30 动脉外膜高倍像（HE）

Fig.10-30　high magnification of artery outer membrane（HE）

中膜—mid membrane　外膜—outer membrane　脂肪细胞—fat cell　自养血管—autotrophy vessel

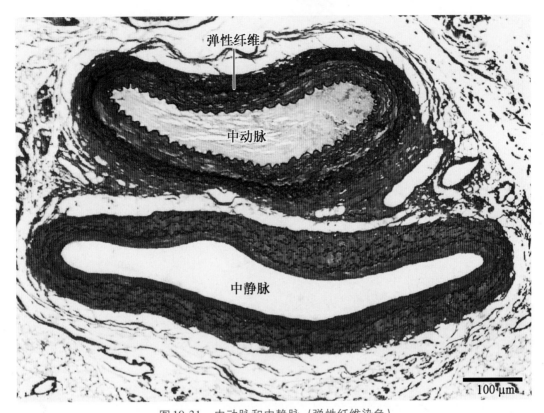

图 10-31 中动脉和中静脉（弹性纤维染色）
Fig.10-31 medium artery and medium vein (elastic fiber stain)
弹性纤维—elastic fiber 中动脉—medium artery 中静脉—medium vein

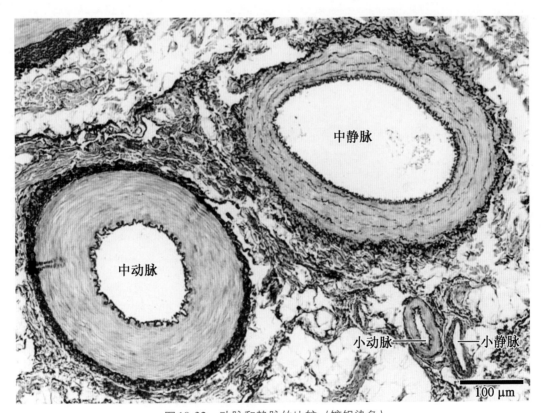

图 10-32 动脉和静脉的比较（镀银染色）
Fig.10-32 comparison of artery and vein (silver stain)
中动脉—medium artery 中静脉—medium vein 小动脉—small artery 小静脉—small vein

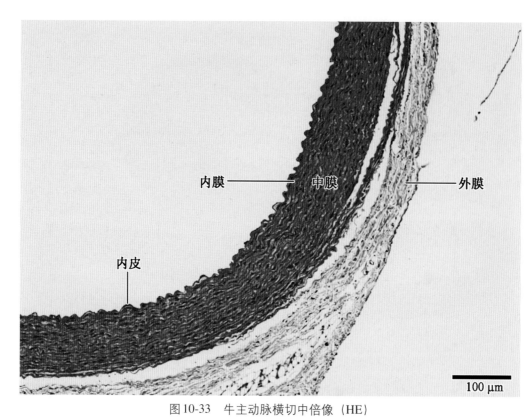

图10-33　牛主动脉横切中倍像（HE）

Fig.10-33　mid magnification of transection of cow aorta（HE）

内皮—endothelium　内膜—internal membrane　中膜—mid membrane　外膜—outer membrane

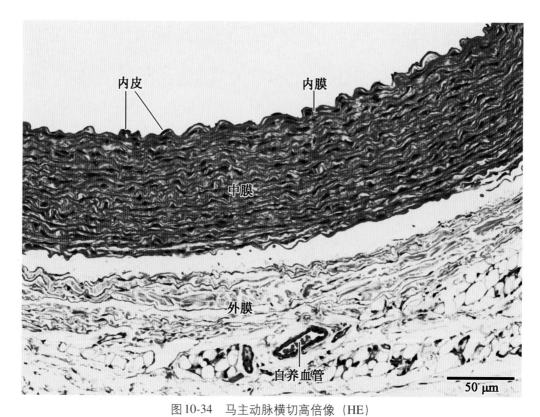

图10-34　马主动脉横切高倍像（HE）

Fig.10-34　high magnification of transection of horse aorta（HE）

内皮—endothelium　内膜—internal membrane　中膜—mid membrane　外膜—outer membrane

自养血管—autotrophic blood vessel

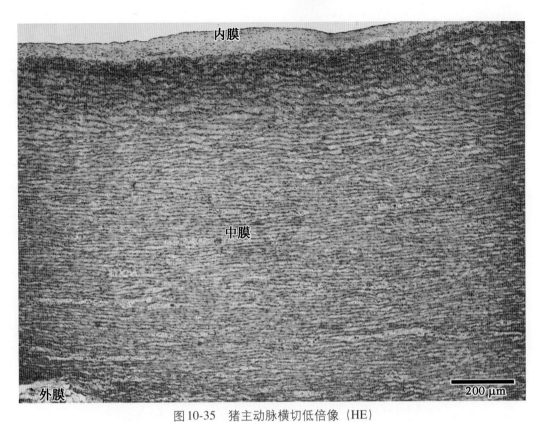

图 10-35 猪主动脉横切低倍像（HE）

Fig.10-35　low magnification of transection of pig aorta（HE）

内膜—internal membrane　中膜—mid membrane　外膜—outer membrane

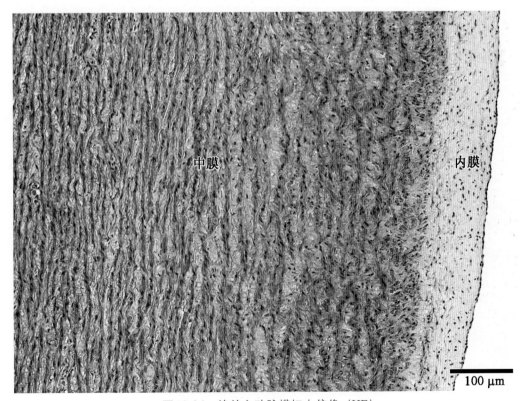

图 10-36 绵羊主动脉横切中倍像（HE）

Fig.10-36　mid magnification of transection of sheep aorta（HE）

内膜—internal membrane　中膜—mid membrane

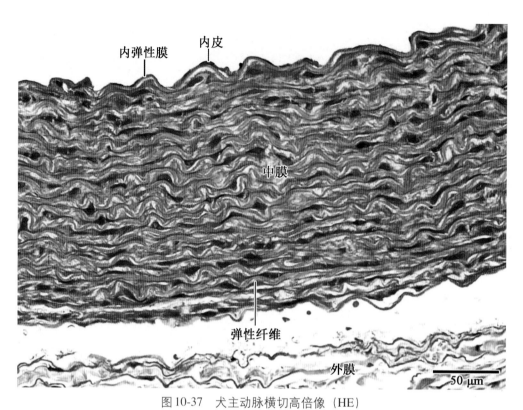

图 10-37 犬主动脉横切高倍像（HE）

Fig.10-37 high magnification of transection of dog aorta（HE）

内弹性膜—internal elastic membrane 内皮—endothelium 中膜—mid membrane

弹性纤维—elastic fiber 外膜—outer membrane

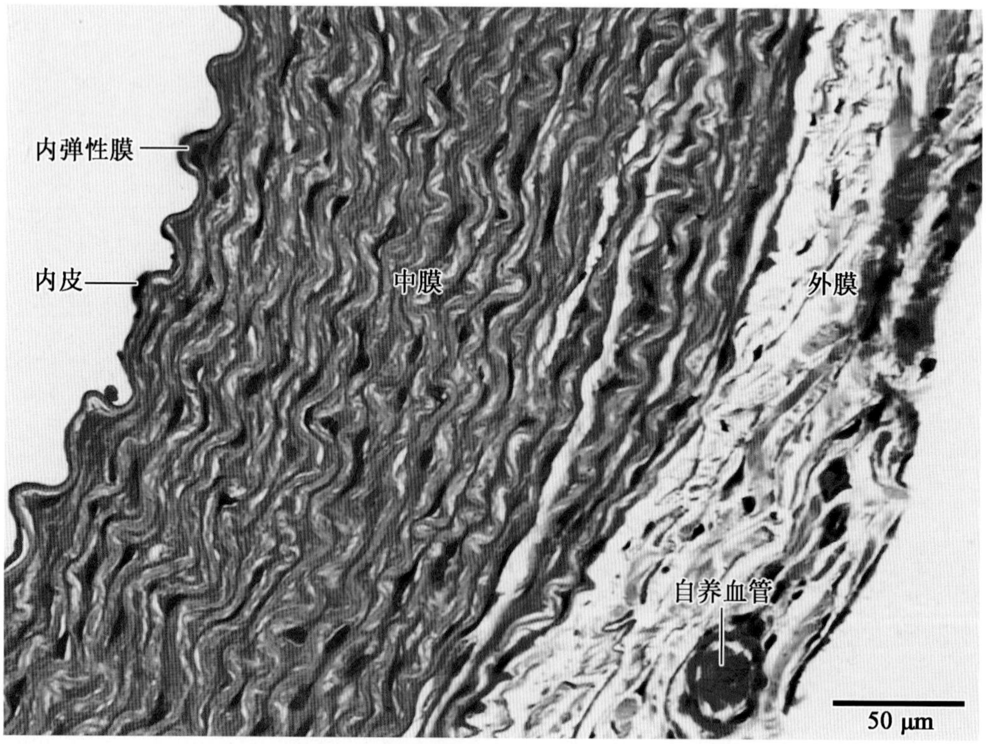

图 10-38 猫主动脉横切高倍像（HE）

Fig.10-38 high magnification of transection of cat aorta（HE）

内弹性膜—internal elastic membrane 内皮—endothelium 中膜—mid membrane

外膜—outer membrane 自养血管—autotrophic blood vessel

第十章 循环系统 Circulatory System

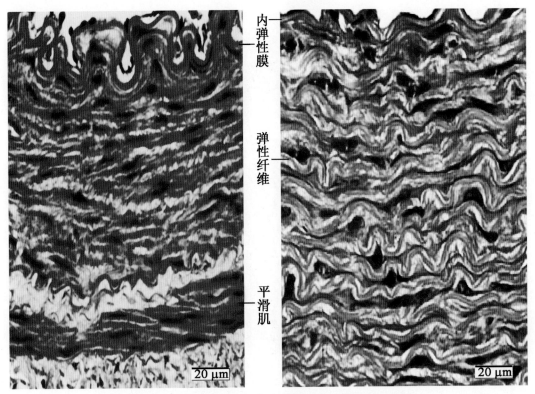

图10-39 中动脉（左）和大动脉（右）的比较（HE）
Fig.10-39 comparison of medium artery (left) and aorta (right)
内弹性膜—internal elastic membrane　弹性纤维—elastic fiber　平滑肌—smooth muscle

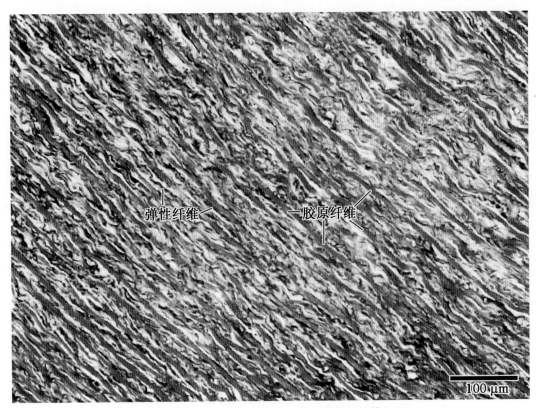

图10-40 牛大动脉中膜高倍像（弹性纤维染色）
Fig.10-40 high magnification of mid-membrane of cow aorta (elastic fiber stain)
弹性纤维—elastic fiber　胶原纤维—collagenous fiber

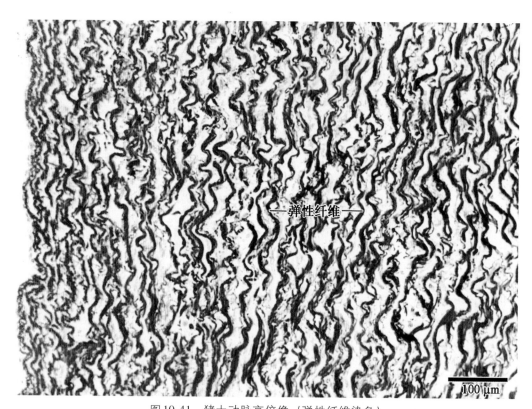

图10-41 猪大动脉高倍像（弹性纤维染色）
Fig.10-41 high magnification of pig aorta (elastic fiber stain)
弹性纤维— elastic fiber

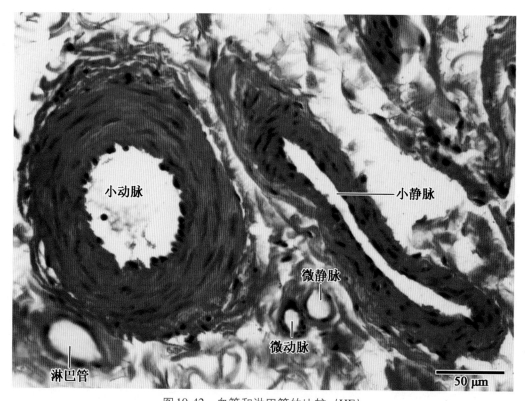

图10-42 血管和淋巴管的比较（HE）
Fig.10-42 comparison of blood and lymph vessels (HE)
小动脉—small artery 小静脉—small vein
微动脉—arteriole 微静脉—venule 淋巴管—lymph vessel

第十章 循环系统 Circulatory System

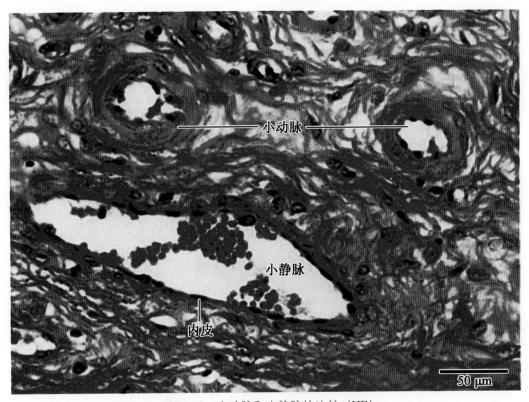

图10-43 小动脉和小静脉的比较（HE）
Fig.10-43 comparison of small artery and small vein（HE）
小动脉—small artery 小静脉—small vein 内皮—endothelium

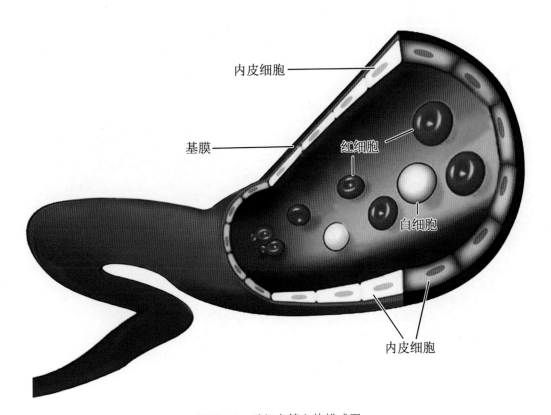

图10-44 毛细血管立体模式图
Fig.10-44 stereogram of capillary
内皮细胞—endothelium 基膜—basal membrane 红细胞—red cell 白细胞—white cell

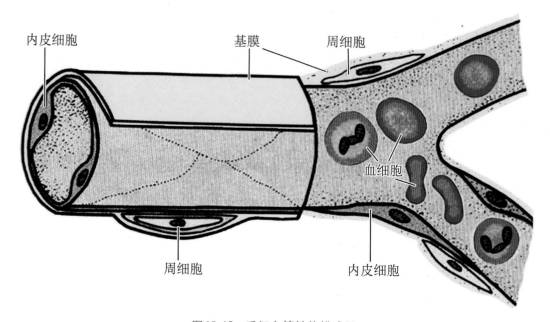

图 10-45 毛细血管结构模式图

Fig.10-45 ideograph of capillary structure

内皮细胞—endothelium　基膜—basal membrane　周细胞—pericyte　血细胞—blood cell

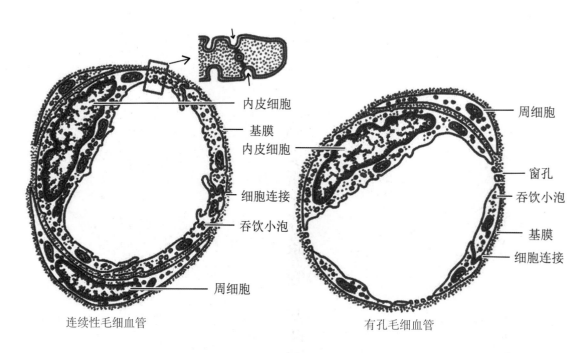

图 10-46 毛细血管横切超微结构模式图

Fig.10-46 ideograph of capillary transection ultrastructure

连续性毛细血管—continuous capillary　有孔毛细血管—perforated capillary　内皮细胞—endothelium
基膜—basal membrane　细胞连接—cell junction　吞饮小泡—pinocytosis vesicle　周细胞—pericyte　窗孔—orifice

第十章 循环系统 Circulatory System

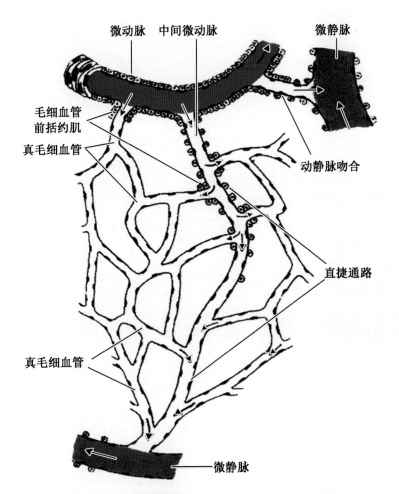

图10-47 微循环血管模式图
Fig.10-47 blood vessel ideograph of microcirculation

微动脉—arteriole
中间微动脉—meta-arteriole
微静脉—venule
毛细血管前括约肌—precapillary sphincter
真毛细血管—true capillary
动静脉吻合—arteriovenous anastomosis
直捷通路—thoroughfare channel

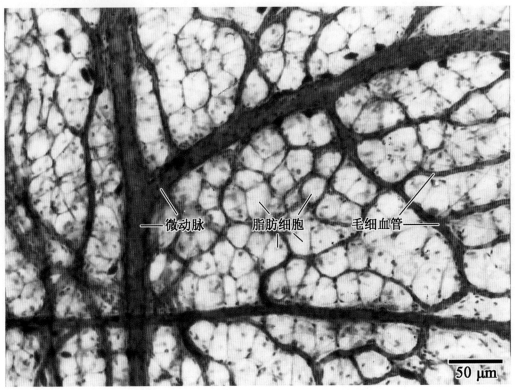

图10-48 脂肪组织内小血管的分布（油红染色）
Fig.10-48 distribution of small blood vessels in adipose tissue（oil red stain）

微动脉— arteriole 脂肪细胞—fat cell 毛细血管—capillary

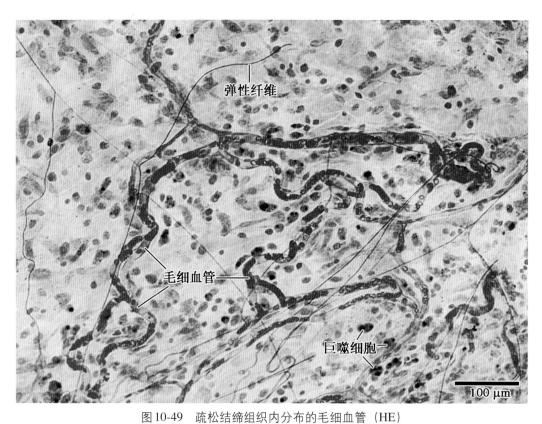

图10-49 疏松结缔组织内分布的毛细血管（HE）

Fig.10-49 distribution of capillaries in loose connective tissue（HE）

弹性纤维—elastic fiber　毛细血管—capillary　巨噬细胞—macrophage

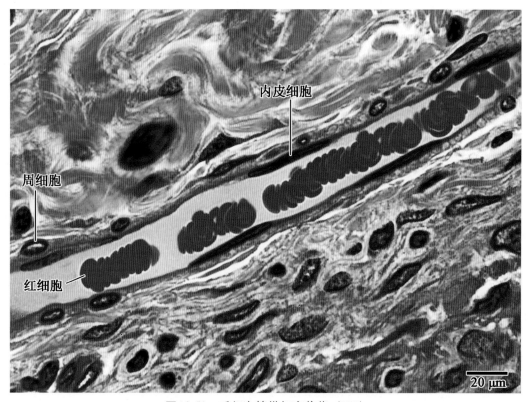

图10-50 毛细血管纵切高倍像（HE）

Fig.10-50 high magnification of capillary longitudinal incision (HE)

内皮细胞—endothelium　周细胞—pericyte　红细胞—red cell

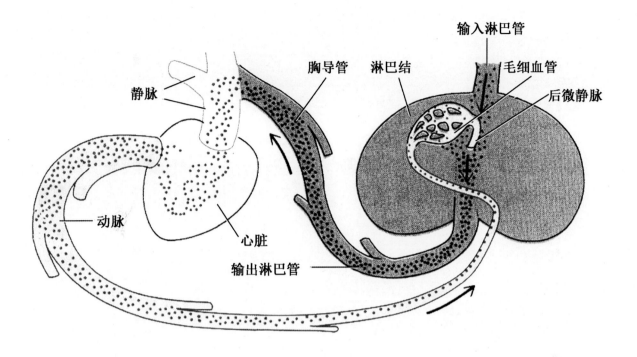

图 10-51　淋巴管与心血管关系模式图

Fig.10-51　relational schema diagram of cardiovascular and lymph vessels

静脉—vein　心脏—heart　动脉—artery　输入淋巴管—afferent lymphatics　淋巴结—lymph node

毛细血管—capillary　后微静脉—postcapillary venules　输出淋巴管—efferent lymphatics　胸导管—thoracic duct

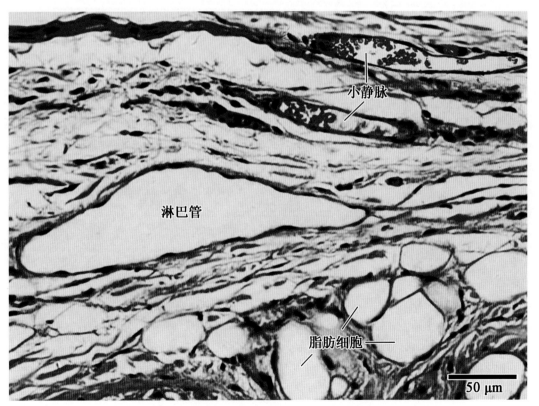

图 10-52　牛纵隔中的淋巴管高倍像

Fig.10-52　high magnification of lymph vessel in cow mediastinum

小静脉—small vein　淋巴管—lymphatic vessel　脂肪细胞—fat cell

第十一章
免疫系统
Immune System

> **Outline**
>
> Immune system consists of lymphoid organs, lymphoid tissues and immune cells. According to their functions, lymphoid organs are classified into primary or central lymphoid organs including thymus, bone marrow and cloacal bursa (in bird), and secondary or peripheral lymphoid organs including spleen, lymph nodes, tonsils, hemal nodes and hemolymph nodes. Lymphoid tissues are divided into lymphoid nodules and diffuse lymphoid tissue. Immune cells include mostly lymphocytes, and other cells including plasma cells, antigen-presenting cells. etc.
>
> Thymus is enclosed by a thin connective tissue capsule that penetrates the parenchyma and divides it into many incomplete lobules, called thymic lobule. Each lobule has a peripheral dark zone known as cortex and a central light zone called medulla. The parenchyma consists of a delicate skeleton of epithelial reticular cells, whose meshes are filled with lymphocytes. Cortex is composed of densely packed T cell precursors (thymocytes) and dispersed epithelial reticular cells. Medulla contains more epithelial reticular cells, mature thymocytes, macrophages and thymic corpuscles. The blood-thymus barrier is present in the cortex, composed of capillaries having a continuous endothelium and a thick basal lamina, perivascular space containing macrophages, and epithelial reticular cells and its basal lamina.
>
> Lymph node is divided into three zones, including peripheral cortex that contains lymphoid nodules with B lymphocytes, paracortical zone that contains T lymphocytes, and medulla comprises medullary cords of lymphoid tissue where many plasma cells reside, and the intervening medullary sinuses.
>
> Spleen has a dense connective tissue capsule containing some smooth muscle fibers. From the capsule, numerous branched trabeculae extend into the parenchyma. The parenchyma has three components: the white pulp, the red pulp and the marginal zone, but no lymphoid vessels. The white pulp consists of densely

> packed lymphoid tissues, which can be further divided into the periarterial lymphatic sheaths and the spleen nodules. The red pulp is composed of the splenic cords and sinusoids. The marginal zone lies between the white pulp and the red pulp, and consists of loose lymphoid tissues.

免疫系统（immune system）由中枢免疫器官（胸腺和骨髓，禽类还有法氏囊）与周围免疫器官（淋巴结、脾和扁桃体等）、淋巴组织（弥散淋巴组织和淋巴小结）及分散于血液、淋巴和其他组织内的淋巴细胞、浆细胞、抗原呈递细胞和它们分泌的免疫球蛋白、补体、各种细胞因子等组成。淋巴细胞在免疫系统中起关键作用。淋巴细胞既能识别病毒、细菌、异体蛋白等抗原物质产生免疫应答，防御和消除病原体的侵害，又可清除体内衰老和损伤的细胞，并监视和销毁体内的突变细胞。

一、免疫器官的发生

（一）胸腺的发生

通常认为胸腺（thymus）起源于胚胎第3、4对咽囊的内胚层和第2、3、4对鳃沟的外胚层。内胚层细胞主要分化形成胸腺髓质的部分上皮细胞；外胚层细胞则分化为皮质和髓质的上皮细胞。咽囊和鳃沟的内、外胚层和神经嵴来源的间充质相互作用，从而启动胸腺的发生。先是第3对咽囊腹侧部的上皮细胞和鳃沟外胚层上皮细胞增生，形成左右两条细胞索。两侧细胞索向胚体尾端伸长，沿胸骨后降入纵隔，与心包膜壁层接触并向中线靠拢、愈合，形成胸腺的原基，细胞索的根部则退化消失。胚胎早期的胸腺实质尚未分成小叶，也无皮质和髓质之分，表面仅包裹一层基膜。胸腺上皮细胞索在其周围的间充质内分支生长，细胞相互连接，形成上皮性网，上皮细胞索间的间充质形成不完整的小隔。随着胸腺的发育，其内出现血管。胚胎中期，淋巴祖细胞不断迁入胸腺，分布于网状上皮细胞的间隙内，并迅速分裂增殖，分化为胸腺细胞。胚胎中期以后，血管和神经伸入分化中的髓质内，巨噬细胞的前体细胞也进入胸腺。此后，胸腺小叶及皮质和髓质分界逐渐清晰。

（二）淋巴结的发生

淋巴结（lymph node）是哺乳动物特有的周围免疫器官，是滤过淋巴和产生免疫应答的重要器官。淋巴结的发生与淋巴管的发生密切相关。在牛、马和灵长类的第7～8周胚体内可观察到毛细淋巴管网；与此同时，局部间充质腔隙也互相融合扩大，形成许多淋巴囊，如颈淋巴囊、髂淋巴囊、乳糜池等，各淋巴囊均与引流一定区域的淋巴管相连接。环绕淋巴囊和大淋巴管周围的细胞不断聚集成群，逐渐形成淋巴结。淋巴结内的淋巴细胞是淋巴祖细胞在肝、骨髓及胸腺内分化后迁移而来的。胎儿淋巴结尚无免疫反应的功能，出生2周后，肠系膜淋巴结内出现浆细胞；出生4周后，可观察到淋巴小结和生发中心。

（三）脾的发生

胚胎时的脾（spleen）是造血器官，但自骨髓开始造血后，脾逐渐成为机体最大的免疫器官，但脾内仍保存少量造血干细胞，在一定条件下可出现造血功能。脾起始于胚胎期胃背系膜内的间充质，其细胞密集成群并凸向腹腔，为腹膜所覆盖。胃背系膜发育成的网膜囊向左突出，脾也被牵向胃的左上方，并参与构成小网膜的边缘。胚胎中期，网膜囊的背叶与体壁黏合，覆盖于左肾上腺及肾的表面。随后，脾血管进入密集的间充质细胞团，并分支形成血窦，即分出脾索和脾窦。卵黄囊血岛的造血干细胞由肝经血流入脾，在血窦周围的网状组织内增生分化为各种造血祖细胞和各种前体细胞。胚胎中期，脾内小动脉周围出现少量T细胞和B细胞，细胞呈小集落状。随着胎龄增长，B细胞集落逐渐增大，形成脾小结。胎儿期后，脾的造血功能活跃，可生成红细胞、粒细胞、血小板和淋巴细胞等。不仅在血窦外可见大小不同的造血灶，血窦内也可见造血集落。与此同时，血窦内皮细胞渐由扁平状变为杆状，还可见巨噬细胞吞噬血细胞现象。此后，密集的淋巴细胞团形成白髓，脾索内的淋巴细胞也增多。脾生成粒细胞和红细胞的功能逐渐被骨髓替代，生成淋巴细胞的功能则保持终生。脾的红髓、白髓日渐分明，淋巴组织不断增多，脾由造血器官逐渐转变为免疫器官。许多T细胞进入小动脉周围的结缔组织内，形成动脉周围淋巴鞘，淋巴鞘内还可见到树突状细胞，脾小体外周出现边缘区，脾的结缔组织逐渐增多，被膜也渐增厚，并出现清晰的小梁。

（四）扁桃体的发生

扁桃体（tonsil）由咽囊内胚层发育而来，内胚层细胞增殖形成细胞索，向下生长伸入间充质内。细胞索的细胞形成扁桃体隐窝上皮，而间充质细胞则围绕隐窝形成网状支架。到胚胎中期，分别由骨髓和胸腺来的B细胞和T细胞进入隐窝上皮内，形成淋巴小结和弥散淋巴组织，深部的间充质形成被膜。扁桃体分为腭扁桃体、舌扁桃体和咽扁桃体，它们与咽黏膜内散在的淋巴组织共同组成咽淋巴环，构成机体的第一道重要防线。

二、免疫器官的组织学结构概述

（一）胸腺（thymus）

胸腺分为胸叶和颈叶，胸叶大，位于心前纵隔内，向前分为左、右颈叶，沿气管两侧分布，前端可达喉部。新生动物的胸腺在生后继续发育，至性成熟期体积达到最大，随后开始退化，直至消失。

1. **被膜**（capsule） 薄层被膜由含弹性纤维的致密结缔组织构成，并伸入实质的小叶间，将胸腺分为不完全分隔的小叶。

2. **皮质**（cortex） 由淡染的上皮网状细胞、巨噬细胞和紧密排列的深染小T淋巴细胞，即胸腺细胞构成，小T淋巴细胞使胸腺皮质染色深暗。上皮网状细胞围绕着毛细血管。毛细血管为皮质中仅有的血管。

3. **髓质**（medulla） 染色浅，相邻的小叶之间是连续的。髓质内有浆细胞、淋巴细胞、巨噬细胞和上皮网状细胞。此外，上皮网状细胞呈同心圆状排列，形成胸腺小体，这是胸腺髓质的组织结构特征。

（二）淋巴结（lymph node）

1. **被膜和小梁** 由含有一些弹性纤维和平滑肌的致密不规则胶原性结缔组织构成，被膜外周常被脂肪组织所包围。输入淋巴管从凸面进入，输出淋巴管和血管从门部通过。

2. **实质** 皮质在外周，髓质在中央，二者分界不明显。猪淋巴结皮质和髓质的位置相反。

（1）皮质 位于被膜下方，分为三部分。

①淋巴小结 紧靠被膜下，在抗原刺激下，淋巴小结发育良好，有明显的暗区、明区和小结帽。淋巴小结内有B细胞、巨噬细胞、滤泡树突细胞、T细胞等。暗区位于基部，大的B细胞分裂分化后，移至明区。小结帽的小淋巴细胞主要是浆细胞的前身，还有一些B记忆细胞。

②副皮质区 位于皮质深部，为厚层弥散淋巴组织，主要含T细胞。区内有许多毛细血管后微静脉，是血液内淋巴细胞进入淋巴结的重要通道。

③皮质淋巴窦 是淋巴结内淋巴流动的通道，包括被膜下淋巴窦和小梁周围淋巴窦。淋巴窦壁衬一层连续内皮细胞、少量网状纤维和扁平网状细胞。窦内有星状的内皮细胞和网状纤维作支架，填充许多巨噬细胞和淋巴细胞。

（2）髓质 位于淋巴结中央，由髓索和髓窦组成。

①髓索 是不规则的淋巴组织索，彼此相连成网，主要含B细胞，还有T细胞、浆细胞、肥大细胞和巨噬细胞等。髓索是产生抗体的部位。髓索中央常有毛细血管后微静脉。

②髓质淋巴窦（髓窦）位于髓索之间，相互连接成网，其结构与皮质淋巴窦相同，有较强的滤过作用。

（三）脾（spleen）

脾由被膜和实质构成，具有造血、滤血、灭血和贮血等作用。

1. **被膜和小梁** 被膜是一层富含平滑肌和弹性纤维的结缔组织，表面被覆间皮。结缔组织伸入脾内分支成许多小梁，互相连接构成脾的支架。

2. **实质** 由白髓、边缘区和红髓组成。

（1）白髓（white pulp）包括脾小结和动脉周围淋巴鞘。脾小结即淋巴小结，主要由B细胞构成。发育良好的脾小结有明区、暗区和小结帽。健康动物的脾小结较少，当受到抗原刺激引起体液免疫应答时，脾小结增多、增大。动脉周围淋巴鞘是围绕中央动脉周围的厚层弥散淋巴组织，由大量T细胞、少量巨噬细胞、交错突细胞等构成，属胸腺依赖区。

（2）边缘区（marginal zone） 在白髓与红髓之间，呈红色。其中淋巴细胞较白髓稀疏，但较红髓密集，主要含B细胞，还有T细胞、巨噬细胞、浆细胞和各种血细胞。中央动脉分支而成的一些毛细血管，其末端在白髓与边缘区之间膨大形成边缘窦，窦的附近有许多的巨噬细胞，能对抗原进行处理。

（3）红髓（red pulp） 分布于被膜下、小梁周围、白髓及边缘区的外侧，含大量血细胞，在新鲜切面上呈红色。红髓包括脾索和脾窦。脾索是富含血细胞的索状淋巴组织，内含T细胞、B细胞、浆细胞、巨噬细胞和其他血细胞。脾索相互连接成网，与脾窦相间排列。脾窦为相互连通的、不规则的静脉窦。窦壁是一层长杆状的内皮细胞，呈纵向平行排列，细胞之间有宽的间隙，脾索内的血细胞可经此穿越进入脾窦，内皮外有不完整的基膜和环行的网状纤维围绕。脾窦如同多孔隙的栅栏，脾收缩时，窦壁的孔隙变窄或消失，脾扩张时孔隙变大。脾窦外侧有较多的巨噬细胞，其突起可通过内皮间隙伸入窦腔内。

（四）扁桃体（tonsil）

扁桃体由淋巴组织构成，位于舌、咽等处上皮下结缔组织中，为重要的防御器官。扁桃体表面的上皮凹陷，形成隐窝。腭扁桃体和舌扁桃体表面为复层扁平上皮；咽扁桃体的上皮为假复层纤毛柱状上皮。上皮下方及隐窝周围密集着淋巴小结及弥散淋巴组织。扁桃体可产生淋巴细胞和抗体，可抗致病微生物。正常情况下，由于扁桃体表面上皮完整和黏液腺不断分泌，可将细菌随同脱落的上皮细胞从隐窝口排出，保持机体健康。当机体抵抗力下降，上皮防御机能减弱，腺体分泌机能降低，扁桃体就会遭受细菌感染而发炎。

三、免疫系统图谱

1. **胸腺** 图11-1～图11-24。
2. **免疫细胞的发生** 图11-25～图11-31。
3. **淋巴结** 图11-32～图11-64。
4. **脾** 图11-65～图11-90。
5. **扁桃体** 图11-91～图11-98。

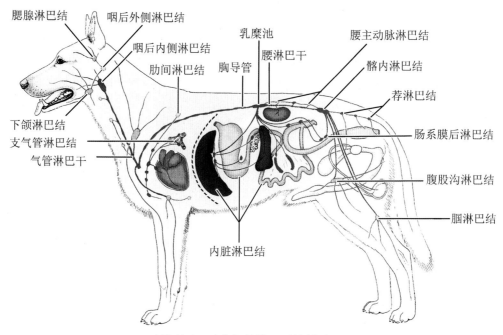

图 11-1　动物机体淋巴系统模式图

Fig.11-1　diagram of animal body lymphatic system

腮腺淋巴结—parotid lymph node　下颌淋巴结—mandibular lymph node　支气管淋巴结—bronchial lymph node
气管淋巴干—tracheal lymph trunk　咽后外侧淋巴结—posterolateral pharyngeal lymph node
咽后内侧淋巴结—posterior medial lymph node　肋间淋巴结—intercostal lymph node　胸导管—thoracic duct
乳糜池—cisterna chyli　腰淋巴干—lumbar trunk　腰主动脉淋巴结—lumbar aortic lymph node
髂内淋巴结—internal iliac lymph node　荐淋巴结—sacral lymph node
肠系膜后淋巴结—posterior mesenteric lymph node　腹股沟淋巴结—inguinal lymph node
腘淋巴结—popliteal lymph node　内脏淋巴结—visceral lymph node

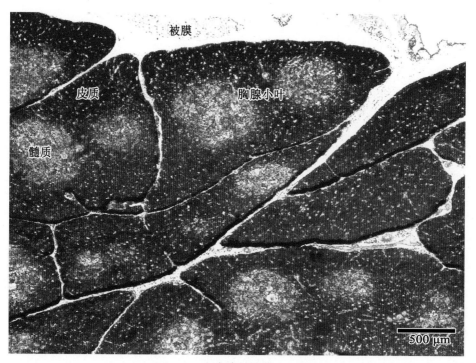

图 11-2　牛胸腺低倍像 1（HE）

Fig.11-2　low magnification of cattle thymus 1（HE）

被膜—capsule　胸腺小叶—thymic lobule　皮质—cortex　髓质—medulla

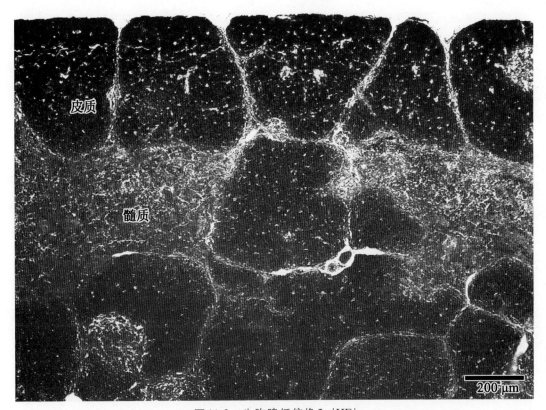

图 11-3 牛胸腺低倍像 2（HE）

Fig.11-3　low magnification of cattle thymus 2（HE）

皮质—cortex　髓质—medulla

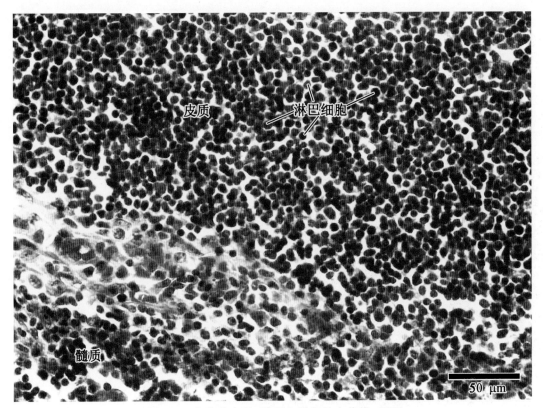

图 11-4 牛胸腺皮质-髓质交界部的中倍像（HE）

Fig.11-4　mid magnification of cattle thymus cortex-medulla junction（HE）

皮质—cortex　髓质—medulla　淋巴细胞—lymphatic cell

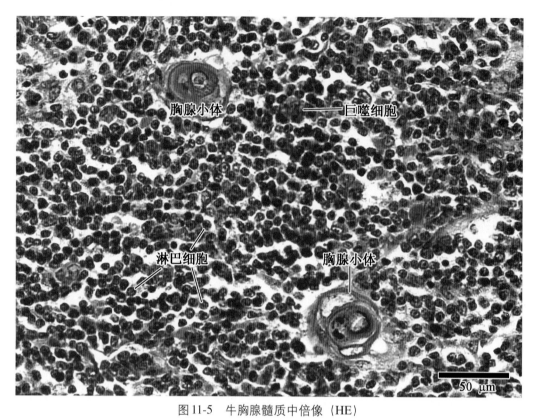

图 11-5 牛胸腺髓质中倍像（HE）

Fig.11-5 mid magnification of cattle thymus medulla（HE）

胸腺小体—thymic corpuscle 淋巴细胞—lymphatic cell 巨噬细胞—macrophage

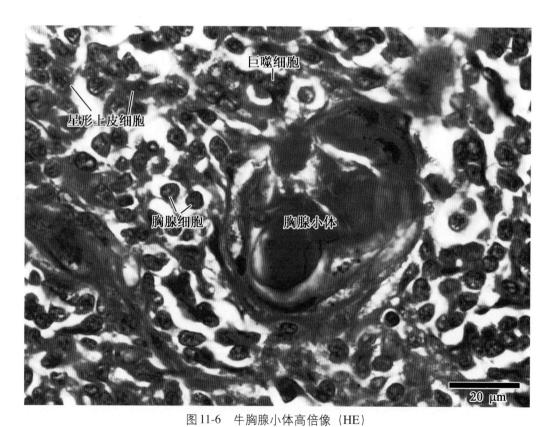

图 11-6 牛胸腺小体高倍像（HE）

Fig.11-6 high magnification of cattle thymic corpuscle（HE）

巨噬细胞—macrophage 星形上皮细胞—astroepithelial cell 胸腺细胞—thymic cell

胸腺小体—thymic corpuscle

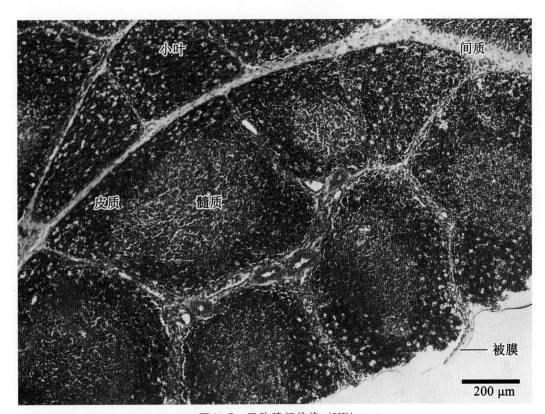

图 11-7 马胸腺低倍像（HE）

Fig.11-7 low magnification of horse thymus（HE）

小叶—lobule 间质—mesenchyme 皮质—cortex 髓质—medulla 被膜—capsule

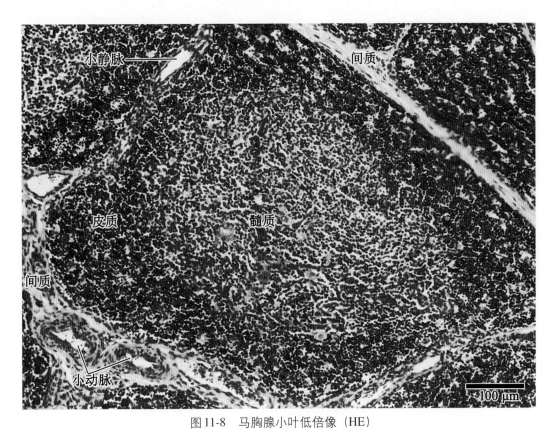

图 11-8 马胸腺小叶低倍像（HE）

Fig.11-8 low magnification of horse thymus lobule（HE）

小静脉—venule 间质—mesenchyme 皮质—cortex 髓质—medulla 小动脉—arteriole

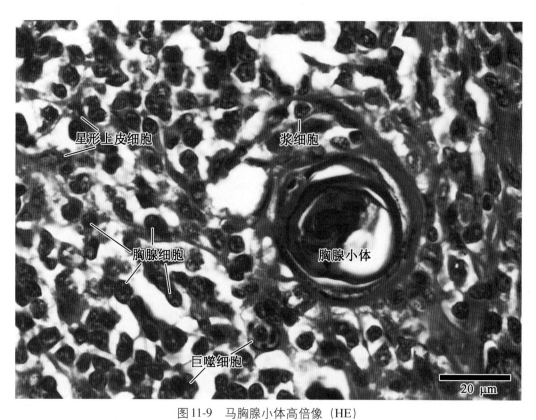

图 11-9 马胸腺小体高倍像（HE）

Fig.11-9　high magnification of horse thymic corpuscle（HE）

星形上皮细胞—astroepithelial cell　浆细胞—plasmocyte　胸腺细胞—thymic cell

胸腺小体—thymic corpuscle　巨噬细胞—macrophage

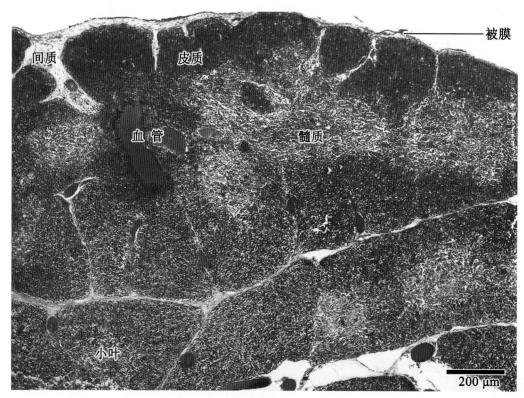

图 11-10　猪胸腺低倍像（HE）

Fig.11-10　low magnification of pig thymus（HE）

间质—mesenchyme　血管—blood vessel　小叶—lobule　皮质—cortex　髓质—medulla　被膜—capsule

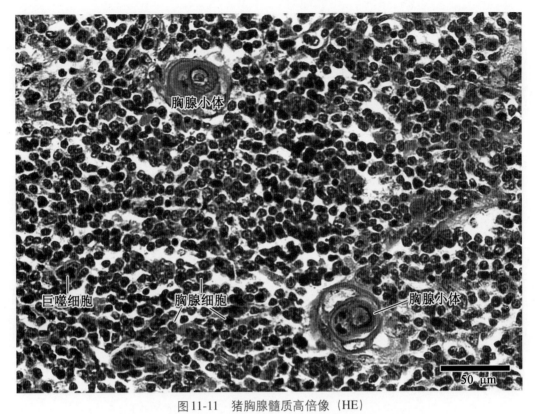

图 11-11 猪胸腺髓质高倍像（HE）

Fig.11-11 high magnification of pig thymus medulla（HE）

胸腺小体—thymic corpuscle 巨噬细胞—macrophage 胸腺细胞—thymic cell

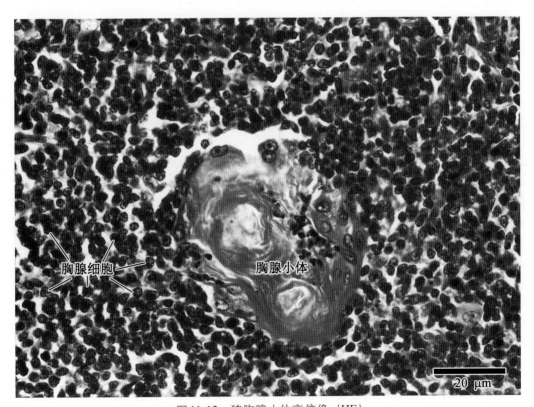

图 11-12 猪胸腺小体高倍像（HE）

Fig.11-12 high magnification of pig thymic corpuscle（HE）

胸腺细胞—thymic cell 胸腺小体—thymic corpuscle

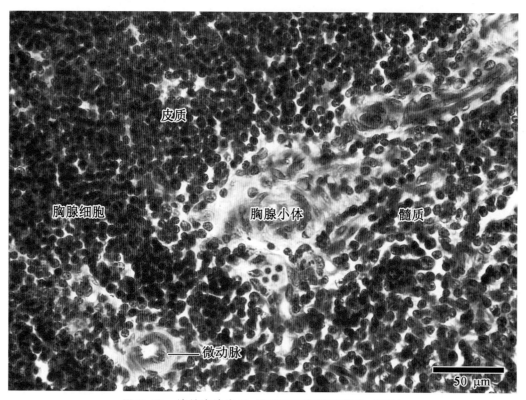

图 11-13 绵羊胸腺皮质-髓质交界部的高倍像（HE）

Fig.11-13 high magnification of sheep thymus cortex-medulla junction（HE）

皮质—cortex　胸腺细胞—thymic cell　胸腺小体—thymic corpuscle　髓质—medulla　微动脉—arteriole

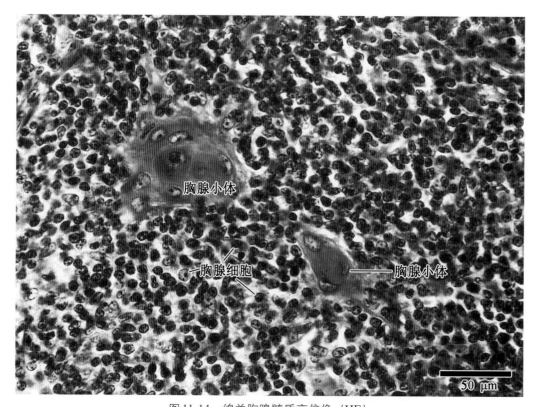

图 11-14 绵羊胸腺髓质高倍像（HE）

Fig.11-14 high magnification of sheep thymus medulla（HE）

胸腺小体—thymic corpuscle　胸腺细胞—thymic cell

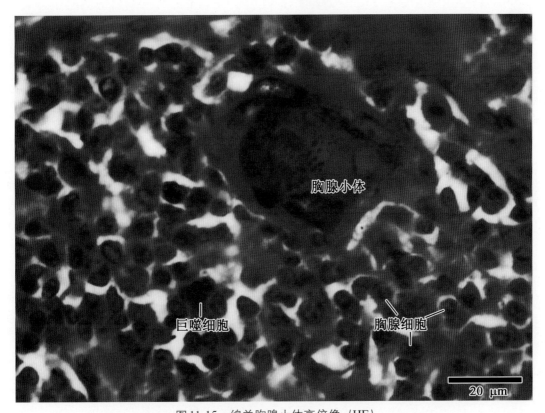

图11-15 绵羊胸腺小体高倍像（HE）

Fig.11-15 high magnification of sheep thymic corpuscle（HE）

胸腺小体—thymic corpuscle 巨噬细胞—macrophage 胸腺细胞—thymic cell

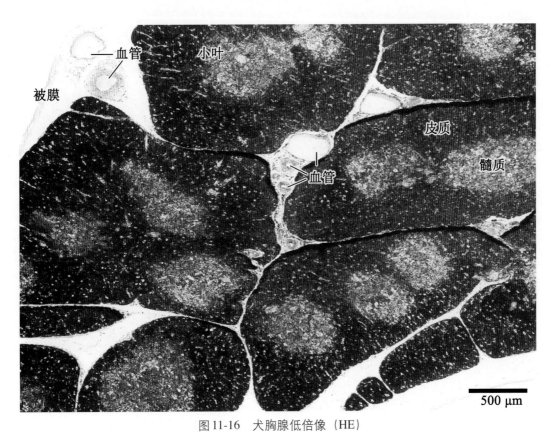

图11-16 犬胸腺低倍像（HE）

Fig.11-16 low magnification of dog thymus（HE）

被膜—capsule 血管—blood vessel 小叶—lobule 皮质—cortex 髓质—medulla

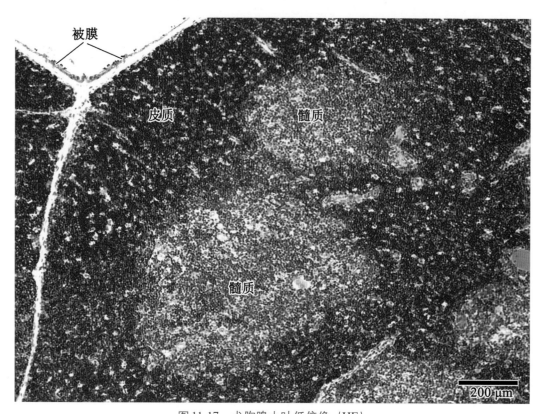

图 11-17　犬胸腺小叶低倍像（HE）

Fig.11-17　low magnification of dog thymus lobule（HE）

被膜—capsule　皮质—cortex　髓质—medulla

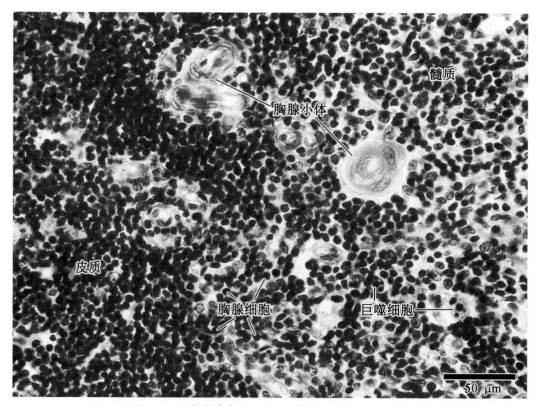

图 11-18　犬胸腺皮质-髓质交界部的高倍像（HE）

Fig.11-18　high magnification of dog thymus cortex-medulla junction（HE）

皮质—capsule　胸腺小体—thymic corpuscle　髓质—medulla　胸腺细胞—thymic cell　巨噬细胞—macrophage

第十一章 免疫系统 Immune System

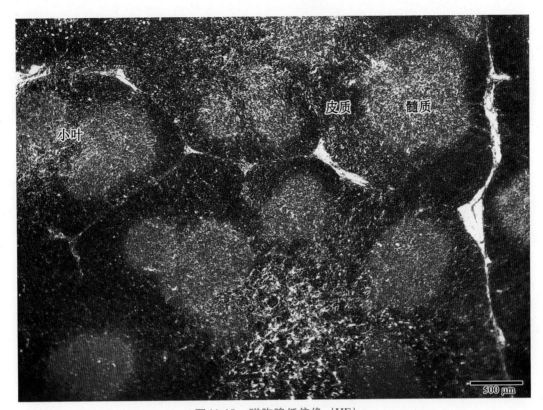

图 11-19 猫胸腺低倍像（HE）

Fig.11-19　low magnification of cat thymus（HE）

小叶—lobule　皮质—cortex　髓质—medulla

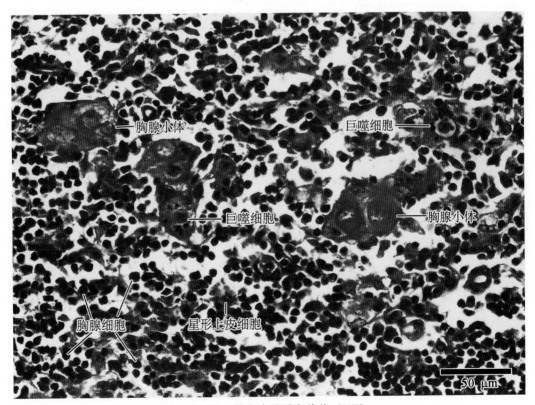

图 11-20 猫胸腺髓质高倍像（HE）

Fig.11-20　high magnification of cat thymus medulla（HE）

胸腺小体—thymic corpuscle　巨噬细胞—macrophage　胸腺细胞—thymic cell　星形上皮细胞—astroepithelial cell

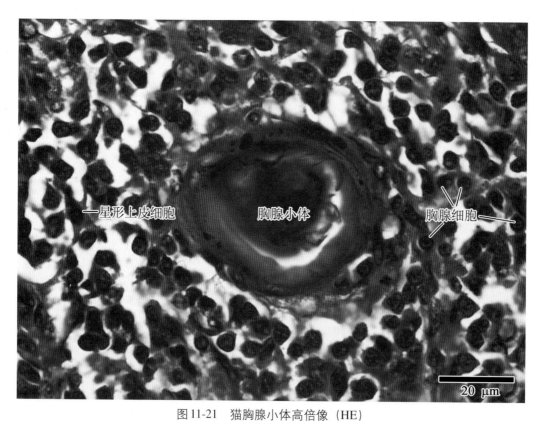

图 11-21 猫胸腺小体高倍像（HE）

Fig.11-21 high magnification of cat thymic corpuscle（HE）

星形上皮细胞—astroepithelial cell　胸腺小体—thymic corpuscle　胸腺细胞—thymic cell

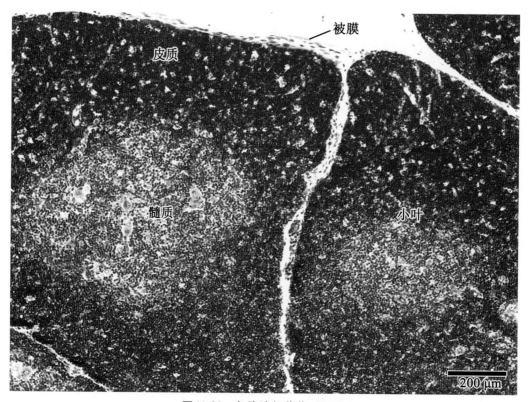

图 11-22 兔胸腺低倍像（HE）

Fig.11-22 low magnification of rabbit thymus（HE）

被膜—capsule　皮质—cortex　髓质—medulla　小叶—lobule

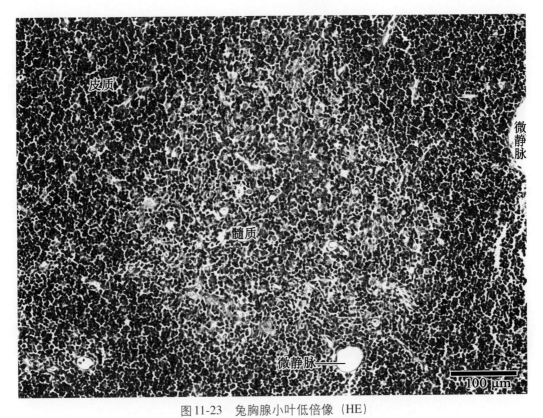

图 11-23　兔胸腺小叶低倍像（HE）
Fig.11-23　low magnification of rabbit thymus lobule（HE）
皮质—cortex　髓质—medulla　微静脉—venule

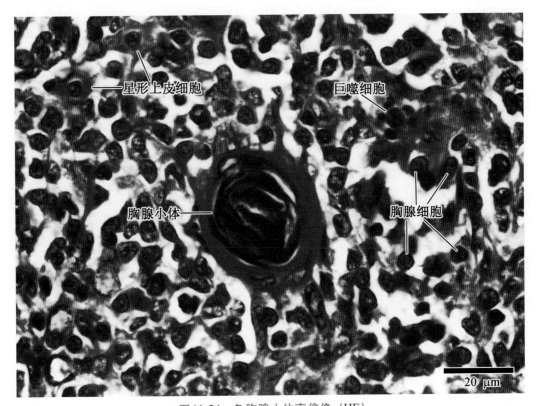

图 11-24　兔胸腺小体高倍像（HE）
Fig.11-24　high magnification of rabbit thymic corpuscle（HE）
星形上皮细胞—astroepithelial cell　巨噬细胞—macrophage　胸腺小体—thymic corpuscle　胸腺细胞—thymic cell

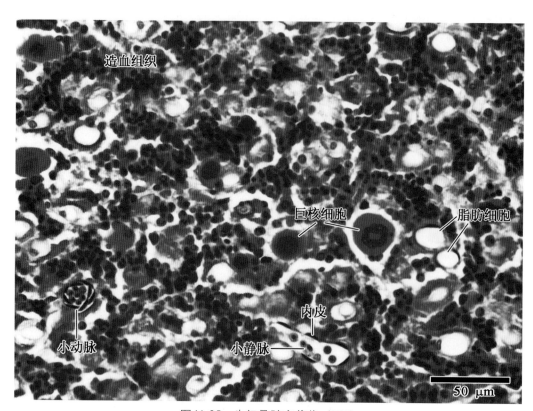

图 11-25 牛红骨髓高倍像（HE）

Fig.11-25 high magnification of cattle red marrow（HE）

造血组织—hemopoietic tissue 巨核细胞—megakaryocyte 脂肪细胞—fat cell

小动脉—arteriole 小静脉—venule 内皮—endothelium

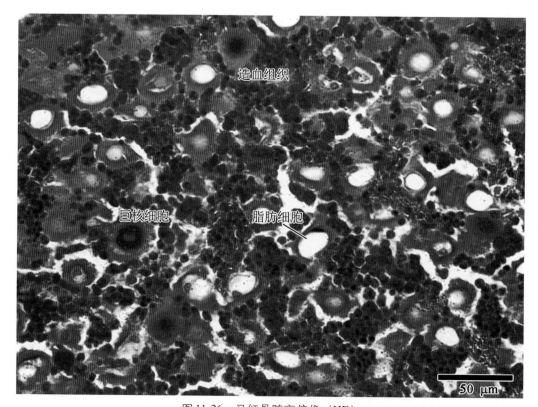

图 11-26 马红骨髓高倍像（HE）

Fig.11-26 high magnification of horse red marrow（HE）

造血组织—hemopoietic tissue 巨核细胞—megakaryocyte 脂肪细胞—fat cell

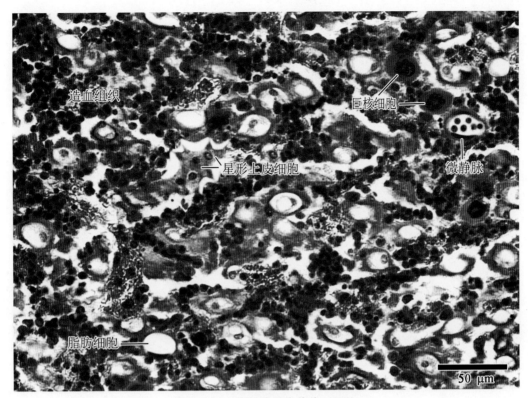

图11-27 猪红骨髓中倍像（HE）

Fig.11-27 mid magnification of pig red marrow （HE）

造血组织—hemopoietic tissue 巨核细胞—megakaryocyte 星形上皮细胞—astroepithelial cell 微静脉—venule
脂肪细胞—fat cell

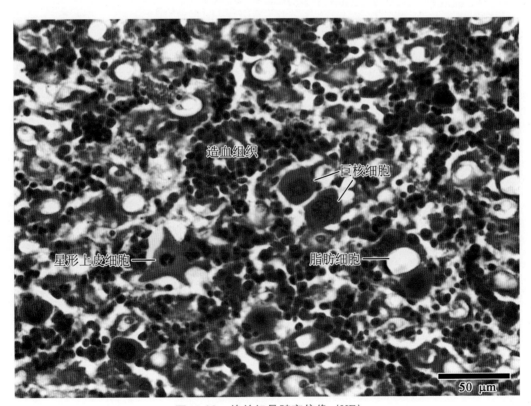

图11-28 绵羊红骨髓高倍像（HE）

Fig.11-28 high magnification of sheep red marrow （HE）

造血组织—hemopoietic tissue 巨核细胞—megakaryocyte 星形上皮细胞—astroepithelial cell 脂肪细胞—fat cell

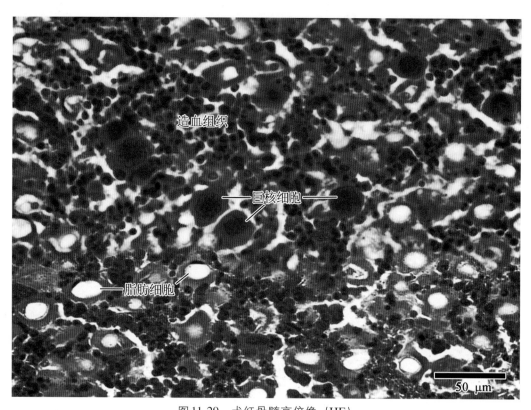

图11-29 犬红骨髓高倍像（HE）

Fig.11-29 high magnification of dog red marrow（HE）

造血组织—hemopoietic tissue　巨核细胞—megakaryocyte　脂肪细胞—fat cell

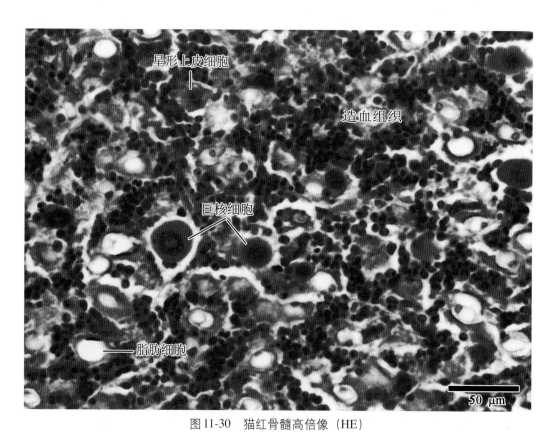

图11-30 猫红骨髓高倍像（HE）

Fig.11-30 high magnification of cat red marrow（HE）

星形上皮细胞—astroepithelial cell　造血组织—hemopoietic tissue　巨核细胞—megakaryocyte　脂肪细胞—fat cell

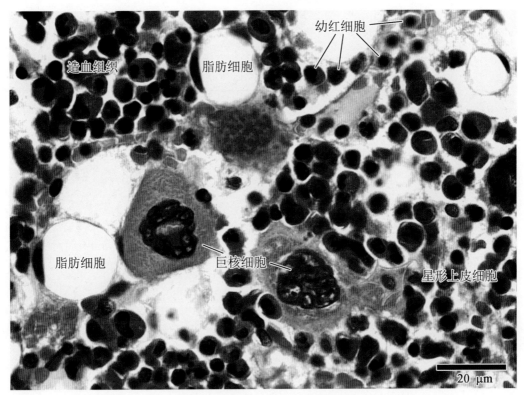

图 11-31 兔红骨髓高倍像（HE）

Fig.11-31 high magnification of rabbit red marrow（HE）

造血组织—hemopoietic tissue 脂肪细胞—fat cell 幼红细胞—normoblast 巨核细胞—megakaryocyte 星形上皮细胞—astroepithelial cell

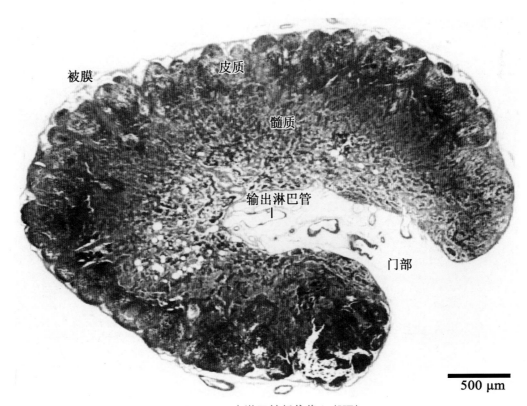

图 11-32 牛淋巴结低倍像 1（HE）

Fig.11-32 low magnification of cattle lymph node 1（HE）

被膜—capsule 皮质—cortex 髓质—medulla 输出淋巴管—efferent duct 门部—portal

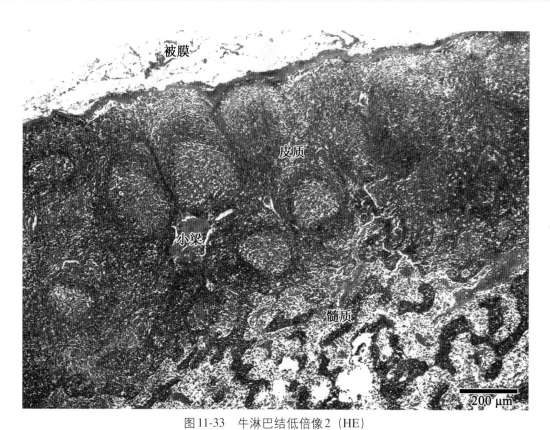

图 11-33 牛淋巴结低倍像 2（HE）

Fig.11-33 low magnification of cattle lymph node 2 （HE）

被膜—capsule 皮质—cortex 小梁—trabecula 髓质—medulla

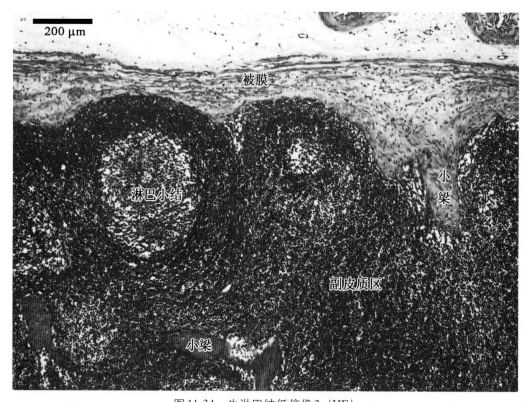

图 11-34 牛淋巴结低倍像 3（HE）

Fig.11-34 low magnification of cattle lymph node 3 （HE）

被膜—capsule 淋巴小结—lymphoid nodule 小梁—trabecula 副皮质区—paracortical zone

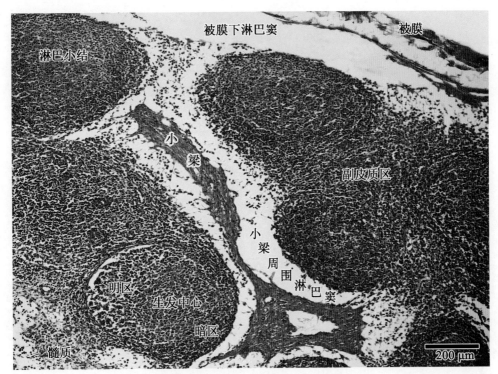

图11-35 牛淋巴结低倍像4（HE）

Fig.11-35 low magnification of cattle lymph node 4 (HE)

被膜—capsule 被膜下淋巴窦—subcapsular sinus 淋巴小结—lymphoid nodule 副皮质区—paracortical zone
小梁—trabecula 小梁周围淋巴窦—peritrabecular sinus 明区—light zone 生发中心—germinal center
暗区—dark zone 髓质—medulla

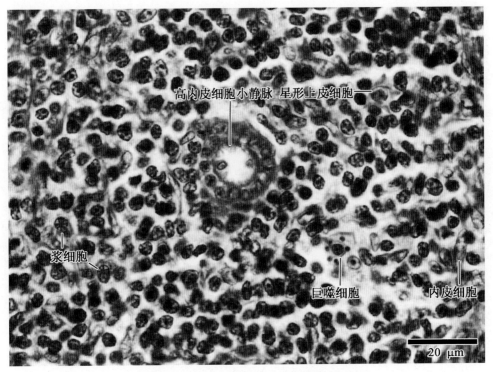

图11-36 牛淋巴结副皮质区高倍像（HE）

Fig.11-36 high magnification of paracortical zone in cattle lymph node (HE)

高内皮细胞小静脉—high endothelial venule 星形上皮细胞—astroepithelial cell 浆细胞—plasmocyte
巨噬细胞—macrophage 内皮细胞—endothelium

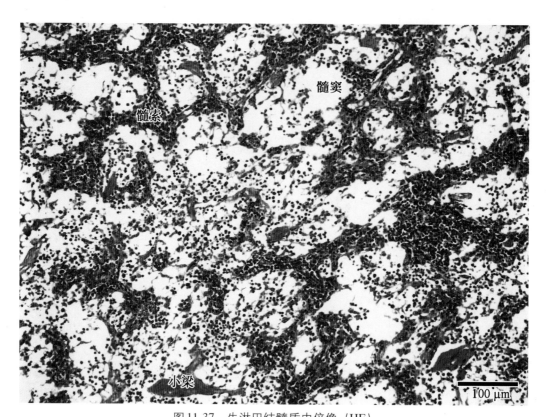

图 11-37　牛淋巴结髓质中倍像（HE）

Fig.11-37　mid magnification of cattle lymph node medulla（HE）

髓索—medullary cord　髓窦—medullary sinus　小梁—trabecula

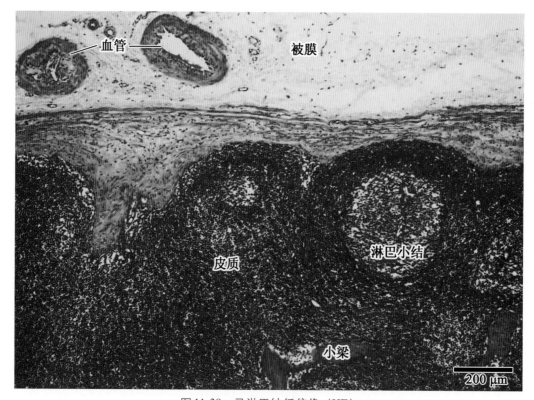

图 11-38　马淋巴结低倍像（HE）

Fig.11-38　low magnification of horse lymph node（HE）

血管—blood vessel　被膜—capsule　皮质—cortex　淋巴小结—lymphoid nodule　小梁—trabecula

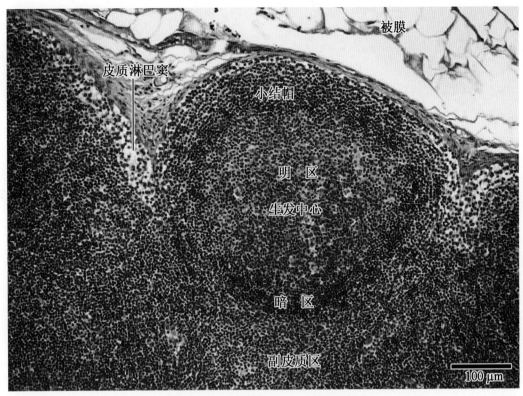

图11-39 马淋巴小结中倍像（HE）

Fig.11-39 mid magnification of horse lymphoid nodule（HE）

被膜—capsule 皮质淋巴窦—cortex sinus 小结帽—nodule cap 明区—light zone 生发中心—germinal center
暗区—dark zone 副皮质区—paracortical zone

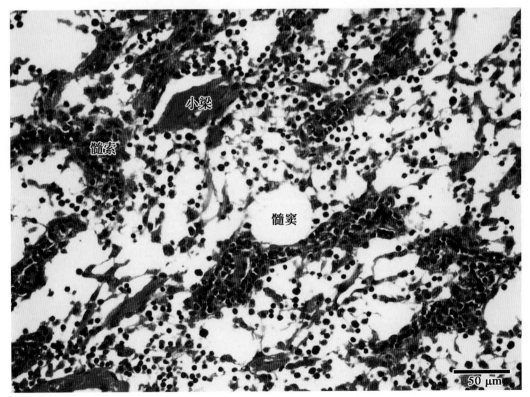

图11-40 马淋巴结髓质高倍像（HE）

Fig.11-40 high magnification of horse lymph node medulla（HE）

小梁—trabecula 髓索—medullary cord 髓窦—medullary sinus

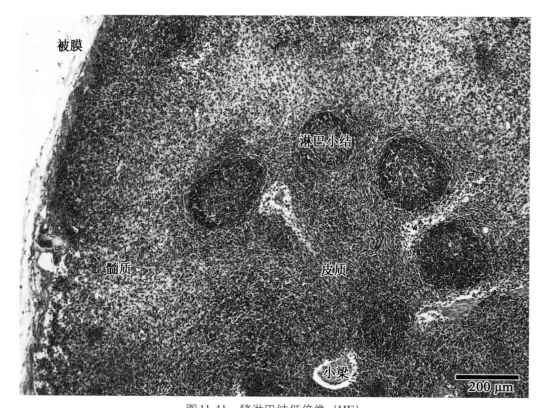

图11-41 猪淋巴结低倍像（HE）

Fig.11-41 low magnification of pig lymph node（HE）

被膜—capsule 淋巴小结—lymphoid nodule 髓质—medulla 皮质—cortex

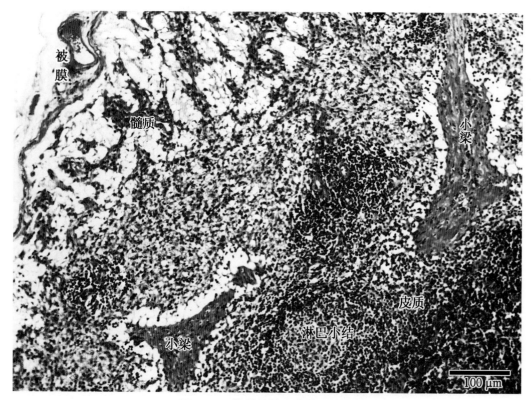

图11-42 猪淋巴结中倍像（HE）

Fig.11-42 mid magnification of pig lymph node（HE）

被膜—capsule 髓质—medulla 小梁—trabecula 淋巴小结—lymphoid nodule 皮质—cortex

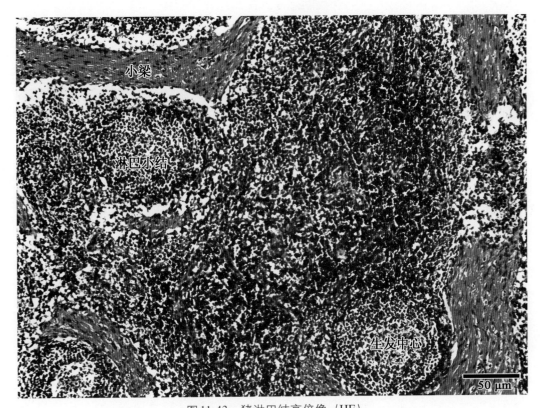

图11-43 猪淋巴结高倍像（HE）

Fig.11-43 high magnification of pig lymph node（HE）

小梁—trabecula 淋巴小结—lymphoid nodule 生发中心—germinal center

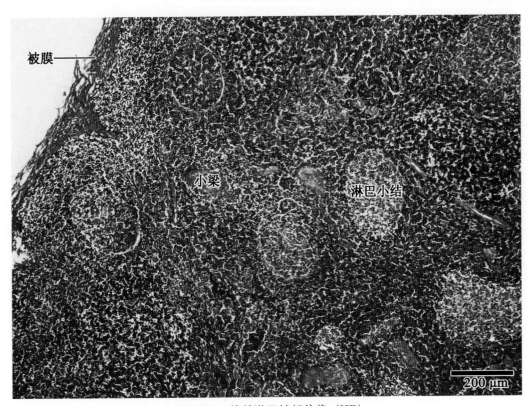

图11-44 绵羊淋巴结低倍像（HE）

Fig.11-44 low magnification of sheep lymph node（HE）

被膜—capsule 小梁—trabecula 淋巴小结—lymphoid nodule

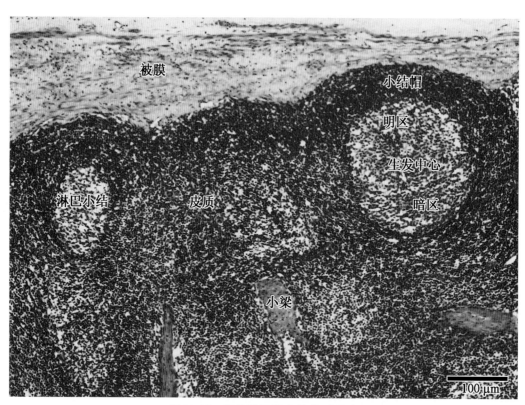

图 11-45　绵羊淋巴结中倍像（HE）

Fig.11-45　mid magnification of sheep lymph node（HE）

被膜—capsule　淋巴小结—lymphoid nodule　皮质—cortex　小梁—trabecula　小结帽—nodule cap　明区—light zone
生发中心—germinal center　暗区—dark zone

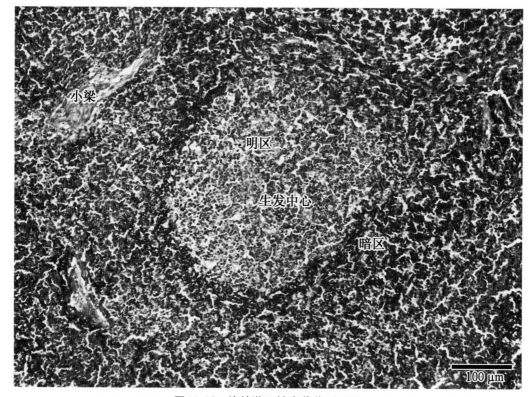

图 11-46　绵羊淋巴结高倍像（HE）

Fig.11-46　high magnification of sheep lymph node（HE）

小梁—trabecula　明区—light zone　生发中心—germinal center　暗区—dark zone

第十一章 免疫系统　Immune System

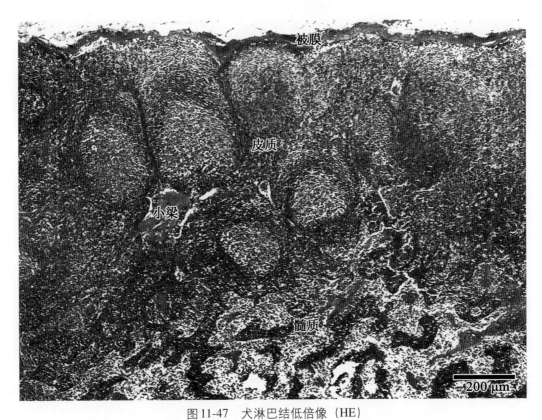

图11-47　犬淋巴结低倍像（HE）
Fig.11-47　low magnification of dog lymph node（HE）
被膜—capsule　皮质—cortex　小梁—trabecula　髓质—medulla

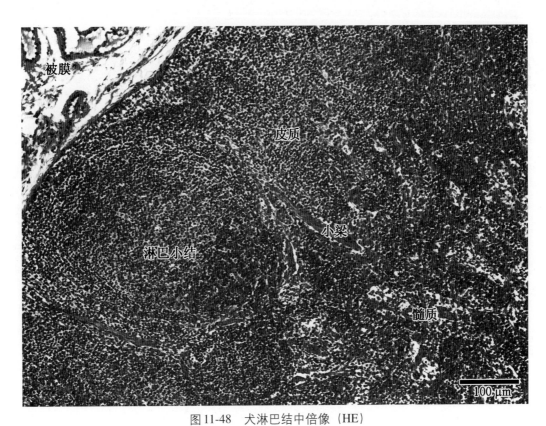

图11-48　犬淋巴结中倍像（HE）
Fig.11-48　mid magnification of dog lymph node（HE）
被膜—capsule　皮质—cortex　淋巴小结—lymphoid nodule　小梁—trabecula　髓质—medulla

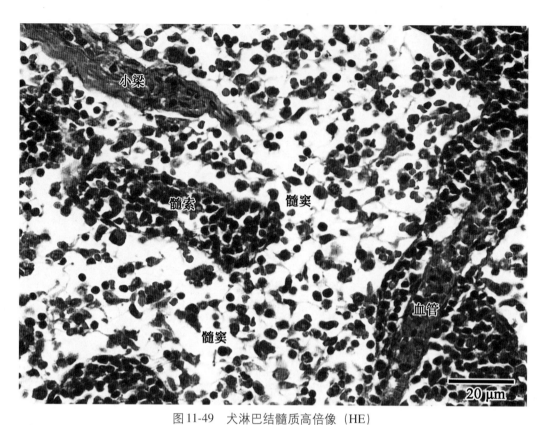

图 11-49 犬淋巴结髓质高倍像（HE）

Fig.11-49 high magnification of dog lymph node medulla (HE)

小梁—trabecula 髓索—medullary cord 髓窦—medullary sinus 血管—blood vessel

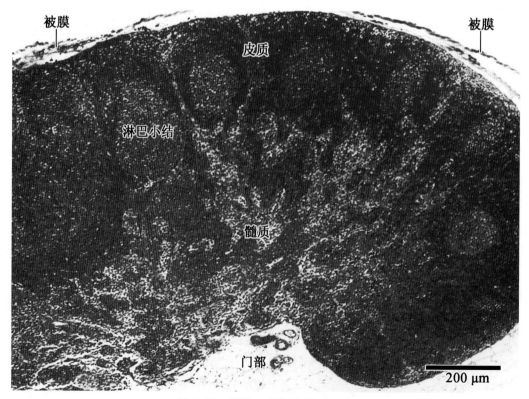

图 11-50 猫淋巴结低倍像（HE）

Fig.11-50 low magnification of cat lymph node (HE)

被膜—capsule 皮质—cortex 淋巴小结—lymphoid nodule 髓质—medulla 门部—portal

第十一章 免疫系统 Immune System

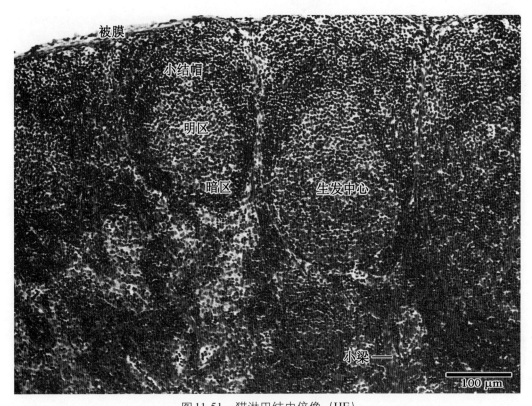

图11-51 猫淋巴结中倍像（HE）
Fig.11-51 mid magnification of cat lymph node（HE）
被膜—capsule 小结帽—nodule cap 明区—light zone 暗区—dark zone
生发中心—germinal center 小梁—trabecula

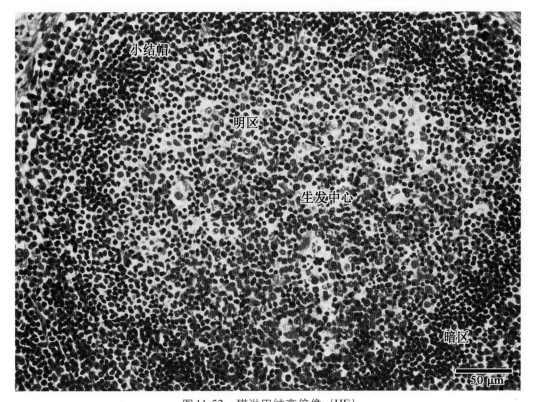

图11-52 猫淋巴结高倍像（HE）
Fig.11-52 high magnification of cat lymph node（HE）
小结帽—nodule cap 明区—light zone 生发中心—germinal center 暗区—dark zone

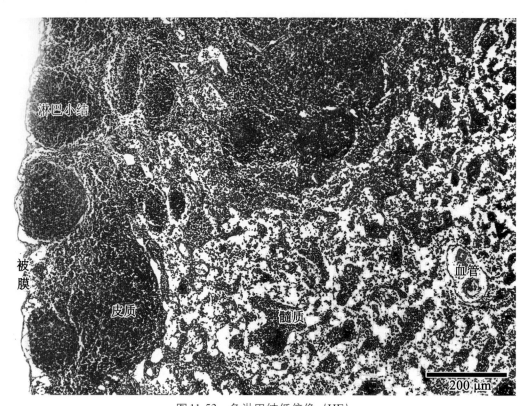

图11-53 兔淋巴结低倍像（HE）

Fig.11-53 low magnification of rabbit lymph node（HE）

淋巴小结—lymphoid nodule 被膜—capsule 皮质—cortex 髓质—medulla 血管—blood vessel

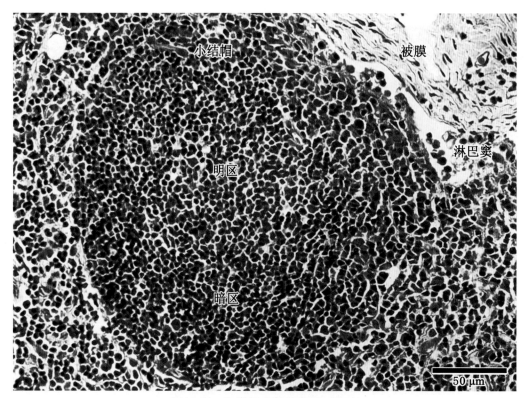

图11-54 兔淋巴小结高倍像（HE）

Fig.11-54 high magnification of rabbit lymphoid nodule（HE）

小结帽—nodule cap 明区—light zone 暗区—dark zone 被膜—capsule 淋巴窦—lymph sinus

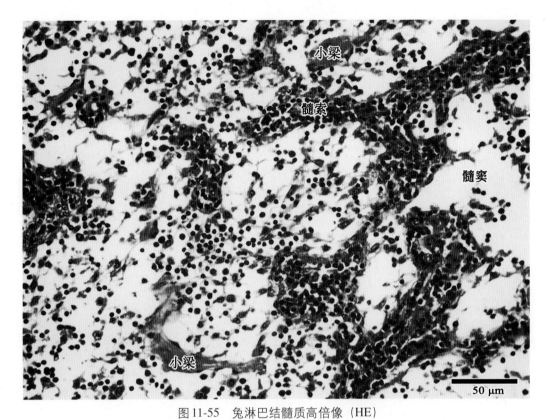

图 11-55　兔淋巴结髓质高倍像（HE）

Fig.11-55　high magnification of rabbit lymph node medulla（HE）

小梁—trabecula　髓索—medullary cord　髓窦—medullary sinus

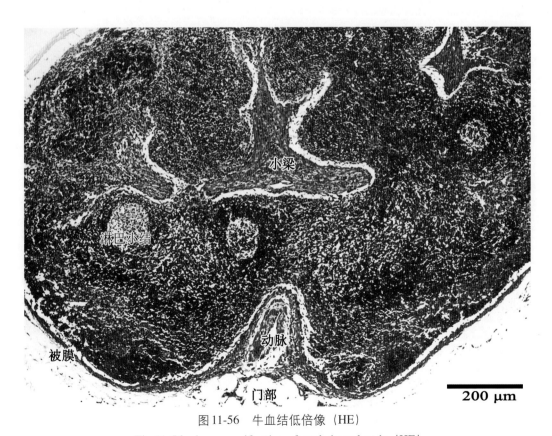

图 11-56　牛血结低倍像（HE）

Fig.11-56　low magnification of cattle hemal node（HE）

被膜—capsule　淋巴小结—lymphoid nodule　小梁—trabecula　动脉—artery　门部—portal

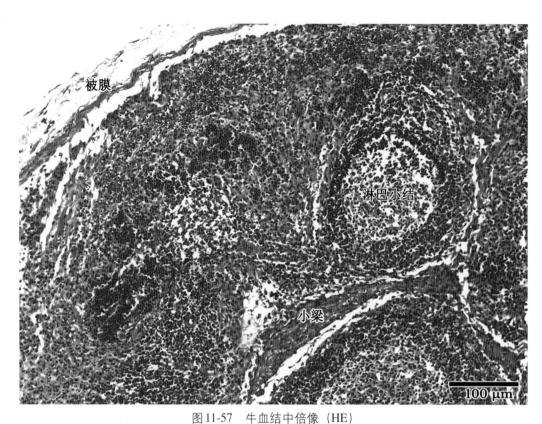

图11-57 牛血结中倍像（HE）

Fig.11-57 mid magnification of cattle hemal node（HE）

被膜—capsule 淋巴小结—lymphoid nodule 小梁—trabecula

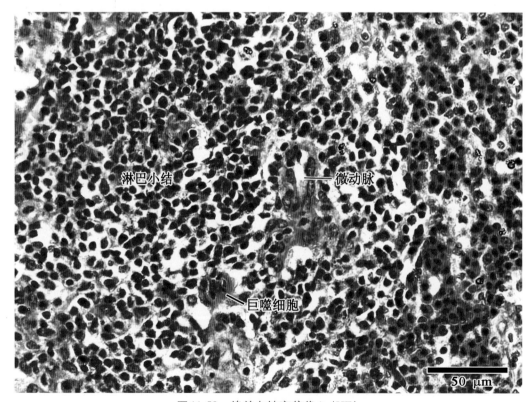

图11-58 绵羊血结高倍像1（HE）

Fig.11-58 high magnification of sheep hemal node 1（HE）

淋巴小结—lymphoid nodule 微动脉—arteriole 巨噬细胞—macrophage

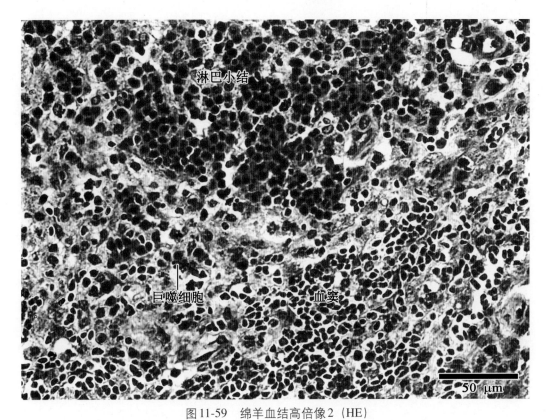

图11-59 绵羊血结高倍像2（HE）

Fig.11-59 high magnification of sheep hemal node 2（HE）

淋巴小结—lymphoid nodule 巨噬细胞—macrophage 血窦—blood sinusoid

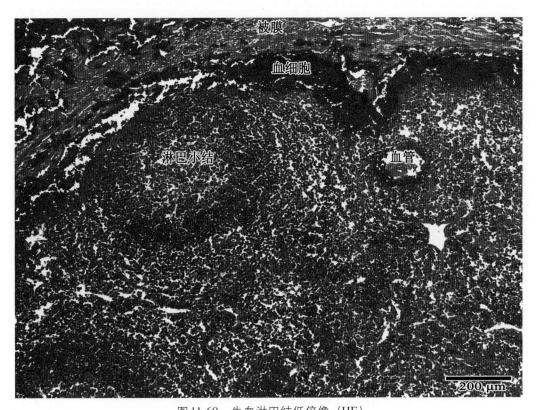

图11-60 牛血淋巴结低倍像（HE）

Fig.11-60 low magnification of cattle haemolymph node（HE）

被膜—capsule 血细胞—blood cell 淋巴小结—lymphoid nodule 血管—blood vessel

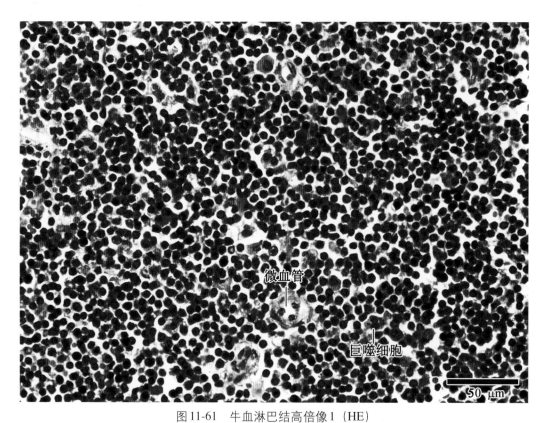

图11-61　牛血淋巴结高倍像1（HE）

Fig.11-61　high magnification of cattle haemolymph node 1 （HE）

微血管—capillaries　巨噬细胞—macrophage

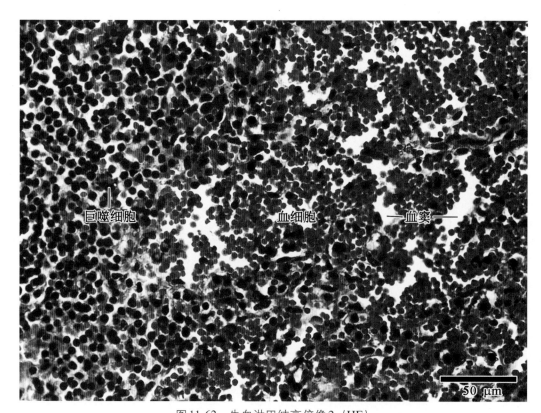

图11-62　牛血淋巴结高倍像2（HE）

Fig.11-62　high magnification of cattle haemolymph node 2 （HE）

巨噬细胞—macrophage　血细胞—blood cell　血窦—blood sinusoid

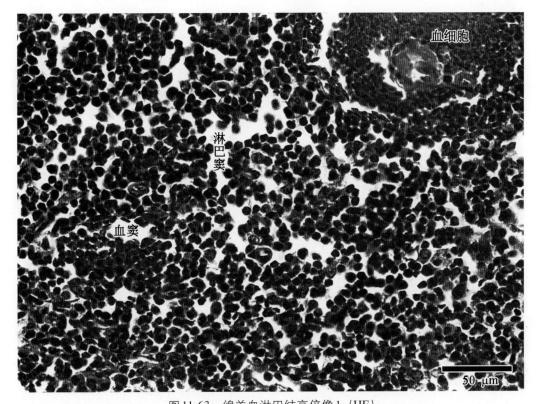

图11-63 绵羊血淋巴结高倍像1（HE）
Fig.11-63 high magnification of sheep haemolymph node 1 （HE）
血窦—blood sinusoid 淋巴窦—lymph sinus 血细胞—blood cell

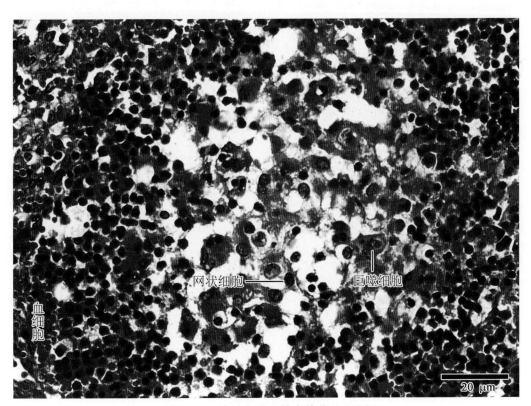

图11-64 绵羊血淋巴结高倍像2（HE）
Fig.11-64 high magnification of sheep haemolymph node 2 （HE）
血细胞—blood cell 网状细胞—reticular cell 巨噬细胞—macrophage

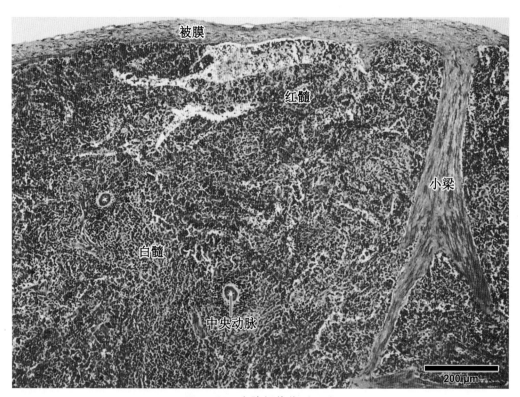

图 11-65　牛脾低倍像（HE）

Fig.11-65　low magnification of cattle spleen（HE）

被膜—capsule　红髓—red pulp　白髓—white pulp　中央动脉—central artery　小梁—trabecula

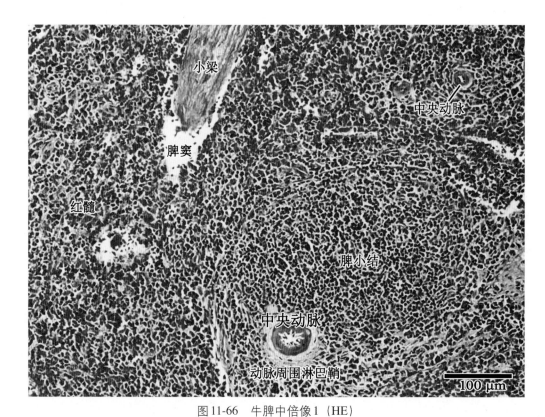

图 11-66　牛脾中倍像1（HE）

Fig.11-66　mid magnification of cattle spleen 1（HE）

小梁—trabecula　脾窦—splenic sinusoid　红髓—red pulp　中央动脉—central artery

脾小结—splenic corpuscle　动脉周围淋巴鞘—periarterial lymphatic sheath

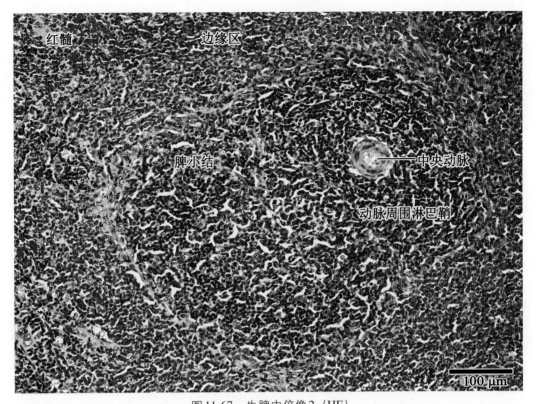

图11-67 牛脾中倍像2（HE）

Fig.11-67 mid magnification of cattle spleen 2（HE）

红髓—red pulp 边缘区—marginal zone 脾小结—splenic corpuscle 中央动脉—central artery

动脉周围淋巴鞘—periarterial lymphatic sheath

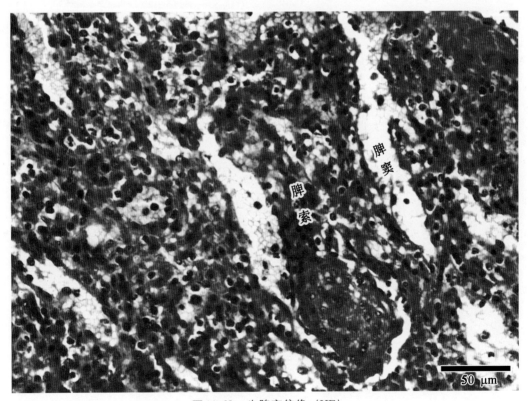

图11-68 牛脾高倍像（HE）

Fig.11-68 high magnification of cattle spleen（HE）

脾索—splenic cord 脾窦—splenic sinusoid

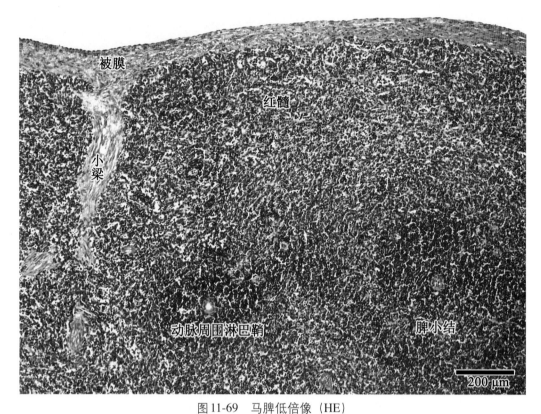

图 11-69　马脾低倍像（HE）

Fig.11-69　low magnification of horse spleen（HE）

被膜—capsule　小梁—trabecula　红髓—red pulp　动脉周围淋巴鞘—periarterial lymphatic sheath

脾小结—splenic corpuscle

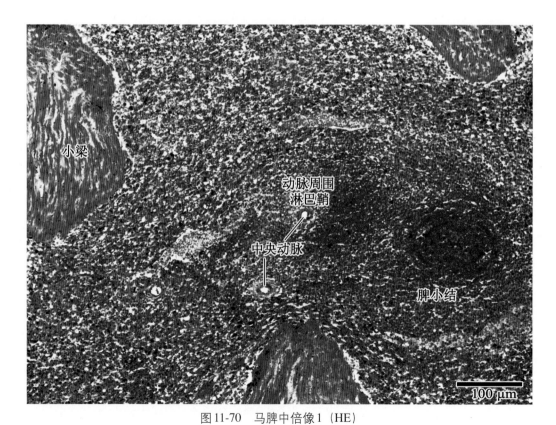

图 11-70　马脾中倍像 1（HE）

Fig.11-70　mid magnification of horse spleen 1（HE）

小梁—trabecula　中央动脉—central artery　动脉周围淋巴鞘—periarterial lymphatic sheath　脾小结—splenic corpuscle

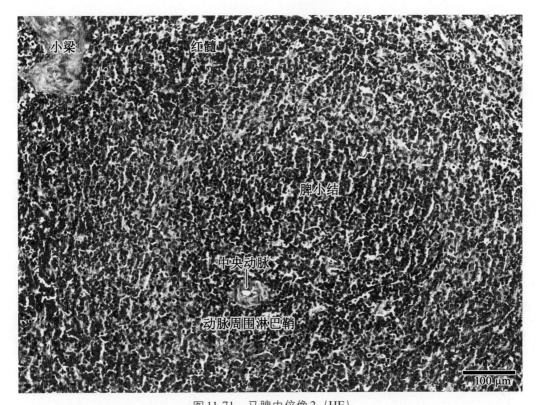

图 11-71 马脾中倍像 2（HE）

Fig.11-71　mid magnification of horse spleen 2（HE）

小梁—trabecula　红髓—red pulp　脾小结—splenic corpuscle　中央动脉—central artery

动脉周围淋巴鞘—periarterial lymphatic sheath

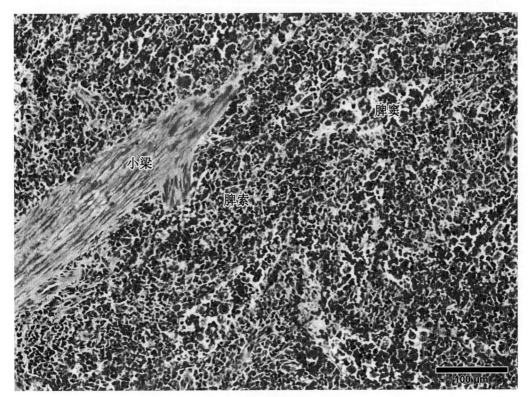

图 11-72 马脾中倍像 3（HE）

Fig.11-72　mid magnification of horse spleen 3（HE）

小梁—trabecula　脾索—splenic cord　脾窦—splenic sinusoid

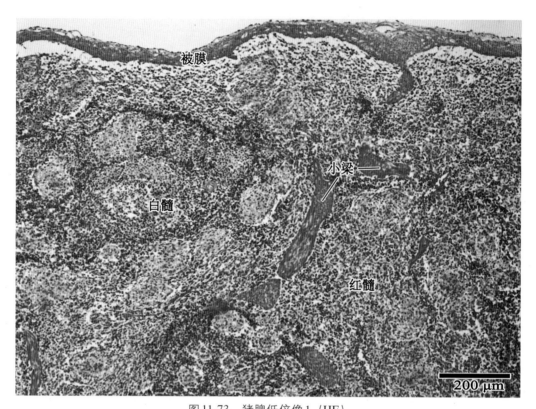

图 11-73　猪脾低倍像1（HE）

Fig.11-73　low magnification of pig spleen 1（HE）

被膜—capsule　白髓—white pulp　小梁—trabecula　红髓—red pulp

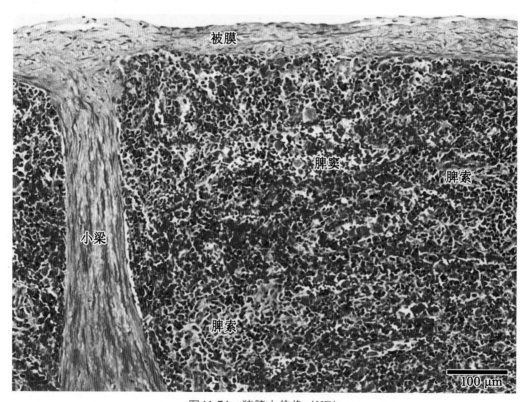

图 11-74　猪脾中倍像（HE）

Fig.11-74　mid magnification of pig spleen（HE）

被膜—capsule　小梁—trabecula　脾窦—splenic sinusoid　脾索—splenic cord

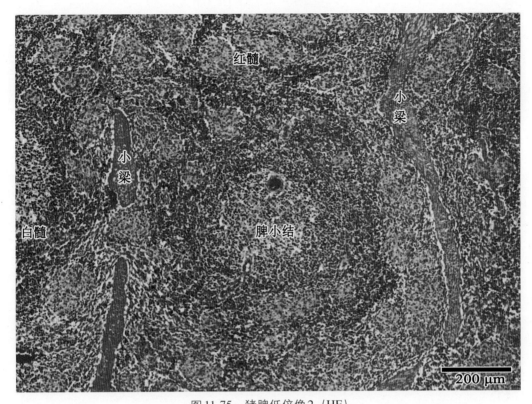

图11-75 猪脾低倍像2（HE）

Fig.11-75 low magnification of pig spleen 2（HE）

白髓—white pulp 小梁—trabecula 红髓—red pulp 脾小结—splenic corpuscle

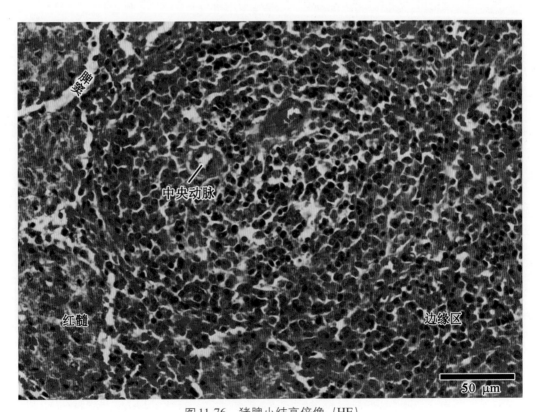

图11-76 猪脾小结高倍像（HE）

Fig.11-76 high magnification of pig splenic corpuscle（HE）

脾窦—splenic sinusoid 中央动脉—central artery 红髓—red pulp 边缘区—marginal zone

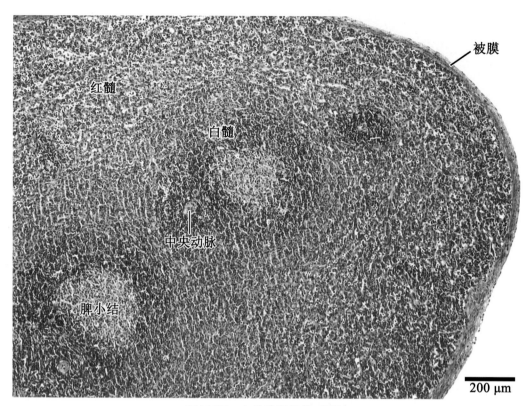

图 11-77　山羊脾低倍像（HE）

Fig.11-77　low magnification of goat spleen（HE）

被膜—capsule　红髓—red pulp　白髓—white pulp　中央动脉—central artery　脾小结—splenic corpuscle

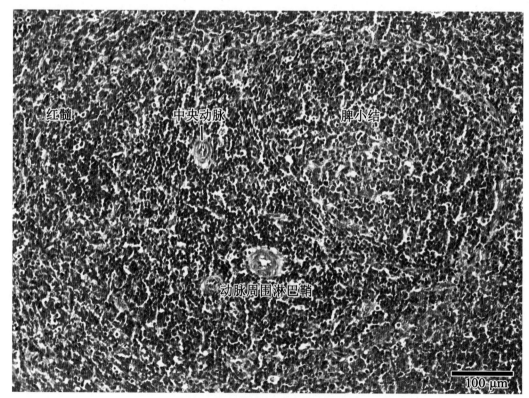

图 11-78　山羊脾白髓中倍像（HE）

Fig.11-78　mid magnification of goat spleen white pulp（HE）

红髓—red pulp　中央动脉—central artery　脾小结—splenic corpuscle　动脉周围淋巴鞘—periarterial lymphatic sheath

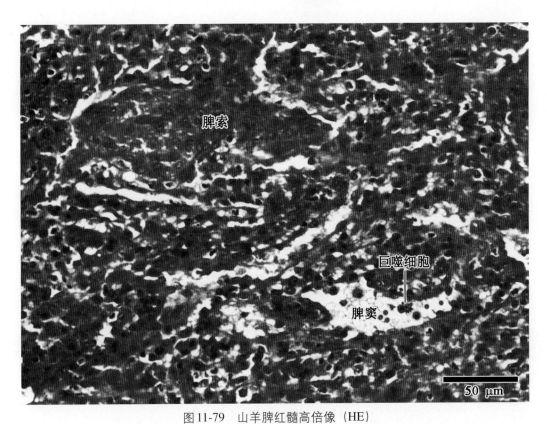

图11-79 山羊脾红髓高倍像（HE）

Fig.11-79 high magnification of goat spleen red pulp（HE）

脾索—splenic cord 脾窦—splenic sinusoid 巨噬细胞—macrophage

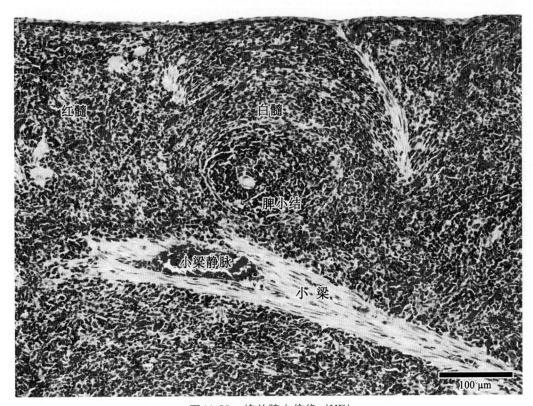

图11-80 绵羊脾中倍像（HE）

Fig.11-80 mid magnification of sheep spleen（HE）

红髓—red pulp 白髓—white pulp 脾小结—splenic corpuscle 小梁静脉—trabecular vein 小梁—trabecula

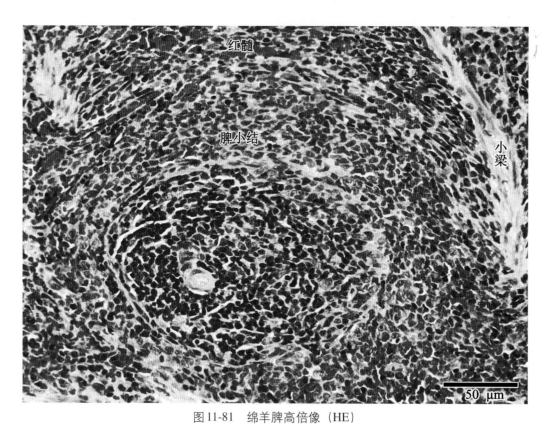

图11-81 绵羊脾高倍像（HE）

Fig.11-81 high magnification of sheep spleen（HE）

红髓—red pulp 脾小结—splenic corpuscle 小梁—trabecula

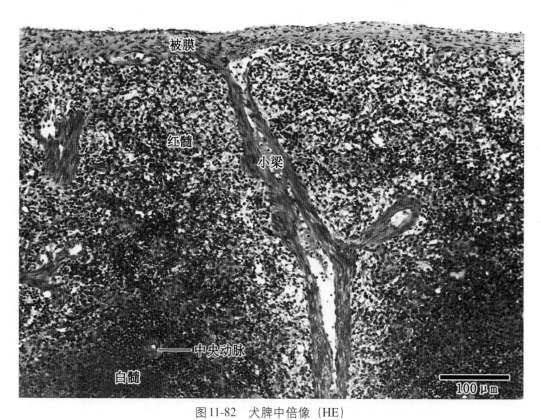

图11-82 犬脾中倍像（HE）

Fig.11-82 mid magnification of dog spleen（HE）

被膜—capsule 红髓—red pulp 小梁—trabecula 中央动脉—central artery 白髓—white pulp

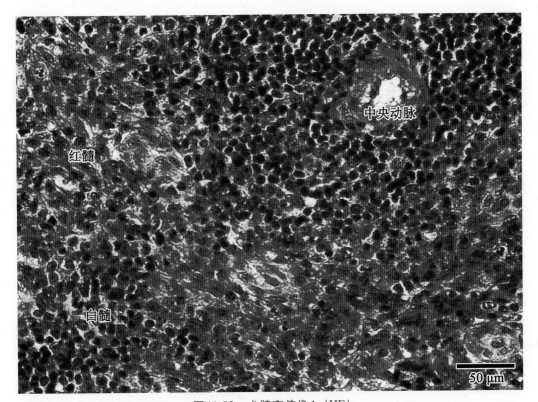

图 11-83　犬脾高倍像 1（HE）

Fig.11-83　high magnification of dog spleen 1（HE）

红髓—red pulp　白髓—white pulp　中央动脉—central artery

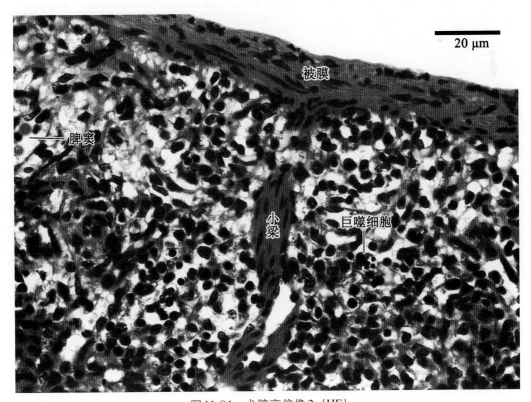

图 11-84　犬脾高倍像 2（HE）

Fig.11-84　high magnification of dog spleen 2（HE）

脾窦—splenic sinusoid　被膜—capsule　小梁—trabecula　巨噬细胞—macrophage

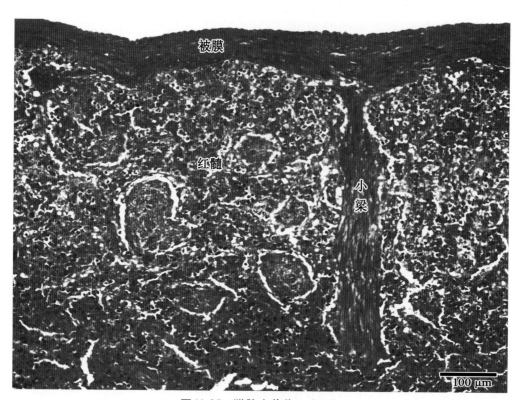

图 11-85　猫脾中倍像 1（HE）

Fig.11-85　mid magnification of cat spleen 1（HE）

被膜—capsule　红髓—red pulp　小梁—trabecula

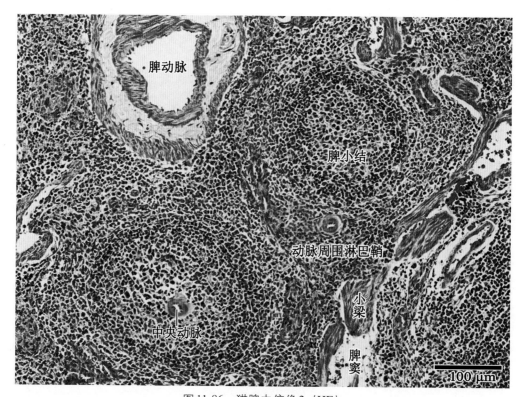

图 11-86　猫脾中倍像 2（HE）

Fig.11-86　mid magnification of cat spleen 2（HE）

脾动脉—splenic artery　中央动脉—central artery　脾小结—splenic corpuscle

动脉周围淋巴鞘—periarterial lymphatic sheath　小梁—trabecula　脾窦—splenic sinusoid

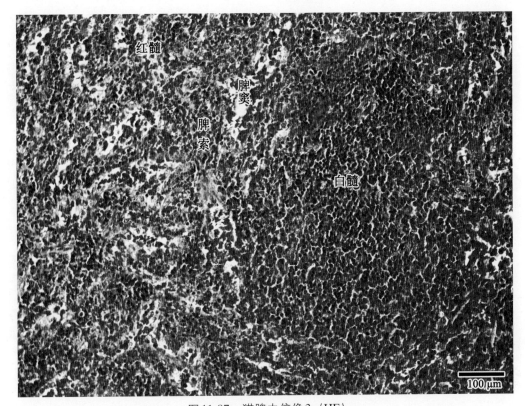

图11-87 猫脾中倍像3（HE）

Fig.11-87 mid magnification of cat spleen 3（HE）

红髓—red pulp 脾索—splenic cord 脾窦—splenic sinusoid 白髓—white pulp

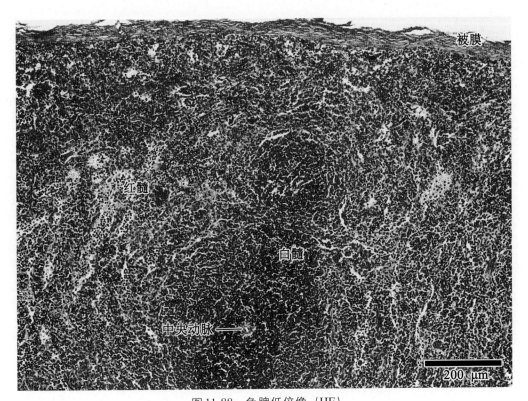

图11-88 兔脾低倍像（HE）

Fig.11-88 low magnification of rabbit spleen（HE）

被膜—capsule 红髓—red pulp 白髓—white pulp 中央动脉—central artery

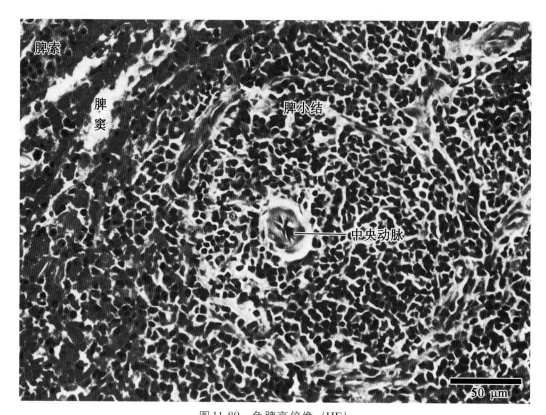

图 11-89 兔脾高倍像（HE）

Fig.11-89　high magnification of rabbit spleen（HE）

脾索—splenic cord　脾窦—splenic sinusoid　脾小结—splenic corpuscle　中央动脉—central artery

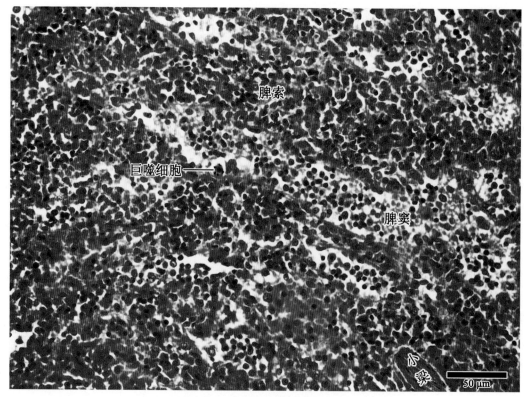

图 11-90 兔脾红髓高倍像（HE）

Fig.11-90　high magnification of rabbit spleen red pulp（HE）

脾索—splenic cord　巨噬细胞—macrophage　脾窦—splenic sinusoid　小梁—trabecula

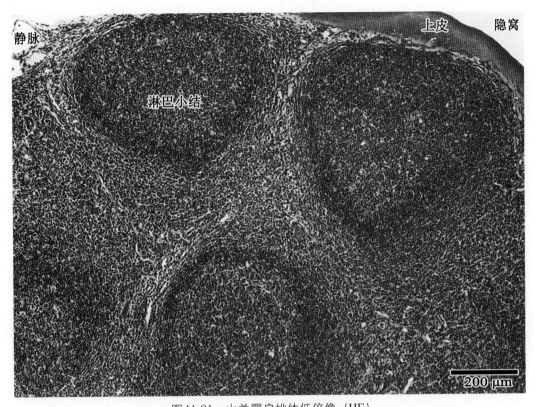

图11-91 山羊腭扁桃体低倍像（HE）

Fig.11-91 low magnification of goat faucial tonsil（HE）

静脉—vein 淋巴小结—lymphoid nodule 上皮—epithelium 隐窝—crypt

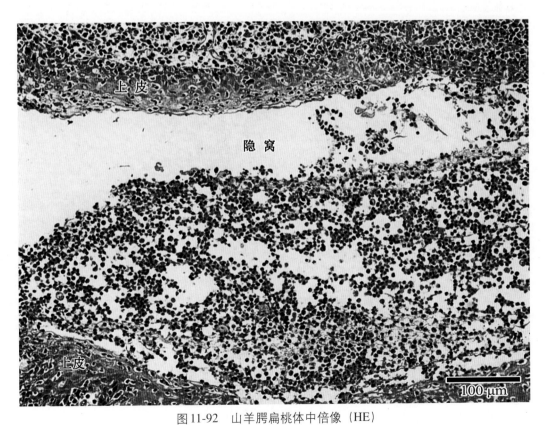

图11-92 山羊腭扁桃体中倍像（HE）

Fig.11-92 mid magnification of goat faucial tonsil（HE）

上皮—epithelium 隐窝—crypt

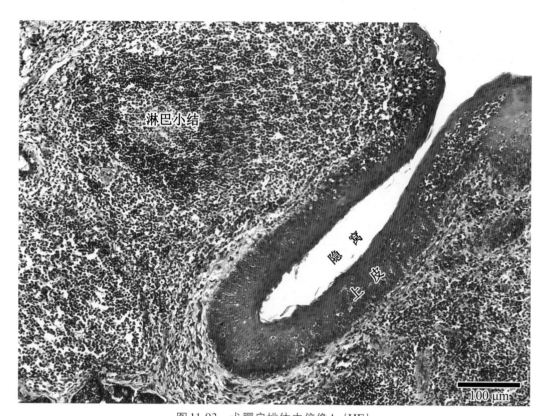

图 11-93 犬腭扁桃体中倍像1（HE）

Fig.11-93 mid magnification of dog faucial tonsil 1（HE）

淋巴小结—lymphoid nodule 隐窝—crypt 上皮—epithelium

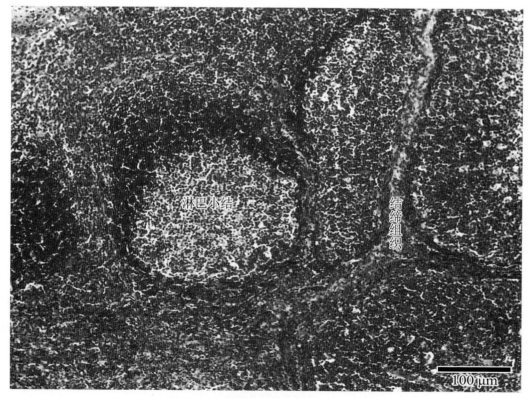

图 11-94 犬腭扁桃体中倍像2（HE）

Fig.11-94 mid magnification of dog faucial tonsil 2（HE）

淋巴小结—lymphoid nodule 结缔组织—connective tissue

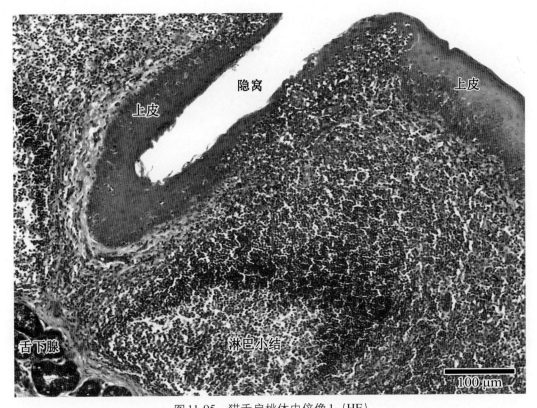

图11-95 猫舌扁桃体中倍像1（HE）

Fig.11-95 mid magnification of cat lingual tonsil 1 （HE）

上皮—epithelium 隐窝—crypt 舌下腺—sublingual gland 淋巴小结—lymphoid nodule

图11-96 猫舌扁桃体中倍像2（HE）

Fig.11-96 mid magnification of cat lingual tonsil 2 （HE）

上皮—epithelium 淋巴小结—lymphoid nodule

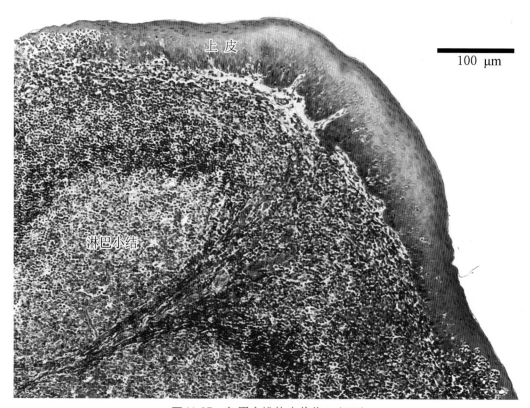

图11-97　兔腭扁桃体中倍像1（HE）

Fig.11-97　mid magnification of rabbit palatine tonsil 1 （HE）

上皮—epithelium　淋巴小结—lymphoid nodule

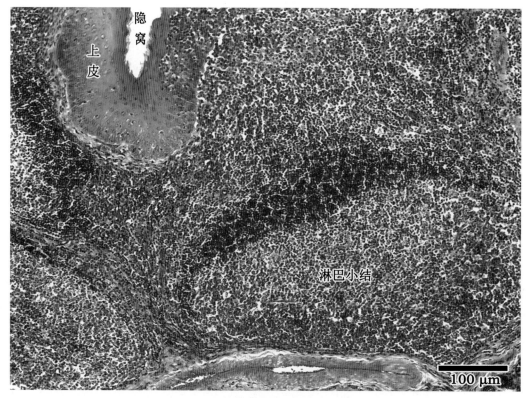

图11-98　兔腭扁桃体中倍像2（HE）

Fig.11-98　mid magnification of rabbit palatine tonsil 2 （HE）

上皮—epithelium　隐窝—crypt　淋巴小结—lymphoid nodule

第十二章
内分泌系统
Endocrine System

> **Outline**
>
> Endocrine system consists of endocrine glands, endocrine tissues and individual endocrine cells that exist in certain organs. Endocrine glands are ductless glands. Individual endocrine cells are distributed in cords, clusters, or follicles surrounded by blood and lymph capillaries. Each kind of endocrine gland can synthesize and secrete one or more types of hormones. Hormones are released and defused into the blood or lymph circulation. They act on target organs and target cells that have specific receptors, and influence their structure and function, regulate their growth, development, breeding, and metabolism of the organism, and maintain the stability of the internal environment. The emphases of this chapter are endocrine glands including hypophysis, adrenal gland and thyroid gland.
>
> Hypophysis is composed of adenohypophysis and neurohypophysis. Adenohypophysis has chromophilic cells (including acidophilic cells and basophilic cells) and chromophobic cells. Chromophilic cells secrete growth hormone, galactin, thyrotropin, gonadotropin and corticotrophin. Neurohypophysis consists of glial cells and unmyelinated nerve fiber, which can transport antidiuretic hormone and oxytocin secreted by supraoptic nucleus and paraventricular nucleus cells.
>
> Adrenal gland is composed of cortex and medulla. In light of morphology and arrangement of the cells, adrenal cortex can be divided into zona multiformis, zona fasciculate and zona reticularis, which can secrete mineralocorticoid, glucocorticoid, androgen and a little amount of estrogen, respectively. Adrenal medulla is composed of medullary cord and the sinusoid and central vein. The cytoplasm of the adrenal medulla cells with abundant secretory granules, which contain adrenalin (epinephrine) and noradrenalin (norepinephrine).
>
> Thyroid gland consists of numerous thyroid follicles. The follicles are lined by a simple cuboidal epithelium and their central cavity contains a gelatinous substance called colloid. The function of the thyroid gland follicles is the secretion of thyroxine. A small population of parafollicular cells is alongside the follicular epithelial cells. Parafollicular cells synthesize and secrete calcitonin.

内分泌系统（endocrine system）是机体重要的调节系统，由独立的内分泌腺、散在的内分泌细胞群以及具有内分泌作用的细胞组成。独立的内分泌腺包括垂体、肾上腺、甲状腺、甲状旁腺、松果体；内分泌细胞群分布很广，如胰岛、肾小球旁器、卵泡、黄体、胎盘、睾丸间质细胞、神经内分泌细胞和消化管的内分泌细胞；兼有内分泌功能的细胞如一部分心肌细胞、肥大细胞等。内分泌系统分泌激素，进入血液循环，并与神经系统共同调节机体的生长、发育、繁殖、代谢和维持内环境的稳定。

一、内分泌系统的发生

（一）垂体的发生

垂体实质由腺垂体和神经垂体两部分组成。腺垂体来自原始口腔，神经垂体来自神经管。胚胎口凹顶的外胚层形成拉克特囊，稍后，间脑的底部神经外胚层形成神经垂体芽。拉克特囊和神经垂体芽逐渐增长并相互接近。一段时间后，拉克特囊的根部退化消失，其远端长大并与神经垂体芽相贴。后来，拉克特囊的前壁形成垂体前叶，在垂体前叶上方形成垂体的结节部；囊的后壁形成垂体的中间部，囊腔只留一裂隙。神经垂体芽的远端膨大，形成神经垂体；其起始部变细，形成漏斗柄。

（二）肾上腺的发生

胚胎生殖腺嵴和肠背系膜之间的腹膜上皮增厚，并向下方伸展形成索状结构，上皮索之间有丰富的血管，形成了原发性皮质。之后腹膜上皮产生嗜碱性细胞，形成了继发性皮质。出生后，原发性皮质退化，继发性皮质扩展并出现分带。肾上腺髓质的发生比皮质晚，交感神经节迁移出来的神经嵴细胞移向皮质中央形成肾上腺髓质，髓质细胞在中央形成嗜铬细胞。

（三）甲状腺的发生

甲状腺起源于内胚层，是胚胎内分泌腺中发生较早的腺体。原始咽底正中处的内胚层细胞增殖，向腹侧形成甲状腺原基。末端分为两个芽突，其根部借甲状舌管与原始咽底壁相连。甲状舌管退化。原基分化，左右芽突细胞增生，芽突演化成两个侧叶，中间形成峡部。有两对咽囊一起形成后鳃体，部分后鳃体迁至甲状腺内，分化为甲状腺滤泡旁细胞。

（四）甲状旁腺的发生

有两对咽囊的背侧上皮细胞增生，形成细胞团，其中一对最初与胸腺原基相连，后脱离移至甲状腺下方，形成了下端一对甲状旁腺；另外一对附着在甲状腺的上端。

（五）松果体的发生

间脑顶部向背侧突出，形成一囊状突起，为松果体原基。囊壁细胞增生，囊腔消失，形成一实质性松果样器官，即松果体。其中的松果体细胞和神经胶质细胞均由神经上皮分化而来。

二、内分泌系统的组织学结构概述

本章重点展示独立的内分泌腺，其他内分泌组织和细胞将在相应的章节展示。

（一）垂体（hypophysis）

腺垂体（adenohypophysis）分为远侧部（distal part）、结节部（tuberal part）、中间部（intermediate part）。远侧部实质由不规则的细胞索组成，这些细胞索被血窦和结缔组织分隔。细胞有两类：嫌色细胞和嗜色细胞。嫌色细胞的胞质少，着色浅；而嗜色细胞的胞质丰富，着色深，又分为嗜酸性细胞和嗜碱性细胞。中间部位于远侧部和神经部之间，以嗜碱性细胞为主。在马，这些区域紧密排列；但在其他家畜，中间部和远侧部被一小的垂体裂分隔开。结节部主要位于漏斗柄周围，由小的嗜碱性细胞索和滤泡组成。

神经垂体（neurohypophysis）包括正中隆起（median eminence）、漏斗柄（infundibular stalk）和神经部（nervous part）。神经垂体含有大量无髓神经纤维，其细胞体位于下丘脑的视上核和室旁核。其轴突形成下丘脑垂体束，经过狭窄的漏斗柄到达神经部。这些细胞的分泌物沿轴突运输，聚集在神经纤维终末区，即赫林体。大量的胶质细胞散在于神经纤维之间，胞核圆形或卵圆形，胞突长。漏斗腔与第三脑室相延续，内衬室管膜细胞；在猫和猪延伸入神经部，而犬和马延伸很短，在反刍动物，漏斗腔延伸不超过漏斗柄。

（二）肾上腺（adrenal gland）

肾上腺表面有致密结缔组织构成的被膜，被膜中含有散在的平滑肌束和未分化的皮质细胞。血管和神经与结缔组织深入实质，将实质分为外周的皮质和中央的髓质。皮质从外至内由多形带、束状带和网状带组成，三带之间无截然界限。在马和猪，肾上腺多形带的细胞为柱状，排列成弓形。在反刍动物，多形带含有多边形的细胞，形成不规则的细胞簇或索。中间带比较小，有的缺如。束状带的细胞质含有大量的脂肪空泡，常呈泡沫状。网状带为皮质最内层，互相吻合的细胞索排列不规则，细胞索周围为窦状隙。肾上腺髓质主要由柱状或多边形的嗜铬细胞组成，细胞形成簇和吻合的细胞索，并被窦状隙隔开。在家畜，髓质常分为外带和内带。

（三）甲状腺（thyroid gland）

甲状腺的每一个腺叶外包结缔组织，并分为小叶。猪和牛的叶间结缔组织丰富。每一小叶由无数大小不同的滤泡组成，滤泡内常常充有胶质。滤泡细胞高低不等，其大小取决于滤泡的活动状态。在静止期，其外观呈鳞片状或矮立方形；在活动期，外观呈立方形或柱状。活动期滤泡细胞顶面附近的胶质周围有空泡；静止期的滤泡，胶质的周围表面较光滑，不出现空泡。滤泡旁细胞存在于甲状腺滤泡细胞之间，或滤泡之间。滤泡旁细胞大，胞质颜色浅，胞核也大，淡染。滤泡旁细胞常单个出现，有时也成群出现。犬的滤泡旁细胞特别丰富。

（四）甲状旁腺（parathyroid gland）

甲状旁腺的实质主要由主细胞簇和索组成。有两种不同功能状态的主细胞，亮主细胞不活动，胞核大而淡染，细胞质嗜碱性；暗主细胞较小，活跃，胞核小而暗，细胞质强嗜碱性。羊的亮细胞位于暗细胞周围；马和牛的甲状旁腺有少量嗜酸性细胞，特别是老龄动物。嗜酸性细胞的胞体大，胞质嗜酸性，核固缩。

（五）松果体（pineal body）

松果体呈卵圆形，表面包有结缔组织被膜，被膜伴随血管深入实质，将实质分成许多小叶，小叶内主要是松果体细胞、神经胶质细胞、无髓神经纤维等。

三、内分泌系统图谱

1. **垂体** 图 12-1～图 12-19。
2. **肾上腺** 图 12-20～图 12-46。
3. **甲状腺** 图 12-47～图 12-63。
4. **甲状旁腺** 图 12-64～图 12-78。
5. **松果体** 图 12-79～图 12-94。

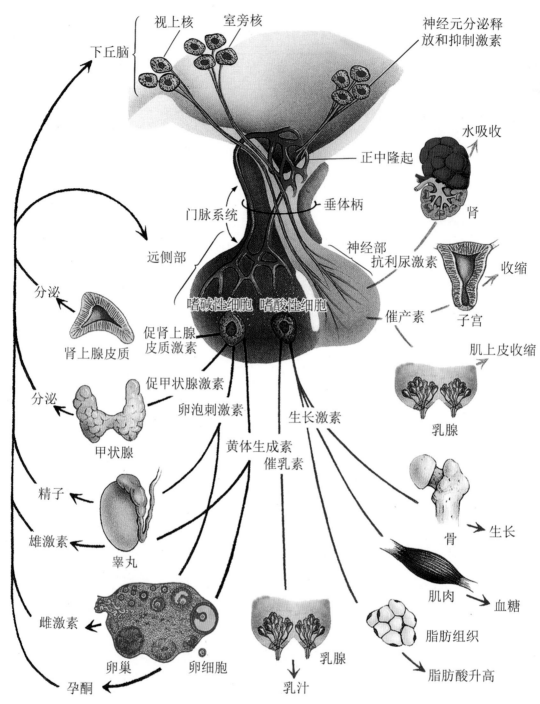

图12-1 内分泌系统模式图

Fig.12-1 ideograph of endocrine system

下丘脑—thalamus 视上核—supraoptic nucleus 室旁核—paraventricular nucleus 神经元分泌释放和抑制激素—neurons secrete releasing and inhibiting hormones 正中隆起—median eminence 门脉系统—portal system 垂体柄—pituitary stalk 远侧部—distal part 神经部—nervous part 嗜碱性细胞—basophil cell 嗜酸性细胞—acidophil cell 抗利尿激素—vasopressin 肾—kidney 水吸收—water absorption 催产素—oxytocin 子宫—uterus 收缩—contract 乳腺—breast 肌上皮收缩—myoepithelial contract 促肾上腺皮质激素—adrenocorticotropic hormone 肾上腺皮质—adrenal cortex 分泌—secretion 促甲状腺激素—thyroid stimulating hormone 甲状腺—thyroid gland 卵泡刺激素—follicle-stimulate hormone 黄体生成素—luteinizing hormone 睾丸—testis 雄激素—testosterone 精子—sperm 卵巢—ovary 卵细胞—egg 孕酮—progesterone 雌激素—estrogen 催乳素—prolactin 乳汁—milk 生长激素—growth hormone 骨—bone 生长—growth 肌肉—muscle 血糖—blood sugar 脂肪组织—fat tissue 脂肪酸升高—higher fatty acids

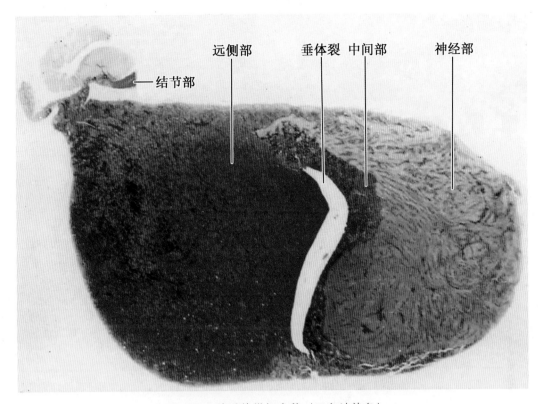

图 12-2 猪垂体纵切全貌（三色法染色）

Fig.12-2 longitudinal view of pig hypophysis（triple stain）

结节部—tuberal part　远侧部—distal part　垂体裂—pituitary crack　中间部—intermediate part　神经部—nervous part

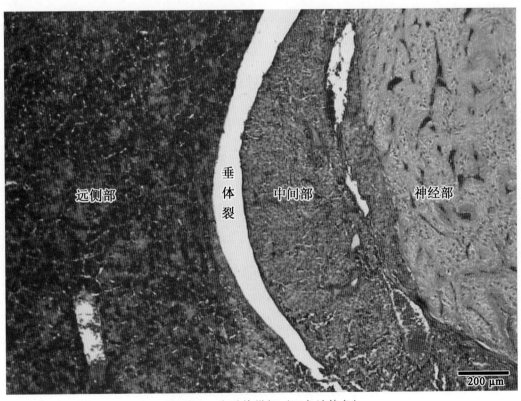

图 12-3 牛垂体纵切（三色法染色）

Fig.12-3 longitudinal view of cow hypophysis（triple stain）

远侧部—distal part　垂体裂—pituitary crack　中间部—intermediate part　神经部—nervous part

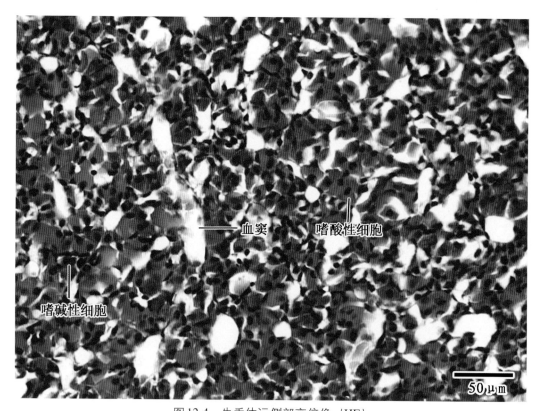

图12-4 牛垂体远侧部高倍像（HE）

Fig.12-4 high magnification of cow hypophysis distal part（HE）

嗜碱性细胞—basophil cell 血窦—blood sinusoid 嗜酸性细胞—acidophil cell

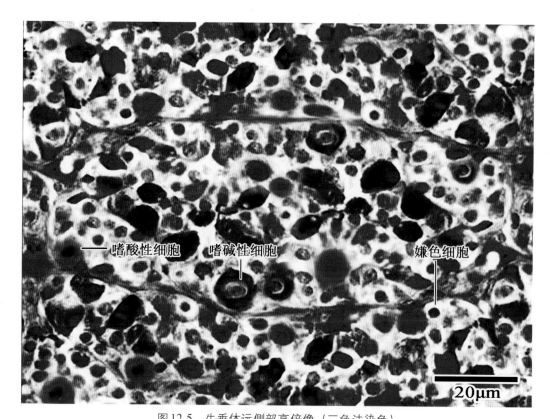

图12-5 牛垂体远侧部高倍像（三色法染色）

Fig.12-5 high magnification of cow hypophysis distal part（triple stain）

嗜酸性细胞—acidophil cell 嗜碱性细胞—basophil cell 嫌色细胞—chromophobe cell

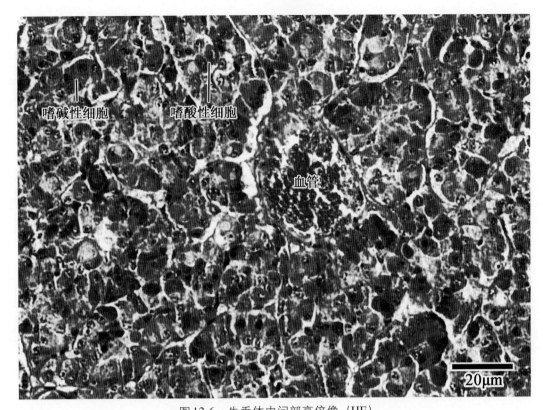

图12-6 牛垂体中间部高倍像（HE）

Fig.12-6 high magnification of cow hypophysis intermediate part（HE）

嗜碱性细胞—basophil cell 嗜酸性细胞—acidophil cell 血管—blood vessel

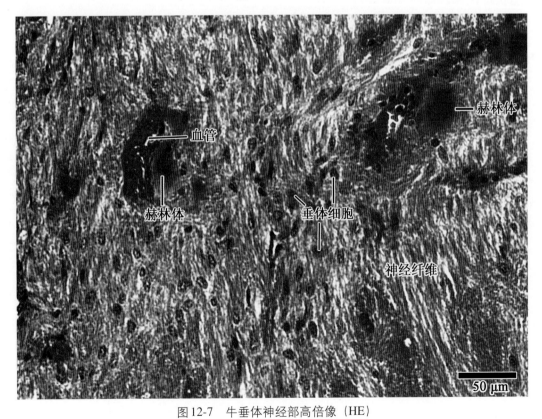

图12-7 牛垂体神经部高倍像（HE）

Fig.12-7 high magnification of cow hypophysis nervous part（HE）

血管—blood vessel 赫林体—Herring body 垂体细胞—pituicyte 神经纤维—nerve fiber

图12-8 马垂体纵切（HE）

Fig.12-8 longitudinal view of horse hypophysis （HE）

远侧部—distal part 中间部—intermediate part 神经部—nervous part 血窦—blood sinusoid

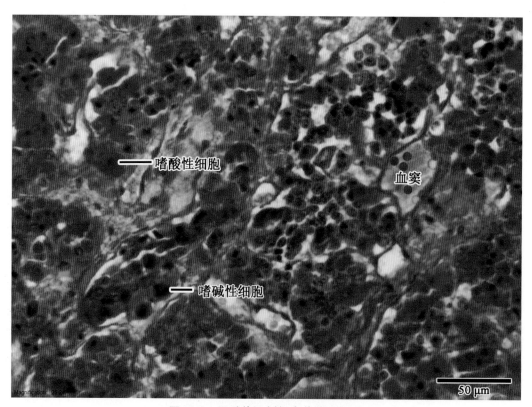

图12-9 马垂体远侧部高倍像（HE）

Fig.12-9 high magnification of horse hypophysis distal part （HE）

嗜酸性细胞—acidophil cell 嗜碱性细胞—basophil cell 血窦—blood sinusoid

第十二章 内分泌系统 Endocrine System

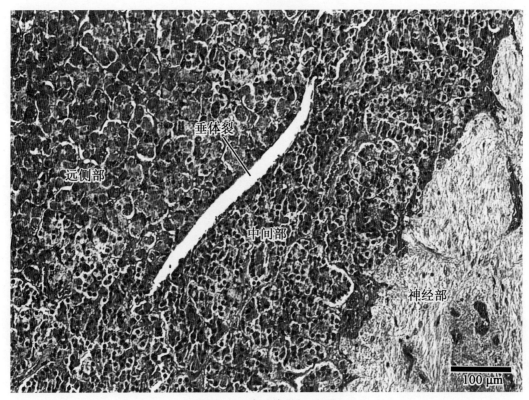

图12-10 猪垂体纵切（HE）

Fig.12-10 longitudinal view of pig hypophysis（HE）

远侧部—distal part　垂体裂—pituitary crack　中间部—intermediate part　神经部—nervous part

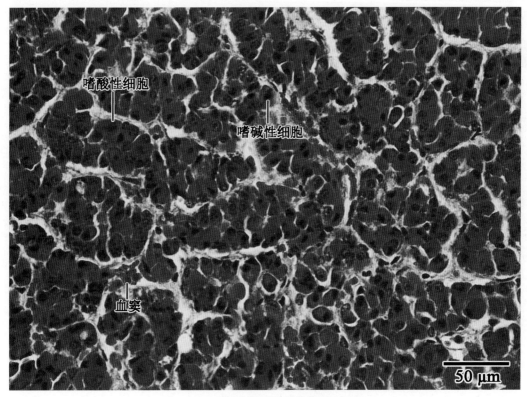

图12-11 猪垂体远侧部高倍像（HE）

Fig.12-11 high magnification of pig hypophysis distal part（HE）

嗜酸性细胞—acidophil cell　嗜碱性细胞—basophil cell　血窦—blood sinusoid

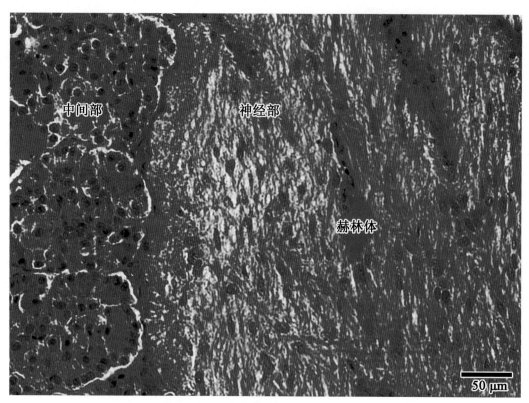

图12-12 猪垂体纵切高倍像（HE）

Fig.12-12　high magnification of pig hypophysis longitudinal view（HE）

中间部—intermediate part　神经部—nervous part　赫林体—Herring body

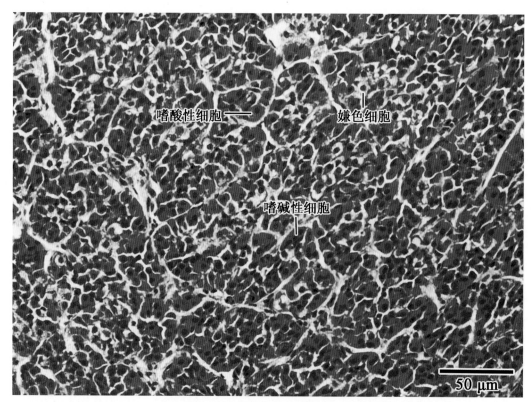

图12-13 绵羊垂体远侧部高倍像（HE）

Fig.12-13　high magnification of sheep hypophysis distal part（HE）

嗜酸性细胞—acidophil cell　嫌色细胞—chromophobe cell　嗜碱性细胞—basophil cell

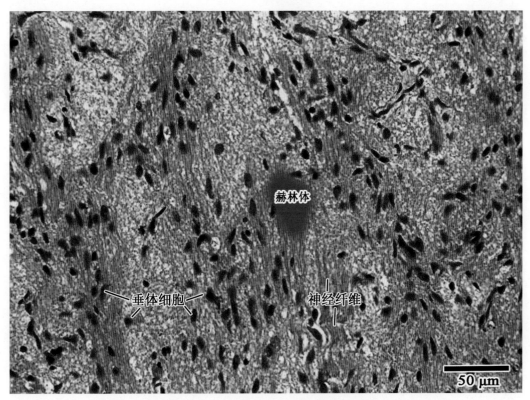

图 12-14 绵羊垂体神经部高倍像（HE）

Fig.12-14 high magnification of sheep hypophysis nervous part（HE）

赫林体—Herring body　垂体细胞—pituicyte　神经纤维—nerve fiber

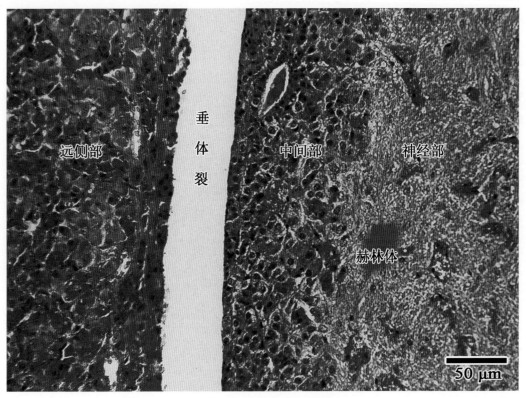

图 12-15 犬垂体纵切（HE）

Fig.12-15 longitudinal view of dog hypophysis（HE）

远侧部—distal part　垂体裂—pituitary crack　中间部—intermediate part　神经部—nervous part　赫林体—Herring body

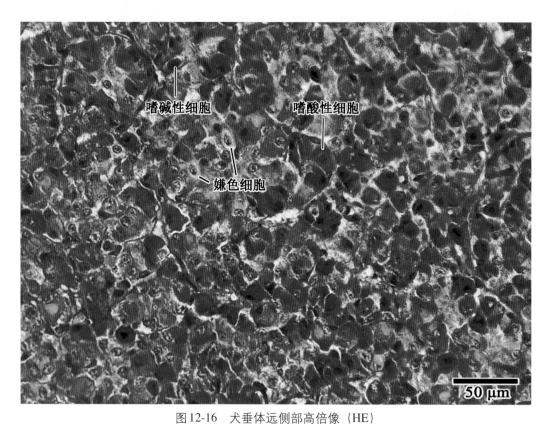

图 12-16　犬垂体远侧部高倍像（HE）

Fig.12-16　high magnification of dog hypophysis distal part（HE）

嗜碱性细胞—basophil cell　嗜酸性细胞—acidophil cell　嫌色细胞—chromophobe cell

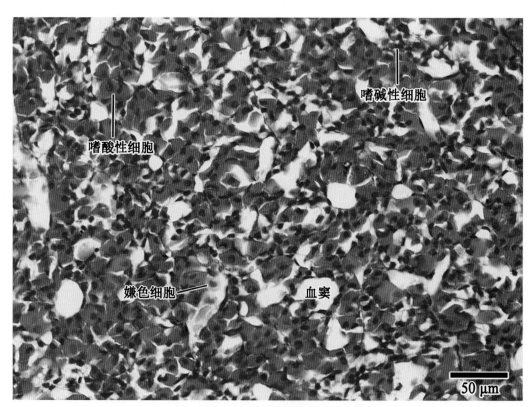

图 12-17　猫垂体远侧部高倍像（HE）

Fig.12-17　high magnification of cat hypophysis distal part（HE）

嗜酸性细胞—acidophil cell　嗜碱性细胞—basophil cell　嫌色细胞—chromophobe cell　血窦—blood sinusoid

第十二章 内分泌系统 Endocrine System

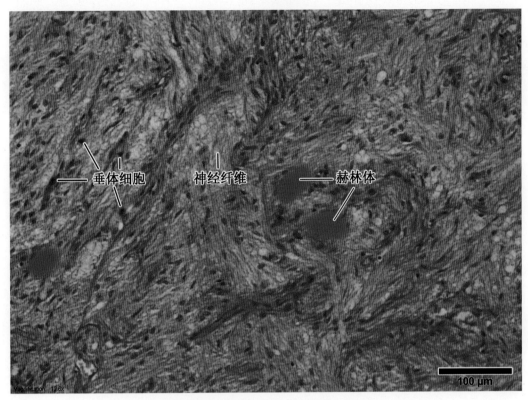

图12-18 猫垂体神经部高倍像（HE）

Fig.12-18 high magnification of cat hypophysis nervous part（HE）

垂体细胞—hypophysis cell 神经纤维—nerve fiber 赫林体—Herring body

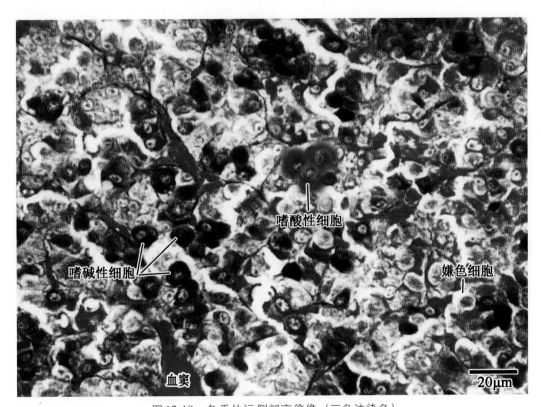

图12-19 兔垂体远侧部高倍像（三色法染色）

Fig.12-19 high magnification of rabbit hypophysis distal part（triple stain）

嗜碱性细胞—basophil cell 嗜酸性细胞—acidophil cell 嫌色细胞—chromophobe cell 血窦—blood sinusoid

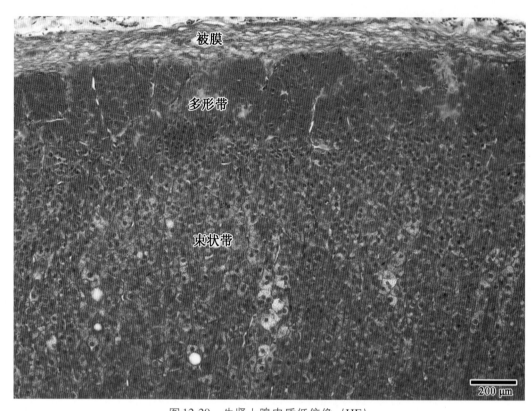

图 12-20　牛肾上腺皮质低倍像（HE）

Fig.12-20　low magnification of cortex in cattle adrenal gland（HE）

被膜—capsule　多形带—multiform zone　束状带—fasciculate zone

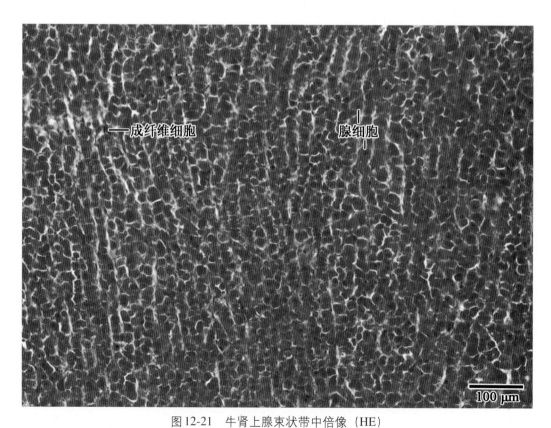

图 12-21　牛肾上腺束状带中倍像（HE）

Fig.12-21　mid magnification of fasciculate zone in cattle adrenal gland（HE）

成纤维细胞—fibroblast　腺细胞—glandular cell

第十二章 内分泌系统 Endocrine System

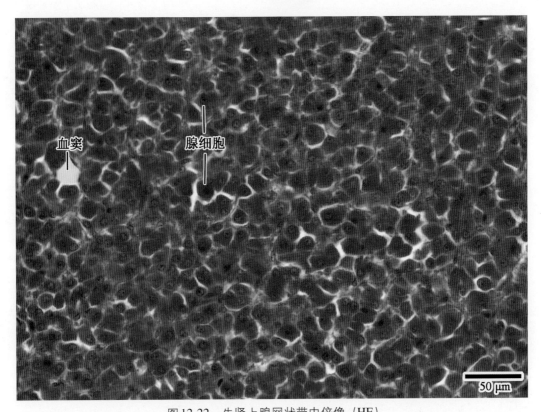

图 12-22 牛肾上腺网状带中倍像（HE）

Fig.12-22 mid magnification of reticular zone in cattle adrenal gland （HE）

血窦—blood sinusoid 腺细胞—glandular cell

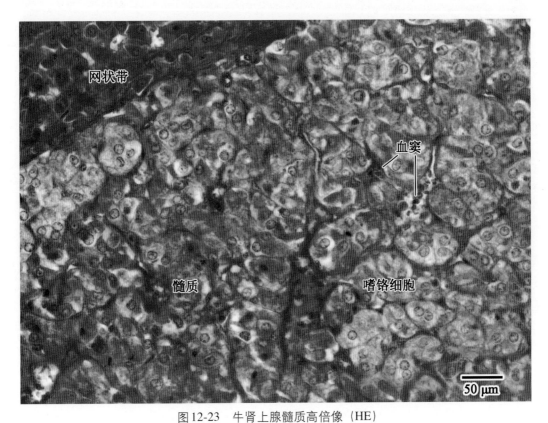

图 12-23 牛肾上腺髓质高倍像（HE）

Fig.12-23 high magnification of medulla in cattle adrenal gland （HE）

网状带—reticular zone 髓质—medulla 血窦—blood sinusoid 嗜铬细胞—chromaffin cell

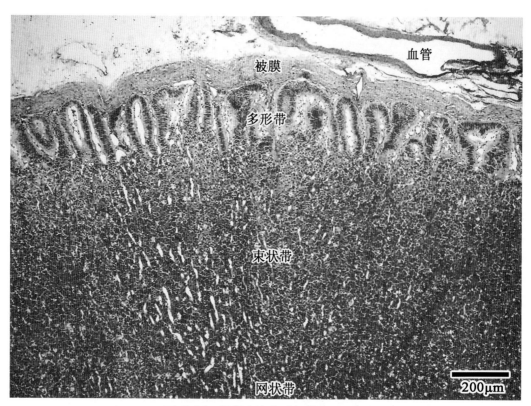

图12-24　马肾上腺皮质低倍像（HE）

Fig.12-24　low magnification of cortex in horse adrenal gland（HE）

被膜—capsule　多形带—multiform zone　束状带—fasciculate zone　网状带—reticular zone　血管—blood vessel

图12-25　马肾上腺束状带中倍像（HE）

Fig.12-25　mid magnification of fasciculate zone in horse adrenal gland（HE）

被膜—capsule　多形带—multiform zone　束状带—fasciculate zone

第十二章 内分泌系统 Endocrine System

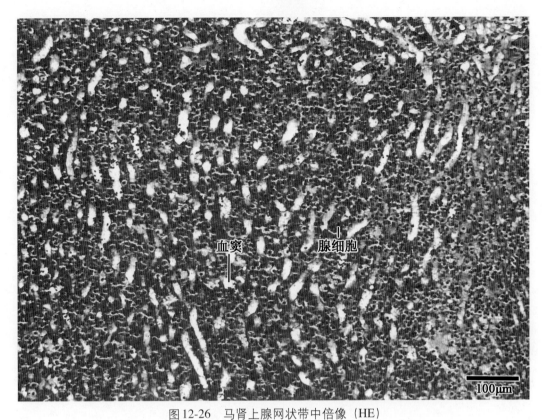

图 12-26 马肾上腺网状带中倍像（HE）
Fig.12-26 mid magnification of reticular zone in horse adrenal gland（HE）
血窦—blood sinusoid 腺细胞—glandular cell

图 12-27 马肾上腺髓质中倍像（HE）
Fig.12-27 mid magnification of medulla in horse adrenal gland（HE）
腺细胞—glandular cell 血窦—blood sinusoid

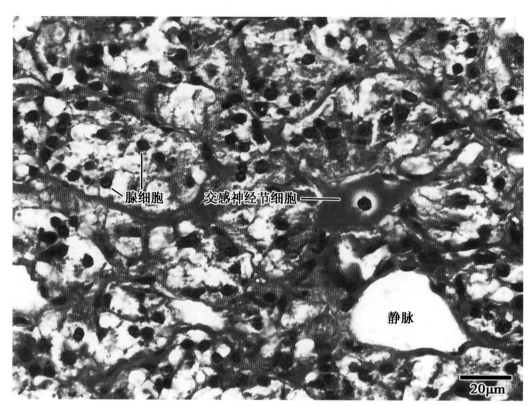

图12-28 马肾上腺髓质高倍像（HE）

Fig.12-28 high magnification of medulla in horse adrenal gland （HE）

腺细胞—glandular cell 交感神经节细胞—sympathetic ganglionic cell 静脉—vein

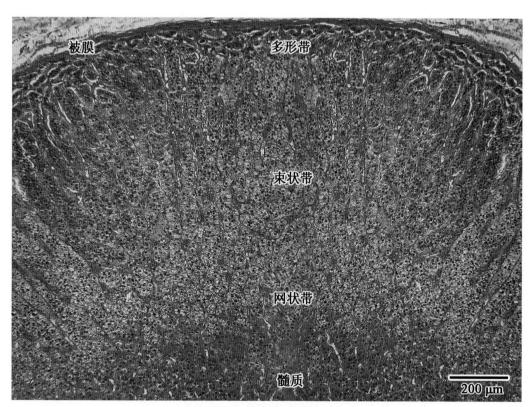

图12-29 猪肾上腺低倍像（HE）

Fig.12-29 low magnification of pig adrenal gland （HE）

被膜—capsule 多形带—multiform zone 束状带—fasciculate zone 网状带—reticular zone 髓质—medulla

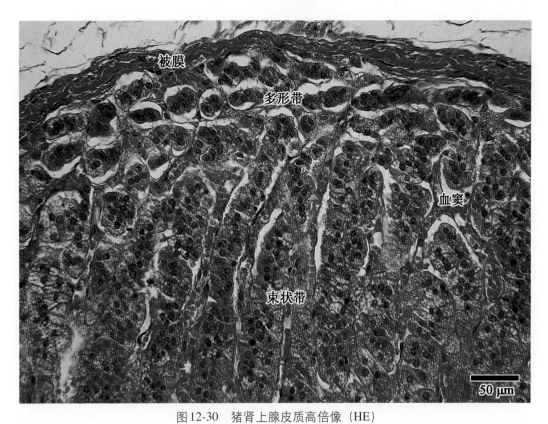

图 12-30 猪肾上腺皮质高倍像（HE）

Fig.12-30 high magnification of cortex in pig adrenal gland（HE）

被膜—capsule 多形带—multiform zone 束状带—fasciculate zone 血窦—blood sinusoid

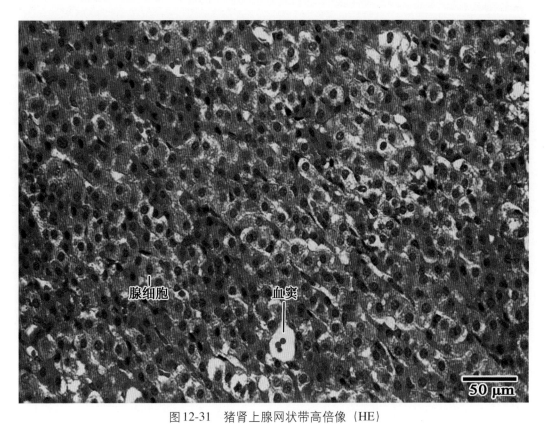

图 12-31 猪肾上腺网状带高倍像（HE）

Fig.12-31 high magnification of reticular zone in pig adrenal gland（HE）

腺细胞—glandular cell 血窦—blood sinusoid

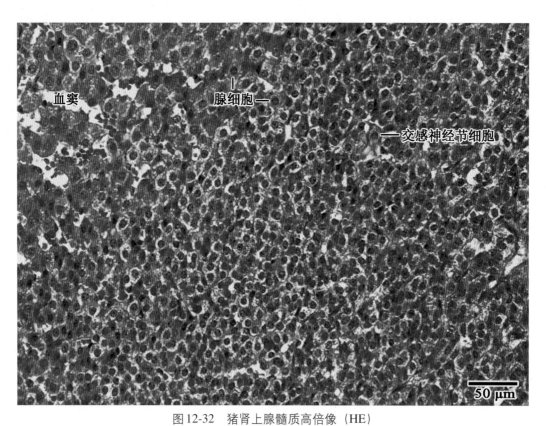

图12-32 猪肾上腺髓质高倍像（HE）

Fig.12-32　high magnification of medulla in pig adrenal gland（HE）

血窦—blood sinusoid　腺细胞—glandular cell　交感神经节细胞—sympathetic ganglionic cell

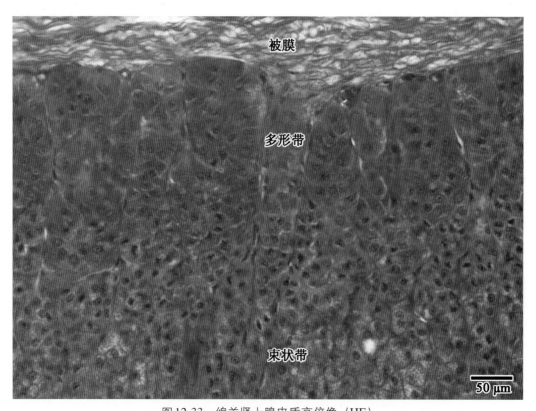

图12-33 绵羊肾上腺皮质高倍像（HE）

Fig.12-33　high magnification of cortex in sheep adrenal gland（HE）

被膜—capsule　多形带—multiform zone　束状带—fasciculate zone

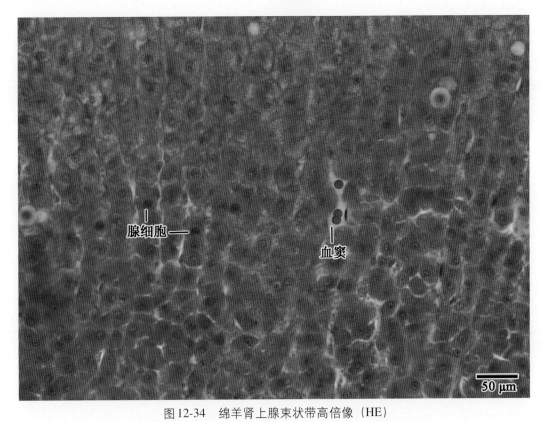

图12-34 绵羊肾上腺束状带高倍像（HE）

Fig.12-34 high magnification of fasciculate zone in sheep adrenal gland（HE）

腺细胞—glandular cell 血窦—blood sinusoid

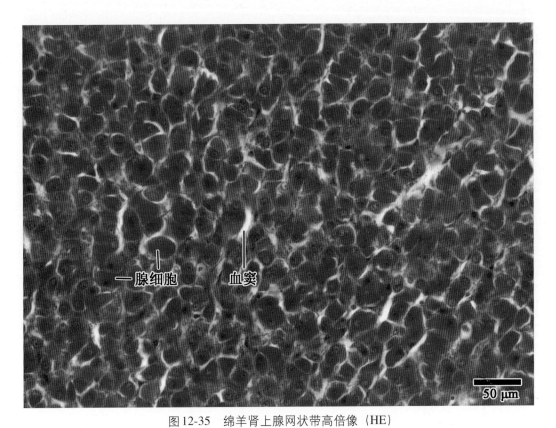

图12-35 绵羊肾上腺网状带高倍像（HE）

Fig.12-35 high magnification of reticular zone in sheep adrenal gland（HE）

腺细胞—glandular cell 血窦—blood sinusoid

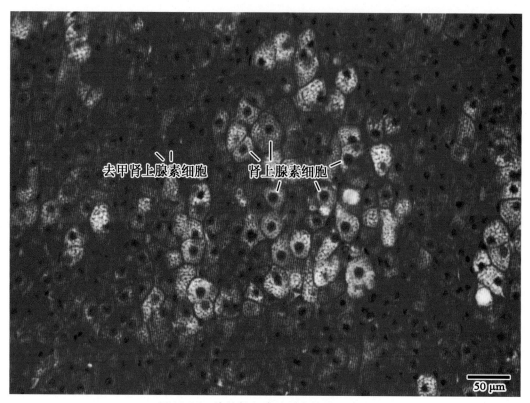

图 12-36 绵羊肾上腺髓质高倍像（HE）

Fig.12-36 high magnification of medulla in sheep adrenal gland（HE）

去甲肾上腺素细胞—noradrenergic cell 肾上腺素细胞—adrenergic cell

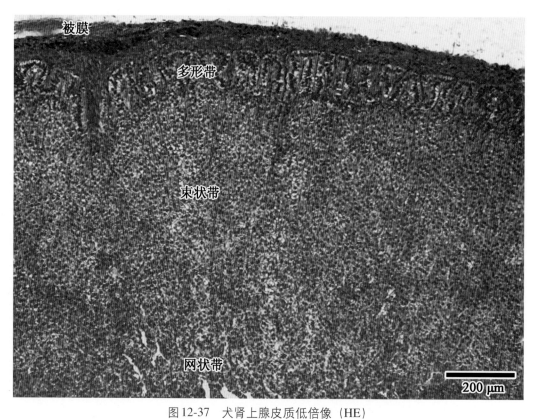

图 12-37 犬肾上腺皮质低倍像（HE）

Fig.12-37 low magnification of cortex in dog adrenal gland（HE）

被膜—capsule 多形带—multiform zone 束状带—fasciculate zone 网状带—reticular zone

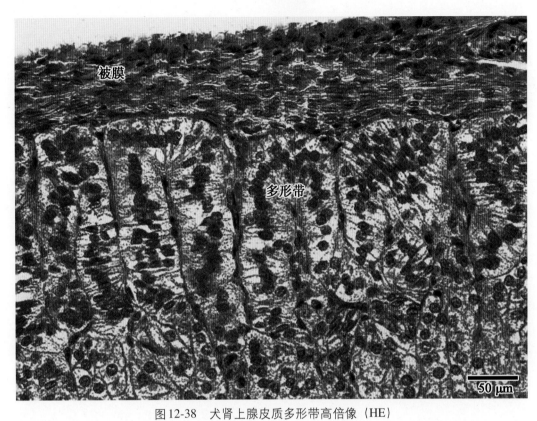

图12-38 犬肾上腺皮质多形带高倍像（HE）
Fig.12-38 high magnification of cortex multiform zone in dog adrenal gland（HE）
被膜—capsule 多形带—multiform zone

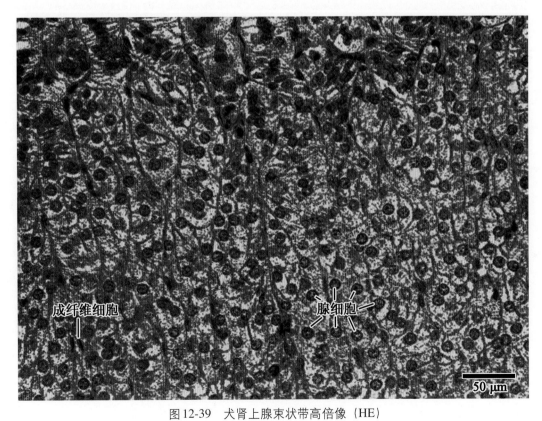

图12-39 犬肾上腺束状带高倍像（HE）
Fig.12-39 high magnification of fasciculate zone in dog adrenal gland（HE）
成纤维细胞—fibroblast 腺细胞—glandular cell

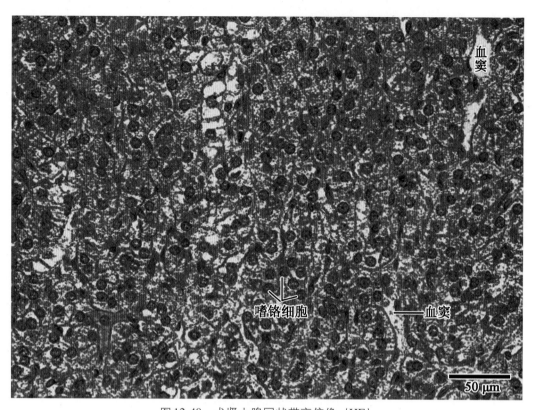

图 12-40 犬肾上腺网状带高倍像（HE）

Fig.12-40 high magnification of reticular zone in dog adrenal gland（HE）

嗜铬细胞—chromaffin cell 血窦—blood sinusoid

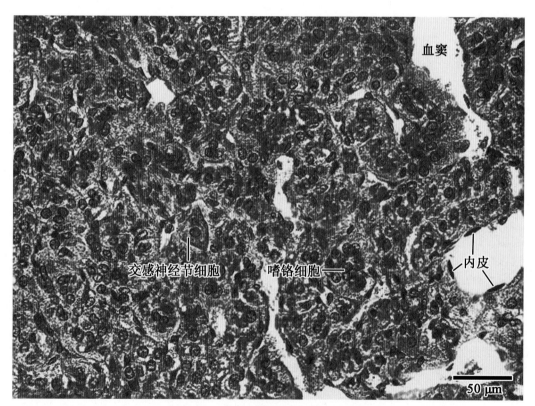

图 12-41 犬肾上腺髓质高倍像（HE）

Fig.12-41 high magnification of medulla in dog adrenal gland（HE）

交感神经节细胞—sympathetic ganglionic cell 嗜铬细胞—chromaffin cell 血窦—blood sinusoid 内皮—endothelium

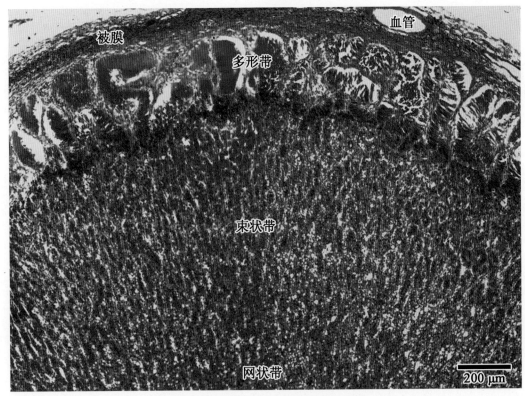

图 12-42 猫肾上腺皮质低倍像（HE）

Fig.12-42 low magnification of cortex in cat adrenal gland（HE）

被膜—capsule 多形带—multiform zone 束状带—fasciculate zone 网状带—reticular zone 血管—blood vessel

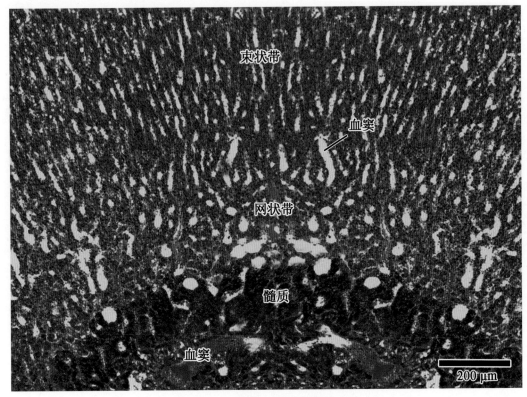

图 12-43 猫肾上腺低倍像（HE）

Fig.12-43 low magnification of cat adrenal gland（HE）

束状带—fasciculate zone 网状带—reticular zone 髓质—medulla 血窦—blood sinusoid

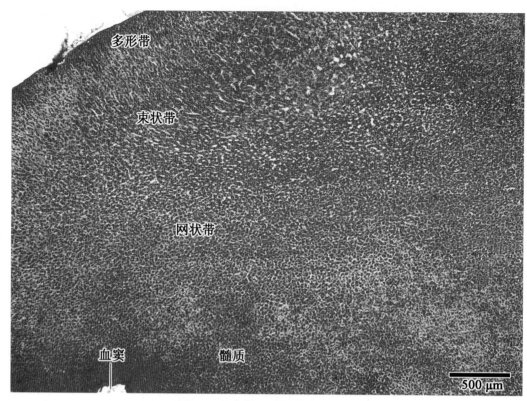

图12-44　兔肾上腺低倍像（HE）

Fig.12-44　low magnification of rabbit adrenal gland（HE）

多形带—multiform zone　束状带—fasciculate zone　网状带—reticular zone　髓质—medulla　血窦—blood sinusoid

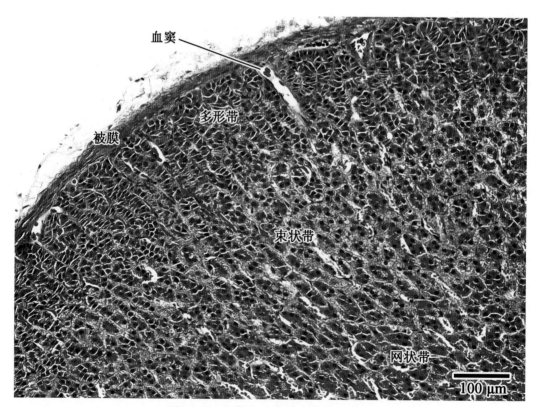

图12-45　兔肾上腺皮质中倍像（HE）

Fig.12-45　mid magnification of cortex in rabbit adrenal gland（HE）

被膜—capsule　血窦—blood sinusoid　多形带—multiform zone　束状带—fasciculate zone　网状带—reticular zone

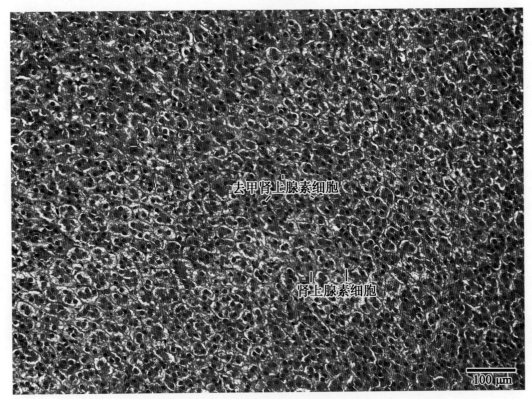

图12-46 兔肾上腺髓质中倍像（HE）

Fig.12-46 mid magnification of medulla in rabbit adrenal gland（HE）

去甲肾上腺素细胞—noradrenaline cell　肾上腺素细胞—adrenergic cell

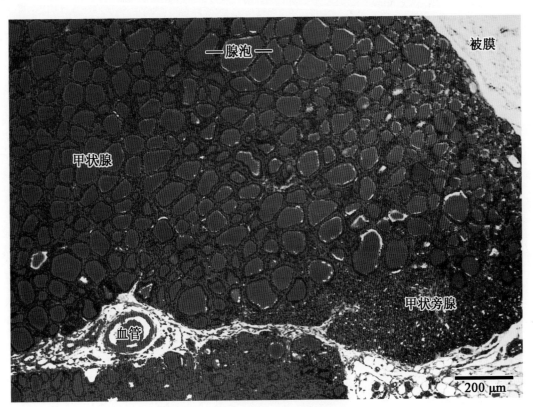

图12-47 牛甲状腺和甲状旁腺低倍像（HE）

Fig.12-47 low magnification of cattle thyroid gland and parathyroid gland（HE）

甲状腺—thyroid gland　腺泡—acinus　血管—blood vessel　被膜—capsule　甲状旁腺—parathyroid gland

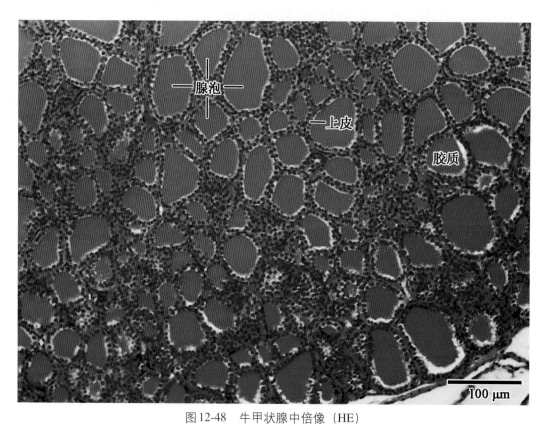

图12-48 牛甲状腺中倍像（HE）

Fig.12-48 mid magnification of cattle thyroid gland（HE）

腺泡—acinus 上皮—epithelium 胶质—colloid

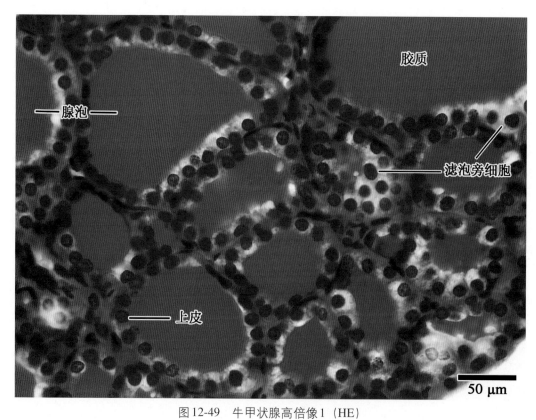

图12-49 牛甲状腺高倍像1（HE）

Fig.12-49 high magnification of cattle thyroid gland 1（HE）

腺泡—acinus 上皮—epithelium 胶质—colloid 滤泡旁细胞—parafollicular cell

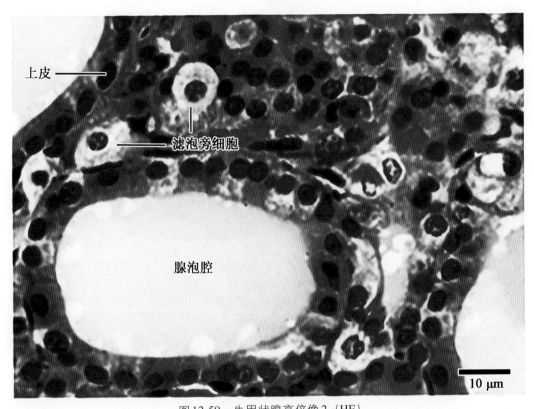

图 12-50 牛甲状腺高倍像 2（HE）

Fig.12-50 high magnification of cattle thyroid gland 2 （HE）

上皮—epithelium 滤泡旁细胞—parafollicular cell 腺泡腔—acinus cavity

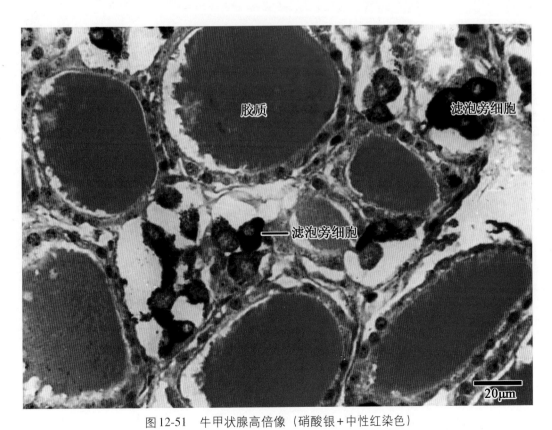

图 12-51 牛甲状腺高倍像（硝酸银+中性红染色）

Fig.12-51 high magnification of cow thyroid gland （silver nitrate+neutral red stain）

胶质—colloid 滤泡旁细胞—parafollicular cell

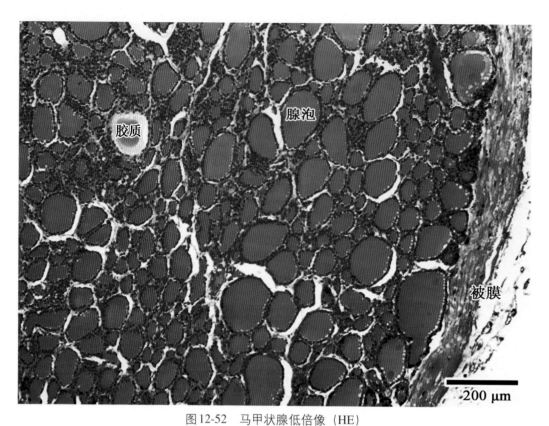

图12-52　马甲状腺低倍像（HE）

Fig.12-52　low magnification of horse thyroid gland（HE）

胶质—colloid　腺泡—acinus　被膜—capsule

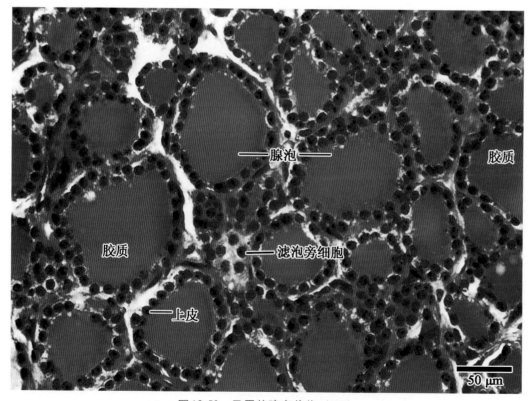

图12-53　马甲状腺高倍像（HE）

Fig.12-53　high magnification of horse thyroid gland（HE）

腺泡—acinus　胶质—colloid　滤泡旁细胞—parafollicular cell　上皮—epithelium

第十二章　内分泌系统　Endocrine System

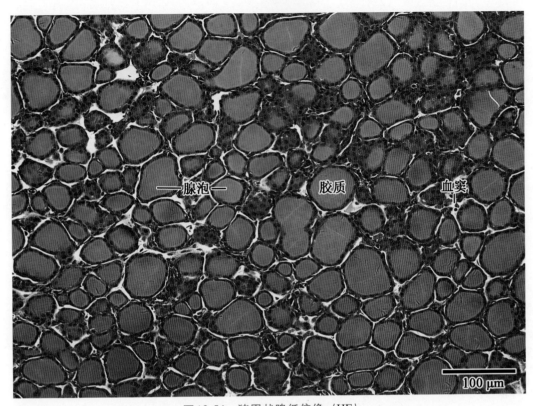

图 12-54　猪甲状腺低倍像（HE）

Fig.12-54　low magnification of pig thyroid gland（HE）

腺泡—acinus　胶质—colloid　血窦—blood sinusoid

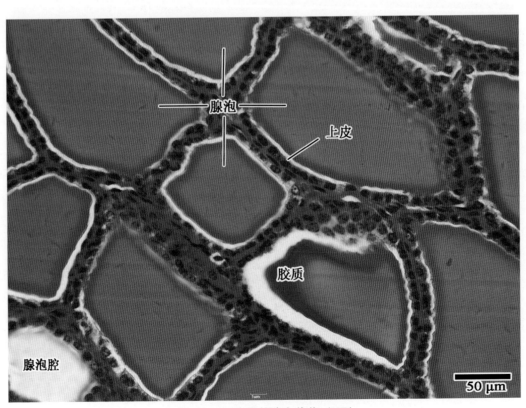

图 12-55　猪甲状腺高倍像（HE）

Fig.12-55　high magnification of pig thyroid gland（HE）

腺泡—acinus　腺泡腔—acinus cavity　上皮—epithelium　胶质—colloid

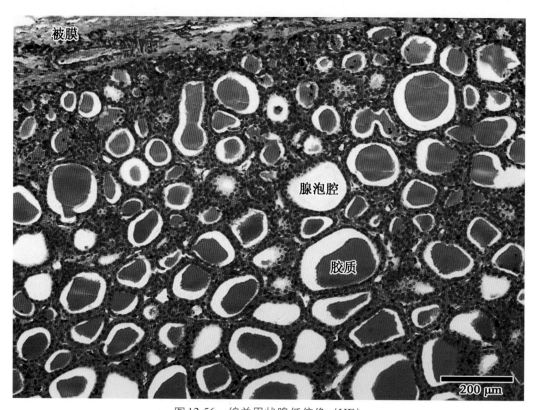

图 12-56 绵羊甲状腺低倍像（HE）

Fig.12-56 low magnification of sheep thyroid gland（HE）

被膜—capsule　腺泡腔—acinus cavity　胶质—colloid

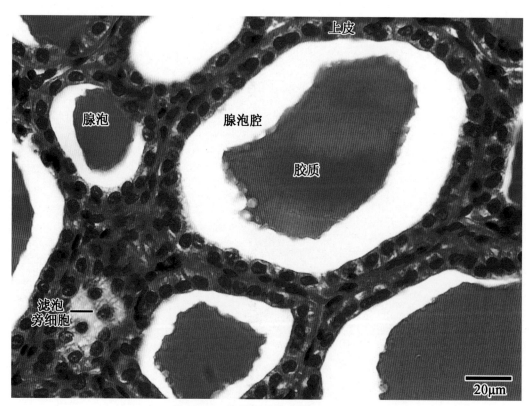

图 12-57 绵羊甲状腺高倍像（HE）

Fig.12-57 high magnification of sheep thyroid gland（HE）

腺泡—acinus　上皮—epithelium　腺泡腔—acinus cavity　胶质—colloid　滤泡旁细胞—parafollicular cell

第十二章　内分泌系统　Endocrine System

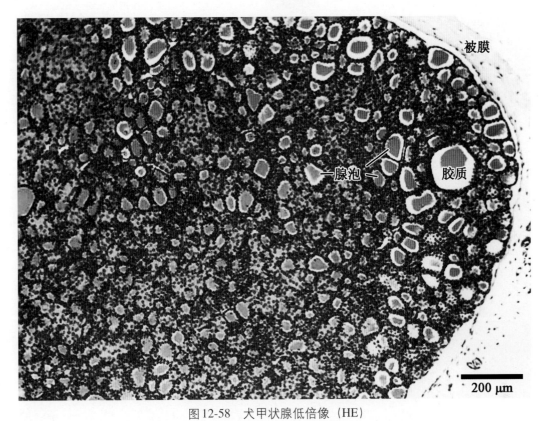

图12-58　犬甲状腺低倍像（HE）
Fig.12-58　low magnification of dog thyroid gland（HE）
被膜—capsule　腺泡—acinus　胶质—colloid

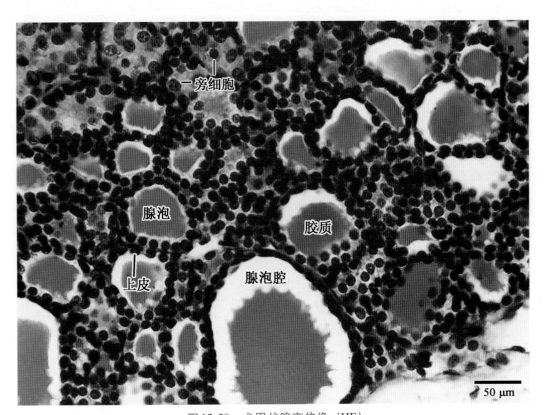

图12-59　犬甲状腺高倍像（HE）
Fig.12-59　high magnification of dog thyroid gland（HE）
旁细胞—parafollicular cell　腺泡—acinus　上皮—epithelium　腺泡腔—acinus cavity　胶质—colloid

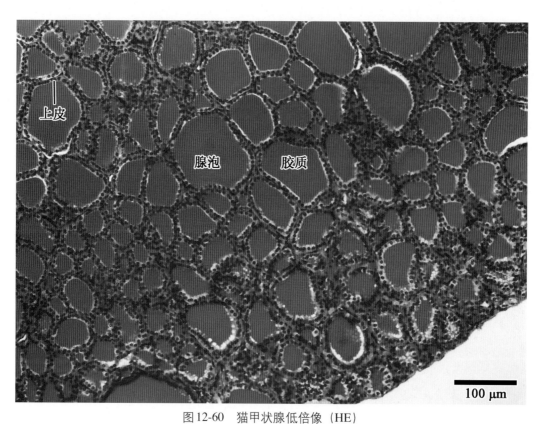

图12-60 猫甲状腺低倍像（HE）

Fig.12-60 low magnification of cat thyroid gland（HE）

上皮—epithelium 腺泡—acinus 胶质—colloid

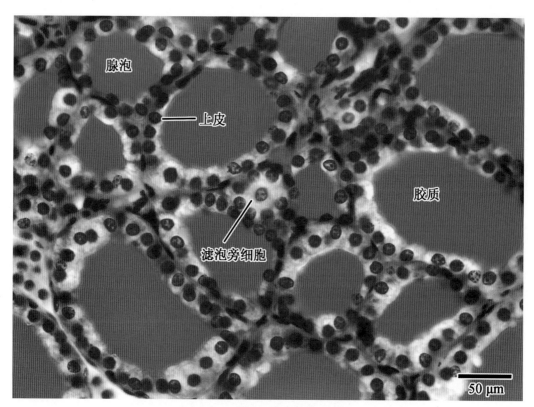

图12-61 猫甲状腺高倍像（HE）

Fig.12-61 high magnification of cat thyroid gland（HE）

腺泡—acinus 上皮—epithelium 滤泡旁细胞—parafollicular cell 胶质—colloid

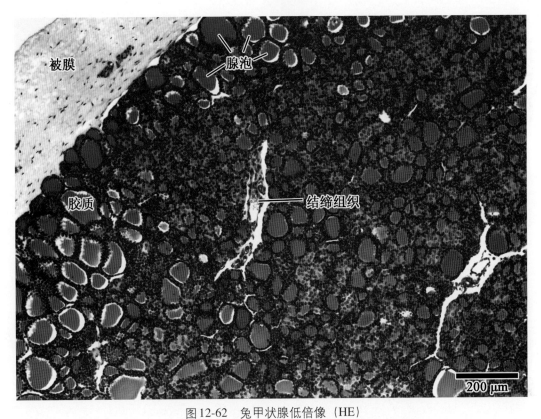

图 12-62 兔甲状腺低倍像（HE）

Fig.12-62　low magnification of rabbit thyroid gland（HE）

被膜—envelop　腺泡—acinus　胶质—colloid　结缔组织—connective tissue

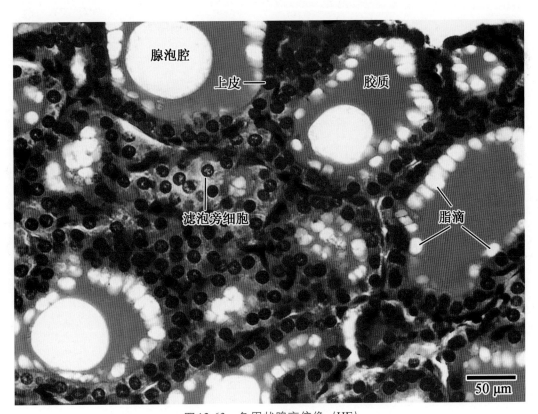

图 12-63 兔甲状腺高倍像（HE）

Fig.12-63　high magnification of rabbit thyroid gland（HE）

腺泡腔—acinus cavity　上皮—epithelium　滤泡旁细胞—parafollicular cell　胶质—colloid　脂滴—lipid droplet

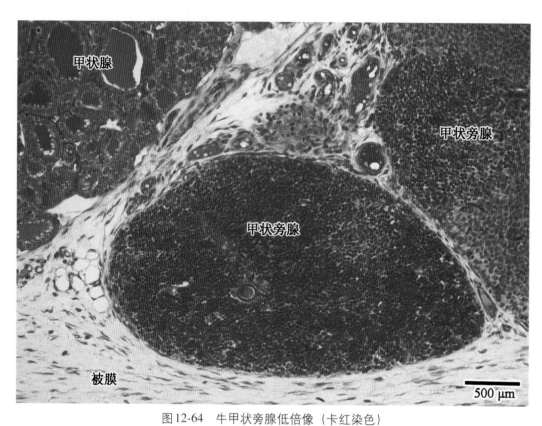

图12-64 牛甲状旁腺低倍像（卡红染色）

Fig.12-64 low magnification of cow parathyroid gland（carmine stain）

甲状腺—thyroid gland 甲状旁腺—parathyroid gland 被膜—capsule

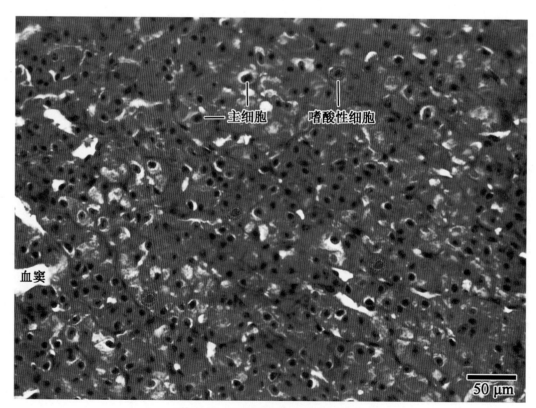

图12-65 牛甲状旁腺中倍像（HE）

Fig.12-65 mid magnification of cow parathyroid gland（HE）

主细胞—chief cell 嗜酸性细胞—oxyphil cell 血窦—blood sinusoid

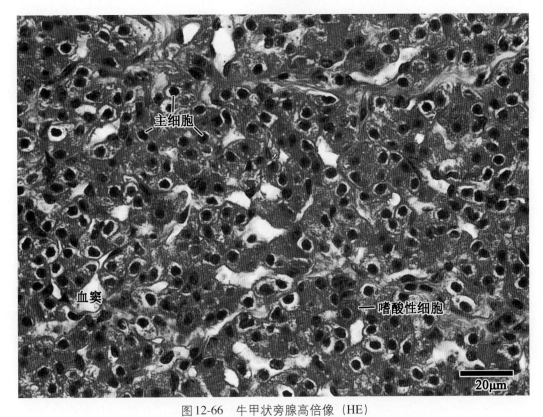

图 12-66 牛甲状旁腺高倍像（HE）
Fig.12-66 high magnification of cow parathyroid gland（HE）
血窦—blood sinusoid 主细胞—chief cell 嗜酸性细胞—oxyphil cell

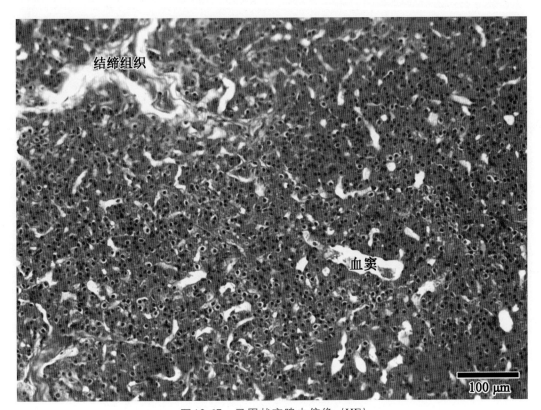

图 12-67 马甲状旁腺中倍像（HE）
Fig.12-67 mid magnification of horse parathyroid gland（HE）
结缔组织—connective tissue 血窦—blood sinusoid

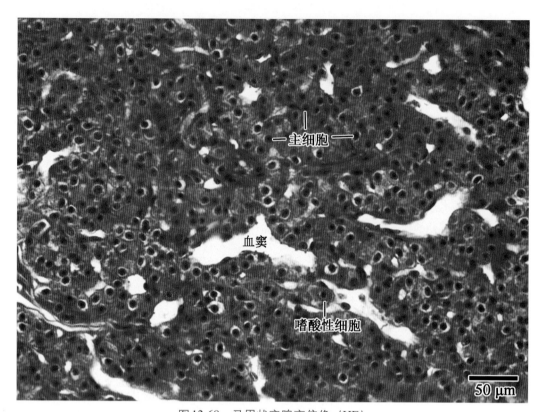

图12-68 马甲状旁腺高倍像（HE）

Fig.12-68 high magnification of horse parathyroid gland（HE）

血窦—blood sinusoid 主细胞—chief cell 嗜酸性细胞—oxyphil cell

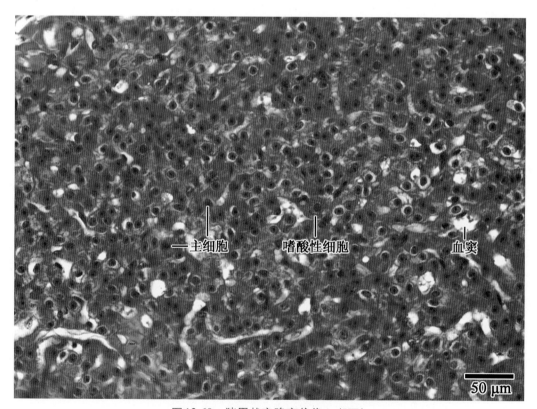

图12-69 猪甲状旁腺高倍像1（HE）

Fig.12-69 high magnification of pig parathyroid gland 1（HE）

主细胞—chief cell 嗜酸性细胞—oxyphil cell 血窦—blood sinusoid

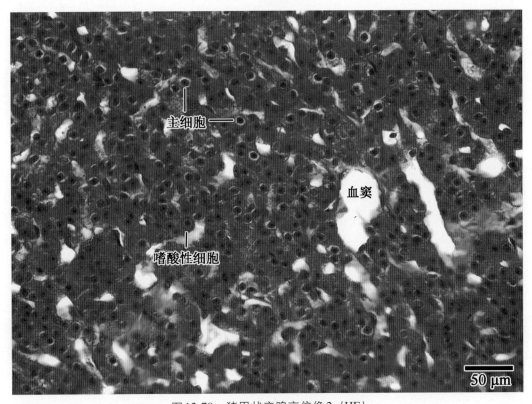

图12-70 猪甲状旁腺高倍像2（HE）

Fig.12-70 high magnification of pig parathyroid gland 2（HE）

主细胞—chief cell 嗜酸性细胞—oxyphil cell 血窦—blood sinusoid

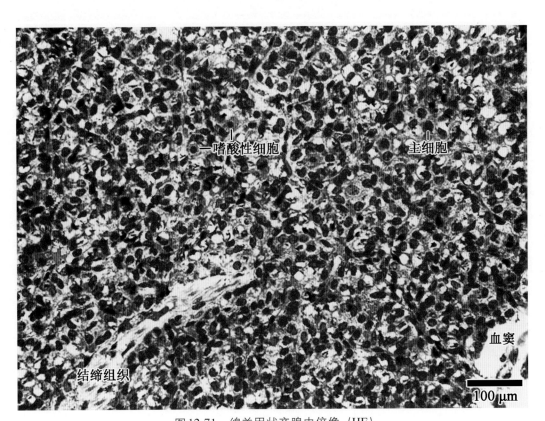

图12-71 绵羊甲状旁腺中倍像（HE）

Fig.12-71 mid magnification of sheep parathyroid gland（HE）

嗜酸性细胞—oxyphil cell 主细胞—chief cell 结缔组织—connective tissue 血窦—blood sinusoid

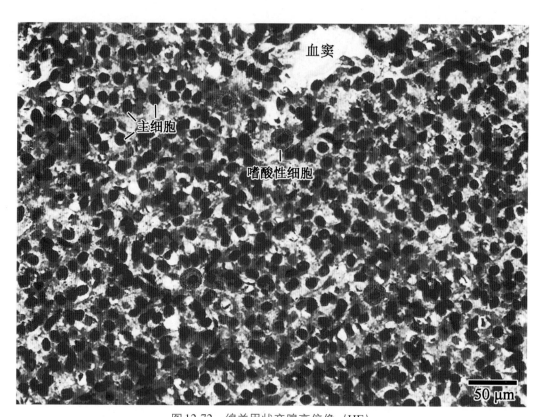

图 12-72 绵羊甲状旁腺高倍像（HE）

Fig.12-72 high magnification of sheep parathyroid gland（HE）

主细胞—chief cell 嗜酸性细胞—oxyphil cell 血窦—blood sinusoid

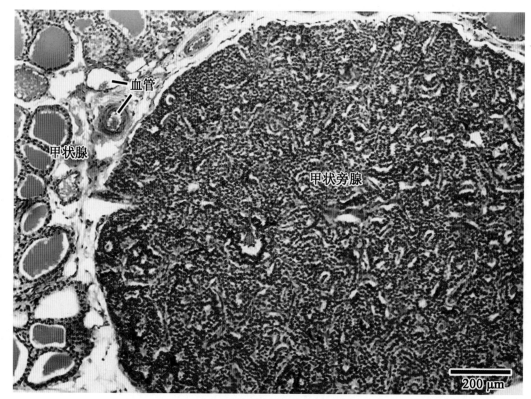

图 12-73 犬甲状旁腺低倍像（HE）

Fig.12-73 low magnification of dog parathyroid gland（HE）

甲状腺—thyroid gland 血管—blood vessel 甲状旁腺—parathyroid gland

第十二章 内分泌系统 Endocrine System

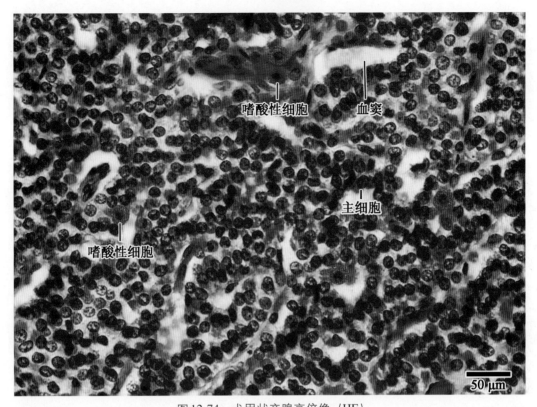

图12-74 犬甲状旁腺高倍像（HE）
Fig.12-74 high magnification of dog parathyroid gland（HE）
嗜酸性细胞—oxyphil cell　血窦—blood sinusoid　主细胞—chief cell

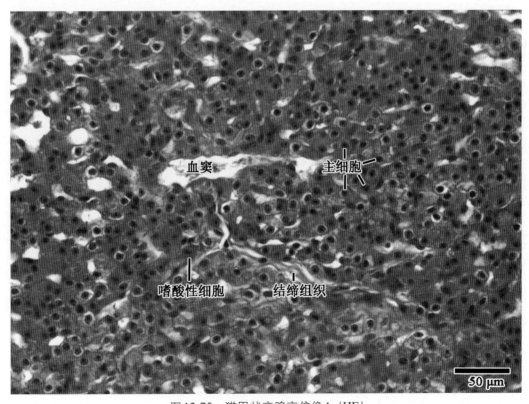

图12-75 猫甲状旁腺高倍像1（HE）
Fig.12-75 high magnification of cat parathyroid gland 1（HE）
血窦—blood sinusoid　主细胞—chief cell　嗜酸性细胞—oxyphil cell　结缔组织—connective tissue

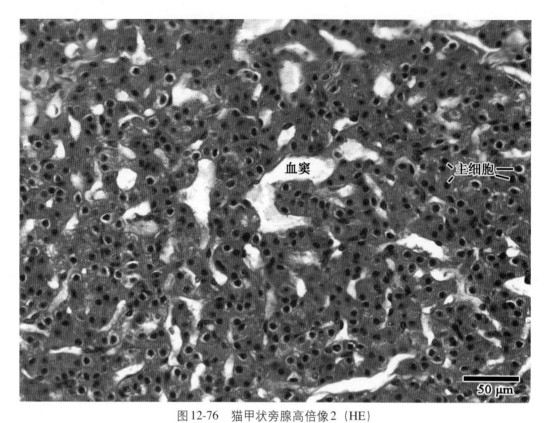

图12-76 猫甲状旁腺高倍像2（HE）

Fig.12-76 high magnification of cat parathyroid gland 2 （HE）

血窦—blood sinusoid 主细胞—chief cell

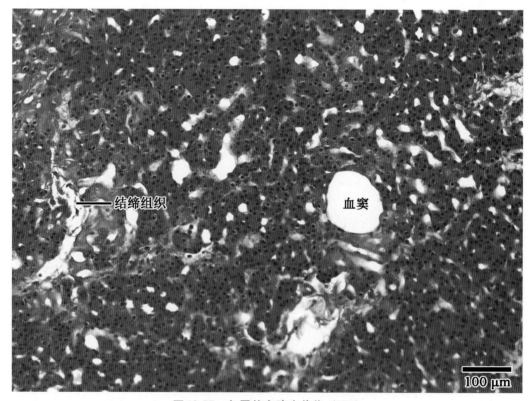

图12-77 兔甲状旁腺中倍像（HE）

Fig.12-77 mid magnification of rabbit parathyroid gland （HE）

结缔组织—connective tissue 血窦—blood sinusoid

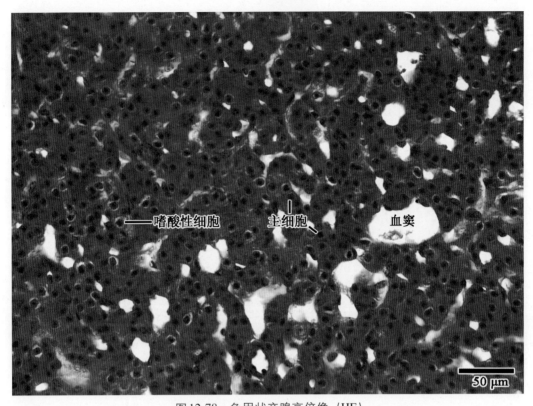

图12-78 兔甲状旁腺高倍像（HE）

Fig.12-78 high magnification of rabbit parathyroid gland（HE）

主细胞—chief cell　嗜酸性细胞—oxyphil cell　血窦—blood sinusoid

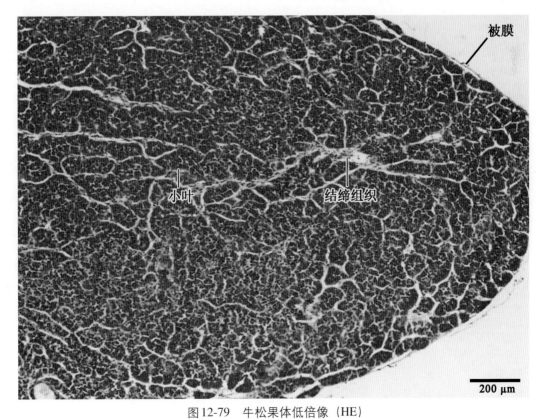

图12-79 牛松果体低倍像（HE）

Fig.12-79 low magnification of cow pineal body（HE）

被膜—capsule　小叶—lobule　结缔组织—connective tissue

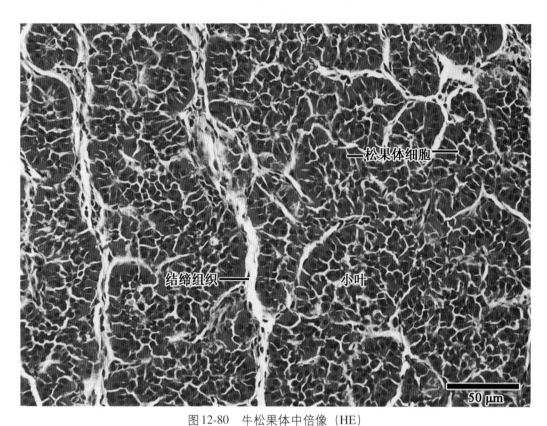

图12-80 牛松果体中倍像（HE）

Fig.12-80 mid magnification of cow pineal body（HE）

结缔组织—connective tissue 小叶—lobule 松果体细胞—pineal cell

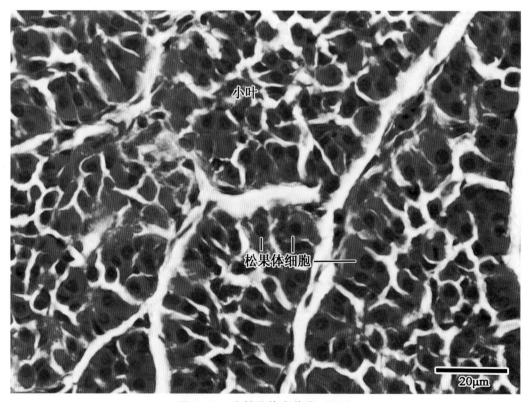

图12-81 牛松果体高倍像（HE）

Fig.12-81 high magnification of cow pineal body（HE）

小叶—lobule 松果体细胞—pineal cell

第十二章 内分泌系统 Endocrine System

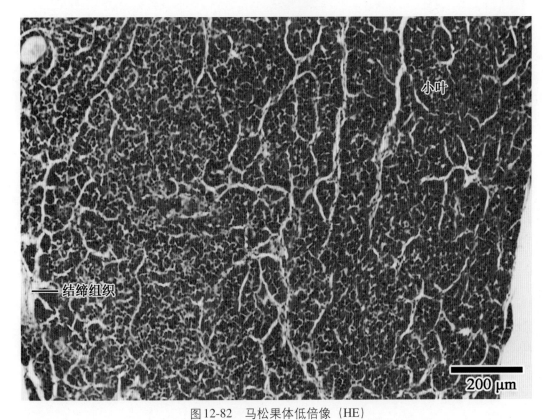

图12-82 马松果体低倍像（HE）
Fig.12-82 low magnification of horse pineal body（HE）
结缔组织—connective tissue　小叶—lobule

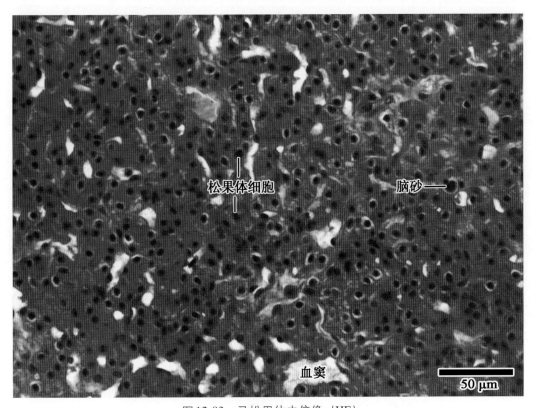

图12-83 马松果体中倍像（HE）
Fig.12-83 mid magnification of horse pineal body（HE）
松果体细胞—pineal cell　脑砂—brain sand　血窦—blood sinusoid

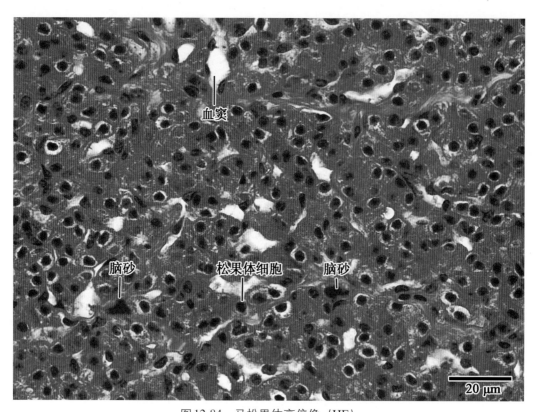

图12-84 马松果体高倍像（HE）

Fig.12-84 high magnification of horse pineal body（HE）

血窦—blood sinusoid 脑砂—brain sand 松果体细胞—pineal cell

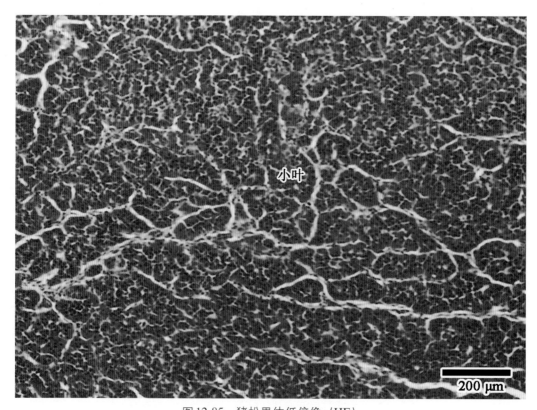

图12-85 猪松果体低倍像（HE）

Fig.12-85 low magnification of pig pineal body（HE）

小叶—lobule

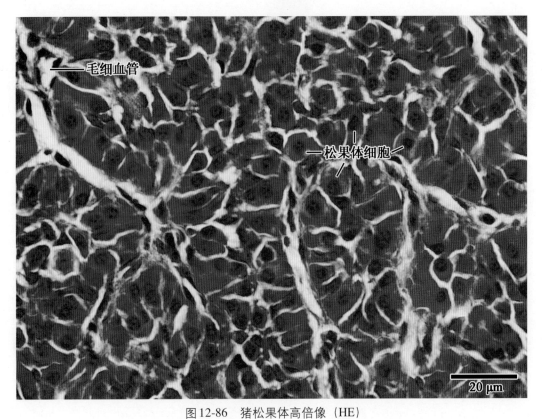

图12-86 猪松果体高倍像（HE）

Fig.12-86 high magnification of pig pineal body （HE）

毛细血管—capillary 松果体细胞—pineal cell

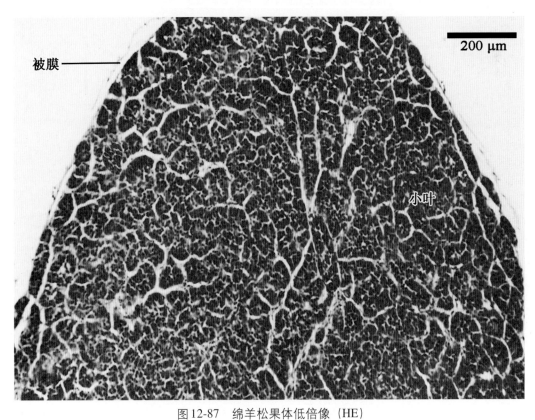

图12-87 绵羊松果体低倍像（HE）

Fig.12-87 low magnification of sheep pineal body （HE）

被膜—capsule 小叶—lobule

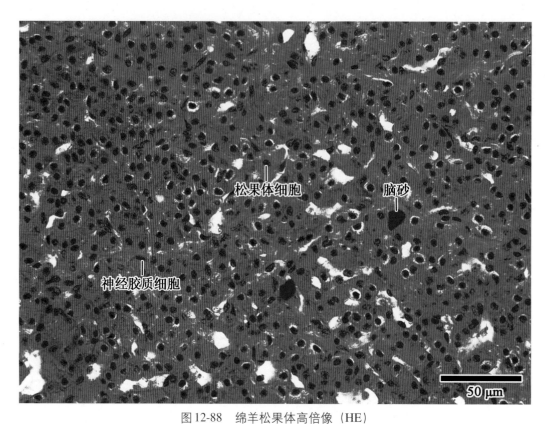

图 12-88 绵羊松果体高倍像（HE）

Fig.12-88 high magnification of sheep pineal body（HE）

松果体细胞—pineal cell　脑砂—brain sand　神经胶质细胞—neurogliocyte

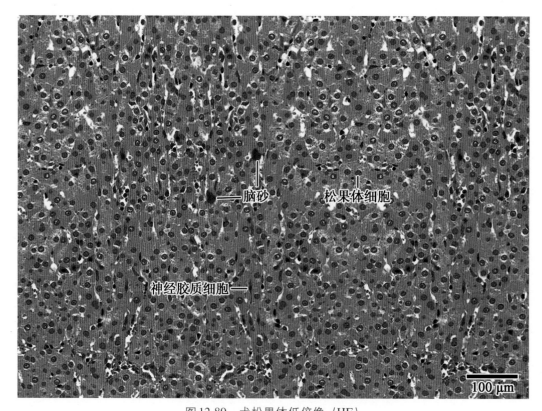

图 12-89 犬松果体低倍像（HE）

Fig.12-89 low magnification of dog pineal body（HE）

脑砂—brain sand　松果体细胞—pineal cell　神经胶质细胞—neurogliocyte

第十二章 内分泌系统 Endocrine System

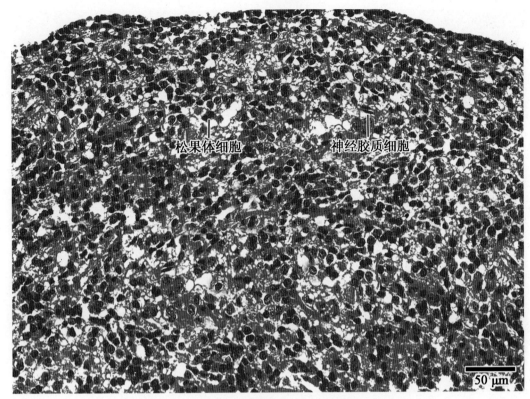

图12-90 犬松果体中倍像（HE）
Fig.12-90 mid magnification of dog pineal body（HE）
松果体细胞—pineal cell 神经胶质细胞—neurogliocyte

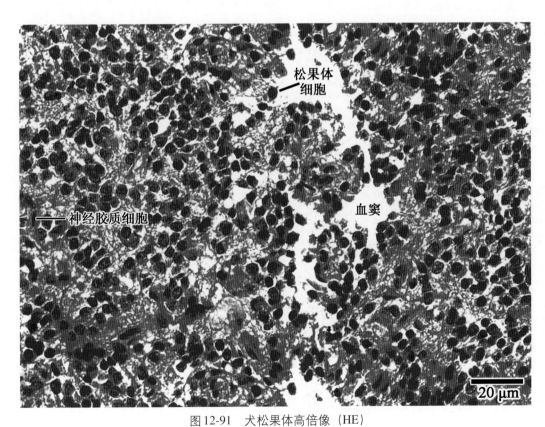

图12-91 犬松果体高倍像（HE）
Fig.12-91 high magnification of dog pineal body（HE）
松果体细胞—pineal cell 神经胶质细胞—neurogliocyte 血窦—blood sinusoid

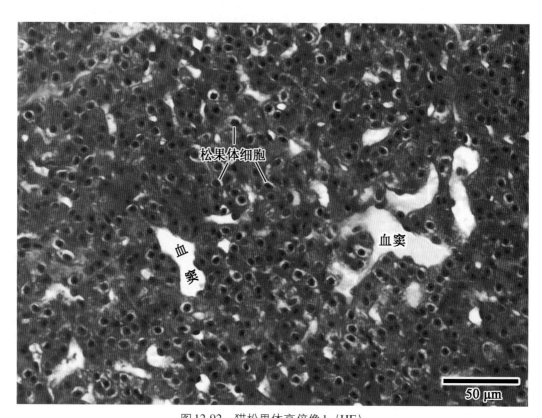

图 12-92 猫松果体高倍像 1 （HE）

Fig.12-92 high magnification of cat pineal body 1 （HE）

松果体细胞—pineal cell 血窦—blood sinusoid

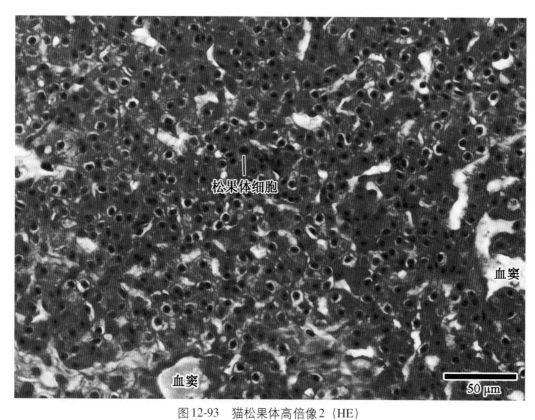

图 12-93 猫松果体高倍像 2 （HE）

Fig.12-93 high magnification of cat pineal body 2 （HE）

松果体细胞—pineal cell 血窦—blood sinusoid

第十二章 内分泌系统 Endocrine System

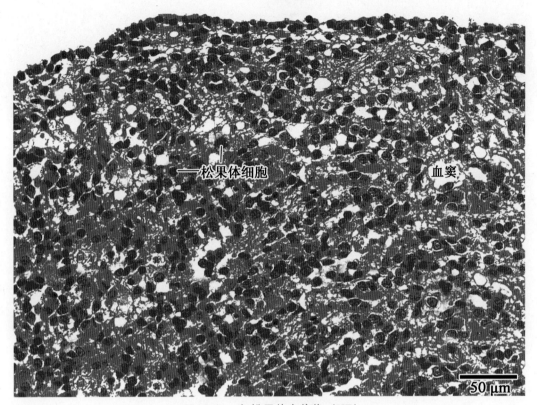

图12-94 兔松果体高倍像（HE）
Fig.12-94 high magnification of rabbit pineal body（HE）
松果体细胞—pineal cell 血窦—blood sinusoid

第十三章
消 化 管
Digestive Tract

> **Outline**
>
> Digestive system consists of digestive tract and digestive glands. Digestive tract consists of: oral cavity, esophagus, stomach, small and large intestines, rectum and anus. Composed of four concentric layers, digestive tract from esophagus through large intestine exhibits considerable regularity. The four layers, named in order from the lumen outward, are mucosa, submucosa, muscularis, and adventitia or serosa. The mucosa, or mucous membrane consists of: superficial epithelium, proper layer and muscularis mucosa. The type of epithelial lining varies based on the segments of digestive tract. The main function of the epithelial lining is to provide a selectively permeable barrier between the contents of the tract and the tissues of the body. Often, large accumulations of typical lymphatic tissue can be found in the stroma. The entire proper layer is composed of loose connective tissue with large number of blood and lymph vessels, sometimes also containing glands and lymphoid tissue. The entire mucosa projects into the lumen as folds or villi. The effective surface of the gut is enlarged extremely by these invaginations and evaginations. In terms of the changing functional activity, mucosa differs considerably from segment to segment of the alimentary tract. Great significance relays on the functional changes of the specialized mucosal cells that are involved in secretion and absorption. As a connective tissue layer, submucosa often contains accumulations of lymphatic tissue as well as glands that extend from the mucosa. Muscularis contains smooth muscle in all parts except the upper esophagus and anal sphincter since there it is composed of skeletal muscle fibers. Musular fibers in muscularis are arranged as inner circular and outer longitudinal layers. Adventitia is composed of several layers of loose connective tissue, alternately collagenous and elastic. Whereas the serosa is a thin layer of loose connective tissue and a simple squamous covering epithelium.
>
> The inner lumen of esophagus is lined by stratified squamous epithelium. The salient feature of which is the esophageal glands in the submucosa.
>
> The characteristics of the stomach are: the gastric mucosa consists of a simple columnar epithelium that

第十三章 消化管 Digestive Tract

> invaginates to various extents into the proper layer, forming gastric pits. The proper layer is filled with branched, tubular gastric glands, which consists of zymogenic cells, oxyntic cells, mucous neck cells, endocrine cells and stem cells. Muscularis consists of three layers of smooth muscle fibers in different directions.
>
> Small intestine includes three segments: duodenum, jejunum and ileum, which are the main portions of the food digestion and nutrient absorption. The mucosa is covered with a simple columnar epithelium, which consists of absorptive cells, goblet cells, endocrine cells and Paneth cells. In duodenum the duodenal glands are present in the submucosa. In ileum, aggregates of lymphoid nodules are present in the proper layer and the submucosa. Large intestine includes three parts: cecum, colon and rectum. The characteristics of which are: the intestinal glands are long, and mingled with a lot of goblet and absorptive cells, and no villi.

消化系统（digestive system）由消化管（digestive tract）和消化腺（digestive glands）组成。消化管从口腔、咽、食管、胃、小肠和大肠直至肛门，为一条衬有上皮的迂曲管道。消化管的功能主要是消化食物、吸收营养和排泄食物残渣。食物中除水、维生素和无机盐可以被消化管上皮直接吸收外，蛋白质、糖类和脂肪等大分子物质需经消化成小分子物质后才能被吸收。此外，消化管黏膜还是机体的重要屏障，黏膜内富有淋巴组织和免疫细胞，对细菌等有害物质具有重要的防御作用。

一、消化管的发生

（一）口腔的发生

1. **口凹的演变**　口腔的发生与颜面发生紧密相连。原始口腔是额鼻突及上颌突、下颌突围成的凹陷。口凹底部是口咽膜，它破裂后，原始口腔便与原始咽相通。在面部发生过程中，上颌突和下颌突向中线生长，左、右下颌突较早愈合，形成下颌和下唇。由额鼻突下缘分化而来的内侧鼻突逐渐愈合，向下方迁移，与同侧的上颌突愈合发育为上唇的正中部分。外侧鼻突与上颌突愈合，两侧上颌突形成上颌和上唇。

2. **腭的发生**　腭来源于腭突。在牛、马胚胎约30天，左、右内侧鼻突愈合后，向原始口腔内长出正中腭突，它形成腭的前部。上颌突向原始口腔内长出外侧腭突，起初在舌的两侧斜向下方生长，以后舌变扁，口腔增大，外侧腭突呈水平方向在舌上方生长，向中线愈合，形成腭的大部分。其前缘与正中腭突愈合。腭前部间充质骨化为硬腭；后部不骨化，为软腭。腭形成后，原始口、鼻腔分隔为永久口腔和鼻腔。

3. **舌的发生**　舌体来源于外侧舌膨大，舌根来自联合突。胚胎下颌突的内侧面细胞增生，形成一对外侧舌膨大，左、右外侧舌膨大生长快，向中线愈合，成为舌体。第2、3、4对腮弓腹侧的间充质增生突向咽腔，称联合突，其前部发育为舌根，后部发育为会厌。舌的上皮来自咽壁内胚层。

4. **齿的发生**　动物一生有乳齿和恒齿。随着年龄的增大，乳齿逐渐脱落而被恒齿代替。齿萌出前，在颌骨内经历了齿胚发生、齿组织分化等过程。齿的原基由原始口腔的外胚层上皮及其下方的间充质发育而成，包括成釉器、齿乳头和齿囊。齿体组织的发生包括釉质、齿本质、齿骨质、齿髓的形成。

（二）原始消化管的形成和分化

三胚层胚盘随着头褶、尾褶和侧褶的形成，由扁平形逐渐卷折为向腹侧弯曲的柱形胚体，此时卵黄囊顶部的内胚层和脏壁中胚层被卷入胚体，形成一条纵行的原始消化管(primitive gut)。它可分成前肠(foregut)、中肠(midgut)和后肠(hindgut)。前肠头端和后肠末端，原先都是盲端，分别被口咽膜和泄殖腔膜封闭，不久该两膜破裂，使前肠、后肠与外界相通。中肠与卵黄囊相连，随着胚体和原肠的生长发育，卵黄囊相对变小，两者连接部分变成细长的卵黄管。

前肠分化为原始咽、食管、胃、胆总管开口处以上的十二指肠及肝、胆囊和胰；此外，除鼻以外的呼吸道也由前肠发生而来。中肠分化为从胆总管开口处以下的小肠以及盲肠、升结肠。后肠分化为降结肠、直肠和肛管。原始消化管内胚层分化为消化管的上皮和腺体，管壁内的结缔组织和肌组织皆由脏壁中胚层分化形成。

（三）食管的发生

胚胎早期食管很短，以后由于心脏位置下降和颈部伸长，食管随之延长。食管腔面的内胚层最初分化为单层上皮，接着细胞迅速增生，上皮增厚，一度使管腔闭塞，后来管腔又重新出现。

（四）胃的发生

食管尾侧的原肠形成梭形膨大，即胃的原基，通过胃的背、腹系膜与胚体后壁和前壁相连。以后由于胃背侧缘发育快，使胃体向背侧扩展，形成胃大弯。由于胃背系膜发育较快，并向左扩展形成网膜囊和大网膜，使胃大弯由背侧转向左侧。胃腹侧缘生长缓慢，形成胃小弯，并由腹侧转向右侧。胃腹系膜生长慢，形成小网膜。

（五）肠的发生

在胃的尾侧形成十二指肠，它生长快，形成凸向腹侧的十二指肠襻。当胃发生旋转时，十二指肠襻就转向右侧，并通过背系膜固定于右侧腹后壁处。随着肠的发生，背系膜增长，而腹系膜则退化消失。十二指肠以下的中肠突向腹侧形成中肠襻。襻顶与卵黄管相连，此时，中肠襻的腹系膜退化消失，背系膜将中肠襻固定于腹顶壁。中肠襻近段发出一囊状的突起，称盲肠突，是盲肠的原基，也是大肠和小肠的分界标志。盲肠突从肝右叶下方逐渐下降到右髂窝处，升结肠随之形成。盲肠突近侧段膨大为盲肠。

二、消化管的组织结构概述

消化管在结构上有一些共性。除口腔和咽外，消化管壁均分为四层，由内向外分别为黏膜、黏膜下层、肌层和外膜。

1. **黏膜**（mucosa） 为消化管壁的最内层，是消化管各段结构差异最大、功能最重要的部位，分上皮、固有层和黏膜肌层三部分。上皮衬于腔面，其类型依部位而异。在口腔、咽、单室胃的无腺部、多室胃的前胃和肛门处为复层扁平上皮，以保护功能为主；单室胃的有腺部、多室胃的皱胃、小肠和大肠为单层柱状上皮，主要功能是消化吸收。固有层（proper layer）为上皮深层的疏松结缔组织，富含毛细血管、毛细淋巴管和神经，有些部位还有大量的腺体、淋巴组织和平滑肌纤维。黏膜肌层（muscularis mucosae）为薄层平滑肌，一般有内环、外纵两层。除口腔及咽外，其余各段均有分布。

2. **黏膜下层**（submucosa） 为疏松结缔组织，含有较大的血管、神经、淋巴管和淋巴组织。在食管和十二指肠的黏膜下层中分别有食管腺和十二指肠腺。有些部位的黏膜和黏膜下层向管腔内突起，形成环行、纵行或不规则的皱襞，扩大黏膜面积。

3. **肌层**（muscular layer） 除口腔、咽、食管（猪和马大部、牛和羊全部）和肛门为骨骼肌外，其他各处均为平滑肌。一般为内环、外纵两层平滑肌，其间有肌间神经丛（myenteric nerve plexus）。

4. **外膜**（adventitia） 是消化管的最外层，分纤维膜和浆膜。纤维膜（fibrosa）仅由结缔组织构成，与相邻器官的结缔组织相连，无明显的界限。食管的颈段和直肠末段的外膜是纤维膜。浆膜（serosa）由薄层结缔组织和间皮构成。浆膜光滑而湿润，可减少器官间的摩擦。

（一）口腔（oral cavity）

口腔壁内没有黏膜肌层、黏膜下层和肌层。其黏膜内有多种腺体，如唇腺、颊腺、舌腺等，分泌浆液性或黏液性的唾液，保持黏膜表面的湿润。

1. **舌**（tongue） 黏膜上皮为复层扁平上皮，角质化程度高，有许多形态和大小不同的舌乳头。丝状乳头、锥状乳头、豆状乳头、菌状乳头、轮廓乳头和叶状乳头。后三者含有味蕾(taste bud)，即味觉感受器。舌肌为骨骼肌，3种走向不同的横行肌、纵行肌和垂直肌交错，收缩时改变舌的形状。

2. **齿**（tooth） 齿分三部分：露在齿龈外的齿冠，埋在齿槽骨内的齿根，二者交界处的齿颈。齿中央有齿髓腔，末端有根尖孔与牙周韧带连接。齿由两类组织构成：一是坚硬的钙化组织，包括釉质、齿本质和齿骨质；

二是松软组织即齿髓。

（二）食管（esophagus）

食管是将食物输送至胃的管道。其腔面有数条纵行皱襞，食物通过时皱襞展平消失。

1.黏膜 上皮为复层扁平上皮，其角化程度随动物种类而异。猪的轻度角化，反刍兽的高度角化，而肉食兽的通常未角化。固有层含淋巴组织，以食管和胃的结合部较多。浅层形成许多乳头，突向上皮。黏膜肌层为散在的纵行平滑肌束，向后逐渐增多，末段形成完整一层。猪和犬的食管前半段无黏膜肌层，后段则很发达。

2.食管其他各层的结构特征

（1）黏膜下层 有分支的管泡状黏液腺或以黏液细胞为主的混合腺，称为食管腺（esophageal gland）。食管腺的分布因动物而异，反刍兽、马和猫仅见于咽和食管的连接部；猪则集中于食管的前半部，向后逐渐减少；犬则在整个黏膜下层内都有。腺导管穿过黏膜层开口于食管腔。分泌的黏液可润滑食管。

（2）肌层 反刍兽和犬的食管肌层全部是骨骼肌；猪食管颈段是骨骼肌，胸段为骨骼肌和平滑肌交错排列，腹段为平滑肌；猫食管的前4/5为骨骼肌，后1/5为平滑肌；马食管的后1/3为平滑肌。肌层多为内环和外纵两层，有时在两层之间出现不规则的副肌层，故食管肌的分层不明显。

（三）胃（stomach）

胃是消化管的膨大部分，前端以贲门接食管，后端以幽门通十二指肠，可暂时储存食物、初步消化和推送食物进入十二指肠。按动物的种类不同分单室胃和多室胃。

1.单室胃（single room stomach） 猪、犬、猫、兔和马属等动物的胃是单室胃。胃黏膜分有腺部和无腺部。无腺部缺胃腺，黏膜苍白，上皮为复层扁平上皮。有腺部黏膜上皮为单层柱状上皮。马胃的无腺部较大；有腺部黏膜富有皱褶，呈红褐色或灰色，内有丰富的贲门腺、胃底腺和幽门腺分布。猪胃黏膜的无腺部很小，仅位于贲门周围，呈苍白色；贲门腺区很大，呈淡灰色；胃底腺区较小，呈棕红色；幽门腺区位于幽门部，呈灰白色。犬、猫等食肉动物胃黏膜全为有腺部；贲门腺区呈环带状，灰白色，较小；胃底腺区大，占胃黏膜面积的2/3，黏膜很厚；幽门腺区黏膜较薄而小。

有腺部黏膜的单层柱状上皮细胞间分布有胃小凹，即胃腺的开口。固有层内有很厚的胃腺。胃腺由主细胞、壁细胞、颈黏液细胞和内分泌细胞组成，分别分泌胃酶、盐酸、黏液和激素。黏膜下层不含腺体，但有血管丛和神经丛。肌层有内斜、中环、外纵行三层平滑肌。中层环行肌形成幽门括约肌。浆膜由薄层结缔组织和间皮构成。

（1）胃底腺（fundic gland） 分布于胃底部的分支管状腺或单管状腺。腺体分为颈部、体部和底部。颈部短而细，与胃小凹相连；体部长；底部稍弯曲，膨大并延伸至黏膜肌层。胃底腺的主细胞、壁细胞、颈黏液细胞和内分泌细胞，可分泌酶类、盐酸、黏液和激素。

（2）贲门腺（cardiac gland） 分布于贲门部，为弯曲的分支管状腺。除猪的外，其他家畜的贲门腺区为一狭长带状。腺体较短，腺腔宽大。腺细胞呈立方形或柱状。犬的贲门腺内有少量的壁细胞，而猪的则可能有散在的主细胞。贲门腺主要分泌黏液。

（3）幽门腺（pyloric gland） 分布于幽门部，为弯曲的分支管状腺。腺细胞呈柱状，胞质染色浅。结构特点是腺体短，分支多且很弯曲，有较多的内分泌细胞。幽门腺主要分泌黏液。

2.多室胃（multicellular stomach） 牛、羊等反刍动物有四个胃，从前到后分别称瘤胃、网胃、瓣胃、皱胃。前三个胃的黏膜衬以复层扁平上皮，浅层细胞角化，黏膜内无腺体，主要贮存食物和分解粗纤维，合称为前胃（forestomach）。皱胃黏膜内分布有消化腺，能分泌胃液，具有化学性消化作用，也称为真胃，结构和功能与单室胃相似。

（1）瘤胃（rumen） 成年时瘤胃最大，占胃总容积的80%，呈前后稍长、左右略扁的椭圆形大囊。瘤胃的前、后两端有较深的前沟和后沟，左、右侧面有较浅的左纵沟和右纵沟，它们围成的环状沟，将瘤胃分为较大的背囊和腹囊。瘤胃以贲门口与食管相连，以瘤网口与网胃相通。瘤胃壁的黏膜呈棕黑色或棕黄色，无腺体，表面有密集的长约1cm的瘤胃乳头，在与瘤胃各沟相对应的内侧面，有光滑的肉柱。在肉柱和瘤胃前庭黏膜上无乳头。

（2）网胃（reticulum） 网胃的容积最小，外形呈梨形，前后稍扁。网胃黏膜呈黑褐色，表面形成许多高低不等的网格状皱褶，形似蜂房。瘤胃和网胃之间的网胃沟黏膜较为平滑，有纵行的皱褶。

（3）瓣胃（omasum） 瓣胃占胃总容积的7%～8%，呈两侧稍扁的球形。瓣胃黏膜形成百余片大小、宽窄不同的瓣叶。瓣叶呈新月形，按宽窄分大、中、小和最小四级，呈有规律地相间排列，横切面很像一叠"百叶"。瓣叶上有许多乳头。

（四）肠（intestine）

肠起自胃的幽门，止于肛门。可分小肠和大肠两部分。草食动物的肠管较长，肉食动物的较短。

1. **小肠**（small intestine） 是细长的管道，是食物消化和吸收的主要场所，前端起于胃的幽门，后端止于盲肠，分为十二指肠、空肠和回肠。小肠的基本结构分为黏膜、黏膜下层、肌层和浆膜。典型结构有肠绒毛、小肠腺、十二指肠腺。

（1）肠绒毛（intestinal villus） 小肠黏膜表面伸向肠腔的指状突起，由黏膜上皮和固有层组成。十二指肠的宽大，空肠的多呈指状，回肠的多呈锥体状。功能是扩大小肠黏膜的表面积。绒毛上皮由吸收细胞、杯状细胞和内分泌细胞组成。

（2）小肠腺（small intestinal gland） 或称肠隐窝（intestinal crypt），分布在肠黏膜固有层的单管腺。由肠绒毛根部的上皮向固有层下陷形成，开口于肠腔。肠腺（intestinal gland）细胞有吸收细胞、杯状细胞、潘氏细胞和内分泌细胞。

①吸收细胞（absorptive cell） 小肠黏膜上皮的主要细胞，呈柱状，细胞游离面有明显的纹状缘，后者由密集而整齐排列的微绒毛构成，可选择性吸收营养物质。

②杯状细胞（goblet cell） 分散在吸收细胞之间。小肠前段较少，后段增多。分泌黏液、润滑和保护肠黏膜。

③潘氏细胞（Paneth cell） 三五成群分布于肠腺底部，呈锥体形。牛、羊的小肠腺内数量多；而猪、犬、猫和兔等动物则无。顶部胞质中有粗大的嗜酸性颗粒，内含肠防御素和溶菌酶等，具有杀灭肠道内有害微生物的作用。

④内分泌细胞（endocrine cell） 分散在胃肠黏膜上皮和腺体内，细胞呈锥形或不规则，胞质中含有很多分泌颗粒，可分泌肽类或胺类激素，调节胃肠蠕动和腺体分泌。

（3）十二指肠腺（duodenal gland） 也称黏膜下腺（submucosal gland）位于十二指肠黏膜下层的分支管泡状腺，其导管穿过黏膜肌层，开口于肠腺底部。分泌含有黏蛋白的碱性液体，保护十二指肠黏膜免受酸性胃液的侵蚀。

2. **大肠**（large intestine） 比小肠短，管径粗，分为盲肠（cecum）、结肠（colon）和直肠（rectum）。草食动物的盲肠特别发达。大肠管壁也分黏膜、黏膜下层、肌层和浆膜。主要功能是吸收水分、无机盐类并进行纤维素的发酵和分解，还分泌黏液、保护和润滑大肠黏膜，以利排便。

三、消化管图谱

1. **消化管的基本结构** 图13-1～图13-2。
2. **口腔** 图13-3～图13-19。
3. **食管** 图13-20～图13-34。
4. **胃** 图13-35～图13-74。
5. **肠** 图13-75～图13-146。

第十三章 消化管 Digestive Tract

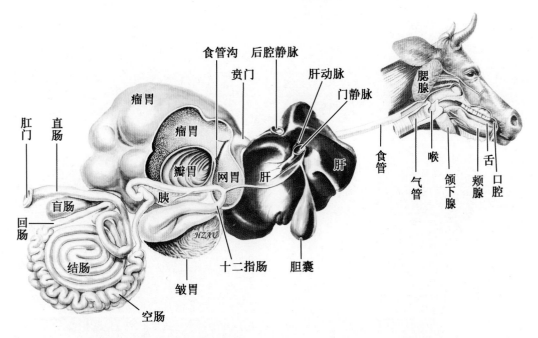

图 13-1　牛消化系统组成示意图

Fig.13-1　composition diagram of cow digestive system

口腔—oral cavity　舌—tongue　颊腺—cheek gland　颌下腺—salivary gland　腮腺—parotid gland　喉—larynx　气管—trachea　食管—esophagus　肝—liver　门静脉—portal vein　肝动脉—liver artery　后腔静脉—postcaval vein　胆囊—gall bladder　贲门—cardia　食管沟—esophageal ditch　瘤胃—rumen　网胃—reticulum　瓣胃—omasum　皱胃—abomasum　十二指肠—duodenum　胰—pancreas　空肠—jejunum　回肠—ileum　盲肠—cecum　结肠—colon　直肠—rectum　肛门—anus

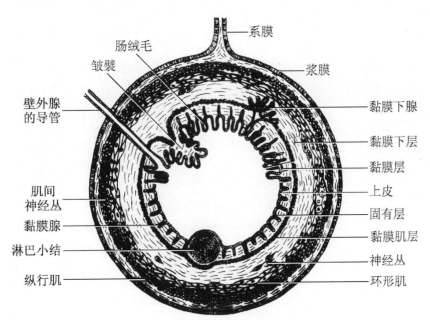

图 13-2　消化管的基本结构模式图

Fig.13-2　basic structure diagram of digestive duct

肠绒毛—intestinal villus　皱襞—plica　壁外腺的导管—duct of outside gland　肌间神经丛—myenteric nervous plexus　黏膜腺—mucosal gland　淋巴小结—lymphoid nodule　纵行肌—longitudinal muscle　系膜—mesentery　浆膜—serosa　黏膜下腺—submucosal gland　黏膜下层—submucosa　黏膜层—mucosa　上皮—epithelium　固有层—proper layer　黏膜肌层—muscularis mucosae　神经丛—nerve plexus　环形肌—circular muscle

409

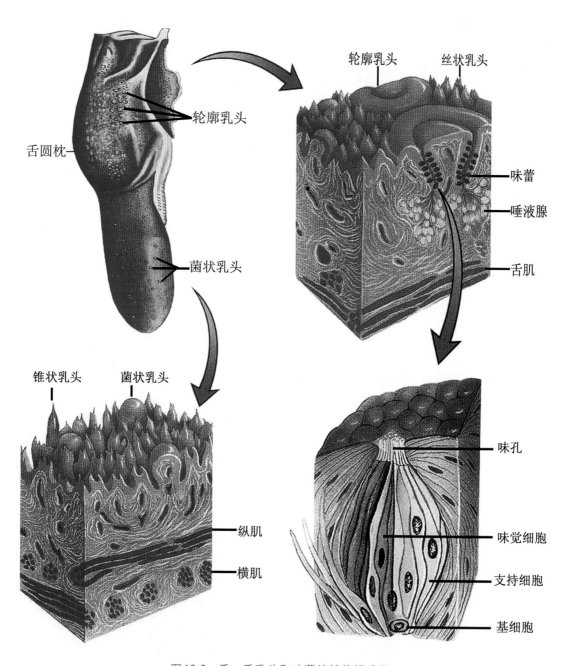

图 13-3 舌、舌乳头和味蕾的结构模式图

Fig.13-3　diagram of tongue, papilla and taste bud

舌圆枕—torus linguae　轮廓乳头—vallatae papilla　菌状乳头—fungiform papilla　锥状乳头—cone papilla
纵肌—longitudinal muscle　横肌—transverse muscle　丝状乳头—filiform papilla　味蕾—taste bud
唾液腺—salivary gland　舌肌—tongue muscle　味孔—taste pore　味觉细胞—taste cell
支持细胞—supporting cell　基细胞—basal cell

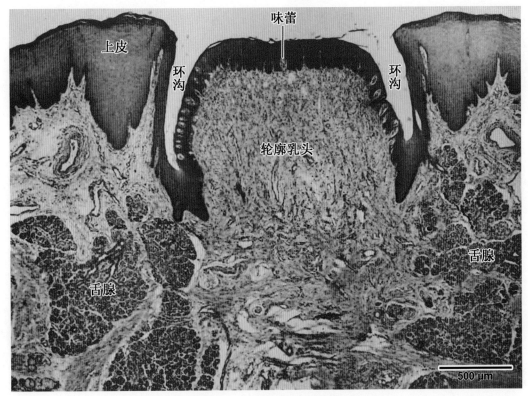

图13-4 牛舌低倍像（HE）

Fig.13-4 low magnification of cow tongue（HE）

上皮—epithelium 环沟—ring groove 味蕾—taste bud 轮廓乳头—vallatae papilla 舌腺—tongue gland

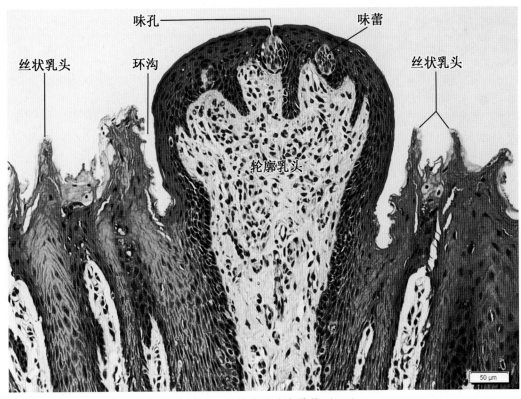

图13-5 牛轮廓乳头高倍像（HE）

Fig.13-5 high magnification of cattle vallatae papilla（HE）

丝状乳头—filiform papilla 环沟—ring groove 轮廓乳头—vallatae papilla 味孔—taste pore 味蕾—taste bud

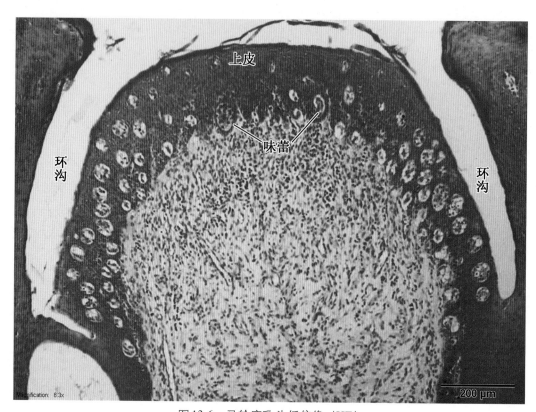

图13-6 马轮廓乳头低倍像（HE）

Fig.13-6　low magnification of horse vallatae papilla（HE）

环沟—ring groove　上皮—epithelium　味蕾—taste bud

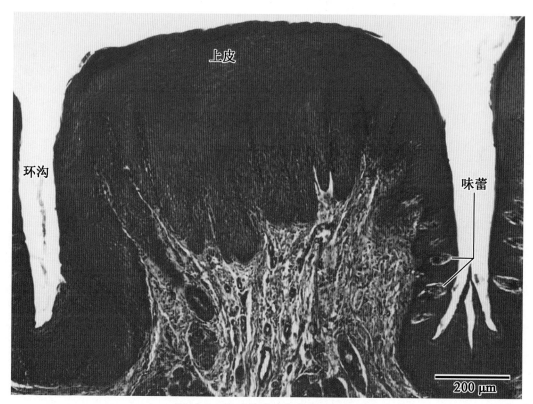

图13-7 马叶状乳头低倍像（HE）

Fig.13-7　low magnification of horse foliate papilla（HE）

环沟—ring groove　上皮—epithelium　味蕾—taste bud

第十三章 消化管 Digestive Tract

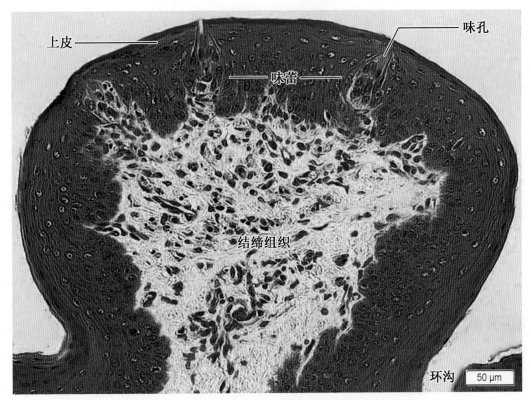

图 13-8 猪菌状乳头高倍像（HE）

Fig.13-8 high magnification of pig fungiform papilla（HE）

上皮—epithelium 味蕾—taste bud 味孔—taste pore 结缔组织—connective tissue 环沟—ring groove

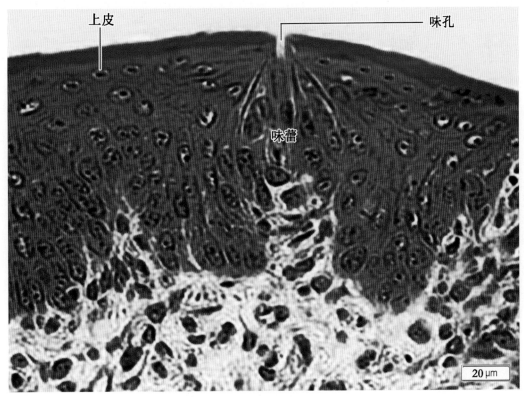

图 13-9 猪味蕾结构的高倍像（HE）

Fig.13-9 high magnification of pig taste bud（HE）

上皮—epithelium 味孔—taste pore 味蕾—taste bud

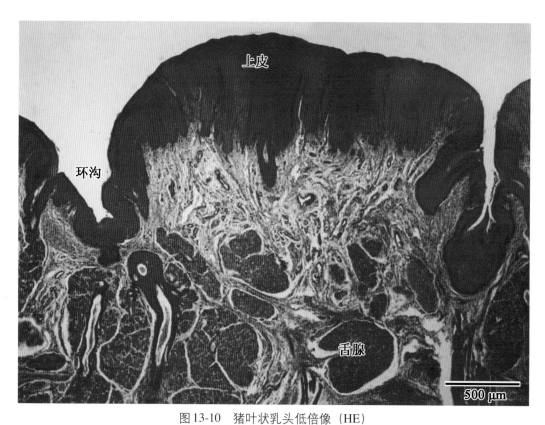

图 13-10 猪叶状乳头低倍像（HE）

Fig.13-10　low magnification of pig foliate papilla（HE）

环沟—ring groove　上皮—epithelium　舌腺—lingual gland

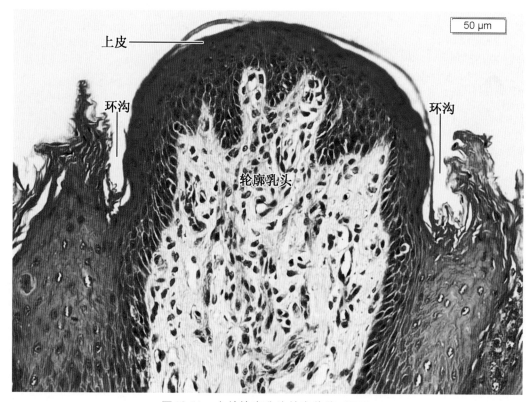

图 13-11 山羊轮廓乳头的高倍像（HE）

Fig.13-11　high magnification of goat circumvallate papilla（HE）

上皮—epithelium　环沟—ring groove　轮廓乳头—vallatae papilla

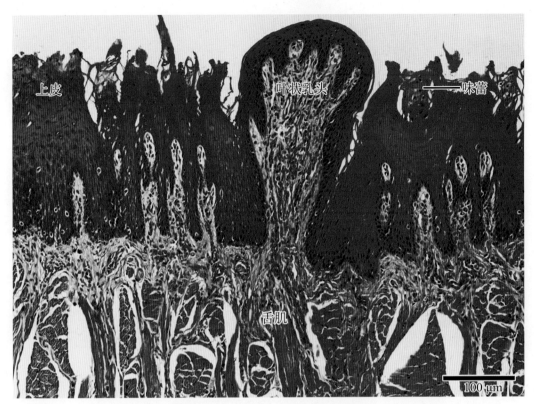

图 13-12 犬叶状乳头中倍像（HE）

Fig.13-12 mid magnification of dog foliate papillae（HE）

上皮—epithelium 叶状乳头—foliate papilla 味蕾—taste bud 舌肌—lingual muscle

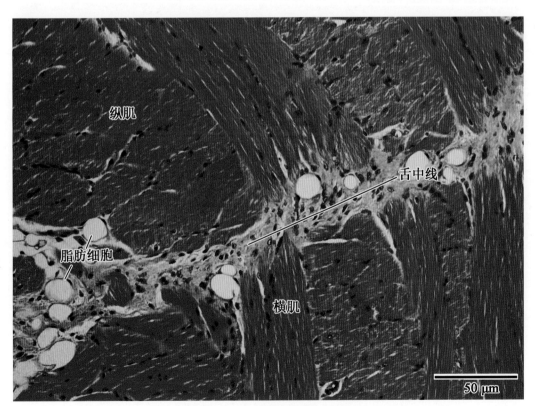

图 13-13 犬舌肌的高倍像（HE）

Fig.13-13 high magnification of dog tongue muscle（HE）

脂肪细胞—fat cell 纵肌—longitudinal muscle 横肌—transverse muscle 舌中线—tongue midline

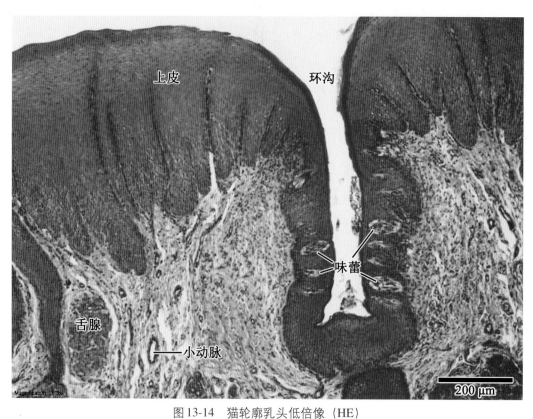

图 13-14 猫轮廓乳头低倍像（HE）

Fig.13-14　low magnification of cat circumvallate papilla（HE）

上皮—epithelium　环沟—ring groove　味蕾—taste bud　舌腺—lingual gland　小动脉—arteriole

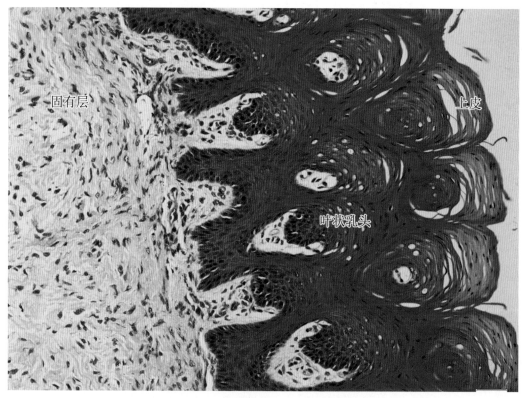

图 13-15　兔叶状乳头中倍像（HE）

Fig.13-15　mid magnification of rabbit foliate papillae（HE）

固有层—proper layer　叶状乳头—foliate papilla　上皮—epithelium

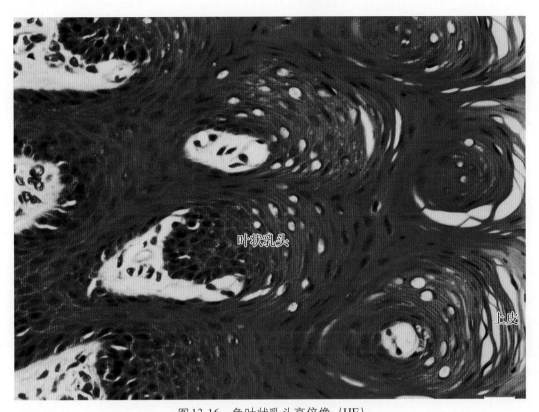

图13-16 兔叶状乳头高倍像（HE）
Fig.13-16 high magnification of rabbit foliate papillae（HE）
叶状乳头—foliate papilla　上皮—epithelium

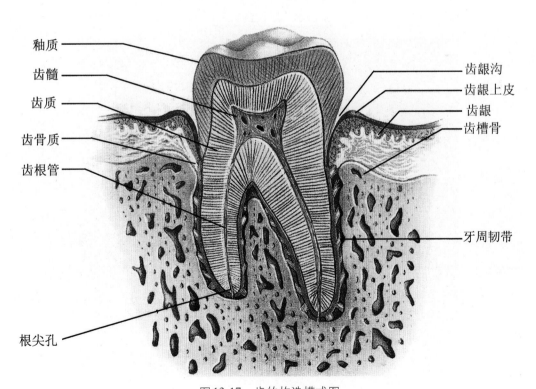

图13-17 齿的构造模式图
Fig.13-17 diagram of tooth structure
釉质—enamel　齿髓—dental pulp　齿质—dentine　齿骨质—crusta petrosa dentis　齿根管—pulp canal
根尖孔—apical foramen　齿龈沟—gingival sulcus　齿龈上皮—gum epithelial　齿龈—gingiva
齿槽骨—alveolar bone　牙周韧带—periodontal ligament

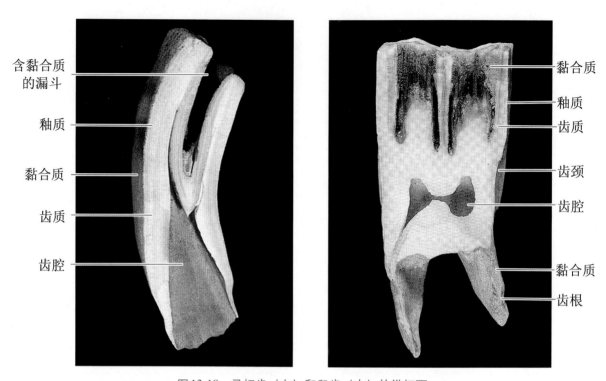

图 13-18 马切齿（左）和臼齿（右）的纵切面

Fig.13-18 longitudinal section of horse incisor(left)and molar(right)

含黏合质的漏斗—funnel containing adhesive material　釉质—enamel　黏合质—cementin　齿质—dentine
齿腔—pulp cavity of a tooth　齿颈— tooth neck　齿根—dedendum

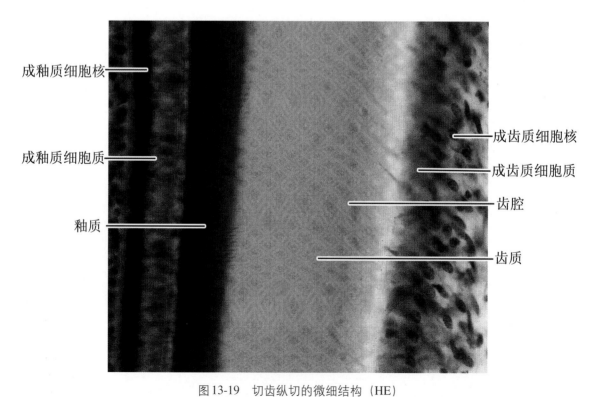

图 13-19 切齿纵切的微细结构（HE）

Fig.13-19 longitudinal section of incisor（HE）

成釉质细胞核—enamel nucleus　成釉质细胞质—into enamel cytoplasm　釉质—enamel　成齿质细胞核—odontoblast nucleus　成齿质细胞质—odontoblast cytoplasm　齿腔—pulp cavity of a tooth　齿质—dentine

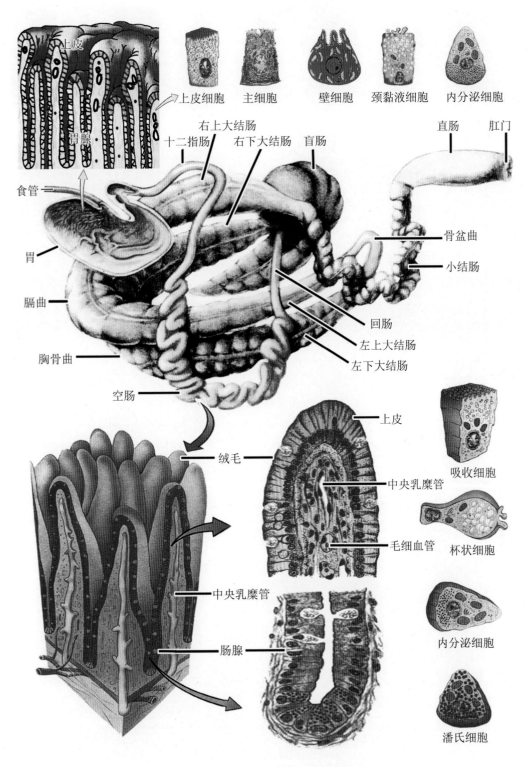

图 13-20　马消化管组成示意图
Fig.13-20　composition diagram of horse digestive system

食管—esophagus　胃—stomach　十二指肠—duodenum　空肠—jejunum　回肠—ileum　盲肠—cecum
右下大结肠—right ventral colon　胸骨曲—flexure sternal　左下大结肠—lower left greater colon　骨盆曲—pelvic bending
左上大结肠— left superior great colon　膈曲—diaphragmatic flexure　右上大结肠—right superior great colon
小结肠—small colon　直肠—rectum　肛门—anus　上皮细胞—epithelial cell　胃腺—gastric gland　主细胞—chief cell
壁细胞—parietal cell　颈黏液细胞—mucous neck cell　内分泌细胞—endocrine cell　绒毛—villi　中央乳糜管—central lacteal
肠腺—intestinal gland　上皮—epithelium　毛细血管—capillary　吸收细胞—absorptive cell　杯状细胞—goblet cell
潘氏细胞—Paneth cell

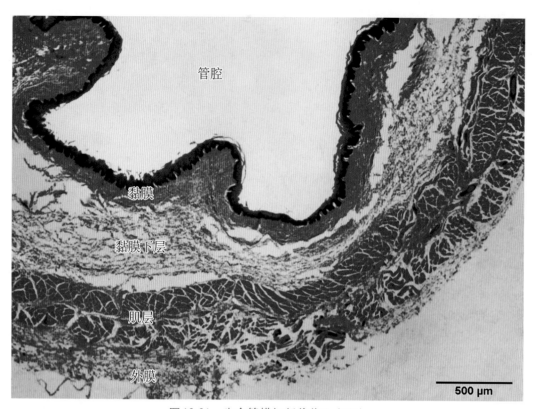

图13-21 牛食管横切低倍像1（HE）

Fig.13-21 low magnification of cow esophagus transaction 1（HE）

管腔—lumen 黏膜—mucosa 黏膜下层—submucosa 肌层—muscular layer 外膜—adventitia

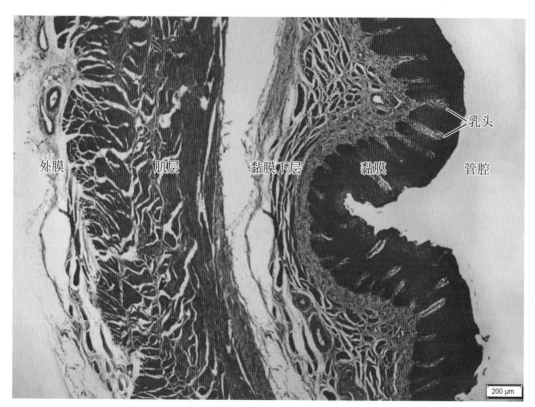

图13-22 牛食管横切低倍像2（HE）

Fig.13-22 low magnification of cow esophagus transaction 2（HE）

外膜—adventitia 肌层—muscular layer 黏膜下层—submucosa 黏膜—mucosa 管腔—lumen 乳头—papilla

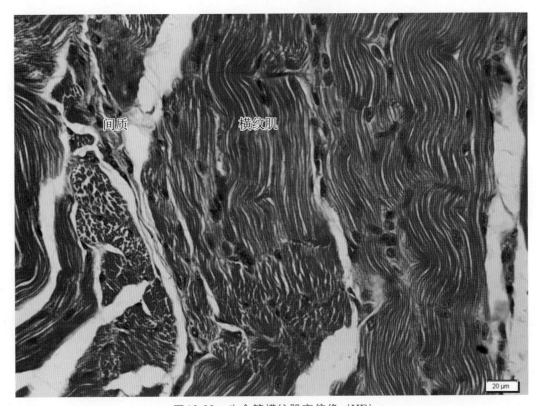

图13-23 牛食管横纹肌高倍像（HE）

Fig.13-23 high magnification of cow esophagus muscle（HE）

间质—mesenchyme 横纹肌—striated muscle

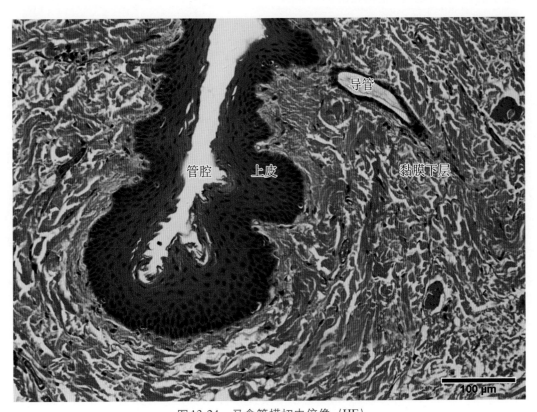

图13-24 马食管横切中倍像（HE）

Fig.13-24 mid magnification of horse esophagus transection（HE）

管腔—lumen 上皮—epithelium 黏膜下层—submucosa 导管—duct

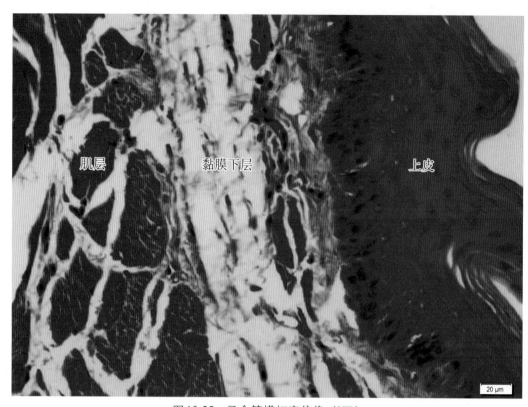

图13-25 马食管横切高倍像（HE）

Fig.13-25 high magnification of horse esophagus transection（HE）

肌层—muscular layer　黏膜下层—submucosa　上皮—epithelium

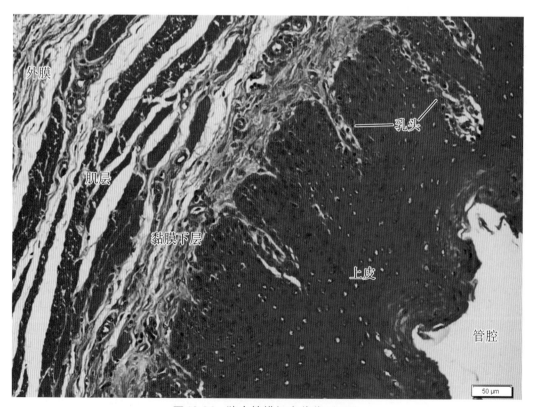

图13-26 猪食管横切高倍像（HE）

Fig.13-26 high magnification of pig esophagus transection（HE）

外膜—adventitia　肌层—muscular layer　黏膜下层—submucosa　上皮—epithelium　管腔—lumen　乳头—papilla

第十三章 消 化 管 Digestive Tract

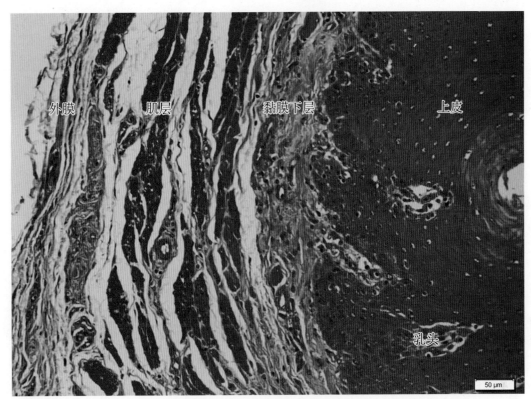

图 13-27 山羊食管横切高倍像（HE）

Fig.13-27 high magnification of goat esophagus transection（HE）

外膜—adventitia 肌层—muscular layer 黏膜下层—submucosa 上皮—epithelium 乳头—papilla

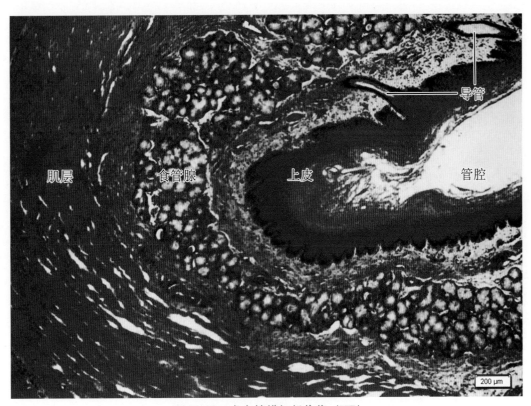

图 13-28 犬食管横切低倍像（HE）

Fig.13-28 low magnification of dog esophagus transection（HE）

肌层—muscular layer 食管腺—esophageal gland 上皮—epithelium 管腔—lumen 导管—duct

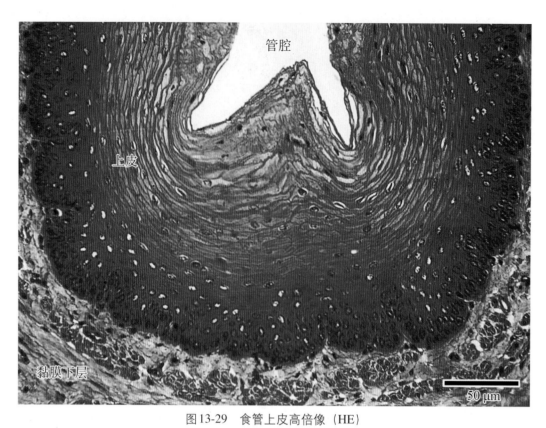

图13-29 食管上皮高倍像（HE）

Fig.13-29 high magnification of esophagus epithelium（HE）

黏膜下层—submucosa 上皮—epithelium 管腔—lumen

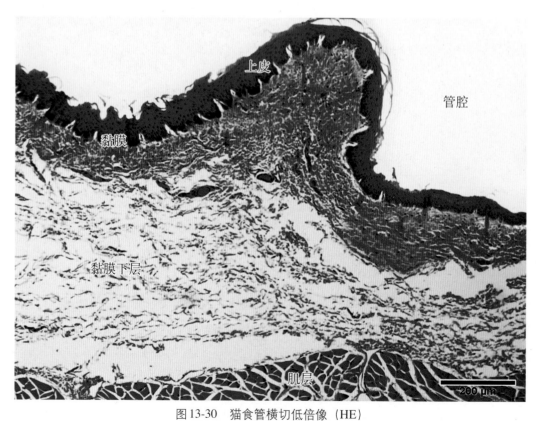

图13-30 猫食管横切低倍像（HE）

Fig.13-30 low magnification of cat esophagus transection（HE）

上皮—epithelium 管腔—lumen 黏膜—mucosa 黏膜下层—submucosa 肌层—muscular layer

第十三章 消化管 Digestive Tract

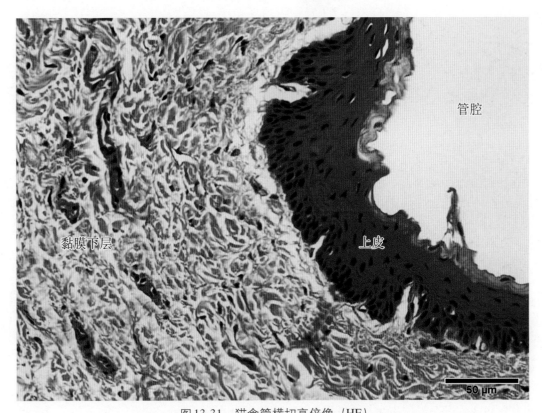

图13-31 猫食管横切高倍像（HE）

Fig.13-31 high magnification of cat esophagus transection（HE）

黏膜下层—submucosa 上皮—epithelium 管腔—lumen

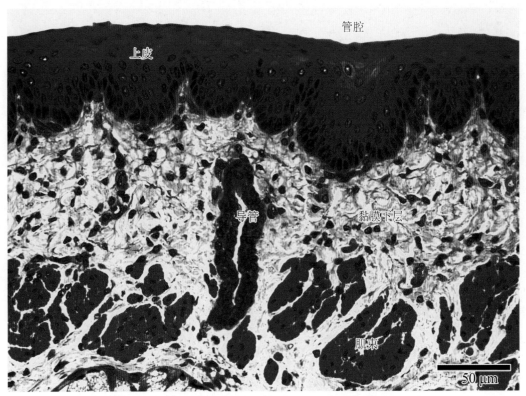

图13-32 兔食管横切高倍像（HE）

Fig.13-32 high magnification of rabbit esophagus transection（HE）

管腔—lumen 上皮—epithelium 导管—duct 黏膜下层—submucosa 肌束—muscle bundle

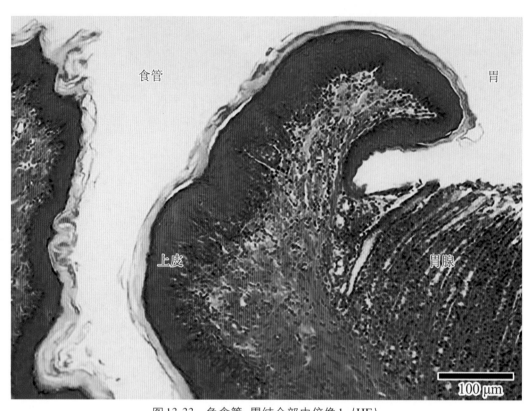

图13-33　兔食管-胃结合部中倍像1（HE）

Fig.13-33　mid magnification of rabbit esophagus-stomach junction 1（HE）

食管—esophagus　胃—stomach　上皮—epithelium　胃腺—gastric gland

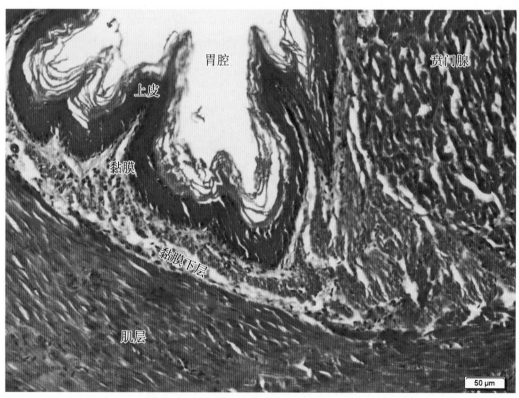

图13-34　兔食管-胃结合部中倍像2（HE）

Fig.13-34　mid magnification of rabbit esophagus-stomach junction 2（HE）

胃腔—gastral cavity　上皮—epithelium　黏膜—mucosa　黏膜下层—submucosa　肌层—muscle　贲门腺—cardiac gland

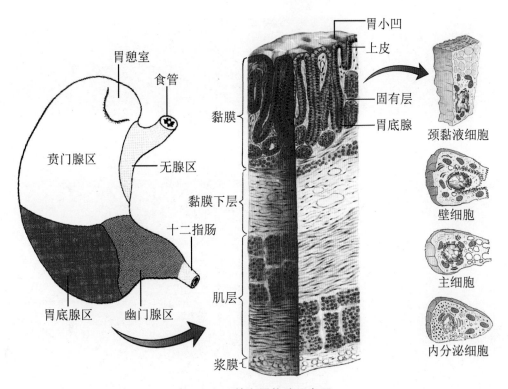

图 13-35　单室胃构造示意图

Fig.13-35　schematic diagram of simple-stomach structure

食管—esophagus　胃憩室—diverticulum　无腺区—no gland area　贲门腺区—cardiac gland region
胃底腺区—fundic gland region　幽门腺区—pyloric gland region　十二指肠—duodenum　黏膜—mucosa
黏膜下层—submucosa　肌层—muscular layer　浆膜—serosa　胃小凹—gastric pit　上皮—epithelium
固有层—proper layer　胃底腺—fundic gland　颈黏液细胞—mucous neck cell　壁细胞—parietal cell　主细胞—chief cell
内分泌细胞—endocrine cell

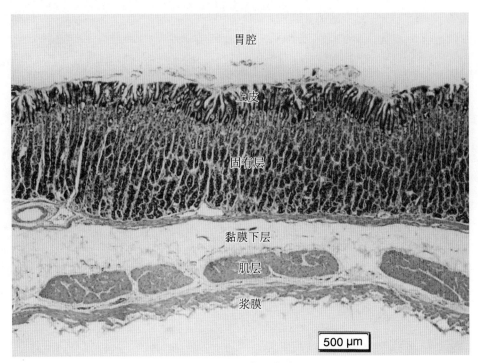

图 13-36　马胃底腺低倍像（HE）

Fig.13-36　low magnification of horse fundic gland（HE）

胃腔—gastral cavity　上皮—epithelium　固有层—proper layer　黏膜下层—submucosa　肌层—muscular layer　浆膜—serosa

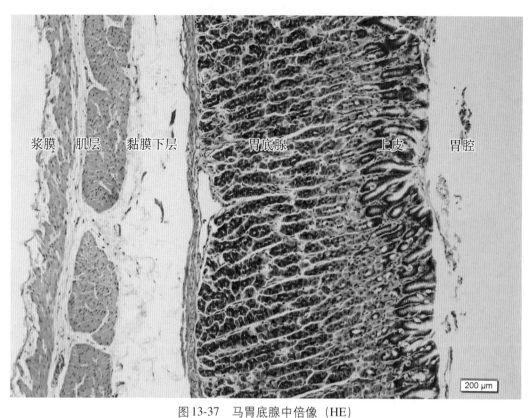

图13-37 马胃底腺中倍像（HE）

Fig.13-37 mid magnification of horse fundic gland （HE）

浆膜—serosa 肌层—muscular layer 黏膜下层—submucosa 胃底腺—fundic gland 上皮—epithelium
胃腔—gastral cavity

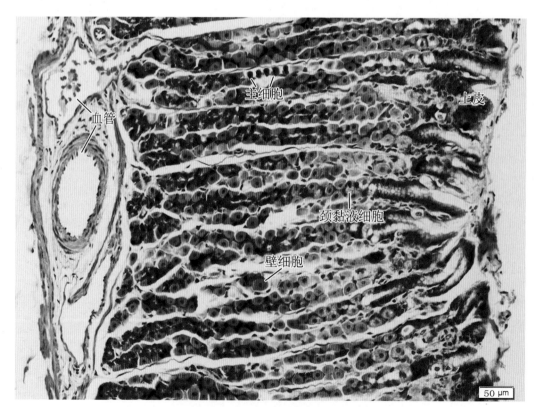

图13-38 马胃底腺高倍像（HE）

Fig.13-38 high magnification of horse fundic gland （HE）

血管—blood vessel 主细胞—chief cell 壁细胞—parietal cell 颈黏液细胞—mucous neck cell 上皮—epithelium

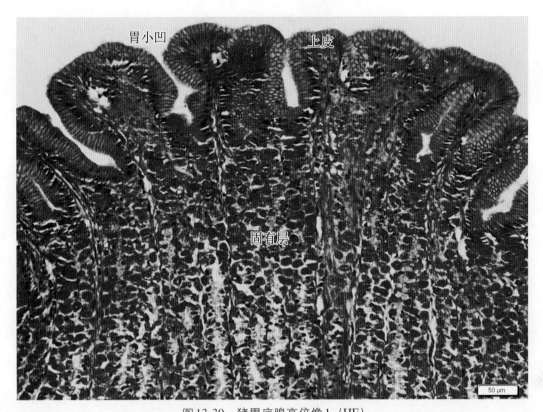

图13-39 猪胃底腺高倍像1（HE）

Fig.13-39　high magnification of pig fundic gland 1（HE）

胃小凹—gastric pit　上皮—epithelium　固有层—proper layer

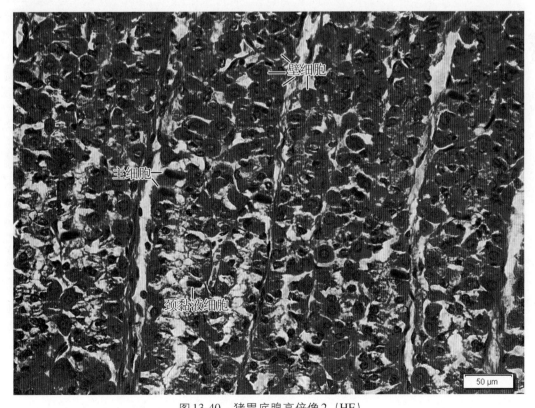

图13-40 猪胃底腺高倍像2（HE）

Fig.13-40　high magnification of pig fundic gland 2（HE）

主细胞—chief cell　壁细胞—parietal cell　颈黏液细胞—mucous neck cell

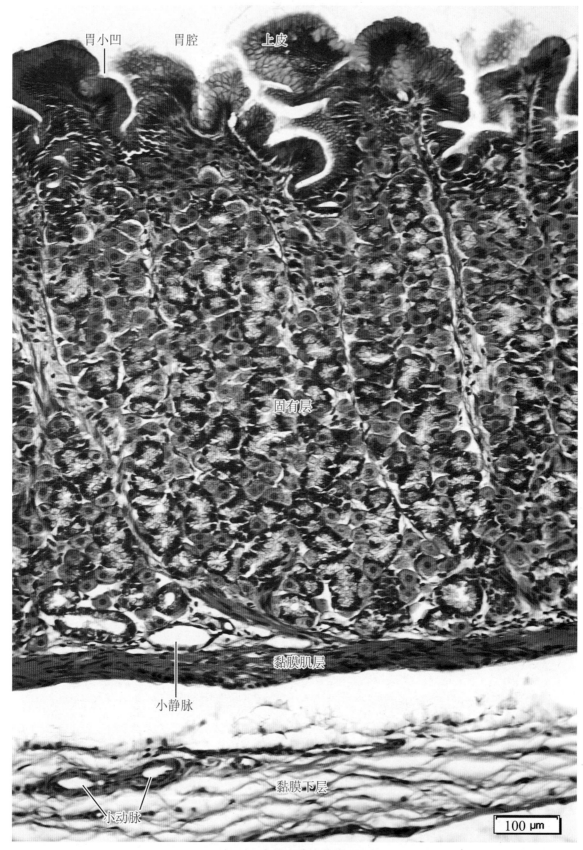

图13-41 犬胃底腺中倍像（HE）

Fig.13-41 mid magnification of dog fundic gland（HE）

胃小凹—gastric pit 胃腔—gastral cavity 上皮—epithelium 固有层—proper layer

小静脉—veinlet 小动脉—arteriole 黏膜肌层—muscularis mucosae 黏膜下层—submucosa

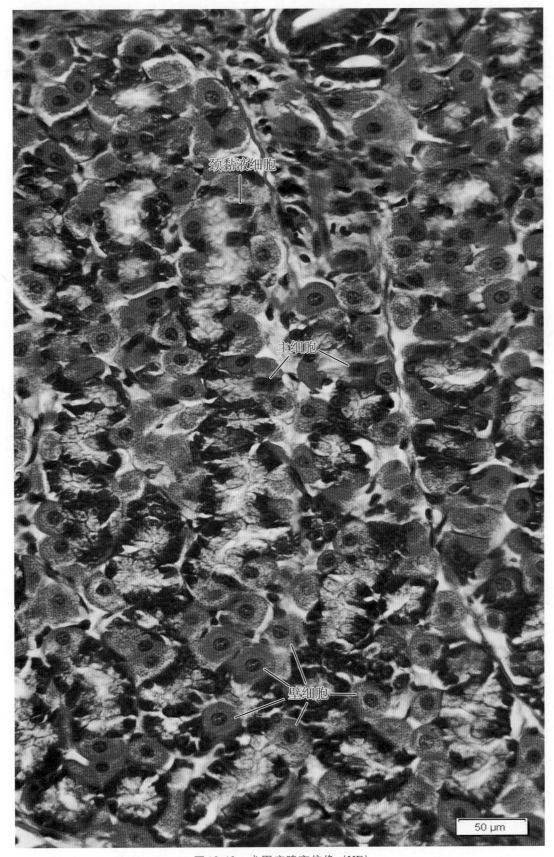

图13-42 犬胃底腺高倍像（HE）

Fig.13-42 high magnification of dog fundic gland（HE）

颈黏液细胞—mucous neck cell 壁细胞—parietal cell 主细胞—chief cell

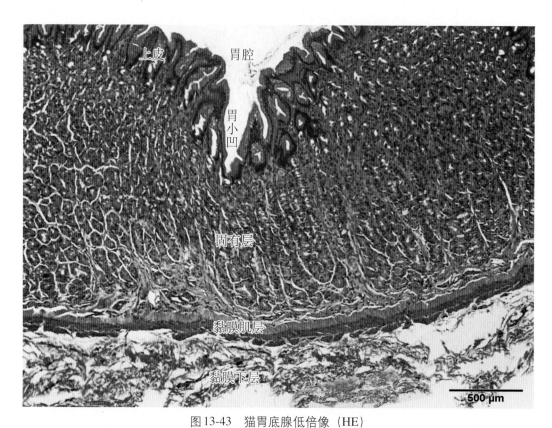

图13-43 猫胃底腺低倍像（HE）

Fig.13-43 low magnification of cat fundic gland （HE）

上皮—epithelium 胃腔—gastral cavity 胃小凹—gastric pit 固有层—proper layer

黏膜肌层—muscularis mucosae 黏膜下层—submucosa

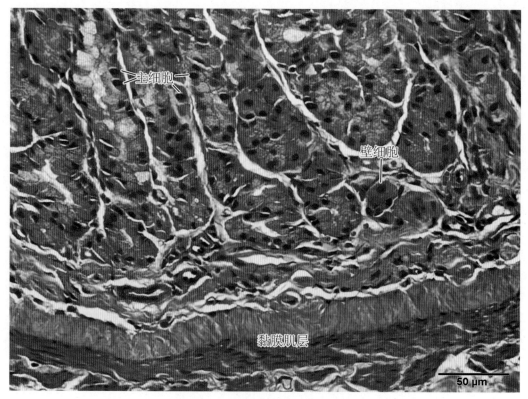

图13-44 猫胃底腺高倍像（HE）

Fig.13-44 high magnification of cat fundic gland （HE）

主细胞—chief cell 壁细胞—parietal cell 黏膜肌层—muscularis mucosae

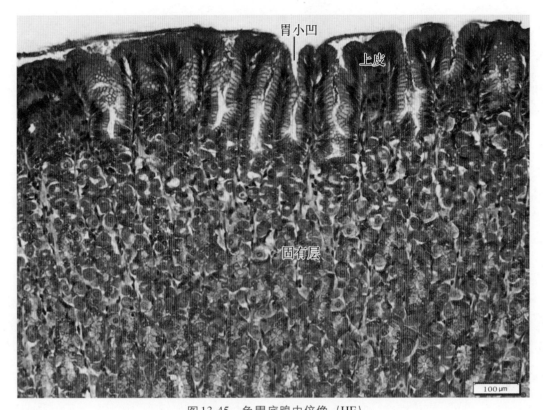

图13-45 兔胃底腺中倍像（HE）

Fig.13-45 mid magnification of rabbit fundic gland（HE）

胃小凹—gastric pit 上皮—epithelium 固有层—proper layer

图13-46 兔胃底腺高倍像（HE）

Fig.13-46 high magnification of rabbit fundic gland（HE）

胃小凹—gastric pit 上皮—epithelium 颈黏液细胞—mucous neck cell 主细胞—chief cell 壁细胞—parietal cell

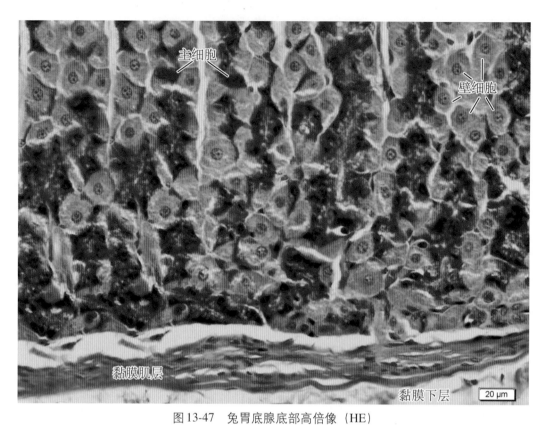

图 13-47 兔胃底腺底部高倍像（HE）

Fig.13-47 high magnification of rabbit fundic gland bottom（HE）

主细胞—chief cell 壁细胞—parietal cell 黏膜肌层—muscularis mucosae 黏膜下层—submucosa

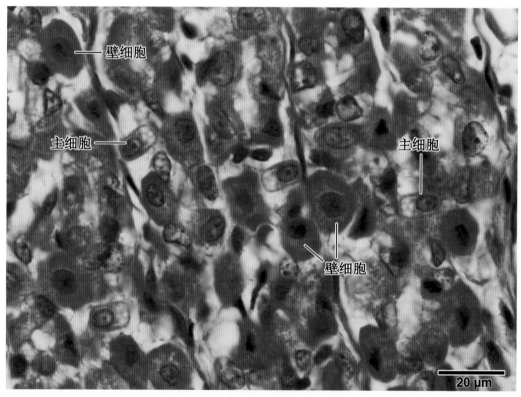

图 13-48 胃底腺体部高倍像（HE）

Fig.13-48 high magnification of fundic gland（HE）

壁细胞—parietal cell 主细胞—chief cell

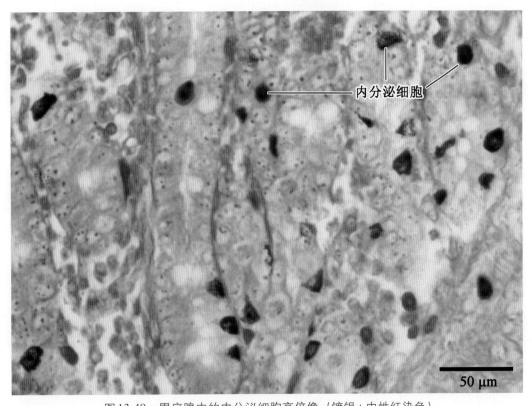

图13-49　胃底腺中的内分泌细胞高倍像（镀银+中性红染色）
Fig.13-49　high magnification of endocrine cells in fundic gland（silver+neutral red stain）
内分泌细胞—endocrine cell

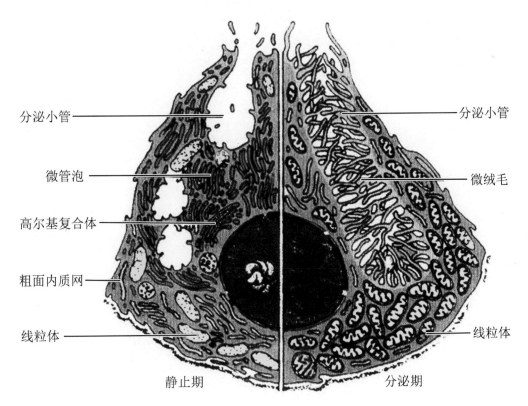

图13-50　壁细胞超微结构模式图
Fig.13-50　ultrastructure images of parietal cell

静止期—resting stage　分泌期—secretory stage　分泌小管—secretory tubule　微管泡—tubulovesicle
高尔基复合体—Golgi complex　粗面内质网—rough endoplasmic reticulum　线粒体—mitochondria　微绒毛—microvilli

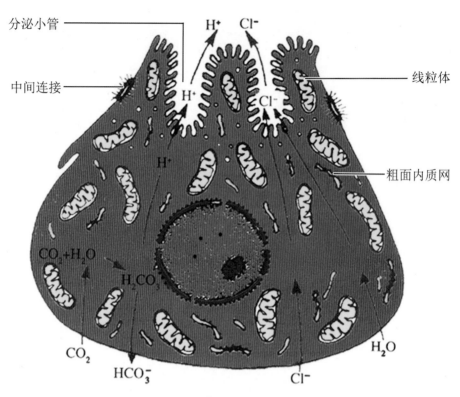

图 13-51 壁细胞合成盐酸示意图
Fig.13-51 schematic diagram of parietal cell synthesis of HCl
分泌小管—secretory tubule 中间连接—intermediate junction 线粒体—mitochondria
粗面内质网—rough endoplasmic reticulum

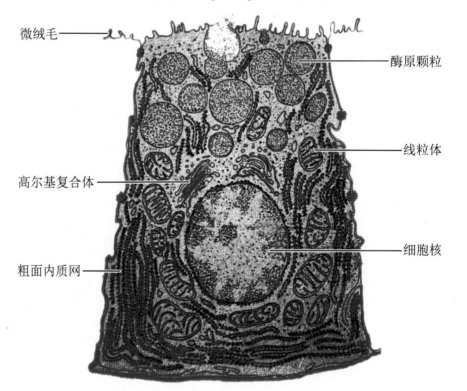

图 13-52 主细胞超微结构模式图
Fig.13-52 ultrastructure images of chief cell
微绒毛—microvilli 高尔基复合体—Golgi complex 粗面内质网—rough endoplasmic reticulum
酶原颗粒—zymogen granule 线粒体—mitochondria 细胞核—nucleus

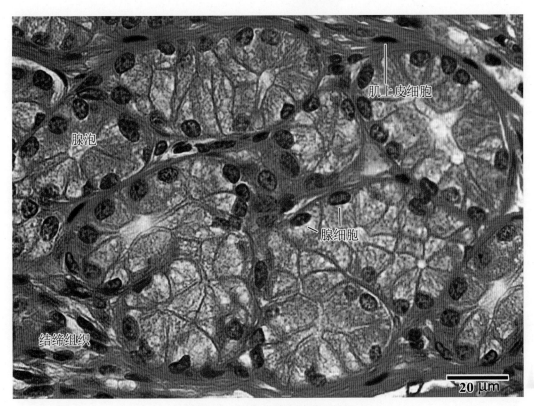

图 13-53 兔贲门腺高倍像（HE）

Fig.13-53 high magnification of rabbit cardiac gland（HE）

腺泡—acinus 结缔组织—connective tissue 肌上皮细胞—myoepithelial cell 腺细胞—gland cell

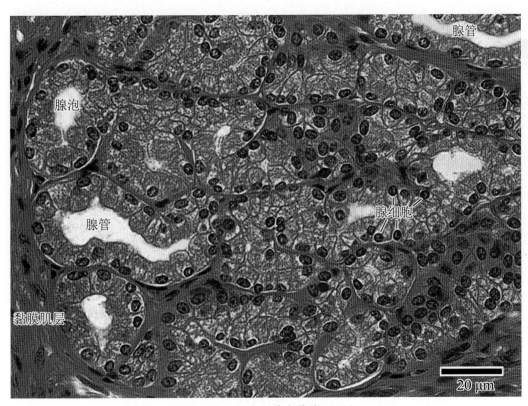

图 13-54 兔幽门腺高倍像（HE）

Fig.13-54 high magnification of rabbit pyloric gland（HE）

腺泡—acinus 腺管—glandular duct 黏膜肌层—muscularis mucosae 腺细胞—gland cell

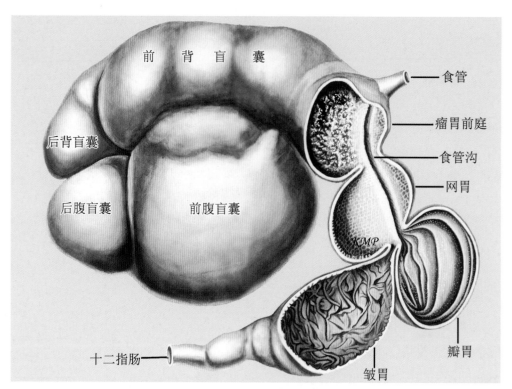

图13-55　牛胃结构示意图

Fig.13-55　schematic diagram of cow stomach structure

前背盲囊—front dorsal sac　后背盲囊—rear dorsal sac　前腹盲囊—front ventral sac
后腹盲囊—posterior ventral sac　食管—esophagus　瘤胃前庭—atrium ventralis　食管沟—esophagus ditch
网胃—reticulum　瓣胃—omasum　皱胃—abomasum　十二指肠—duodenum

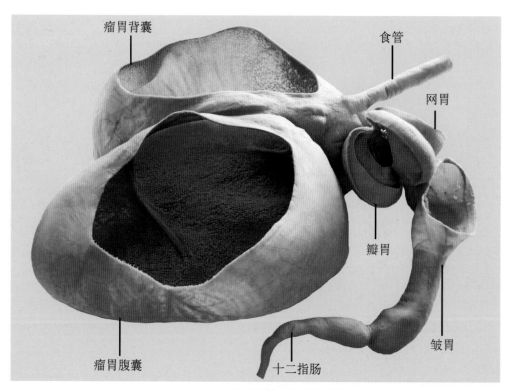

图13-56　山羊胃结构（塑化标本）

Fig.13-56　structure of goat stomach（plastic specimen）

瘤胃背囊—rumen dorsal sac　瘤胃腹囊—rumen ventral sac　食管—esophagus　网胃—reticulum　瓣胃—omasum
皱胃—abomasum　十二指肠—duodenum

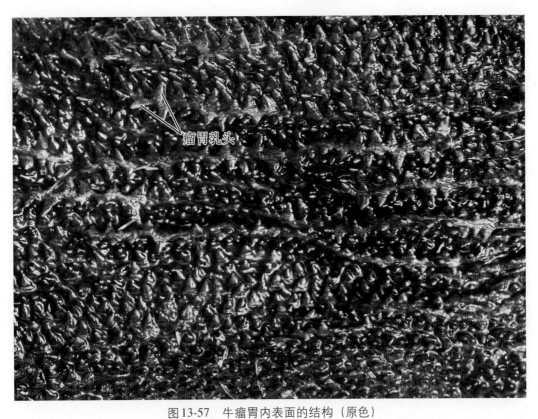

图13-57 牛瘤胃内表面的结构（原色）
Fig.13-57 structure of cow rumen internal surface（original color）
瘤胃乳头—rumen papillae

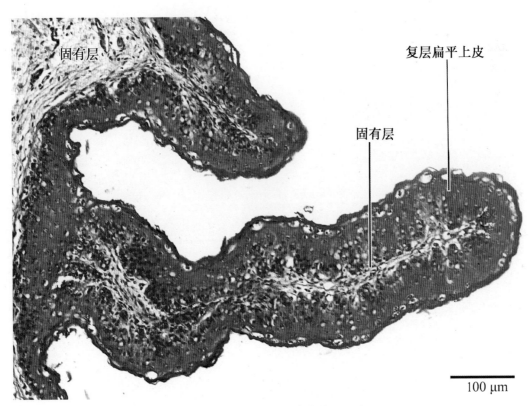

图13-58 牛瘤胃乳头中倍像（HE）
Fig.13-58 mid magnification of cow rumen papillae（HE）
固有层—proper layer 复层扁平上皮— stratified squamous epithelium

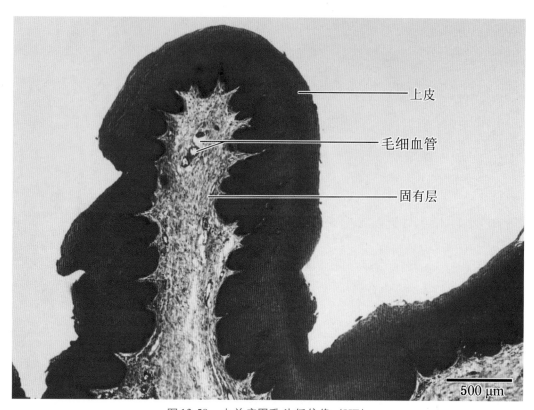

图13-59 山羊瘤胃乳头低倍像（HE）

Fig.13-59 low magnification of goat rumen papillae（HE）

上皮—epithelium 毛细血管—capillary 固有层—proper layer

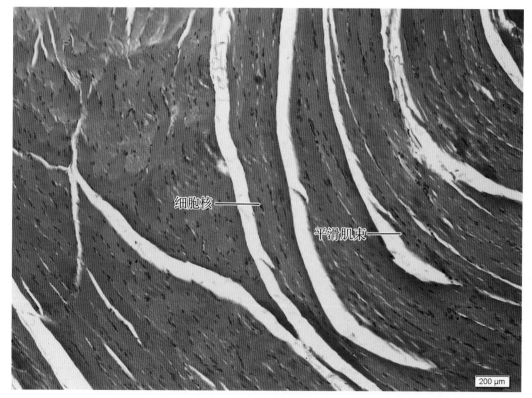

图13-60 牛瘤胃肌层低倍像（HE）

Fig.13-60 low magnification of cow rumen muscular layer（HE）

细胞核—nucleus 平滑肌束—smooth muscle bundle

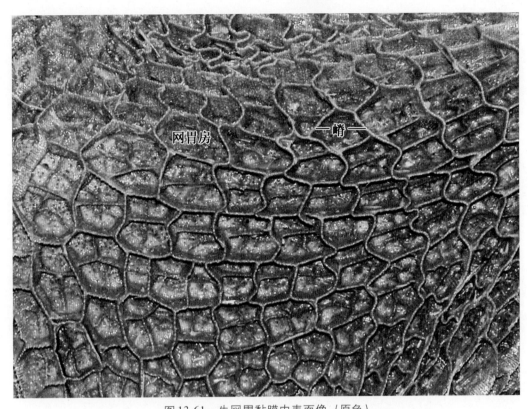

图13-61 牛网胃黏膜内表面像（原色）
Fig.13-61 internal surface of cow reticulum mucosa（original color）
网胃房—reticulum cell 嵴—ridge

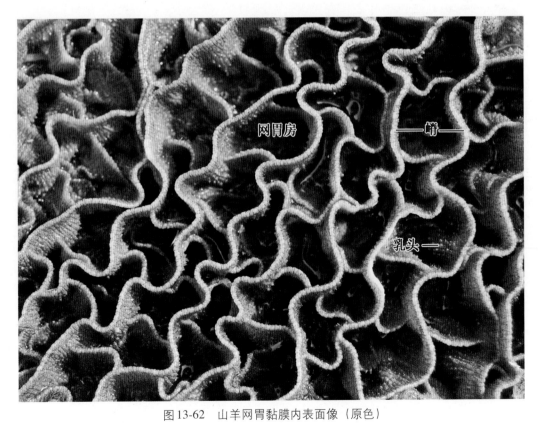

图13-62 山羊网胃黏膜内表面像（原色）
Fig.13-62 internal surface of goat reticulum mucosa（original color）
网胃房—reticulum cell 嵴—ridge 乳头—papilla

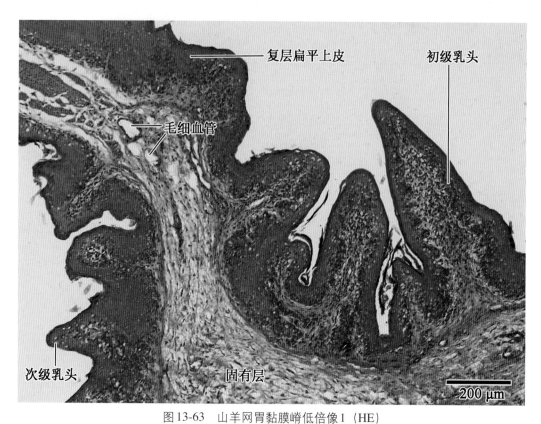

图13-63 山羊网胃黏膜嵴低倍像1（HE）

Fig.13-63 low magnification of goat reticulum mucosa ridge 1（HE）

复层扁平上皮—stratified squamous epithelium　毛细血管—capillary
初级乳头—primary papilla　次级乳头—secondary papilla　固有层—proper layer

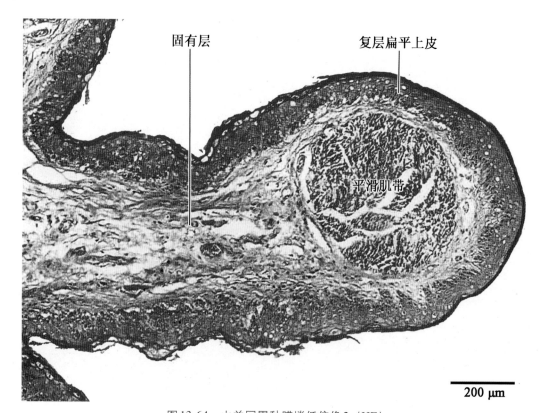

图13-64 山羊网胃黏膜嵴低倍像2（HE）

Fig.13-64 low magnification of goat reticulum mucosa ridge 2（HE）

固有层—proper layer　复层扁平上皮—stratified squamous epithelium　平滑肌带—smooth muscle bundle

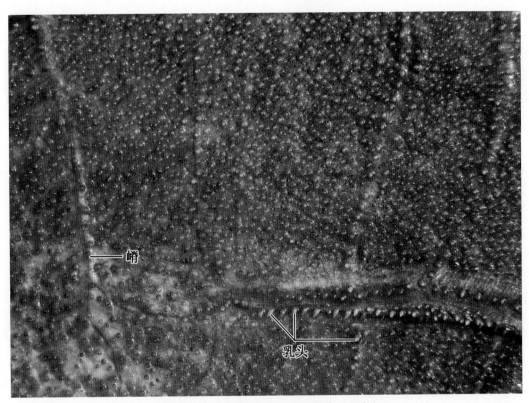

图13-65　牛瓣胃叶片黏膜表面像（原色）

Fig.13-65　surface of cow omasum lamina（original color）

嵴—ridge　乳头—papilla

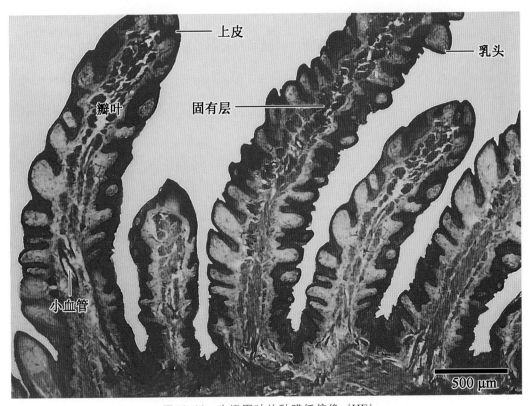

图13-66　牛瓣胃叶片黏膜低倍像（HE）

Fig.13-66　low magnification of cow omasum lamina（HE）

上皮—epithelium　瓣叶—omasum lamina　固有层—proper layer　乳头—papilla　小血管—small blood vessel

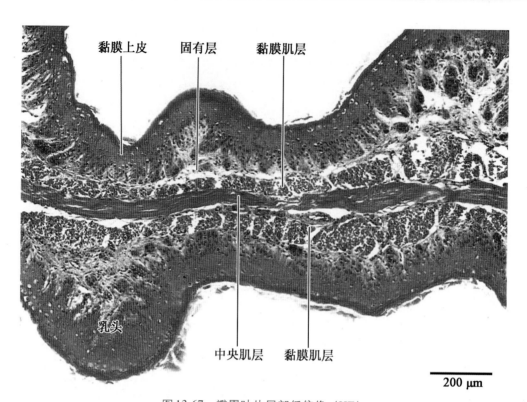

图13-67 瓣胃叶片局部低倍像（HE）

Fig.13-67 low magnification of omasum lamina（HE）

黏膜上皮—mucosa epithelium 固有层—proper layer 黏膜肌层—muscularis mucosae
乳头—papilla 中央肌层—middle muscular layer

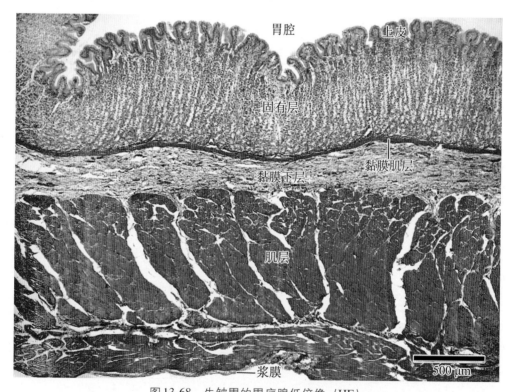

图13-68 牛皱胃的胃底腺低倍像（HE）

Fig.13-68 low magnification of cow abomasum fundic gland（HE）

胃腔—gastral cavity 上皮—epithelium 固有层—proper layer 黏膜下层—submucosa
黏膜肌层—muscularis mucosae 肌层—muscular layer 浆膜—serosa

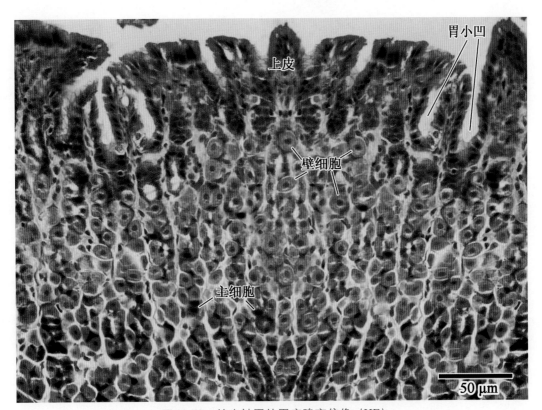

图13-69 犊牛皱胃的胃底腺高倍像（HE）

Fig.13-69 high magnification of calf abomasum fundic gland（HE）

上皮—epithelium 胃小凹—gastric pit 壁细胞—parietal cell 主细胞—chief cell

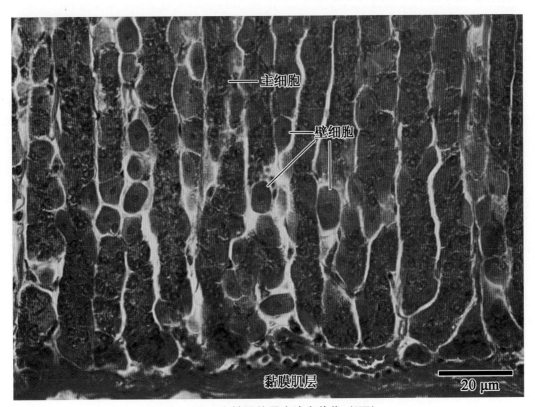

图13-70 牛皱胃的胃底腺高倍像（HE）

Fig.13-70 high magnification of cow abomasum fundic gland（HE）

主细胞—chief cell 壁细胞—parietal cell 黏膜肌层—muscularis mucosae

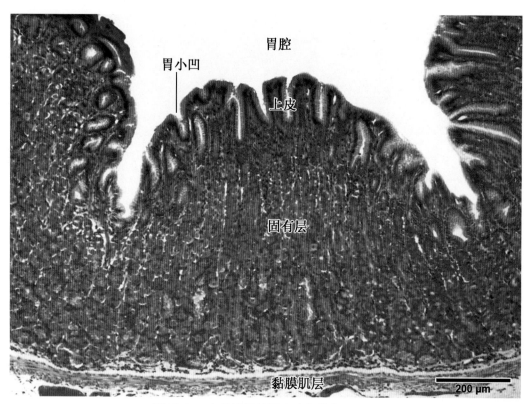

图13-71 山羊皱胃的胃底腺低倍像（HE）

Fig.13-71 low magnification of goat abomasum fundic gland（HE）

胃小凹—gastric pit　胃腔—gastral cavity　上皮—epithelium　固有层—proper layer　黏膜肌层—muscularis mucosae

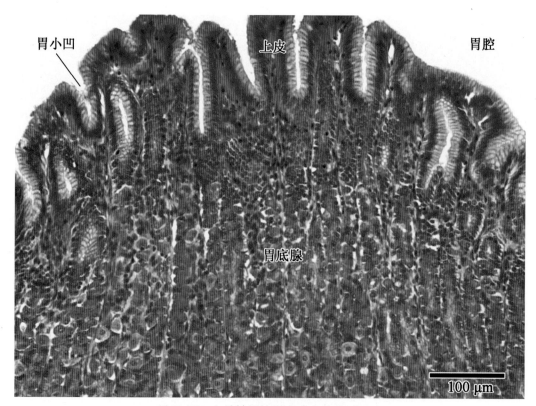

图13-72 山羊皱胃的胃底腺中倍像（HE）

Fig.13-72 mid magnification of goat abomasum fundic gland（HE）

胃小凹—gastric pit　上皮—epithelium　胃腔—gastral cavity　胃底腺—fundic gland

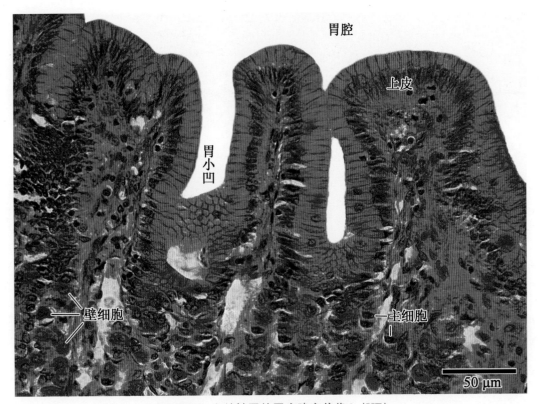

图13-73 山羊皱胃的胃底腺高倍像1（HE）

Fig.13-73 high magnification of goat abomasum fundic gland 1 （HE）

胃腔—gastral cavity 胃小凹—gastric pit 上皮—epithelium 壁细胞—parietal cell 主细胞—chief cell

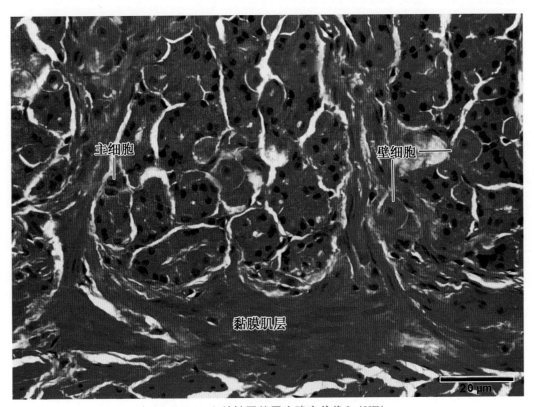

图13-74 山羊皱胃的胃底腺高倍像2（HE）

Fig.13-74 high magnification of goat abomasum fundic gland 2 （HE）

主细胞—chief cell 壁细胞—parietal cell 黏膜肌层—muscularis mucosae

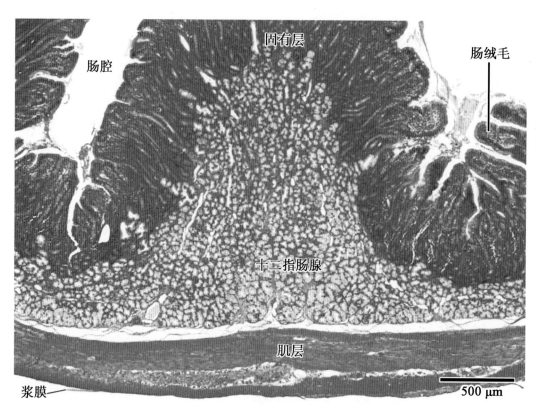

图13-75　马十二指肠横切低倍像（HE）

Fig.13-75　low magnification of horse duodenum transection（HE）

肠腔—intestinal lumen　固有层—proper layer　十二指肠腺—duodenal gland　肌层—muscle layer　浆膜—serosa

肠绒毛—intestinal villus

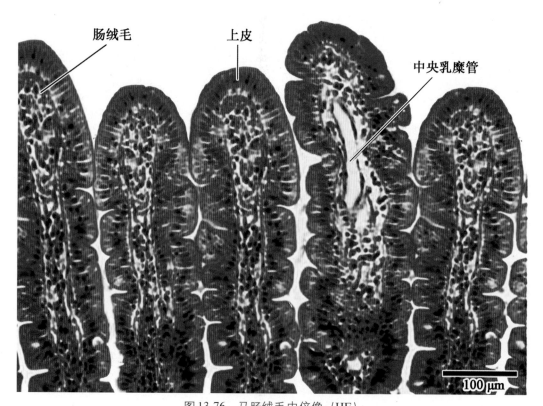

图13-76　马肠绒毛中倍像（HE）

Fig.13-76　mid magnification of horse intestinal villus（HE）

肠绒毛—intestinal villus　上皮—epithelium　中央乳糜管—central lacteal

图 13-77　肠绒毛横切高倍像（HE）

Fig.13-77　high magnification of intestinal villus transection（HE）

中央乳糜管—central lacteal　吸收细胞—absorptive cell　杯状细胞—goblet cell

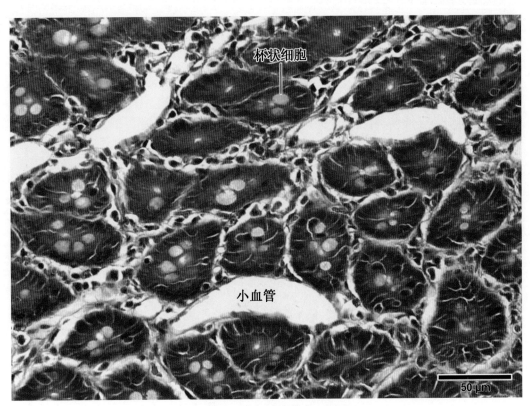

图 13-78　马肠腺高倍像（HE）

Fig.13-78　high magnification of horse intestinal gland（HE）

杯状细胞—goblet cell　小血管—small blood vessel

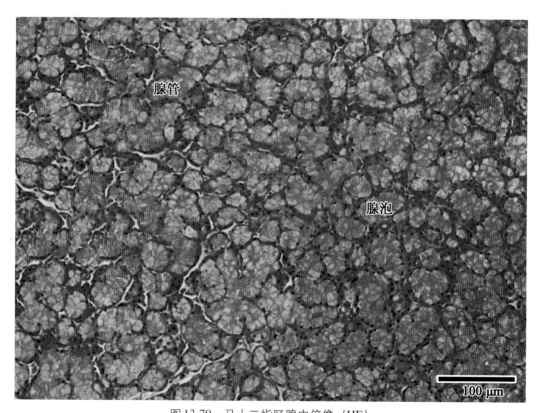

图13-79 马十二指肠腺中倍像（HE）

Fig.13-79 mid magnification of horse duodenal gland（HE）

腺管—glandular duct 腺泡—acinus

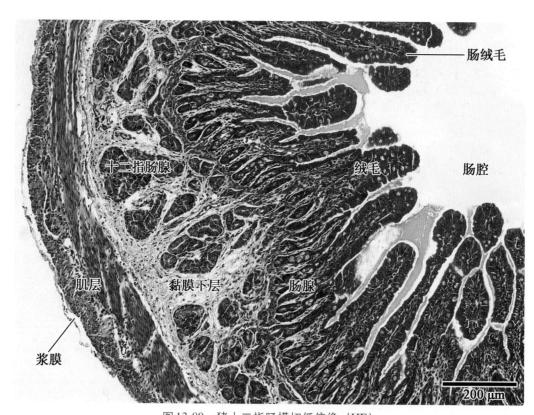

图13-80 猪十二指肠横切低倍像（HE）

Fig.13-80 low magnification of pig duodenum transection（HE）

浆膜—serosa 肌层—muscle layer 十二指肠腺—duodenal gland 黏膜下层—submucosa 肠腺—intestinal gland 肠绒毛—intestinal villus 肠腔—intestinal lumen

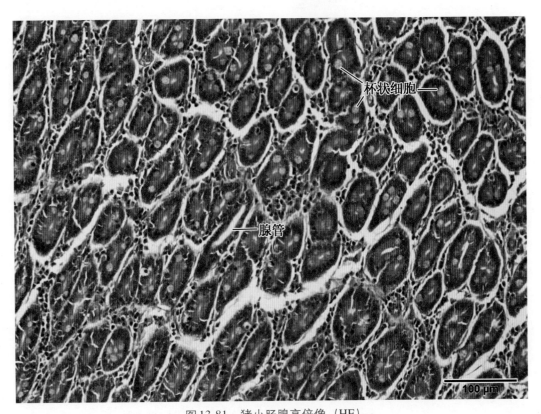

图 13-81 猪小肠腺高倍像（HE）

Fig.13-81 high magnification of pig small intestinal gland（HE）

腺管—glandular duct 杯状细胞—goblet cell

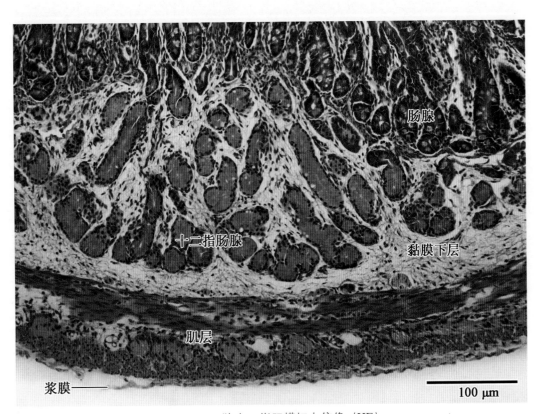

图 13-82 猪十二指肠横切中倍像（HE）

Fig.13-82 mid magnification of pig duodenum transection（HE）

肠腺—intestinal gland 十二指肠腺—duodenal gland 黏膜下层—submucosa 肌层—muscle layer 浆膜—serosa

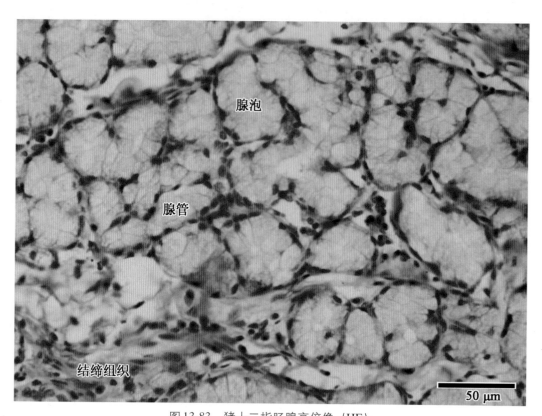

图 13-83 猪十二指肠腺高倍像（HE）

Fig.13-83 high magnification of pig duodenal gland（HE）

腺泡—acinus 腺管—glandular duct 结缔组织—connective tissue

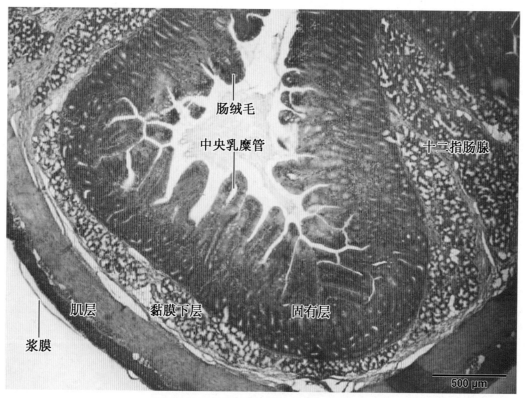

图 13-84 犬十二指肠横切低倍像（HE）

Fig.13-84 low magnification of dog duodenum transection（HE）

浆膜—serosa 肌层—muscle layer 黏膜下层—submucosa 肠绒毛—intestinal villus 中央乳糜管—central lacteal
固有层—proper layer 十二指肠腺—duodenal gland

第十三章 消化管 Digestive Tract

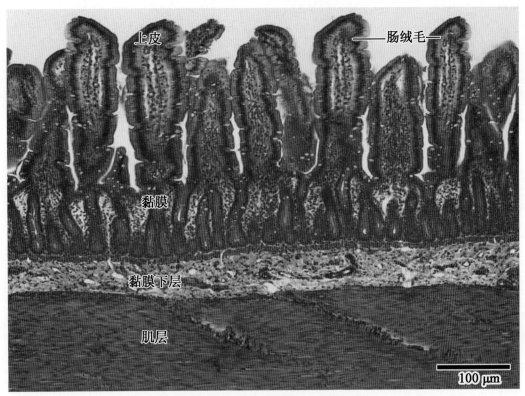

图13-85 犬十二指肠横切中倍像（HE）

Fig.13-85 mid magnification of dog duodenum transection（HE）

上皮—epithelium 肠绒毛—intestinal villus 黏膜—mucosa 黏膜下层—submucosa 肌层—muscle layer

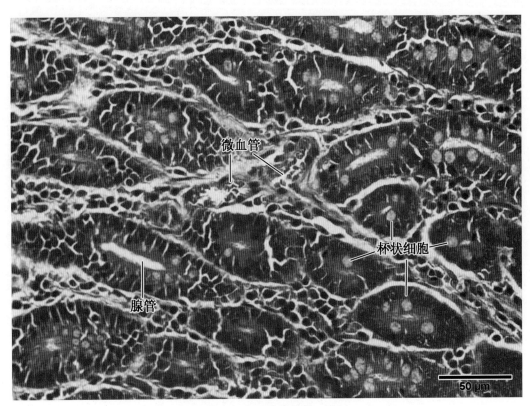

图13-86 犬肠腺高倍像（HE）

Fig.13-86 high magnification of dog intestinal gland（HE）

微血管—capillary 腺管—glandular duct 杯状细胞—goblet cell

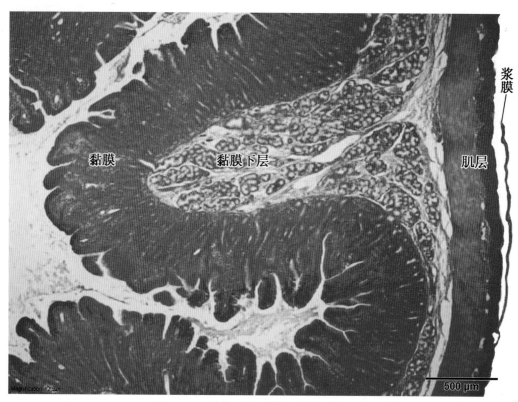

图13-87　猫十二指肠横切低倍像1（HE）

Fig.13-87　low magnification of cat duodenum transaction 1 （HE）

黏膜—mucosa　黏膜下层—submucosa　肌层—muscle layer　浆膜—serosa

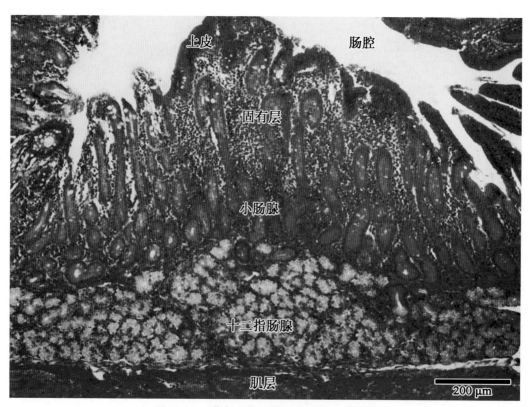

图13-88　猫十二指肠横切低倍像2（HE）

Fig.13-88　low magnification of cat duodenum transaction 2 （HE）

上皮—epithelium　固有层—proper layer　小肠腺—small intestinal gland　十二指肠腺—duodenal gland　肌层—muscle layer　肠腔—enteric cavity

第十三章 消化管 Digestive Tract

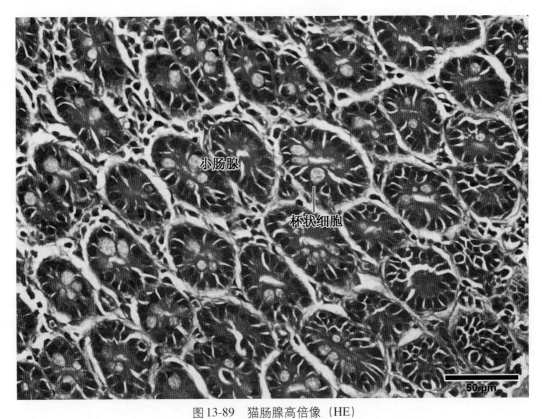

图 13-89 猫肠腺高倍像（HE）

Fig.13-89 high magnification of cat intestinal gland（HE）

小肠腺—small intestinal gland　杯状细胞—goblet cell

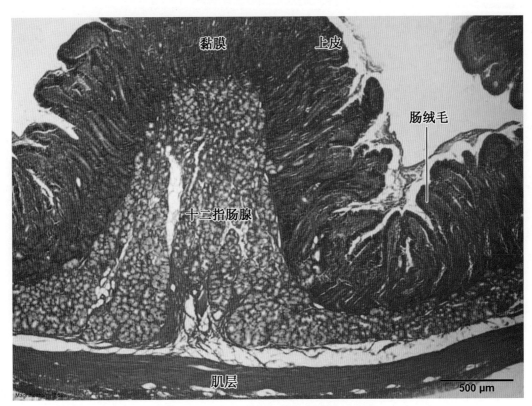

图 13-90 兔十二指肠横切低倍像（HE）

Fig.13-90 low magnification of rabbit duodenum transection（HE）

黏膜—mucosa　上皮—epithelium　十二指肠腺—duodenal gland　肠绒毛—intestinal villus　肌层—muscle layer

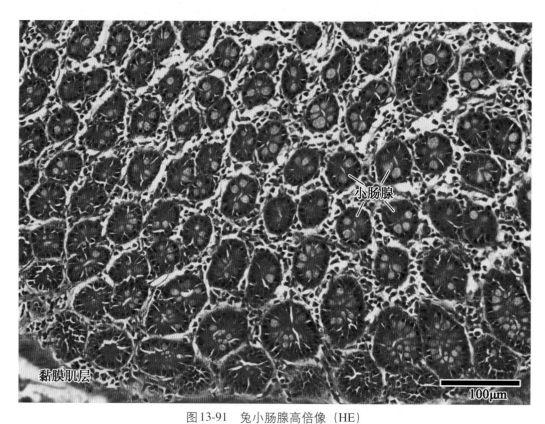

图 13-91　兔小肠腺高倍像（HE）

Fig.13-91　high magnification of rabbit small intestinal gland（HE）

小肠腺—small intestinal gland　黏膜肌层—muscularis mucosae

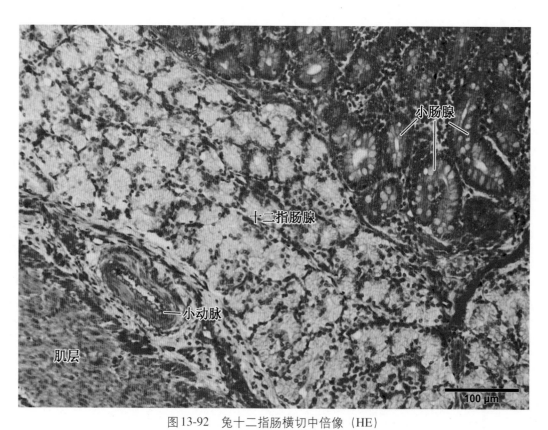

图 13-92　兔十二指肠横切中倍像（HE）

Fig.13-92　mid magnification of rabbit duodenum transection（HE）

小肠腺—small intestinal gland　十二指肠腺—duodenal gland　小动脉—arteriole　肌层—muscular layer

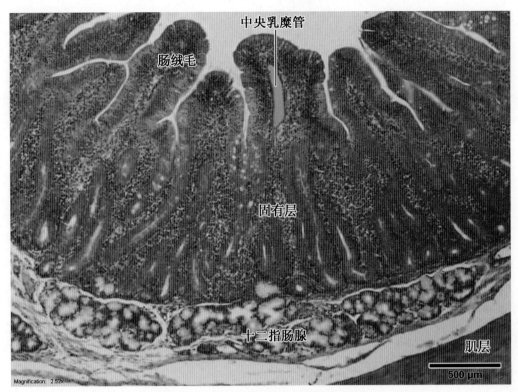

图13-93 牛十二指肠横切低倍像1（HE）

Fig.13-93 low magnification of cow duodenum transaction 1（HE）

肠绒毛—intestinal villus 中央乳糜管—central lacteal 固有层— proper layer 十二指肠腺—duodenal gland

肌层—muscle layer

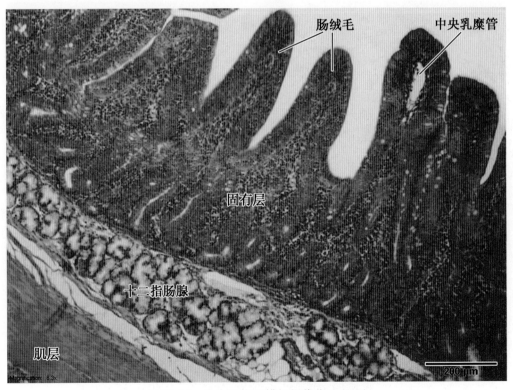

图13-94 牛十二指肠横切低倍像2（HE）

Fig.13-94 low magnification of cow duodenum transaction 2（HE）

肌层—muscle layer 十二指肠腺—duodenal gland 固有层— proper layer 肠绒毛—intestinal villus

中央乳糜管—central lacteal

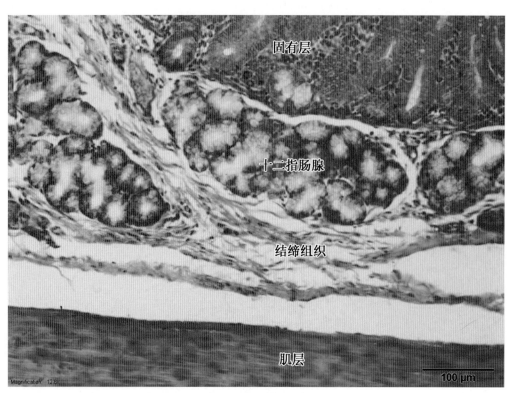

图13-95 牛十二指肠横切中倍像（HE）

Fig.13-95 mid magnification of cow duodenum transection（HE）

固有层— proper layer 十二指肠腺—duodenal gland 结缔组织—connective tissue 肌层—muscle layer

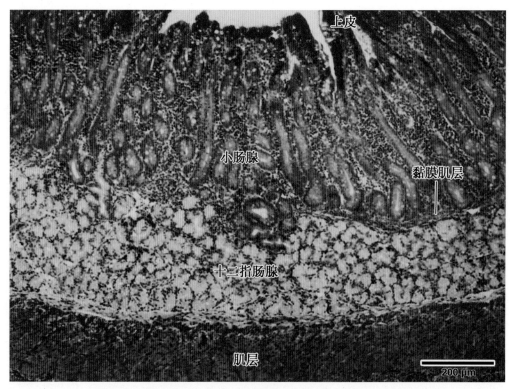

图13-96 山羊十二指肠横切低倍像1（HE）

Fig.13-96 low magnification of goat duodenum transaction 1（HE）

上皮—epithelium 小肠腺—small intestinal gland 黏膜肌层—muscularis mucosae 十二指肠腺—duodenal gland
肌层—muscular layer

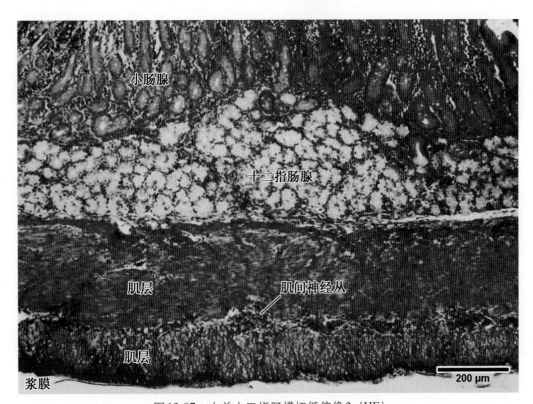

图13-97 山羊十二指肠横切低倍像2（HE）

Fig.13-97 low magnification of goat duodenum transaction 2 （HE）

小肠腺—small intestinal gland 十二指肠腺—duodenal gland 肌层—muscular layer

肌间神经丛—myenteric nerve plexus 浆膜—serosa

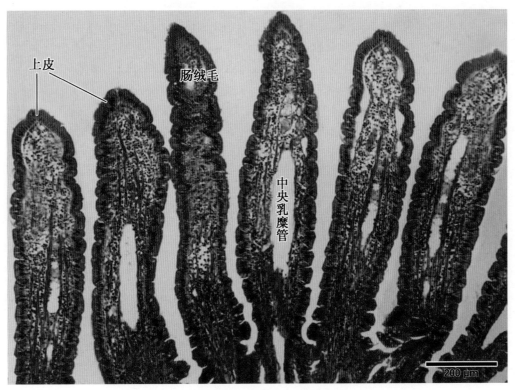

图13-98 绵羊肠绒毛低倍像（HE）

Fig.13-98 low magnification of sheep intestinal villus （HE）

上皮—epithelium 肠绒毛—intestinal villus 中央乳糜管—central lacteal

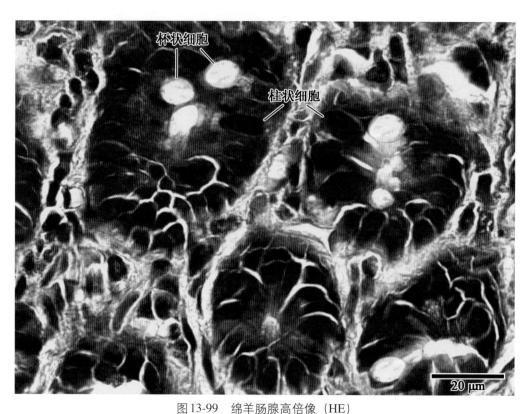

图 13-99 绵羊肠腺高倍像（HE）
Fig.13-99 high magnification of sheep intestinal gland（HE）
杯状细胞—goblet cell 柱状细胞—columnar cell

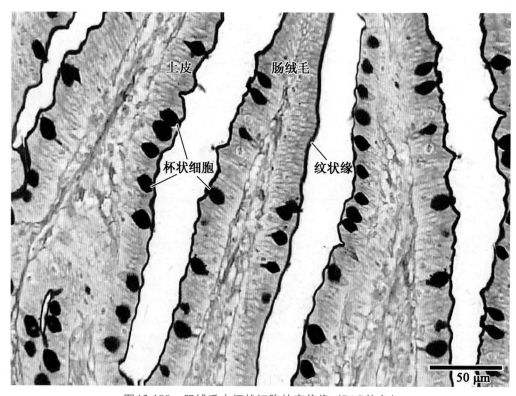

图 13-100 肠绒毛中杯状细胞的高倍像（PAS 染色）
Fig.13-100 high magnification of intestinal villus goblet cells（PAS stain）
上皮—epithelium 肠绒毛—intestinal villus 杯状细胞—goblet cell 纹状缘—striated border

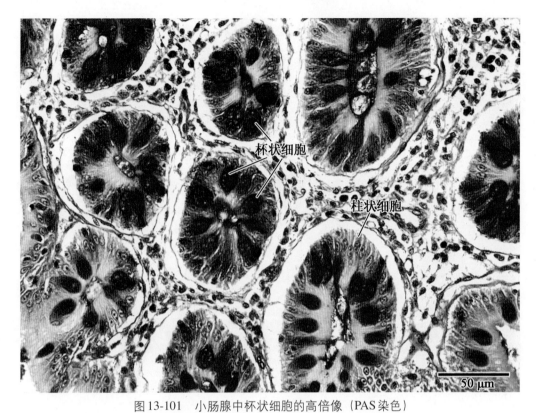

图13-101　小肠腺中杯状细胞的高倍像（PAS染色）

Fig.13-101　high magnification of small intestinal gland goblet cells（PAS stain）

杯状细胞—goblet cell　柱状细胞—columnar cell

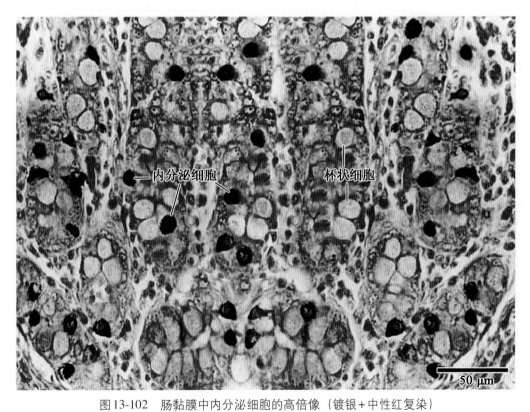

图13-102　肠黏膜中内分泌细胞的高倍像（镀银+中性红复染）

Fig.13-102　high magnification of intestinal mucosa endocrine cells（silver+neutral red）

内分泌细胞—endocrine cell　杯状细胞—goblet cell

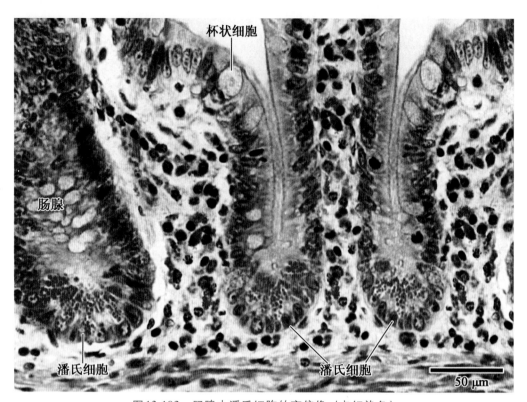

图13-103 肠腺中潘氏细胞的高倍像（卡红染色）
Fig.13-103 high magnification of intestinal glandular Paneth cells（carmine red stain）
肠腺—intestinal gland 潘氏细胞—Paneth cell 杯状细胞—goblet cell

图13-104 肠肌间神经丛的高倍像（硝酸银+中性红染色）
Fig.13-104 high magnification of intermyenteric plexus（silver nitrate+neutral red stain）
神经元—neuron 肌层—muscular layer

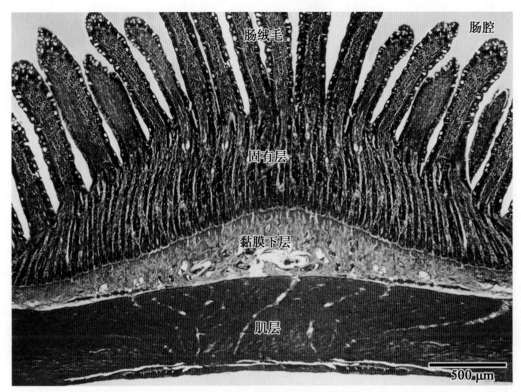

图13-105　马空肠横切低倍像1（HE）

Fig.13-105　low magnification of horse jejunum transaction 1（HE）

肠绒毛—intestinal villus　固有层—proper layer　黏膜下层—submucosa　肌层—muscle layer　肠腔—enteric cavity

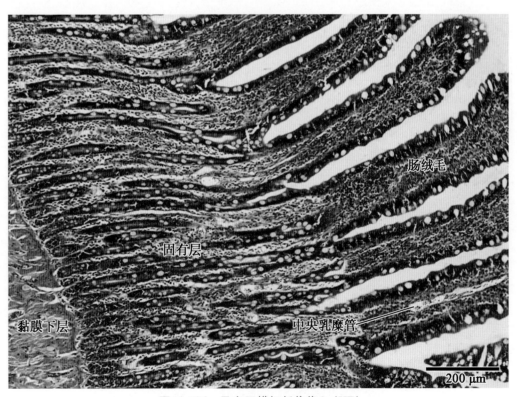

图13-106　马空肠横切低倍像2（HE）

Fig.13-106　low magnification of horse jejunum transaction 2（HE）

黏膜下层—submucosa　固有层—proper layer　肠绒毛—intestinal villus　中央乳糜管—central lacteal

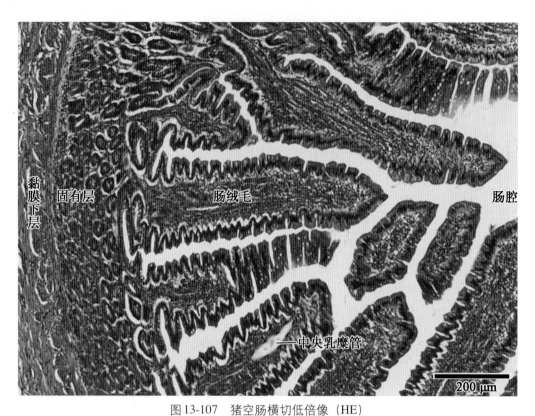

图13-107 猪空肠横切低倍像（HE）

Fig.13-107 low magnification of pig jejunum transection（HE）

黏膜下层—submucosa 固有层—proper layer 肠绒毛—intestinal villus 肠腔—intestinal lumen
中央乳糜管—central lacteal

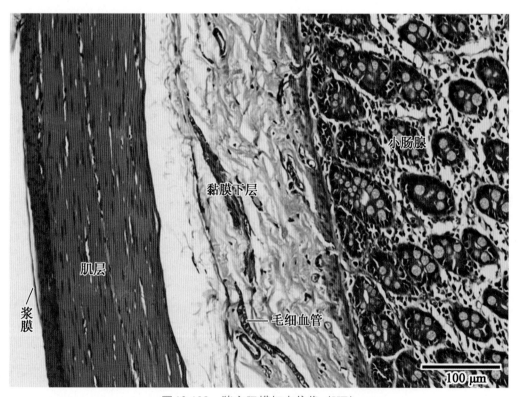

图13-108 猪空肠横切中倍像（HE）

Fig.13-108 mid magnification of pig jejunum transection（HE）

浆膜—serosa 肌层—muscle layer 黏膜下层—submucosa 小肠腺—small intestinal gland 毛细血管—capillary

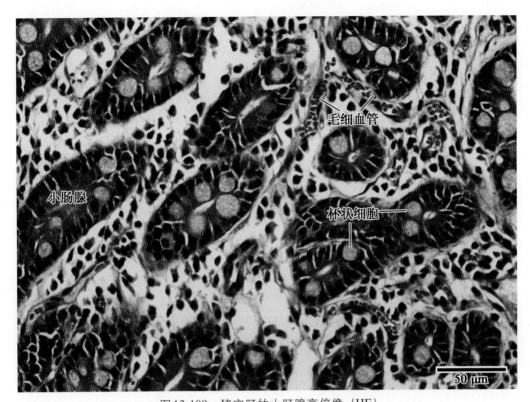

图13-109 猪空肠的小肠腺高倍像（HE）

Fig.13-109 high magnification of small intestinal glands in pig jejunum（HE）

小肠腺—small intestinal gland　毛细血管—capillary　杯状细胞—goblet cell

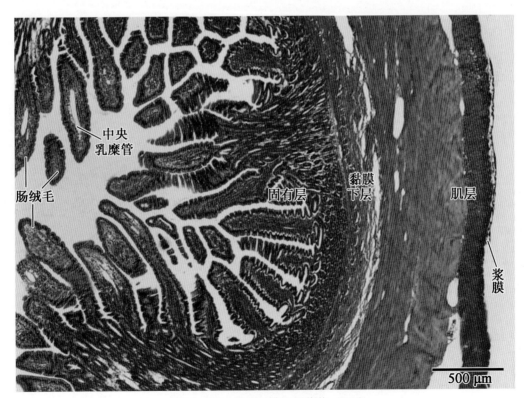

图13-110 犬空肠横切低倍像（HE）

Fig.13-110 low magnification of dog jejunum transection（HE）

肠绒毛—intestinal villus　中央乳糜管—central lacteal　固有层—proper layer　黏膜下层—submucosa

肌层—muscle layer　浆膜—serosa

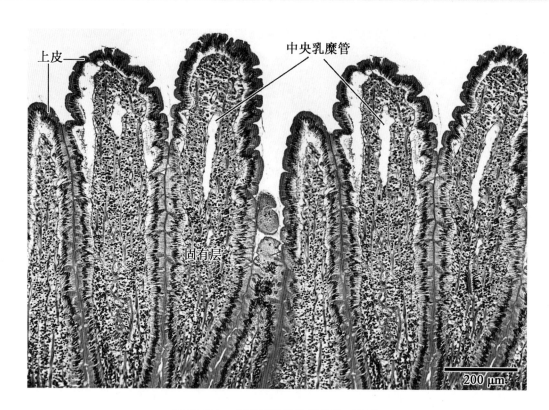

图13-111 空肠绒毛纵切低倍像（HE）

Fig.13-111 low magnification of jejunum longitudinal section（HE）

上皮—epithelium 中央乳糜管—central lacteal 固有层—proper layer

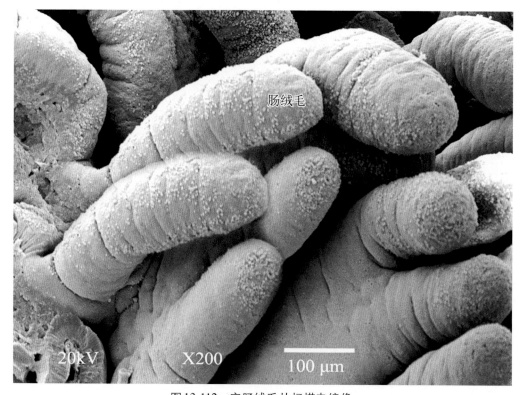

图13-112 空肠绒毛丛扫描电镜像

Fig.13-112 scanning electrical microscope image of jejunum villi floccus

肠绒毛—intestinal villus

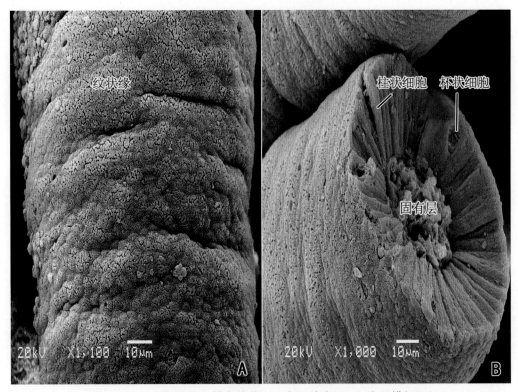

图 13-113　空肠绒毛扫描电镜像（A 表示外表面，B 表示横切面）

Fig.13-113　scanning electrical microscope image of jejunum villi（A indicating outside surface, B indicating transection）

纹状缘—striated border　柱状细胞—columnar cell　杯状细胞—goblet cell

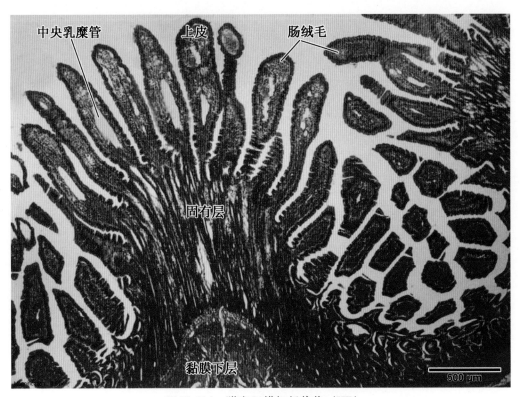

图 13-114　猫空肠横切低倍像（HE）

Fig.13-114　low magnification of cat jejunum transection（HE）

中央乳糜管—central lacteal　上皮—epithelium　肠绒毛—intestinal villus　固有层—proper layer　黏膜下层—submucosa

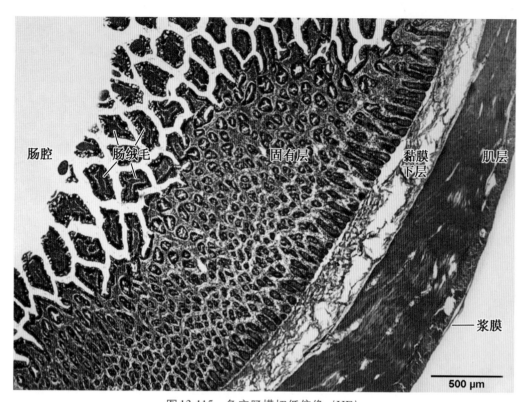

图 13-115　兔空肠横切低倍像（HE）

Fig.13-115　low magnification of rabbit jejunum transection（HE）

肠腔—enteric cavity　肠绒毛—intestinal villus　固有层—proper layer　黏膜下层—submucosa

肌层—muscle layer　浆膜—serosa

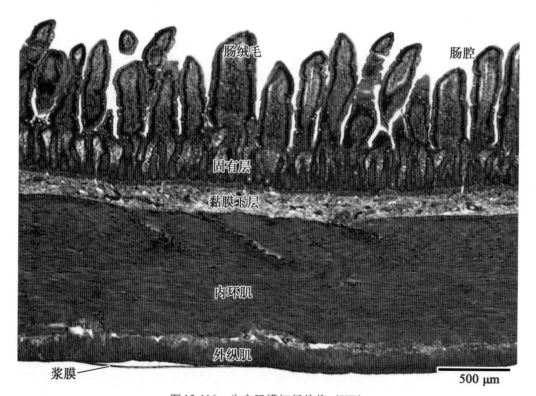

图 13-116　牛空肠横切低倍像（HE）

Fig.13-116　low magnification of cow jejunum transection（HE）

肠绒毛—intestinal villus　固有层—proper layer　黏膜下层—submucosa　内环肌—inner ring muscle

外纵肌—longitudinal muscle　浆膜—serosa　肠腔—enteric cavity

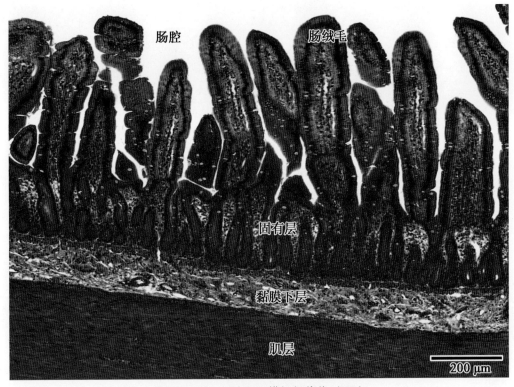

图 13-117　山羊空肠横切低倍像（HE）

Fig.13-117　low magnification of goat jejunum transection（HE）

肠腔—enteric cavity　肠绒毛—intestinal villus　固有层— proper layer　黏膜下层—submucosa　肌层—muscle layer

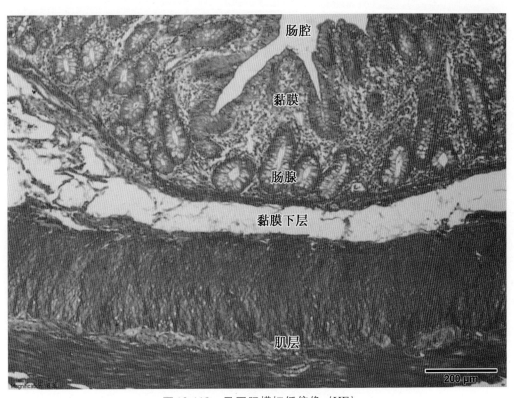

图 13-118　马回肠横切低倍像（HE）

Fig.13-118　low magnification of horse ileum transection（HE）

肠腔—enteric cavity　黏膜—mucosa　肠腺—intestinal gland　黏膜下层—submucosa　肌层—muscle layer

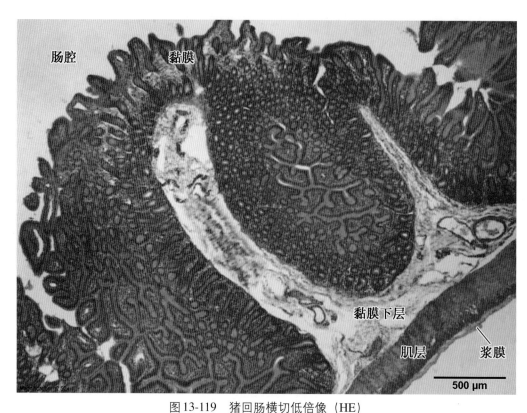

图13-119 猪回肠横切低倍像（HE）

Fig.13-119 low magnification of pig ileum transection（HE）

肠腔—enteric cavity　黏膜—mucosa　黏膜下层—submucosa　肌层—muscle layer　浆膜—serosa

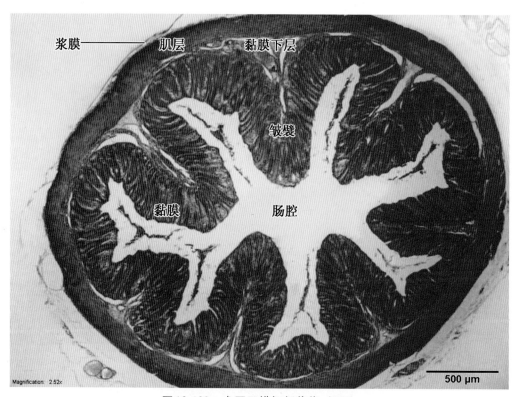

图13-120 犬回肠横切低倍像（HE）

Fig.13-120 low magnification of dog ileum transection（HE）

浆膜—serosa　肌层—muscle layer　黏膜下层—submucosa　皱襞—plica　黏膜—mucosa　肠腔—enteric cavity

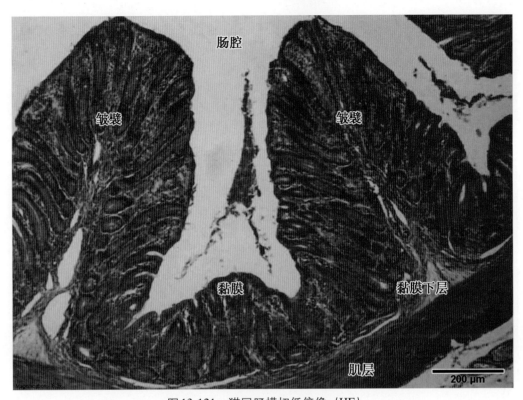

图13-121 猫回肠横切低倍像（HE）
Fig.13-121 low magnification of cat ileum transection（HE）
肠腔—enteric cavity 皱襞—plica 黏膜—mucosa 黏膜下层—submucosa 肌层—muscle layer

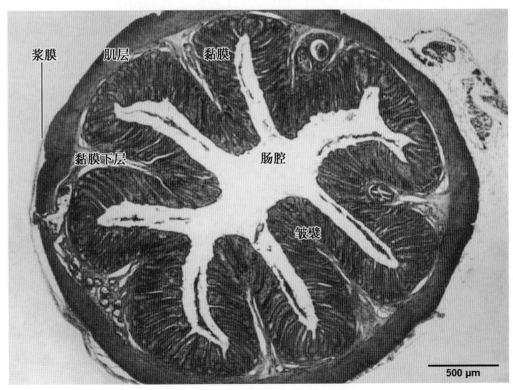

图13-122 兔回肠横切低倍像（HE）
Fig.13-122 low magnification of rabbit ileum transection（HE）
浆膜—serosa 肌层—muscle layer 黏膜—mucosa 黏膜下层—submucosa 肠腔—enteric cavity 皱襞—plica

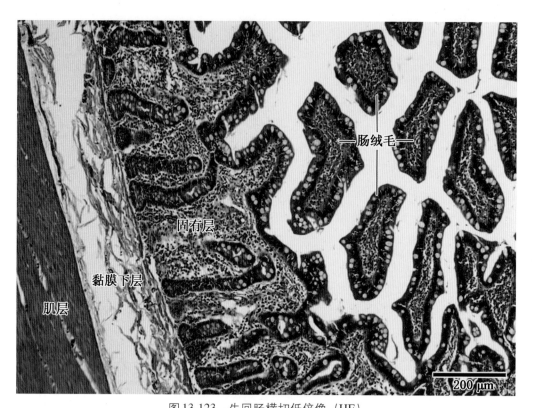

图13-123　牛回肠横切低倍像（HE）

Fig.13-123　low magnification of cow ileum transection（HE）

肌层—muscle layer　黏膜下层—submucosa　固有层— proper layer　肠绒毛—intestinal villus

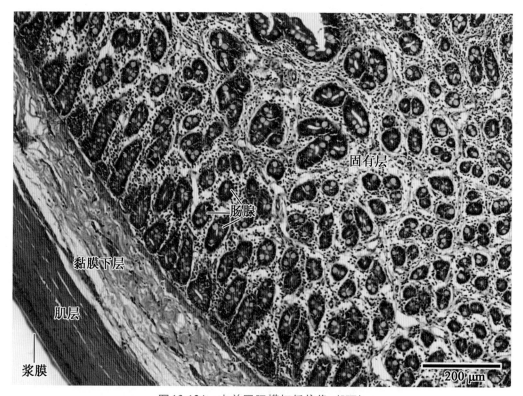

图13-124　山羊回肠横切低倍像（HE）

Fig.13-124　low magnification of goat ileum transection（HE）

浆膜—serosa　肌层—muscle layer　黏膜下层—submucosa　肠腺—intestinal gland　固有层— proper layer

第十三章 消化管 Digestive Tract

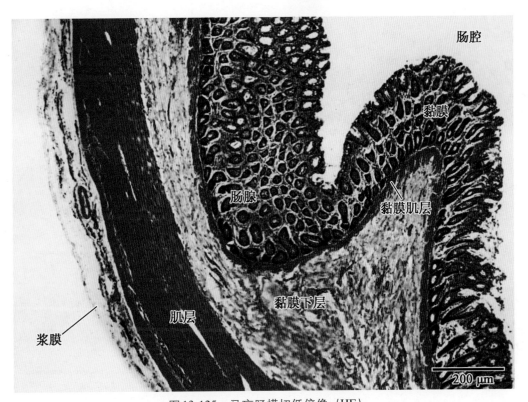

图13-125 马盲肠横切低倍像（HE）

Fig.13-125 low magnification of horse cecum transection（HE）

浆膜—serosa 肌层—muscle layer 肠腺—intestinal gland 黏膜下层—submucosa
黏膜肌层—muscularis mucosae 黏膜—mucosa 肠腔—enteric cavity

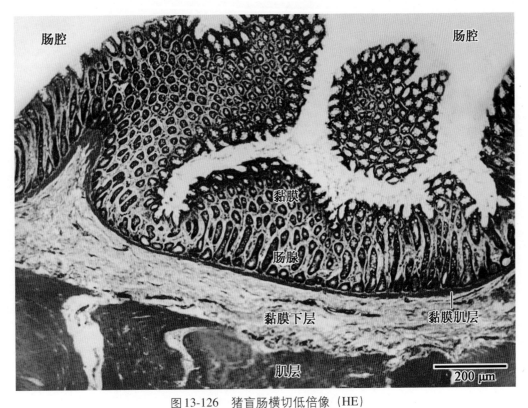

图13-126 猪盲肠横切低倍像（HE）

Fig.13-126 low magnification of pig cecum transection（HE）

肠腔—enteric cavity 黏膜—mucosa 肠腺—intestinal gland 黏膜下层—submucosa
黏膜肌层—muscularis mucosae 肌层—muscle layer

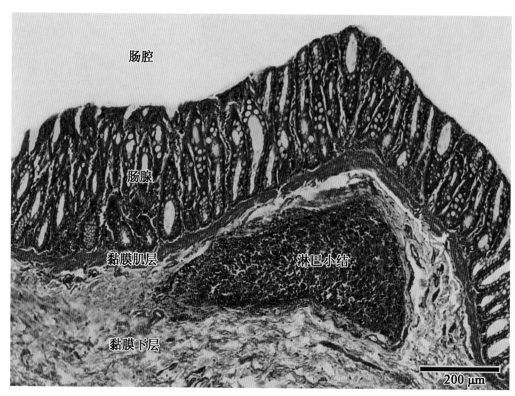

图 13-127　犬盲肠皱襞中倍像（HE）

Fig.13-127　mid magnification of dog cecum plica（HE）

肠腔—enteric cavity　肠腺—intestinal gland　黏膜肌层—muscularis mucosae　淋巴小结—lymphoid nodule
黏膜下层—submucosa

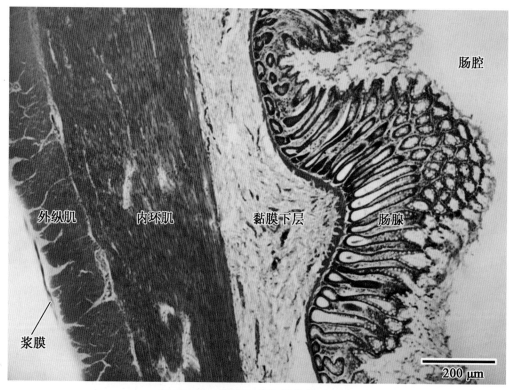

图 13-128　猫盲肠横切低倍像（HE）

Fig.13-128　low magnification of cat cecum transection（HE）

浆膜—serosa　外纵肌—longitudinal muscle　内环肌—ring muscle　黏膜下层—submucosa
肠腺—intestinal gland　肠腔—enteric cavity

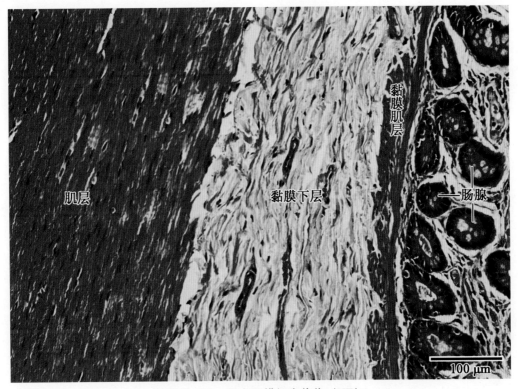

图13-129 兔盲肠横切中倍像（HE）

Fig.13-129 mid magnification of rabbit cecum transection（HE）

肌层—muscle layer   黏膜下层—submucosa   黏膜肌层—muscularis mucosae   肠腺—intestinal gland

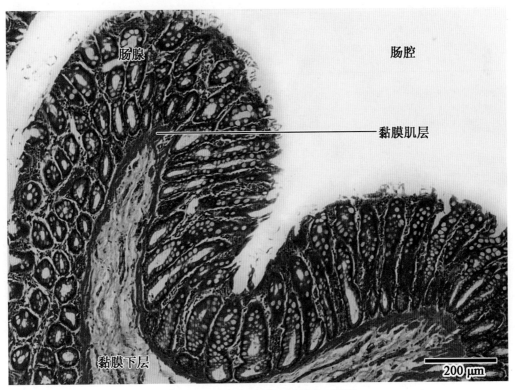

图13-130 牛盲肠横切低倍像（HE）

Fig.13-130 low magnification of cow cecum transection（HE）

肠腺—intestinal gland   肠腔—enteric cavity   黏膜肌层—muscularis mucosae   黏膜下层—submucosa

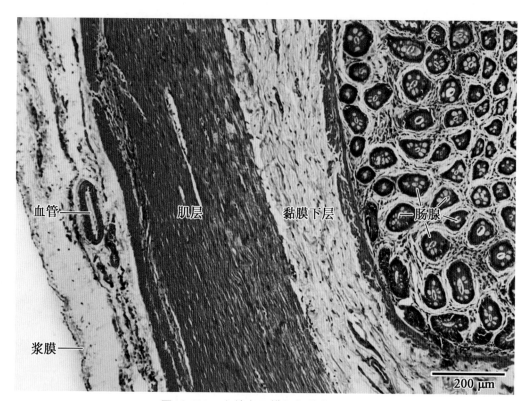

图13-131　山羊盲肠横切低倍像（HE）

Fig.13-131　low magnification of goat cecum transection（HE）

浆膜—serosa　血管—blood vessel　肌层—muscle layer　黏膜下层—submucosa　肠腺—intestinal gland

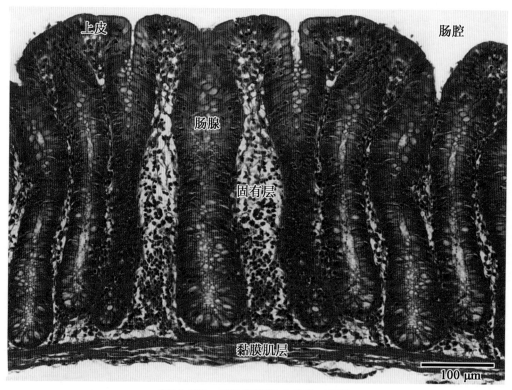

图13-132　马结肠横切中倍像（HE）

Fig.13-132　mid magnification of horse colon transection（HE）

上皮—epithelium　肠腔—enteric cavity　肠腺—intestinal gland　固有层—proper layer　黏膜肌层—muscularis mucosae

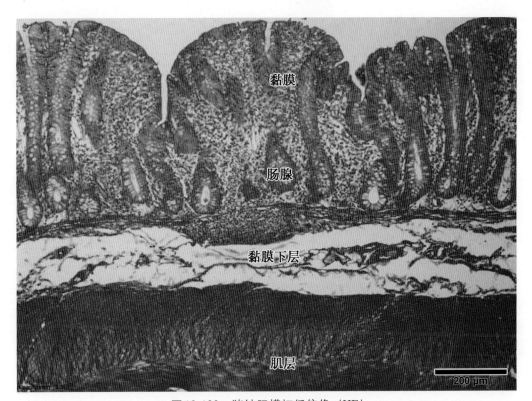

图13-133 猪结肠横切低倍像（HE）

Fig.13-133 low magnification of pig colon transection（HE）

黏膜—mucosa 肠腺—intestinal gland 黏膜下层—submucosa 肌层—muscle layer

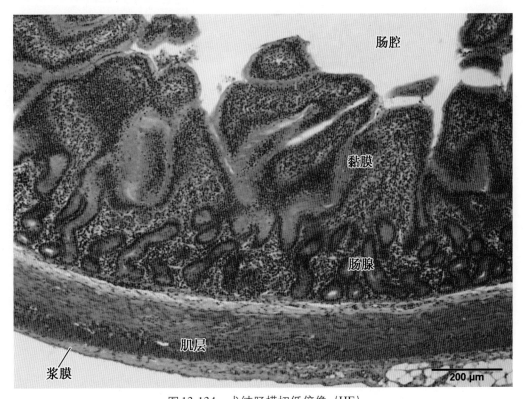

图13-134 犬结肠横切低倍像（HE）

Fig.13-134 low magnification of dog colon transection（HE）

肠腔—enteric cavity 黏膜—mucosa 肠腺—intestinal gland 肌层—muscle layer 浆膜—serosa

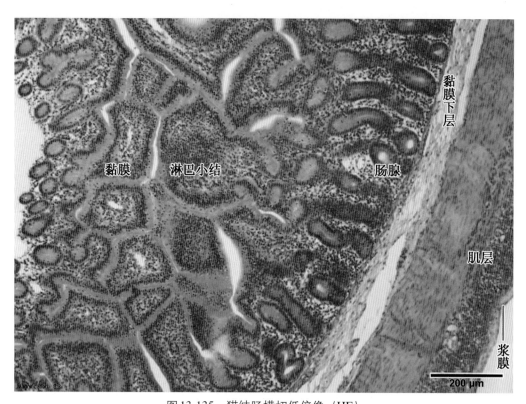

图 13-135 猫结肠横切低倍像（HE）

Fig.13-135 low magnification of cat colon transection（HE）

黏膜—mucosa 淋巴小结—lymphoid nodule 肠腺—intestinal gland 黏膜下层—submucosa
肌层—muscle layer 浆膜—serosa

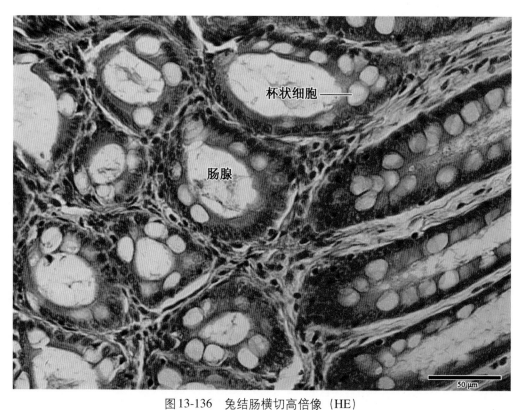

图 13-136 兔结肠横切高倍像（HE）

Fig.13-136 high magnification of rabbit colon transection（HE）

肠腺—intestinal gland 杯状细胞—goblet cell

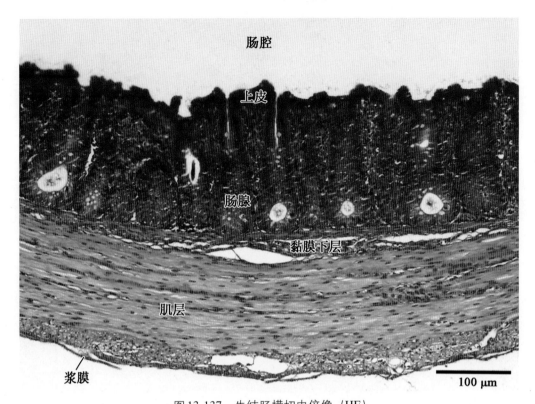

图13-137　牛结肠横切中倍像（HE）

Fig.13-137　mid magnification of cow colon transection（HE）

肠腔—enteric cavity　上皮—epithelium　肠腺—intestinal gland　黏膜下层—submucosa

肌层—muscle layer　浆膜—serosa

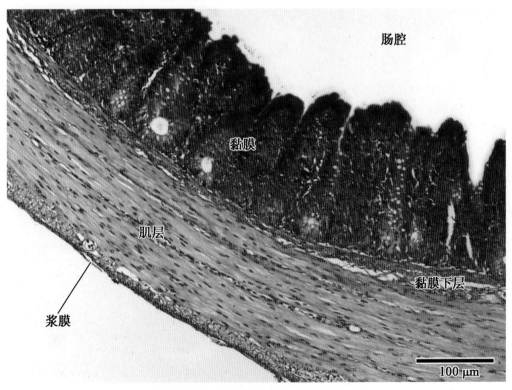

图13-138　山羊结肠横切中倍像（HE）

Fig.13-138　mid magnification of goat colon transection（HE）

肠腔—enteric cavity　黏膜—mucosa　黏膜下层—submucosa　肌层—muscle layer　浆膜—serosa

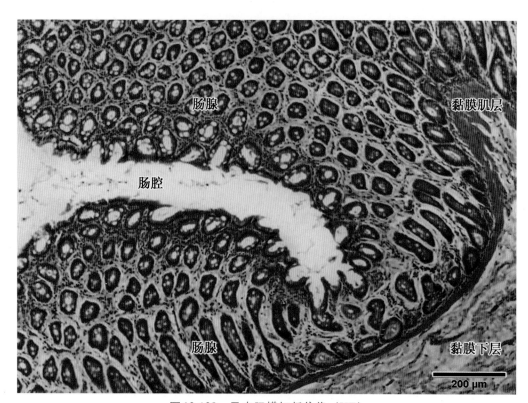

图13-139 马直肠横切低倍像（HE）

Fig.13-139 low magnification of horse rectum transection（HE）

肠腔—enteric cavity　肠腺—intestinal gland　黏膜肌层—muscularis mucosae　黏膜下层—submucosa

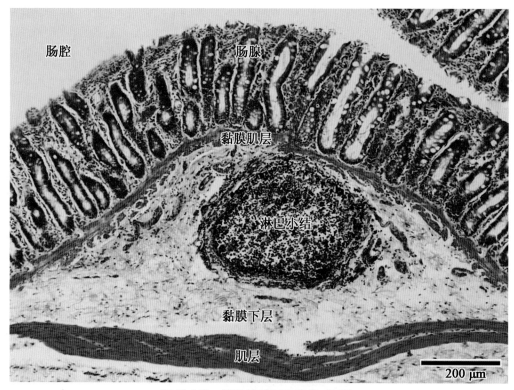

图13-140 猪直肠横切高倍像（HE）

Fig.13-140 high magnification of pig rectum transection（HE）

肠腔—enteric cavity　肠腺—intestinal gland　黏膜肌层—muscularis mucosae　淋巴小结—lymphoid nodule

黏膜下层—submucosa　肌层—muscle layer

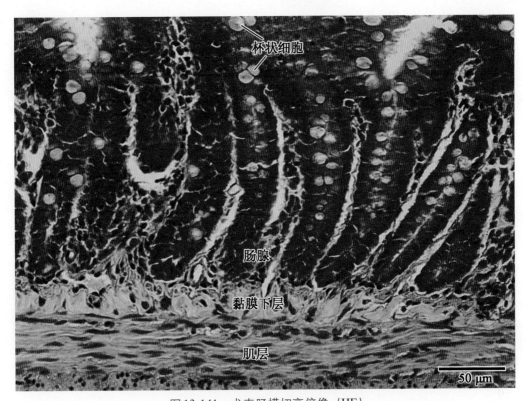

图13-141 犬直肠横切高倍像（HE）

Fig.13-141 high magnification of dog rectum transection（HE）

杯状细胞—goblet cell 肠腺—intestinal gland 黏膜下层—submucosa 肌层—muscle layer

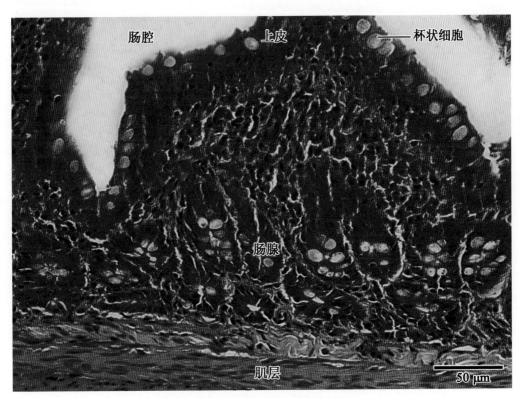

图13-142 猫直肠横切高倍像（HE）

Fig.13-142 high magnification of cat rectum transection（HE）

肠腔—enteric cavity 上皮—epithelium 杯状细胞—goblet cell 肠腺—intestinal gland 肌层—muscle layer

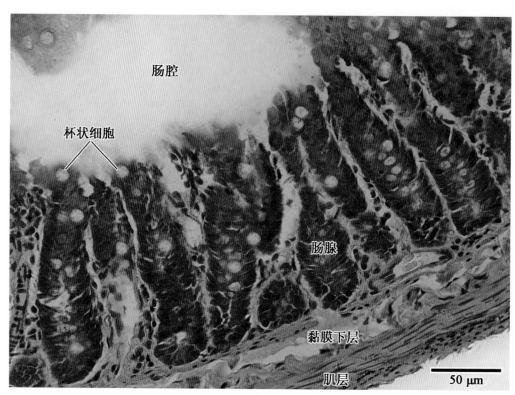

图13-143 兔直肠横切高倍像（HE）

Fig.13-143　high magnification of rabbit rectum transection（HE）

肠腔—enteric cavity　杯状细胞—goblet cell　肠腺—intestinal gland　黏膜下层—submucosa　肌层—muscle layer

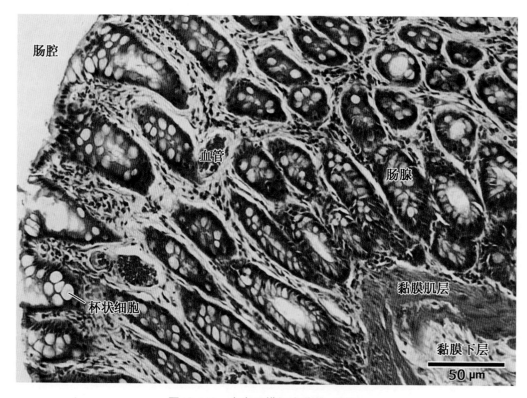

图13-144 牛直肠横切高倍像（HE）

Fig.13-144　high magnification of cow rectum transection（HE）

肠腔—enteric cavity　杯状细胞—goblet cell　血管—blood vessel　肠腺—intestinal gland

黏膜肌层—muscularis mucosae　黏膜下层—submucosa

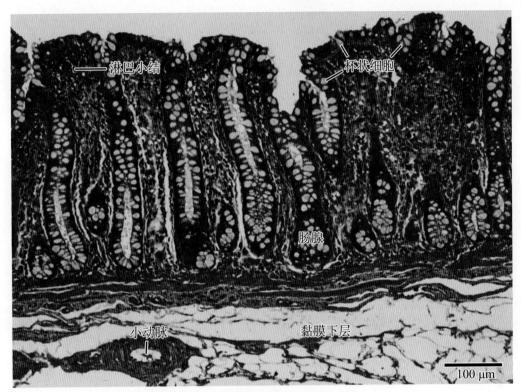

图13-145　绵羊直肠横切高倍像（HE）

Fig.13-145　high magnification of sheep rectum transection（HE）

淋巴小结—lymphoid nodule　杯状细胞—goblet cell　肠腺—intestinal gland　小动脉—arteriole　黏膜下层—submucosa

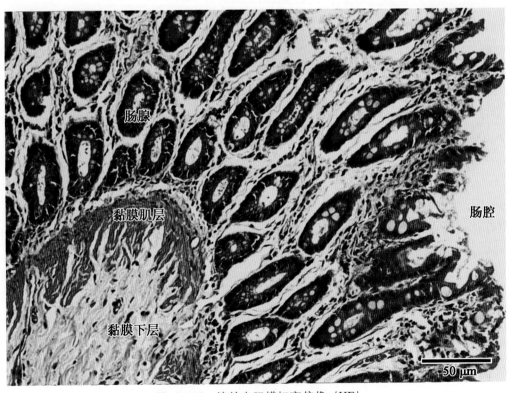

图13-146　绵羊直肠横切高倍像（HE）

Fig.13-146　high magnification of sheep rectum transection（HE）

肠腺—intestinal gland　黏膜肌层—muscularis mucosae　黏膜下层—submucosa　肠腔—enteric cavity

第十四章
消 化 腺
Digestive Glands

Outline

Digestive glands consist of several components, namely, salivary gland, liver and pancreas, which have excretory ducts that open into the digestive tract.

Salivary glands serve to wet and lubricate the oral cavity and its contents, to initiate digestion of food and to promote excretion of certain substances such as urea. The acini of salivary gland are composed of serous acini, mucous acini and mixed acini. The acini are rounded or oval and made up of a single layer of pyramidal epithelial cells with their apices pointing towards a centrally located lumen. The acini contain serous or mucous cells.

Liver is mainly composed of the hepatic lobules. Liver lobule is the structural and functional unit of the liver. It includes hepatic plate, hepatic sinusoid, perisinusoidal space, bile canaliculi and central vein. The liver lobule is a prism with six sides, with portal regions at the periphery. The hepatic plates are made up of hepatocytes, which radiate from the central vein, and are separated by vascular sinusoids. Endothelial cells of sinusoids are separated from the underlying hepatocytes by a subendothelial space known as Disse space. Sinusoids contain phagocytic cells known as Kupffer cells. Portal regions are located in the corners of the lobules, containing connective tissue, an interlobular arteries, an interlobular venule and an interlobular bile ducts. The bile canaliculi, the first portions of the bile duct system, are formed by the surfaces of the adjoining hepatocytes.

Pancreas consists of both exocrine and endocrine glands that produces digestive enzymes and hormones. The exocrine portion contains a lot of rounded acini, which consist of a single layer of pyramidal epithelial cells with their apices pointing towards a centrally located lumen. The endocrine cells of the pancreas form the pancreatic islets and contain four types of principal cells, each secreting a different hormone: A-cell secreting glucagons, B-cell secreting insulin, D-cell secreting somatostatin and PP-cell secreting pancreatic polypeptide.

第十四章 消化腺 Digestive Glands

消化腺（digestive glands）分为两大类：分布于消化管壁内的称为小消化腺，如唇腺、颊腺、胃腺和肠腺等；独立存在于消化管壁外的称为大消化腺，如腮腺、颌下腺、舌下腺、肝和胰等。消化腺的分泌物经由导管输入消化管。

一、消化腺的发生

（一）唾液腺的发生

三对大唾液腺（salivary glands）即腮腺、颌下腺、舌下腺，起源于胚胎的原始口腔上皮。上皮细胞增殖，形成实心细胞索，以出芽的方式向深层的间充质内生长，反复分支，远端膨大、分化形成腺泡，上皮周围的间充质分化为结缔组织。

（二）肝与胆的发生

肝与胆囊及胆管都起源于同一个胚胎原基——肝憩室。

1. 肝憩室的发生与演变 原肠胚期，在前肠尾部近卵黄囊处的腹侧内胚层细胞增殖并向腹侧生长，形成一囊状突起称肝憩室（hepatic diverticulum）。肝憩室迅速生长延伸，长入心脏与卵黄囊之间的间充质即原始横膈内。肝憩室生长增大，其末端分为头支与尾支，头支为肝的原基，发育为肝实质和肝内胆管及肝管；尾支发育为胆囊和胆囊管。头支和尾支与原始消化管连接的部分，分化为胆总管。头支的供血丰富，生长快，肝突入腹腔并占据腹腔的大部。肝周围的原始横膈间充质分化为肝的被膜。尾支的远端膨大，形成胆囊（马等无胆囊动物除外）；近端细长，形成胆囊管。肝憩室的基部演变为胆总管，其开口处最初位于十二指肠的腹侧壁，以后随十二指肠的发育和旋转而转向背侧壁。

2. 肝的发生 肝憩室刚形成时为单层柱状上皮和薄层间充质构成的盲囊，以后头支上皮细胞迅速增殖，在原始横膈内反复分支成肝细胞索，相互连接成网，将经过横膈内的左右卵黄静脉和脐静脉分割成许多相互吻合的毛细血管，它们与横膈间充质内的毛细血管共同发育为肝血窦，分布于肝细胞索之间。随后肝细胞索内的肝细胞间出现小腔为原始胆小管，肝内胆管网和中央静脉逐渐形成，肝细胞索与肝血窦分别围绕中央静脉形成肝小叶，随着胎龄增长，肝小叶不断增多。肝细胞索相连形成肝板，胚胎后期的肝板仍较厚，由3～5层肝细胞组成，出生后才逐渐形成单层肝细胞板。

3. 胆囊的发生 肝憩室的尾支远端膨大呈囊状，伸入胃腹系膜内分化形成胆囊。胆囊最初为实心细胞团，后来才出现腔，腔面衬以由内胚层分化来的单层柱状上皮。胃腹系膜内的间充质分化为胆囊的结缔组织和肌层。胆囊管和肝外胆管起初也为内胚层形成的实心细胞索，以后经过管腔重建才出现管腔。

（三）胰的发生

胰起源于胚胎原基的背胰芽（dorsal pancreatic bud）和腹胰芽（ventral pancreatic bud）。原肠胚末期，在前肠末端腹侧靠近肝憩室的尾侧，内胚层上皮增生，向肠壁外突出形成腹胰芽。背胰芽由腹胰芽对侧的内胚层上皮细胞增生，突出肠壁而成，体积较大。背胰芽和腹胰芽的上皮细胞不断增殖，形成的细胞索反复分支，其末端形成腺泡，而与原始消化管上皮相连的分支形成各级导管，于是背、腹两个胰芽分化成了背胰（dorsal pancreas）和腹胰（ventral pancreas）。在背胰和腹胰的中轴线上各有一条贯穿腺体全长的总导管，分别称背胰管和腹胰管。由于胃和十二指肠向右旋转和肠壁的不均等生长，致使腹胰和腹胰管的开口转至背侧，并与背胰融合，形成一个胰。腹胰管与背胰管的远段连通，形成胰腺的主胰导管，与胆总管会合后共同开口于十二指肠乳头。背胰管的近段退化或保留形成副胰管，开口于十二指肠副乳头。

胰腺的实质来源于原始消化管的内胚层。胰腺导管上皮内的未分化细胞，即胚胎早期的干细胞可分化为胰腺细胞和胰岛细胞。最初上皮细胞排列呈条索状，并分支成胰管系统，细胞索的末端细胞增生成团，后来分化为胰的各级导管及腺泡。胰腺小导管的部分上皮细胞仍可增生，逐渐与上皮细胞分离，向管壁外突出，聚集成团，最终脱离管壁，形成独立的胰岛，并分化形成A、B、D等细胞。细胞排列成不规则的团索状，其间有丰富的毛细血管，形成典型的内分泌腺结构。

二、消化腺的组织学结构概述

（一）唾液腺（salivary glands）

唾液腺为复管泡状腺，被膜较薄，腺实质分为许多小叶，由分支的导管和腺泡组成。

1. **腺泡**（alveoli） 呈泡状或管泡状，由单层立方或锥形腺细胞组成，为腺的分泌部。腺细胞与基膜之间以及部分导管上皮与基膜之间有肌上皮细胞，细胞扁平、有突起。肌上皮细胞的收缩有助于分泌物排出。腺泡分浆液性、黏液性和混合性3种类型。

2. **导管** 是反复分支的上皮性管道，是腺的排泄部，末端与腺泡相连。唾液腺导管可分为闰管（intercalated duct）、纹状管（striated duct）、小叶间导管和总导管。

（二）肝（liver, hepar）

肝是最大的消化腺，可分泌胆汁促进脂肪的分解与吸收，并参与多种物质的合成、贮存、代谢、转化和分解。因此，肝又是一个极其重要的物质代谢器官，其生理作用远远超过消化腺的范畴。肝被膜表面有浆膜，结缔组织在肝门处随门静脉、肝动脉、肝管的分支及淋巴管和神经等伸入肝实质内，将其分成若干个肝小叶。

1. **肝小叶**（hepatic lobule） 肝的基本结构和功能单位。呈多面棱柱体，横断面呈不规则的多边形。小叶中央有一条纵贯长轴的中央静脉，其外周是放射状相间排列的肝板和肝血窦。

（1）**肝板**（hepatic plate） 单行肝细胞互相连接，排列而成的细胞板。相邻肝板间有分支相吻合，凹凸不平，形成立体网格结构。肝板以中央静脉为中心，呈放射状排列。肝板的切面呈索状。

（2）**肝血窦**（hepatic sinusoid） 肝板之间吻合成网状的毛细血管。窦腔大而不规则，窦壁由一层内皮细胞构成。肝血窦内有枯否细胞。枯否细胞即肝血窦内的巨噬细胞，属于单核吞噬细胞系统。窦周间隙也称迪塞间隙，是血窦内皮细胞与肝细胞之间的微小缝隙。贮脂细胞存在于窦周间隙，形态不规则，胞体小而有突起，胞质内有大小不等的脂滴。可贮存脂肪和维生素A，产生基质和网状纤维。

（3）**胆小管**（bile canaliculus） 由相邻肝细胞间局部细胞膜凹陷而成，肝细胞分泌的胆汁直接排入胆小管。胆小管在肝板内相互沟通成网，并在肝小叶边缘汇集成小叶内胆管，汇入小叶间胆管。

2. **门管区**（portal area） 门管区也称汇管区，是相邻几个肝小叶之间的区域，结缔组织中有小叶间动脉、小叶间静脉和小叶间胆管的横断面。这三种管道分别来源于肝动脉、肝门静脉和肝管的分支。

（三）胆囊（gall bladder）

胆囊是贮存、浓缩和酸化胆汁的囊状器官。胆囊壁由黏膜、肌层和外膜三层组成。

1. **黏膜** 空虚时，黏膜形成许多高而有分支的皱襞；充盈时，皱襞变小或消失。皱襞间的上皮向固有层内凹陷形成黏膜憩室。上皮为单层柱状，其游离面的微绒毛形成不明显的纹状缘。牛胆囊的柱状细胞之间常夹有杯状细胞和散在的内分泌细胞。固有层为薄层疏松结缔组织，内含淋巴小结和盘曲的管状腺。牛的腺体较多，猪和食肉类动物的较少。根据动物种类、个体以及在黏膜上所处的位置不同，腺体可能是浆液性的，或者是黏液性的。

2. **肌层** 较薄，主要为环行平滑肌。

3. **外膜** 除与肝的连接部为纤维膜外，其余均为浆膜。

（四）胰（pancreas）

胰包括外分泌部和内分泌部两部分。外分泌部是重要的消化腺，分泌胰液，起消化作用；内分泌部分泌激素，参与调节体内的糖代谢。胰表面有薄层结缔组织被膜，并伸入实质将胰分隔成许多不明显的小叶。

1. **外分泌部**（exocrine portion） 外分泌部构成胰实质的绝大部分，为浆液性的复管泡状腺，由腺泡和导管组成。腺泡呈管状或泡状，由一层锥形的腺泡细胞围成。腺泡细胞核大而圆，位于细胞基部。基部胞质内含有丰富的粗面内质网和核糖体，显较强的嗜碱性。胞质顶部有许多圆形或卵圆形的酶原颗粒，呈嗜酸性。腺泡腔很小，腔内有一些延伸到其中的闰管上皮细胞。该细胞扁平状，胞质淡染，胞核卵圆形，称泡心细胞（centroacinar cell）。导管是输送胰液至十二指肠的管道，包括闰管、小叶内导管、小叶间导管和胰管。闰管长而细，由单层扁平上皮围成，起始于腺泡。小叶内导管变粗，为单层立方上皮。小叶间导管会合成粗大的胰管，开口于十二指肠，管壁上皮也由单层低柱状变为高柱状。导管上皮能分泌大量的水和钠、钾、重碳酸盐等电解质。

2.内分泌部（endocrine portion） 内分泌部即胰岛（pancreas islet），是分散在外分泌部腺泡之间的内分泌细胞团，胰尾部较多。小胰岛仅几个细胞，大胰岛有数百个细胞。胰岛细胞之间有丰富的毛细血管，利于激素的通过。胰岛细胞在HE染色标本中着色浅，各类细胞不易区分。用特殊染色法（如Mallory-Azan法）或电镜观察，可显示A、B、D、PP等多种细胞。

三、消化腺图谱

1.**唾液腺** 图14-1～图14-52。

2.**肝** 图14-53～图14-87。

3.**胆囊** 图14-88～图14-94。

4.**胰** 图14-95～图14-120。

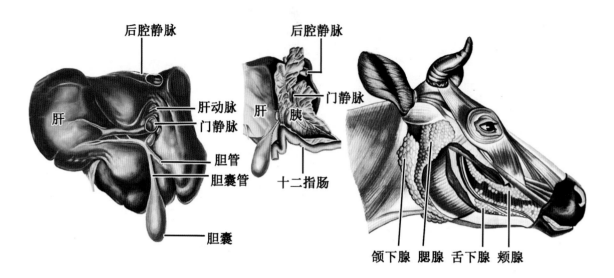

图 14-1 消化腺示意图

Fig.14-1 diagram of digestive glands

肝—liver 后腔静脉—postcaval vein 肝动脉—liver artery 门静脉—portal vein 胆管—bile duct
胆囊管—cystic gall duct 胆囊—gallbladder 胰—pancreas 十二指肠—duodenum
颌下腺—salivary gland 腮腺—parotid gland 舌下腺—sublingual gland 颊腺—cheek gland

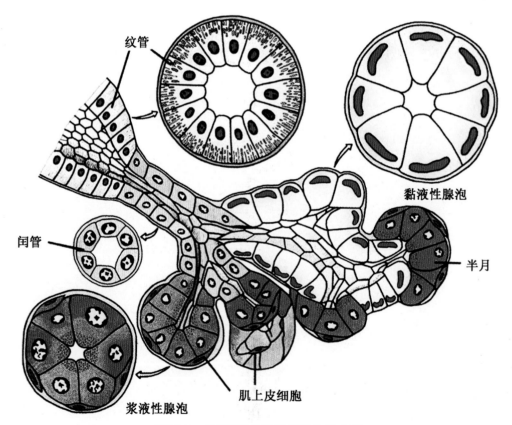

图 14-2 唾液腺腺泡和导管结构模式图

Fig.14-2 structural pattern diagram of salivary glands and ducts

纹管—striated duct 闰管—intercalated duct 浆液性腺泡—serous acinus 肌上皮细胞—serous acinus
黏液性腺泡—mucous acinus 半月—demilune

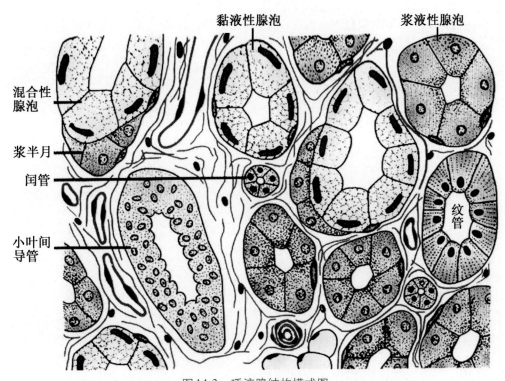

图 14-3 唾液腺结构模式图

Fig.14-3 structural pattern diagram of salivary glands

黏液性腺泡—mucous acinus 浆液性腺泡—serous acinus 混合性腺泡—mixed acinus 浆半月—serous demilune
闰管—intercalated duct 小叶间导管—interlobular duct 纹管—striated duct

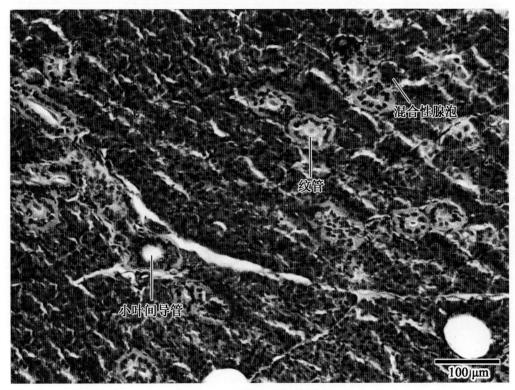

图 14-4 牛腮腺中倍像（HE）

Fig.14-4 mid magnification of cattle parotid gland（HE）

混合性腺泡—mixed acinus 纹管—striated duct 小叶间导管—interlobular duct

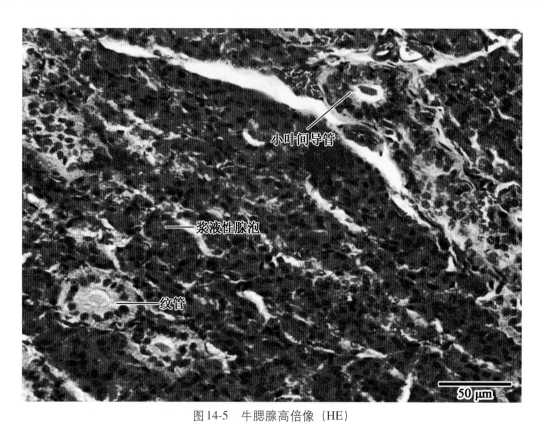

图14-5 牛腮腺高倍像（HE）

Fig.14-5 high magnification of cattle parotid gland（HE）

纹管—striated duct 浆液性腺泡—serous acinus 小叶间导管—interlobular duct

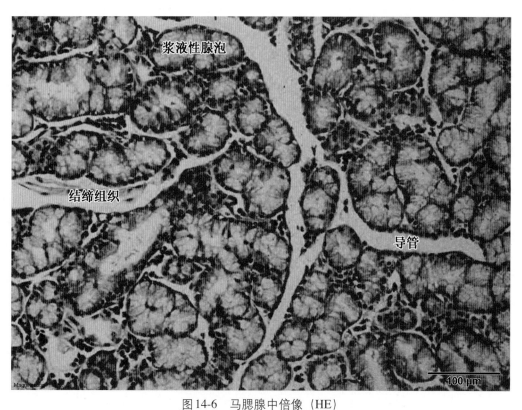

图14-6 马腮腺中倍像（HE）

Fig.14-6 mid magnification of horse parotid gland（HE）

浆液性腺泡—serous acinus 结缔组织—connective tissue 导管—duct

第十四章 消化腺 Digestive Glands

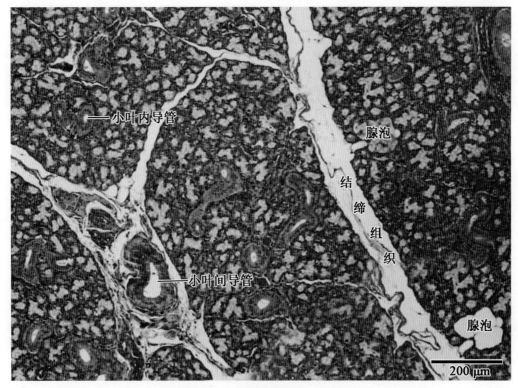

图 14-7 猪腮腺低倍像（HE）

Fig.14-7 low magnification of pig parotid gland（HE）

小叶内导管—lobular duct 小叶间导管—interlobular duct 腺泡—acinus 结缔组织—connective tissue

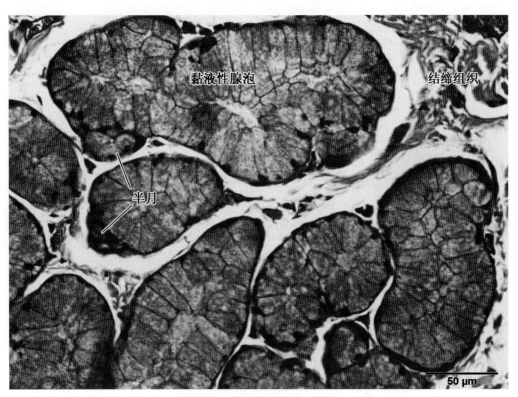

图 14-8 猪腮腺高倍像（HE）

Fig.14-8 high magnification of pig parotid gland（HE）

黏液性腺泡—mucous acinus 结缔组织—connective tissue 半月—demilune

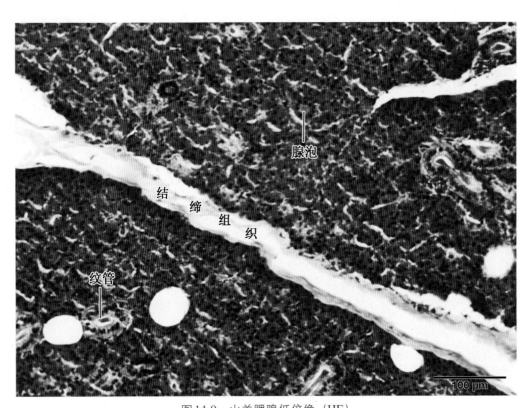

图 14-9 山羊腮腺低倍像（HE）

Fig.14-9 low magnification of goat parotid gland （HE）

腺泡—acinus 结缔组织—connective tissue 纹管—striated duct

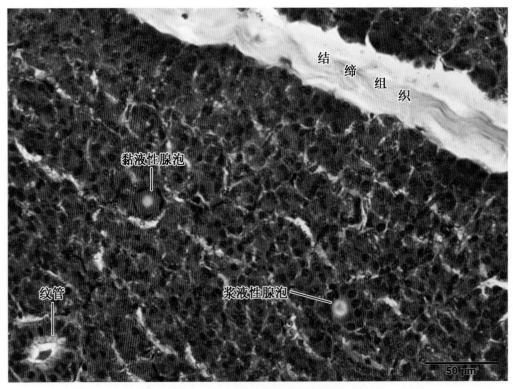

图 14-10 山羊腮腺高倍像（HE）

Fig.14-10 high magnification of goat parotid gland （HE）

纹管—striated duct 黏液性腺泡—mucous acinus 浆液性腺泡—serous acinus 结缔组织—connective tissue

第十四章 消化腺 Digestive Glands

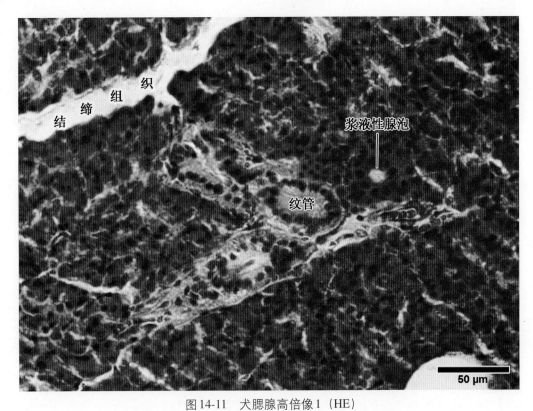

图 14-11　犬腮腺高倍像 1（HE）

Fig.14-11　high magnification of dog parotid gland 1（HE）

结缔组织—connective tissue　纹管—striated duct　浆液性腺泡—serous acinus

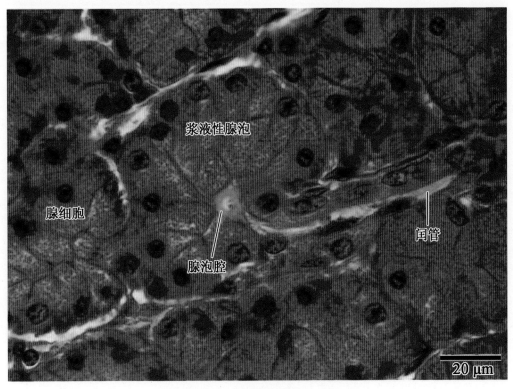

图 14-12　犬腮腺高倍像 2（HE）

Fig.14-12　high magnification of dog parotid gland 2（HE）

腺细胞—glandular cell　浆液性腺泡—serous acinus　腺泡腔—acinus cavity　闰管—intercalated duct

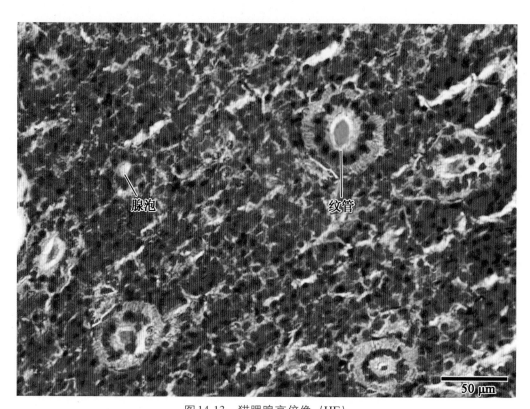

图14-13 猫腮腺高倍像（HE）
Fig.14-13 high magnification of cat parotid gland（HE）
腺泡—acinus 纹管—striated duct

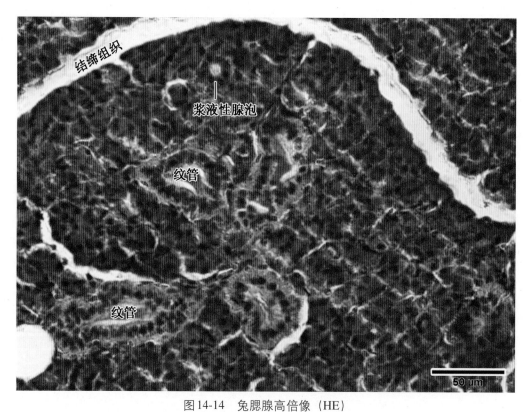

图14-14 兔腮腺高倍像（HE）
Fig.14-14 high magnification of rabbit parotid gland（HE）
结缔组织—connective tissue 浆液性腺泡—serous acinus 纹管—striated duct

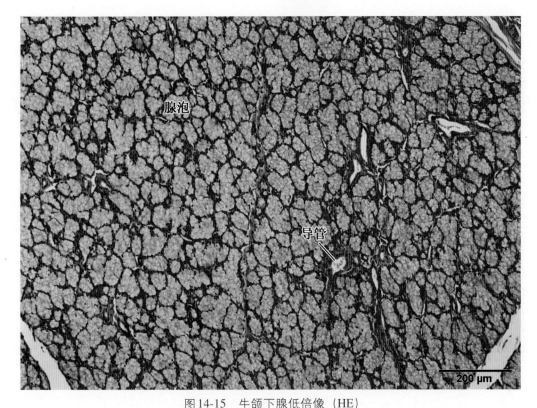

图 14-15 牛颌下腺低倍像（HE）

Fig.14-15 low magnification of cattle submaxillary gland（HE）

腺泡—acinus 导管—duct

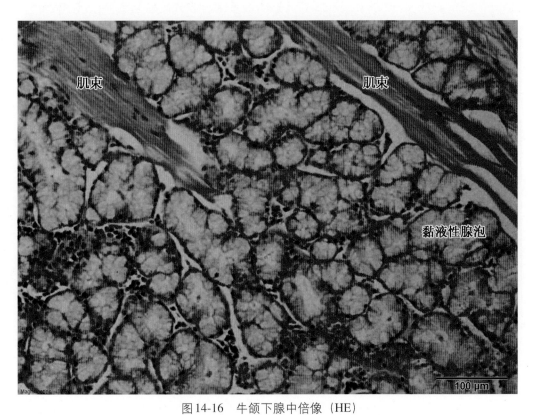

图 14-16 牛颌下腺中倍像（HE）

Fig.14-16 mid magnification of cattle submaxillary gland（HE）

肌束—muscle bundle 黏液性腺泡—mucous acinus

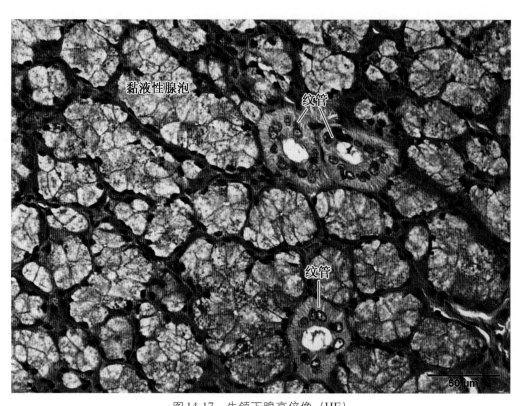

图 14-17 牛颌下腺高倍像（HE）

Fig.14-17 high magnification of cattle submaxillary gland（HE）

黏液性腺泡—mucous acinus 纹管—striated duct

图 14-18 马颌下腺低倍像（HE）

Fig.14-18 low magnification of horse submaxillary gland（HE）

腺泡—acinus 导管—duct 结缔组织—connective tissue

第十四章 消化腺 Digestive Glands

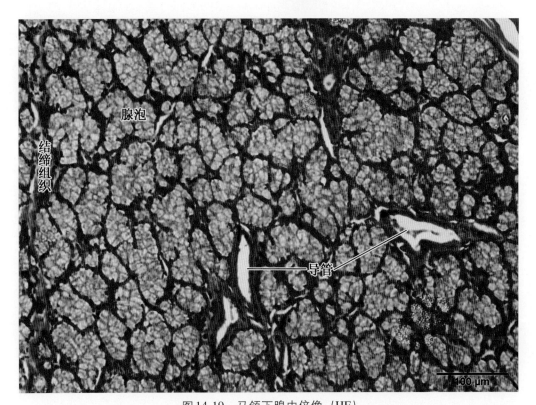

图14-19 马颌下腺中倍像（HE）

Fig.14-19 mid magnification of horse submaxillary gland（HE）

腺泡—acinus 结缔组织—connective tissue 导管—duct

图14-20 马颌下腺高倍像（HE）

Fig.14-20 high magnification of horse submaxillary gland（HE）

黏液性腺泡—mucous acinus 导管—duct

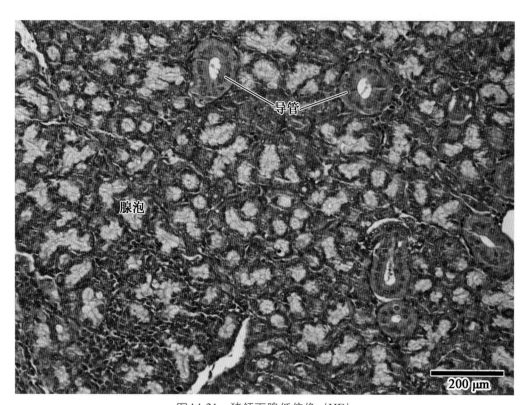

图14-21 猪颌下腺低倍像（HE）

Fig.14-21 low magnification of pig submaxillary gland（HE）

腺泡—acinus 导管—duct

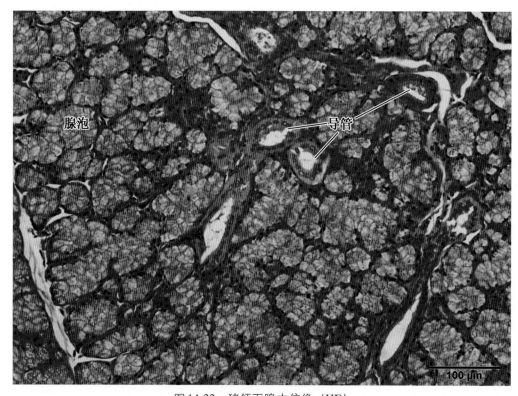

图14-22 猪颌下腺中倍像（HE）

Fig.14-22 mid magnification of pig submaxillary gland（HE）

腺泡—acinus 导管—duct

第十四章 消化腺　Digestive Glands

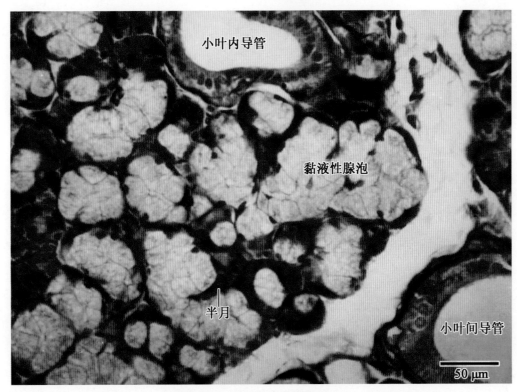

图14-23　猪颌下腺高倍像（HE）

Fig.14-23　high magnification of pig submaxillary gland（HE）

小叶内导管—lobular duct　黏液性腺泡—mucous acinus　半月—demilune　小叶间导管—interlobular duct

图14-24　山羊颌下腺低倍像（HE）

Fig.14-24　low magnification of goat submaxillary gland（HE）

腺泡—acinus　导管—duct　结缔组织—connective tissue

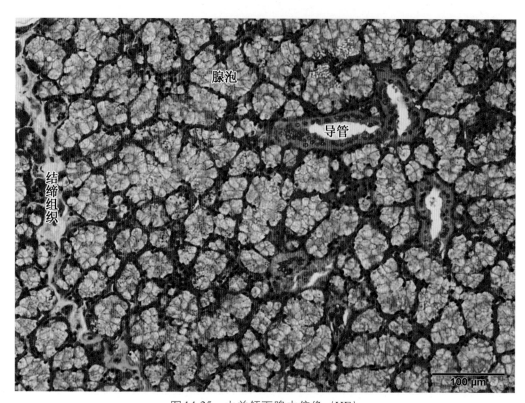

图 14-25 山羊颌下腺中倍像（HE）

Fig.14-25 mid magnification of goat submaxillary gland（HE）

结缔组织—connective tissue 腺泡—acinus 导管—duct

图 14-26 山羊颌下腺高倍像（HE）

Fig.14-26 high magnification of goat submaxillary gland（HE）

黏液性腺泡—mucous acinus 半月—demilune 纹管—striated duct

第十四章 消化腺 Digestive Glands

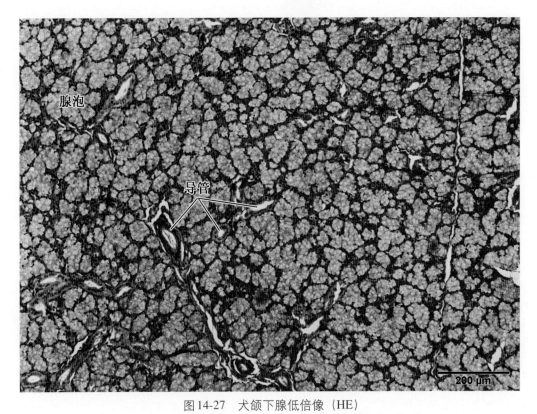

图14-27 犬颌下腺低倍像（HE）

Fig.14-27 low magnification of dog submaxillary gland（HE）

腺泡—acinus 导管—duct

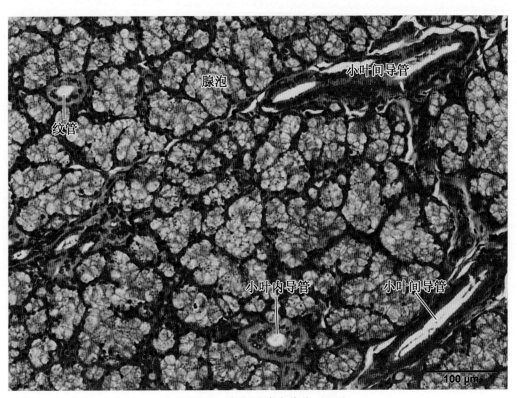

图14-28 犬颌下腺中倍像（HE）

Fig.14-28 mid magnification of dog submaxillary gland（HE）

纹管—striated duct 腺泡—acinus 小叶内导管—lobular duct 小叶间导管—interlobular duct

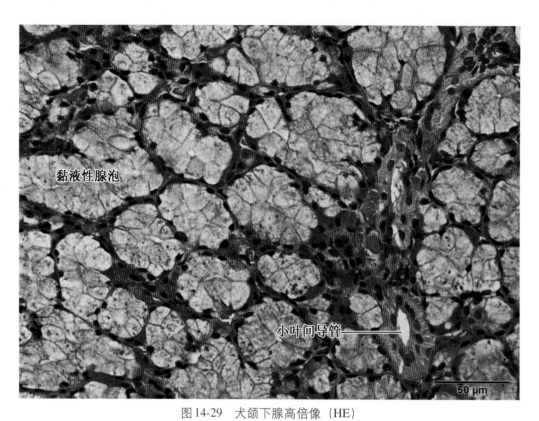

图 14-29 犬颌下腺高倍像（HE）

Fig.14-29 high magnification of dog submaxillary gland（HE）

黏液性腺泡—mucous acinus　小叶间导管—interlobular duct

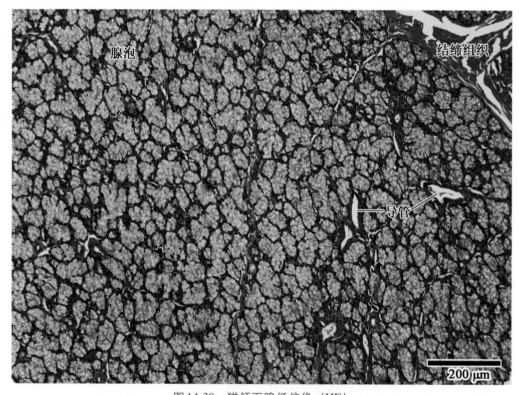

图 14-30 猫颌下腺低倍像（HE）

Fig.14-30 low magnification of cat submaxillary gland（HE）

腺泡—acinus　导管—duct　结缔组织—connective tissue

第十四章 消化腺 Digestive Glands

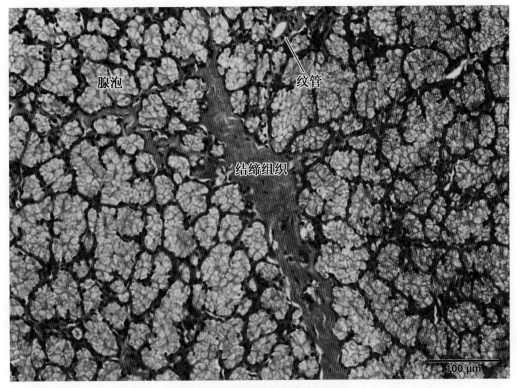

图14-31 猫颌下腺中倍像（HE）
Fig.14-31 mid magnification of cat submaxillary gland（HE）
腺泡—acinus 纹管—striated duct 结缔组织—connective tissue

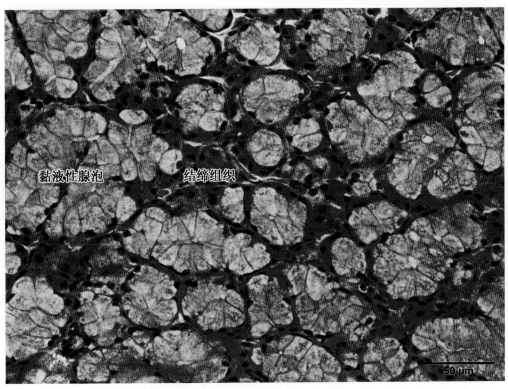

图14-32 猫颌下腺高倍像（HE）
Fig.14-32 high magnification of cat submaxillary gland（HE）
黏液性腺泡—mucous acinus 结缔组织—connective tissue

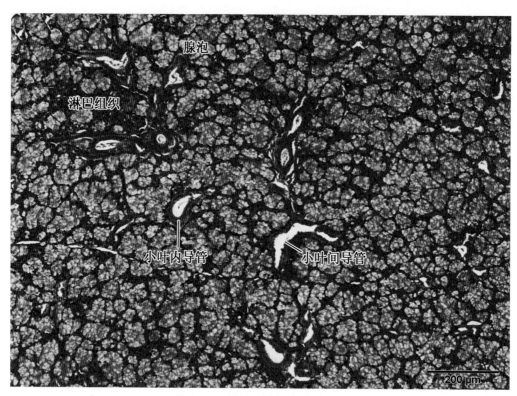

图14-33 兔颌下腺低倍像（HE）

Fig.14-33　low magnification of rabbit submaxillary gland（HE）

腺泡—acinus　淋巴组织—lymph tissue　小叶内导管—lobular duct　小叶间导管—interlobular duct

图14-34 兔颌下腺中倍像（HE）

Fig.14-34　mid magnification of rabbit submaxillary gland（HE）

腺泡—acinus　小叶间导管—interlobular duct

第十四章 消化腺 Digestive Glands

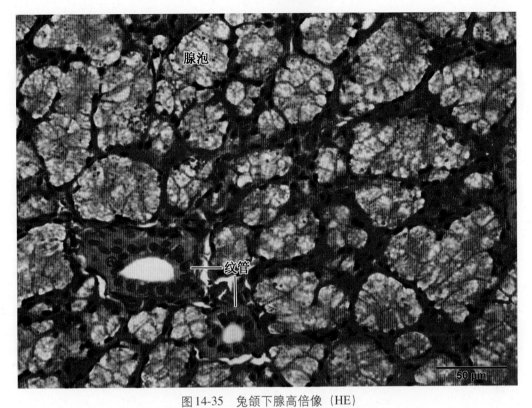

图14-35 兔颌下腺高倍像（HE）

Fig.14-35 high magnification of rabbit submaxillary gland（HE）

腺泡—acinus 纹管—striated duct

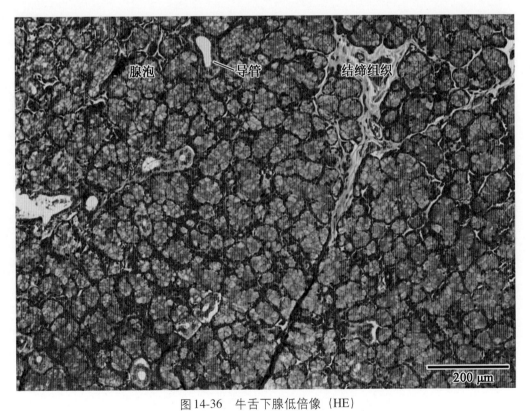

图14-36 牛舌下腺低倍像（HE）

Fig.14-36 low magnification of cattle sublingual gland（HE）

腺泡—acinus 导管—duct 结缔组织—connective tissue

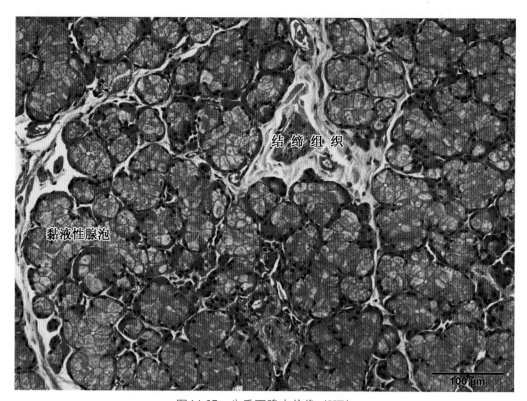

图14-37　牛舌下腺中倍像（HE）
Fig.14-37　mid magnification of cattle sublingual gland（HE）
黏液性腺泡—mucous acinus　结缔组织—connective tissue

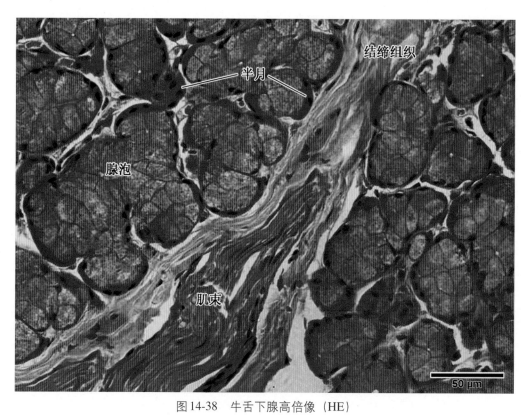

图14-38　牛舌下腺高倍像（HE）
Fig.14-38　high magnification of cattle sublingual gland（HE）
腺泡—acinus　半月—demilune　结缔组织—connective tissue　肌束—muscular bundle

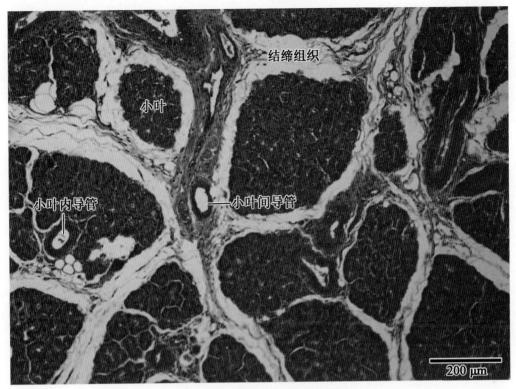

图14-39 马舌下腺低倍像（HE）

Fig.14-39 low magnification of horse sublingual gland（HE）

小叶—lobule　结缔组织—connective tissue　小叶内导管—lobular duct　小叶间导管—interlobular duct

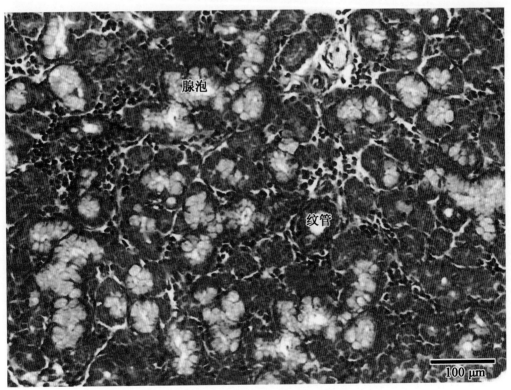

图14-40 马舌下腺中倍像（HE）

Fig.14-40 mid magnification of horse sublingual gland（HE）

腺泡—acinus　纹管—striated duct

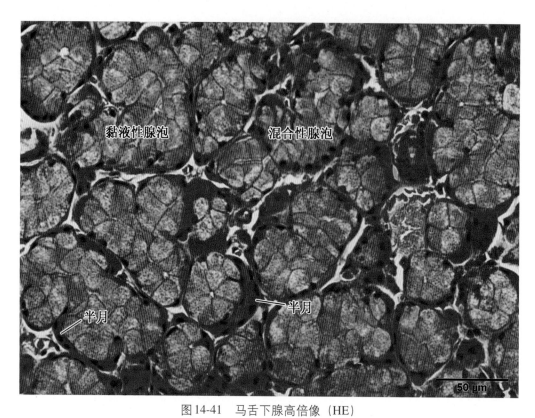

图14-41　马扬下腺高倍像（HE）

Fig.14-41　high magnification of horse sublingual gland（HE）

黏液性腺泡—mucous acinus　混合性腺泡—mixed acinus　半月—demilune

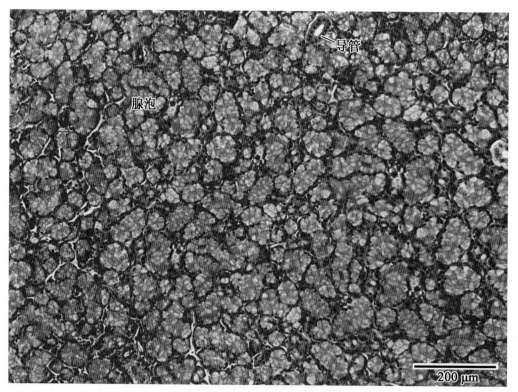

图14-42　猪舌下腺低倍像（HE）

Fig.14-42　low magnification of pig sublingual gland（HE）

腺泡—acinus　导管—duct

第十四章 消化腺　Digestive Glands

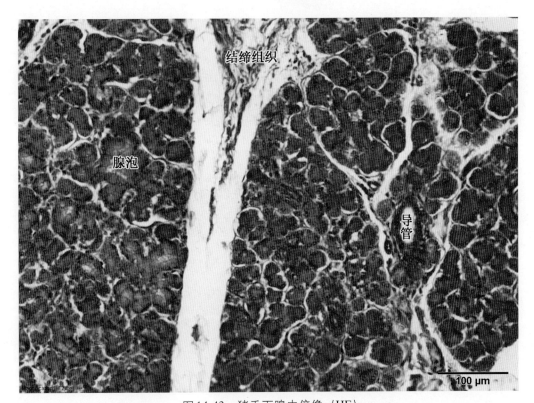

图 14-43　猪舌下腺中倍像（HE）
Fig.14-43　mid magnification of pig sublingual gland（HE）
腺泡—acinus　结缔组织—connective tissue　导管—duct

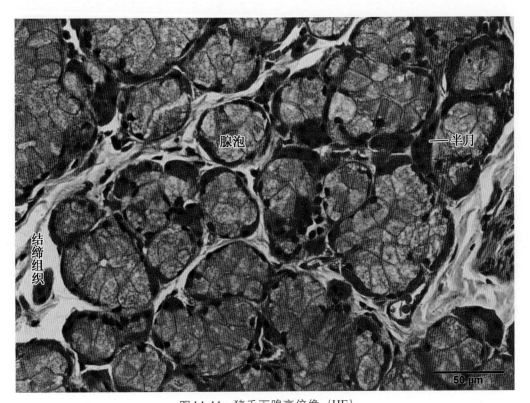

图 14-44　猪舌下腺高倍像（HE）
Fig.14-44　high magnification of pig sublingual gland（HE）
腺泡—acinus　结缔组织—connective tissue　半月—demilune

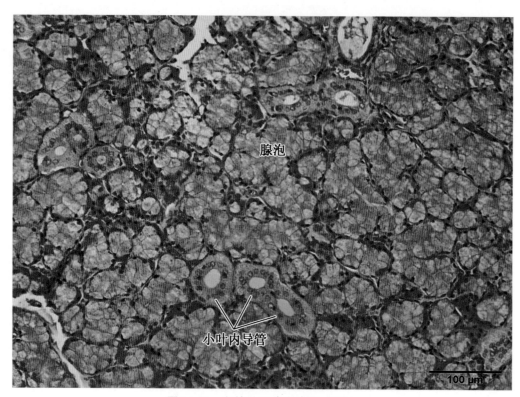

图14-45 山羊舌下腺中倍像（HE）
Fig.14-45 mid magnification of goat sublingual gland（HE）
腺泡—acinus 小叶内导管—lobular duct

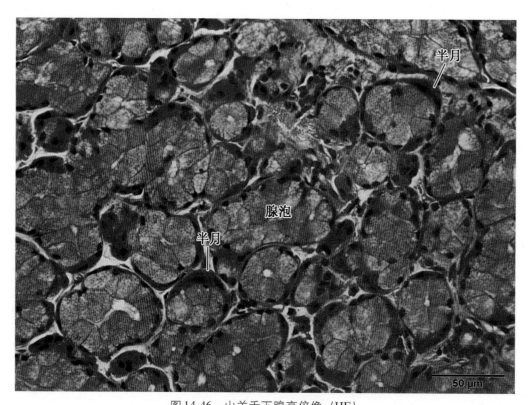

图14-46 山羊舌下腺高倍像（HE）
Fig.14-46 high magnification of goat sublingual gland（HE）
腺泡—acinus 半月—demilune

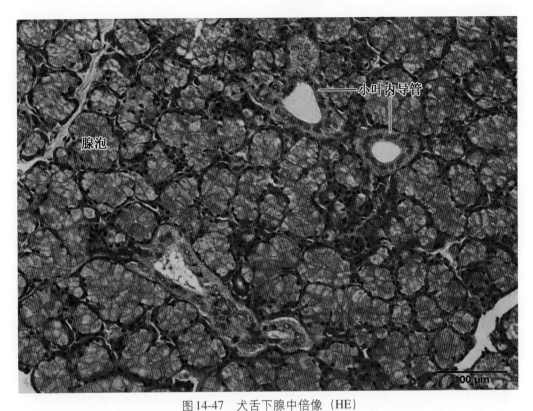

图 14-47 犬舌下腺中倍像（HE）
Fig.14-47 mid magnification of dog sublingual gland（HE）
腺泡—acinus 小叶内导管—lobular duct

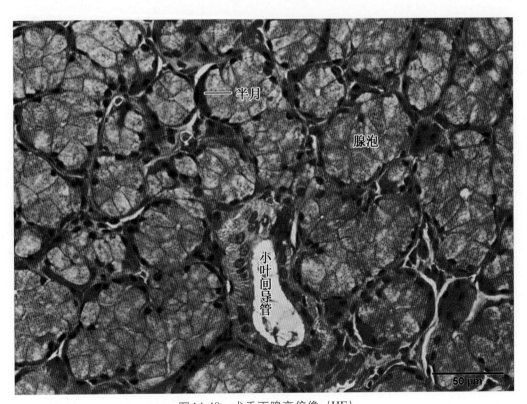

图 14-48 犬舌下腺高倍像（HE）
Fig.14-48 high magnification of dog sublingual gland（HE）
半月—demilune 腺泡—acinus 小叶间导管—interlobular duct

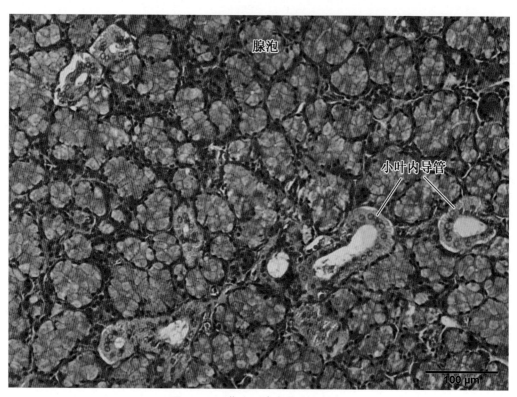

图14-49 猫舌下腺中倍像（HE）
Fig.14-49 mid magnification of cat sublingual gland（HE）
腺泡—acinus 小叶内导管—lobular duct

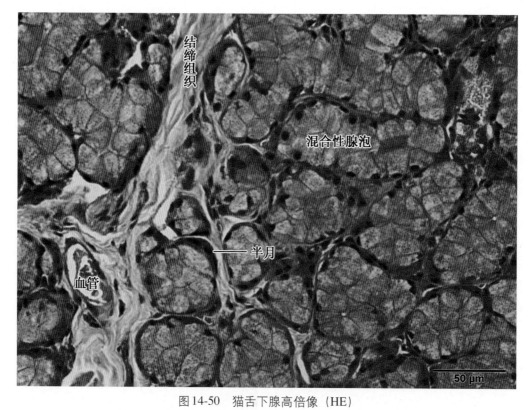

图14-50 猫舌下腺高倍像（HE）
Fig.14-50 high magnification of cat sublingual gland（HE）
结缔组织—connective tissue 血管—blood vessel 混合性腺泡—mixed acinus 半月—demilune

第十四章 消化腺 Digestive Glands

图 14-51　兔舌下腺中倍像（HE）
Fig.14-51　mid magnification of rabbit sublingual gland（HE）
腺泡—acinus　闰管—intercalated duct　结缔组织—connective tissue

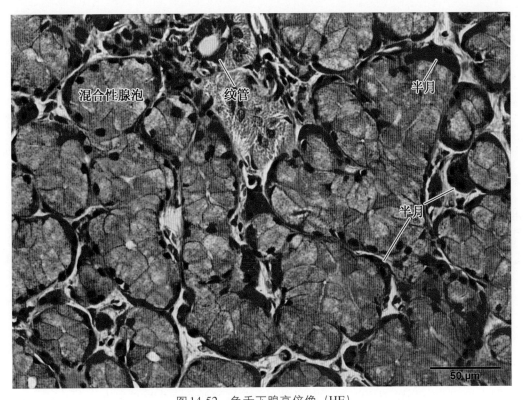

图 14-52　兔舌下腺高倍像（HE）
Fig.14-52　high magnification of rabbit sublingual gland（HE）
混合性腺泡—mixed acinus　纹管—striated duct　半月—demilune

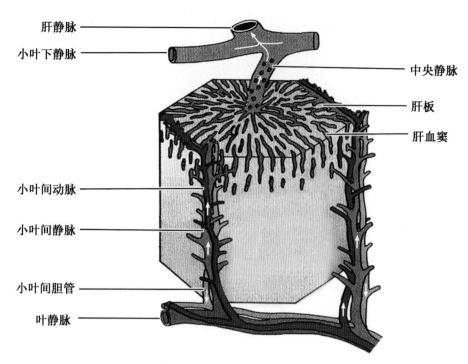

图 14-53　肝小叶立体模式图
Fig.14-53　stereogram of liver lobule

肝静脉—liver vein　小叶下静脉—sublobular vein　小叶间动脉—interlobular artery　小叶间静脉—interlobular vein
小叶间胆管—interlobular bile duct　叶静脉—lobular vein　中央静脉—central vein　肝板—liver plate
肝血窦—hepatic sinusoid

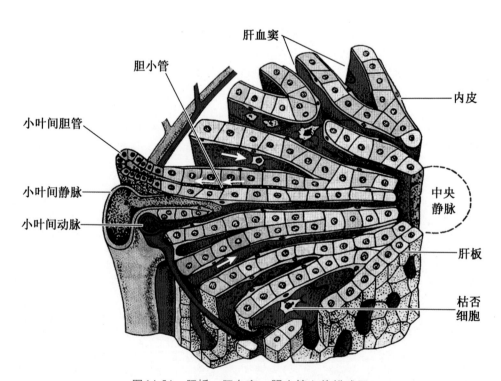

图 14-54　肝板、肝血窦、胆小管立体模式图
Fig.14-54　stereogram of liver plate, hepatic sinusoid and bile canaliculus

小叶间胆管—interlobular bile duct　小叶间静脉—interlobular vein　小叶间动脉—interlobular artery
胆小管—bile canaliculus　肝血窦—hepatic sinusoid　内皮—endothelium　中央静脉—central vein　肝板—liver plate
枯否细胞—Kupffer cell

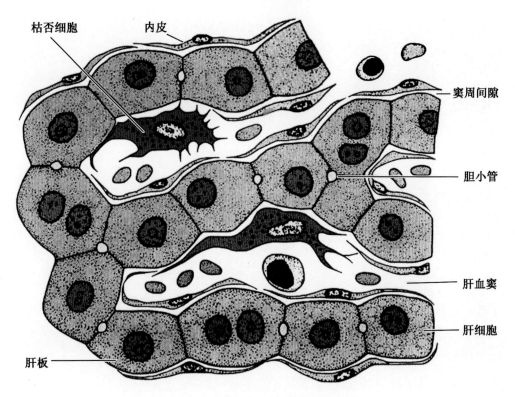

图 14-55 肝板、肝血窦高倍模式图

Fig.14-55 high magnification ideograph of liver plate and hepatic sinusoid

内皮—endothelium　枯否细胞—Kupffer cell　肝板—liver plate　窦周间隙—perisinusoidal space
胆小管—bile canaliculus　肝血窦—hepatic sinusoid　肝细胞—hepatocyte

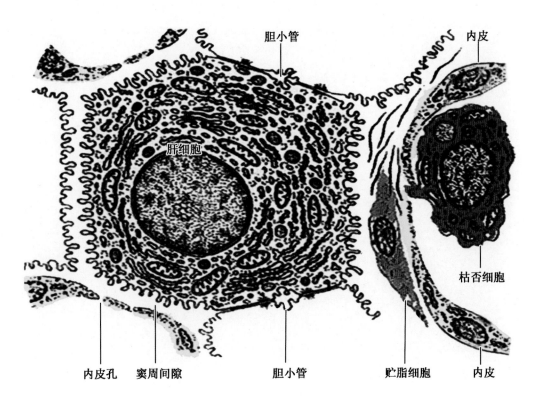

图 14-56 肝细胞、肝血窦和胆小管的关系

Fig.14-56 relationship among hepatocyte, hepatic sinusoid and bile canaliculus

胆小管—bile canaliculus　内皮—endothelium　肝细胞—hepatocyte　内皮孔—endothelium pore
窦周间隙—perisinusoidal space　贮脂细胞—fat store cell　枯否细胞—Kupffer cell

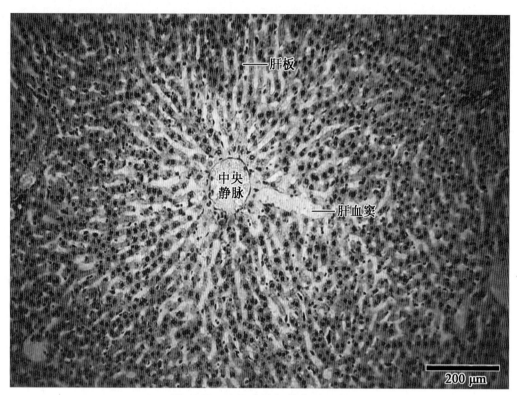

图 14-57　牛肝小叶低倍像（HE）

Fig.14-57　low magnification of cattle liver lobule（HE）

肝板—liver plate　中央静脉—central vein　肝血窦—hepatic sinusoid

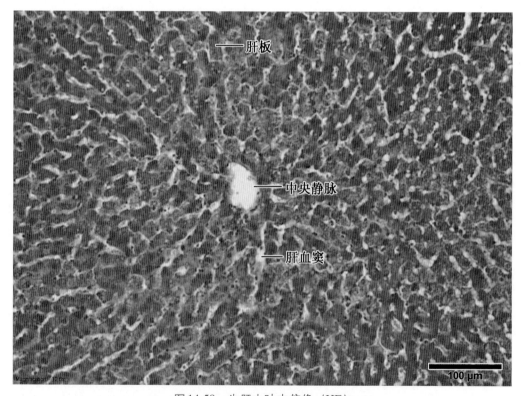

图 14-58　牛肝小叶中倍像（HE）

Fig.14-58　mid magnification of cattle liver lobule（HE）

肝板—liver plate　中央静脉—central vein　肝血窦—hepatic sinusoid

第十四章 消 化 腺　Digestive Glands

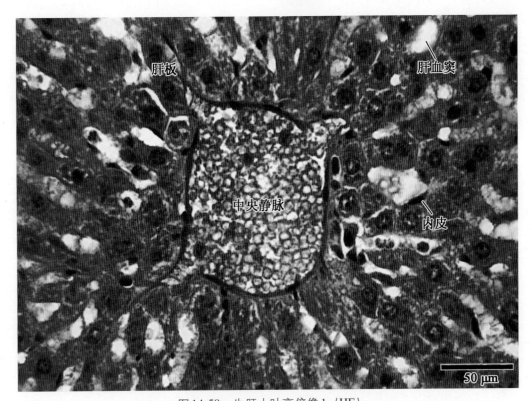

图 14-59　牛肝小叶高倍像1（HE）

Fig.14-59　high magnification of cattle liver lobule 1（HE）

肝板—liver plate　中央静脉—central vein　肝血窦—hepatic sinusoid　内皮—endothelium

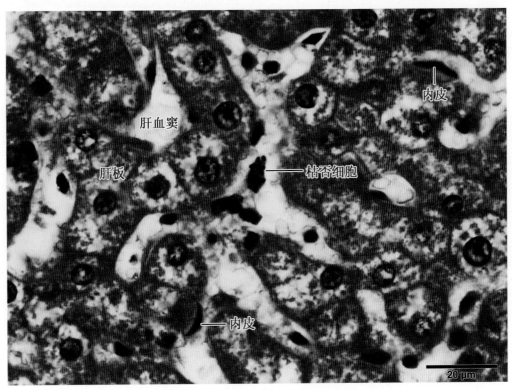

图 14-60　牛肝小叶高倍像2（HE）

Fig.14-60　high magnification of cattle liver lobule 2（HE）

肝血窦—hepatic sinusoid　肝板—liver plate　枯否细胞—Kupffer cell　内皮—endothelium

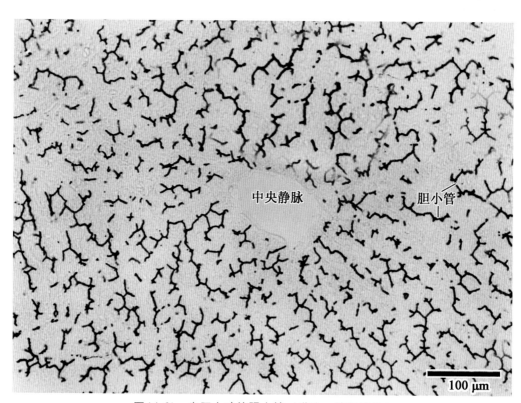

图 14-61　牛肝小叶的胆小管（灌注，镀银染色）
Fig.14-61　bile canaliculus in cattle hepatic lobule（injection，silver stain）
中央静脉—central vein　胆小管—bile canaliculus

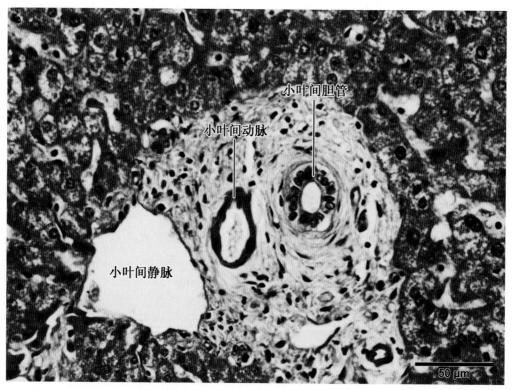

图 14-62　牛肝门管区高倍像（HE）
Fig.14-62　high magnification of cattle liver portal area（HE）
小叶间静脉—interlobular vein　小叶间动脉—interlobular artery　小叶间胆管—interlobular bile duct

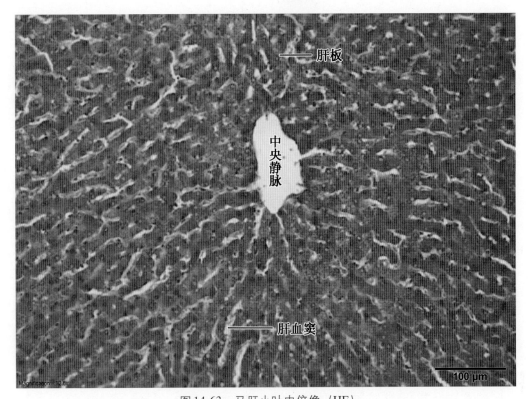

图14-63 马肝小叶中倍像（HE）
Fig.14-63 mid magnification of horse liver lobule（HE）
肝板—liver plate 中央静脉—central vein 肝血窦—hepatic sinusoid

图14-64 马肝小叶高倍像（HE）
Fig.14-64 high magnification of horse liver lobule（HE）
内皮—endothelium 肝板—liver plate 中央静脉—central vein 肝血窦—hepatic sinusoid

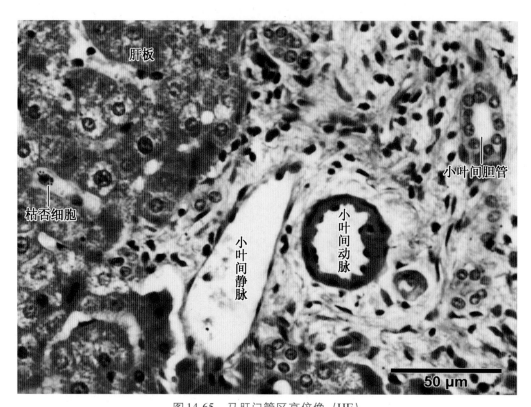

图 14-65 马肝门管区高倍像（HE）

Fig.14-65 high magnification of horse liver portal area（HE）

肝板—liver plate 枯否细胞—Kupffer cell 小叶间静脉—interlobular vein

小叶间动脉—interlobular artery 小叶间胆管—interlobular bile duct

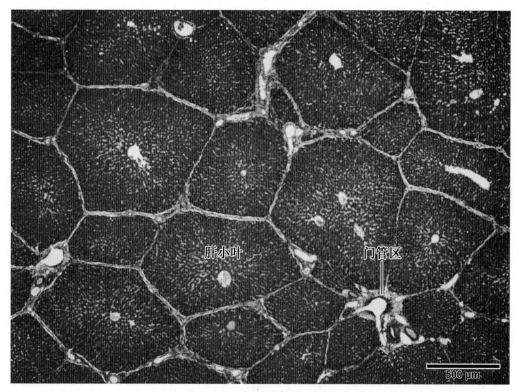

图 14-66 猪肝低倍像（HE）

Fig.14-66 low magnification of pig liver（HE）

肝小叶—liver lobule 门管区—portal area

第十四章 消化腺 Digestive Glands

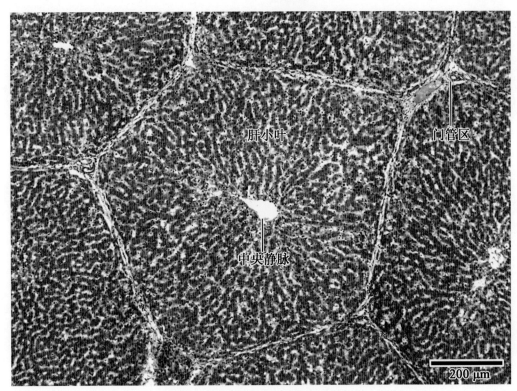

图 14-67 猪肝小叶低倍像（HE）
Fig.14-67 low magnification of pig liver lobule（HE）
肝小叶—liver lobule　门管区—portal area　中央静脉—central vein

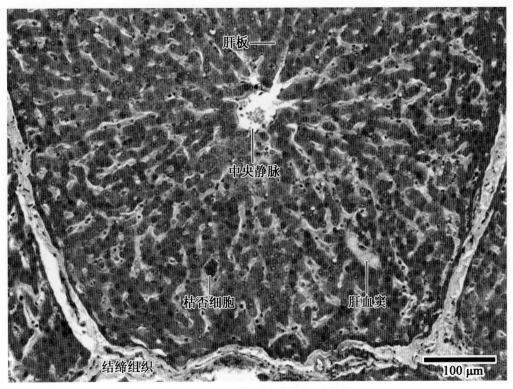

图 14-68 猪肝小叶中倍像（HE）
Fig.14-68 mid magnification of pig liver lobule（HE）
肝板—liver plate　中央静脉—central vein　枯否细胞—Kupffer cell　肝血窦—hepatic sinusoid　结缔组织—connective tissue

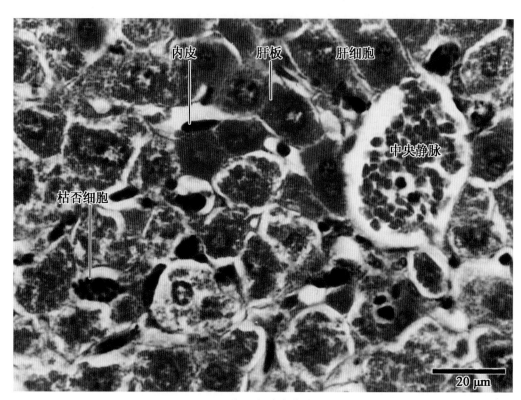

图14-69　猪肝小叶高倍像（HE）

Fig.14-69　high magnification of pig liver lobule（HE）

枯否细胞—Kupffer cell　内皮—endothelium　肝板—liver plate　肝细胞—hepatocyte　中央静脉—central vein

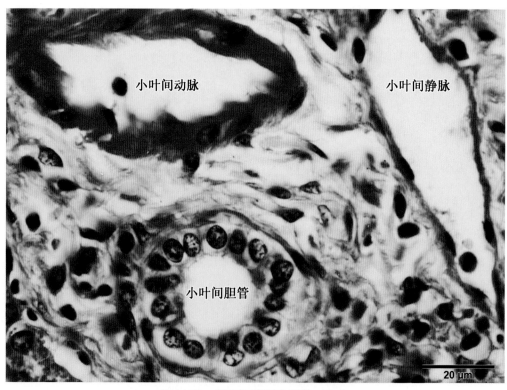

图14-70　猪肝门管区高倍像（HE）

Fig.14-70　high magnification of pig liver portal area（HE）

小叶间动脉—interlobular artery　小叶间胆管—interlobular bile duct　小叶间静脉—interlobular vein

第十四章 消化腺 Digestive Glands

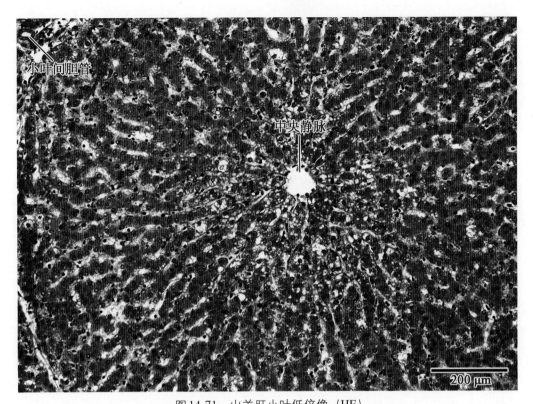

图 14-71 山羊肝小叶低倍像（HE）
Fig.14-71 low magnification of goat liver lobule（HE）
小叶间胆管—interlobular bile duct　中央静脉—central vein

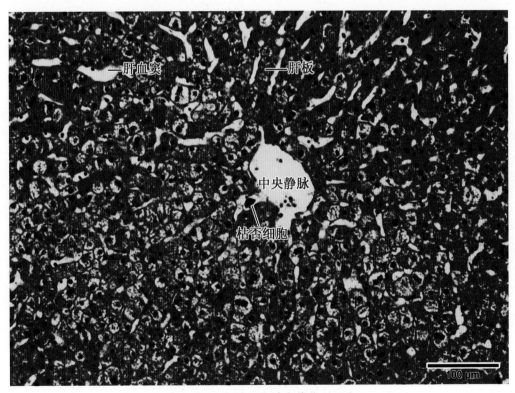

图 14-72 山羊肝小叶中倍像（HE）
Fig.14-72 mid magnification of goat liver lobule（HE）
肝血窦—hepatic sinusoid　肝板—liver plate　中央静脉—central vein　枯否细胞—Kupffer cell

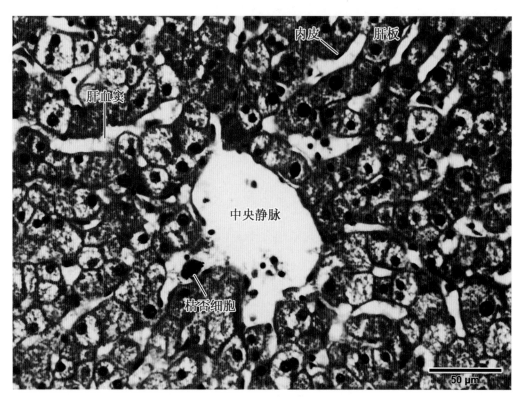

图 14-73　山羊肝小叶高倍像（HE）

Fig.14-73　high magnification of goat liver lobule（HE）

肝血窦—hepatic sinusoid　内皮—endothelium　肝板—liver plate　中央静脉—central vein　枯否细胞—Kupffer cell

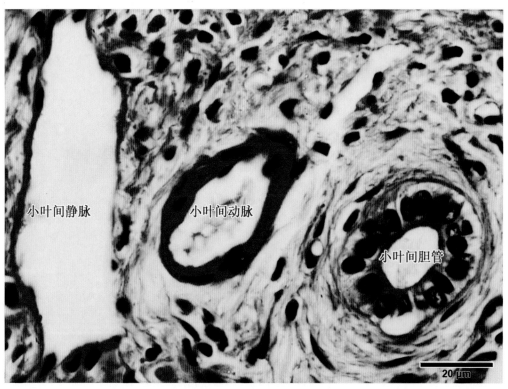

图 14-74　山羊肝门管区高倍像（HE）

Fig.14-74　high magnification of goat liver portal area（HE）

小叶间静脉—interlobular vein　小叶间动脉—interlobular artery　小叶间胆管—interlobular bile duct

第十四章 消 化 腺 Digestive Glands

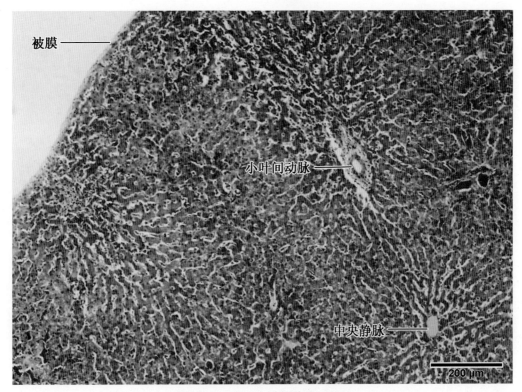

图14-75 犬肝小叶低倍像（HE）
Fig.14-75 low magnification of dog liver lobule（HE）
被膜—capsule 小叶间动脉—interlobular artery 中央静脉—central vein

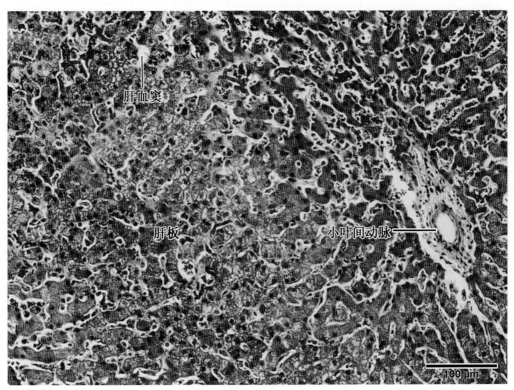

图14-76 犬肝小叶中低倍像（HE）
Fig.14-76 mid magnification of dog liver lobule（HE）
肝血窦—hepatic sinusoid 肝板—liver plate 小叶间动脉—interlobular artery

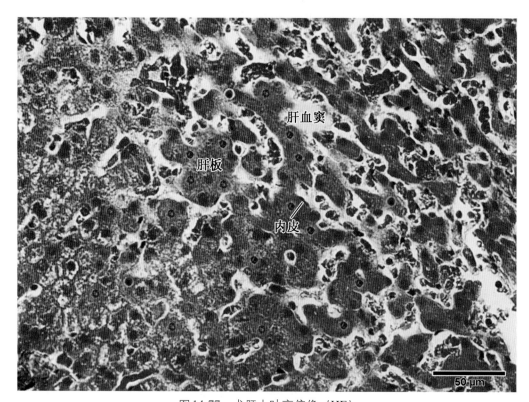

图14-77 犬肝小叶高倍像（HE）

Fig.14-77　high magnification of dog liver lobule（HE）

肝血窦—hepatic sinusoid　肝板—liver plate　内皮—endothelium

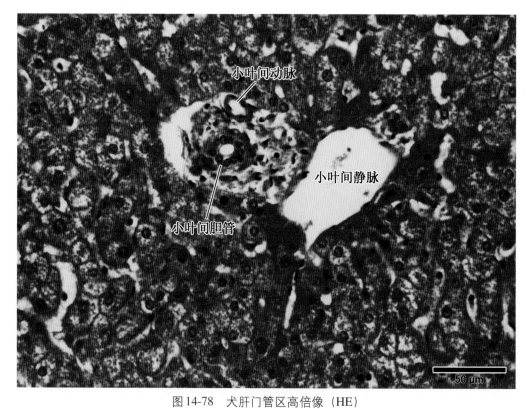

图14-78 犬肝门管区高倍像（HE）

Fig.14-78　high magnification of dog liver portal area（HE）

小叶间动脉—interlobular artery　小叶间胆管—interlobular bile duct　小叶间静脉—interlobular vein

第十四章 消化腺 Digestive Glands

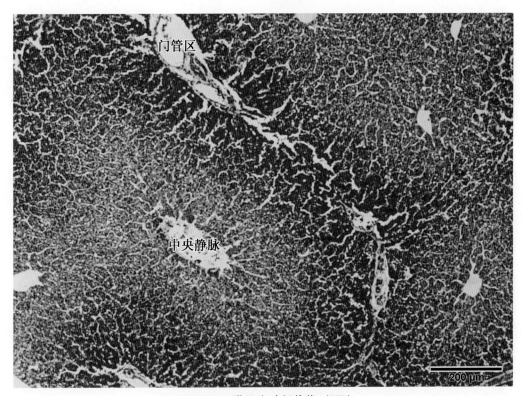

图14-79 猫肝小叶低倍像（HE）
Fig.14-79 low magnification of cat liver lobule（HE）
门管区—portal area 中央静脉—central vein

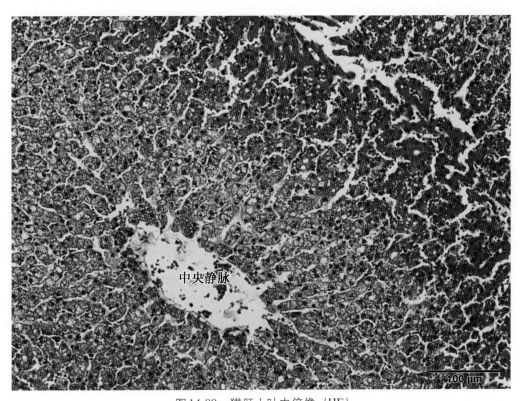

图14-80 猫肝小叶中倍像（HE）
Fig.14-80 mid magnification of cat liver lobule（HE）
中央静脉—central vein

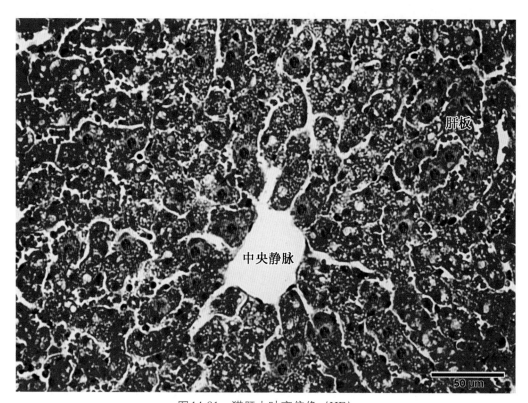

图14-81 猫肝小叶高倍像（HE）

Fig.14-81 high magnification of cat liver lobule（HE）

肝板—liver plate 中央静脉—central vein

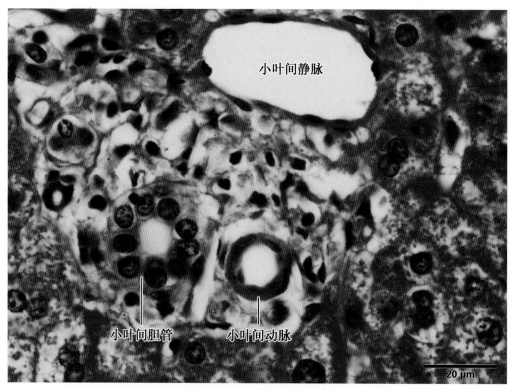

图14-82 猫肝门管区高倍像（HE）

Fig.14-82 high magnification of cat liver portal area（HE）

小叶间静脉—interlobular vein 小叶间胆管—interlobular bile duct 小叶间动脉—interlobular artery

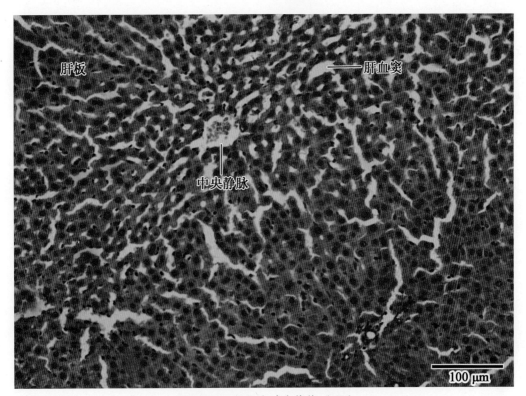

图14-83 兔肝小叶中倍像（HE）

Fig.14-83 mid magnification of rabbit liver lobule（HE）

肝板—liver plate　肝血窦—hepatic sinusoid　中央静脉—central vein

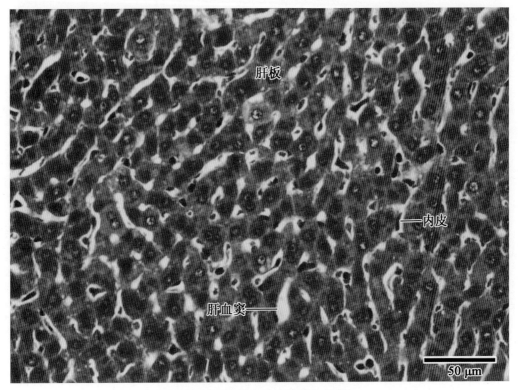

图14-84 兔肝小叶高倍像1（HE）

Fig.14-84 high magnification of rabbit liver lobule 1（HE）

肝板—liver plate　肝血窦—hepatic sinusoid　内皮—endothelium

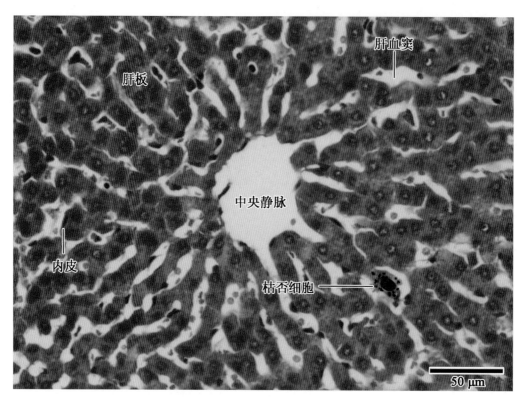

图14-85 兔肝小叶高倍像2（HE）

Fig.14-85 high magnification of rabbit liver lobule 2 （HE）

肝板—liver plate 肝血窦—hepatic sinusoid 内皮—endothelium 中央静脉—central vein 枯否细胞—Kupffer cell

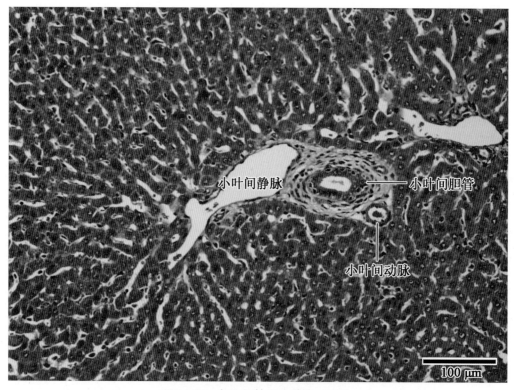

图14-86 兔肝门管区中倍像（HE）

Fig.14-86 mid magnification of rabbit liver portal area （HE）

小叶间静脉—interlobular vein 小叶间胆管—interlobular bile duct 小叶间动脉—interlobular artery

第十四章 消化腺　Digestive Glands

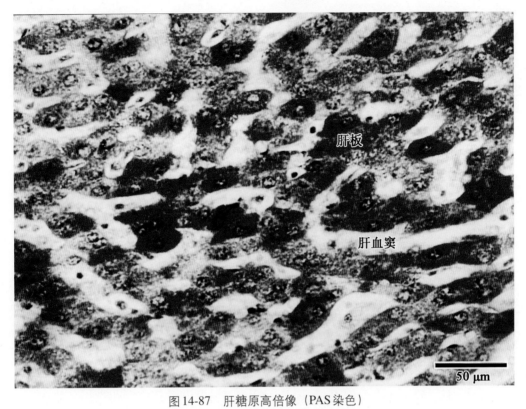

图 14-87　肝糖原高倍像（PAS染色）

Fig.14-87　high magnification of hepatic glycogen（PAS stain）

肝板—liver plate　肝血窦—hepatic sinusoid　糖原（图中的红色颗粒）—hepatic glycogens（red particles）

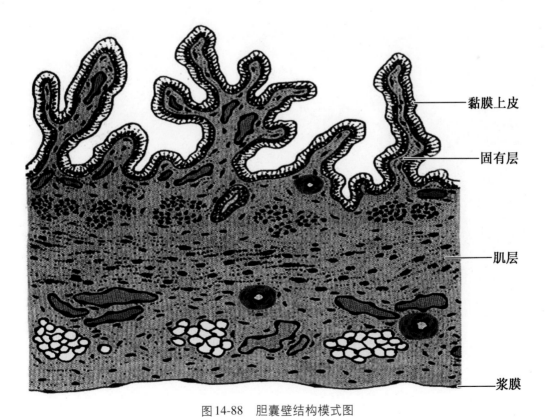

图 14-88　胆囊壁结构模式图

Fig.14-88　structural ideograph of gallbladder wall

黏膜上皮—mucosa epithelium　固有层—lamina propria　肌层—muscular layer　浆膜—serosa

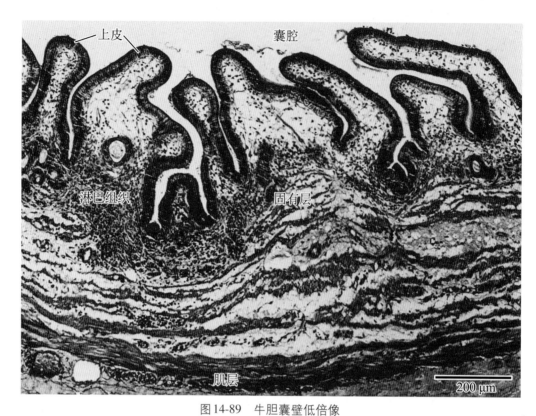

图 14-89　牛胆囊壁低倍像

Fig.14-89　low magnification of cow gallbladder wall

上皮—epithelium　囊腔—lumen　淋巴组织—lymph tissue　固有层—proper layer　肌层—muscular layer

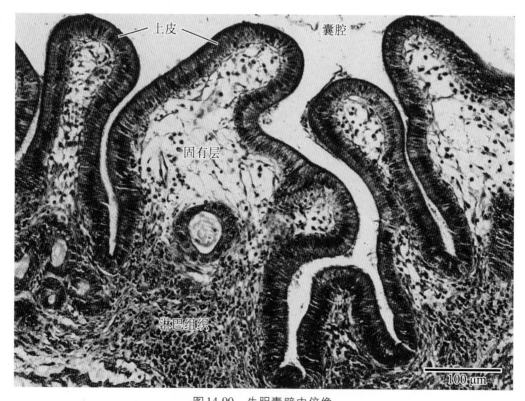

图 14-90　牛胆囊壁中倍像

Fig.14-90　mid magnification of cow gallbladder wall

上皮—epithelium　囊腔—lumen　固有层—proper layer　淋巴组织—lymph tissue

第十四章 消 化 腺 Digestive Glands

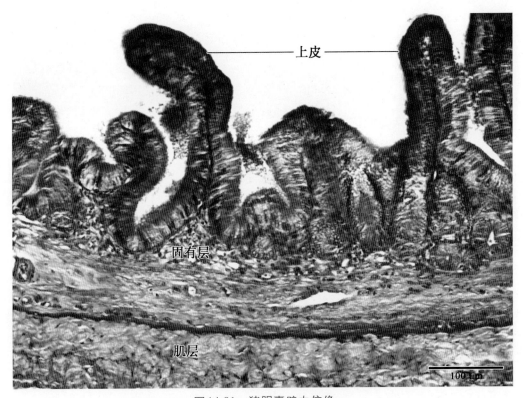

图 14-91　猪胆囊壁中倍像

Fig.14-91　mid magnification of pig gallbladder wall

上皮—epithelium　固有层—lamina propria　肌层—muscular layer

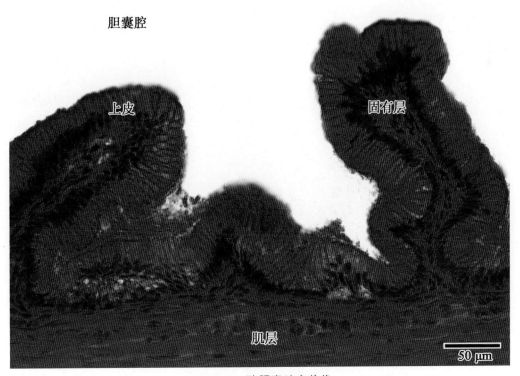

图 14-92　猪胆囊壁高倍像

Fig.14-92　high magnification of pig gallbladder wall

胆囊腔—cystic cavity　上皮—epithelium　固有层—lamina propria　肌层—muscular layer

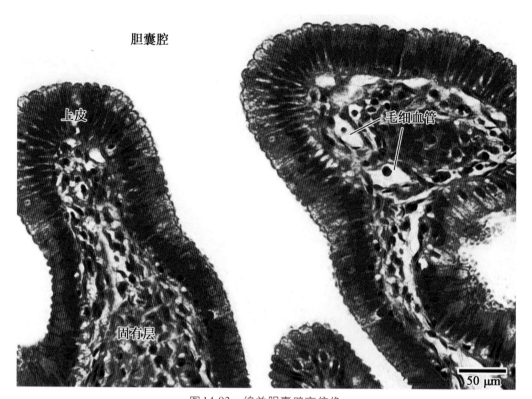

图14-93　绵羊胆囊壁高倍像

Fig.14-93　high magnification of sheep gallbladder wall

胆囊腔—cystic cavity　上皮—epithelium　固有层—lamina propria　毛细血管—capillary

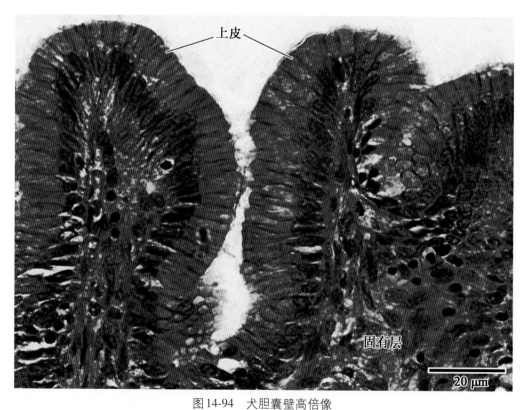

图14-94　犬胆囊壁高倍像

Fig.14-94　high magnification of dog gallbladder wall

上皮—epithelium　固有层—lamina propria

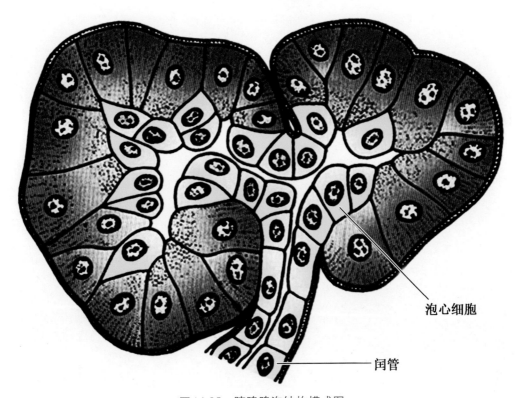

图 14-95 胰腺腺泡结构模式图
Fig.14-95 structural ideograph of pancreatic acini
闰管—intercalated duct　泡心细胞—centroacinar cell

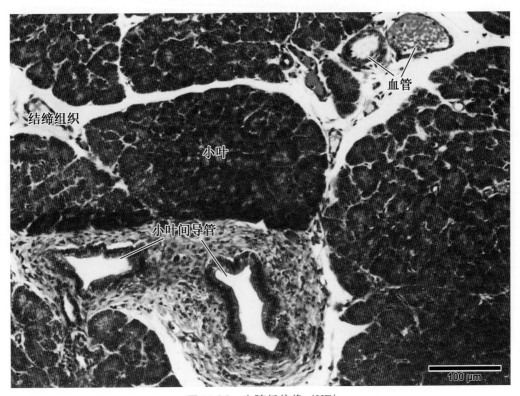

图 14-96 牛胰低倍像（HE）
Fig.14-96 low magnification of cattle pancreas（HE）
结缔组织—connective tissue　小叶—lobule　血管—blood vessel　小叶间导管—interlobular duct

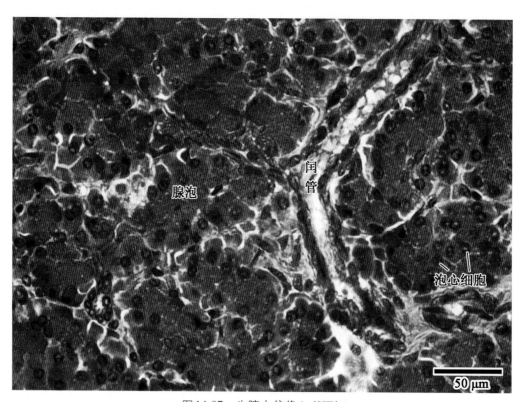

图 14-97　牛胰中倍像 1（HE）

Fig.14-97　mid magnification of cattle pancreas 1（HE）

腺泡—acinus　闰管—intercalated duct　泡心细胞—centroacinar cell

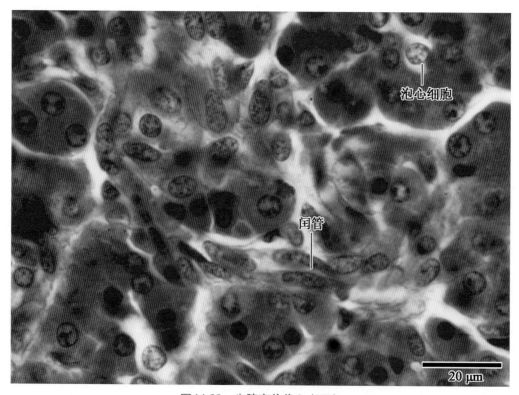

图 14-98　牛胰高倍像 2（HE）

Fig.14-98　high magnification of cattle pancreas 2（HE）

闰管—intercalated duct　泡心细胞—centroacinar cell

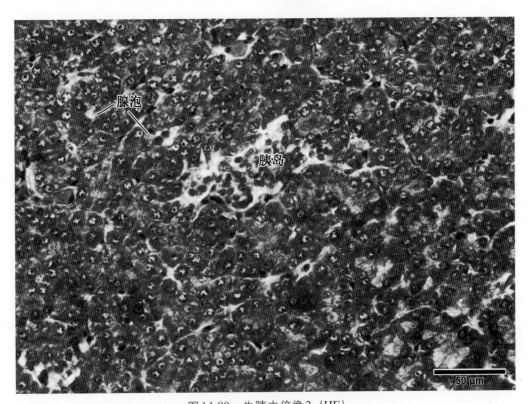

图14-99 牛胰中倍像2（HE）

Fig.14-99 mid magnification of cattle pancreas 2（HE）

腺泡—acinus 胰岛—pancreas islet

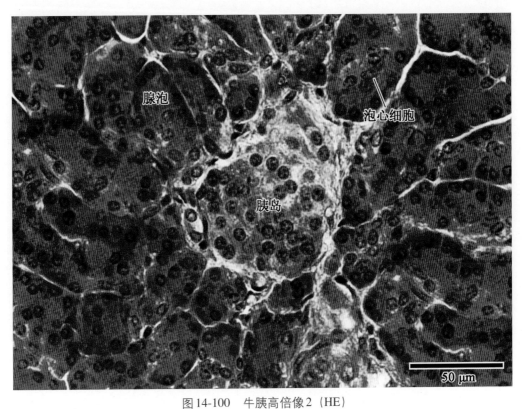

图14-100 牛胰高倍像2（HE）

Fig.14-100 high magnification of cattle pancreas 2（HE）

腺泡—acinus 胰岛—pancreas islet 泡心细胞—centroacinar cell

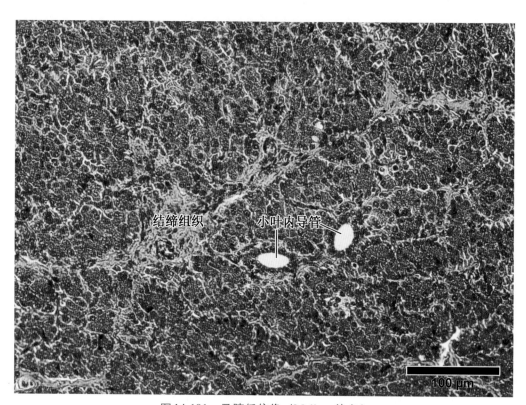

图 14-101　马胰低倍像（Mallory 染色）

Fig.14-101　low magnification of horse pancreas（Mallory stain）

结缔组织—connective tissue　小叶内导管—interlobular duct

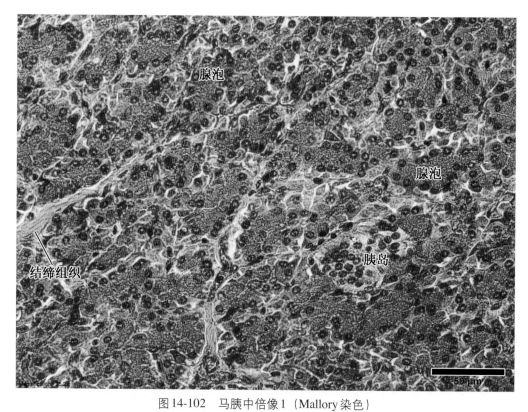

图 14-102　马胰中倍像 1（Mallory 染色）

Fig.14-102　mid magnification of horse pancreas 1（Mallory stain）

结缔组织—connective tissue　腺泡—acinus　胰岛—pancreas islet

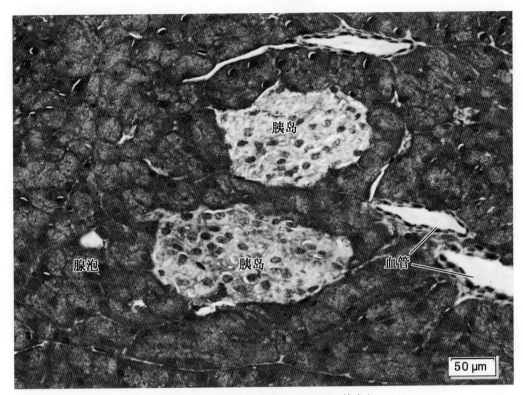

图14-103 马胰中倍像2（Mallory染色）

Fig.14-103 mid magnification of horse pancreas 2（Mallory stain）

腺泡—acinus 胰岛—pancreas islet 血管—blood vessel

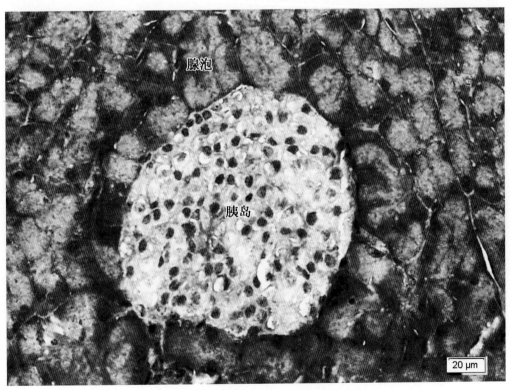

图14-104 马胰高倍像（Mallory染色）

Fig.14-104 high magnification of horse pancreas（Mallory stain）

腺泡—acinus 胰岛—pancreas islet

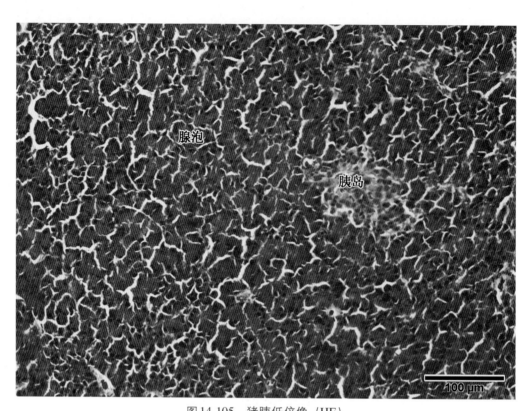

图 14-105　猪胰低倍像（HE）

Fig.14-105　low magnification of pig pancreas（HE）

腺泡—acinus　胰岛—pancreas islet

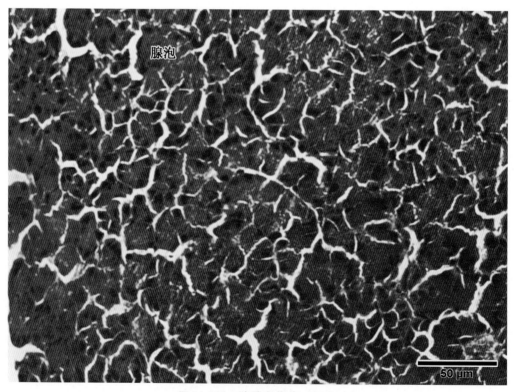

图 14-106　猪胰高倍像（HE）

Fig.14-106　high magnification of pig pancreas（HE）

腺泡—acinus

图 14-107 猪胰高倍像（Mallory-Azan 染色）
Fig.14-107 high magnification of pig pancreas（Mallory-Azan stain）
腺泡—acinus 胰岛—pancreas islet

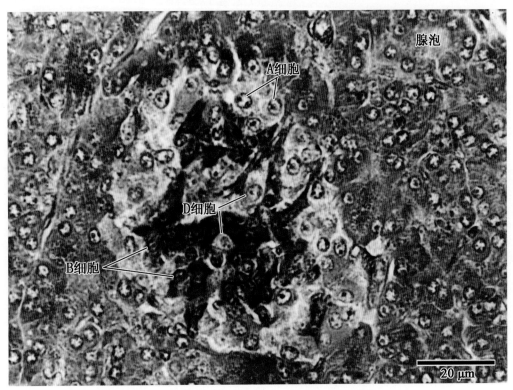

图 14-108 猪胰岛高倍像（Mallory-Azan 染色）
Fig.14-108 high magnification of pig pancreas islet（Mallory-Azan stain）
腺泡—acinus A 细胞—A cell B 细胞—B cell D 细胞—D cell

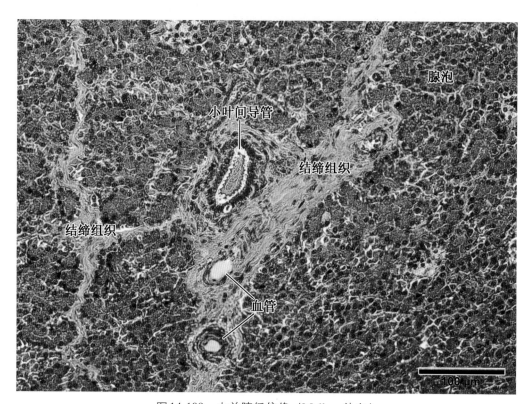

图14-109　山羊胰低倍像（Mallory染色）

Fig.14-109　low magnification of goat pancreas（Mallory stain）

结缔组织—connective tissue　小叶间导管—interlobular duct　腺泡—acinus　血管—blood vessel

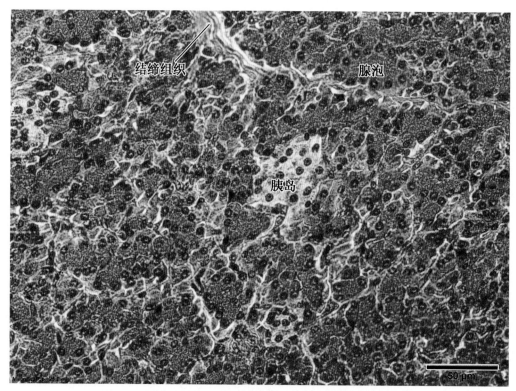

图14-110　山羊胰高倍像1（Mallory染色）

Fig.14-110　high magnification of goat pancreas 1（Mallory stain）

结缔组织—connective tissue　腺泡—acinus　胰岛—pancreas islet

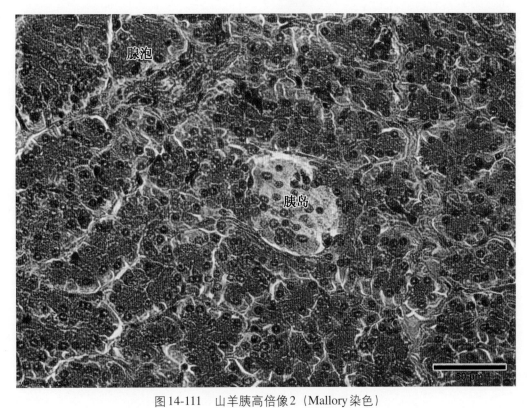

图 14-111　山羊胰高倍像 2（Mallory 染色）
Fig.14-111　high magnification of goat pancreas 2（Mallory stain）
腺泡—acinus　胰岛—pancreas islet

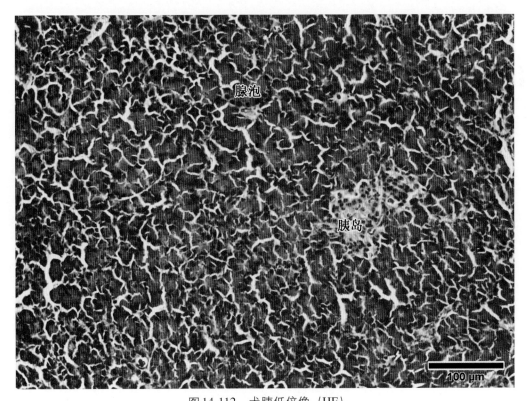

图 14-112　犬胰低倍像（HE）
Fig.14-112　low magnification of dog pancreas（HE）
腺泡—acinus　胰岛—pancreas islet

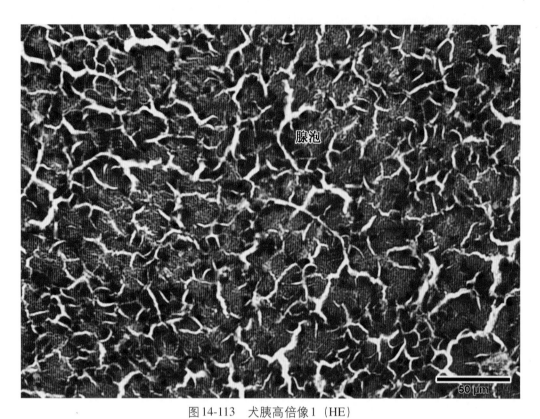

图14-113 犬胰高倍像1（HE）

Fig.14-113 high magnification of dog pancreas 1（HE）

腺泡—acinus

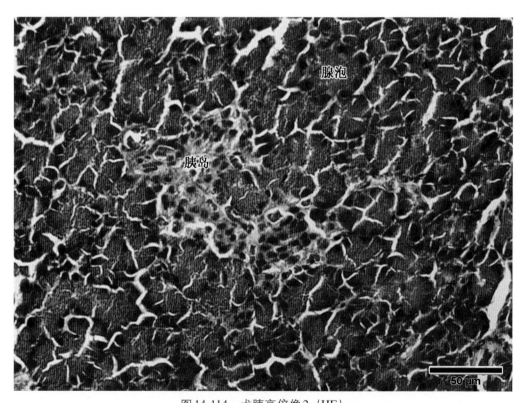

图14-114 犬胰高倍像2（HE）

Fig.14-114 high magnification of dog pancreas 2（HE）

腺泡—acinus 胰岛—pancreas islet

第十四章 消化腺 Digestive Glands

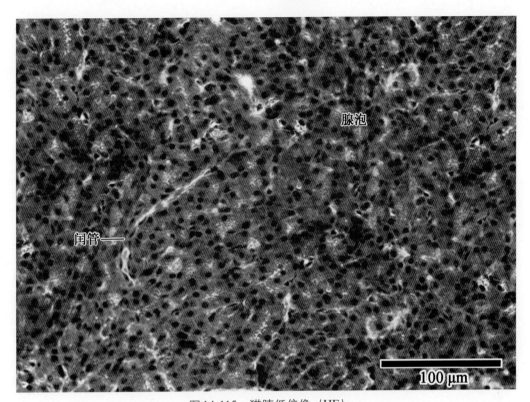

图14-115 猫胰低倍像（HE）

Fig.14-115 low magnification of cat pancreas（HE）

腺泡—acinus 闰管—intercalated duct

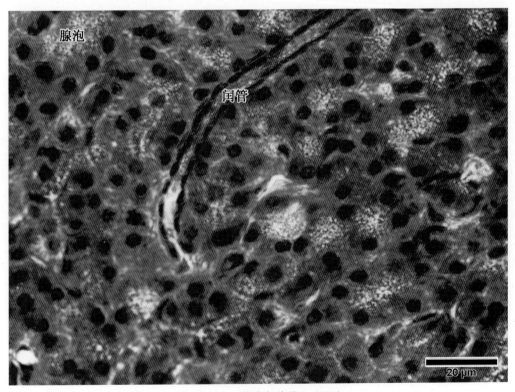

图14-116 猫胰高倍像1（HE）

Fig.14-116 high magnification of cat pancreas 1（HE）

腺泡—acinus 闰管—intercalated duct

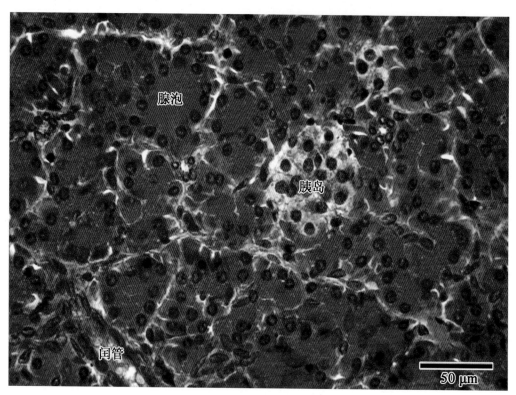

图14-117 猫胰高倍像2（HE）

Fig.14-117 high magnification of cat pancreas 2 (HE)

腺泡—acinus 胰岛—pancreas islet 闰管—intercalated duct

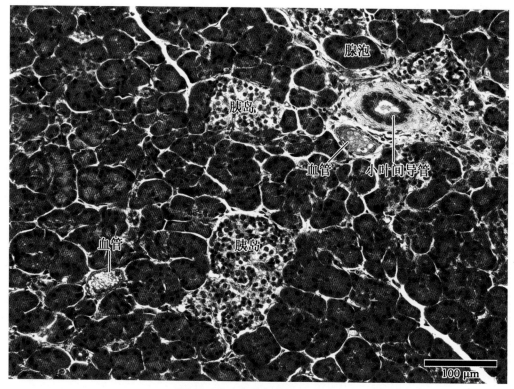

图14-118 兔胰低倍像（HE）

Fig.14-118 low magnification of rabbit pancreas (HE)

腺泡—acinus 胰岛—pancreas islet 血管—blood vessel 小叶间导管—interlobular duct

第十四章 消化腺 Digestive Glands

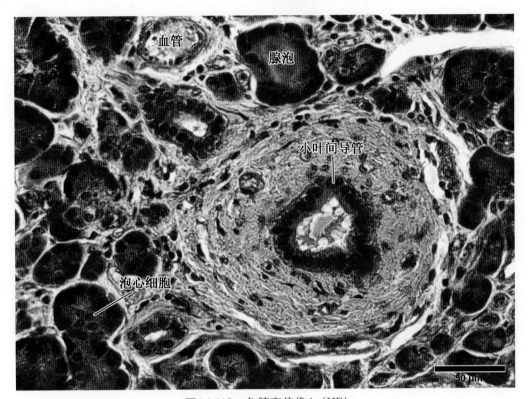

图 14-119　兔胰高倍像 1（HE）

Fig.14-119　high magnification of rabbit pancreas 1 （HE）

血管—blood vessel　腺泡—acinus　泡心细胞—centroacinar cell　小叶间导管—interlobular duct

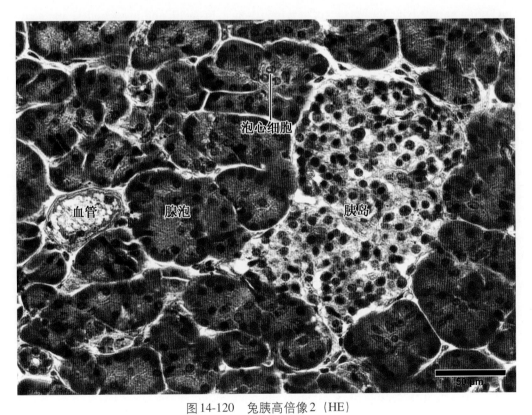

图 14-120　兔胰高倍像 2（HE）

Fig.14-120　high magnification of rabbit pancreas 2 （HE）

血管—blood vessel　腺泡—acinus　泡心细胞—centroacinar cell　胰岛—pancreas islet

第十五章
呼吸系统
Respiratory System

Outline

Respiratory system is composed of the lung and a series of tracts linking the pulmonary tissue with the external environment. This system is customarily divided into two principal subdivisions; a conducting portion, including nasal cavity, nasopharynx, larynx, trachea, bronchi, and bronchioles; and a respiratory portion, including alveoli and their associated structures. The important function of the respiratory system is gas exchange.

Nasal cavity mucosa is composed of epidermis and lamina propria. Nasal cavity can be separated into vestibular portion, respiratory region and olfactory region. The first portion is the intumescence of ingressing nose. Respiratory region is the main portion. The mucosa is pink in the normal physiological condition and the behind anodic is olfactory region.

The walls of trachea and bronchus become arranged in concentric layers of mucosa, submucosa and adventitia. The mucosa is lined with ciliated pseudostratified columnar epithelium that contains lots of goblet cells. The submucosa is composed of loose connective tissue. The adventitia consists of hyaline cartilage and fibers.

The outer surface of lung is serosa named ectoptygma or visceral pleura. The lung can be separated into parenchyma and interstitium. The parenchyma consists of branch and terminal alveoli pulmonis. The interstitium includes connective tissue, blood vessel, lymphatic vessel, nerves and so on. The parenchyma includes two subdivisions: air-conducting portion and respiratory portion. The air-conducting portion contains bronchus, segmental bronchus, small bronchus, bronchiole and terminal bronchiole. The respiratory portion is composed of respiratory bronchiole, alveolar duct, alveolar sac and pulmonary alveoli. In order to adapt the respiratory function, the epithelium of bronchus has a great change. In the larger bronchioles, the epithelium is ciliated pseudostratified columnar, which decreases in height and complexity to become ciliated simple columnar or cuboidal epithelium in the smaller terminal bronchioles.

第十五章 呼吸系统 Respiratory System

Alveoli are specialized sac-like structures that make up the greater part of the lung. Only within them do the O_2 and CO_2 exchange between the air and the blood. Alveolar epithelium includes type I cells and type II cells. The former, squamous alveolar cells, make up 97% of the alveolar surfaces. The latter, cuboidal cells, make up the remaining 3% of the alveolar surfaces. They can secret surfactant which decreases alveolar surface tension. The alveolar septum serves as the site of the blood-air barrier. The blood-air barrier refers to the cells and cell products. Across them, gases diffuse between the alveolar compartment and the capillary compartment.

呼吸系统（respiratory system）包括鼻、咽、喉、气管、支气管和肺。从鼻腔到肺终末细支气管主要是输送气体、温暖和净化空气。肺内呼吸性细支气管到肺泡进行气体交换。此外，鼻有嗅觉功能，喉与发音有关，肺还参与生物活性物质的合成与代谢过程。

一、呼吸系统的发生

呼吸系统由原肠分化而成，鼻腔上皮来自表面外胚层，其他部分的上皮由原始消化管内胚层分化而来。喉气管憩室的上端开口于咽的部分发育为喉，其余部分发育为气管。憩室的末端膨大发育为肺芽，是支气管和肺的原基。肺芽迅速生长并成树状分支，形成肺叶支气管并发育为肺段支气管，最终形成终末细支气管和有气体交换功能的呼吸性细支气管、肺泡管和肺泡囊。随着肺泡数量增多，肺泡上皮出现Ⅰ型细胞和Ⅱ型细胞并分泌表面活性物质。喉气管憩室和肺芽周围的间充质分化为喉、气管和各级支气管壁的结缔组织、软骨和平滑肌，并分化为肺内间质中的结缔组织。

二、呼吸器官的组织学结构概述

（一）鼻腔（nasal cavity）

鼻前庭是鼻腔入口处的膨大部，由皮肤覆盖，黏膜层有复层扁平上皮，固有层富有皮脂腺和汗腺。鼻呼吸部黏膜内有丰富的静脉丛。假复层柱状纤毛上皮间夹杂杯状细胞，节律性摆动排除异物，固有层血管和腺体发达，分泌鼻液保证鼻腔的湿润。嗅区黏膜为假复层纤毛柱状上皮，由嗅细胞、支持细胞、基底细胞组成。固有层内含分泌浆液的嗅腺，嗅细胞为双极神经元，其中央轴突汇集多数嗅细胞嗅丝，穿过筛板达嗅球，周围轴突凸出上皮表面，成为细长的嗅毛。

（二）咽（pharynx）

咽覆有假复层柱状纤毛上皮并有杯状细胞分布，固有层内有管状混合腺，弹性纤维将黏膜层和基层分开，肌层有环形肌和纵形肌。富含弹性纤维的结缔组织将肌层和疏松结缔组织外膜隔开。

（三）喉（larynx）

喉的内表面覆有假复层纤毛柱状上皮。喉的支架由甲状软骨、环状软骨和会厌软骨、杓状软骨构成。固有层内分布有丰富的弹性纤维，固有层和黏膜下层均有腺体分布。喉肌分为内、外两组。喉的神经均为迷走神经分支。

（四）气管和支气管（trachea and bronchus）

气管和支气管结构相似，管壁由内向外为黏膜、黏膜下层和外膜。黏膜由上皮和固有层组成，上皮为假复层纤毛柱状上皮，包含纤毛柱状细胞、杯状细胞、基细胞、刷细胞、小颗粒细胞；由疏松结缔组织构成的固有层与上皮细胞之间有明显的基膜，富含淋巴组织，是机体局部免疫的关键部位；固有层深部和黏膜下层有腺体分布；透明软骨环和纤维结缔组织构成的外膜调节气管的管径，保持气管畅通。

（五）肺（lung）

肺表面附有浆膜，组织分为实质和间质。肺实质导气部从叶支气管、段支气管、小支气管经细支气管到达终末细支气管均为肺内气体通道。终末细支气管继续分支为呼吸性细支气管（respiratory bronchiole）、肺泡管（alveolar duct）、肺泡囊（alveolar sac）和肺泡（alveolus），进行气体交换的呼吸部。

叶支气管到小支气管，上皮为假复层纤毛柱状上皮，杯状细胞和腺体逐渐减少，固有层含有较多弹性纤维、淋巴组织和环形平滑肌束，外膜的软骨层逐渐变为软骨片并减少。细支气管，上皮由假复层纤毛柱状上皮变为单层纤毛柱状上皮，杯状细胞、腺体、软骨片均减少至消失，环形肌和黏膜皱襞增多。终末细支气管，上皮为单层纤毛柱状上皮，具有完整的基层和黏膜皱襞，无腺体分布。

呼吸性细支气管，管腔小，管壁有少量肺泡，上皮由单层纤毛柱状上皮逐渐变为单层柱状或单层立方上皮。上皮下结缔组织有少量环形肌，皱襞消失。肺泡管，管壁不完整，有结节状膨大，上皮为单层立方上皮或单层扁平上皮，结缔组织较薄。肺泡囊是若干肺泡共同开口围成的囊状结构，切片中看不到完整的球形肺泡，无平滑肌。肺泡表面以扁平上皮细胞为主，胶原纤维和弹性纤维与相邻肺泡的扁平上皮组成肺泡隔将相邻肺泡隔开。

三、呼吸系统图谱

1. **鼻腔**　图 15-1 ～图 15-6。
2. **咽**　图 15-7。
3. **气管**　图 15-8 ～图 15-13。
4. **支气管**　图 15-14 ～图 15-16。
5. **肺**　图 15-17 ～图 15-40。

第十五章 呼吸系统　Respiratory System

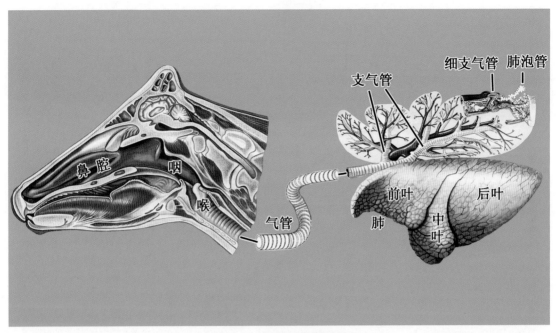

图 15-1　呼吸系统模式图

Fig.15-1　ideograph of respiratory system

鼻腔—nasal cavity　咽—pharynx　喉—throat　气管—trachea　支气管—bronchus　细支气管—bronchiole
肺泡管—alveolar duct　肺—lung　前叶—anterior lobe　中叶—mid lobe　后叶—posterior lobe

图 15-2　牛鼻前庭高倍像（HE）

Fig.15-2　high magnification of cow nasal vestibule（HE）

角化的上皮—keratinized epithelium　复层扁平上皮—stratified squamous epithelium　皮下组织—hypodermis

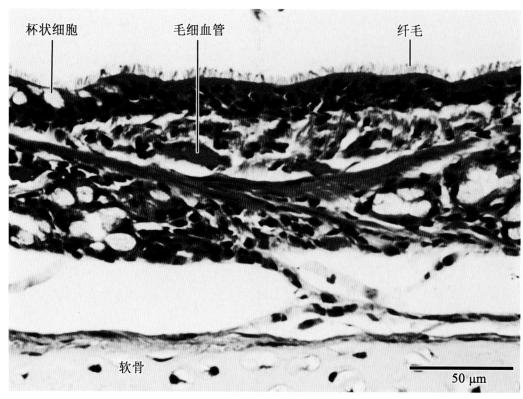

图 15-3　牛固有鼻腔高倍像（HE）

Fig.15-3　high magnification of cow proper nasal cavity（HE）

杯状细胞—goblet cell　毛细血管—capillary　纤毛—cilium　软骨—cartilage

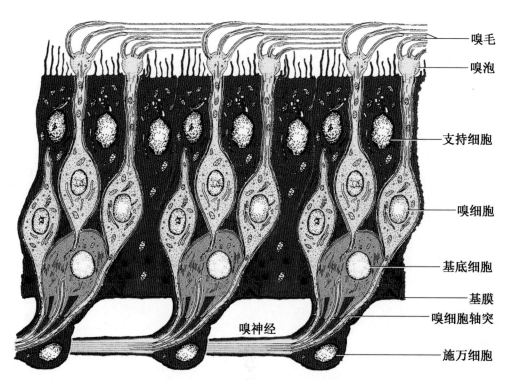

图 15-4　鼻腔嗅区黏膜结构模式图

Fig.15-4　mucosa structure ideograph of nasal cavity olfactory region

嗅毛—olfactory cilium　嗅泡—olfactory vesicle　支持细胞—supporting cell　嗅细胞—olfactory cell
基底细胞—basal cell　基膜—basement membrane　嗅细胞轴突—olfactory cell axon　施万细胞—Schwann cell
嗅神经—olfactory nerve

第十五章 呼吸系统 Respiratory System

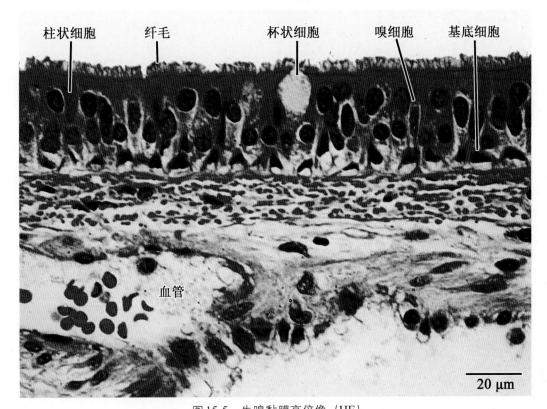

图 15-5 牛嗅黏膜高倍像（HE）

Fig.15-5 high magnification of cow olfactory mucosa（HE）

柱状细胞—columnar cell 纤毛—cilium 杯状细胞—goblet cell 嗅细胞—olfactory cell
基底细胞—basal cell 血管—blood vessel

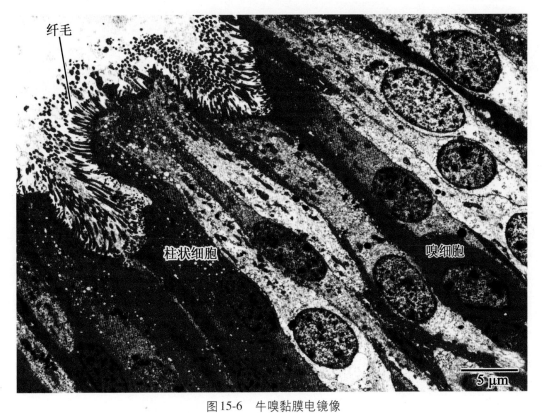

图 15-6 牛嗅黏膜电镜像

Fig.15-6 electron microscope photo of cow olfactory mucosa

纤毛—cilium 柱状细胞—columnar cell 嗅细胞—olfactory cell

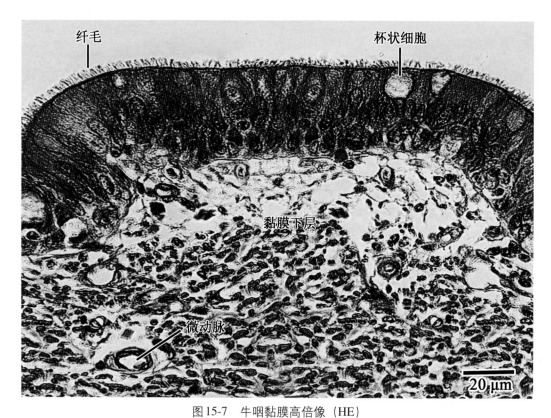

图 15-7 牛咽黏膜高倍像（HE）

Fig.15-7 high magnification of cow pharyngeal mucosa（HE）

纤毛—cilium 杯状细胞—goblet cell 黏膜下层—submucosa 微动脉—arteriole

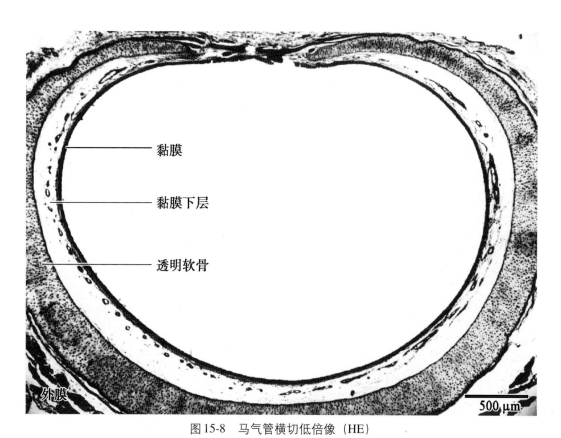

图 15-8 马气管横切低倍像（HE）

Fig.15-8 low magnification of horse tracheal cross section（HE）

黏膜—mucosa 黏膜下层—submucosa 透明软骨—hyaline cartilage 外膜—adventitia

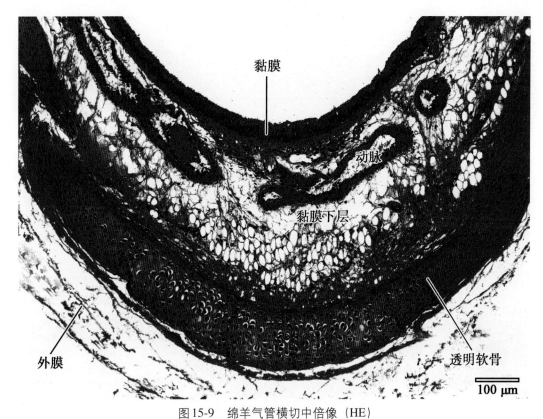

图15-9 绵羊气管横切中倍像（HE）

Fig.15-9 mid magnification of sheep tracheal cross section（HE）

黏膜—mucosa 动脉—artery 黏膜下层—submucosa 外膜—adventitia 透明软骨—hyaline cartilage

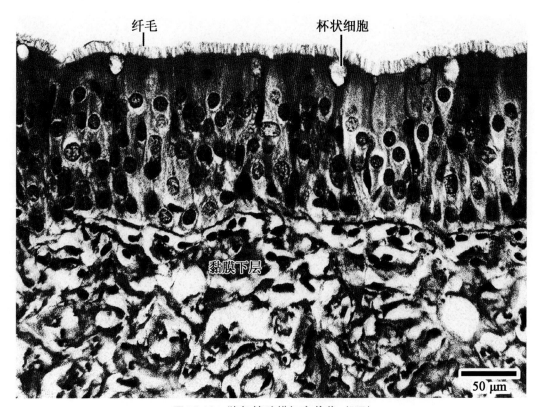

图15-10 猪气管壁横切高倍像（HE）

Fig.15-10 high magnification of pig tracheal cross section（HE）

纤毛—cilium 杯状细胞—goblet cell 黏膜下层—submucosa

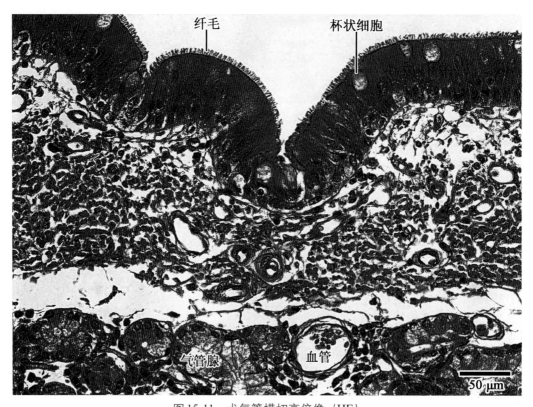

图 15-11 犬气管横切高倍像（HE）

Fig.15-11　high magnification of dog tracheal cross section（HE）

纤毛—cilium　杯状细胞—goblet cell　气管腺—tracheal gland　血管—blood vessel

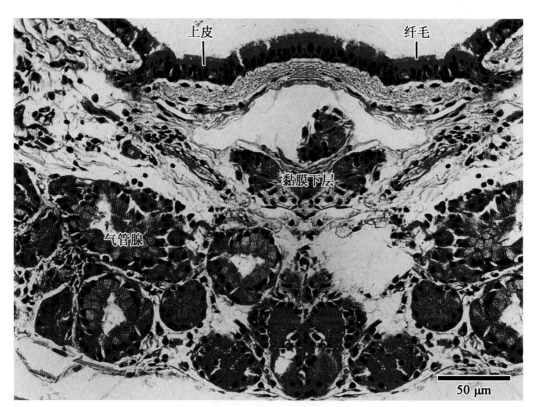

图 15-12 猫气管横切高倍像（HE）

Fig.15-12　high magnification of cat tracheal cross section（HE）

上皮—epithelium　纤毛—cilium　黏膜下层—submucosa　气管腺—tracheal gland

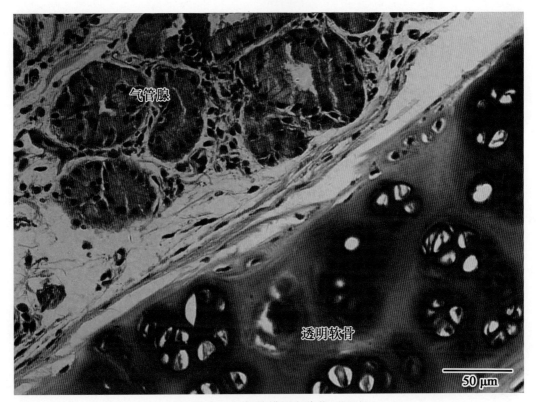

图15-13 兔气管横切高倍像（HE）

Fig.15-13 high magnification of rabbit tracheal cross section（HE）

气管腺—tracheal gland 透明软骨—hyaline cartilage

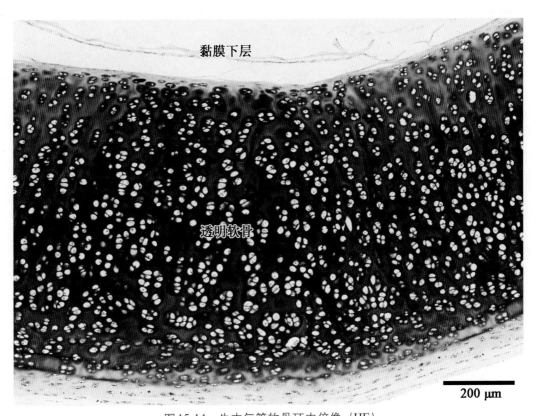

图15-14 牛支气管软骨环中倍像（HE）

Fig.15-14 mid magnification of cow bronchus cartilage（HE）

黏膜下层—submucosa 透明软骨—hyaline cartilage

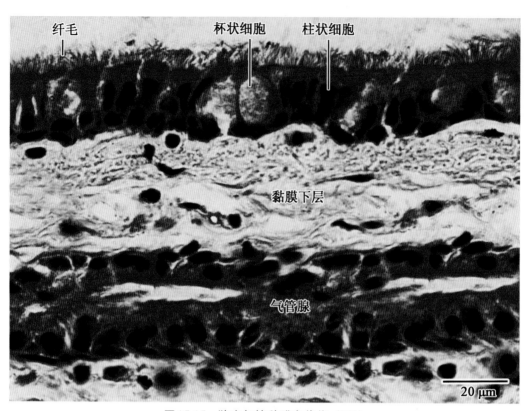

图15-15 猪支气管黏膜高倍像（HE）

Fig.15-15 high magnification of pig bronchus mucosa（HE）

纤毛—cilium　杯状细胞—goblet cell　柱状细胞—columnar cell　黏膜下层—submucosa　气管腺—tracheal gland

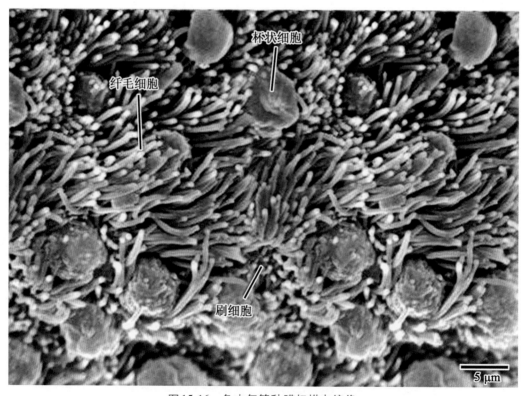

图15-16 兔支气管黏膜扫描电镜像

Fig.15-16 scanning electron microscope photo of rabbit bronchus mucosa

纤毛细胞—ciliated cell　杯状细胞—goblet cell　刷细胞—brush cell

第十五章 呼吸系统 Respiratory System

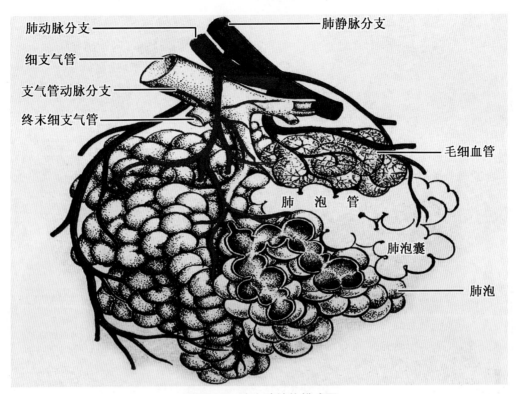

图15-17 肺小叶结构模式图

Fig.15-17 ideograph of pulmonary lobule

肺动脉分支—pulmonary branch 细支气管—bronchiole 支气管动脉分支—bronchial artery branch 终末细支气管—bronchiole terminals 肺静脉分支—pulmonary vein branch 毛细血管—capillary 肺泡管—alveolar duct 肺泡囊—alveolar sac 肺泡—alveolar

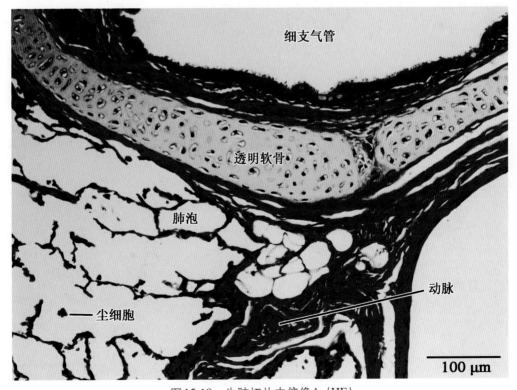

图15-18 牛肺切片中倍像1（HE）

Fig.15-18 mid magnification of cow lung section 1（HE）

细支气管—bronchiole 透明软骨—hyaline cartilage 肺泡—alveolus 尘细胞—dust cell 动脉—artery

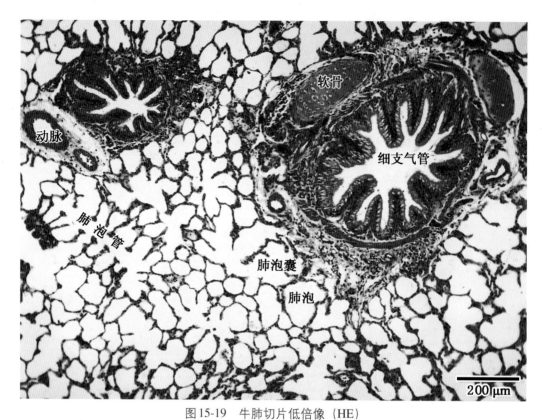

图15-19　牛肺切片低倍像（HE）

Fig.15-19　low magnification of cow lung section （HE）

动脉—artery　软骨—cartilage　细支气管—bronchiole　肺泡管—alveolar duct

肺泡囊—alveolar sac　肺泡—alveolus

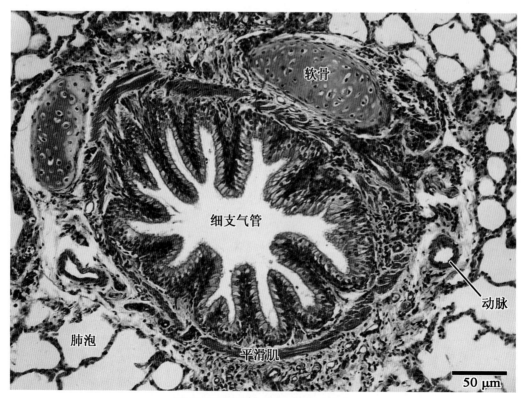

图15-20　牛肺切片高倍像（HE）

Fig.15-20　high magnification of cow lung section （HE）

软骨—cartilage　细支气管—bronchiole　肺泡—alveolus　平滑肌—smooth muscle　动脉—artery

第十五章 呼吸系统　Respiratory System

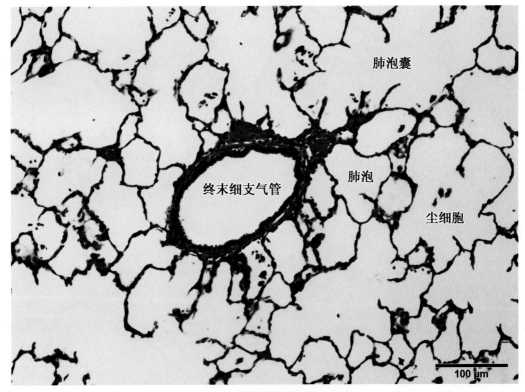

图 15-21　牛肺切片中倍像 2（HE）

Fig.15-21　mid magnification of cow lung section 2（HE）

终末细支气管—bronchiole terminals　肺泡囊—alveolar sac　肺泡—alveolus　尘细胞—dust cell

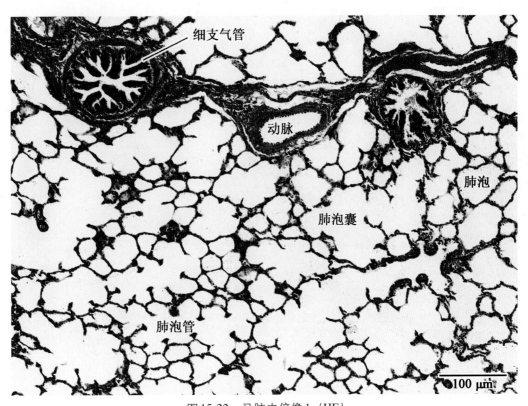

图 15-22　马肺中倍像 1（HE）

Fig.15-22　mid magnification of horse lung 1（HE）

细支气管—bronchiole　动脉—artery　肺泡管—alveolar duct　肺泡囊—alveolar sac　肺泡—alveolus

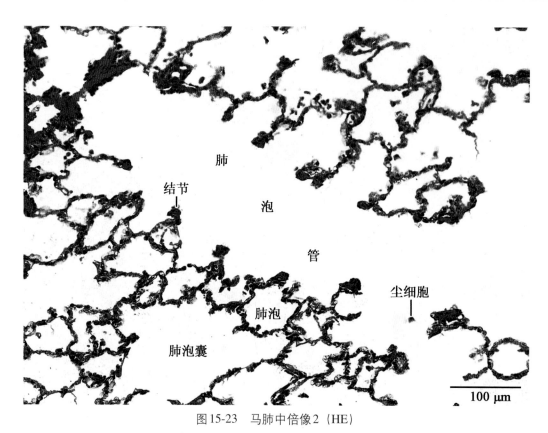

图15-23　马肺中倍像2（HE）

Fig.15-23　mid magnification of horse lung 2（HE）

肺泡管—alveolar duct　肺泡囊—alveolar sac　肺泡—alveolus　尘细胞—dust cell　结节—tuber

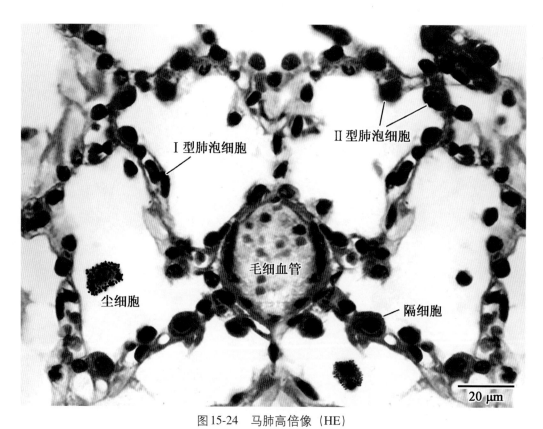

图15-24　马肺高倍像（HE）

Fig.15-24　high magnification of horse lung（HE）

Ⅰ型肺泡细胞—type Ⅰ alveolar cell　Ⅱ型肺泡细胞—type Ⅱ alveolar cell　尘细胞—dust cell
毛细血管—capillary　隔细胞—septum cell

第十五章 呼吸系统　Respiratory System

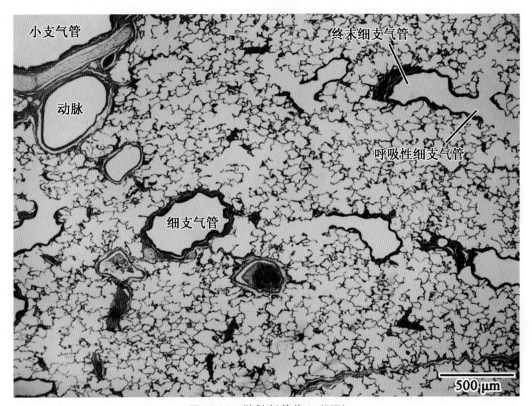

图15-25　猪肺低倍像1（HE）

Fig.15-25　low magnification of pig lung 1（HE）

小支气管—bronchium　动脉—artery　细支气管—bronchiole　终末细支气管—bronchiole terminals
呼吸性细支气管—respiratory bronchiole

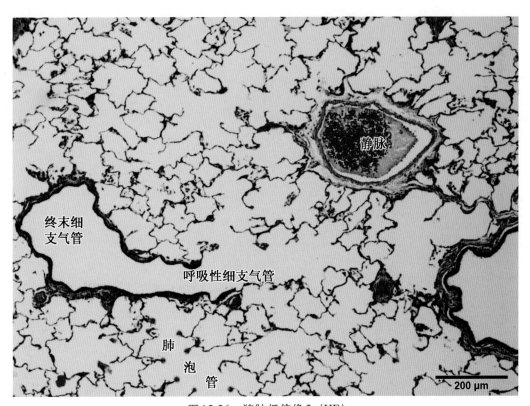

图15-26　猪肺低倍像2（HE）

Fig.15-26　low magnification of pig lung 2（HE）

静脉—vein　终末细支气管—bronchiole terminals　呼吸性细支气管—respiratory bronchiole　肺泡管—alveolar duct

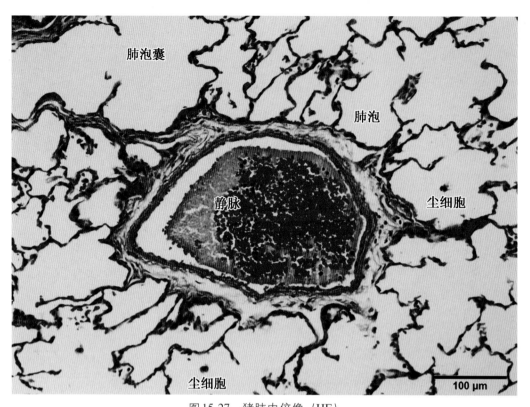

图15-27 猪肺中倍像（HE）

Fig.15-27 mid magnification of pig lung（HE）

肺泡囊—alveolar sac 静脉—vein 肺泡—alveolus 尘细胞—dust cell

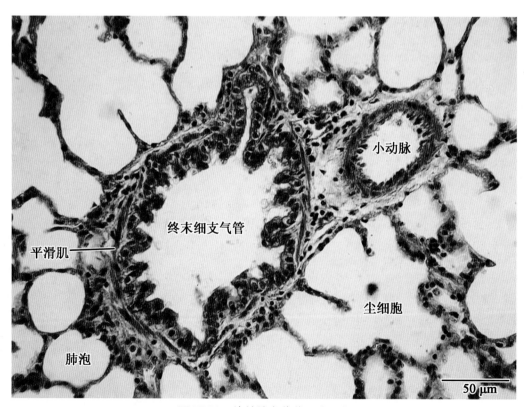

图15-28 绵羊肺高倍像1（HE）

Fig.15-28 high magnification of sheep lung 1（HE）

平滑肌—smooth muscle 肺泡—alveolus 终末细支气管—bronchiole terminals 小动脉—small artery 尘细胞—dust cell

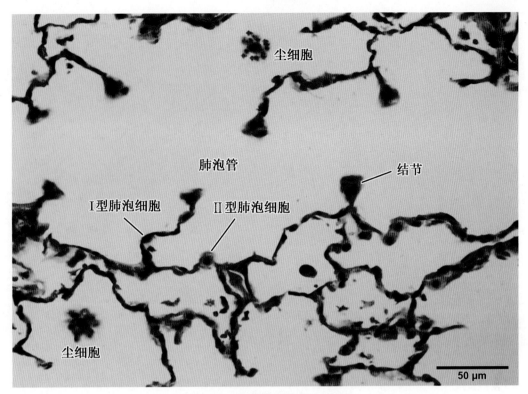

图 15-29　绵羊肺高倍像 2（HE）

Fig.15-29　high magnification of sheep lung 2（HE）

尘细胞—dust cell　肺泡管—alveolar duct　Ⅰ型肺泡细胞—type Ⅰ alveolar cell　Ⅱ型肺泡细胞—type Ⅱ alveolar cell
结节—tuber

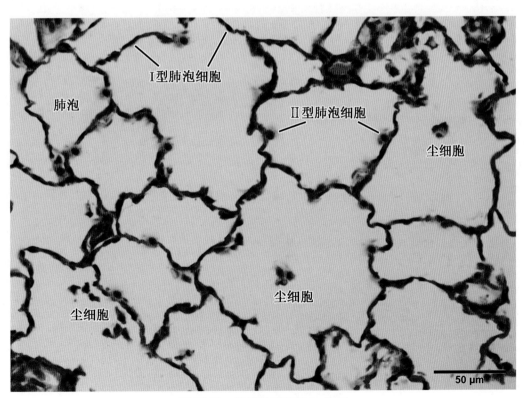

图 15-30　绵羊肺高倍像 3（HE）

Fig.15-30　high magnification of sheep lung 3（HE）

肺泡—alveolus　Ⅰ型肺泡细胞—type Ⅰ alveolar cell　Ⅱ型肺泡细胞—type Ⅱ alveolar cell　尘细胞—dust cell

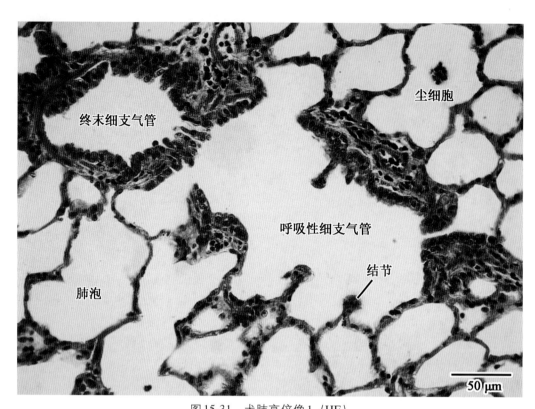

图15-31 犬肺高倍像1（HE）

Fig.15-31 high magnification of dog lung 1 (HE)

终末细支气管—bronchiole terminals 呼吸性细支气管—respiratory bronchiole

肺泡—alveolus 尘细胞—dust cell 结节—tuber

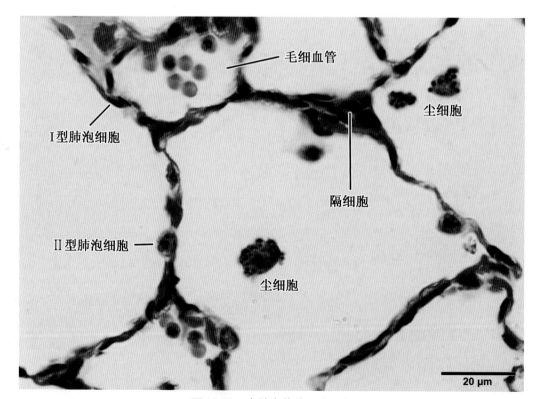

图15-32 犬肺高倍像2（HE）

Fig.15-32 high magnification of dog lung 2 (HE)

Ⅰ型肺泡细胞—type Ⅰ alveolar cell Ⅱ型肺泡细胞—type Ⅱ alveolar cell 毛细血管—capillary

尘细胞—dust cell 隔细胞—septum cell

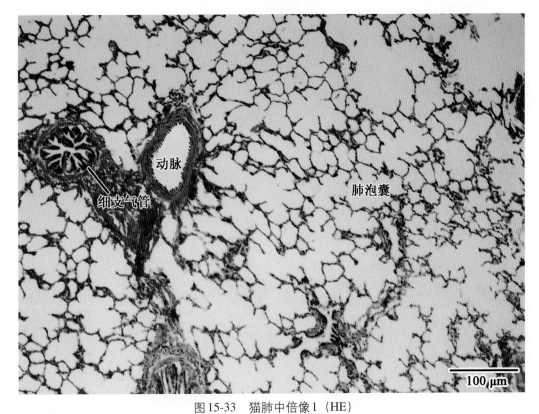

图 15-33　猫肺中倍像1（HE）

Fig.15-33　mid magnification of cat lung 1 （HE）

细支气管—bronchiole　动脉—artery　肺泡囊—alveolar sac

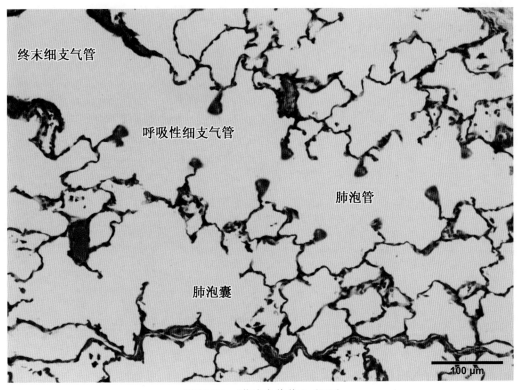

图 15-34　猫肺中倍像2（HE）

Fig.15-34　mid magnification of cat lung 2 （HE）

终末细支气管—bronchiole terminals　呼吸性细支气管—respiratory bronchiole　肺泡管—alveolar duct　肺泡囊—alveolar sac

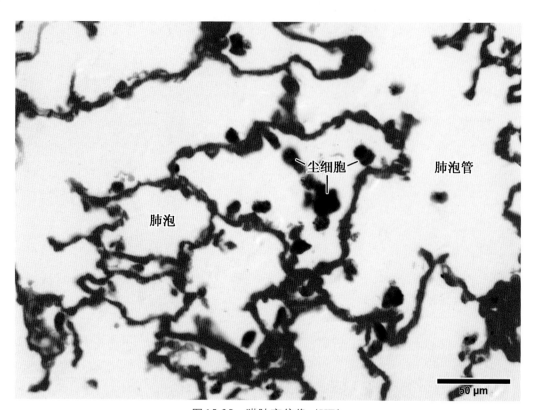

图15-35 猫肺高倍像（HE）

Fig.15-35 high magnification of cat lung （HE）

肺泡—alveolus　尘细胞—dust cell　肺泡管—alveolar duct

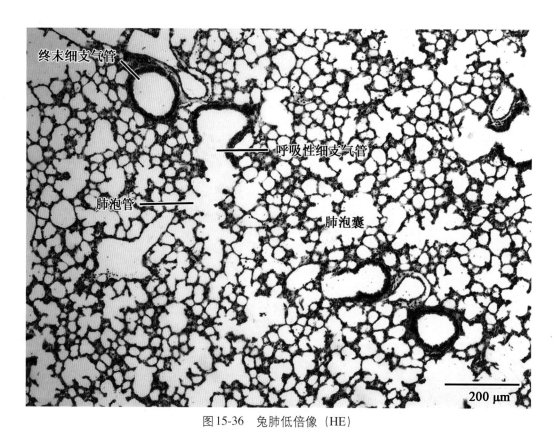

图15-36 兔肺低倍像（HE）

Fig.15-36 low magnification of rabbit lung （HE）

终末细支气管—bronchiole terminals　呼吸性细支气管—respiratory bronchiole　肺泡管—alveolar duct　肺泡囊—alveolar sac

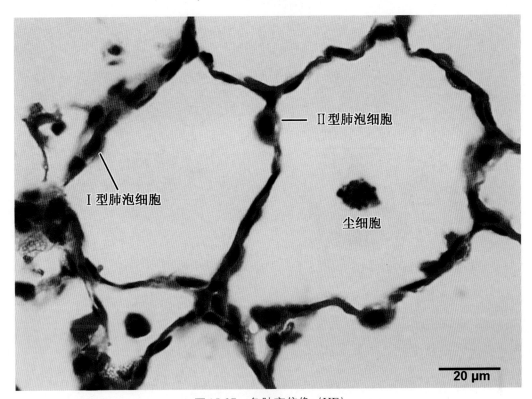

图15-37 兔肺高倍像（HE）
Fig.15-37 high magnification of rabbit lung（HE）
Ⅰ型肺泡细胞—type Ⅰ alveolar cell   Ⅱ型肺泡细胞—type Ⅱ alveolar cell   尘细胞—dust cell

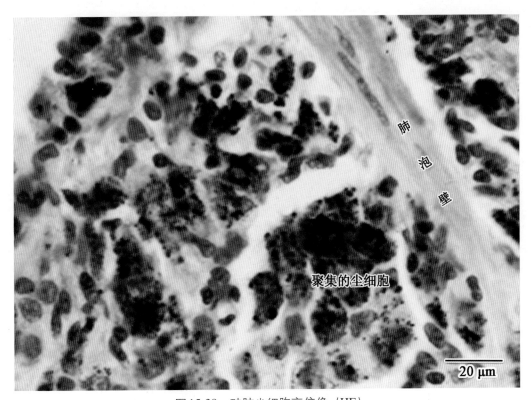

图15-38 硅肺尘细胞高倍像（HE）
Fig.15-38 high magnification of dust cells in silicosis lung（HE）
肺泡壁—alveolar wall   聚集的尘细胞—collective dust cell

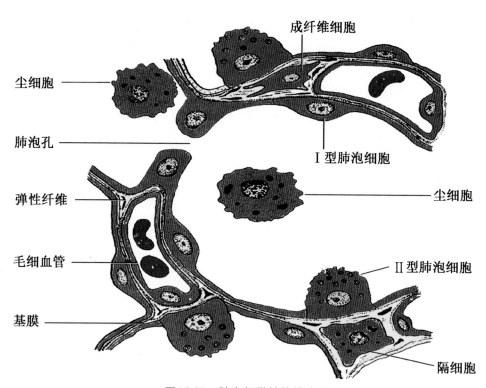

图15-39 肺泡超微结构模式图

Fig.15-39 ideograph of alveolus ultrastructure

尘细胞—dust cell　肺泡孔—alveolar pore　弹性纤维—elastic fiber　毛细血管—capillary　基膜—basal membrane
成纤维细胞—fibroblast　Ⅰ型肺泡细胞—type Ⅰ alveolar cell　Ⅱ型肺泡细胞—type Ⅱ alveolar cell　隔细胞—septum cell

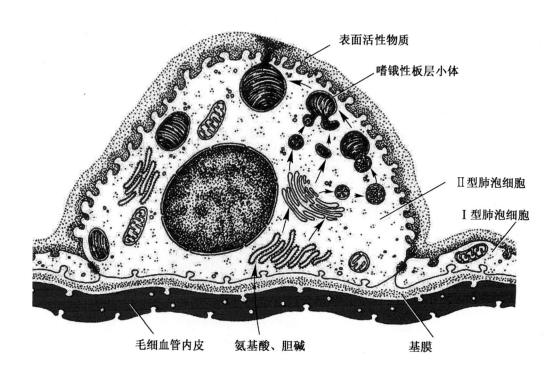

图15-40 肺泡细胞超微结构模式图

Fig.15-40 ultrastructure ideograph of type Ⅱ alveolar cell

表面活性物质—surfactant　嗜锇性板层小体—osmiophilic multilamellar body　Ⅱ型肺泡细胞—type Ⅱ alveolar cell
Ⅰ型肺泡细胞—type Ⅰ alveolar cell　毛细血管内皮—capillary endothelium　氨基酸、胆碱—amino acid and choline
基膜—basal membrane

第十六章
泌尿系统
Urinary System

Outline

The components of the urinary system include two kidneys, two ureters, a bladder, and a urethra. Kidneys are the important organ of the urinary system, which elaborate a fluid product called urine; ureters, two fibromuscular tubes, conduct the urine to a single urinary bladder where the fluid accumulates for periodic evacuation via the single urethra that connects the bladder to the exterior.

Kidneys are respectively wrapped by a capsule of dense connective tissue. The parenchyma of kidney consists of an outer cortex and an inner medulla. The functional unit of kidney is uriniferous tubule. Uriniferous tubule is composed of a long convoluted portion called nephron and a system of intrarenal collecting ducts. Nephron is composed of several regions of diversified morphology, but all of them are characterized by cells that have an elaborate shape with numerous lateral interdigitating processes. The blind end of nephron is indented by a network of capillaries and supporting cells to form a filtering body called the renal corpuscle. In addition, nephron consists of a proximal convoluted tubule, a straight region of the proximal tubule, a thin limb, a straight region of the distal tubule, a macula densa region of the distal tubule, and a distal convoluted tubule. Nephrons are situated within the kidney in a characteristic position with the renal corpuscle and proximal convoluted tubules located in the cortex. The straight portion of the proximal tubule, the thin limb segment, and the straight portion of the distal tubule form a looping structure called the loop of Henle, which enters into medullary pyramid by way of a medullary ray, forms a hairpin loop within the medulla, and returns to the cortex via the same medullary ray. Nephrons are classified as superficial or juxtamedullary by the position of their renal corpuscles within the superficial or juxtamedullary region of the cortex. Nephron begins with a renal corpuscle located in the cortex and is roughly oval in shape. Each renal corpuscle consists of tufts of capillaries and their supporting cells, which have developed within a double-walled capsule in one end of the developing renal tubule. The outer wall of the capsule is called the parietal layer; the inner wall is the visceral (podocyte) layer. The space between the two walls of the capsule is called Bowman's space. The epithelium of this visceral wall covers the tufts of capillaries much like a glove covers each fingers of a hand. Between the epithelium and capillaries is an extracellular layer, the glomerular basement membrane. The parietal layer of capsular epithelium is continuous with the epithelium of the neck of the tubule. Bowman's space is therefore continuous with the lumen of the remaining nephron, so that fluid formed by filtration within the renal corpuscle enters the lumen of the proximal convoluted tubule. Nephrons empty into a complex system of collecting ducts. Urine is conveyed from the kidney to the bladder where it is stored. The Juxtaglomerular complex is the structure adjacent to the renal corpuscle, which consists of juxtaglomerular cells, macula densa and extraglomerular mesangial cells. The function of these apparatuses relates to the maintenance of blood pressure by producing and secreting hormones.

The walls of ureter, bladder and urethra are similar in their basic structure, being composed of an inner mucosal layer, a middle muscular layer, and an external adventitial coat of connective tissue that binds the structure to the surrounding connective tissue.

一、泌尿系统的发生

泌尿系统（urinary system）由肾、输尿管、膀胱和尿道组成，产生、贮存和排出尿液。动物在泌尿器官的胚胎发育过程中，前后产生3代位置不同的肾，即前肾、中肾和后肾。这3种肾在某些脊椎动物的成体中都可看到，因此，胚胎泌尿器官的演化过程，反映了系统发育的真实情况。

中段中胚层演化为生肾节后，其前部的生肾节形成许多独立的小管，称为前肾小管。前肾小管与其侧面的前肾管连接起来，向胚体后部延伸到泄殖腔；前肾小管的近端有开口开放于体腔。家畜的前肾没有泌尿作用，很快退化。当前肾还未完全消失，前肾管尚未到达泄殖腔时，中肾已开始发育。中部的生肾节产生中肾小管，小管的一端内陷成为肾小囊，包在由毛细血管形成的肾小球外；小管的另一端连到前肾管上，此时前肾管改名为中肾管。中肾是胚胎早期和中期的排泄器官，只在一定时期内起作用，以后即失去作用，被后肾代替。后肾在中肾还未退化时就已开始发生，在中肾管的后端形成一个输尿管芽或称后肾管，这个管以后继续向前延伸，深入后肾的生肾组织中，成为后肾中集合小管以及排泄管道。另外，在后肾形成后不久，中段中胚层即分出一些密集的细胞包围后肾管盲端的膨大部，这些细胞团逐渐裂出腔隙形成肾小管。肾小管延伸、弯曲而分成不同的段落，前端与血管形成肾小体，后段以弓形集合小管与集合小管相接。另一些间充质细胞则形成肾的间质。后肾是家畜永久性泌尿器官。

泄殖腔被尿直肠隔分隔为背侧的直肠和腹侧的尿生殖窦。膀胱和尿道主要由尿生殖窦演变而来。尿生殖窦分为3部分：上段的膨大部发育成膀胱，其顶端最初与尿囊相通，以后尿囊闭锁，退化成韧带；中段的狭窄部演变成雌性尿道或雄性尿道的前列腺部；下段的扁平部演变成雌性的阴道前庭或雄性尿道的海绵体部。

二、泌尿器官的组织学结构概述

（一）肾（kidney）

肾实质由大量泌尿小管构成，其间有少量结缔组织和血管。泌尿小管包括肾单位和集合小管。

1. **肾单位（nephron）** 由肾小体和肾小管组成。

（1）肾小体（renal corpuscle）球形，由血管球和肾小囊组成。血管球（肾小球）是一团球状的动脉毛细血管，其管壁有孔，无隔膜，血管内皮外有一层基膜。肾小囊为近端小管起始端凹陷的双层杯状结构，包围血管球，外层为壁层，为单层扁平上皮；内层为脏层，称足细胞，体积大，从胞体上伸出大的初级突起后，再分出

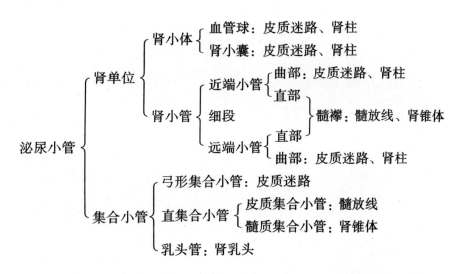

许多次级突起，紧包在毛细血管基膜上，形成滤过膜。壁层和脏层之间为肾小囊腔，原尿滤过首先进入肾小囊腔。

（2）肾小管（renal tubule） 分近端小管、细段和远端小管。近端小管与肾小囊相连并盘曲于肾小体附近，离开皮质进入髓质。近端小管曲部长而弯曲，管壁由单层立方细胞组成，管腔小。小管上皮细胞侧面有许多侧突，与相邻细胞的侧突交错，故细胞界限不清。近端小管直部的细胞变矮，侧突不明显。细段管径小，由单层扁平上皮构成，有核部分凸向管腔。远端小管包括直部和曲部。直部经髓质又返回所属肾小体附近的皮质，盘曲形成远曲小管。其上皮为单层立方上皮，细胞矮小，着色浅，核圆并位于细胞中央。

2.**集合小管**（collecting tubule） 分弓形集合小管、直集合小管和乳头管。弓形集合小管与远曲小管延续，进入髓放线与直集合小管连接。直集合小管在髓放线和髓质内下行，至肾乳头处改称乳头管。集合小管内衬单层立方上皮，细胞界线清晰，管腔大，细胞着色浅。靠近肾乳头开口处，上皮转为变移上皮。

3.**球旁复合体**（juxtaglomerular complex） 在肾小体血管极，由球旁细胞、致密斑和球外系膜细胞组成，也称肾小球旁器，具有内分泌和调节功能。

（二）排尿管道（micturition pipeline）

主要包括输尿管、膀胱和尿道。除尿道外，其他各部均分为黏膜、肌层和外膜三层。

1.**输尿管**（ureter） 细而长，起于肾盂，止于膀胱。黏膜有纵走皱襞，衬以变移上皮。肌层平滑肌可分为内纵、中环和外纵三层。外膜为疏松结缔组织。

2.**膀胱**（bladder） 贮尿器官，黏膜形成许多不规则的皱襞，上皮为变移上皮，其厚度随尿量不同而异。固有层富含弹性纤维，常有淋巴小结。肌层特别发达，分层不太规则，中层最厚，在膀胱颈部形成括约肌。外膜随部位不同而异，膀胱体和膀胱顶部为浆膜，膀胱颈为疏松结缔组织外膜。

3.**尿道**（urethra） 雌、雄两性尿道差异大。

（1）公畜尿道 兼有排尿和排精的功能，又称尿生殖道。分为骨盆部和阴茎部。

①骨盆部 是指自膀胱颈到坐骨弓的一段。管壁分为黏膜、血管层、前列腺组织层和肌层。

②阴茎部 为骨盆部尿道的延续，位于阴茎的尿道海绵体内。分为黏膜、海绵体层及白膜三层。

（2）母畜尿道 较短，管壁由黏膜、肌层和外膜组成。黏膜衬以复层扁平上皮，固有层疏松结缔组织含有淋巴小结、尿道腺（牛）和稠密的静脉丛，形成海绵状结构。肌层很薄。分内环、外纵两层。外膜疏松结缔组织含有丰富的血管和神经丛。

三、泌尿系统图谱

1.**肾** 图16-1～图16-46。

2.**排尿管道** 图16-47～图16-70。

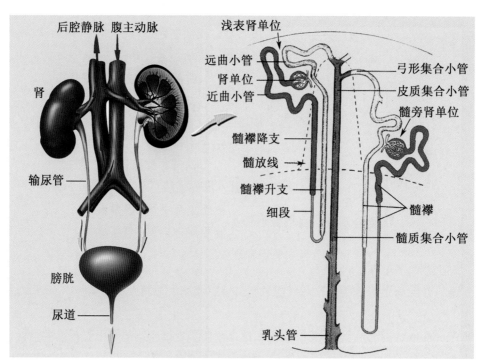

图 16-1 泌尿系统结构模式图

Fig.16-1 ideograph of urinary system structure

后腔静脉—posterior vena cava 腹主动脉—abdominal aorta 肾—kidney 输尿管—ureter 膀胱—urinary bladder 尿道—urethra 浅表肾单位—superficial nephron 远曲小管—distal convoluted tubule 肾单位—nephron 近曲小管—proximal convoluted tubule 髓袢降支—descending limb 髓放线—medullary ray 髓袢升支—ascending limb 细段—thin segment 乳头管—papillary ducts 弓形集合小管—arched collecting tubule 皮质集合小管—cortical collecting ducts 髓旁肾单位—juxtamedullary nephron 髓袢—medullary loops 髓质集合小管—medullary collecting ducts

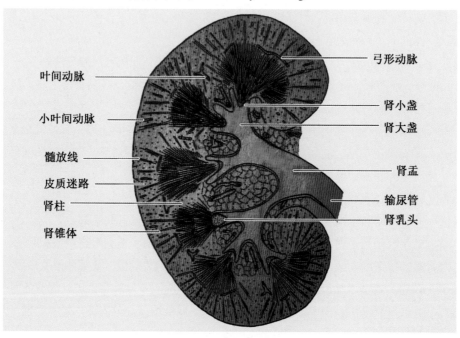

图 16-2 肾结构模式图

Fig.16-2 ideograph of kidney structure

叶间动脉—interlobar artery 小叶间动脉—interlobular artery 髓放线—medullary ray 皮质迷路—cortical labyrinth 肾柱—renal column 肾锥体—renal pyramid 弓形动脉—arcuate artery 肾小盏—minor renal calices 肾大盏—greater renal calices 肾盂—renal pelvis 输尿管—ureter 肾乳头—renal papillae

第十六章 泌尿系统 Urinary System

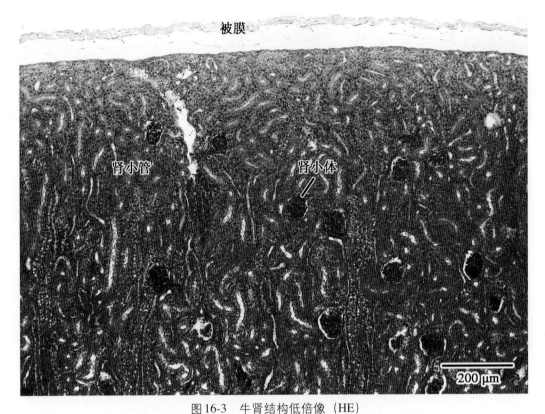

图16-3 牛肾结构低倍像（HE）

Fig.16-3 low magnification of cow kidney structure（HE）

被膜—capsule 肾小管—kidney tubules 肾小体—renal corpuscle

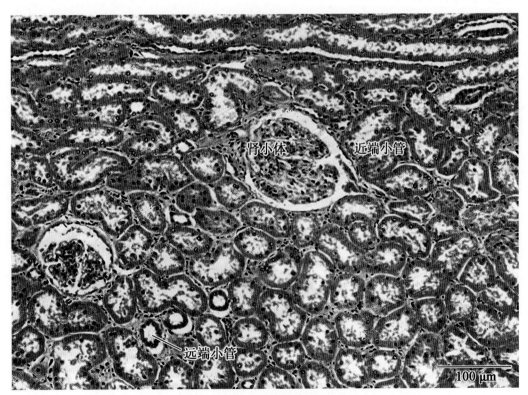

图16-4 牛肾皮质结构中倍像（HE）

Fig.16-4 mid magnification of cow kidney cortex structure（HE）

肾小体—renal corpuscle 近端小管—proximal tubule 远端小管—distal tubule

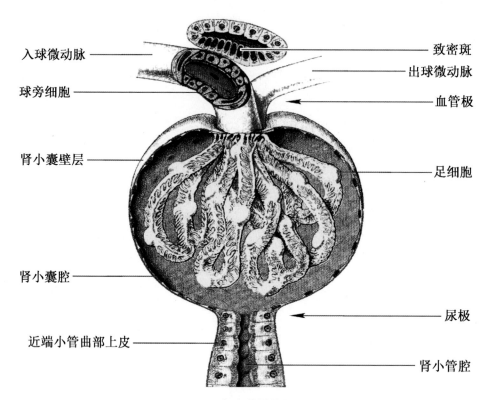

图16-5　肾小体结构模式图

Fig.16-5　ideograph of renal corpuscle

入球微动脉—afferent arteriole　球旁细胞—juxtaglomerular cell　肾小囊壁层—parietal renal capsule
肾小囊腔—capsular space　近端小管曲部上皮—proximal convoluted tubule epithelium　致密斑—macula densa
出球微动脉—efferent arteriole　血管极—vascular pole　足细胞—podocyte　尿极—urinary pole
肾小管腔—renal tubular cavity

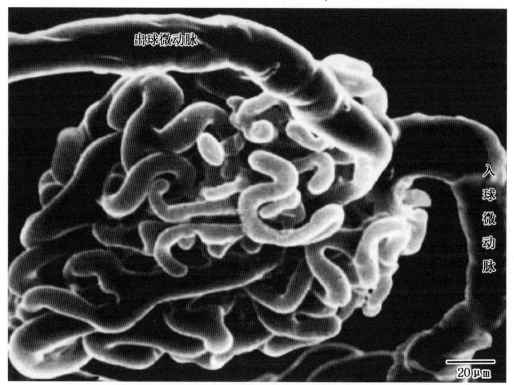

图16-6　肾小球铸型扫描电镜像

Fig.16-6　scanning electron microscope photo of renal glomerulus perfusion

入球微动脉—afferent arteriole　出球微动脉—efferent arteriole

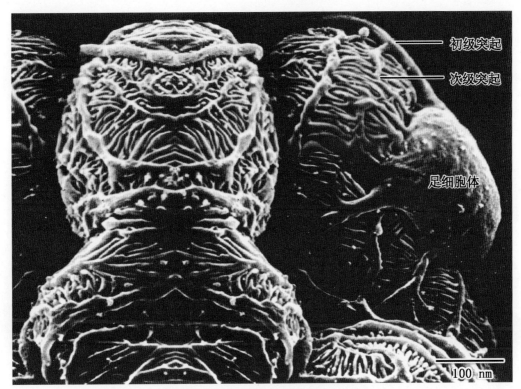

图 16-7 肾小球毛细血管扫描电镜像

Fig.16-7 scanning electron microscope photo of renal glomerulus capillary

初级突起—primary process 次级突起—secondary process 足细胞体—podocyte body

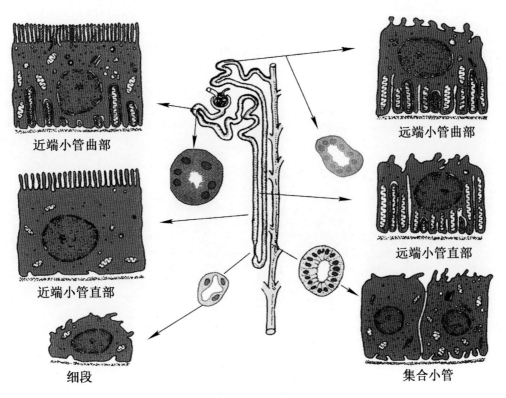

图 16-8 肾小管结构模式图

Fig.16-8 ideograph of renal tubule

近端小管曲部—proximal convoluted tubule 近端小管直部—proximal straight tubule 细段—thin segment
远端小管曲部—distal convoluted tubule 远端小管直部—distal straight tubule 集合小管—collecting tubule

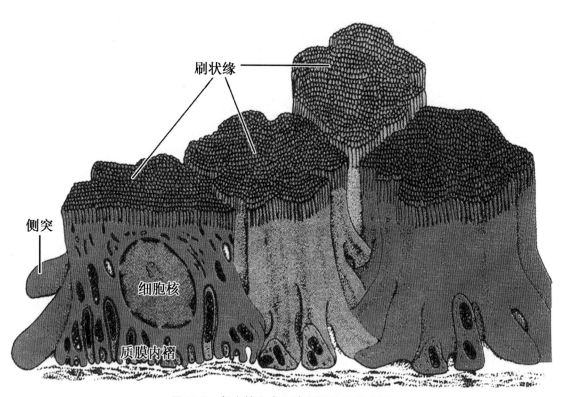

图16-9 肾小管上皮细胞超微结构模式图

Fig.16-9 ultrastructure ideograph of renal tubule epithelium

刷状缘—brush border 侧突—lateral process 细胞核—nucleus 质膜内褶—plasma membrane infolding

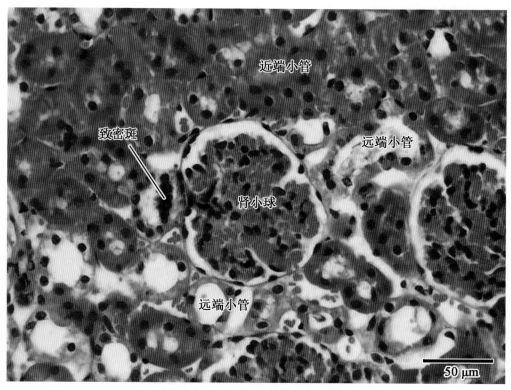

图16-10 牛肾结构高倍像1（HE）

Fig.16-10 high magnification of cow kidney structure 1（HE）

致密斑—macula densa 近端小管—proximal tubule 肾小球—renal glomerulus 远端小管—distal tubule

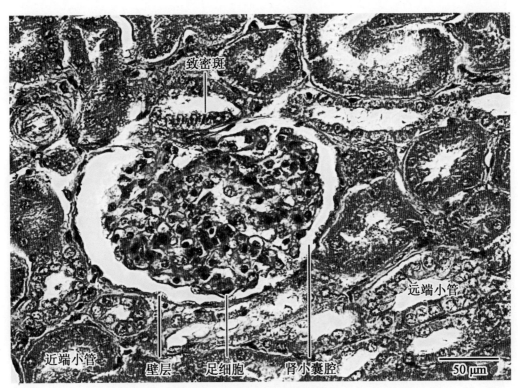

图16-11 牛肾结构高倍像2（HE）

Fig.16-11 high magnification of cow kidney structure 2（HE）

致密斑—macula densa 近端小管—proximal tubule 壁层—parietal renal capsule
足细胞—podocyte 肾小囊腔—capsular space 远端小管—distal tubule

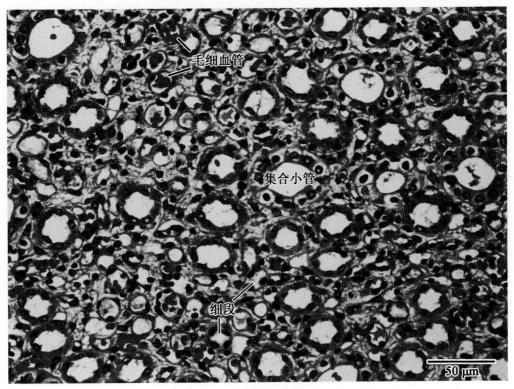

图16-12 牛肾髓质高倍像1（HE）

Fig.16-12 high magnification of cow renal medulla 1（HE）

毛细血管—capillary 集合小管—collecting tubule 细段—thin segment

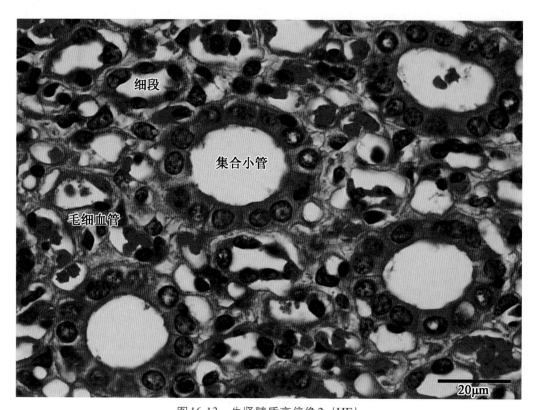

图 16-13　牛肾髓质高倍像 2（HE）

Fig.16-13　high magnification of cow renal medulla 2（HE）

细段—thin segment　集合小管—collecting tubule　毛细血管—capillary

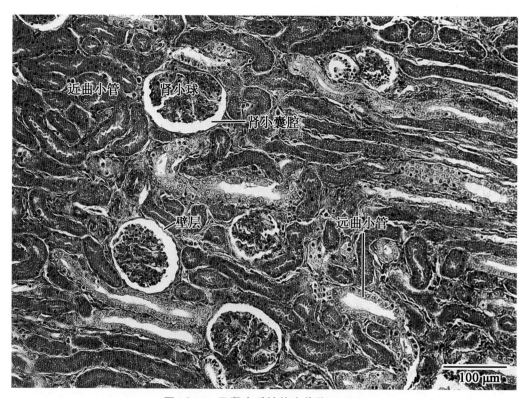

图 16-14　马肾皮质结构中倍像（HE）

Fig.16-14　mid magnification of horse kidney cortex structure（HE）

近曲小管—proximal convoluted tubule　肾小球—renal glomerulus　肾小囊腔—capsular space

壁层—wall layer　远曲小管—distal convoluted tubule

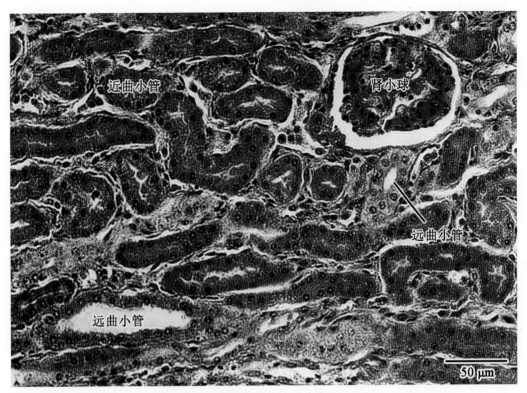

图 16-15　马肾皮质结构高倍像 1（HE）

Fig.16-15　high magnification of horse kidney cortex structure 1（HE）

近曲小管—proximal convoluted tubule　远曲小管—distal convoluted tubule　肾小球—renal glomerulus

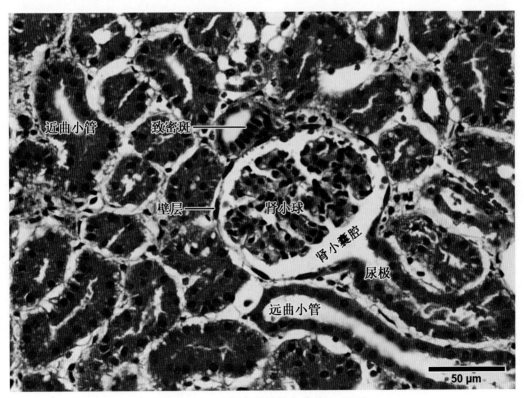

图 16-16　马肾皮质结构高倍像 2（HE）

Fig.16-16　high magnification of horse kidney cortex structure 2（HE）

近曲小管—proximal convoluted tubule　致密斑—macula densa　壁层—wall layer　肾小球—renal glomerulus

肾小囊腔—capsular space　尿极—urinary pole　远曲小管—distal convoluted tubule

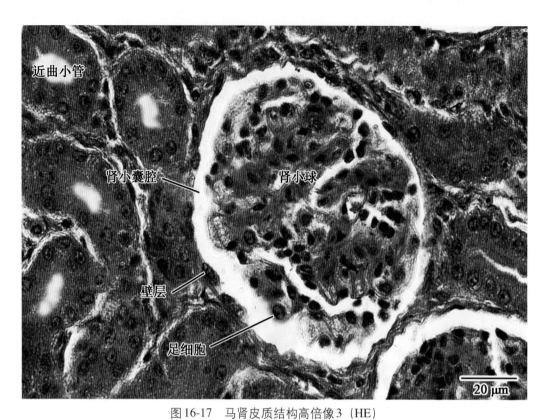

图 16-17　马肾皮质结构高倍像 3（HE）

Fig.16-17　high magnification of horse kidney cortex structure 3（HE）

近曲小管—proximal convoluted tubule　肾小囊腔—capsular space　壁层—wall layer

肾小球—renal glomerulus　足细胞—podocyte

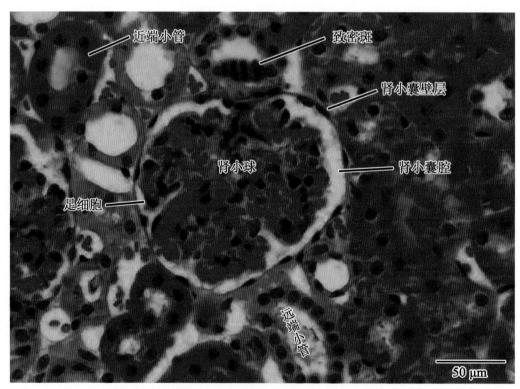

图 16-18　马肾皮质结构高倍像 4（HE）

Fig.16-18　high magnification of horse kidney cortex structure 4（HE）

近端小管—proximal tubule　致密斑—macula densa　肾小囊壁层—renal capsule wall layer　肾小球—renal glomerulus

肾小囊腔—capsular space　足细胞—podocyte　远端小管—distal tubule

第十六章 泌尿系统 Urinary System

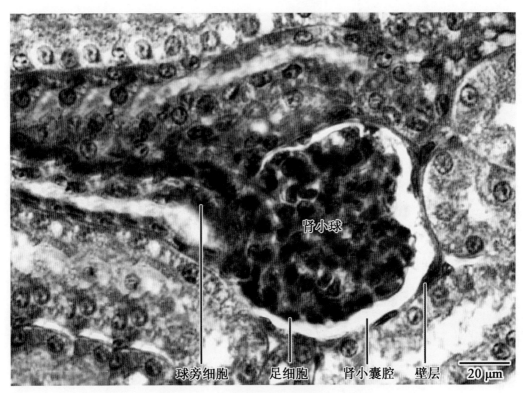

图 16-19 马肾皮质结构高倍像 5（HE）

Fig.16-19 high magnification of horse kidney cortex structure 5（HE）

肾小球—renal glomerulus 球旁细胞—juxtaglomerular cell 足细胞—podocyte

肾小囊腔—capsular space 壁层—wall layer

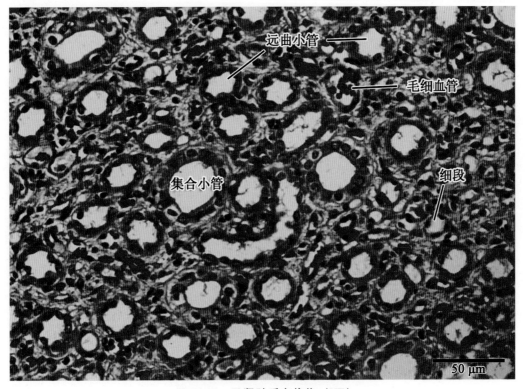

图 16-20 马肾髓质高倍像（HE）

Fig.16-20 high magnification of horse renal medulla（HE）

远曲小管—distal convoluted tubule 毛细血管—capillary 集合小管—collecting tubule 细段—thin segment

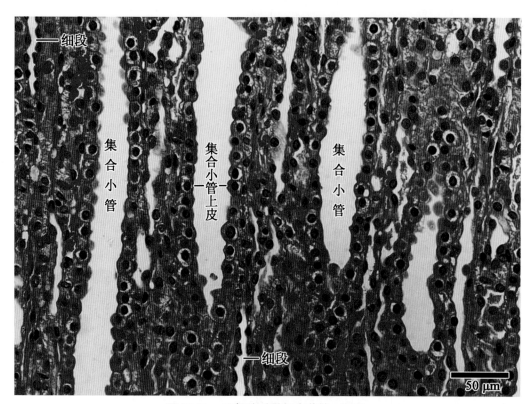

图16-21 马肾髓质纵切高倍像（HE）

Fig.16-21 high magnification of horse kidney medulla length cutting（HE）

细段—thin segment 集合小管—collecting tubule 集合小管上皮—collecting tubule epithelium

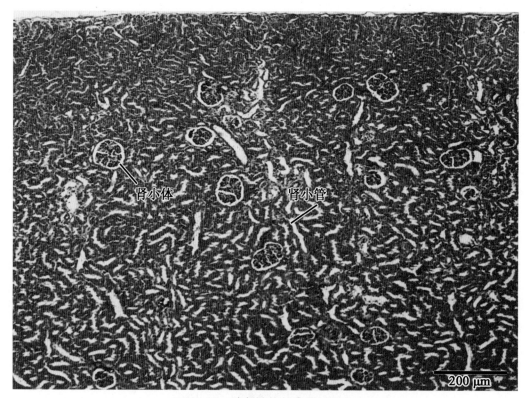

图16-22 猪肾结构低倍像（HE）

Fig.16-22 low magnification of pig kidney structure（HE）

肾小体— renal corpuscle 肾小管—kidney tubules

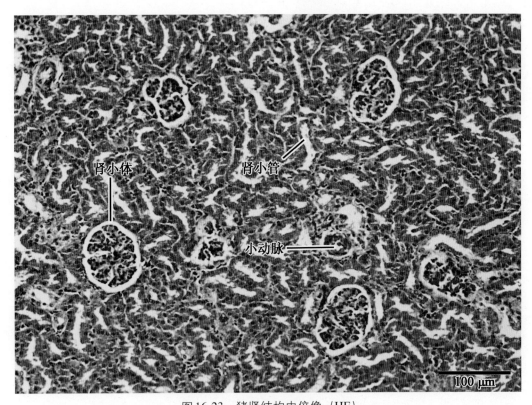

图16-23 猪肾结构中倍像（HE）

Fig.16-23 mid magnification of pig kidney structure（HE）

肾小体— renal corpuscle 肾小管—kidney tubules 小动脉—arteriole

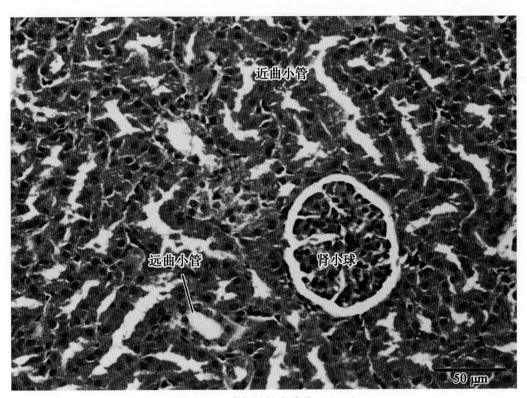

图16-24 猪肾结构高倍像1（HE）

Fig.16-24 high magnification of pig kidney structure 1（HE）

近曲小管—proximal convoluted tubule 远曲小管—distal convoluted tubule 肾小球—renal glomerulus

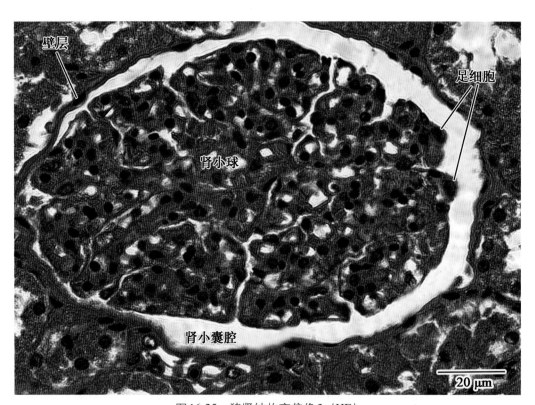

图 16-25 猪肾结构高倍像 2（HE）

Fig.16-25　high magnification of pig kidney structure 2（HE）

壁层—wall layer　　肾小球—renal glomerulus　　足细胞—podocyte　　肾小囊腔—capsular space

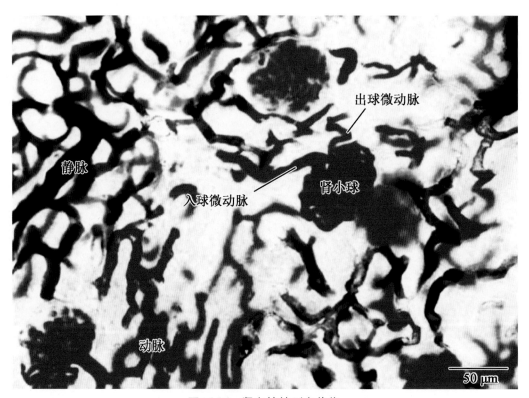

图 16-26 肾血管铸型高倍像

Fig.16-26　high magnification of renal perfusion

静脉—vein　　动脉—artery　　入球微动脉—afferent arteriole　　出球微动脉—efferent arteriole　　肾小球—renal glomerulus

第十六章 泌尿系统 Urinary System

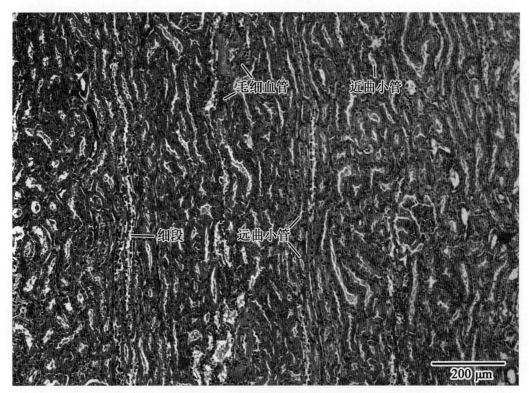

图 16-27 猪肾髓质低倍像（HE）

Fig.16-27 low magnification of pig renal medulla（HE）

毛细血管—capillary 近曲小管—proximal convoluted tubule 细段—thin segment 远曲小管—distal convoluted tubule

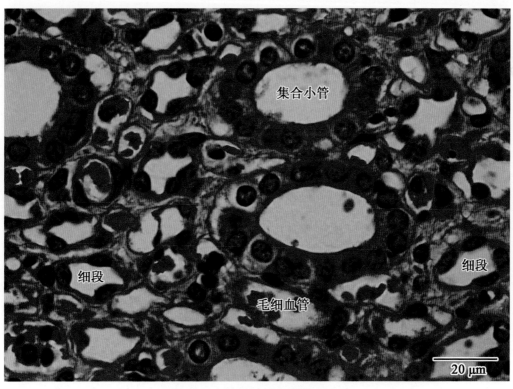

图 16-28 猪肾髓质高倍像（HE）

Fig.16-28 high magnification of pig renal medulla（HE）

细段—thin segment 集合小管—collecting tubule 毛细血管—capillary

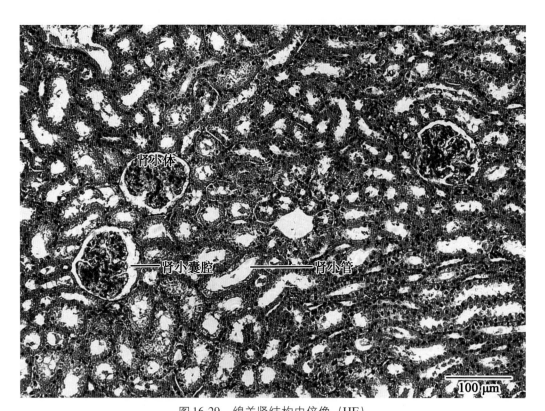

图 16-29　绵羊肾结构中倍像（HE）

Fig.16-29　mid magnification of sheep kidney structure（HE）

肾小体— renal corpuscle　肾小囊腔—capsular space　肾小管—kidney tubules

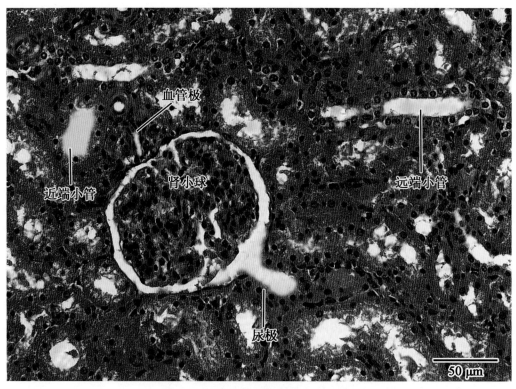

图 16-30　绵羊肾结构高倍像 1（HE）

Fig.16-30　high magnification of sheep kidney structure 1（HE）

近端小管—proximal tubule　血管极—vascular pole　肾小球—renal glomerulus　尿极—urinary pole　远端小管—distal tubule

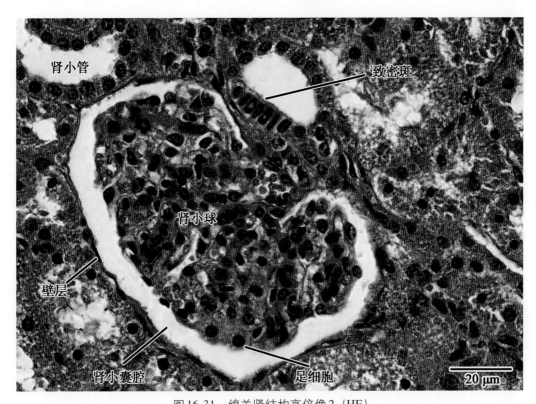

图16-31 绵羊肾结构高倍像2（HE）

Fig.16-31 high magnification of sheep kidney structure 2（HE）

肾小管—kidney tubules 致密斑—macula densa 肾小球—renal glomerulus 壁层—wall layer
肾小囊腔—capsular space 足细胞—podocyte

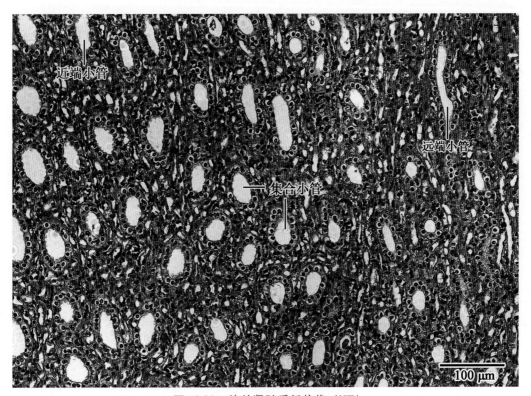

图16-32 绵羊肾髓质低倍像（HE）

Fig.16-32 low magnification of sheep renal medulla（HE）

近端小管—proximal tubule 集合小管—collecting tubule 远端小管—distal tubule

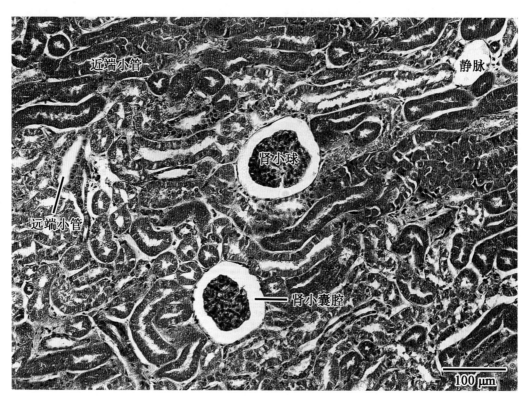

图 16-33　犬肾皮质中倍像（HE）

Fig.16-33　mid magnification of dog kidney cortex（HE）

近端小管—proximal tubule　远端小管—distal tubule　肾小球—renal glomerulus　肾小囊腔—capsular space　静脉—vein

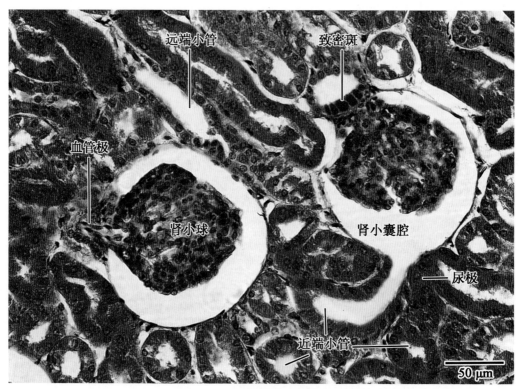

图 16-34　犬肾皮质高倍像 1（HE）

Fig.16-34　high magnification of dog kidney cortex 1（HE）

远端小管—distal tubule　血管极—vascular pole　肾小球—renal glomerulus　致密斑—macula densa
肾小囊腔—capsular space　近端小管—proximal tubule　尿极—urinary pole

第十六章 泌尿系统 Urinary System

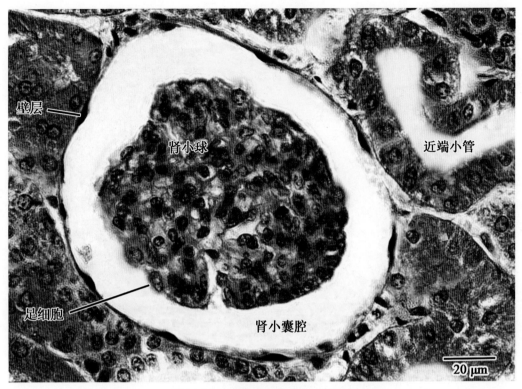

图 16-35　犬肾皮质高倍像2（HE）

Fig.16-35　high magnification of dog kidney cortex 2（HE）

壁层—wall layer　足细胞—podocyte　肾小球—renal glomerulus　肾小囊腔—capsular space　近端小管—proximal tubule

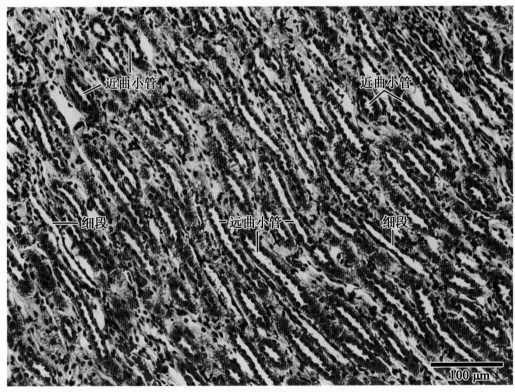

图 16-36　犬肾髓质纵切低倍像（HE）

Fig.16-36　low magnification of dog renal medulla length cutting（HE）

近曲小管—proximal convoluted tubule　细段—thin segment　远曲小管—distal convoluted tubule

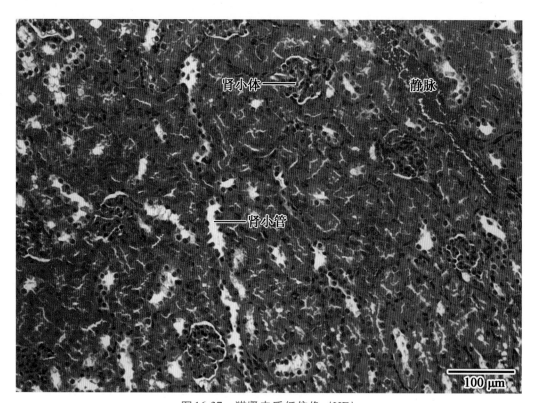

图 16-37　猫肾皮质低倍像（HE）

Fig.16-37　low magnification of cat kidney cortex（HE）

肾小体—renal corpuscle　肾小管—renal tubule　静脉—vein

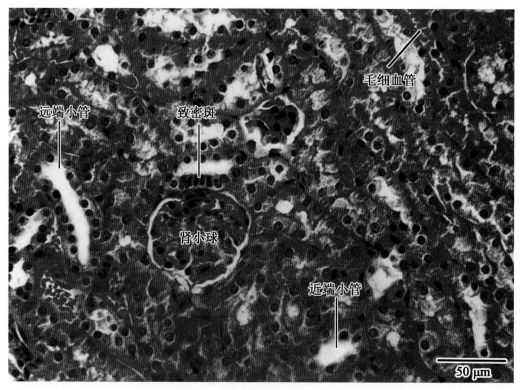

图 16-38　猫肾皮质高倍像 1（HE）

Fig.16-38　high magnification of cat kidney cortex 1（HE）

远端小管—distal tubule　致密斑—macula densa　肾小球—renal glomerulus　毛细血管—capillary　近端小管—proximal tubule

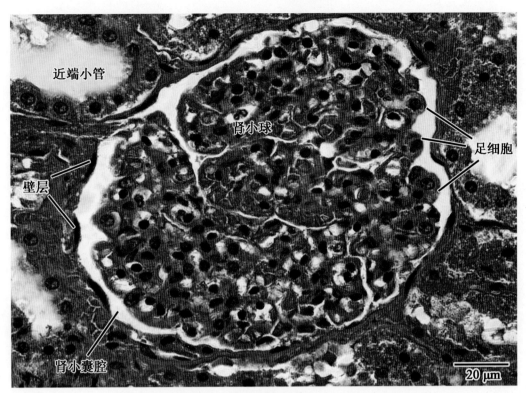

图 16-39　猫肾皮质高倍像 2（HE）

Fig.16-39　high magnification of cat kidney cortex 2（HE）

近端小管—proximal tubule　壁层—wall layer　肾小囊腔—capsular space　肾小球—renal glomerulus　足细胞—podocyte

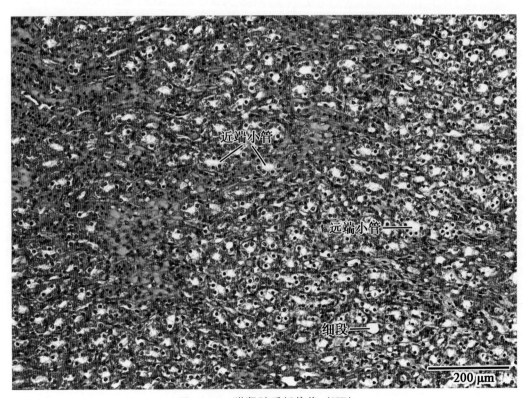

图 16-40　猫肾髓质低倍像（HE）

Fig.16-40　low magnification of cat renal medulla（HE）

近端小管—proximal tubule　远端小管—distal tubule　细段—thin segment

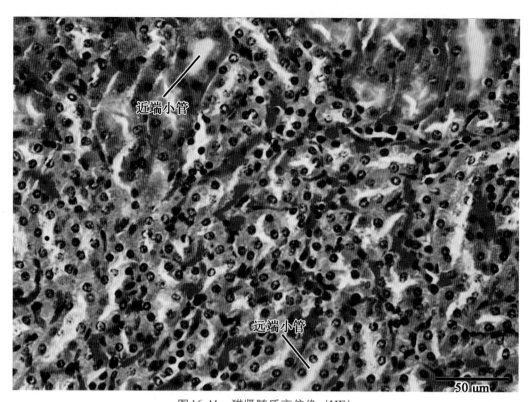

图16-41 猫肾髓质高倍像（HE）

Fig.16-41 high magnification of cat renal medulla（HE）

近端小管—proximal tubule 远端小管—distal tubule

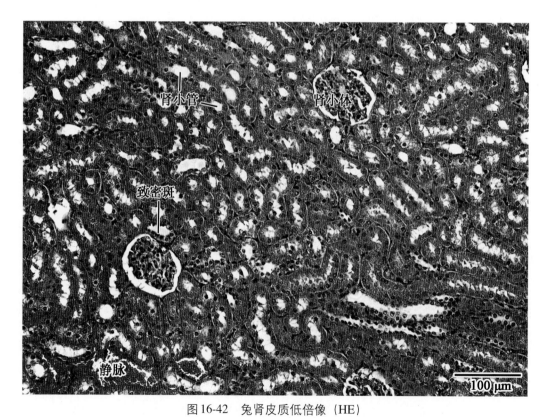

图16-42 兔肾皮质低倍像（HE）

Fig.16-42 low magnification of rabbit kidney cortex（HE）

肾小管—renal tubule 肾小体—renal corpuscle 致密斑—macula densa 静脉—vein

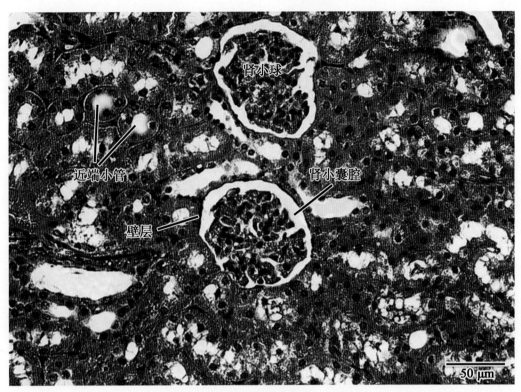

图16-43　兔肾皮质高倍像1（HE）

Fig.16-43　high magnification of rabbit kidney cortex 1 （HE）

肾小球—renal glomerulus　近端小管—proximal tubule　壁层—wall layer　肾小囊腔—capsular space

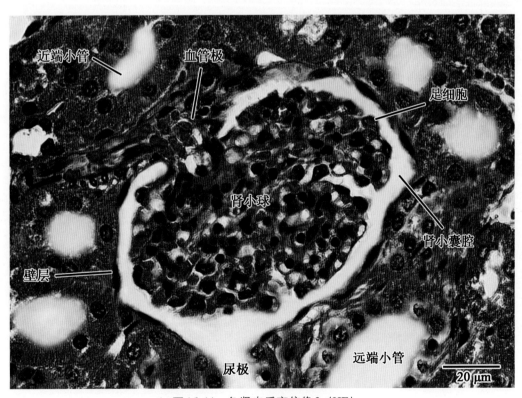

图16-44　兔肾皮质高倍像2（HE）

Fig.16-44　high magnification of rabbit kidney cortex 2 （HE）

近端小管—proximal tubule　壁层—wall layer　血管极—vascular pole　肾小球—renal glomerulus
足细胞—podocyte　肾小囊腔—capsular space　远端小管—distal tubule　尿极—urinary pole

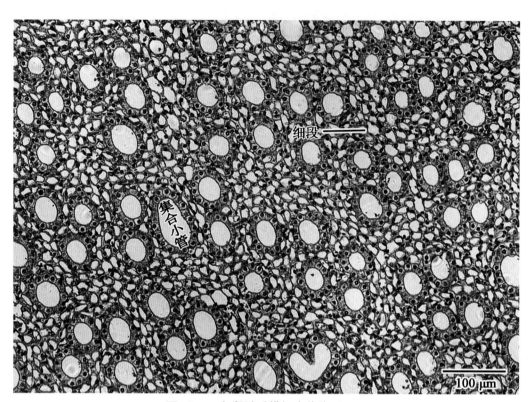

图16-45 兔肾髓质横切中倍像（HE）
Fig.16-45 mid magnification of rabbit renal medulla cross cutting（HE）
集合小管—collecting tubule 细段—thin segment

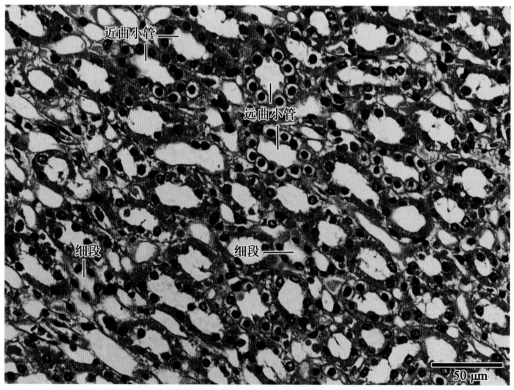

图16-46 兔肾髓质纵切高倍像（HE）
Fig.16-46 high magnification of rabbit renal medulla length cutting（HE）
近曲小管—proximal convoluted tubule 远曲小管—distal convoluted tubule 细段—thin segment

第十六章 泌尿系统 Urinary System

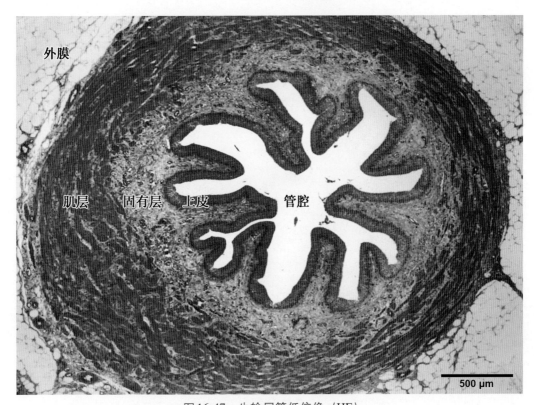

图16-47　牛输尿管低倍像（HE）

Fig.16-47　low magnification of cow ureter（HE）

外膜—adventitia　肌层—muscular layer　固有层—proper layer　上皮—epithelium　管腔—lumen

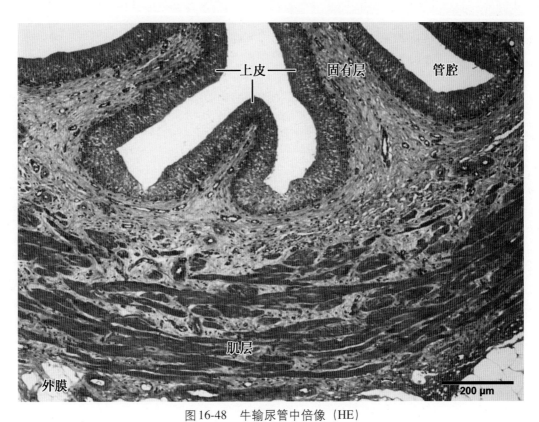

图16-48　牛输尿管中倍像（HE）

Fig.16-48　mid magnification of cow ureter（HE）

上皮—epithelium　固有层—proper layer　管腔—lumen　肌层—muscular layer　外膜—adventitia

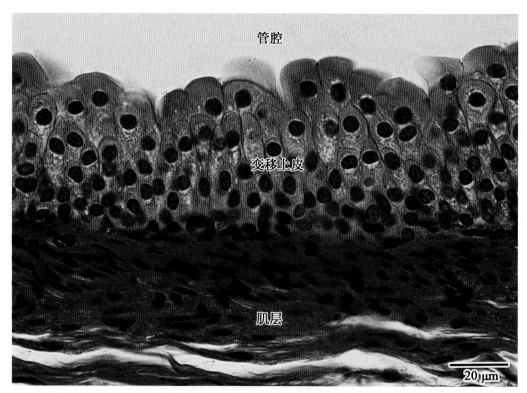

图16-49 牛输尿管高倍像（HE）

Fig.16-49 high magnification of cow ureter（HE）

管腔—lumen 变移上皮—transitional epithelium 肌层—muscular layer

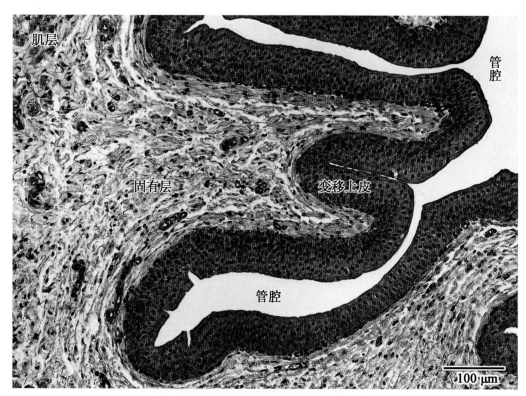

图16-50 马输尿管中倍像（HE）

Fig.16-50 mid magnification of horse ureter（HE）

肌层—muscular layer 固有层—proper layer 变移上皮—transitional epithelium 管腔—lumen

第十六章 泌尿系统 Urinary System

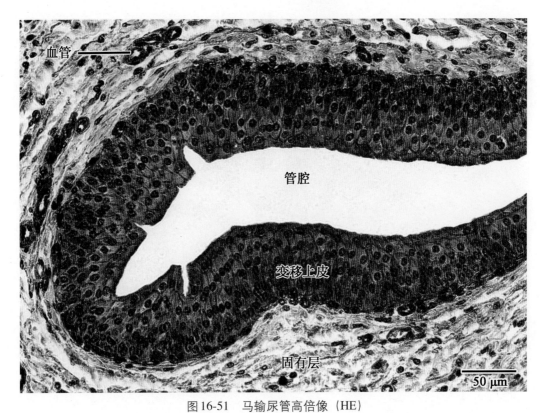

图16-51 马输尿管高倍像（HE）

Fig.16-51 high magnification of horse ureter（HE）

血管—blood vessel　管腔—lumen　变移上皮—transitional epithelium　固有层—proper layer

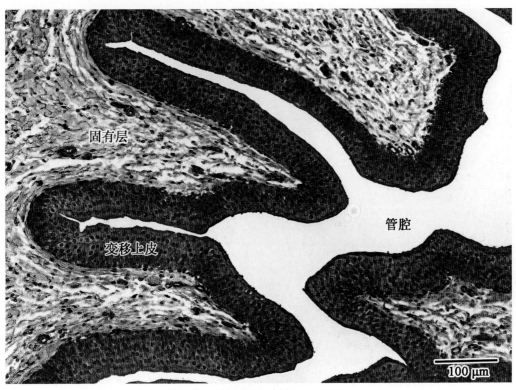

图16-52 猪输尿管中倍像（HE）

Fig.16-52 mid magnification of pig ureter（HE）

固有层—proper layer　变移上皮—transitional epithelium　管腔—lumen

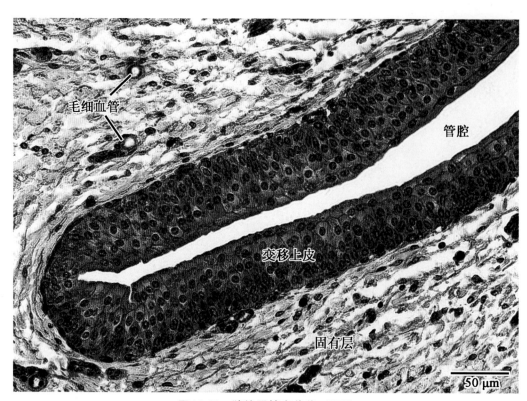

图 16-53 猪输尿管高倍像（HE）

Fig.16-53 high magnification of pig ureter（HE）

毛细血管—capillary 管腔—lumen 变移上皮—transitional epithelium 固有层—proper layer

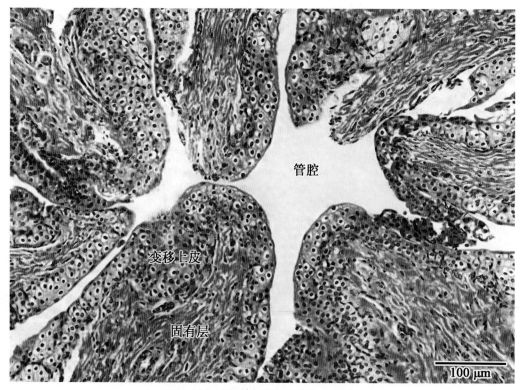

图 16-54 绵羊输尿管中倍像（HE）

Fig.16-54 mid magnification of sheep ureter（HE）

管腔—lumen 变移上皮—transitional epithelium 固有层—proper layer

第十六章 泌尿系统 Urinary System

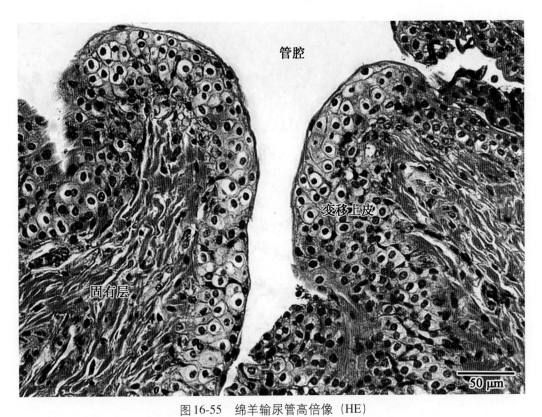

图 16-55 绵羊输尿管高倍像（HE）
Fig.16-55 high magnification of sheep ureter（HE）
管腔—lumen 变移上皮—transitional epithelium 固有层—proper layer

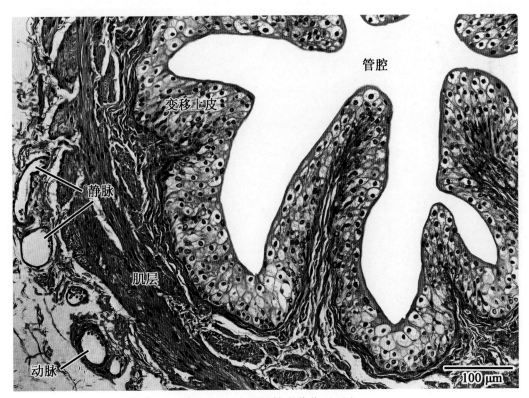

图 16-56 犬输尿管中倍像（HE）
Fig.16-56 mid magnification of dog ureter（HE）
静脉—vein 动脉—artery 肌层—muscular layer 变移上皮—transitional epithelium 管腔—lumen

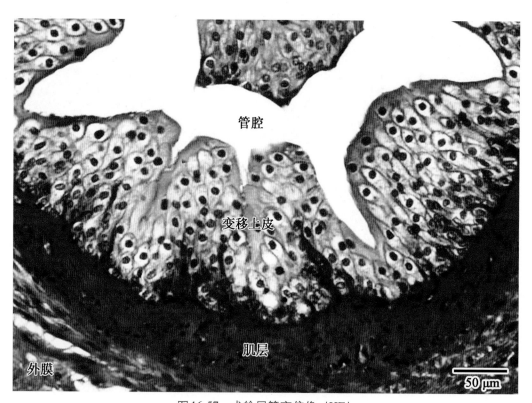

图 16-57　犬输尿管高倍像（HE）

Fig.16-57　high magnification of dog ureter（HE）

管腔—lumen　变移上皮—transitional epithelium　肌层—muscular layer　外膜—adventitia

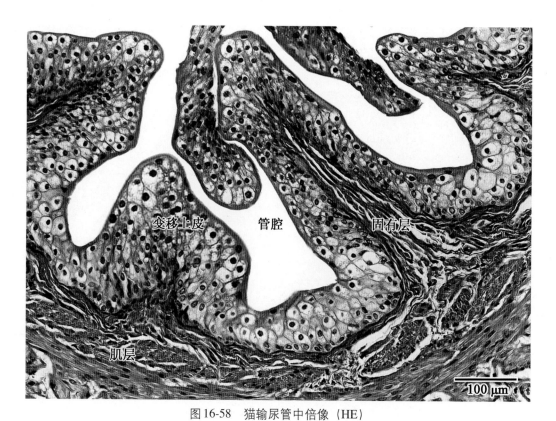

图 16-58　猫输尿管中倍像（HE）

Fig.16-58　mid magnification of cat ureter（HE）

管腔—lumen　变移上皮—transitional epithelium　肌层—muscular layer　固有层—proper layer

第十六章 泌尿系统 Urinary System

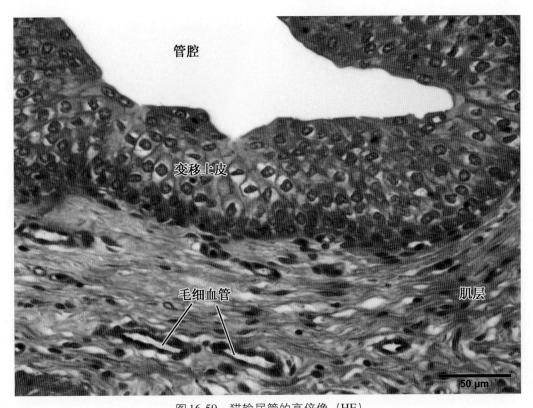

图16-59 猫输尿管的高倍像（HE）

Fig.16-59 high magnification of cat ureter（HE）

管腔—lumen　变移上皮—transitional epithelium　毛细血管—capillary　肌层—muscular layer

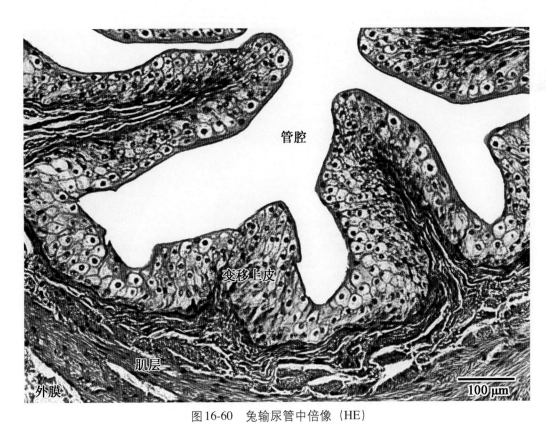

图16-60 兔输尿管中倍像（HE）

Fig.16-60 mid magnification of rabbit ureter（HE）

管腔—lumen　变移上皮—transitional epithelium　肌层—muscular layer　外膜—adventitia

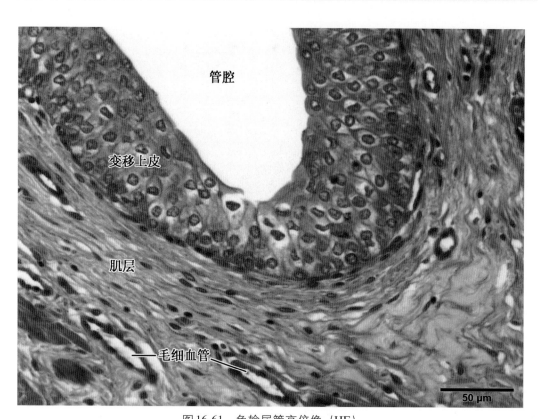

图 16-61 兔输尿管高倍像（HE）

Fig.16-61 high magnification of rabbit ureter（HE）

管腔—lumen 变移上皮—transitional epithelium 肌层—muscular layer 毛细血管—capillary

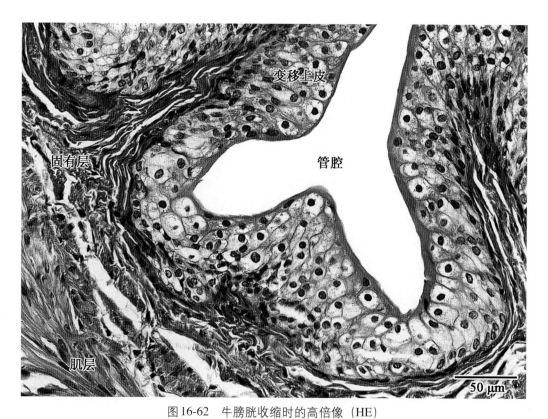

图 16-62 牛膀胱收缩时的高倍像（HE）

Fig.16-62 high magnification of bladder contractions in cow（HE）

变移上皮—transitional epithelium 固有层—proper layer 管腔—lumen 肌层—muscular layer

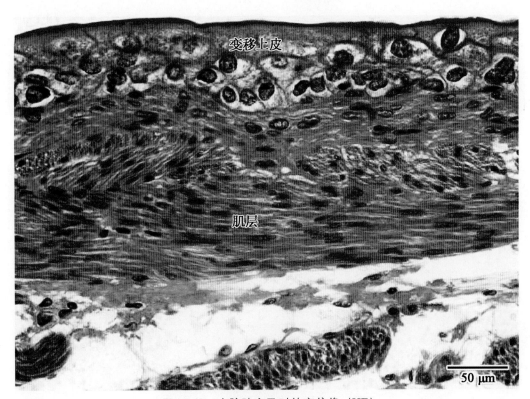

图16-63 牛膀胱充盈时的高倍像（HE）

Fig.16-63　high magnification of cow bladder filling（HE）

变移上皮—transitional epithelium　肌层—muscular layer

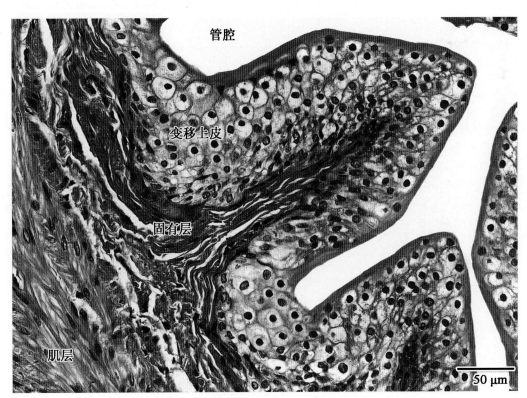

图16-64 马膀胱收缩时的高倍像（HE）

Fig.16-64　high magnification of bladder contractions in horse（HE）

管腔—lumen　变移上皮—transitional epithelium　固有层—proper layer　肌层—muscular layer

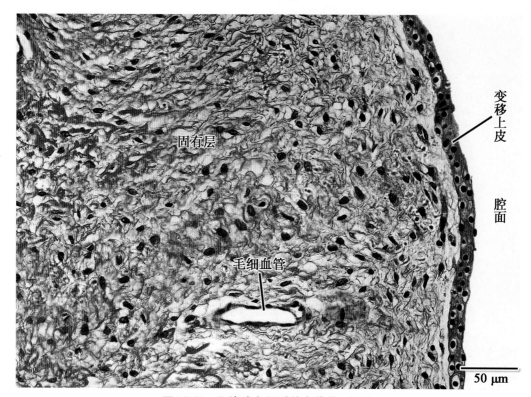

图16-65 马膀胱充盈时的高倍像（HE）

Fig.16-65　high magnification of horse bladder filling（HE）

固有层—proper layer　毛细血管—capillary　变移上皮—transitional epithelium　腔面—cavosurface

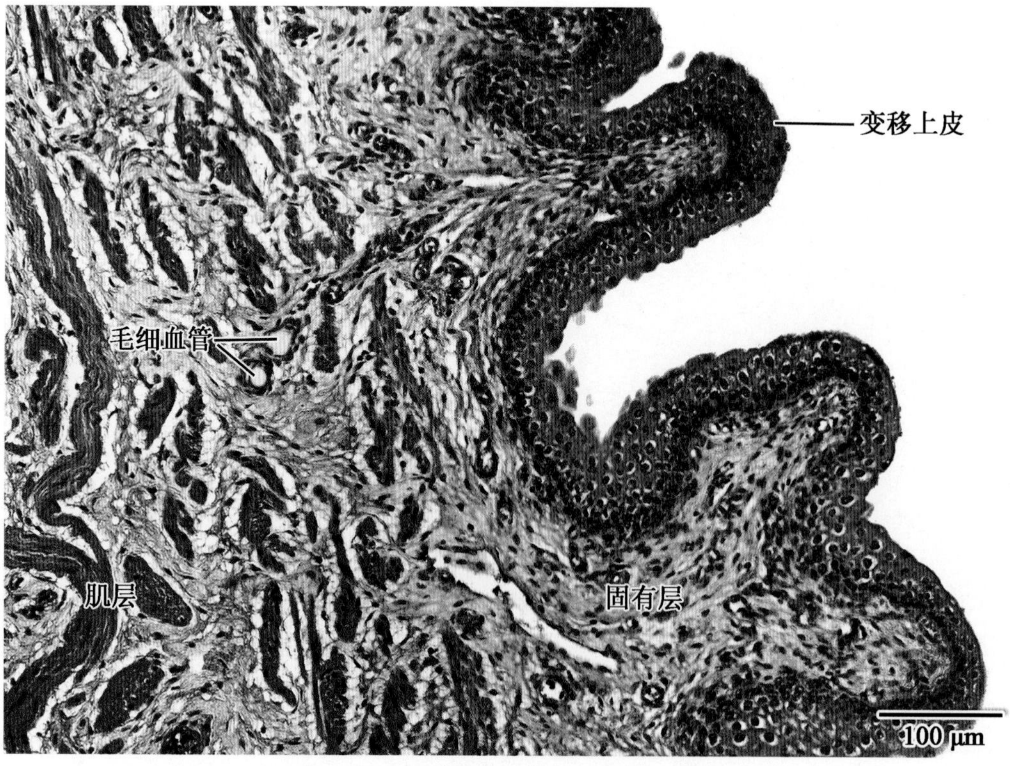

图16-66　猪膀胱收缩时的中倍像（HE）

Fig.16-66　mid magnification of bladder contractions in pig（HE）

肌层—muscular layer　毛细血管—capillary　变移上皮—transitional epithelium　固有层—proper layer

第十六章 泌尿系统 Urinary System

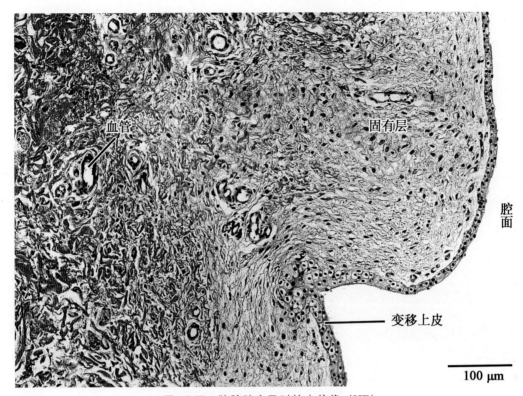

图 16-67　猪膀胱充盈时的中倍像（HE）
Fig.16-67　mid magnification of pig bladder filling（HE）

血管—blood vessel　固有层—proper layer　腔面—cavosurface　变移上皮—transitional epithelium

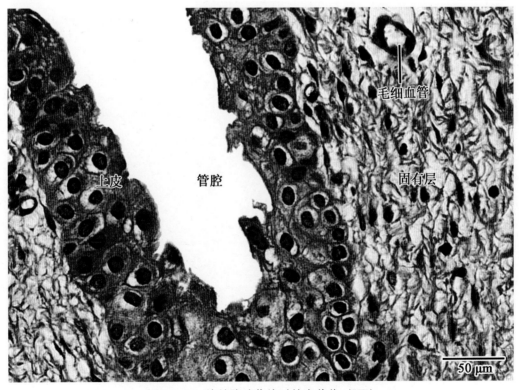

图 16-68　绵羊膀胱收缩时的高倍像（HE）
Fig.16-68　high magnification of sheep bladder contractions（HE）

上皮—epithelium　管腔—lumen　毛细血管—capillary　固有层—proper layer

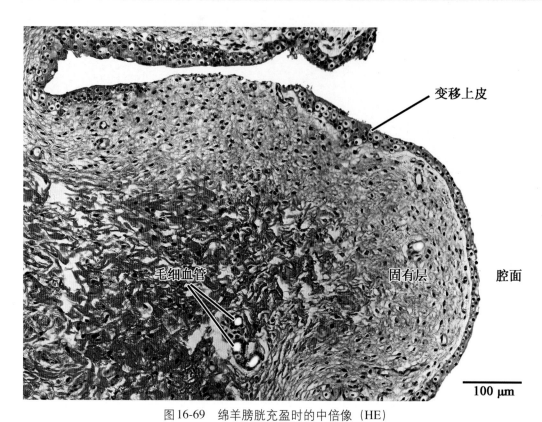

图16-69 绵羊膀胱充盈时的中倍像（HE）
Fig.16-69 mid magnification of sheep bladder filling（HE）

毛细血管—capillary　变移上皮—transitional epithelium　固有层—proper layer　腔面—cavosurface

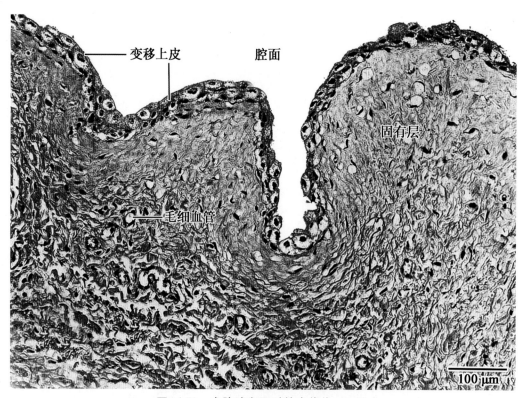

图16-70 犬膀胱充盈时的中倍像（HE）
Fig.16-70 mid magnification of dog bladder filling（HE）

变移上皮—transitional epithelium　腔面—cavosurface　固有层—proper layer　毛细血管—capillary

第十七章
雌性生殖系统
Female Reproductive System

Outline

Female reproductive system of domestic animals consists of two ovaries, two oviducts, uterus, vagina, and external genitalia.

Ovary is the female genital gland which produces ova and estrogen during reproductive period. There is a simple squamous or cuboidal epithelium, named germinal epithelium, wrapping the ovary surface. The dense connective tissue, named tunica albuginea, locates the next layer. The parenchyma can be divided into the peripheral cortex region and the central medulla region. In the peripheral cortex, there are a lot of different ovarian follicles, corpus luteum, atretic follicles, which are embedded in stroma. The principal component of the medulla is loose connective tissue with plentiful blood vessels, lymphatic vessels and nerves. All the follicles, in the light of their structure, can be divided into primordial, growing and mature ones, and the growing follicles can be further classified into the primary and secondary follicles. When follicles matured, they deliver ova and the remaining develop into corpus luteum. The process that sexually mature mammals release their ova, is called ovulation. Corpus luteum comprises granulosa lutein cells and theca lutein cells from the theca interna of the ovulated follicle to form a temporary endocrine gland with rich blood capillaries. Cells of corpus luteum secrete progesterone and estrogens. Atretic follicles originate from the degeneration of follicles in various developing stages. Both ovulation and ovarian hormone production are controlled by the cyclical release of luteinizing hormone (LH) and follicle stimulating hormone (FSH) from the anterior pituitary of the gonadotrophic hormones.

Uterus is the pregnant organ which divides into corner, body and cervix. The wall of uterus consists of three layers: endometrium, myometrium and perimetrium. Uterine glands are in the lamina propria of endometrium. The structure of endometrium can change periodically following the estrous cycle.

雌性生殖系统（female reproductive system）包括卵巢、输卵管、子宫、阴道和外生殖器。卵巢产生的卵子在输卵管中受精后，输送到子宫。子宫是孕育胎儿的器官。

一、生殖系统的发生

生殖系统中的生殖腺和主要管道，都是由中胚层发生的。当中肾还是胚胎主要排泄器官的时候，生殖腺即开始在中肾腹内侧发育。此处的脏中胚层细胞，分裂成为多层的上皮细胞团，向中肾组织内突入，称为生殖嵴，外层为间皮及原始生殖细胞构成的生殖上皮，内部为生殖上皮产生的索状细胞团和间充质。在生殖嵴中，有些大的细胞称为原始生殖细胞。这是未分性别时期的生殖腺结构。不论生殖腺将来发育的方向如何，生殖嵴表面生殖上皮的细胞都伸向内面的间充质中，并形成细胞索。如果胚胎发育成雄性个体，则细胞索发育并出现腔隙，而成为生精小管，生殖上皮深部的间充质发育成为白膜，生殖上皮转化为间质细胞。如果胚胎发育成雌性个体，则细胞索分裂成许多细胞团，内包原始生殖细胞，将来分化成卵原细胞，并形成原始卵泡。原始卵泡在胚胎期可进一步发育，但到个体娩出后，则又逐步退化。到性成熟时，未发育的原始卵泡开始逐步发育。

生殖管的发育和泌尿器官的发育密切相关。性别未分化时期的胚胎，中肾管的外侧由脏中胚层内凹形成两条米勒管（Müllerian duct），向后延伸通至泄殖腔。随后肾的发育，输尿管的末端与中肾管分开，并开口在尿囊末端的膀胱；中肾管与米勒管则开口于泄殖腔。伴随性别的分化，后两套管道相应地退化或发育。当胚胎发育为雄性，一部分中肾小管与睾丸相连成为附睾，一部分中肾管转变为输精管，开口于骨盆部尿道；米勒管退化，在成体留有遗迹，称为雄性子宫。如胚胎发育成雌性，中肾管退化，一部分中肾小管退化形成卵巢冠和卵巢旁体；米勒管发育成输卵管、子宫和阴道。

二、雌性生殖器官的组织学结构概述

（一）卵巢（ovary）

卵巢的结构依动物的种类、年龄、生殖周期的阶段而有所不同。卵巢由被膜和实质组成。

1.**被膜** 包括生殖上皮和白膜。卵巢表面有单层扁平或立方形的生殖上皮，下方是结缔组织构成的白膜。马卵巢的生殖上皮仅位于排卵窝处，其余部分均被覆浆膜。

2.**实质** 外周为皮质，内部为髓质。

（1）皮质 位于白膜内侧，由基质、卵泡（follicle）和黄体（corpus luteum）构成。基质中主要是紧密排列的幼稚结缔组织细胞，呈菱形，细胞核长杆状。基质中胶原纤维少，网状纤维多。皮质中的卵泡大小、形态各不相同，是卵泡发育的不同阶段。通常外周的卵泡较小而多，朝向髓质的较大。未能发育成熟的退化成闭锁卵泡（atretic follicle）。幼年期的卵巢含许多小卵泡，性成熟后卵泡发育，出现不同发育阶段的卵泡。卵泡由中央的一个卵母细胞及其周围的卵泡细胞组成。根据卵泡的发育特点，将卵泡分为原始卵泡、生长卵泡（初级卵泡、次级卵泡）和成熟卵泡。

成熟卵泡排卵后，卵泡壁塌陷，卵泡膜血管破裂出血，血液充满卵泡腔，形成血体（红体）。同时，残留在卵泡壁的颗粒细胞和内膜细胞向腔内侵入，胞体增大，胞质内出现脂质颗粒，形成黄体。黄体是内分泌腺，主要分泌孕酮及雌激素，保证胚胎的附植和发育。黄体发育程度和存在时间，取决于卵细胞是否受精。如未妊娠，黄体则逐渐退化。如果已妊娠，黄体在整个妊娠期维持其大小和分泌功能。黄体完成其功能后即退化成为结缔组织瘢痕，称为白体。

（2）髓质 位于卵巢中部，面积小。疏松结缔组织中有许多血管、神经及淋巴管。在近卵巢门处有少量的平滑肌，血管、淋巴管及神经由门部进入卵巢。

（二）输卵管（oviduct）

输卵管为输送卵子和受精的管道。分为漏斗部、壶腹部、峡部。管壁组织结构分黏膜、肌层、浆膜三层。黏膜的表面有许多纵行皱襞。上皮为单层柱状。猪及反刍动物有的部分是假复层柱状上皮。上皮细胞有两种类型，即游离面带有可动纤毛的柱状细胞及不带有纤毛的柱状细胞。

（三）子宫（uterus）

子宫是胎儿附植及孕育的地方。子宫壁的结构分为内膜、肌层和外膜三层。

1. 子宫内膜 由上皮和固有层构成。上皮随动物种类和发情周期而不同，猪和反刍动物为单层柱状或假复层柱状上皮，猫等为单层柱状上皮。上皮有分泌功能，游离面有暂时的纤毛。固有层的浅层有较多的细胞成分及子宫腺导管。细胞以梭形或星形的胚性结缔组织细胞为主，细胞突起相互连接。还有巨噬细胞、肥大细胞、淋巴细胞、白细胞和浆细胞等。固有层的深层中细胞少，布满了分支管状的子宫腺及其导管（肉阜处除外）。子宫腺分泌物为富含糖原等营养物质的浓稠黏液。

子宫肉阜（caruncle）是反刍动物固有层深部形成的圆形加厚部分，有数十个至上百个，内含丰富的成纤维细胞和大量的血管。羊的子宫肉阜中心凹陷，牛的子宫肉阜为圆形隆突。子宫肉阜参与胎盘的形成，属胎盘的母体部分。

2. 肌层和外膜 子宫肌层由发达的内环、外纵两层平滑肌组成。在两层间或内层深部存在大的血管及淋巴管，这些血管主要是供应子宫内膜营养，在反刍动物子宫肉阜处特别发达。

三、雌性生殖系统图谱

1. **卵巢** 图 17-1 ~ 图 17-43。
2. **输卵管** 图 17-44 ~ 图 17-54。
3. **子宫** 图 17-55 ~ 图 17-76。
4. **阴道** 图 17-77 ~ 图 17-90。

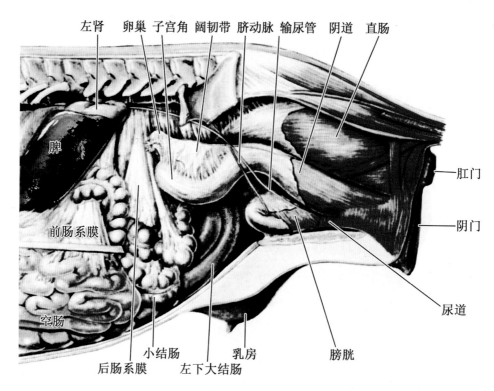

图 17-1　母马生殖器官的位置关系

Fig.17-1　Positional relationship of female horse reproductive organs

左肾—left kidney　卵巢—ovary　子宫角—cornua uteri　阔韧带—broad ligament　脐动脉—umbilical artery
输尿管—ureter　阴道—vagina　直肠—rectum　脾—spleen　前肠系膜—front mesentery　空肠—jejunum
后肠系膜—later mesentery　小结肠—small colon　左下大结肠—left ventral colon　乳房—breast　膀胱—bladder
尿道—urethra　阴门—pussy　肛门—anus

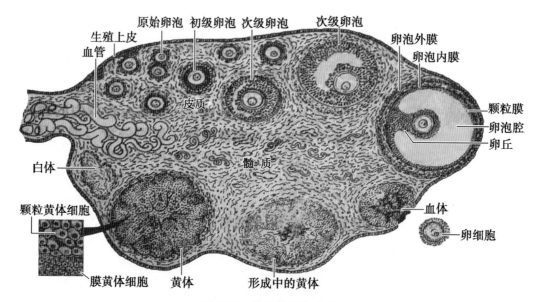

图 17-2　卵巢结构模式图

Fig.17-2　Ideograph of ovary structure

血管—blood vessel　生殖上皮—germinal epithelium　原始卵泡—primordial follicle　初级卵泡—primary follicle
次级卵泡—secondary follicle　卵泡外膜—follicular outer membrane　卵泡内膜—follicular inner membrane
颗粒膜—membrana granulosa　卵泡腔—follicular cavity　卵丘—germ hillock　血体—corpus hemorrhagicum
卵细胞—oocyte　形成中的黄体—growing corpus luteum　黄体—corpus luteum　膜黄体细胞—theca lutein cell
颗粒黄体细胞—granulosa lutein cell　白体—corpus albicans　皮质—cortex　髓质—medulla

第十七章 雌性生殖系统 Female Reproductive System

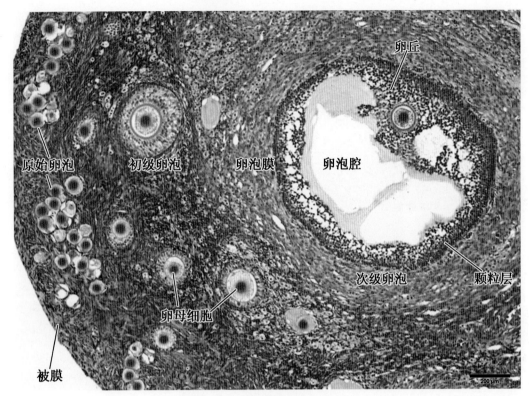

图 17-3 牛卵巢低倍像 1（HE）

Fig.17-3 low magnification of cattle ovary 1（HE）

被膜—capsule　原始卵泡—primordial follicle　初级卵泡—primary follicle　次级卵泡—secondary follicle
卵泡膜—follicular membrane　颗粒层—granular layer　卵泡腔—follicular cavity　卵丘—germ hillock　卵母细胞—oocyte

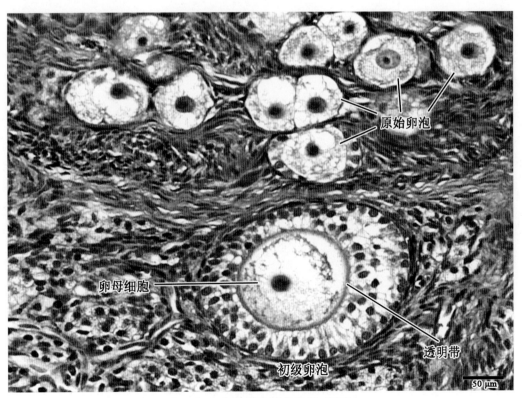

图 17-4 牛卵巢皮质高倍像（HE）

Fig.17-4 high magnification of cattle ovary cortex（HE）

原始卵泡—primordial follicle　初级卵泡—primary follicle　卵母细胞—oocyte　透明带—zona pellucida

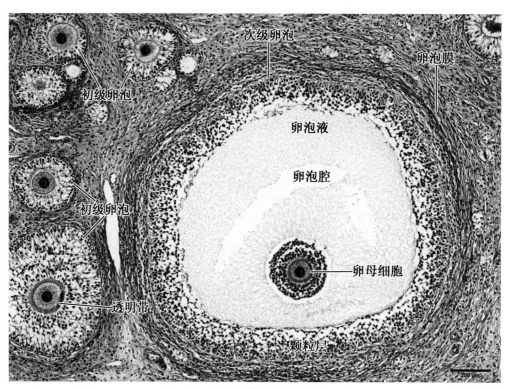

图 17-5　牛卵巢皮质低倍像（HE）

Fig.17-5　low magnification of cattle ovary cortex （HE）

初级卵泡—primary follicle　次级卵泡—secondary follicle　卵泡液—follicular fluid　卵泡腔—follicular cavity　卵母细胞—oocyte　卵泡膜—follicular membrane　颗粒层—granular layer　透明带—zona pellucida

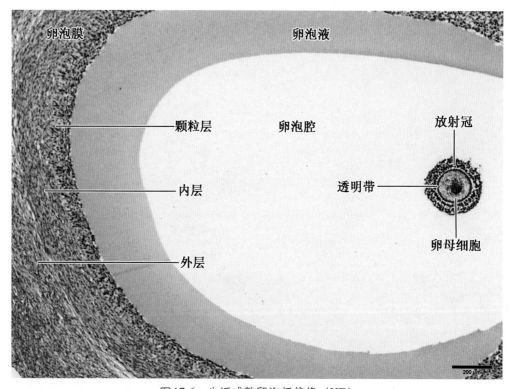

图 17-6　牛近成熟卵泡低倍像（HE）

Fig.17-6　low magnification of cattle maturing follicle （HE）

卵泡膜—follicular membrane　颗粒层—granular layer　内层—inner layer　外层—outer layer　卵泡液—follicular fluid　卵泡腔—follicular cavity　放射冠—corona radiata　透明带—zona pellucida　卵母细胞—oocyte

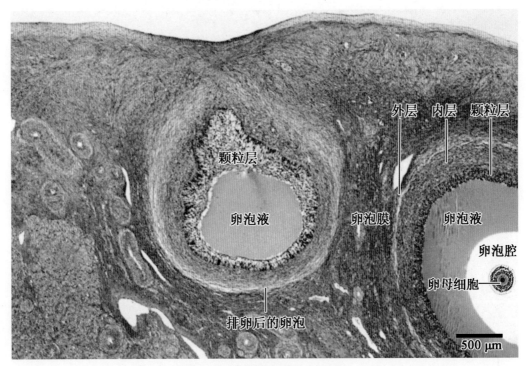

图17-7 牛卵巢低倍像2（HE）

Fig.17-7 low magnification of cattle ovary 2（HE）

颗粒层—granular layer 卵泡液—follicular fluid 排卵后的卵泡—ovulated follicle 卵泡膜—follicular membrane
外层—outer layer 内层—inner layer 卵泡腔—follicular cavity 卵母细胞—oocyte

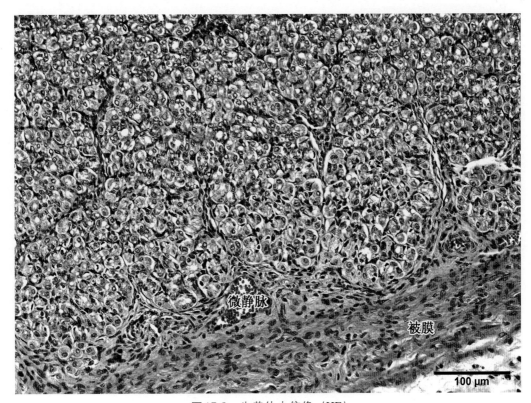

图17-8 牛黄体中倍像（HE）

Fig.17-8 mid magnification of cattle corpus luteum（HE）

微静脉—venule 被膜—capsule

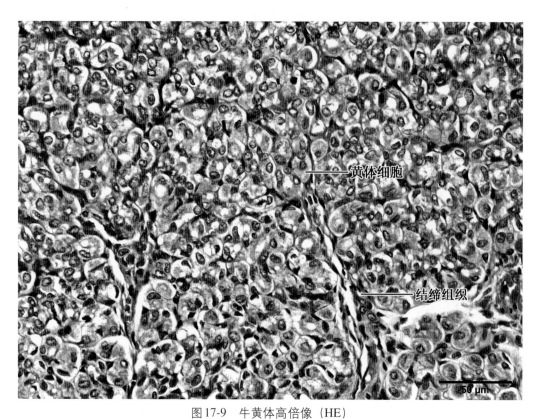

图17-9　牛黄体高倍像（HE）

Fig.17-9　high magnification of cattle corpus luteum（HE）

黄体细胞—luteal cell　结缔组织—connective tissue

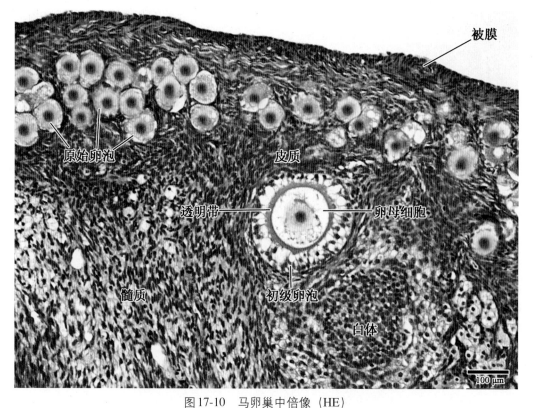

图17-10　马卵巢中倍像（HE）

Fig.17-10　mid magnification of horse ovary（HE）

被膜—capsule　皮质—cortex　原始卵泡—primordial follicle　初级卵泡—primary follicle

透明带—zona pellucida　卵母细胞—oocyte　髓质—medulla　白体—corpus albicans

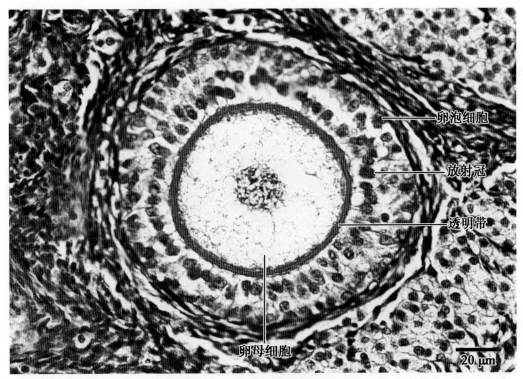

图 17-11　马初级卵泡高倍像（HE）

Fig.17-11　high magnification of horse primary follicle （HE）

卵泡细胞—follicular cell　放射冠—corona radiata　透明带—zona pellucida　卵母细胞—oocyte

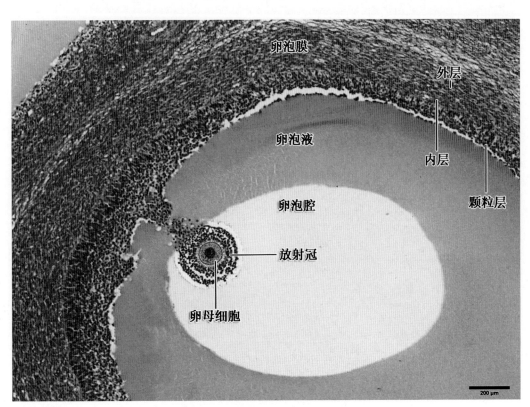

图 17-12　马成熟卵泡低倍像（HE）

Fig.17-12　low magnification of horse mature follicle （HE）

卵泡膜—follicular membrane　外层—outer layer　内层—inner layer　颗粒层—granular layer　卵泡液—follicular fluid　卵泡腔—follicular cavity　放射冠—corona radiata　卵母细胞—oocyte

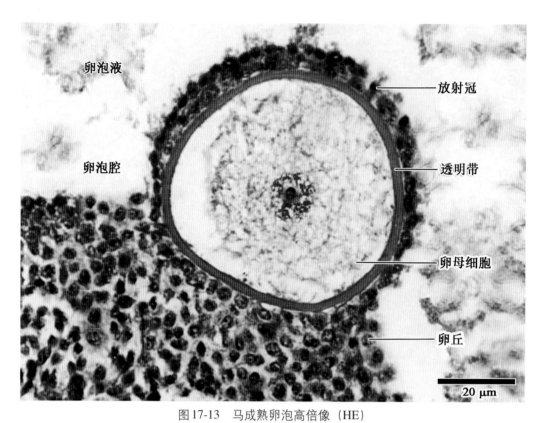

图 17-13 马成熟卵泡高倍像（HE）

Fig.17-13 high magnification of horse mature follicle（HE）

卵泡液—follicular fluid 卵泡腔—follicular cavity 放射冠—corona radiata 透明带—zona pellucida

卵母细胞—oocyte 卵丘—germ hillock

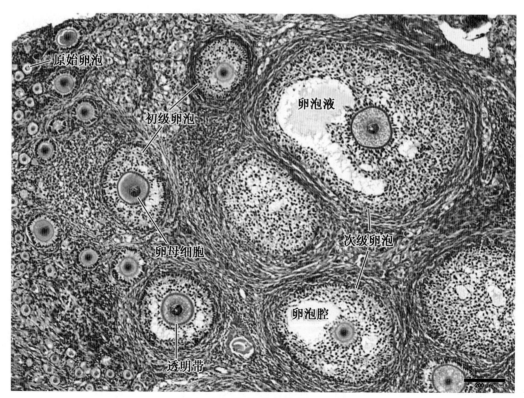

图 17-14 猪卵巢低倍像（HE）

Fig.17-14 low magnification of pig ovary（HE）

原始卵泡—primordial follicle 初级卵泡—primary follicle 次级卵泡—secondary follicle 卵泡液—follicular fluid

卵泡腔—follicular cavity 卵母细胞—oocyte 透明带—zona pellucida

第十七章 雌性生殖系统 Female Reproductive System

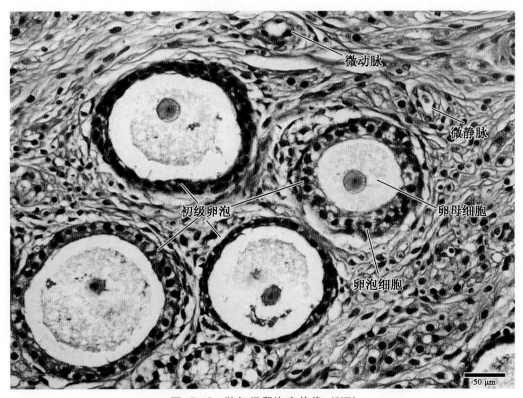

图 17-15　猪初级卵泡高倍像（HE）

Fig.17-15　high magnification of pig primary follicle（HE）

微动脉—arteriole　微静脉—venule　初级卵泡—primary follicle　卵母细胞—oocyte　卵泡细胞—follicular cell

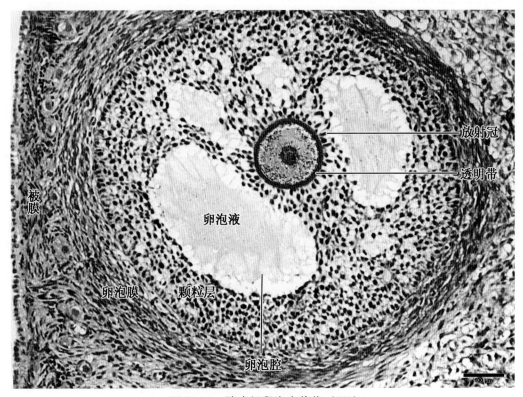

图 17-16　猪次级卵泡中倍像（HE）

Fig.17-16　mid magnification of pig secondary follicle（HE）

被膜—capsule　卵泡膜—follicular membrane　颗粒层—granular layer　卵泡液—follicular fluid
卵泡腔—follicular cavity　放射冠—corona radiata　透明带—zona pellucida

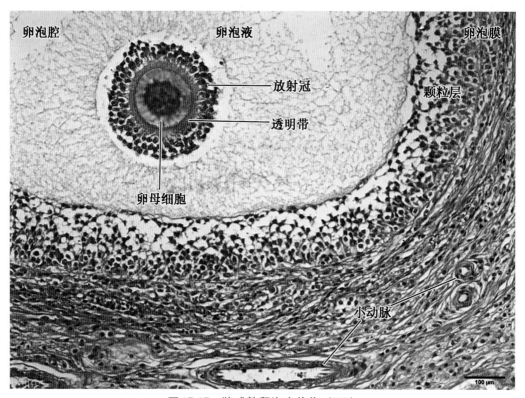

图17-17 猪成熟卵泡中倍像（HE）

Fig.17-17 mid magnification of pig mature follicle（HE）

卵泡腔—follicular cavity 卵泡液—follicular fluid 放射冠—corona radiata 透明带—zona pellucida
卵母细胞—oocyte 卵泡膜—follicular membrane 颗粒层—granular layer 小动脉—arteriole

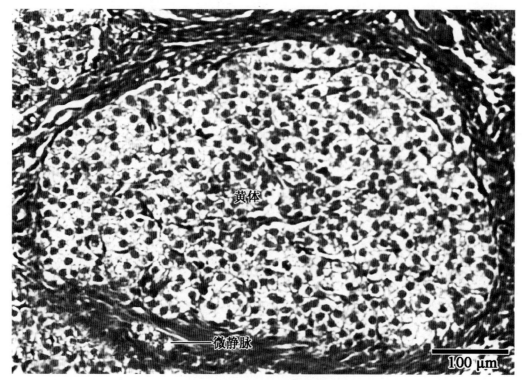

图17-18 猪黄体中倍像（HE）

Fig.17-18 mid magnification of pig corpus luteum（HE）

黄体—corpus luteum 微静脉—venule

第十七章 雌性生殖系统 Female Reproductive System

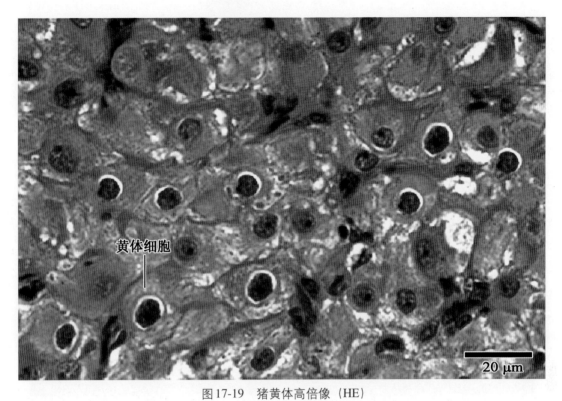

图17-19 猪黄体高倍像（HE）
Fig.17-19 high magnification of pig corpus luteum（HE）
黄体细胞—luteal cell

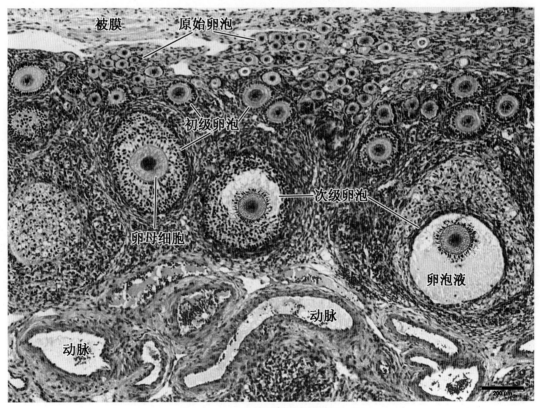

图17-20 绵羊卵巢低倍像（HE）
Fig.17-20 low magnification of sheep ovary（HE）
被膜—capsule 原始卵泡—primordial follicle 初级卵泡—primary follicle 次级卵泡—secondary follicle
卵泡液—follicular fluid 卵母细胞—oocyte 动脉—artery

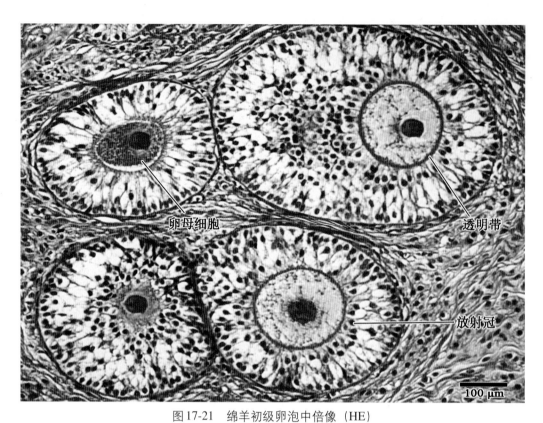

图 17-21　绵羊初级卵泡中倍像（HE）

Fig.17-21　mid magnification of sheep primary follicle（HE）

卵母细胞—oocyte　透明带—zona pellucida　放射冠—corona radiata

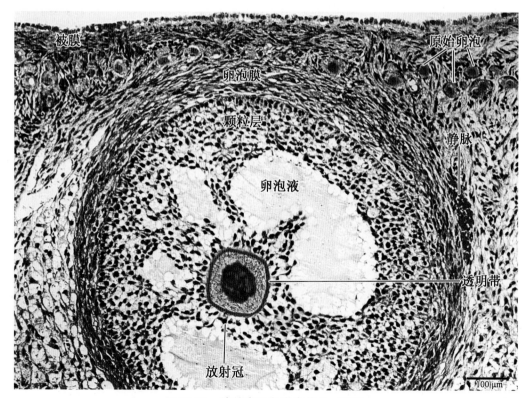

图 17-22　绵羊次级卵泡中倍像（HE）

Fig.17-22　mid magnification of sheep secondary follicle（HE）

被膜—capsule　原始卵泡—primordial follicle　卵泡膜—follicular membrane　颗粒层—granular layer
卵泡液—follicular fluid　放射冠—corona radiata　透明带—zona pellucida　静脉—venule

第十七章 雌性生殖系统　Female Reproductive System

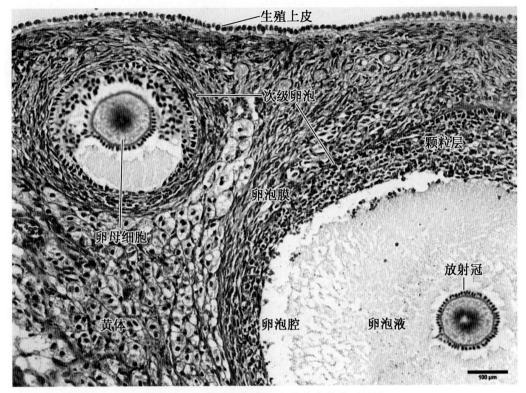

图17-23　绵羊次级卵泡和黄体中倍像（HE）

Fig.17-23　mid magnification of sheep secondary follicle and corpus luteum（HE）

生殖上皮—germinal epithelium　次级卵泡—secondary follicle　卵泡膜—follicular membrane　颗粒层—granular layer
卵泡腔—follicular cavity　卵泡液—follicular fluid　放射冠—corona radiata　卵母细胞—oocyte　黄体—corpus luteum

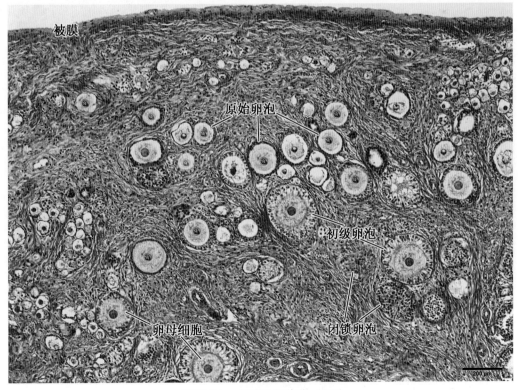

图17-24　犬卵巢低倍像（HE）

Fig.17-24　low magnification of dog ovary（HE）

被膜—capsule　原始卵泡—primordial follicle　初级卵泡—primary follicle　闭锁卵泡—atretic follicle　卵母细胞—oocyte

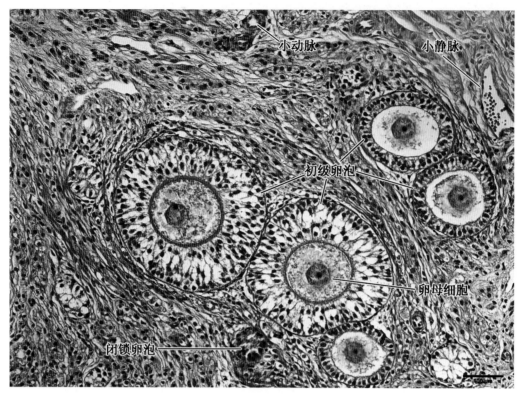

图 17-25 犬卵巢皮质中倍像（HE）

Fig.17-25 mid magnification of dog ovary cortex（HE）

小动脉—arteriole 小静脉—venule 初级卵泡—primordial follicle 闭锁卵泡—atretic follicle 卵母细胞—oocyte

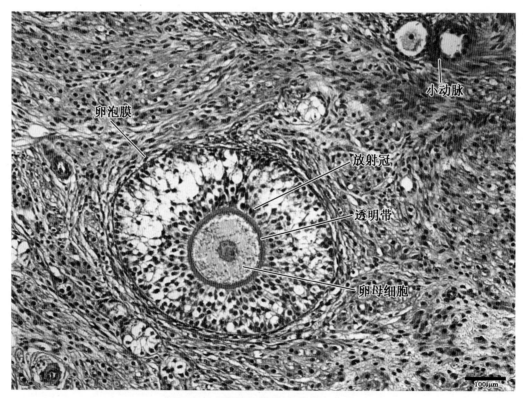

图 17-26 犬初级卵泡中倍像（HE）

Fig.17-26 mid magnification of dog primary follicle（HE）

小动脉—arteriole 卵泡膜—follicular membrane 放射冠—corona radiata 透明带—zona pellucida 卵母细胞—oocyte

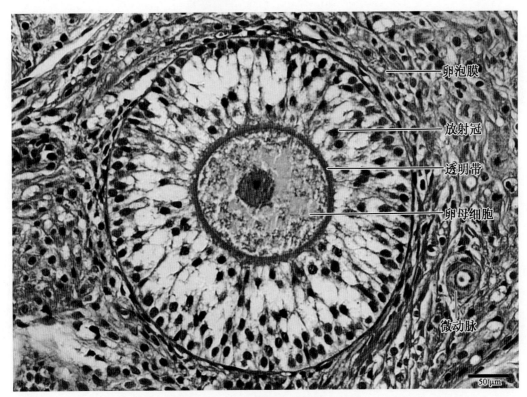

图17-27 犬初级卵泡高倍像（HE）

Fig.17-27 high magnification of dog primary follicle（HE）

卵泡膜—follicular membrane 放射冠—corona radiata 透明带—zona pellucida 卵母细胞—oocyte 微动脉—arteriole

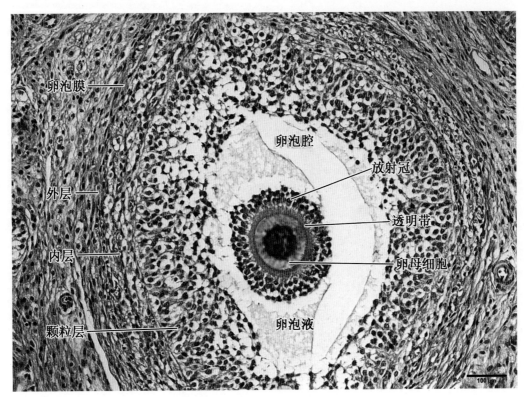

图17-28 犬次级卵泡中倍像（HE）

Fig.17-28 mid magnification of dog secondary follicle（HE）

卵泡膜—follicular membrane 外层—outer layer 内层—inner layer 颗粒层—granular layer 卵泡腔—follicular cavity 卵泡液—follicular fluid 放射冠—corona radiata 透明带—zona pellucida 卵母细胞—oocyte

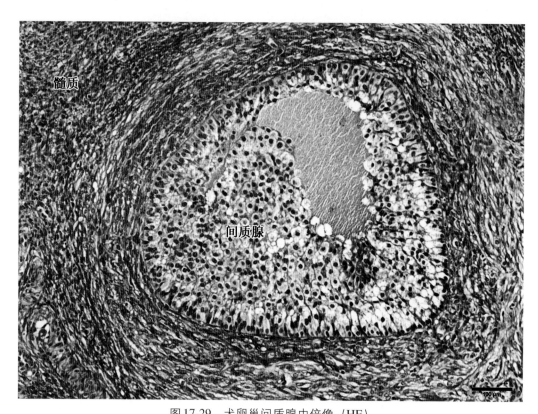

图17-29 犬卵巢间质腺中倍像（HE）

Fig.17-29 mid magnification of dog interstitial gland（HE）

髓质—medulla 间质腺—interstitial gland

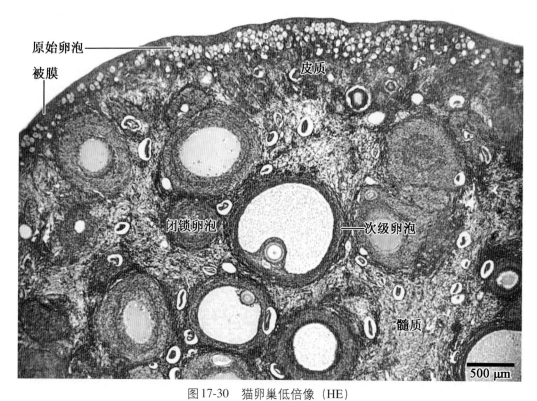

图17-30 猫卵巢低倍像（HE）

Fig.17-30 low magnification of cat ovary（HE）

被膜—capsule 原始卵泡—primordial follicle 皮质—cortex 闭锁卵泡—atretic follicle

次级卵泡—secondary follicle 髓质—medulla

第十七章 雌性生殖系统 Female Reproductive System

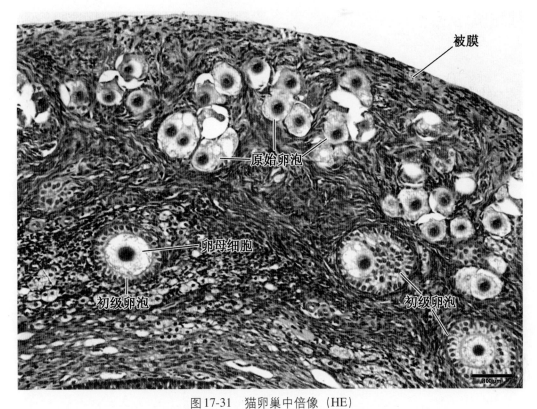

图 17-31 猫卵巢中倍像（HE）

Fig.17-31 mid magnification of cat ovary（HE）

被膜—capsule 原始卵泡—primordial follicle 初级卵泡—primary follicle 卵母细胞—oocyte

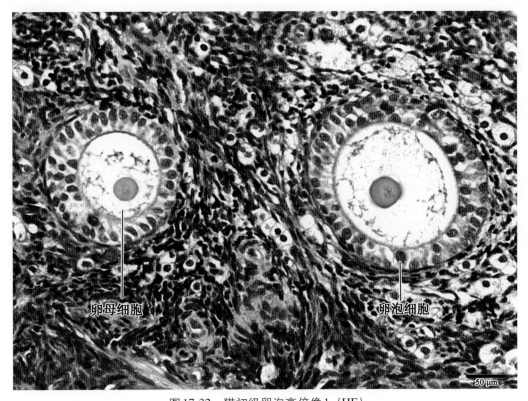

图 17-32 猫初级卵泡高倍像 1（HE）

Fig.17-32 high magnification of cat primary follicle 1（HE）

卵母细胞—oocyte 卵泡细胞—follicular cell

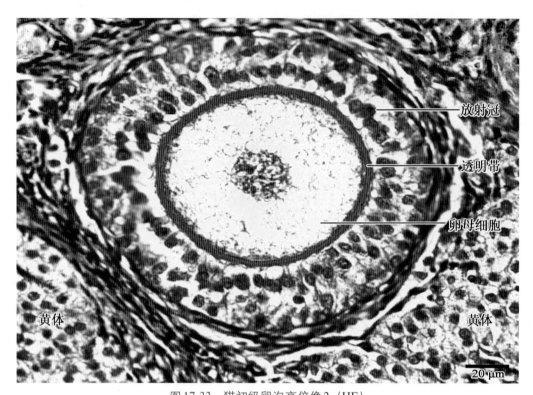

图 17-33 猫初级卵泡高倍像 2（HE）

Fig.17-33 high magnification of cat primary follicle 2（HE）

放射冠—corona radiata 透明带—zona pellucida 卵母细胞—oocyte 黄体—corpus luteum

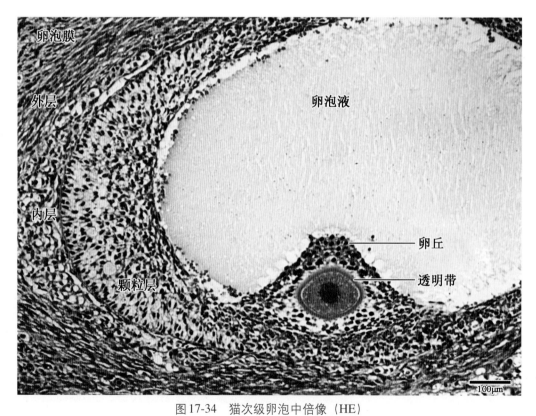

图 17-34 猫次级卵泡中倍像（HE）

Fig.17-34 mid magnification of cat secondary follicle（HE）

卵泡膜—follicular membrane 外层—outer layer 内层—inner layer 颗粒层—granular layer

卵泡液—follicular fluid 卵丘—germ hillock 透明带—zona pellucida

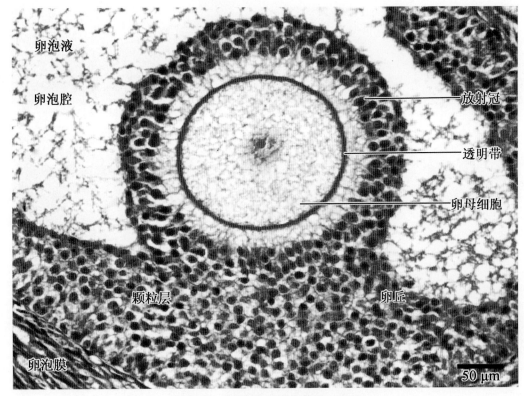

图17-35 猫次级卵泡高倍像（HE）

Fig.17-35 high magnification of cat secondary follicle（HE）

卵泡液—follicular fluid　卵泡腔—follicular cavity　卵泡膜—follicular membrane　颗粒层—granular layer　放射冠—corona radiata　透明带—zona pellucida　卵母细胞—oocyte　卵丘—germ hillock

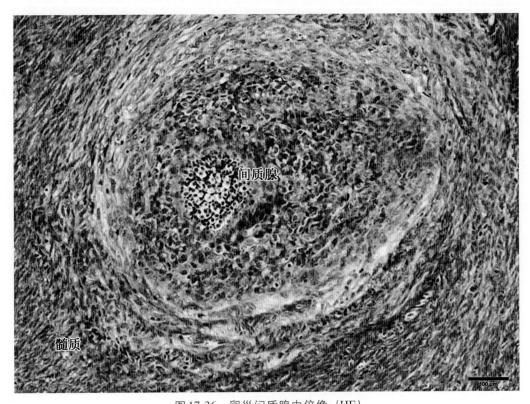

图17-36 卵巢间质腺中倍像（HE）

Fig.17-36 mid magnification of interstitial gland（HE）

髓质—medulla　间质腺—interstitial gland

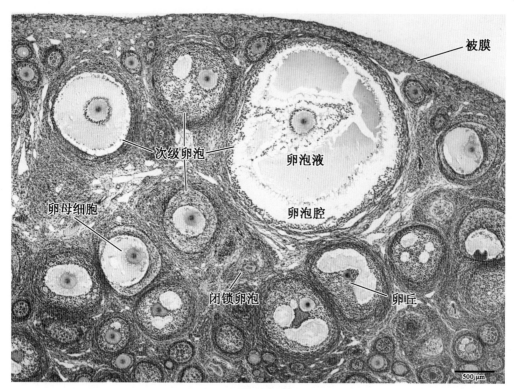

图 17-37　兔卵巢低倍像（HE）

Fig.17-37　low magnification of rabbit ovary（HE）

被膜—capsule　次级卵泡—secondary follicle　卵泡液—follicular fluid　卵泡腔—follicular cavity
卵母细胞—oocyte　闭锁卵泡—atretic follicle　卵丘—germ hillock

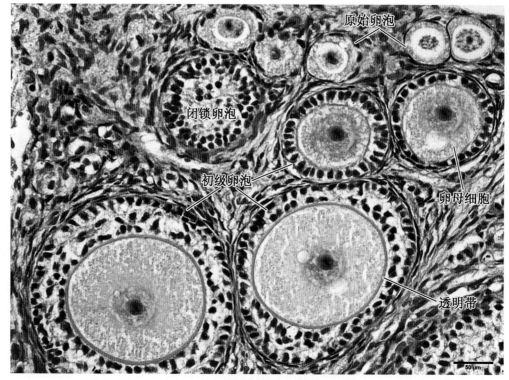

图 17-38　兔卵巢皮质高倍像1（HE）

Fig.17-38　high magnification of rabbit ovary cortex 1（HE）

原始卵泡—primordial follicle　闭锁卵泡—atretic follicle　初级卵泡—primary follicle　卵母细胞—oocyte
透明带—zona pellucida

第十七章 雌性生殖系统 Female Reproductive System

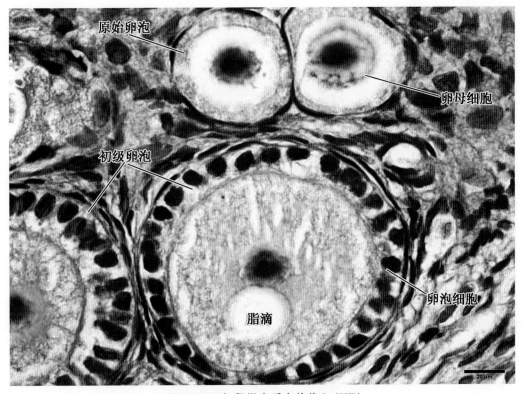

图17-39 兔卵巢皮质高倍像2（HE）

Fig.17-39 high magnification of rabbit ovary cortex 2（HE）

原始卵泡—primordial follicle　初级卵泡—primary follicle　卵母细胞—oocyte　卵泡细胞—follicular cell　脂滴—lipid droplet

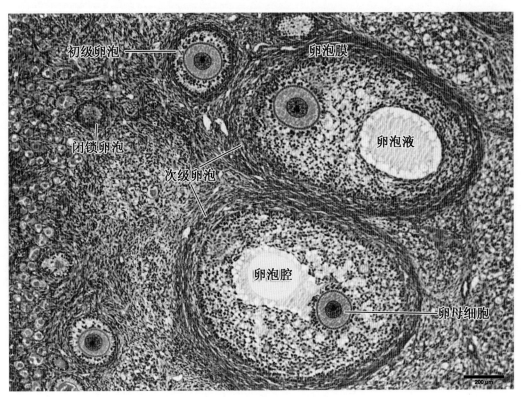

图17-40 兔次级卵泡低倍像1（HE）

Fig.17-40 low magnification of rabbit secondary follicle 1（HE）

初级卵泡—primary follicle　闭锁卵泡—atretic follicle　次级卵泡—secondary follicle　卵泡膜—follicular membrane　卵泡液—follicular fluid　卵泡腔—follicular cavity　卵母细胞—oocyte

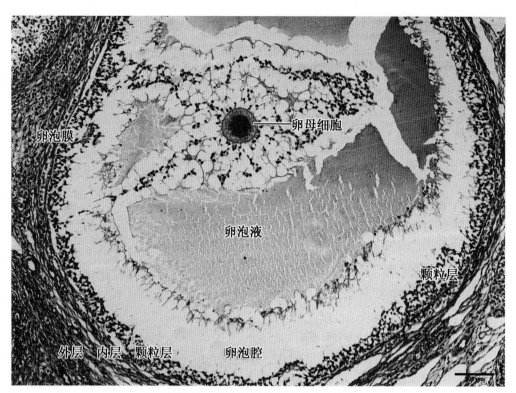

图17-41　兔次级卵泡低倍像2（HE）

Fig.17-41　low magnification of rabbit secondary follicle 2（HE）

卵泡膜—follicular membrane　外层—outer layer　内层—inner layer　颗粒层—granular layer　卵母细胞—oocyte
卵泡液—follicular fluid　卵泡腔—follicular cavity

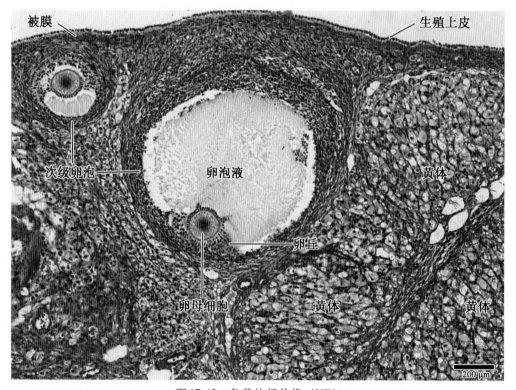

图17-42　兔黄体低倍像（HE）

Fig.17-42　low magnification of rabbit corpus luteum（HE）

被膜—capsule　生殖上皮—germinal epithelium　次级卵泡—secondary follicle　卵泡液—follicular fluid
卵母细胞—oocyte　卵丘—germ hillock　黄体—corpus luteum

第十七章 雌性生殖系统 Female Reproductive System

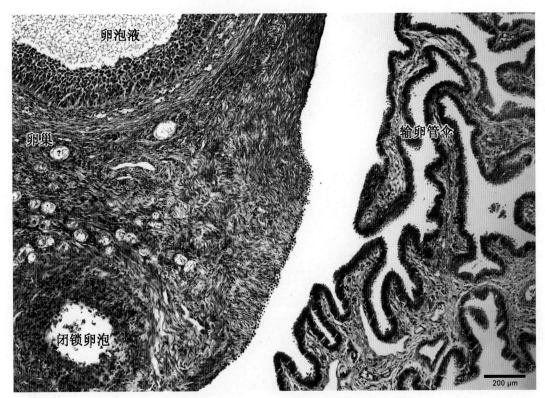

图17-43 牛卵巢和输卵管伞低倍像（HE）

Fig.17-43 low magnification of cattle ovary and fimbriae of uterine tube（HE）

卵巢—ovary　卵泡液—follicular fluid　闭锁卵泡—atretic follicle　输卵管伞—fimbriae of uterine tube

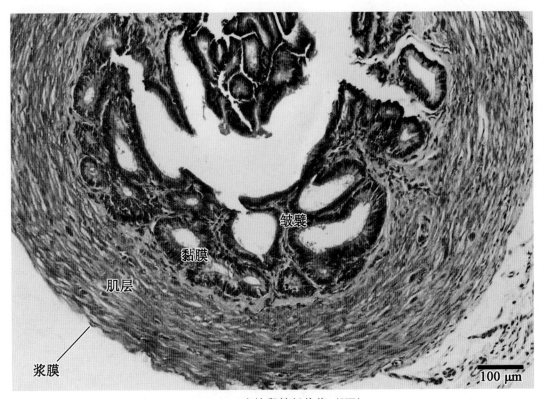

图17-44 牛输卵管低倍像（HE）

Fig.17-44 low magnification of cattle oviduct（HE）

浆膜—serosa　肌层—muscular layer　黏膜—mucosa　皱襞—plica

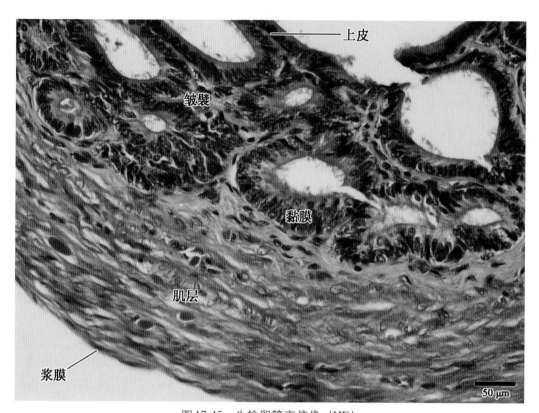

图17-45　牛输卵管高倍像（HE）

Fig.17-45　high magnification of cattle oviduct（HE）

浆膜—serosa　肌层—muscular layer　黏膜—mucosa　皱襞—plica　上皮—epithelium

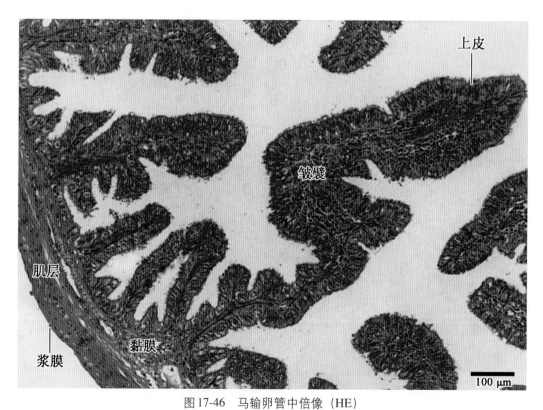

图17-46　马输卵管中倍像（HE）

Fig.17-46　mid magnification of horse oviduct（HE）

浆膜—serosa　肌层—muscular layer　黏膜—mucosa　皱襞—plica　上皮—epithelium

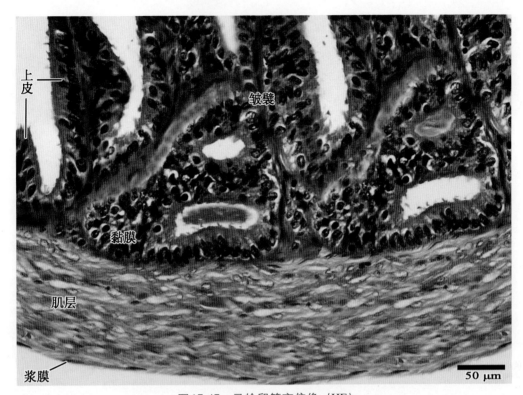

图17-47 马输卵管高倍像（HE）

Fig.17-47 high magnification of horse oviduct（HE）

浆膜—serosa 肌层—muscular layer 黏膜—mucosa 皱襞—plica 上皮—epithelium

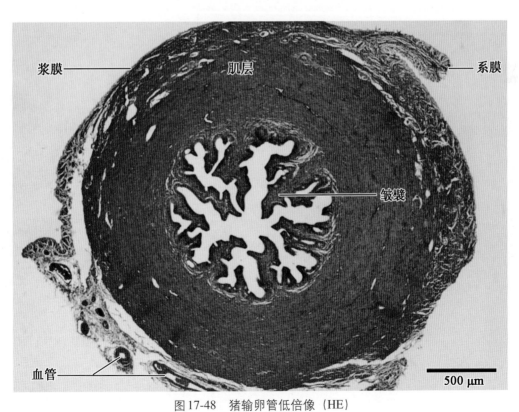

图17-48 猪输卵管低倍像（HE）

Fig.17-48 low magnification of pig oviduct（HE）

浆膜—serosa 肌层—muscular layer 系膜—mesentery 皱襞—plica 血管—blood vessel

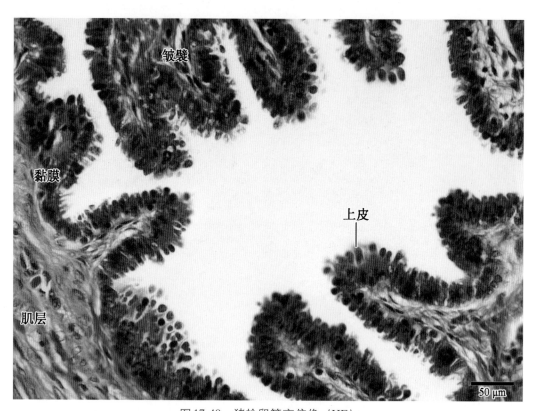

图17-49 猪输卵管高倍像（HE）

Fig.17-49 high magnification of pig oviduct（HE）

肌层—muscular layer 黏膜—mucosa 皱襞—plica 上皮—epithelium

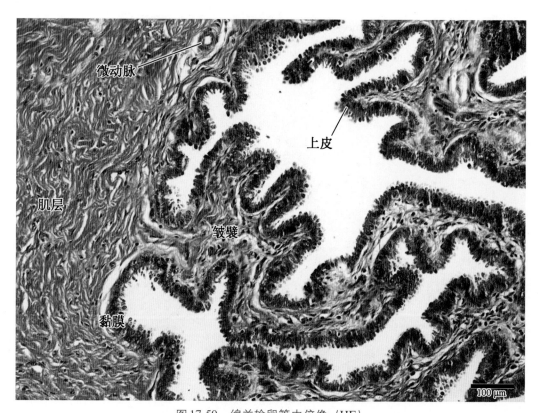

图17-50 绵羊输卵管中倍像（HE）

Fig.17-50 mid magnification of sheep oviduct（HE）

微动脉—arteriole 肌层—muscular layer 黏膜—mucosa 皱襞—plica 上皮—epithelium

第十七章 雌性生殖系统 Female Reproductive System

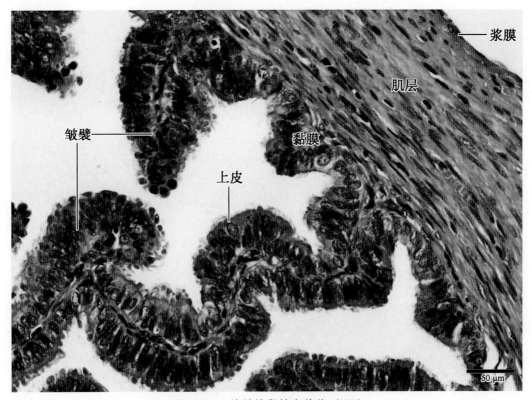

图17-51 绵羊输卵管高倍像（HE）

Fig.17-51 high magnification of sheep oviduct（HE）

浆膜—serosa　肌层—muscular layer　黏膜—mucosa　皱襞—plica　上皮—epithelium

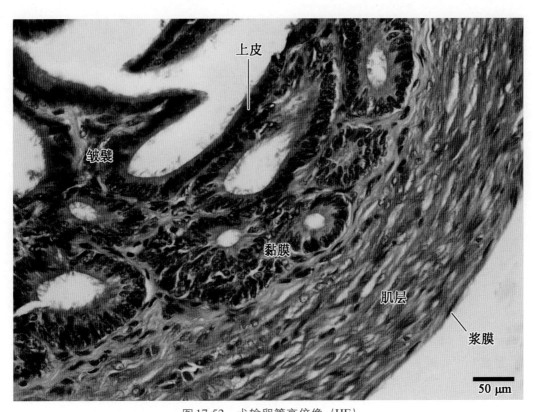

图17-52 犬输卵管高倍像（HE）

Fig.17-52 high magnification of dog oviduct（HE）

上皮—epithelium　皱襞—plica　黏膜—mucosa　肌层—muscular layer　浆膜—serosa

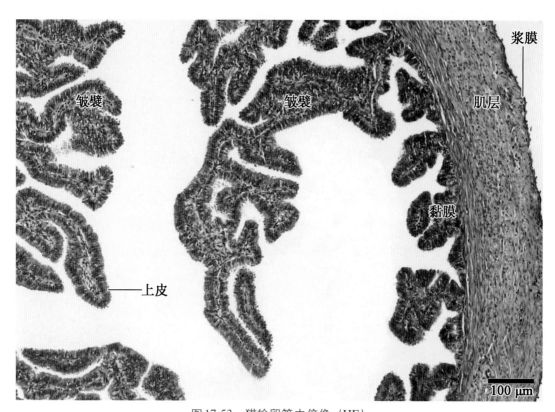

图 17-53 猫输卵管中倍像（HE）

Fig.17-53 mid magnification of cat oviduct（HE）

皱襞—plica 上皮—epithelium 黏膜—mucosa 肌层—muscular layer 浆膜—serosa

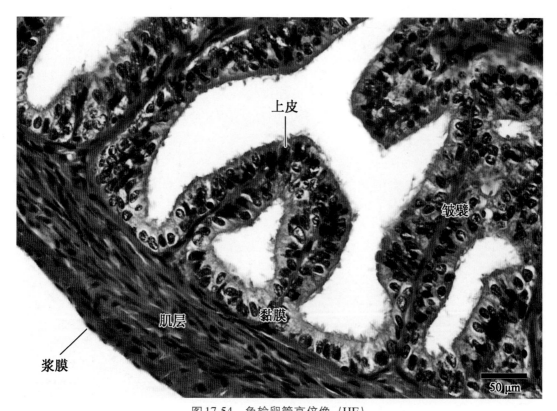

图 17-54 兔输卵管高倍像（HE）

Fig.17-54 high magnification of rabbit oviduct（HE）

浆膜—serosa 肌层—muscular layer 黏膜—mucosa 上皮—epithelium 皱襞—plica

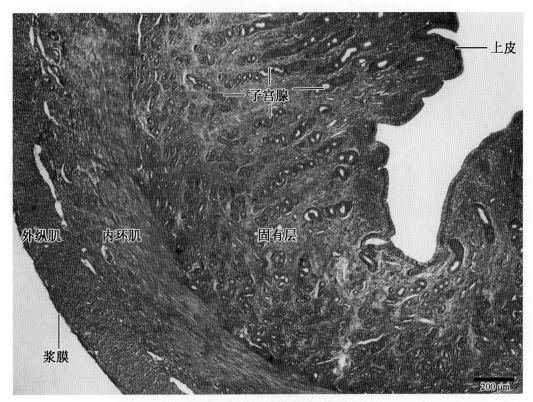

图17-55 牛子宫低倍像（HE）

Fig.17-55 low magnification of cattle uterus（HE）

浆膜—serosa 外纵肌—outer longitudinal muscle 内环肌—inner ring muscle 固有层—lamina propria 子宫腺—uterus gland 上皮—epithelium

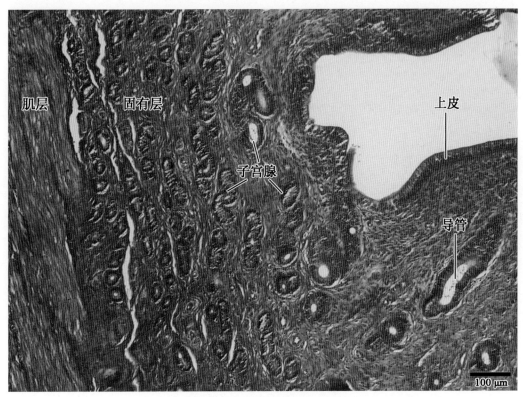

图17-56 牛子宫中倍像（HE）

Fig.17-56 mid magnification of cattle uterus（HE）

肌层—muscular layer 固有层—lamina propria 子宫腺—uterus gland 上皮—epithelium 导管—duct

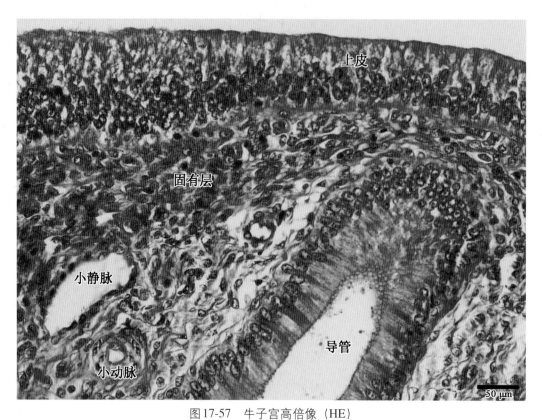

图17-57　牛子宫高倍像（HE）

Fig.17-57　high magnification of cow uterus（HE）

小静脉—venule　小动脉—arteriole　固有层—lamina propria　上皮—epithelium　导管—duct

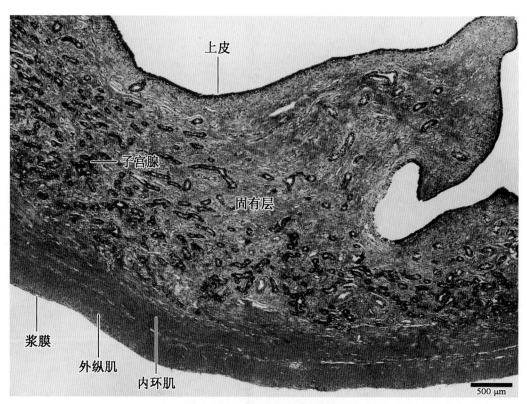

图17-58　马子宫低倍像（HE）

Fig.17-58　low magnification of horse uterus（HE）

浆膜—serosa　外纵肌—outer longitudinal muscle　内环肌—inner ring muscle

固有层—lamina propria　子宫腺—uterus gland　上皮—epithelium

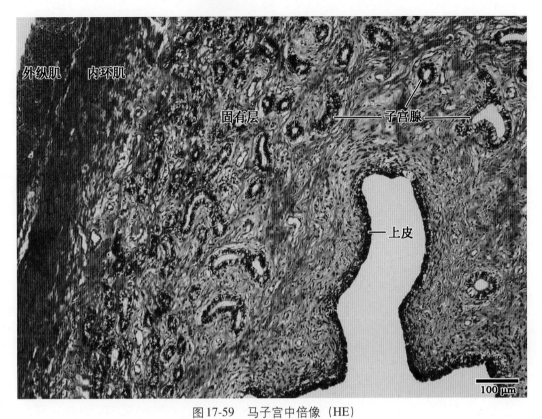

图 17-59 马子宫中倍像（HE）

Fig.17-59 mid magnification of horse uterus （HE）

外纵肌—outer longitudinal muscle　内环肌—inner ring muscle　固有层—lamina propria

子宫腺—uterus gland　上皮—epithelium

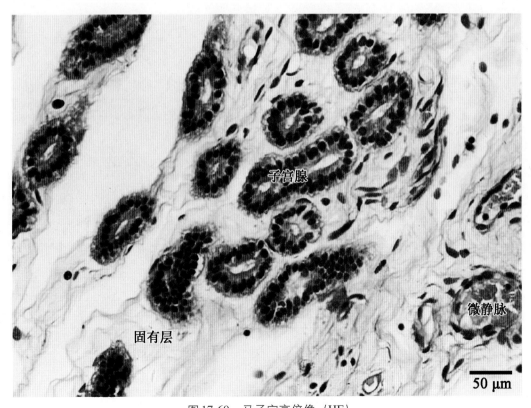

图 17-60 马子宫高倍像（HE）

Fig.17-60 high magnification of horse uterus （HE）

固有层—lamina propria　子宫腺—uterus gland　微静脉—venule

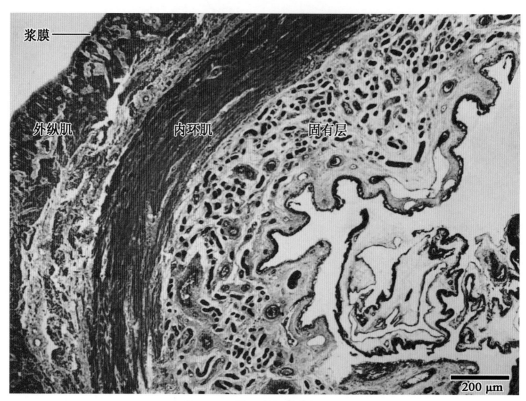

图 17-61 猪子宫低倍像（HE）

Fig.17-61 low magnification of pig uterus（HE）

浆膜—serosa 外纵肌—outer longitudinal muscle 内环肌—inner ring muscle 固有层—lamina propria

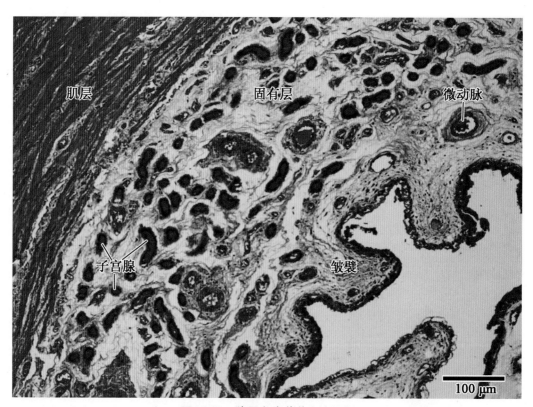

图 17-62 猪子宫中倍像1（HE）

Fig.17-62 mid magnification of pig uterus 1（HE）

肌层—muscular layer 固有层—lamina propria 微动脉—arteriole 子宫腺—uterus gland 皱襞—plica

第十七章 雌性生殖系统 Female Reproductive System

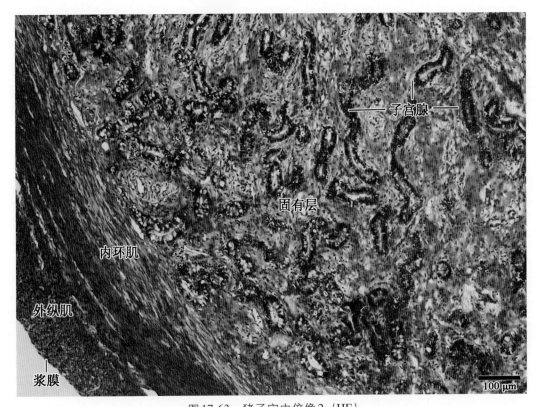

图 17-63 猪子宫中倍像 2（HE）

Fig.17-63 mid magnification of pig uterus 2（HE）

浆膜—serosa 外纵肌—outer longitudinal muscle 内环肌—inner ring muscle

固有层—lamina propria 子宫腺—uterus gland

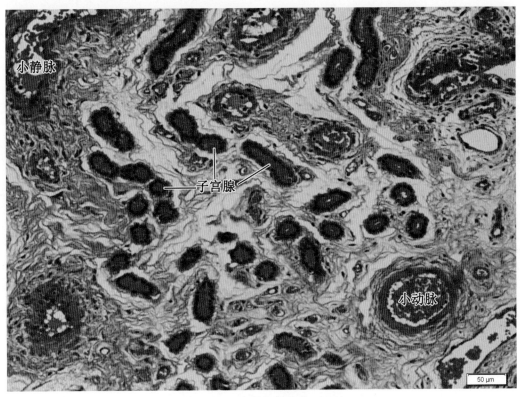

图 17-64 猪子宫高倍像（HE）

Fig.17-64 high magnification of pig uterus（HE）

小静脉—venule 子宫腺—uterus gland 小动脉—arteriole

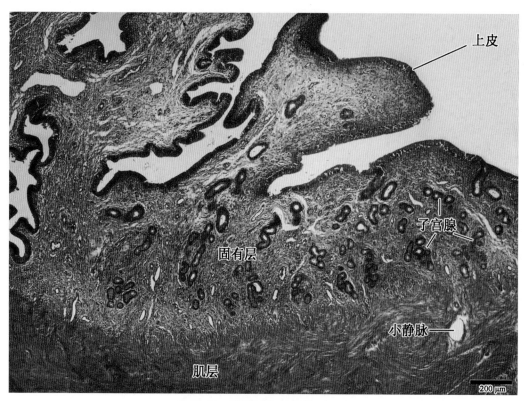

图17-65 绵羊子宫低倍像（HE）

Fig.17-65 low magnification of sheep uterus（HE）

肌层—muscular layer　固有层—lamina propria　子宫腺—uterus gland　小静脉—venule　上皮—epithelium

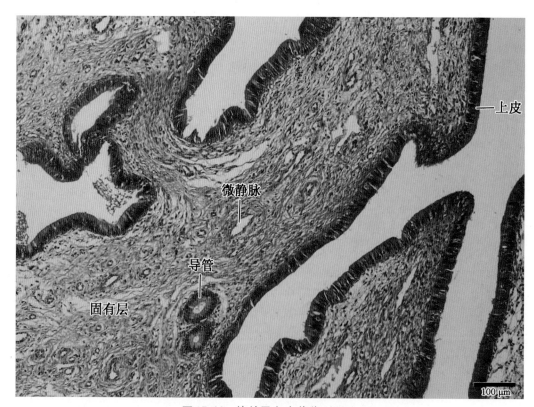

图17-66 绵羊子宫中倍像（HE）

Fig.17-66 mid magnification of sheep uterus（HE）

固有层—lamina propria　导管—duct　微静脉—venule　上皮—epithelium

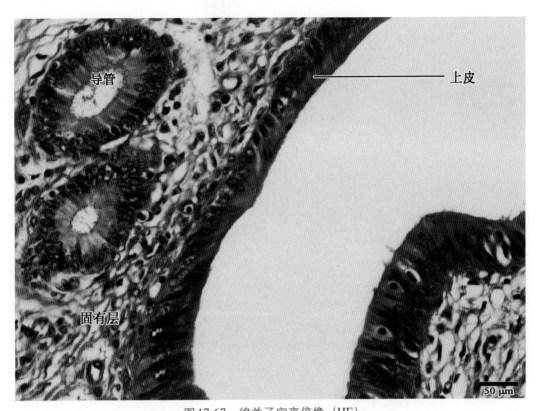

图17-67 绵羊子宫高倍像（HE）

Fig.17-67 high magnification of sheep uterus（HE）

固有层—lamina propria 导管—duct 上皮—epithelium

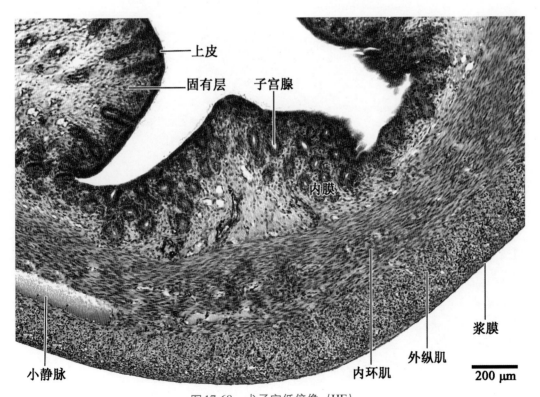

图17-68 犬子宫低倍像（HE）

Fig.17-68 low magnification of dog uterus（HE）

上皮—epithelium 固有层—lamina propria 子宫腺—uterus gland 内膜—inner membrane
小静脉—venule 内环肌—inner ring muscle 外纵肌—outer longitudinal muscle 浆膜—serosa

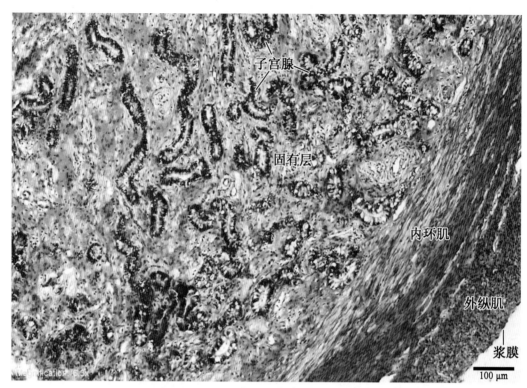

图17-69 犬子宫中倍像（HE）

Fig.17-69 mid magnification of dog uterus（HE）

子宫腺—uterus gland 固有层—lamina propria 内环肌—inner ring muscle 外纵肌—outer longitudinal muscle 浆膜—serosa

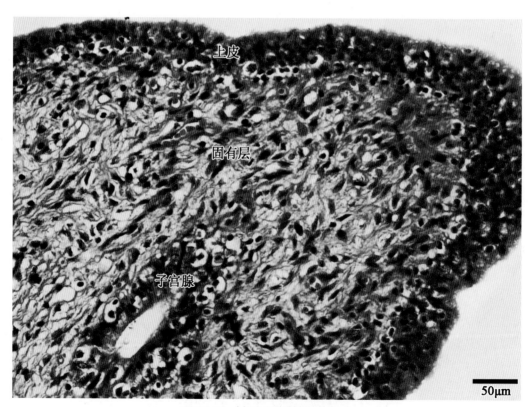

图17-70 犬子宫高倍像（HE）

Fig.17-70 high magnification of dog uterus（HE）

上皮—epithelium 固有层—lamina propria 子宫腺—uterus gland

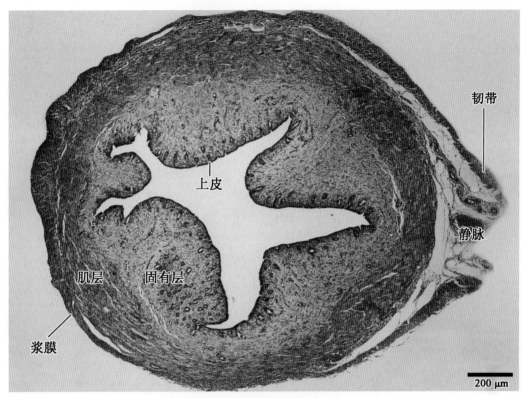

图17-71 猫子宫低倍像（HE）

Fig.17-71 low magnification of cat uterus（HE）

浆膜—serosa 肌层—muscular layer 固有层—lamina propria 上皮—epithelium 韧带—ligament 静脉—venule

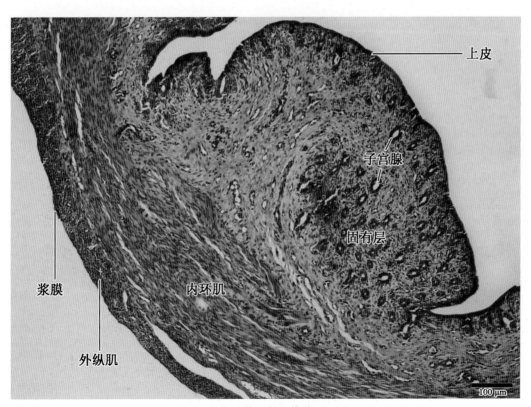

图17-72 猫子宫中倍像（HE）

Fig.17-72 mid magnification of cat uterus（HE）

浆膜—serosa 外纵肌—outer longitudinal muscle 内环肌—inner ring muscle
固有层—lamina propria 子宫腺—uterus gland 上皮—epithelium

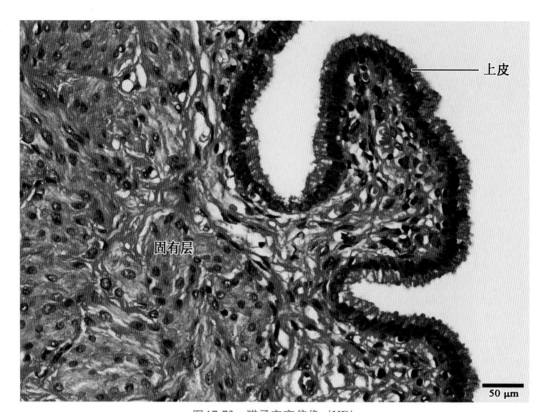

图17-73 猫子宫高倍像（HE）

Fig.17-73 high magnification of cat uterus（HE）

固有层—lamina propria 上皮—epithelium

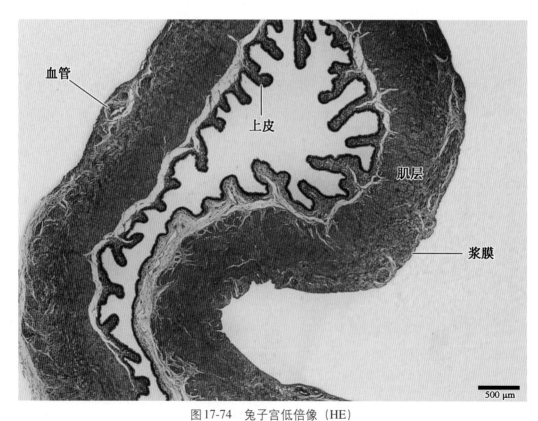

图17-74 兔子宫低倍像（HE）

Fig.17-74 low magnification of rabbit uterus（HE）

血管—blood vessel 上皮—epithelium 肌层—muscular layer 浆膜—serosa

第十七章 雌性生殖系统 Female Reproductive System

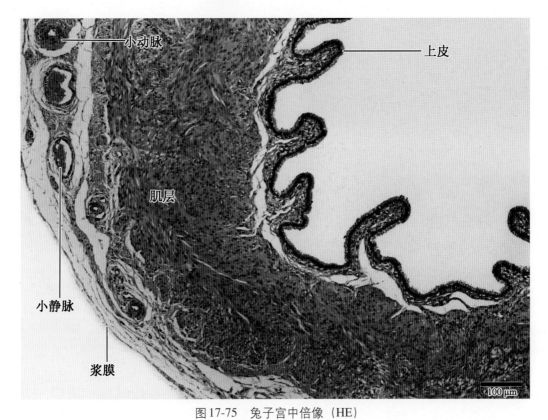

图17-75 兔子宫中倍像（HE）
Fig.17-75 mid magnification of rabbit uterus（HE）
小动脉—arteriole 小静脉—venule 浆膜—serosa 肌层—muscular layer 上皮—epithelium

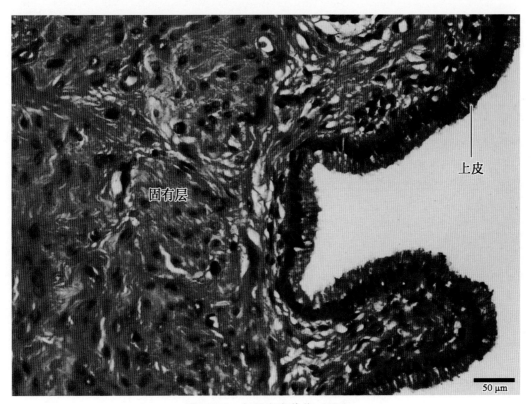

图17-76 兔子宫高倍像（HE）
Fig.17-76 high magnification of rabbit uterus（HE）
固有层—lamina propria 上皮—epithelium

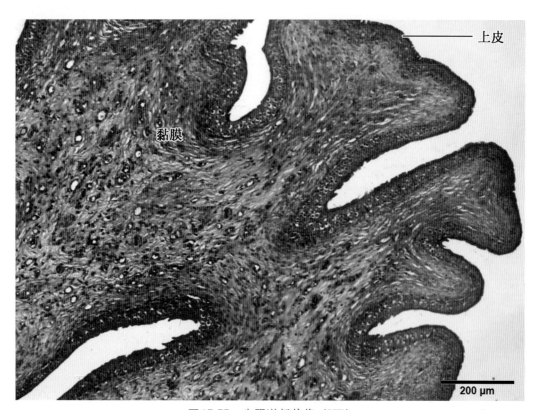

图17-77　牛阴道低倍像（HE）

Fig.17-77　low magnification of cattle vagina（HE）

黏膜—mucosa　上皮—epithelium

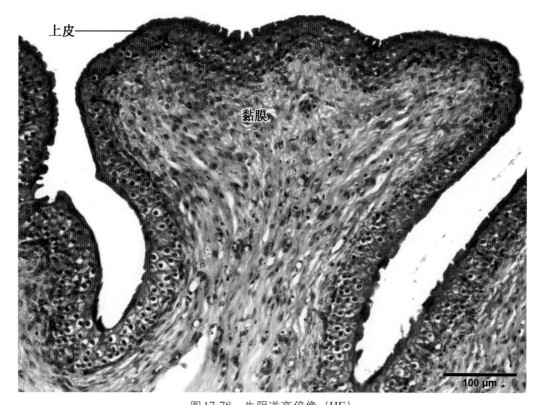

图17-78　牛阴道高倍像（HE）

Fig.17-78　high magnification of cattle vagina（HE）

上皮—epithelium　黏膜—mucosa

第十七章 雌性生殖系统 Female Reproductive System

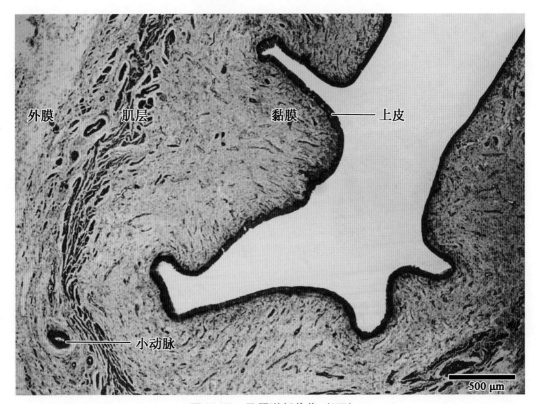

图 17-79 马阴道低倍像（HE）

Fig.17-79 low magnification of horse vagina（HE）

外膜—adventitia 肌层—muscular layer 黏膜—mucosa 上皮—epithelium 小动脉—arteriole

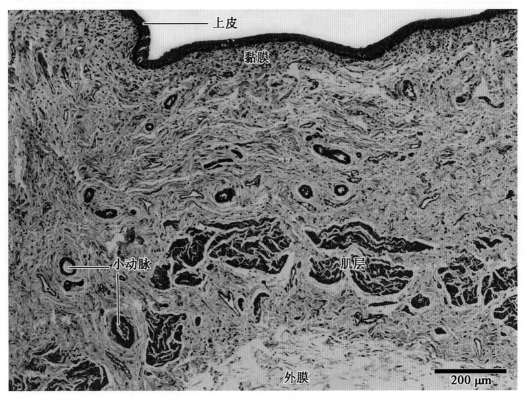

图 17-80 马阴道中倍像（HE）

Fig.17-80 mid magnification of horse vagina（HE）

小动脉—arteriole 上皮—epithelium 黏膜—mucosa 肌层—muscular layer 外膜—adventitia

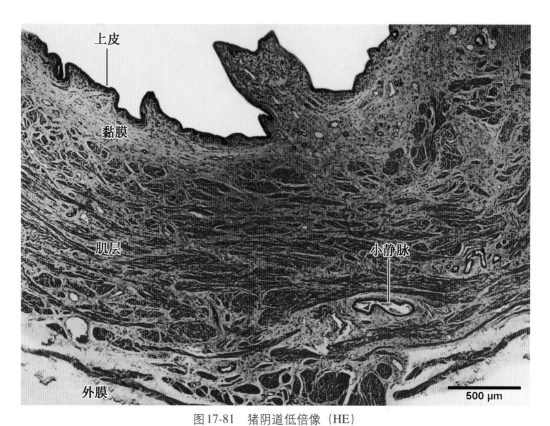

图17-81 猪阴道低倍像（HE）

Fig.17-81 low magnification of pig vagina（HE）

上皮—epithelium 黏膜—mucosa 肌层—muscular layer 外膜—adventitia 小静脉—venule

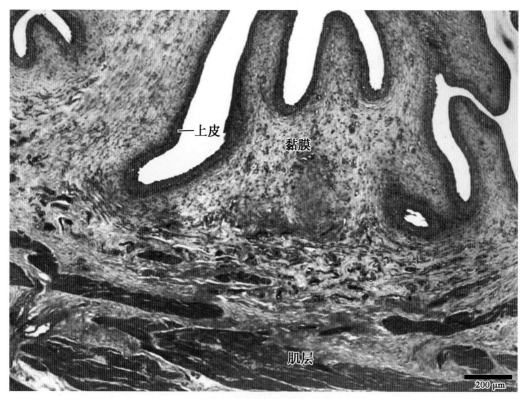

图17-82 猪阴道中倍像（HE）

Fig.17-82 mid magnification of pig vagina（HE）

上皮—epithelium 黏膜—mucosa 肌层—muscular layer

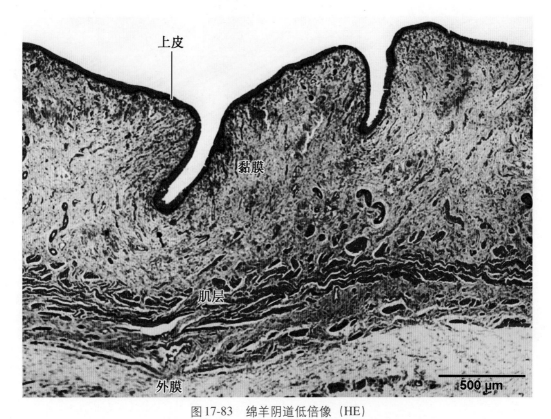

图17-83 绵羊阴道低倍像（HE）

Fig.17-83 low magnification of sheep vagina （HE）

上皮—epithelium 黏膜—mucosa 肌层—muscular layer 外膜—adventitia

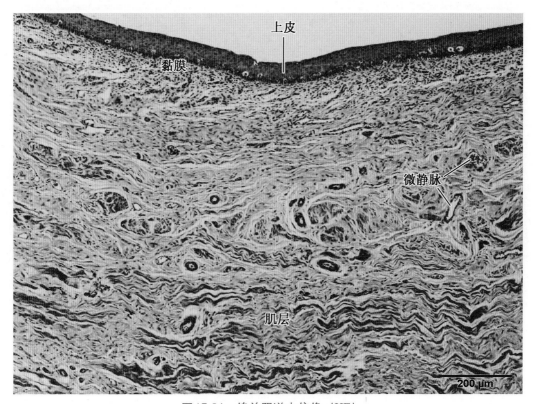

图17-84 绵羊阴道中倍像（HE）

Fig.17-84 mid magnification of sheep vagina （HE）

上皮—epithelium 黏膜—mucosa 肌层—muscular layer 微静脉—venule

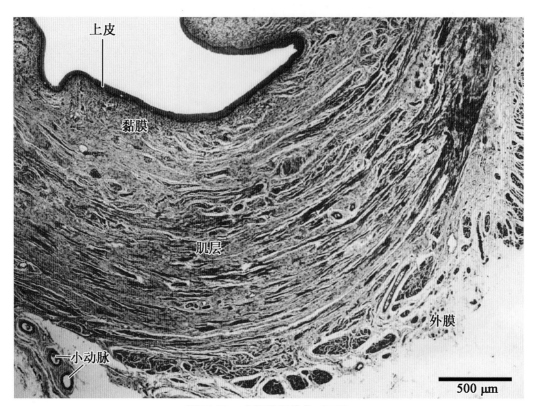

图17-85 犬阴道低倍像（HE）

Fig.17-85 low magnification of dog vagina（HE）

上皮—epithelium 黏膜—mucosa 肌层—muscular layer 外膜—adventitia 小动脉—arteriole

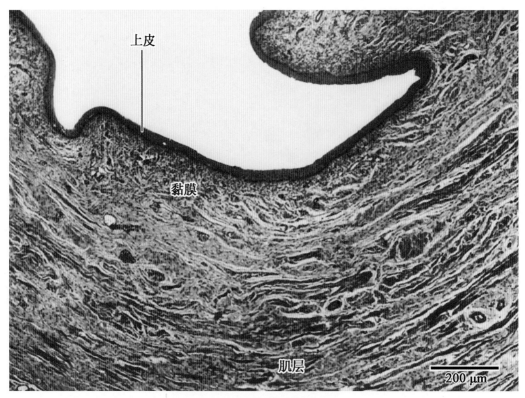

图17-86 犬阴道中倍像（HE）

Fig.17-86 mid magnification of dog vagina（HE）

上皮—epithelium 黏膜—mucosa 肌层—muscular layer

第十七章 雌性生殖系统 Female Reproductive System

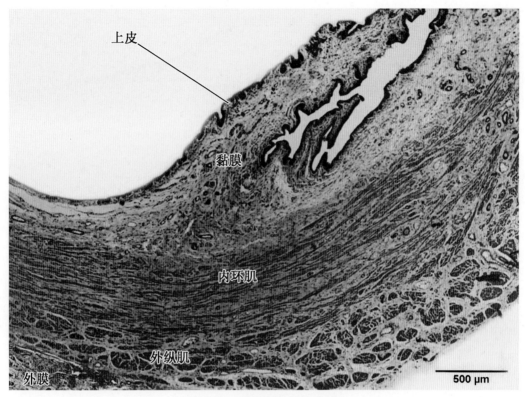

图17-87 猫阴道低倍像（HE）

Fig.17-87 low magnification of cat vagina（HE）

上皮—epithelium 黏膜—mucosa 内环肌—inner ring muscle 外纵肌—outer longitudinal muscle 外膜—adventitia

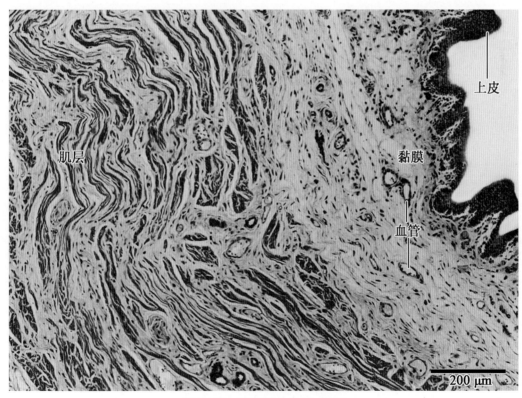

图17-88 猫阴道中倍像（HE）

Fig.17-88 mid magnification of cat vagina（HE）

肌层—muscular layer 黏膜—mucosa 上皮—epithelium 血管—blood vessel

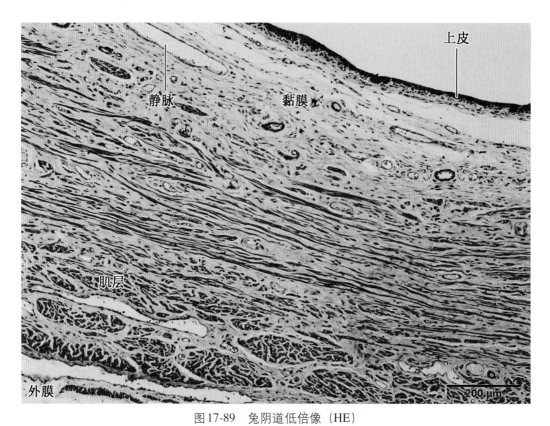

图17-89　兔阴道低倍像（HE）

Fig.17-89　low magnification of rabbit vagina（HE）

静脉—vein　上皮—epithelium　黏膜—mucosa　肌层—muscular layer　外膜—adventitia

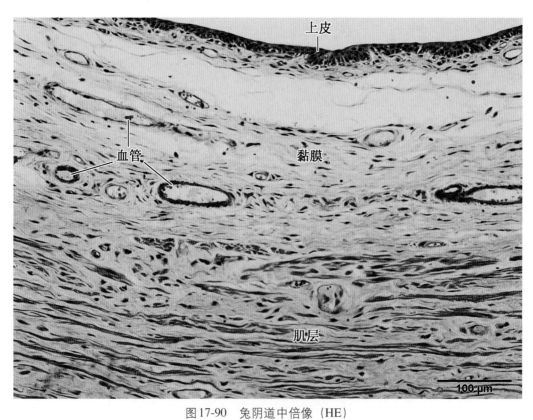

图17-90　兔阴道中倍像（HE）

Fig.17-90　mid magnification of rabbit vagina（HE）

上皮—epithelium　血管—blood vessel　黏膜—mucosa　肌层—muscular layer

第十八章 雄性生殖系统
Male Reproductive System

Outline

Male reproductive system consists of testes, genital ducts, accessory sex glands and external genital organs. Testes or male gonads are paired organs that lie in the scrotal sac, have a dual function for production of male gametes, spermatozoa, and secretion of male sex hormones, principally testosterone. Testes are composed of parenchyma and interstitium. They are wrapped by a thick capsule of dense connective tissue, namely the tunica albuginea, from which the connective tissue forms the mediastinum testis and septa. Septa separate the parenchyma into many testicular lobules, which is occupied by 1-4 long, thin and coiled seminiferous tubules. Near the mediastinum, coiled tubules transform into the straight seminiferous tubules and enter the mediastinum testis and form the rete testis. The parenchyma of the testis consists of the convoluted, straight seminiferous tubules and rete testis. The stroma includes the connective tissue filling between the parenchyma. Seminiferous tubules are the site of spermatogenesis and composed of the spermatogenous epithelium, which comprises the spermatogonia, primary and second spermatocytes, spermatids and spermatozoa. The processes of spermatozoa arising from spermatogonia are referred to as spermatogenesis. Interstitial cells of the testis distribute in the stroma and produce androgen.

Epididymis attaches on the one side of testis and can be separated into three parts: the head, body and tail, covered with tunica albuginea consisting of dense connective tissue. The parenchyma comprises the efferent ductules of testis and duct of epididymis. The latter is a long, convoluted tube connected with ductus deferens at the end. Secretions of the epithelium lining in the duct play a fundamental role for the development and maturation of spermatozoa. Ampullae and accessory sex glands i.e, seminal vesicles, prostate, bulbourethral glands, contribute to the formation of seminal plasma.

雄性生殖系统（male reproductive system）由睾丸、生殖管道、副性腺和外生殖器组成。睾丸产生精子和分泌雄性激素。生殖管道包括附睾和输精管，促进精子成熟以及营养、贮存和输送精子。副性腺包括精囊腺、前列腺和尿道球腺，分泌精清，与精子构成精液。

一、雄性生殖器官的发生

雄性和雌性的生殖系统发生与泌尿系统的发生关系密切，其主要器官均起源于中胚层。睾丸和卵巢由三种胚胎组织，即中胚层上皮、间充质和原始生殖细胞发育形成。

（一）睾丸的发生

胚胎细胞的Y染色体短臂上 SRY 基因指导未分化性腺向睾丸分化。睾丸决定因子诱导初级性索浓集并伸入未分化性腺的髓质内，形成生精小管索。靠近门部的初级性索相互吻合形成睾丸网，位于表面上皮与髓质之间的间充质分化为一层较厚的结缔组织白膜。此后，生精小管索分化为长襻状的生精小管。从胎儿期至出生后青春期之前，生精小管是没有管腔的细胞索，由来自原始生殖细胞的精原细胞和来自初级性索的支持细胞构成。靠近睾丸网的生精小管分化为直精小管。生精小管之间的间充质细胞分化为间质细胞，分泌雄激素睾酮和雄烯二酮，诱导中肾管和外生殖器向雄性分化。随后，胎盘产生的绒毛膜促性腺激素达到高峰，刺激间质细胞分泌雄激素。

（二）生殖管道的发生

雄雌两性胚胎早期都有两对生殖管道：一对称中肾管，发育为雄性生殖管道；另一对称中肾旁管，发育为雌性生殖管道。胚胎早期，生殖系统处于未分化阶段，两对生殖管道同时存在。若未分化性腺分化成睾丸，间质细胞分泌的睾酮开始促进中肾管发育，而支持细胞分泌的抗中肾旁管激素则抑制中肾旁管的发育，使其逐渐退化。中肾管的头段高度弯曲，形成附睾管，与中肾管演变而成的输出小管相连，后者又与睾丸网相连。中肾管的中段形成输精管，尾段形成射精管和精囊。中肾管最初开口于尿生殖窦，随着尿生殖窦发育为膀胱，射精管开口移向下方，最终开口于尿道。

（三）副性腺的发生

中肾管的尾端向侧面分支生长，形成精囊腺。位于尿道前列腺部的内胚层上皮细胞增生，向周围的间充质内生长，分化形成前列腺上皮，上皮周围的间充质分化成为前列腺的基质和平滑肌。尿道球腺由海绵体部的尿道上皮细胞增生而形成，其周围的间充质分化形成基质和平滑肌。

（四）外生殖器的发生

胚胎早期，雄雌两性的尿生殖膜上方的间充质增生，产生一个生殖结节。尿生殖膜的两侧各形成两条隆起，内侧的较小，为尿生殖褶；外侧的较大，为阴唇阴囊隆起。尿生殖褶之间的凹陷为尿道沟，沟底覆有尿生殖膜。胚胎早期，外生殖腺处于未分化状态，雄雌两性的外生殖器相似；中期以后，性别特征呈现明显区别。在二氢睾酮（DHT）的诱导下，外生殖器向雄性分化。生殖结节伸长形成阴茎，两侧的尿生殖褶沿阴茎的腹侧面，由后向前合并成管，形成尿道海绵体。后来龟头处形成一个皮肤反褶，称为包皮。左右阴唇阴囊隆起移向尾侧，并相互靠拢在中线处愈合成阴囊。

二、雄性生殖器官的组织学结构概述

（一）睾丸（testis）

睾丸表面被覆一层浆膜，其深面为致密结缔组织白膜。白膜沿纵轴伸向尾端形成睾丸纵隔，并呈放射状分出许多睾丸小隔，将睾丸实质分成许多睾丸小叶，每个小叶中有1～4条精小管。精小管分为生精小管和直精小管两段。生精小管以盲端起自小叶边缘，在小叶内盘曲折叠，管壁细胞分为生精细胞和支持细胞。生精细胞包括精原细胞、初级精母细胞、次级精母细胞、精子细胞及精子。它们依次由生精小管的基底部向管腔排列。生精小管上皮外有一薄层基膜，基膜外为一层肌样细胞。支持细胞呈高柱状或锥状。细胞底部附着在基膜上，顶部伸达腔面。在相邻支持细胞的侧面之间，镶嵌有各级生精细胞。在游离端，多个变态中的精子细胞以头部嵌附其上。生精小管末端变为短而直的直精小管。直精小管细，管壁无生精细胞，仅由单层立方或柱状的支持细胞组成。直精小管在睾丸纵隔中相互吻合形成睾丸网，最后汇合成6～12条较粗的睾丸输出管，从睾丸头端走出进入附睾头。分布在生精小管间的疏松结缔组织称为睾丸间质，间质中有一种特殊的内分泌细胞，称睾丸间质细胞；其分布于生精小管之间，胞体较大，呈圆形或不规则状，胞质强嗜酸性，其主要作用是分泌雄性激素即睾酮。

（二）附睾（epididymis）

附睾是贮存精子和精子成熟的地方，分附睾头、体和尾。附睾管长而弯曲，黏膜上皮为假复层柱状，由两类细胞组成。一类称纤毛细胞，数量多，高柱状，游离面有成簇的静纤毛，其主要功能是吞饮吸收破碎的精子，分泌甘油磷酸胆碱、唾液酸蛋白等。另一类细胞称基细胞，位于上皮细胞的基部，胞体小而呈椭圆形，胞质染色淡，可分裂增生补充纤毛细胞。

（三）输精管（seminiferous duct）

输精管是附睾管的延续，为肌性管道，管腔小而壁厚，管壁由黏膜、肌层和被膜构成。黏膜表面有很多纵行皱襞，黏膜上皮由假复层柱状过渡到单层柱状。肌层为发达的平滑肌，牛、马、猪的分为内环、中斜、外纵三层，羊的只有内环、外纵两层。

（四）副性腺（accessory glands）

1. 精囊腺（vesicular gland） 为分叶状的分支管状腺或复管泡状腺。腺上皮为假复层柱状上皮，由高柱状细胞及小而圆的基底细胞构成。叶内导管和排泄管衬以单层立方上皮。肉食动物无精囊腺，猪的精囊腺发达，马属动物的呈囊状。精囊腺的分泌物呈弱碱性，含丰富果糖，有营养精子和稀释精液的作用。

2. 前列腺（prostate gland） 为复管泡状腺（反刍类）或复管状腺。腺体外包以较厚的结缔组织被膜，被膜伸入实质形成小梁。被膜及小梁内均有平滑肌纤维。前列腺分腺体部和扩散部。马、犬、猫的腺体部大，扩散部小；牛、猪的则相反；羊无腺体部。腺上皮呈单层扁平、立方、柱状或假复层柱状，与腺体分泌状态有关。导管部上皮为单层柱状或扁平，最大导管的开口部为变移上皮，开口于尿生殖道内。

3. 尿道球腺（bulbourethral gland） 为复管状腺（猪、猫）或复管泡状腺（马、牛、羊），外被结缔组织被膜，并伸入实质分出若干小叶。腺泡衬以单层柱状上皮，腺内导管衬以假复层柱状上皮，腺导管则衬以变移上皮。

（五）阴茎（penis）

阴茎表面被覆无毛皮肤，内部的背侧有两条阴茎海绵体，腹侧有一条尿道海绵体。每条海绵体周围包有致密结缔组织白膜。海绵体是勃起组织，内含大量血窦和小梁。平时血窦中血液很少，海绵体呈柔软状态。阴茎勃起时，小梁中的动脉平滑肌舒张，大量血液涌入血窦内，血窦充血膨胀，白膜下的静脉受压，血液回流受阻海绵体变硬。

三、雄性生殖系统图谱

1. **睾丸** 图 18-1 ～图 18-28。
2. **精子** 图 18-29 ～图 18-31。
3. **附睾** 图 18-32 ～图 18-48。
4. **输精管** 图 18-49 ～图 18-60。
5. **副性腺** 图 18-61 ～图 18-84。
6. **阴茎** 图 18-85 ～图 18-86。

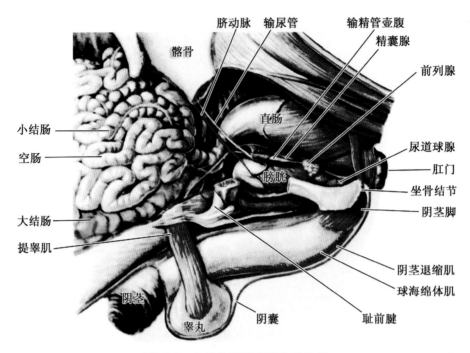

图 18-1　公马生殖器官的位置关系

Fig.18-1　the positional relationship of male horse reproductive organs

髂骨—ilium　脐动脉—umbilical artery　输尿管—ureter　直肠—rectum　膀胱—bladder
输精管壶腹—ampulla ductus deferentis　精囊腺—seminal vesicle　前列腺—prostate
尿道球腺—bulbourethral gland　肛门—anus　坐骨结节—ischial tuberosity　阴茎脚—penis root
阴茎退缩肌—penile retract　球海绵体肌—bulbocavernosus muscle　耻前腱—prepubic tendon
阴囊—scrotum　睾丸—testis　阴茎—penis　提睾肌—cremaster　大结肠—big colon　空肠—jejunum　小结肠—microcolon

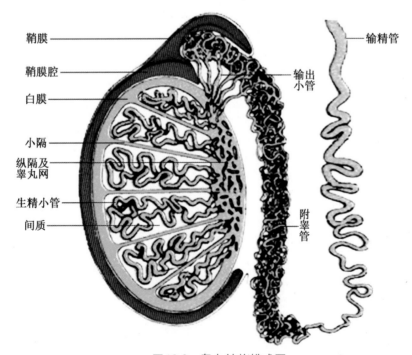

图 18-2　睾丸结构模式图

Fig.18-2　ideograph of testis structure

鞘膜—tunica vaginalis　鞘膜腔—vagina tunic cavity　白膜—albuginea　小隔—septulum
纵隔及睾丸网—mediastinum and rete testis　生精小管—seminiferous tubule　间质—mesenchyme
输出小管—efferent ductules　附睾管—ductus epididymidis　输精管—spermaduct

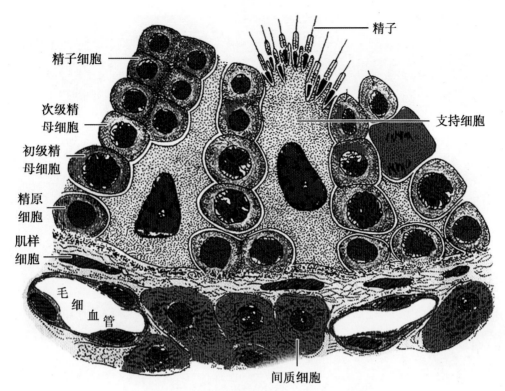

图18-3 生精小管上皮和睾丸间质结构模式图

Fig.18-3 ideograph of testis structure

毛细血管—capillary 肌样细胞—myoid cell 精原细胞—spermatogonium 初级精母细胞—primary spermatocyte 次级精母细胞—secondary spermatocyte 精子细胞—spermatoblast 精子—sperm 支持细胞—supporting cell 间质细胞—interstitial cell

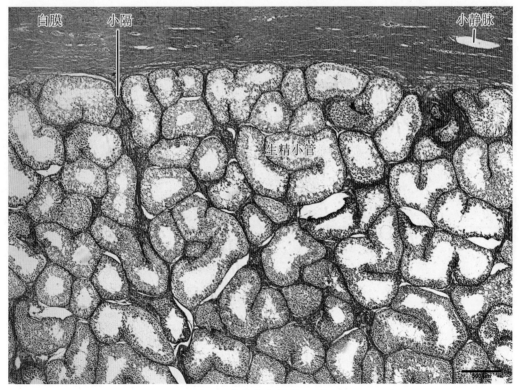

图18-4 牛睾丸低倍像1（HE）

Fig.18-4 low magnification of cattle testis structure 1（HE）

白膜—albuginea 小隔—septulum 小静脉—veinlet 生精小管—seminiferous tubule

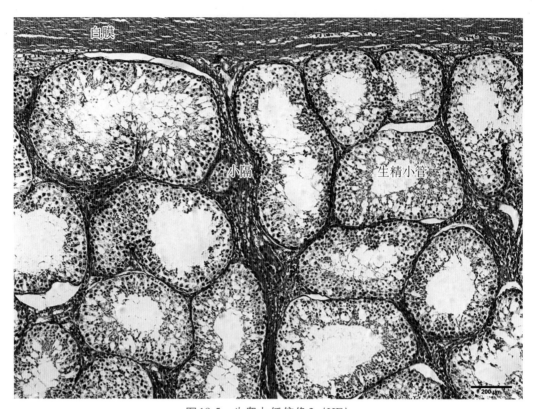

图18-5 牛睾丸低倍像2（HE）

Fig.18-5 low magnification of cattle testis structure 2 （HE）

白膜—albuginea 小隔—septulum 生精小管—seminiferous tubule

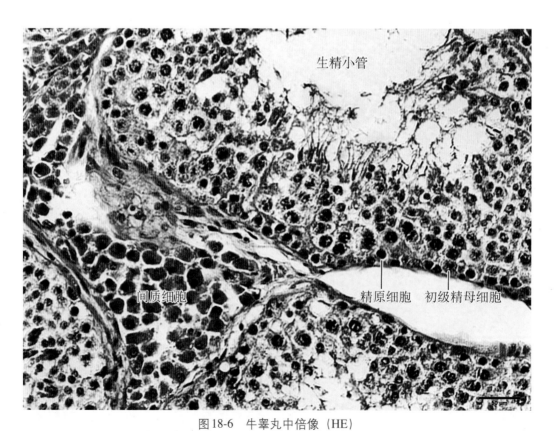

图18-6 牛睾丸中倍像（HE）

Fig.18-6 mid magnification of cattle testis structure （HE）

生精小管—seminiferous tubule 间质细胞—interstitial cell 精原细胞—spermatogonium

初级精母细胞—primary spermatocyte

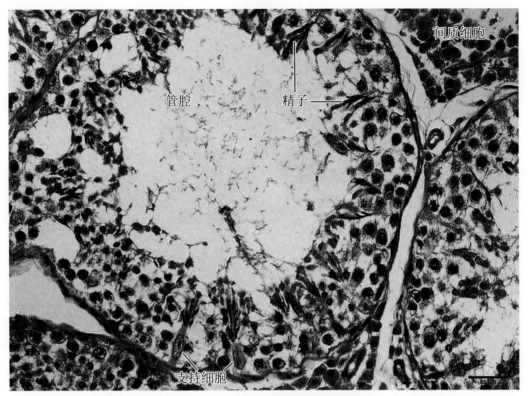

图18-7 牛生精小管高倍像1（HE）

Fig.18-7 high magnification of cattle seminiferous tubule 1 （HE）

管腔—lumen 精子—sperm 间质细胞—interstitial cell 支持细胞—supporting cell

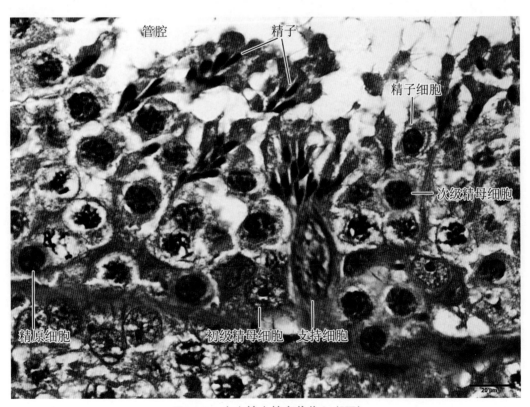

图18-8 牛生精小管高倍像2（HE）

Fig.18-8 high magnification of cattle seminiferous tubule 2 （HE）

管腔—lumen 精子—sperm 精子细胞—spermatoblast 精原细胞—spermatogonium 初级精母细胞—primary spermatocyte

次级精母细胞—secondary spermatocyte 支持细胞—supporting cell

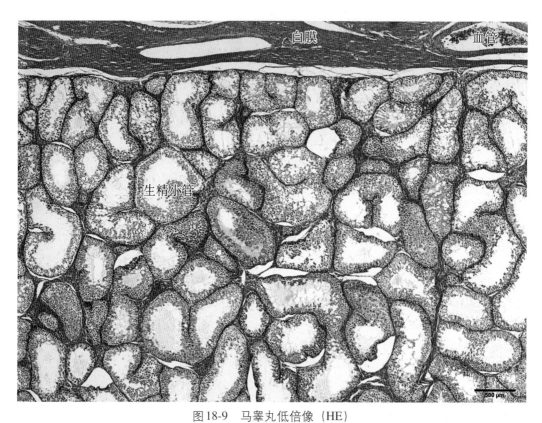

图18-9 马睾丸低倍像（HE）

Fig.18-9 low magnification of horse testis（HE）

白膜—albuginea 血管—blood vessel 生精小管—seminiferous tubule

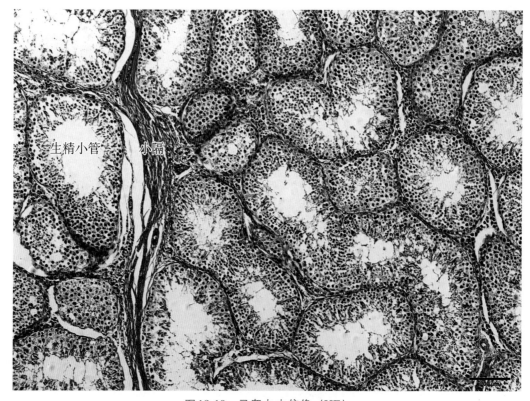

图18-10 马睾丸中倍像（HE）

Fig.18-10 mid magnification of horse testis（HE）

生精小管—seminiferous tubule 小隔—septulum

第十八章 雄性生殖系统 Male Reproductive System

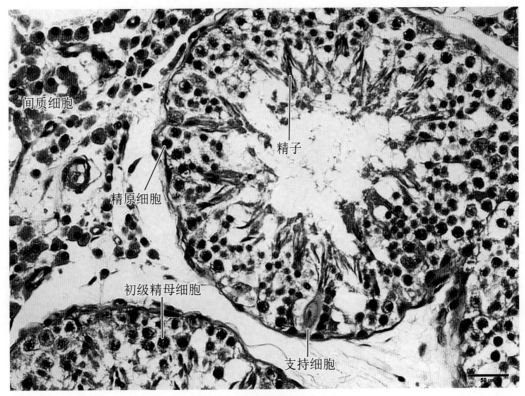

图 18-11 马睾丸高倍像（HE）

Fig.18-11 high magnification of horse testis（HE）

间质细胞—interstitial cell 精原细胞—spermatogonium 初级精母细胞—primary spermatocyte 精子—sperm

支持细胞—supporting cell

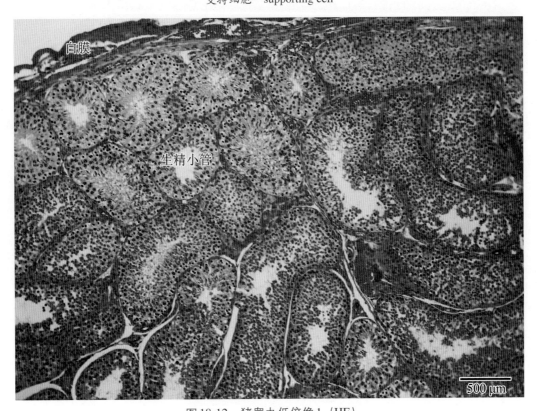

图 18-12 猪睾丸低倍像 1（HE）

Fig.18-12 low magnification of pig testis 1（HE）

白膜—albuginea 生精小管—seminiferous tubule

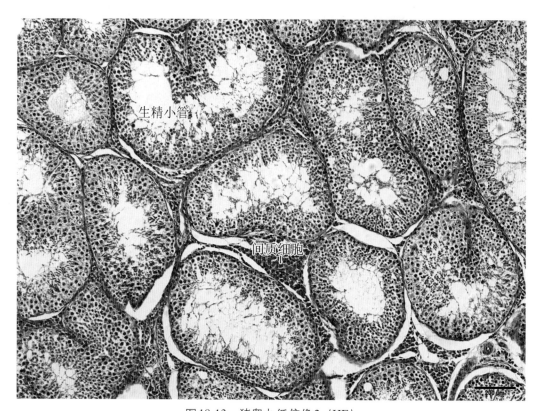

图 18-13　猪睾丸低倍像 2（HE）

Fig.18-13　low magnification of pig testis 2（HE）

生精小管—seminiferous tubule　间质细胞—interstitial cell

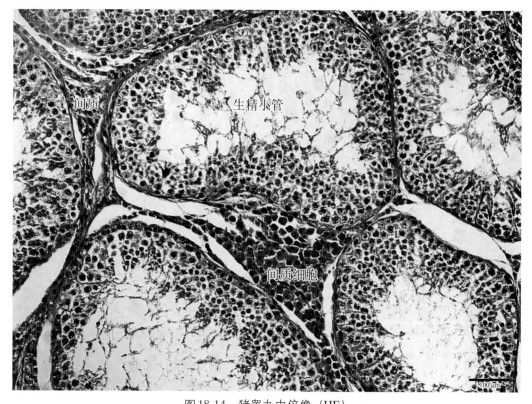

图 18-14　猪睾丸中倍像（HE）

Fig.18-14　mid magnification of pig testis（HE）

间质—mesenchyme　生精小管—seminiferous tubule　间质细胞—interstitial cell

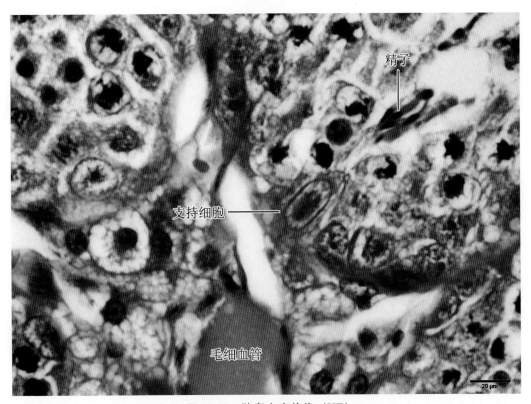

图 18-15 猪睾丸高倍像（HE）

Fig.18-15　high magnification of pig testis（HE）

精子—sperm　支持细胞—supporting cell　毛细血管—capillary

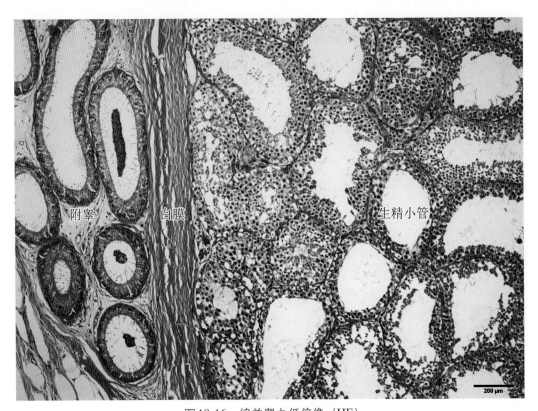

图 18-16　绵羊睾丸低倍像（HE）

Fig.18-16　low magnification of sheep testis（HE）

附睾—epididymis　白膜—albuginea　生精小管—seminiferous tubule

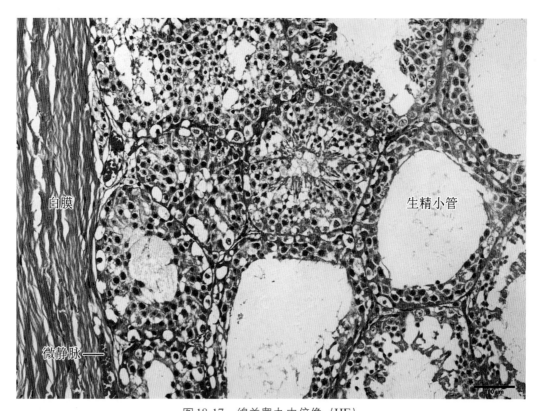

图 18-17　绵羊睾丸中倍像（HE）

Fig.18-17　mid magnification of sheep testis（HE）

白膜—albuginea　生精小管—seminiferous tubule　微静脉—venule

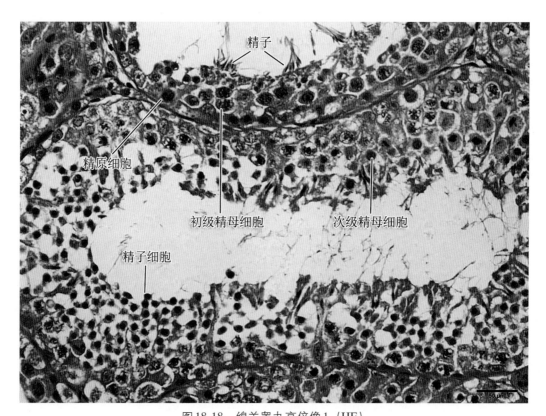

图 18-18　绵羊睾丸高倍像 1（HE）

Fig.18-18　high magnification of sheep testis 1（HE）

精原细胞—spermatogonium　初级精母细胞—primary spermatocyte

次级精母细胞—secondary spermatocyte　精子细胞—spermatoblast　精子—sperm

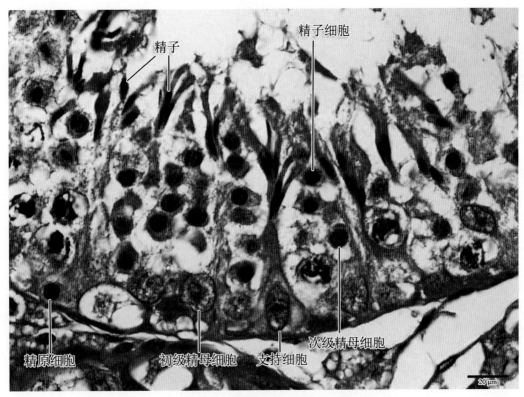

图 18-19　绵羊睾丸高倍像 2（HE）

Fig.18-19　high magnification of sheep testis 2（HE）

精子—sperm　精子细胞—spermatoblast　精原细胞—spermatogonium　初级精母细胞—primary spermatocyte

次级精母细胞—secondary spermatocyte　支持细胞—supporting cell

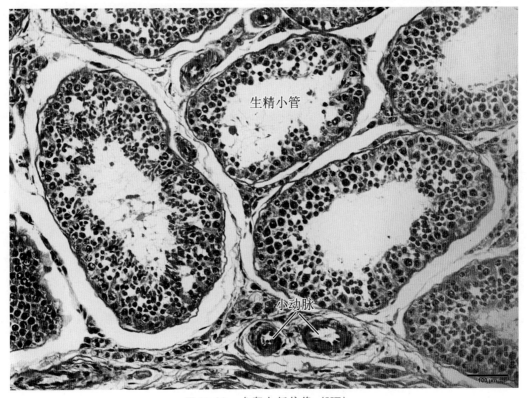

图 18-20　犬睾丸低倍像（HE）

Fig.18-20　low magnification of dog testis（HE）

生精小管—seminiferous tubule　小动脉—arteriole

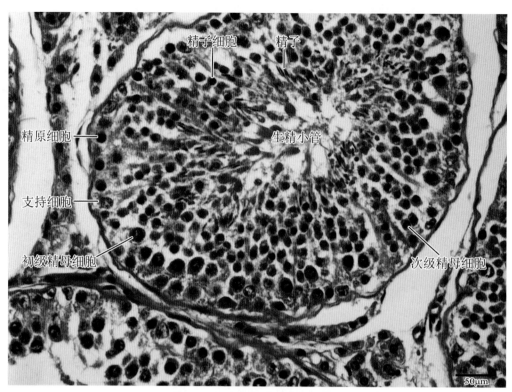

图18-21　犬睾丸高倍像1（HE）

Fig.18-21　high magnification of dog testis 1（HE）

生精小管—seminiferous tubule　精子细胞—spermatoblast　精子—sperm　精原细胞—spermatogonium
初级精母细胞—primary spermatocyte　次级精母细胞—secondary spermatocyte　支持细胞—supporting cell

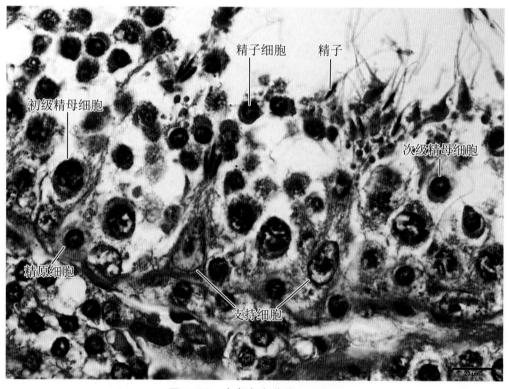

图18-22　犬睾丸高倍像2（HE）

Fig.18-22　high magnification of dog testis 2（HE）

精子细胞—spermatoblast　精子—sperm　精原细胞—spermatogonium　初级精母细胞—primary spermatocyte
次级精母细胞—secondary spermatocyte　支持细胞—supporting cell

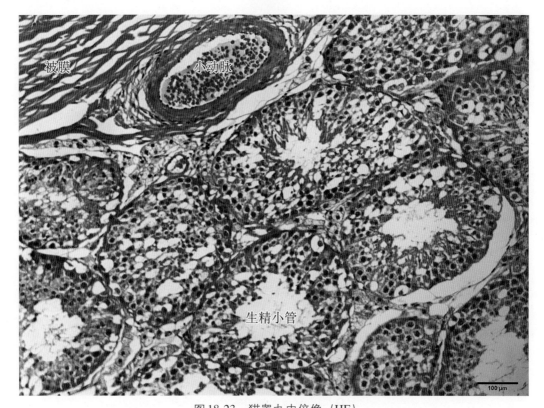

图18-23 猫睾丸中倍像（HE）

Fig.18-23　mid magnification of cat testis（HE）

被膜—capsule　小动脉—arteriole　生精小管—seminiferous tubule

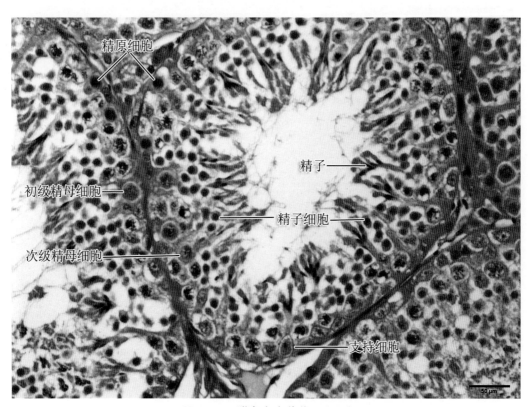

图18-24 猫睾丸高倍像1（HE）

Fig.18-24　high magnification of cat testis 1（HE）

精原细胞—spermatogonium　初级精母细胞—primary spermatocyte　次级精母细胞—secondary spermatocyte　精子—sperm

精子细胞—spermatoblast　支持细胞—supporting cell

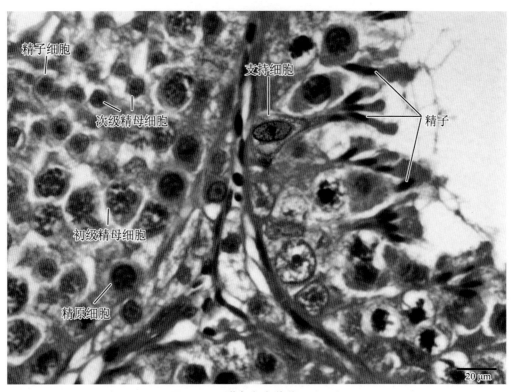

图 18-25 猫睾丸高倍像 2（HE）

Fig.18-25 high magnification of cat testis 2（HE）

精原细胞—spermatogonium 初级精母细胞—primary spermatocyte 次级精母细胞—secondary spermatocyte

精子细胞—spermatoblast 精子—sperm 支持细胞—supporting cell

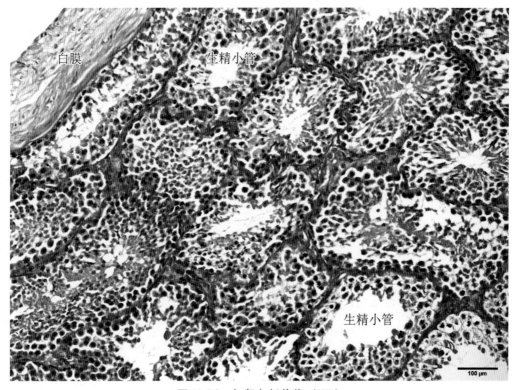

图 18-26 兔睾丸低倍像（HE）

Fig.18-26 low magnification of rabbit testis（HE）

白膜—albuginea 生精小管—seminiferous tubule

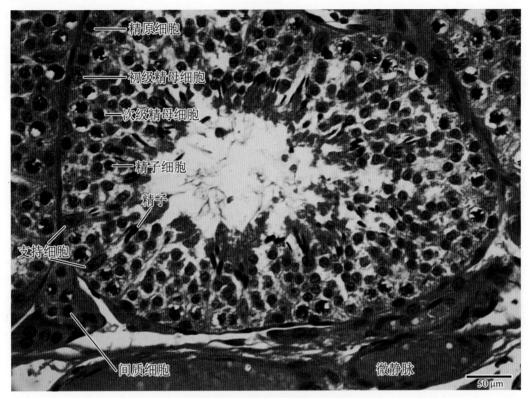

图18-27 兔睾丸高倍像1（HE）

Fig.18-27 high magnification of rabbit testis 1（HE）

精原细胞—spermatogonium 初级精母细胞—primary spermatocyte 次级精母细胞—secondary spermatocyte
精子细胞—spermatoblast 精子—sperm 支持细胞—supporting cell 间质细胞—interstitial cell 微静脉—venule

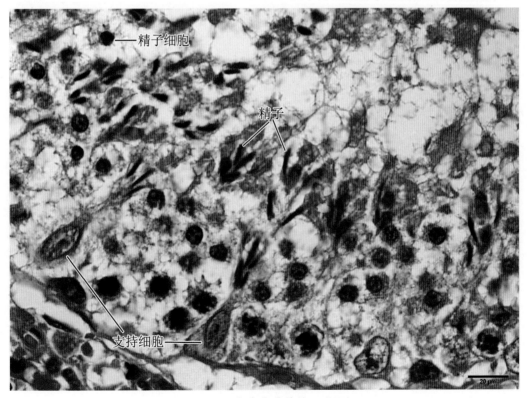

图18-28 兔睾丸高倍像2（HE）

Fig.18-28 high magnification of rabbit testis 2（HE）

精子细胞—spermatoblast 精子—sperm 支持细胞—supporting cell

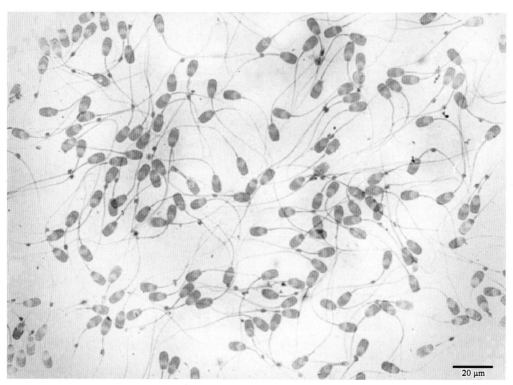

图18-29 猪精子高倍像1（涂片法，铁苏木精染色）
Fig.18-29 high magnification of pig sperm 1（smear, iron hematoxylin stain）

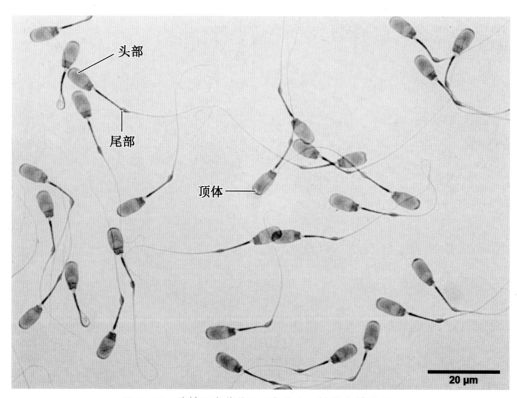

图18-30 猪精子高倍像2（涂片法，铁苏木精染色）
Fig.18-30 high magnification of pig sperm 2（smear, iron hematoxylin stain）
头部—head 尾部—tail 顶体—acrosome

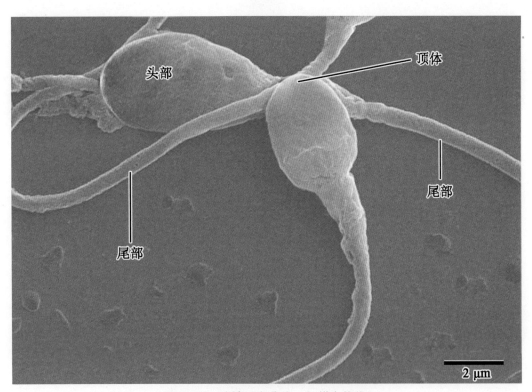

图18-31 猪精子高倍像（扫描电镜像）
Fig.18-31 high magnification of pig sperm（SEM）
头部—head 尾部—tail 顶体—acrosome

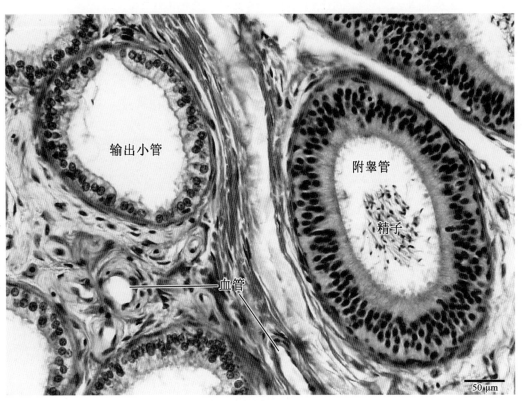

图18-32 牛附睾高倍像（HE）
Fig.18-32 high magnification of cattle epididymis（HE）
输出小管—efferent duct 附睾管—epididymal duct 精子—sperm 血管—blood vessel

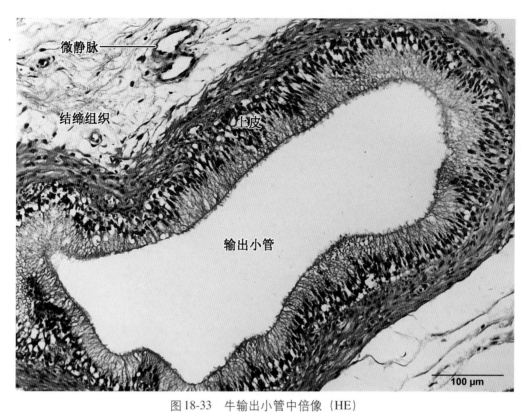

图18-33　牛输出小管中倍像（HE）

Fig.18-33　mid magnification of cattle efferent duct（HE）

微静脉—venule　结缔组织—connective tissue　输出小管—efferent duct　上皮—epithelium

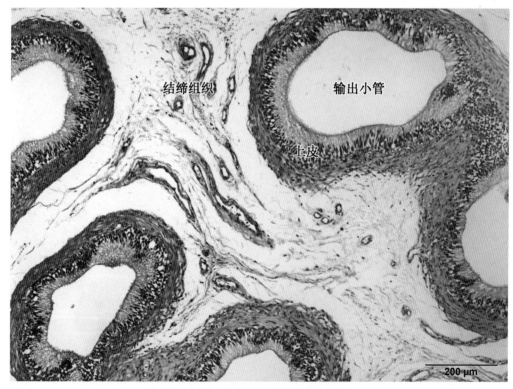

图18-34　马输出小管低倍像（HE）

Fig.18-34　low magnification of horse efferent duct（HE）

结缔组织—connective tissue　输出小管—efferent duct　上皮—epithelium

第十八章 雄性生殖系统　Male Reproductive System

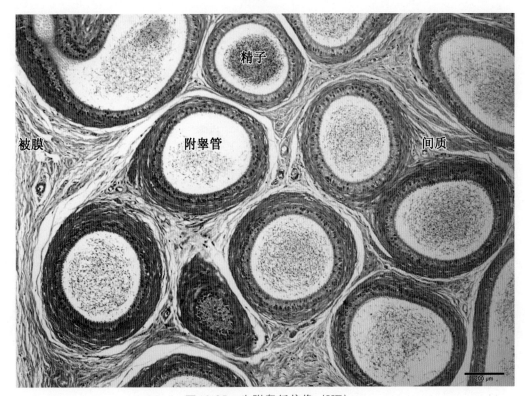

图 18-35　牛附睾低倍像（HE）

Fig.18-35　low magnification of cattle epididymis（HE）

被膜—capsule　附睾管—epididymal duct　精子—sperm　间质—mesenchyme

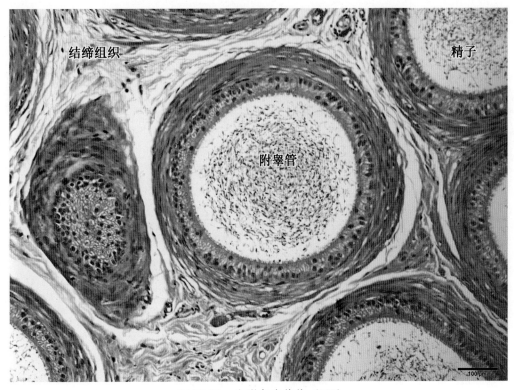

图 18-36　牛附睾中倍像（HE）

Fig.18-36　mid magnification of cattle epididymis（HE）

结缔组织—connective tissue　附睾管—epididymal duct　精子—sperm

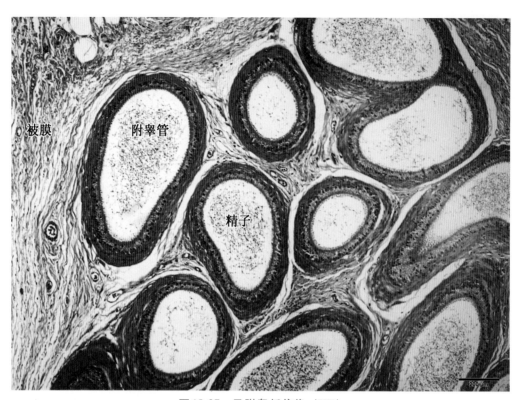

图18-37　马附睾低倍像（HE）

Fig.18-37　low magnification of horse epididymis（HE）

被膜—capsule　附睾管—epididymal duct　精子—sperm

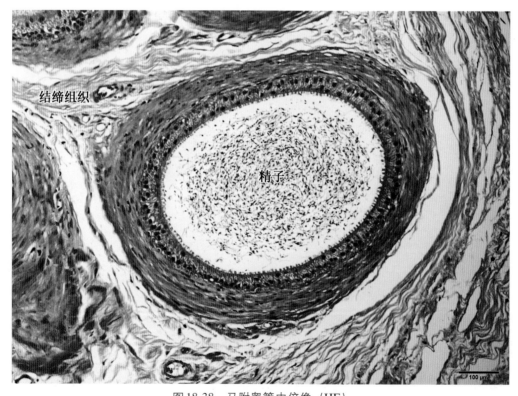

图18-38　马附睾管中倍像（HE）

Fig.18-38　mid magnification of horse epididymal duct（HE）

结缔组织—connective tissue　精子—sperm

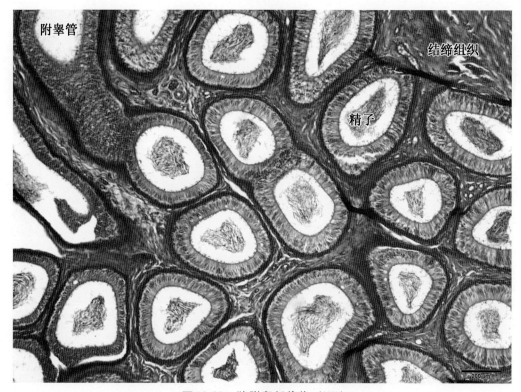

图18-39 猪附睾低倍像（HE）

Fig.18-39 low magnification of pig epididymis（HE）

附睾管—epididymal duct 精子—sperm 结缔组织—connective tissue

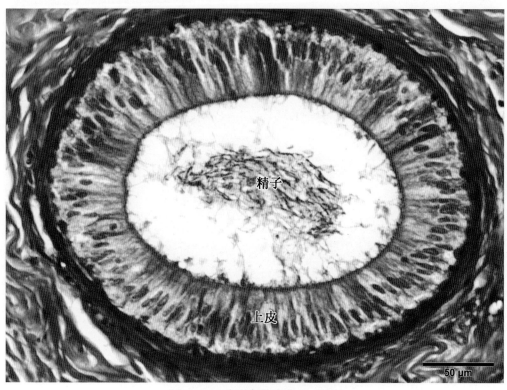

图18-40 猪附睾管高倍像（HE）

Fig.18-40 high magnification of pig epididymal duct（HE）

精子—sperm 上皮—epithelium

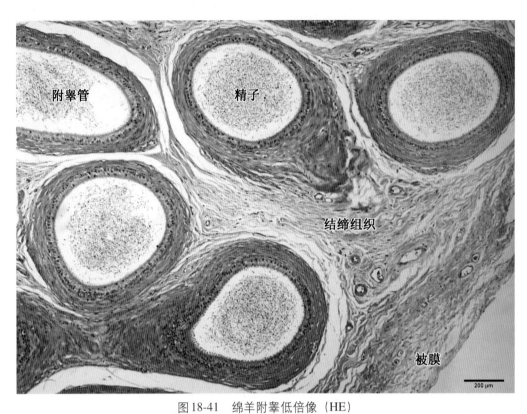

图 18-41　绵羊附睾低倍像（HE）

Fig.18-41　low magnification of sheep epididymis（HE）

附睾管—epididymal duct　精子—sperm　结缔组织—connective tissue　被膜—capsule

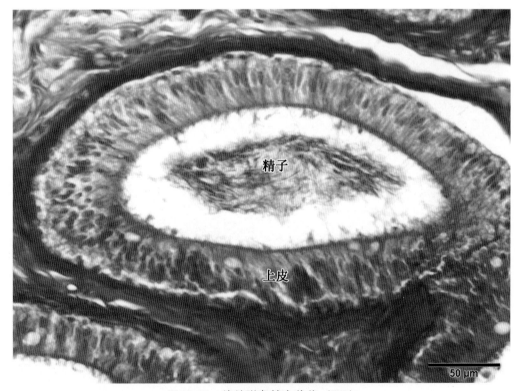

图 18-42　绵羊附睾管高倍像（HE）

Fig.18-42　high magnification of sheep epididymal duct（HE）

精子—sperm　上皮—epithelium

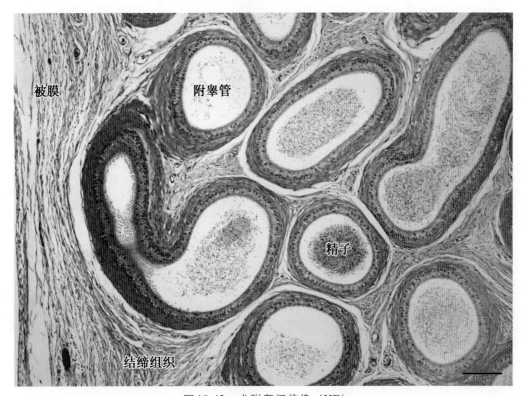

图18-43 犬附睾低倍像（HE）

Fig.18-43 low magnification of dog epididymis （HE）

被膜—capsule 附睾管—epididymal duct 精子—sperm 结缔组织—connective tissue

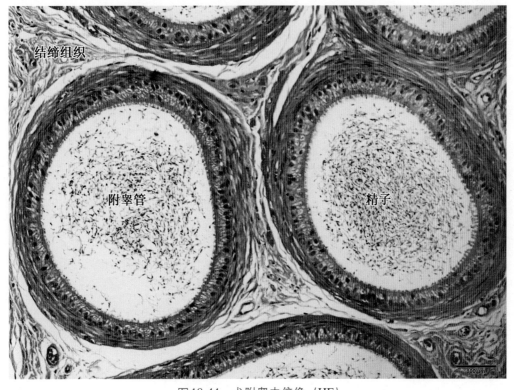

图18-44 犬附睾中倍像（HE）

Fig.18-44 mid magnification of dog epididymis （HE）

结缔组织—connective tissue 附睾管—epididymal duct 精子—sperm

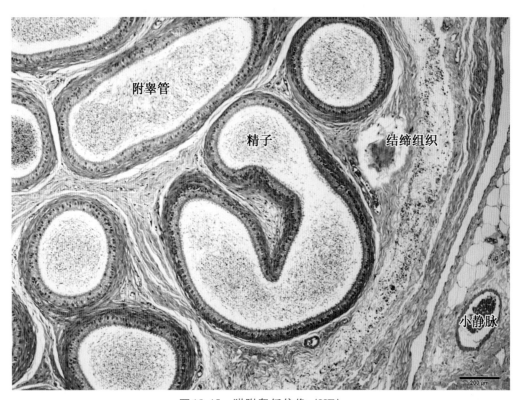

图 18-45　猫附睾低倍像（HE）

Fig.18-45　low magnification of cat epididymis（HE）

附睾管—epididymal duct　精子—sperm　结缔组织—connective tissue　小静脉—venule

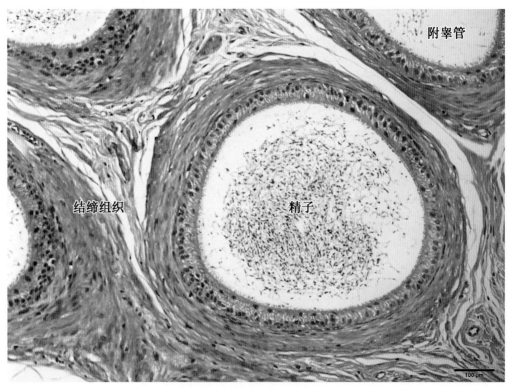

图 18-46　猫附睾中倍像（HE）

Fig.18-46　mid magnification of cat epididymis（HE）

附睾管—epididymal duct　结缔组织—connective tissue　精子—sperm

第十八章 雄性生殖系统 Male Reproductive System

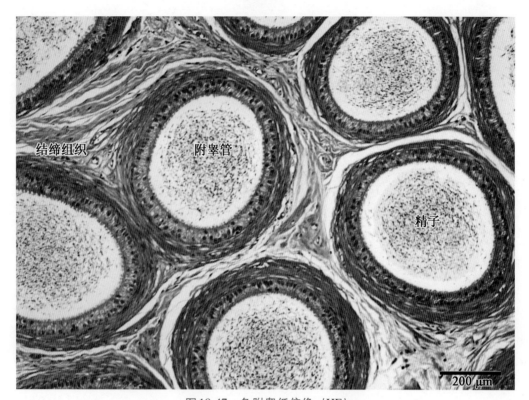

图18-47 兔附睾低倍像（HE）

Fig.18-47 low magnification of rabbit epididymis（HE）

结缔组织—connective tissue 附睾管—epididymal duct 精子—sperm

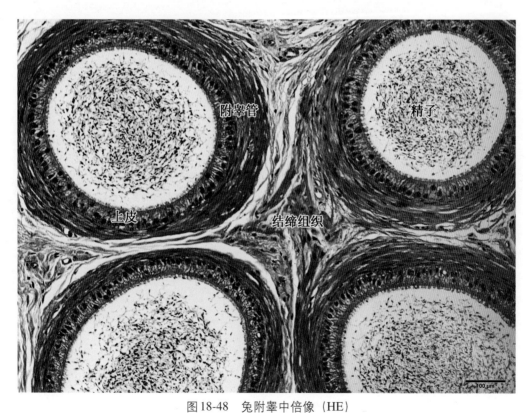

图18-48 兔附睾中倍像（HE）

Fig.18-48 mid magnification of rabbit epididymis（HE）

附睾管—epididymal duct 精子—sperm 上皮—epithelium 结缔组织—connective tissue

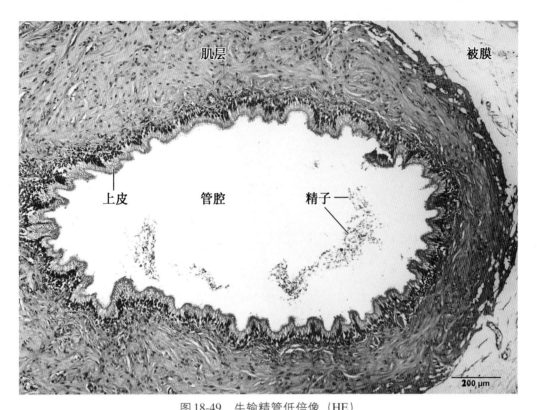

图 18-49　牛输精管低倍像（HE）

Fig.18-49　low magnification of cattle spermaduct（HE）

肌层—muscular layer　被膜—capsule　上皮—epithelium　管腔—lumen　精子—sperm

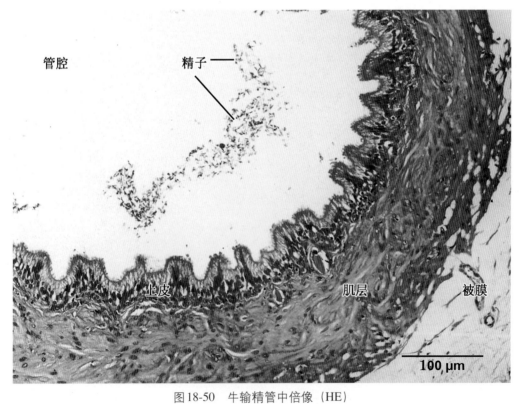

图 18-50　牛输精管中倍像（HE）

Fig.18-50　mid magnification of cattle spermaduct（HE）

管腔—lumen　精子—sperm　上皮—epithelium　肌层—muscular layer　被膜—capsule

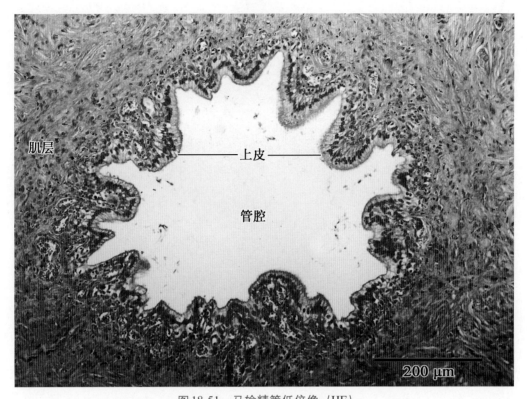

图 18-51　马输精管低倍像（HE）
Fig.18-51　low magnification of horse spermaduct（HE）
肌层—muscular layer　上皮—epithelium　管腔—lumen

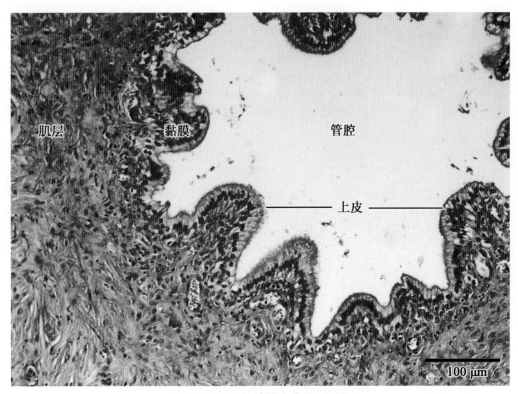

图 18-52　马输精管中倍像（HE）
Fig.18-52　mid magnification of horse spermaduct（HE）
肌层—muscular layer　黏膜—mucosa　管腔—lumen　上皮—epithelium

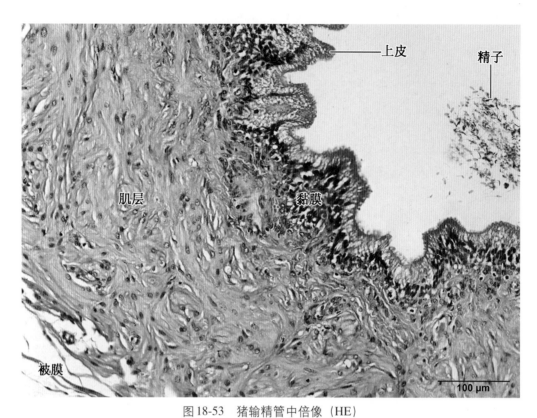

图 18-53 猪输精管中倍像（HE）

Fig.18-53 mid magnification of pig spermaduct（HE）

被膜—capsule 肌层—muscular layer 黏膜—mucosa 上皮—epithelium 精子—sperm

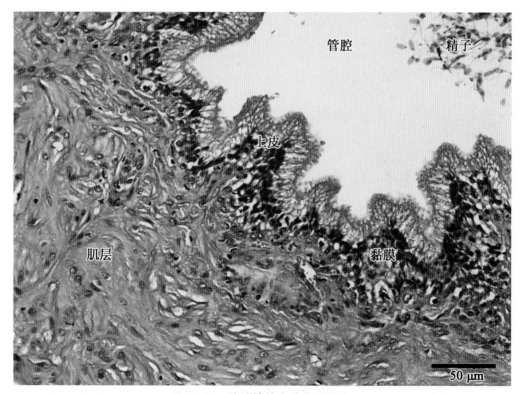

图 18-54 猪输精管高倍像（HE）

Fig.18-54 high magnification of pig spermaduct（HE）

肌层—muscular layer 黏膜—mucosa 上皮—epithelium 管腔—lumen 精子—sperm

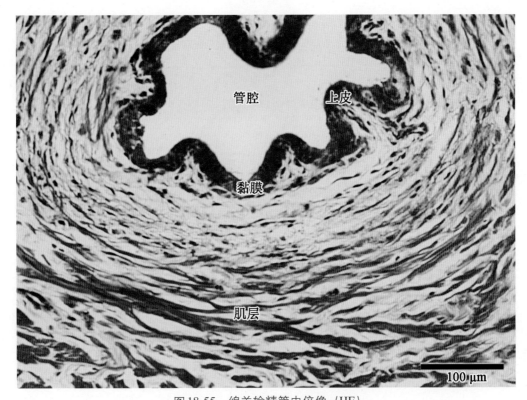

图18-55 绵羊输精管中倍像（HE）

Fig.18-55 mid magnification of sheep spermaduct（HE）

管腔—lumen 黏膜—mucosa 肌层—muscular layer 上皮—epithelium

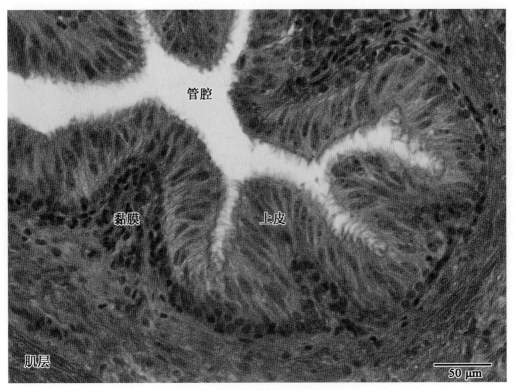

图18-56 绵羊输精管高倍像（HE）

Fig.18-56 high magnification of sheep spermaduct（HE）

管腔—lumen 黏膜—mucosa 肌层—muscular layer 上皮—epithelium

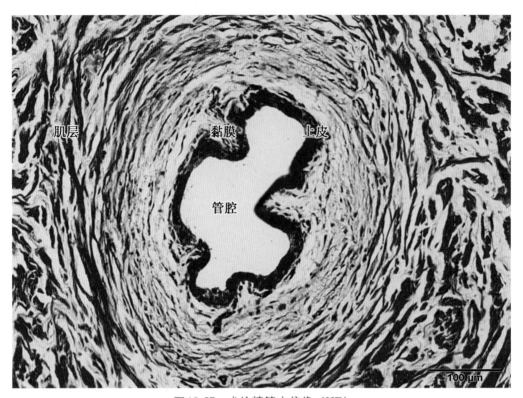

图 18-57　犬输精管中倍像（HE）

Fig.18-57　mid magnification of dog spermaduct（HE）

肌层—muscular layer　黏膜—mucosa　上皮—epithelium　管腔—lumen

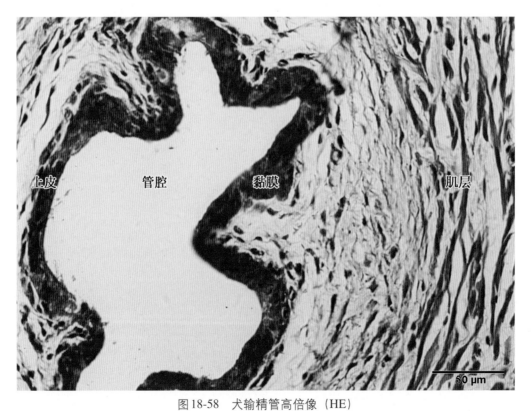

图 18-58　犬输精管高倍像（HE）

Fig.18-58　high magnification of dog spermaduct（HE）

上皮—epithelium　管腔—lumen　黏膜—mucosa　肌层—muscular layer

第十八章 雄性生殖系统　Male Reproductive System

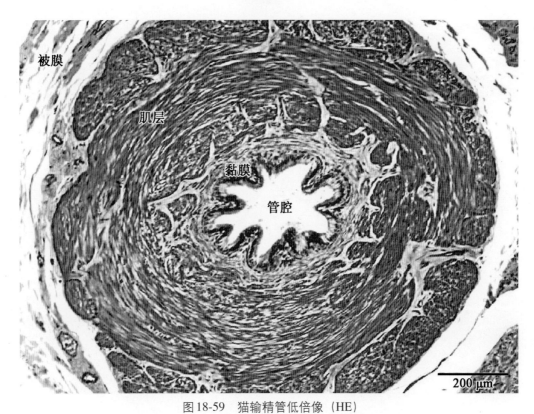

图 18-59　猫输精管低倍像（HE）

Fig.18-59　low magnification of cat spermaduct（HE）

被膜—capsule　肌层—muscular layer　黏膜—mucosa　管腔—lumen

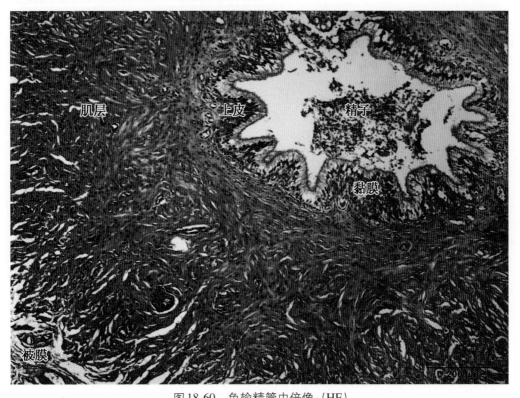

图 18-60　兔输精管中倍像（HE）

Fig.18-60　mid magnification of rabbit spermaduct（HE）

被膜—capsule　肌层—muscular layer　黏膜—mucosa　上皮—epithelium　精子—sperm

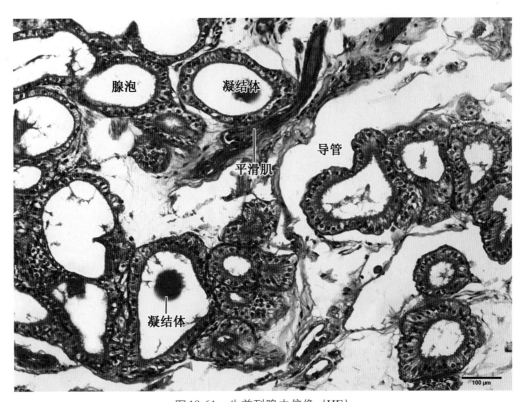

图 18-61　牛前列腺中倍像（HE）

Fig.18-61　mid magnification of cattle prostate gland（HE）

腺泡—acinus　凝结体—concretion　平滑肌—smooth muscle　导管—duct

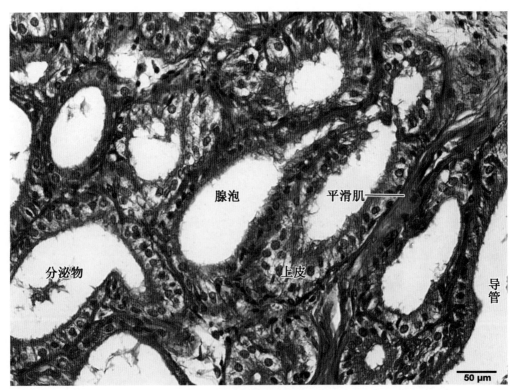

图 18-62　牛前列腺高倍像（HE）

Fig.18-62　high magnification of cattle prostate gland（HE）

腺泡—acinus　平滑肌—smooth muscle　分泌物—secretion　上皮—epithelium　导管—duct

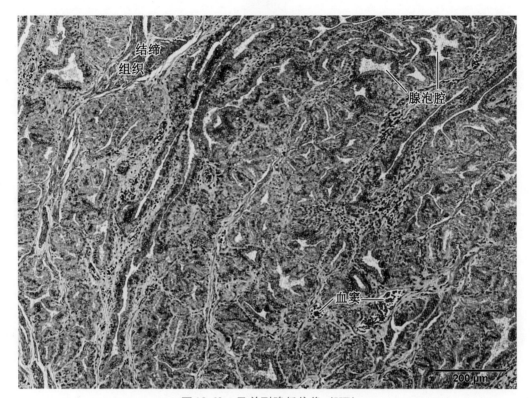

图18-63 马前列腺低倍像（HE）

Fig.18-63 low magnification of horse prostate gland（HE）

结缔组织—connective tissue 腺泡腔—acinus cavity 血窦—blood sinusoid

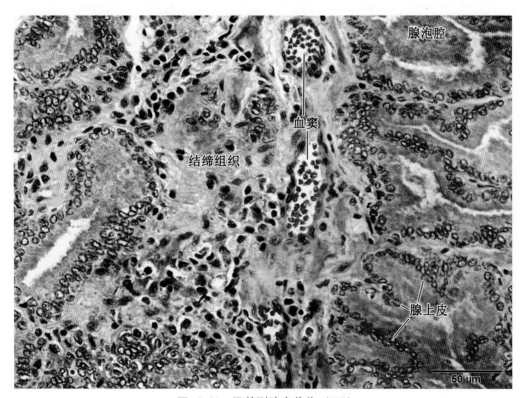

图18-64 马前列腺高倍像（HE）

Fig.18-64 high magnification of horse prostate gland（HE）

结缔组织—connective tissue 血窦—blood sinusoid 腺泡腔—acinus cavity 腺上皮—glandular epithelium

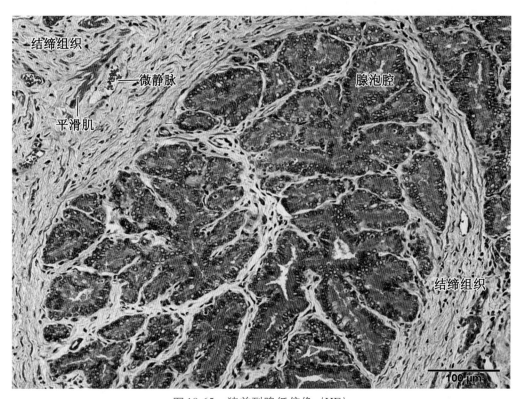

图 18-65 猪前列腺低倍像（HE）

Fig.18-65 low magnification of pig prostate gland（HE）

结缔组织—connective tissue 平滑肌—smooth muscle 微静脉—venule 腺泡腔—acinus cavity

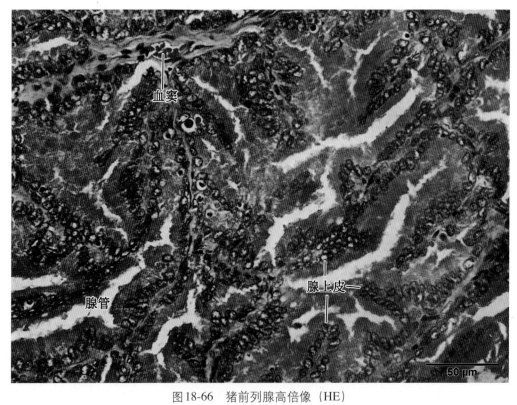

图 18-66 猪前列腺高倍像（HE）

Fig.18-66 high magnification of pig prostate gland（HE）

血窦—blood sinusoid 腺管—glandular duct 腺上皮—glandular epithelium

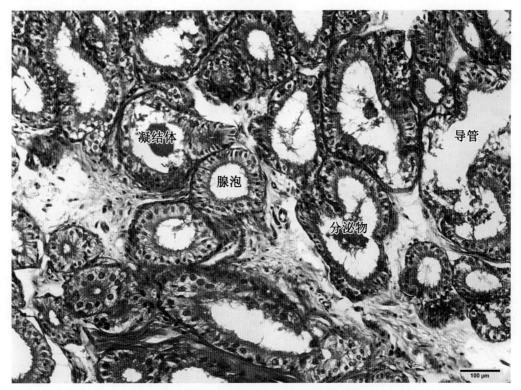

图 18-67 绵羊前列腺中倍像（HE）

Fig.18-67 mid magnification of sheep prostate gland（HE）

凝结体—concretion 腺泡—acinus 分泌物—secretion 导管—duct

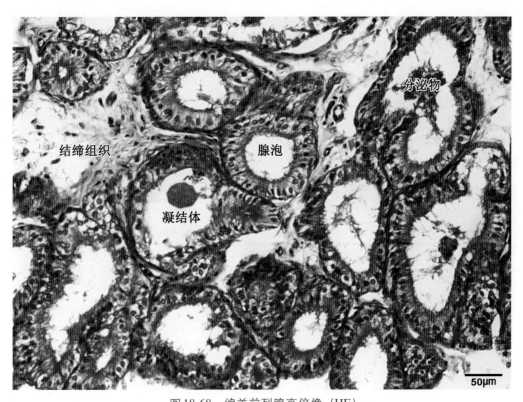

图 18-68 绵羊前列腺高倍像（HE）

Fig.18-68 high magnification of sheep prostate gland（HE）

结缔组织—connective tissue 凝结体—concretion 腺泡—acinus 分泌物—secretion

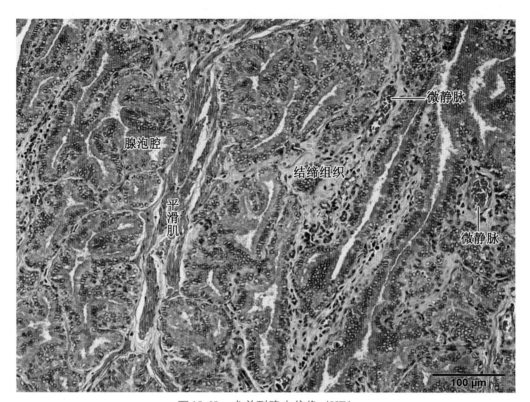

图 18-69 犬前列腺中倍像（HE）

Fig.18-69 mid magnification of dog prostate gland（HE）

腺泡腔—acinus cavity　平滑肌—smooth muscle　结缔组织—connective tissue　微静脉—venule

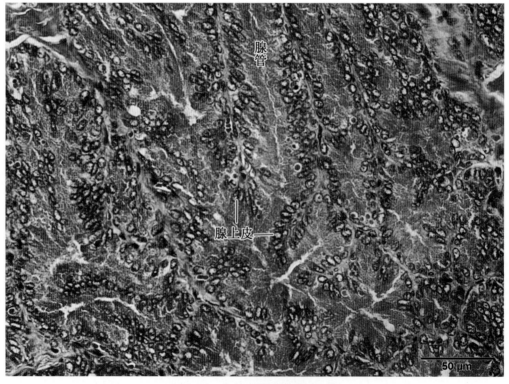

图 18-70 犬前列腺高倍像（HE）

Fig.18-70 high magnification of dog prostate gland（HE）

腺管—glandular duct　腺上皮—glandular epithelium

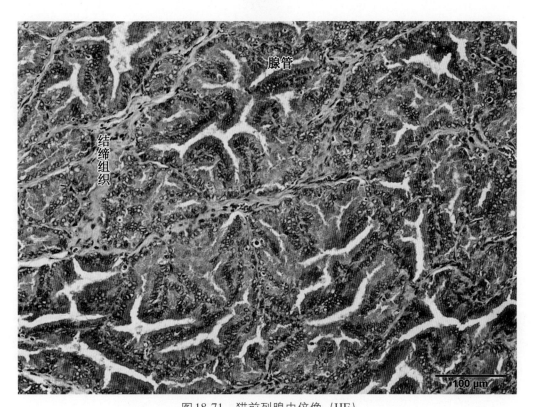

图18-71 猫前列腺中倍像（HE）

Fig.18-71 mid magnification of cat prostate gland（HE）

结缔组织—connective tissue 腺管—glandular duct

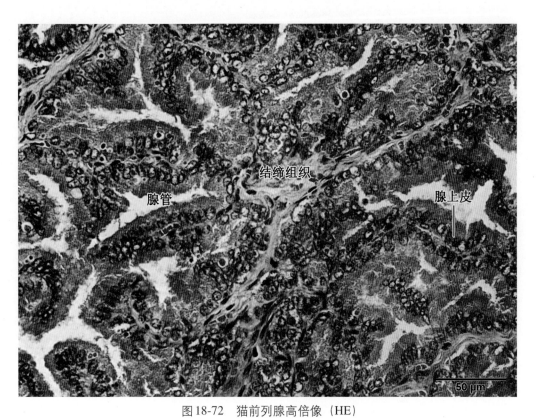

图18-72 猫前列腺高倍像（HE）

Fig.18-72 high magnification of cat prostate gland（HE）

腺管—glandular duct 结缔组织—connective tissue 腺上皮—glandular epithelium

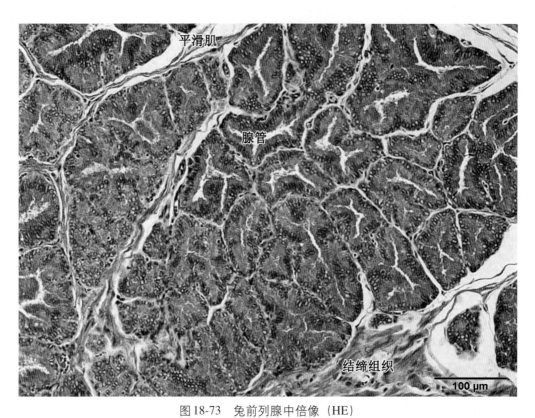

图 18-73　兔前列腺中倍像（HE）

Fig.18-73　mid magnification of rabbit prostate gland（HE）

平滑肌—smooth muscle　腺管—glandular duct　结缔组织—connective tissue

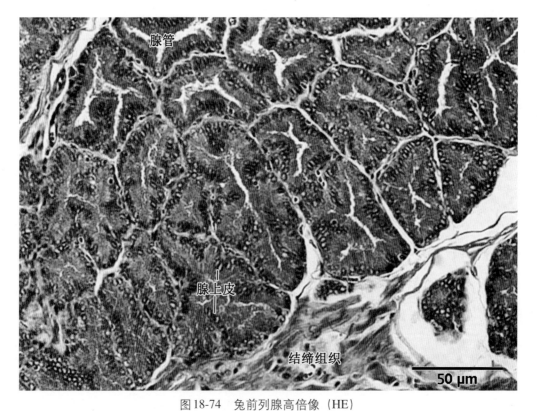

图 18-74　兔前列腺高倍像（HE）

Fig.18-74　high magnification of rabbit prostate gland（HE）

腺管—glandular duct　腺上皮—glandular epithelium　结缔组织—connective tissue

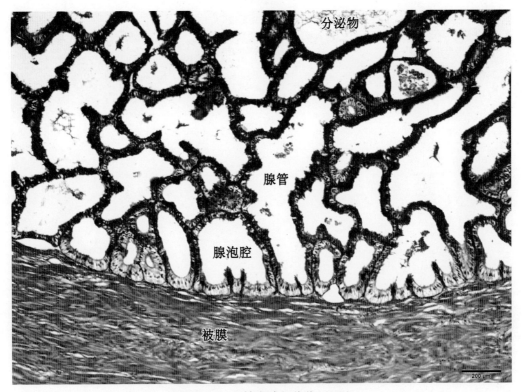

图18-75 牛精囊腺低倍像（HE）

Fig.18-75 low magnification of cattle vesicular gland（HE）

被膜—capsule 腺泡腔—acinus cavity 腺管—glandular duct 分泌物—secretion

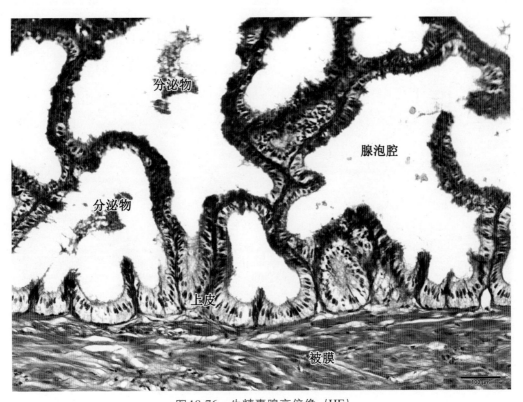

图18-76 牛精囊腺高倍像（HE）

Fig.18-76 high magnification of cattle vesicular gland（HE）

被膜—capsule 上皮—epithelium 分泌物—secretion 腺泡腔—acinus cavity

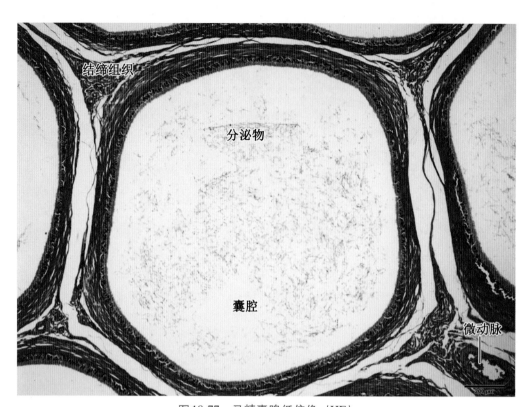

图 18-77 马精囊腺低倍像（HE）

Fig.18-77 low magnification of horse vesicular gland（HE）

结缔组织—connective tissue　分泌物—secretion　囊腔—lumen　微动脉—arteriole

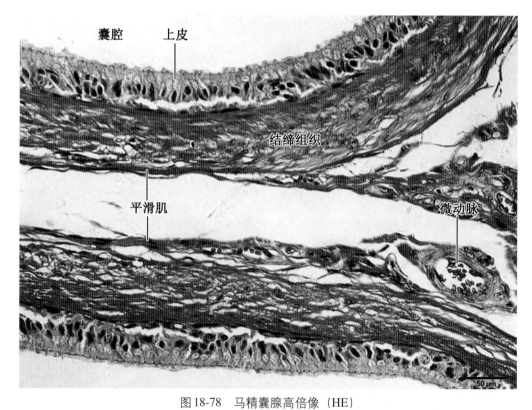

图 18-78 马精囊腺高倍像（HE）

Fig.18-78 high magnification of horse vesicular gland（HE）

囊腔—lumen　上皮—epithelium　结缔组织—connective tissue

平滑肌—smooth muscle　微动脉—arteriole

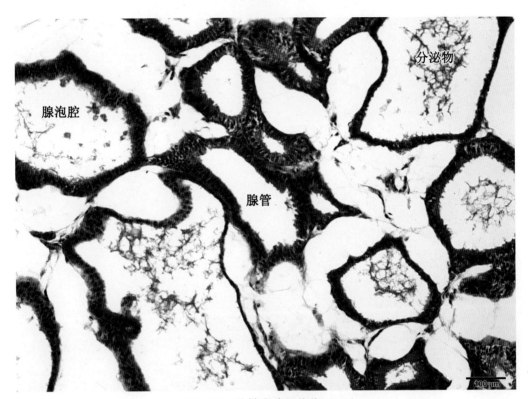

图 18-79　猪精囊腺低倍像（HE）

Fig.18-79　low magnification of pig vesicular gland（HE）

腺泡腔—acinus cavity　腺管—glandular duct　分泌物—secretion

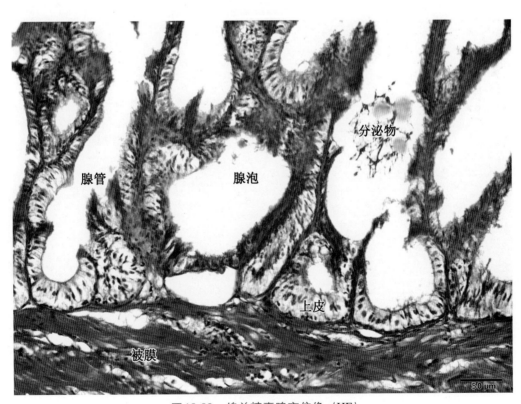

图 18-80　绵羊精囊腺高倍像（HE）

Fig.18-80　high magnification of sheep vesicular gland（HE）

腺管—glandular duct　腺泡—acinus　分泌物—secretion　上皮—epithelium　被膜—capsule

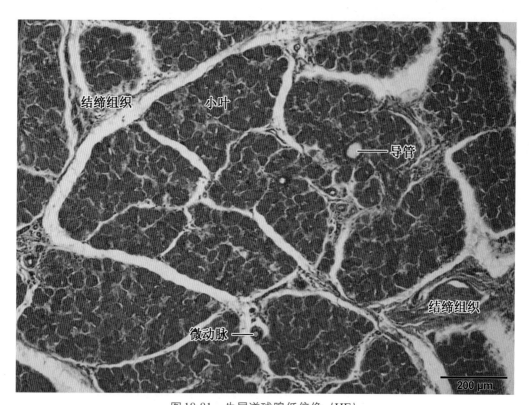

图 18-81　牛尿道球腺低倍像（HE）

Fig.18-81　low magnification of cattle bulbourethral gland（HE）

结缔组织—connective tissue　小叶—lobule　导管—duct　微动脉—arteriole

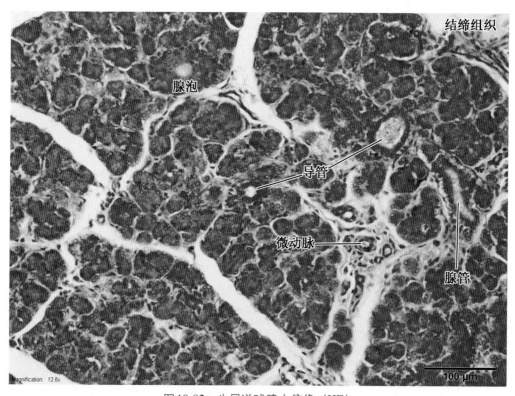

图 18-82　牛尿道球腺中倍像（HE）

Fig.18-82　mid magnification of cattle bulbourethral gland（HE）

结缔组织—connective tissue　腺泡—acinus　导管—duct　微动脉—arteriole　腺管—glandular duct

第十八章 雄性生殖系统　Male Reproductive System

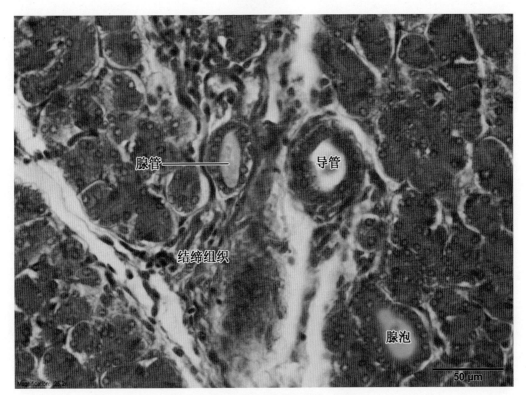

图18-83　马尿道球腺高倍像（HE）

Fig.18-83　high magnification of horse bulbourethral gland（HE）

腺管—glandular duct　导管—duct　结缔组织—connective tissue　腺泡—acinus

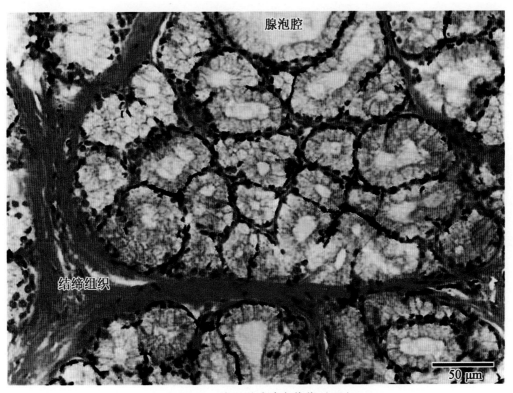

图18-84　猪尿道球腺高倍像（HE）

Fig.18-84　high magnification of pig bulbourethral gland（HE）

结缔组织—connective tissue　腺泡腔—acinus cavity

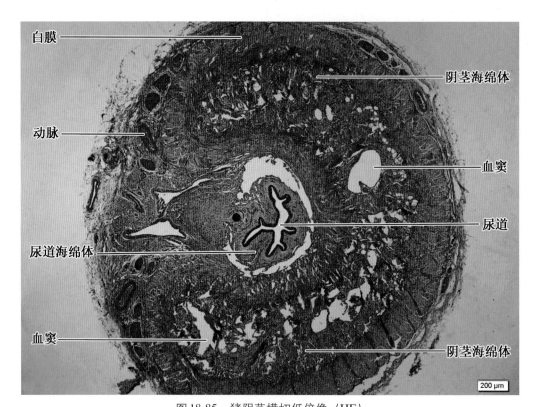

图18-85 猪阴茎横切低倍像（HE）

Fig.18-85 low magnification of pig penis cross section （HE）

白膜—albuginea 动脉—artery 尿道海绵体—corpus spongiosum urethra

血窦—blood sinusoid 阴茎海绵体—corpus cavernosum 尿道—urethra

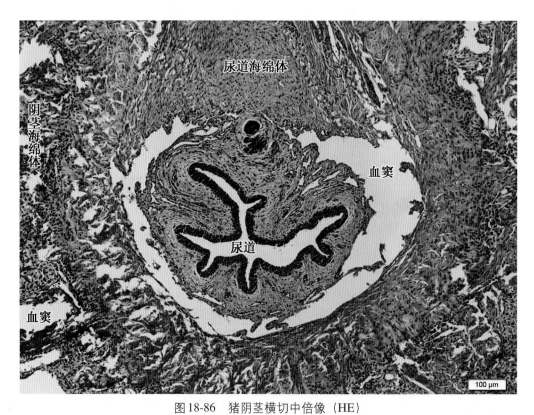

图18-86 猪阴茎横切中倍像（HE）

Fig.18-86 mid magnification of pig penis cross section （HE）

阴茎海绵体—corpus cavernosum 尿道海绵体—corpus spongiosum urethra

血窦—blood sinusoid 尿道—urethra

第十九章
被皮系统
Integumentary System

Outline

Integumentary system consists of the skin and epidermal derivatives, including hairs, feather, horn, beak, crest, squama and sebaceous, sweat glands and mammary glands in different outer surfaces of the animal body.

Skin is the largest organ of the creature. It consists of epidermis, surface epithelial layer, and dermis, the subjacent layer of connective tissue. Hypodermis, a loose connective tissue and adipose tissue lies beneath the dermis and binds skin loosely to the subjacent tissue. Epidermis consists of a stratified squamous keratinized epithelium. Cells of the epidermis can be classified into two types: keratinocytes and nonkeratinocytes. Keratinocytes are main cells of the epidermis. Nonkeratinocytes are less abundant and found between keratinocytes. They include melanocytes, Langerhans cells and Merkel cells. The thickness of epidermis varies in light of the region of the body. Epidermis of the palms and soles is the thickest and has the most typical structure, in which five layers can be distinguished: stratum basale, stratum spinosum, stratum granulosum, stratum lucidum and stratum corneum. Other epidermises of the body are thin, in which the stratum lucidum is not always present. Dermis can be subdivided into two layers: the papillary layer and the reticular layer. Papillary layer which is thinner, consists of loose connective tissue and has many dermal papillae. Reticular layer is thicker and composed of irregular dense connective tissue. Its bundles of coarse collagenous fibers and elastic fibers run in various directions and form a fiber network. The network is responsible for the flexibility and elasticity of the skin. Skin possesses many functions, such as protection, sensory, absorption, excretion, thermoregulation and participates in immune responses and material metabolism of the body.

被皮系统（integumentary system）由皮肤及其衍生物，包括被毛、皮脂腺、汗腺、乳腺、蹄、枕、角等组成。

一、被皮系统的发生

（一）皮肤的发生

浅层的表皮来源于体壁外胚层，为角化的复层扁平上皮；深层的真皮主要由中胚层形成，为致密结缔组织；皮下组织是来源于中胚层的疏松结缔组织或脂肪组织。表皮由两类细胞组成。一类是角质形成细胞，为表皮的主要成分，参与表皮角化；另一类是非角质形成细胞，数量少，分散在角质形成细胞之间，如黑素细胞、朗格汉斯细胞和梅克尔细胞。

（二）皮肤主要附属器的发生

1. **毛的发生** 表皮生发层细胞向真皮内增生，形成一实心的上皮细胞柱，为毛芽基，并斜向下生长，末端膨大成毛球；其底部的间充质突入毛球，称毛乳头。毛芽基中央的细胞经角质化成为毛根，周边的细胞分化为毛囊的内根鞘。当毛分化时，斜行的毛囊的钝角侧细胞增生，为皮脂腺原基，在其下方的间充质衍化出一束平滑肌，为竖毛肌。

2. **皮脂腺的发生** 皮脂腺多由毛囊外根鞘一侧的细胞增生突入间充质形成，这些上皮突出物分支形成若干腺泡的原基及与其相通的导管。腺泡中央的细胞变大，胞质内出现小脂滴，周边的一层细胞小为基细胞。充满脂滴的中央细胞解体，排入毛囊至皮肤表面，构成皮脂。

3. **汗腺的发生** 汗腺分两种，一种为常见的外泌小汗腺，另一种为顶浆分泌的大汗腺。外泌汗腺原基开始于掌、足底和指（趾）腹面，然后出现于其他部位。汗腺源自表皮生发层细胞增殖，向真皮内长出的圆柱状细胞索，细胞索逐渐延长，末端弯曲并卷绕成球状，形成分泌部的原基，与表皮相连的部分则为导管的原基。大汗腺发生于妊娠中期，遍布在身体大部分皮肤中，是从毛囊的一侧增生而成，其导管开口于毛囊或皮肤表面。

二、被皮系统的组织学结构概述

（一）皮肤（skin）

皮肤是机体最大的器官，被覆于整个身体表面，在唇、肛门、鼻和眼睑边缘及泌尿生殖管道外口与黏膜相连续。皮肤具有多种功能，如保护机体免受物理、化学及生物的侵害；构成防止体内水分丢失的屏障，吸收紫外线防止辐射的损伤及合成维生素D；分泌汗液调节体温；通过各种神经末梢监测外部环境变化以及免疫防御功能。

皮肤由表面角化的复层扁平上皮及深部的结缔组织层，即表皮和真皮构成。表皮与真皮相互交错，构成表皮嵴和真皮乳头，二者间由基膜隔开。指端由嵴构成的指纹即是表皮嵴与真皮乳头二者相互交错的证据。在皮肤与深部组织结构之间是皮下组织。

1. **表皮（epidermis）** 按角质层的厚度，皮肤分为厚皮肤与薄皮肤。厚皮肤的表皮由五层组成。最深层是基底层（表皮生发层），为单层立方至柱状细胞，负责细胞的更新。棘层最厚，细胞呈多边形，特征是以大量的突起与周围棘细胞的突起形成桥粒。颗粒层细胞不断积累透明角质颗粒，并最终充满整个细胞，破坏细胞核及细胞器。第四层为透明层，较薄，只在掌部和暴露于日光下皮肤的颗粒层与角质层间，呈现薄的透明区域。透明层细胞无细胞核也无细胞器，但充满密集排列的角蛋白丝和角母蛋白。最外层是角质层，由多层干硬的死细胞组成。表皮由四种细胞构成：角质形成细胞、黑素细胞、朗格汉斯细胞和梅克尔细胞。薄皮肤仅有三层或四层结构。在薄皮肤内，透明层常缺失，角质层、颗粒层、棘层均较薄。

2. **真皮（dermis）** 真皮由不规则致密胶原性结缔组织组成，主要含Ⅰ型胶原蛋白和许多弹性纤维，其功能是将皮肤固定于皮下组织。真皮可分为松散交织的乳头层和深部粗大致密的网状层。乳头层是与表皮嵴相互交错的表浅区域。乳头层与网状层无明确分界。

（二）皮肤衍生物（skin derivatives）

皮肤的衍生物包括毛、角、皮脂腺、汗腺及蹄匣。这些结构从表皮向下长入真皮及皮下组织，且与外界相连通。每根毛由角质细胞形成的毛干和包在毛囊中的毛根和毛球组成，毛囊与皮脂腺相连，皮脂腺将皮脂分泌于毛囊颈部。竖毛肌附着于毛囊，支撑皮脂腺，并嵌入皮肤的表层。汗腺为弯曲的单管状腺，分泌汗液，通过长导管排至皮肤表面。肌上皮细胞包绕着汗腺分泌部。

（三）乳腺（mammary gland）

乳腺为复管泡状腺。由被膜、间质和实质组成。被膜是富有脂肪的结缔组织膜。被膜的结缔组织深入实质，将其分为许多小叶。腺实质包括分泌部与导管部。

1. 分泌部 由腺泡组成。腺泡呈卵圆形或球形。单层腺上皮细胞的形态随分泌活动而变化。细胞内聚集脂滴和蛋白颗粒时，细胞呈高柱状或锥状，顶端突入腺泡腔内，胞核为球形，多位于细胞的基部，此时腺泡腔较小。随着分泌物排出，细胞变成立方形，腺泡腔增大，并充满分泌物。在腺上皮细胞与基膜之间有肌上皮细胞，其收缩有助于分泌和乳汁的排出。在一个腺小叶内的各腺泡，其分泌活动并不完全一致。因此，可见某些腺泡的上皮细胞为高柱状，而另一些细胞为立方形。

2. 导管部 自小叶内导管开始，其上皮为单层立方上皮，有肌上皮细胞。小叶内导管与很多腺泡相连，进入叶间结缔组织后，汇入小叶间导管，管壁为单层柱状上皮或双层立方上皮。小叶间导管集合成输乳管，输导整个腺叶的乳汁。输乳管等大导管的管壁为双层矮柱状上皮，并有纵行的平滑肌纤维。乳头管的上皮为复层扁平上皮。

乳汁主要含有蛋白质、脂肪、乳糖及无机盐等。在分娩后不久乳腺分泌的乳汁叫初乳。初乳的成分及生物学特性均与哺乳期乳汁有所不同，特别是乳糖及脂肪含量较少，蛋白质（尤其是球蛋白）、维生素A含量丰富。初乳中还有初乳小体、酶、溶菌素等。初乳小体为圆形或卵圆形的有核细胞，胞质中充满脂滴。

3. 乳腺的生长发育和泌乳的变化 乳腺的结构一直处在变化过程中，主要表现在分泌部的新生或吸收，间质的增多或减少。动物体在性成熟前和两个泌乳期之间的乳腺静止期中，乳腺内主要是结缔组织、分散的输乳管和一些萎缩塌陷的腺泡或细胞索。腺泡和小的导管均为单层立方上皮或扁平上皮。性成熟后，乳腺开始发育。特别是在妊娠期间，在激素的影响下，腺组织的发育尤为迅速，腺泡显著增生，间质相对明显减少。分娩后，进入哺乳时期，乳腺发育达到全面活动期。到哺乳后期，腺组织逐渐缩小以至停止分泌活动，结缔组织和脂肪增多，乳腺转入静止期。

三、被皮系统图谱

1. **皮肤** 图19-1 ～图19-17。
2. **毛** 图19-18 ～图19-26。
3. **汗腺** 图19-27 ～图19-31。
4. **皮脂腺** 图19-32 ～图19-38。
5. **乳腺** 图19-39 ～图19-54。
6. **蹄** 图19-55 ～图19-59。
7. **趾垫** 图19-60。

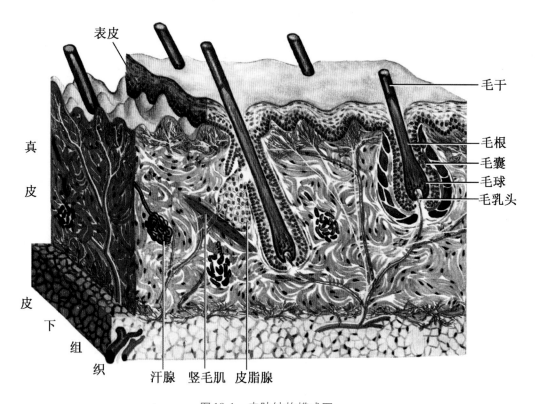

图 19-1　皮肤结构模式图

Fig.19-1　ideograph of skin structure

表皮—epidermis　真皮—dermis　皮下组织—subcutaneous tissue　毛干—hair shaft　毛根—hair root　毛囊—hair follicle　毛球—hair bulb　毛乳头—hair papilla　汗腺—sweat gland　竖毛肌—arrector pilli muscle　皮脂腺—sebaceous gland

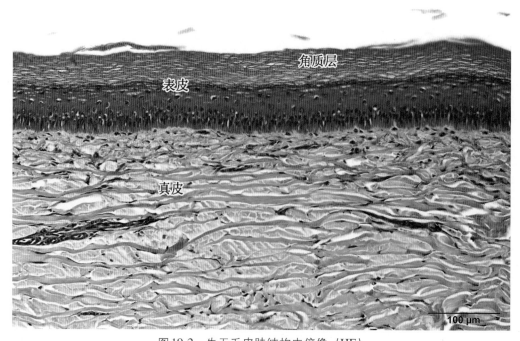

图 19-2　牛无毛皮肤结构中倍像（HE）

Fig.19-2　mid magnification of cattle hairless skin structure（HE）

角质层—horny layer　表皮—epidermis　真皮—dermis

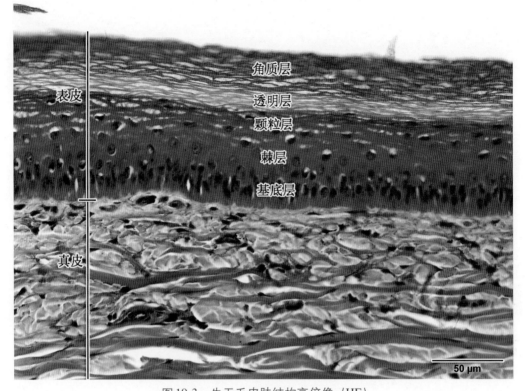

图19-3 牛无毛皮肤结构高倍像（HE）

Fig.19-3 high magnification of cattle hairless skin structure（HE）

表皮—epidermis 真皮—dermis 角质层—horny layer 透明层—hyaline layer
颗粒层—granular layer 棘层—spinosum layer 基底层—basal layer

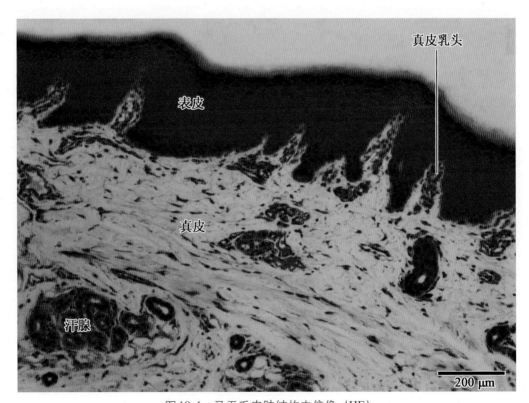

图19-4 马无毛皮肤结构中倍像（HE）

Fig.19-4 mid magnification of horse hairless skin structure（HE）

表皮—epidermis 真皮—dermis 真皮乳头—dermal papilla 汗腺—sweat gland

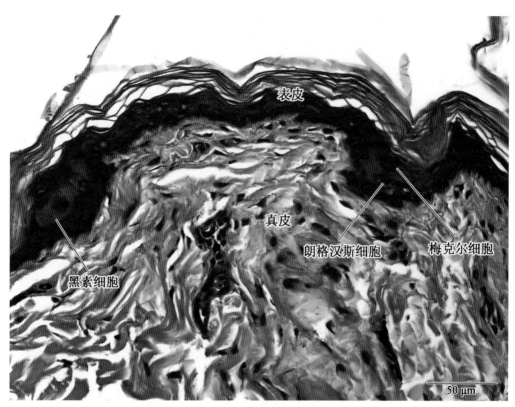

图 19-5 猫无毛皮肤结构高倍像（HE）

Fig.19-5 high magnification of cat hairless skin structure（HE）

表皮—epidermis 真皮—dermis 黑素细胞—melanocyte 朗格汉斯细胞—Langerhans cell 梅克尔细胞—Merkel cell

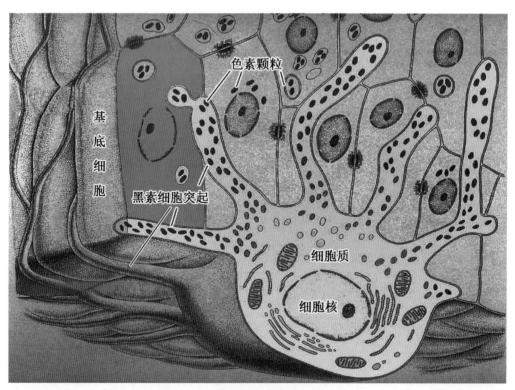

图 19-6 黑素细胞超微结构模式图

Fig.19-6 ultrastructure ideograph of melanocyte

基底细胞—basal cell 黑素细胞突起—melanocyte process 色素颗粒—pigment granules

细胞质—cytoplasm 细胞核—cell nucleus

第十九章 被皮系统 Integumentary System

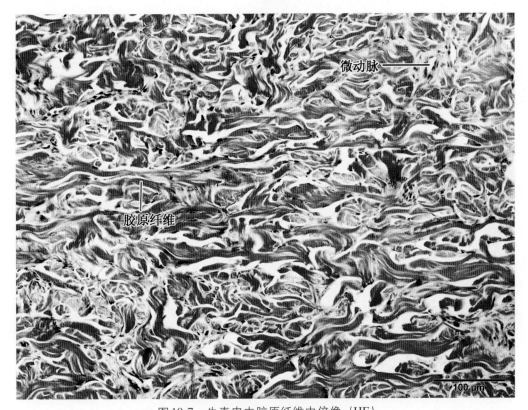

图 19-7　牛真皮中胶原纤维中倍像（HE）
Fig.19-7　mid magnification of the collagenous fiber in cattle dermis（HE）
胶原纤维—collagenous fiber　微动脉—arteriole

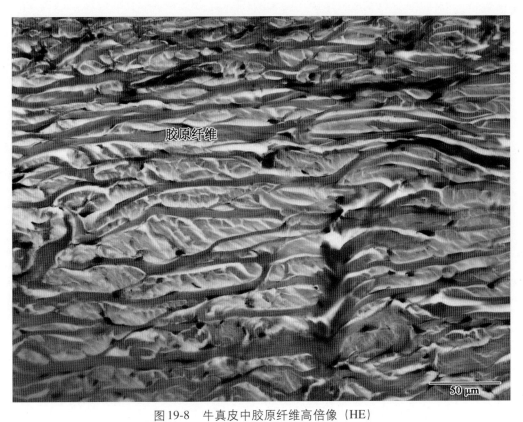

图 19-8　牛真皮中胶原纤维高倍像（HE）
Fig.19-8　high magnification of the collagenous fiber in cattle dermis（HE）
胶原纤维—collagenous fiber

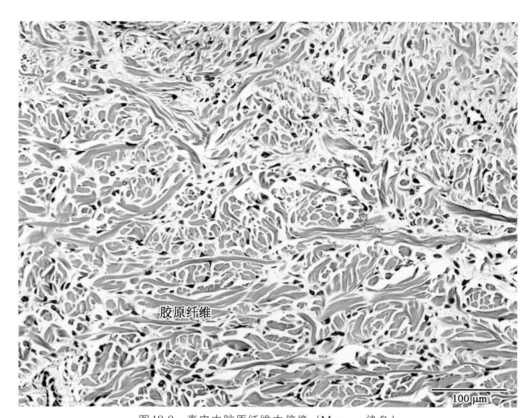

图19-9 真皮中胶原纤维中倍像（Masson 染色）
Fig.19-9 mid magnification of collagenous fiber in dermis（Masson stain）
胶原纤维—collagenous fiber

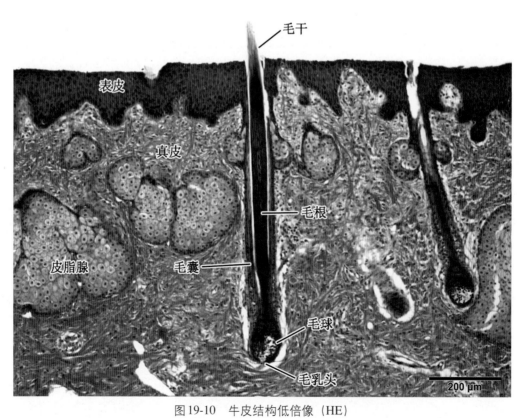

图19-10 牛皮结构低倍像（HE）
Fig.19-10 low magnification of cattle skin structure（HE）
表皮—epidermis 真皮—dermis 皮脂腺—sebaceous gland 毛干—hair shaft 毛根—hair root
毛囊—hair follicle 毛球—hair bulb 毛乳头—hair papilla

第十九章 被皮系统 Integumentary System

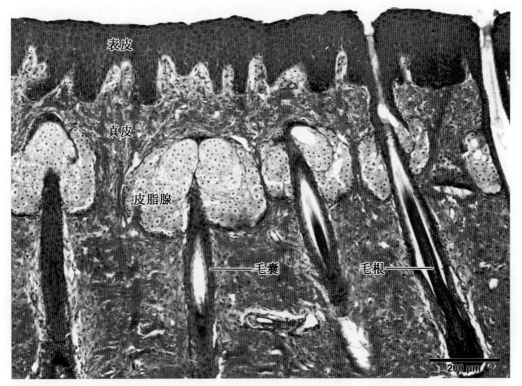

图19-11 马皮结构低倍像（HE）

Fig.19-11 low magnification of horse skin structure（HE）

表皮—epidermis 真皮—dermis 皮脂腺—sebaceous gland 毛囊—hair follicle 毛根—hair root

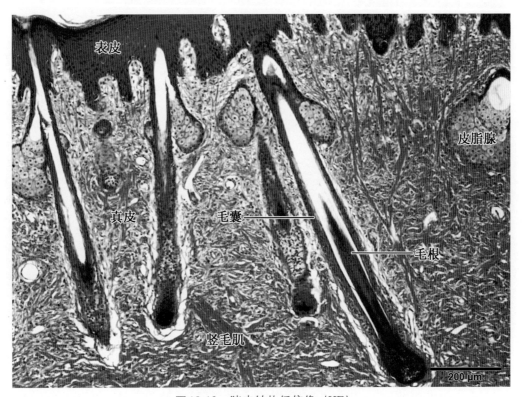

图19-12 猪皮结构低倍像（HE）

Fig.19-12 low magnification of pig skin structure（HE）

表皮—epidermis 真皮—dermis 毛囊—hair follicle 毛根—hair root
皮脂腺—sebaceous gland 竖毛肌—arrector pilli muscle

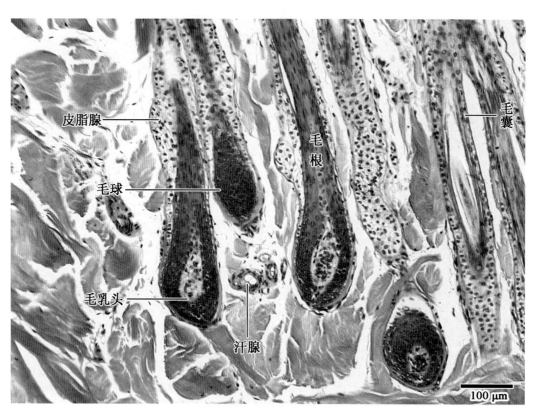

图 19-13　绵羊皮结构中倍像（HE）

Fig.19-13　mid magnification of sheep skin structure（HE）

皮脂腺—sebaceous gland　毛球—hair bulb　毛乳头—hair papilla　毛根—hair root　毛囊—hair follicle　汗腺—sweat gland

图 19-14　山羊皮结构低倍像（HE）

Fig.19-14　low magnification of goat skin structure（HE）

表皮—epidermis　真皮—dermis　毛孔—pore　皮脂腺—sebaceous gland　毛囊—hair follicle

第十九章 被皮系统 Integumentary System

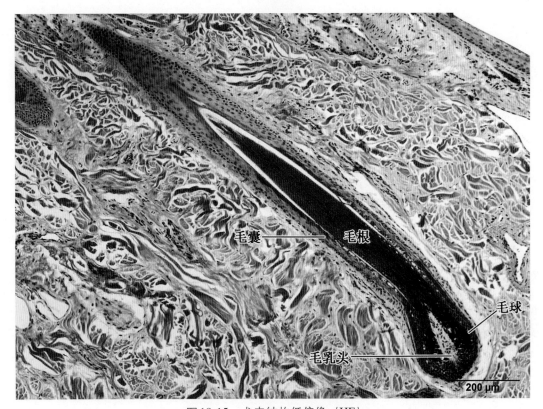

图19-15 犬皮结构低倍像（HE）
Fig.19-15 low magnification of dog skin structure（HE）
毛囊—hair follicle　毛根—hair root　毛球—hair bulb　毛乳头—hair papilla

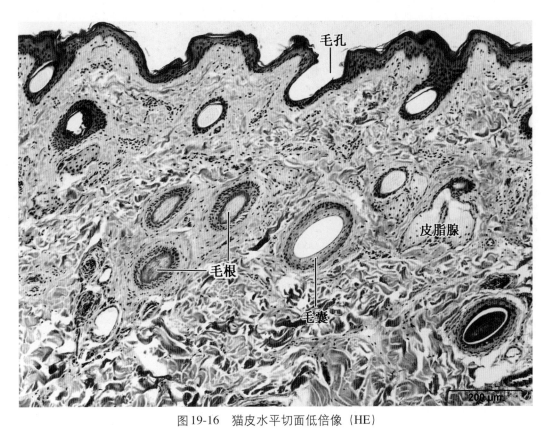

图19-16 猫皮水平切面低倍像（HE）
Fig.19-16 low magnification of cat skin horizontal section（HE）
毛孔—pore　毛根—hair root　毛囊—hair follicle　皮脂腺—sebaceous gland

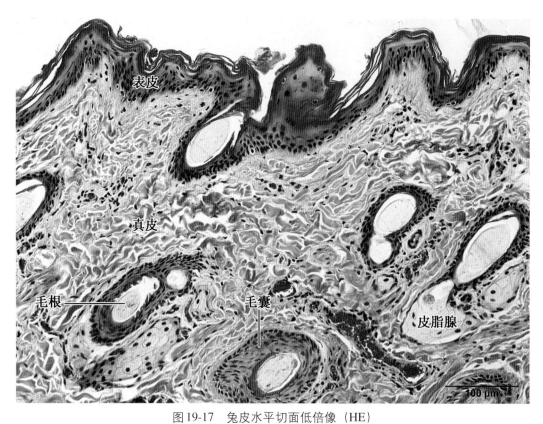

图 19-17　兔皮水平切面低倍像（HE）

Fig.19-17　low magnification of rabbit skin horizontal section（HE）

表皮—epidermis　真皮—dermis　毛根—hair root　毛囊—hair follicle　皮脂腺—sebaceous gland

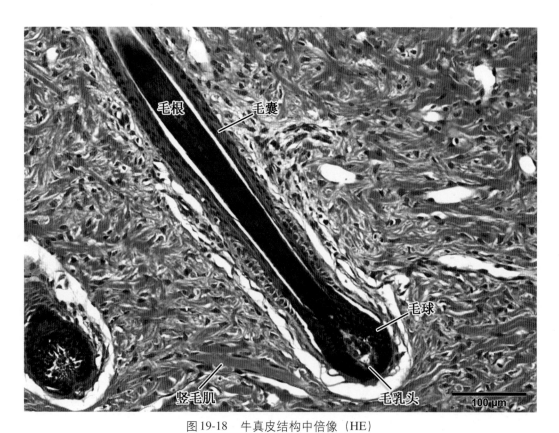

图 19-18　牛真皮结构中倍像（HE）

Fig.19-18　mid magnification of cattle dermis（HE）

毛根—hair root　毛囊—hair follicle　毛球—hair bulb　毛乳头—hair papilla　竖毛肌—arrector pilli muscle

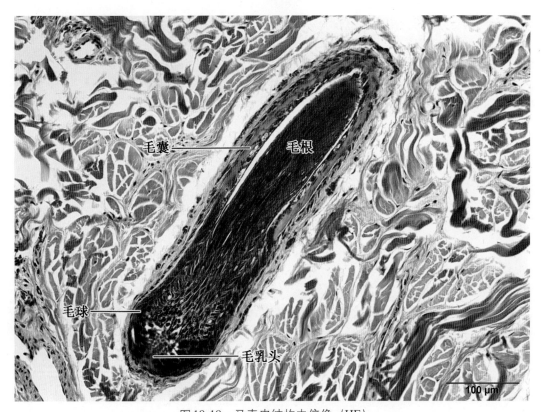

图19-19 马真皮结构中倍像（HE）

Fig.19-19 mid magnification of horse dermis（HE）

毛囊—hair follicle 毛根—hair root 毛球—hair bulb 毛乳头—hair papilla

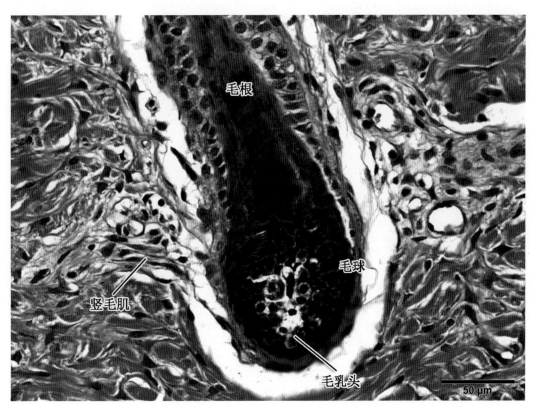

图19-20 猪真皮结构高倍像（HE）

Fig.19-20 high magnification of pig dermis（HE）

毛根—hair root 毛球—hair bulb 毛乳头—hair papilla 竖毛肌—arrector pilli muscle

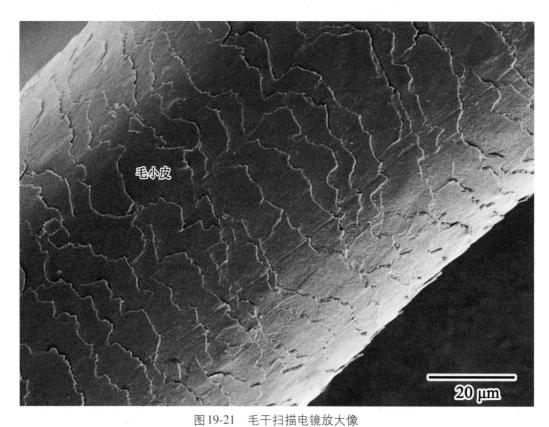

图19-21 毛干扫描电镜放大像

Fig.19-21 high magnification of hair shaft on scanning electron microscope

毛小皮—hair cuticle

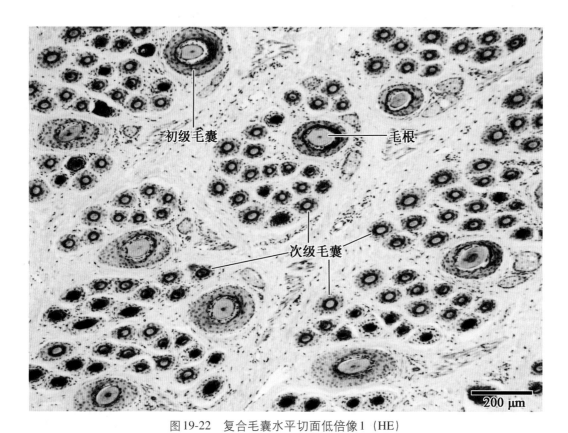

图19-22 复合毛囊水平切面低倍像1（HE）

Fig.19-22 low magnification of composite hair follicles horizontal section 1 （HE）

初级毛囊—primary hair follicle 次级毛囊—secondary hair follicle 毛根—hair root

第十九章 被皮系统 Integumentary System

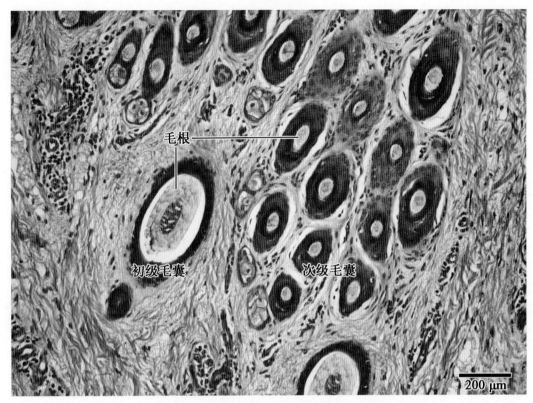

图 19-23 复合毛囊水平切面低倍像 2 （HE）

Fig.19-23 low magnification of composite hair follicles horizontal section 2 (HE)

毛根—hair root　初级毛囊—primary hair follicle　次级毛囊—secondary hair follicle

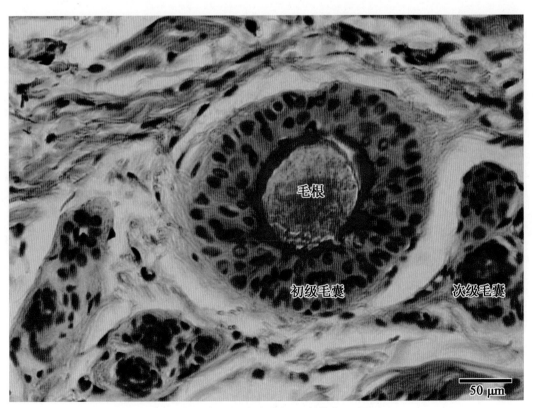

图 19-24 复合毛囊水平切面高倍像 1 （HE）

Fig.19-24 high magnification of composite hair follicles horizontal section 1 (HE)

毛根—hair root　初级毛囊—primary hair follicle　次级毛囊—secondary hair follicle

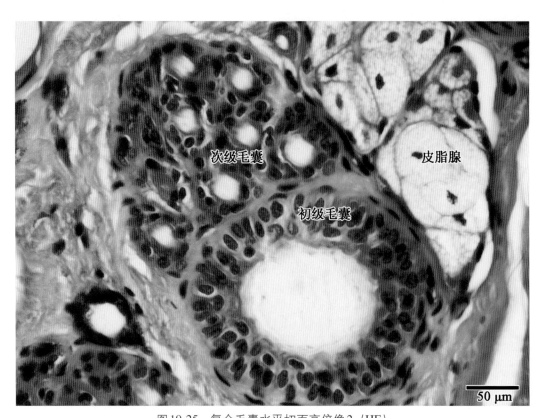

图19-25 复合毛囊水平切面高倍像2（HE）

Fig.19-25 high magnification of composite hair follicles horizontal section 2 （HE）

次级毛囊—secondary hair follicle　初级毛囊—primary hair follicle　皮脂腺—sebaceous gland

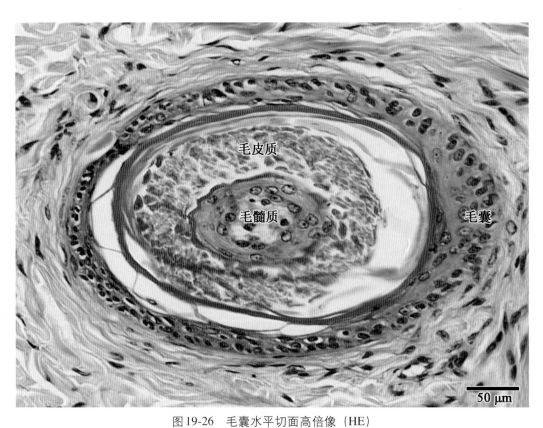

图19-26 毛囊水平切面高倍像（HE）

Fig.19-26 high magnification of hair follicle horizontal section （HE）

毛皮质—hair cortex　毛髓质—hair medulla　毛囊—hair follicle

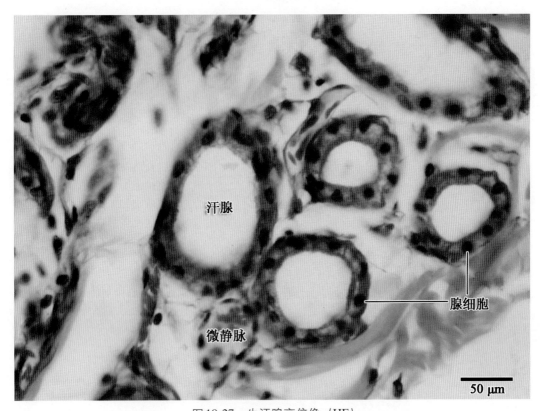

图 19-27 牛汗腺高倍像（HE）

Fig.19-27 high magnification of cattle sweat gland（HE）

汗腺—sweat gland 腺细胞—glandular cell 微静脉—venule

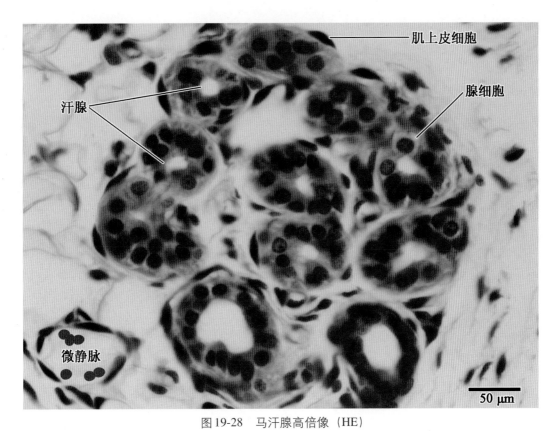

图 19-28 马汗腺高倍像（HE）

Fig.19-28 high magnification of horse sweat gland（HE）

汗腺—sweat gland 腺细胞—glandular cell 肌上皮细胞—myoepithelial cell 微静脉—venule

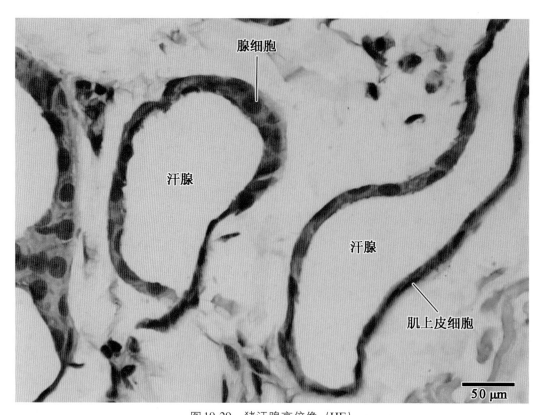

图19-29 猪汗腺高倍像（HE）

Fig.19-29 high magnification of pig sweat gland（HE）

汗腺—sweat gland 腺细胞—glandular cell 肌上皮细胞—myoepithelial cell

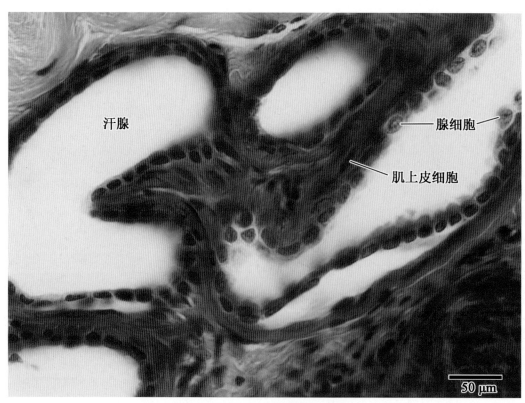

图19-30 山羊汗腺高倍像（HE）

Fig.19-30 high magnification of goat sweat gland（HE）

汗腺—sweat gland 腺细胞—glandular cell 肌上皮细胞—myoepithelial cell

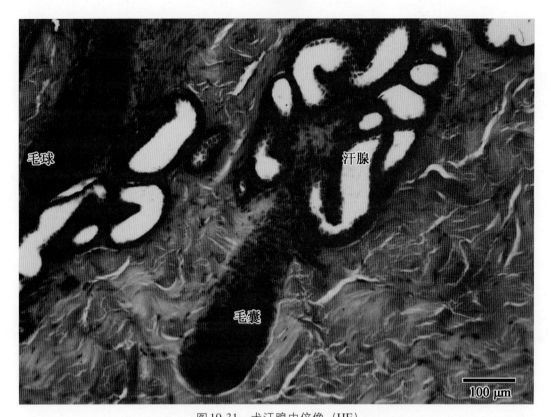

图19-31 犬汗腺中倍像（HE）

Fig.19-31　mid magnification of dog sweat gland（HE）

汗腺—sweat gland　　毛球—hair bulb　　毛囊—hair follicle

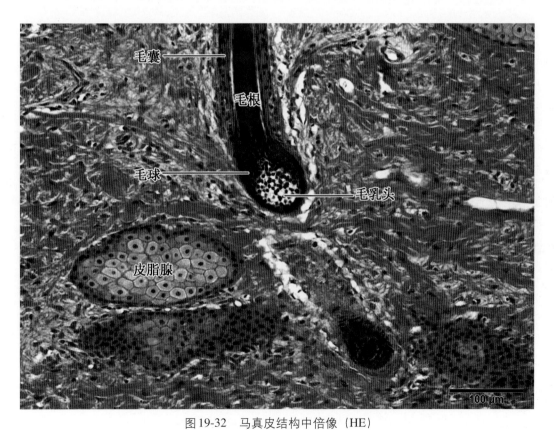

图19-32 马真皮结构中倍像（HE）

Fig.19-32　mid magnification of horse dermis structure（HE）

毛囊—hair follicle　　毛根—hair root　　毛球—hair bulb　　毛乳头—hair papilla　　皮脂腺—sebaceous gland

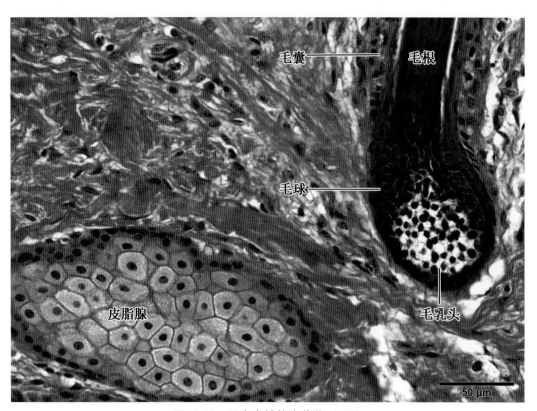

图 19-33 马真皮结构高倍像（HE）

Fig.19-33 high magnification of horse dermis structure（HE）

毛囊—hair follicle 毛根—hair root 毛球—hair bulb 毛乳头—hair papilla 皮脂腺—sebaceous gland

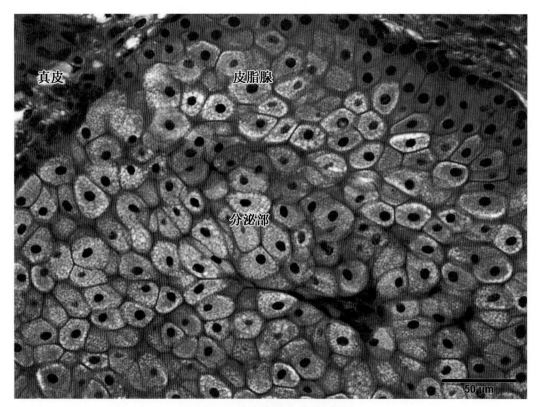

图 19-34 猪皮脂腺高倍像（HE）

Fig.19-34 high magnification of pig sebaceous gland（HE）

真皮—dermis 皮脂腺—sebaceous gland 分泌部—secretory portion

第十九章　被皮系统　Integumentary System

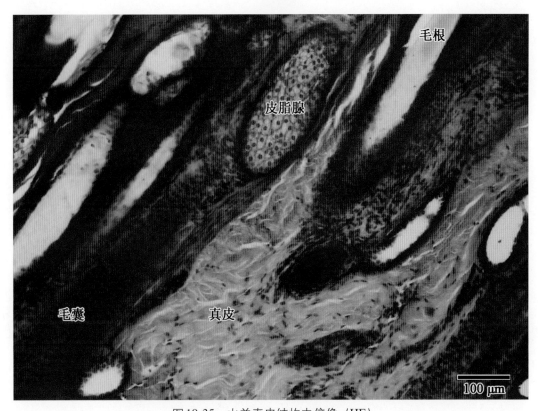

图19-35　山羊真皮结构中倍像（HE）

Fig.19-35　mid magnification of goat dermis structure（HE）

皮脂腺—sebaceous gland　毛根—hair root　毛囊—hair follicle　真皮—dermis

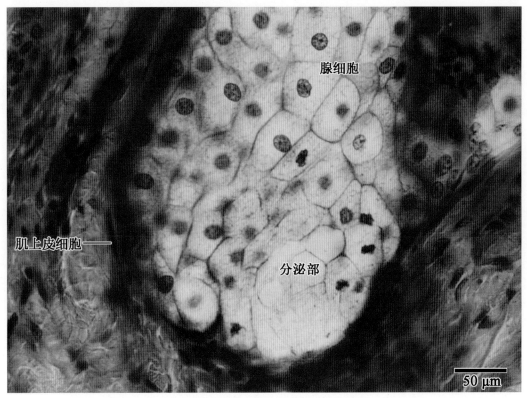

图19-36　山羊真皮结构高倍像（HE）

Fig.19-36　high magnification of goat dermis structure（HE）

肌上皮细胞—myoepithelial cell　腺细胞—glandular cell　分泌部—secretory portion

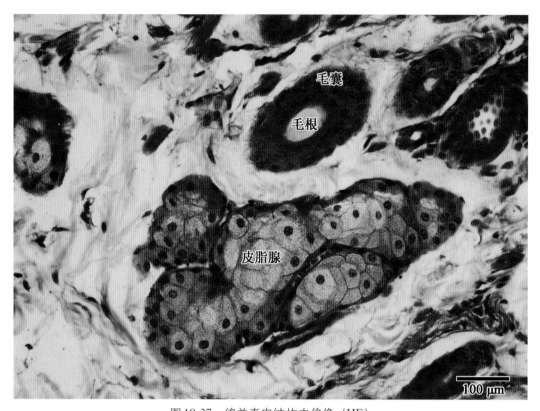

图 19-37　绵羊真皮结构中倍像（HE）

Fig.19-37　mid magnification of sheep dermis structure（HE）

毛囊—hair follicle　毛根—hair root　皮脂腺—sebaceous gland

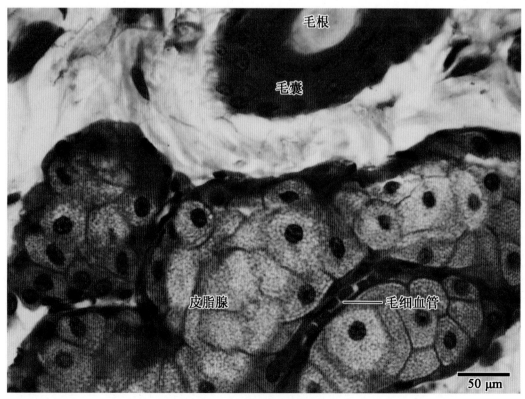

图 19-38　绵羊真皮结构高倍像（HE）

Fig.19-38　high magnification of sheep dermis structure（HE）

毛根—hair root　毛囊—hair follicle　皮脂腺—sebaceous gland　毛细血管—capillary

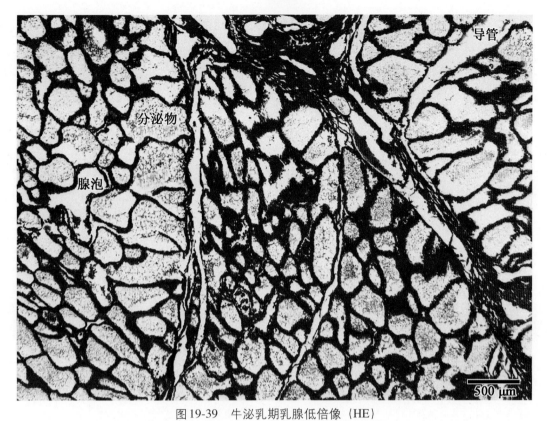

图19-39 牛泌乳期乳腺低倍像（HE）

Fig.19-39 low magnification of cow breast in lactation period（HE）

腺泡—acinus 分泌物—secretion 导管—duct

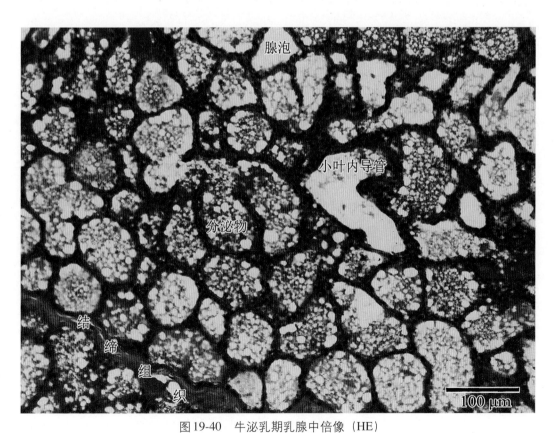

图19-40 牛泌乳期乳腺中倍像（HE）

Fig.19-40 mid magnification of cow breast in lactation period（HE）

腺泡—acinus 分泌物—secretion 小叶内导管—intralobular duct 结缔组织—connective tissue

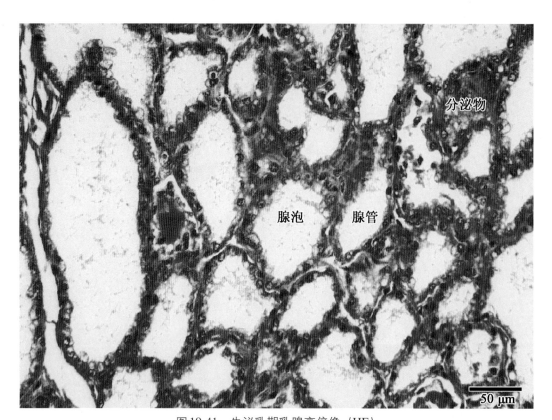

图 19-41 牛泌乳期乳腺高倍像（HE）

Fig.19-41 high magnification of cow breast in lactation period（HE）

分泌物—secretion 腺泡—acinus 腺管—glandular duct

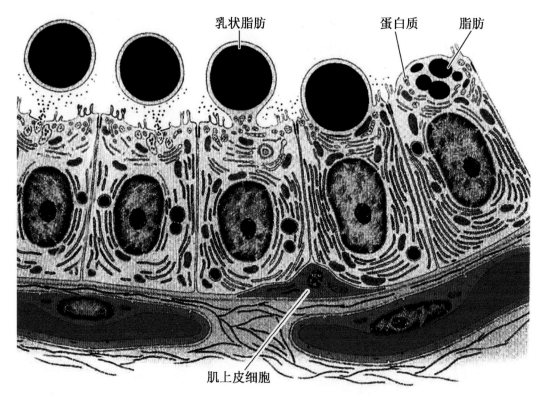

图 19-42 泌乳期乳腺细胞超微结构模式图

Fig.19-42 ultrastructure ideograph of mammary glandular cell in lactation

乳状脂肪—fat emulsion 蛋白质—protein 脂肪—fat 肌上皮细胞—myoepithelial cell

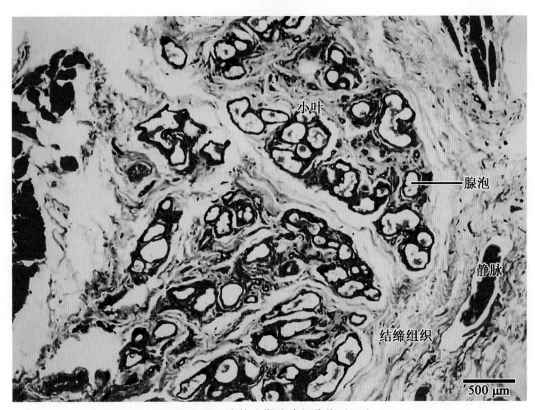

图 19-43 牛静止期乳腺低倍像（HE）

Fig.19-43 low magnification of cow breast in resting period（HE）

小叶—lobule 腺泡—acinus 结缔组织—connective tissue 静脉—vein

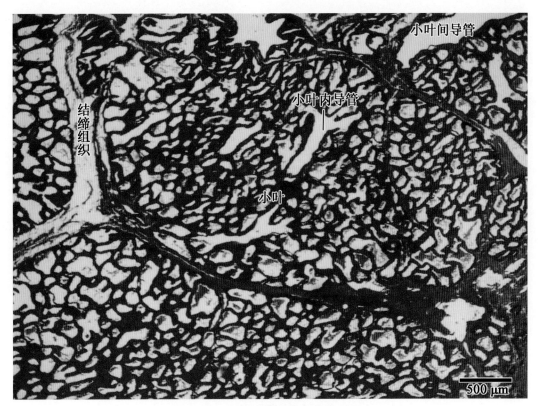

图 19-44 马泌乳期乳腺低倍像（HE）

Fig.19-44 low magnification of horse breast in lactation period（HE）

结缔组织—connective tissue 小叶—lobule 小叶内导管—intralobular duct 小叶间导管—interlobular duct

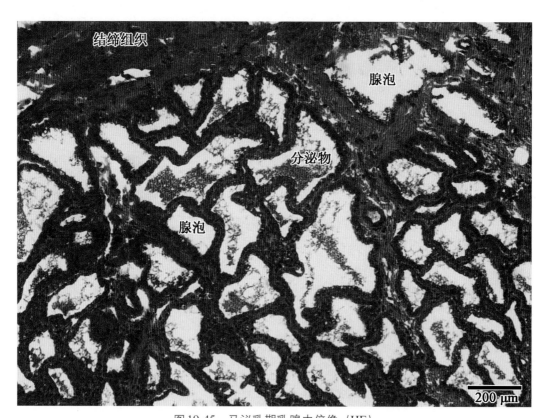

图19-45　马泌乳期乳腺中倍像（HE）

Fig.19-45　mid magnification of horse breast in lactation period（HE）

结缔组织—connective tissue　腺泡—acinus　分泌物—secretion

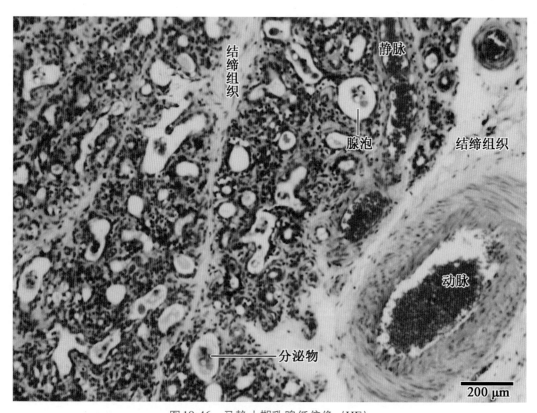

图19-46　马静止期乳腺低倍像（HE）

Fig.19-46　low magnification of horse breast in resting period（HE）

结缔组织—connective tissue　腺泡—acinus　静脉—vein　动脉—artery　分泌物—secretion

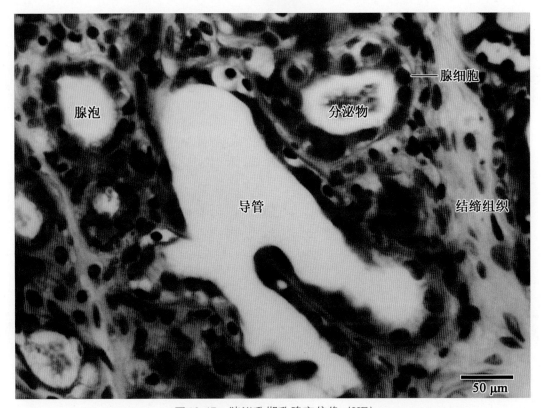

图19-47 猪泌乳期乳腺高倍像（HE）

Fig.19-47　high magnification of pig breast in lactation period（HE）

腺泡—acinus　导管—duct　分泌物—secretion　腺细胞—gland cell　结缔组织—connective tissue

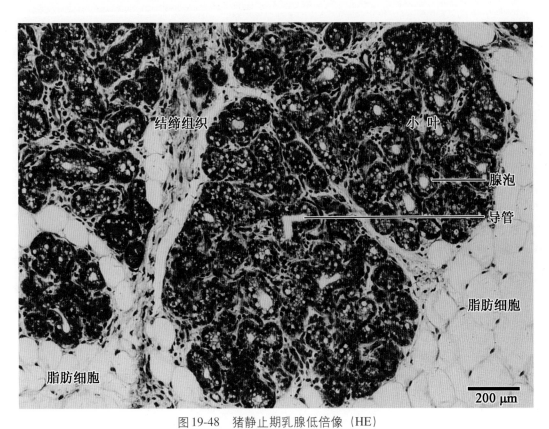

图19-48 猪静止期乳腺低倍像（HE）

Fig.19-48　low magnification of pig breast in resting period（HE）

结缔组织—connective tissue　小叶—lobule　腺泡—acinus　导管—duct　脂肪细胞—fat cell

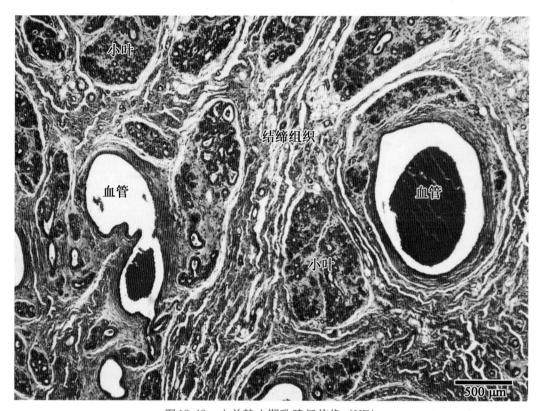

图19-49 山羊静止期乳腺低倍像（HE）
Fig.19-49 low magnification of goat breast in resting period（HE）
小叶—lobule 结缔组织—connective tissue 血管—blood vessel

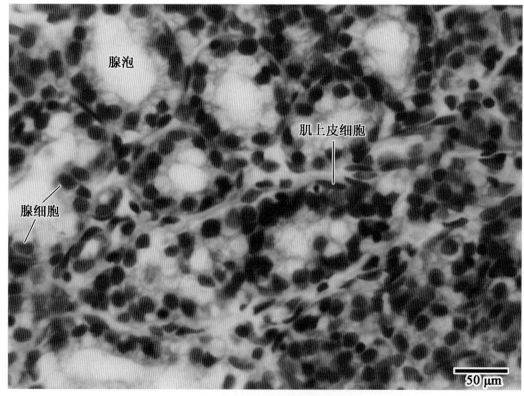

图19-50 犬泌乳期乳腺高倍像（HE）
Fig.19-50 high magnification of dog breast in lactation period（HE）
腺泡—acinus 腺细胞—gland cell 肌上皮细胞—myoepithelial cell

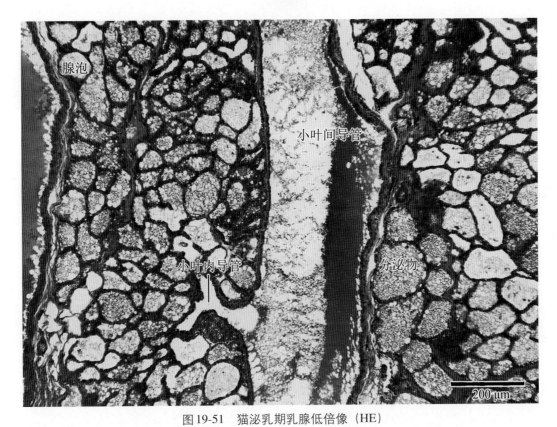

图 19-51 猫泌乳期乳腺低倍像（HE）

Fig.19-51 low magnification of cat breast in lactation period（HE）

腺泡—acinus 小叶间导管—interlobular duct 小叶内导管—intralobular duct 分泌物—secretion

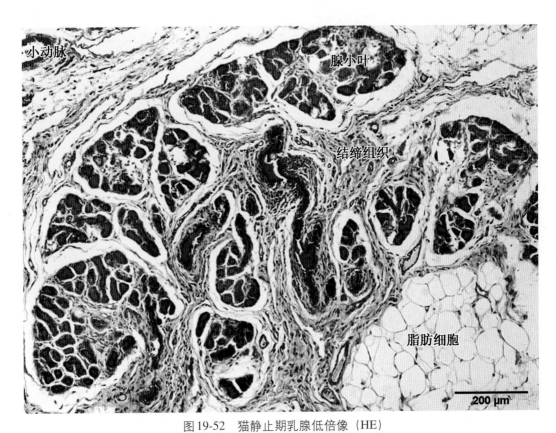

图 19-52 猫静止期乳腺低倍像（HE）

Fig.19-52 low magnification of cat breast in resting period（HE）

小动脉—arteriole 腺小叶—lobule 结缔组织—connective tissue 脂肪细胞—fat cell

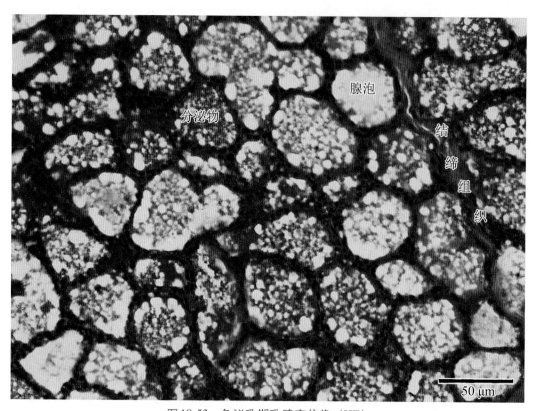

图 19-53 兔泌乳期乳腺高倍像（HE）

Fig.19-53　high magnification of rabbit breast in lactation period（HE）

分泌物—secretion　腺泡—acinus　结缔组织—connective tissue

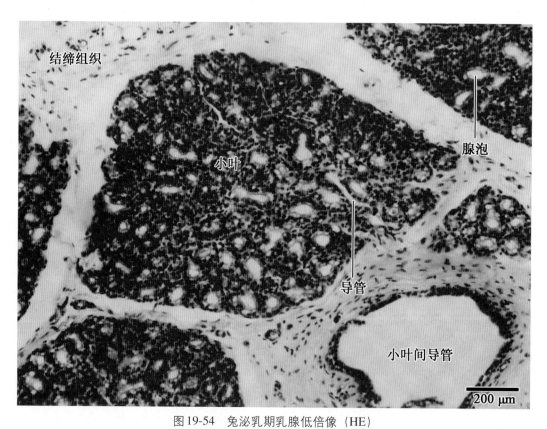

图 19-54 兔泌乳期乳腺低倍像（HE）

Fig.19-54　low magnification of rabbit breast in lactation period（HE）

结缔组织—connective tissue　小叶—lobule　腺泡—acinus　导管—duct　小叶间导管—interlobular duct

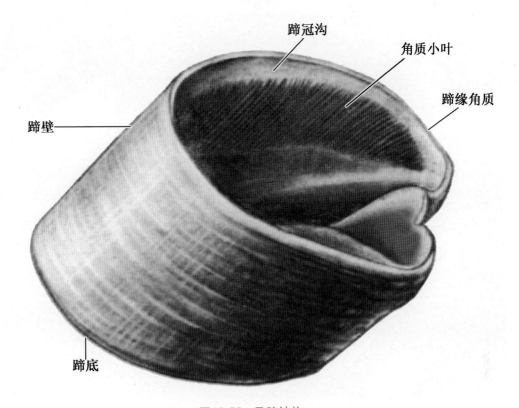

图 19-55　马蹄结构

Fig.19-55　structure of horse hoof

蹄冠沟—coronet ditch　　角质小叶—cutin lobule　　蹄缘角质—hoof edge cutin　　蹄壁—hoof wall　　蹄底—hoof bottom

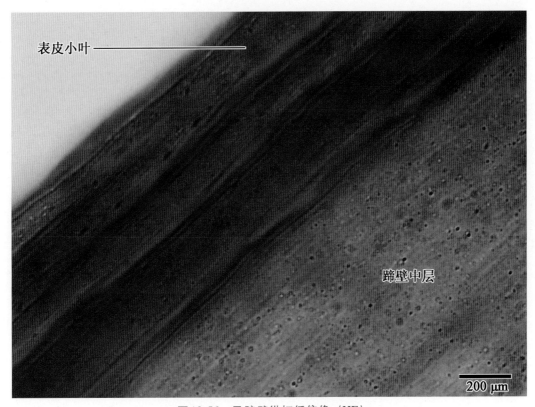

图 19-56　马蹄壁纵切低倍像（HE）

Fig.19-56　low magnification of horse hoof wall lengthwise section（HE）

表皮小叶—cuticle lobule　　蹄壁中层—hoof wall middle

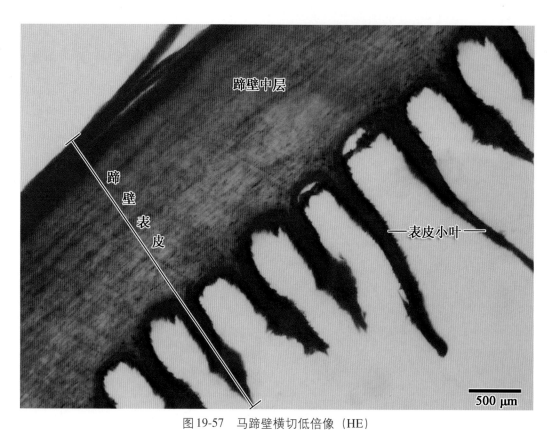

图 19-57 马蹄壁横切低倍像（HE）

Fig.19-57 low magnification of horse hoof wall transection（HE）

蹄壁中层—hoof wall middle　表皮小叶—cuticle lobule　蹄壁表皮—hoof wall epidermis

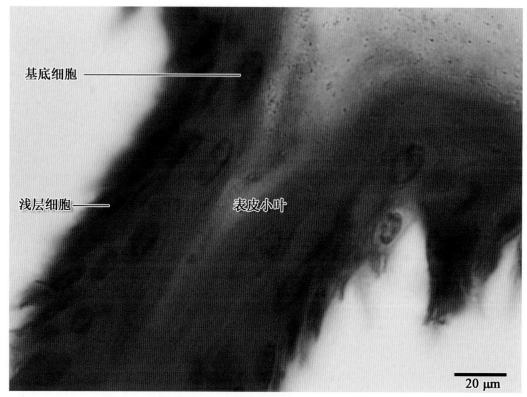

图 19-58 马蹄壁小叶横切高倍像（HE）

Fig.19-58 high magnification of horse hoof wall lobule transection（HE）

基底细胞—basal cell　浅层细胞—shallow cell　表皮小叶—cuticle lobule

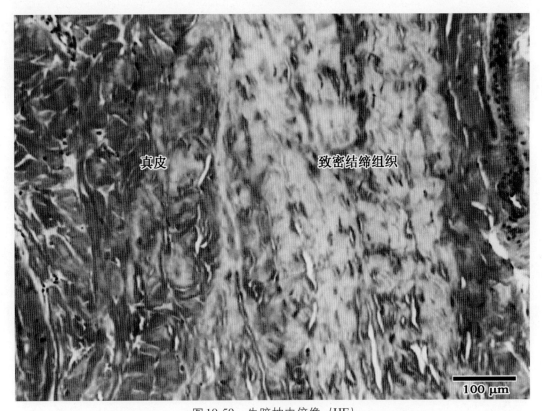

图19-59 牛蹄枕中倍像（HE）

Fig.19-59 mid magnification of cattle hoof pillow（HE）

真皮—dermis 致密结缔组织—dense connective tissue

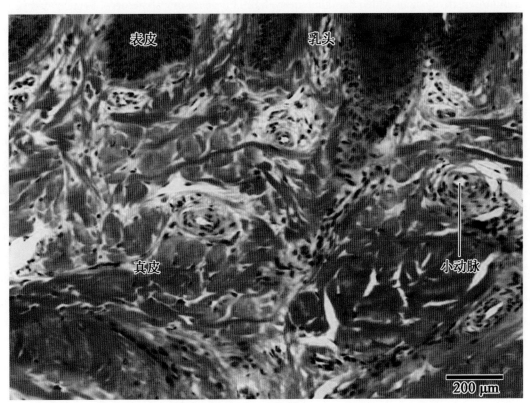

图19-60 犬趾垫低倍像（HE）

Fig.19-60 low magnification of dog toe pad（HE）

表皮—epidermis 真皮—dermis 乳头—dermal papilla 小动脉—arteriole

第二十章 感觉器官
Sensory Organs

Outline

Information about the external world is conveyed to the central nervous system by sensory units called receptor. This chapter shows the special organs responsible for the reception of light and sound waves.

Eye is a photosensitive organ which contains an eyeball and accessory structures. Vertebrate eye has the resembling basic structure though difference in shape and structure. The eyeball is composed of three outer tunics enclosing the refracting media. The three tunics are: tunica fibrosa, tunica vascularis posterior five-sixths—the sclera, and the retina. Judging from the transverse section of the colorless and transparent cornea, it contains five layers: epithelium, anterior limiting lamina, stroma, posterior limiting lamina and endothelium. Tunica vascularis has three components: iris, ciliary body and choroids. From the exterior to the interior, retina consists of four layers of cell: ①pigment epithelium; ②photoreceptor cell, of which there are two types: rods and cones that differ in function. Responding to illumination of low intensity, rods are useful in night vision. In contrary, cones are sensitive to bright and color; ③Bipolar cells. The bipolar neuron connects the rods and cones to the ganglion cells; ④Ganglion cells. They contact with the bipolar cells through their dendrites and send to the brain, axons that converge at the optic papilla forming the optic nerve.

Ear is responsible for creature's equilibrium and hearing. This organ comprises three components: the external ear, the middle ear and the internal ear. The external ear includes auricle, external auditory meatus, which receives auditory stimuli, sound waves. The middle ear includes tympanic cavity and pharyngotympanic tube, in which sound waves are transmitted from air to the internal ear. The internal ear of mammalian comprises two labyrinths, the osseous labyrinth and the membranous labyrinth. The former includes vestibule, osseous semicircular ducts and cochlea, the latter consists of three subcompartments: utricle and saccule, membranous semicircular ducts and membranous cochlea. The membranous labyrinth is permeated inner lymph. The outer lymph is filled between the osseous labyrinth and the membranous labyrinth. The function of the lymph is to supply nutrition for internal ear and convey sound waves.

一、眼和耳的发生

（一）眼的发生

1. 眼球的发生 眼球的发生有三种来源：①神经外胚层形成视网膜和视神经；②表皮外胚层形成晶状体和角膜上皮；③来自神经嵴的中外胚层(mesectoderm)形成角膜的其他成分及巩膜与血管膜等。中外胚层的结构和间充质相似，故称外胚间充质(ectomesenchyme)。

2. 眼附属结构的发生 视杯前方的体表外胚层形成睑褶，即眼睑的原基。睑褶的外层分化成眼睑皮肤，内层形成结膜并与球结膜和角膜上皮相连续。睑褶中部的间充质，发育成睑板、结缔组织和肌组织。上下睑黏合缘外侧，上皮细胞分化成毛囊，长出睫毛。同时，在黏合缘的内侧，上皮呈管状内陷形成睑板腺。泪腺由结膜外侧体表外层上皮发生、分化而来。眶部泪腺出现较早。

（二）耳的发生

耳郭是由胚胎时期第1对鳃沟周围的6个小结节融合形成的。外耳道起源于第1对鳃沟。第一对咽囊内侧部形成咽鼓管，外侧部为盲端，扩大成鼓室。鼓室周围的间充质分化成3块听小骨，它们彼此形成微动关节，以利于声波的传导。胚胎神经孔未闭合之前，菱脑诱导两侧的外胚层上皮增厚形成耳板，继而内陷，最后与外胚层分离，形成耳泡。耳泡的内侧发生一个囊管，形成内淋巴管，囊管的盲端膨大成内淋巴囊，逐渐伸达硬脑膜，并被硬脑膜包围，囊底与硬脑膜紧密接触，形成内耳膜迷路内淋巴液的排出通道。早期耳泡由薄膜组成膜迷路；以后前部发育形成半规管及椭圆囊和球囊，后部形成蜗管的原基，逐渐分化出毛细胞和支持细胞，神经末梢与毛细胞构成突触，发育成位觉感受器。

二、眼和耳的组织学结构概述

（一）眼（eye）

眼由眼球和附属器官组成。眼球由眼球壁和内容物组成。眼球内容物包括房水、晶状体和玻璃体。这些结构清澈透明并有屈光作用。本节重点展示眼球壁的组织结构。眼球壁构成眼球的外壳，由外向内依次为纤维膜、血管膜和视网膜三层。

1. 眼球壁（eyeball wall）

（1）纤维膜（fibrous tunic） 为眼球的最外层。主要由致密结缔组织构成，保护眼球内部结构和维持眼球形态。纤维膜前方1/6为角膜，后5/6为巩膜，两者交界处成角膜缘。

（2）血管膜（vascular tunic） 位于纤维膜的内侧，是眼球壁的中间层，由疏松结缔组织、丰富的血管和色素细胞构成，故又称色素膜（uvea），司营养功能。血管膜自前向后分为虹膜、睫状体和脉络膜。

（3）视网膜（retina） 是眼球壁的最内层，柔软而透明。衬于睫状体和虹膜内面者没有感光作用，称视网膜盲部；衬于脉络膜内面者有感光作用，称视网膜视部。两部分在锯齿缘相移行。一般所说的视网膜就是指视部。

视网膜属高度特化的神经组织。在HE染色切片上可分为十层，而这十层结构，是由四种细胞形成，即色素上皮细胞、视细胞、双极细胞和节细胞。

2. 眼的附属器官

（1）眼睑（eyelid） 是眼的保护器官，其外为皮肤，内面为黏膜，又称为睑结膜，中间为睑板。眼睑的皮肤薄，有许多细毛及不发达的皮脂腺和汗腺。眼睑的边缘有一系列的睫毛，下眼睑的睫毛短，肉食动物和猪的下眼睑完全没有睫毛。马和肉食动物的睑结膜衬以复层柱状上皮；猪和反刍动物则为变移上皮，并含有杯状细胞。

（2）睑板（tarsus） 由致密结缔组织构成，内有分支的复管泡状腺，称睑板腺（tarsal gland）。腺中央有一条直行的排泄管，开口于睑缘。上眼睑的睑板腺比较发达，但猪的很小。睑板腺分泌睑脂，以润泽眼睑。在睑板的外侧面有薄层的横纹肌，为睑匝肌，可闭眼。

(3) 泪腺（lacrimal gland） 位于眼眶上外侧，多为浆液性复管泡状腺，但猪以黏液性腺为主。腺细胞类似浆液腺细胞，核圆形，位于细胞中央。腺体有较长的闰管与排泄管相连。闰管和排泄管的上皮分别为单层立方和柱状上皮，较大的排泄管可由2层上皮细胞组成。猪的大排泄管尚有软骨环绕。

（二）耳（ear）

耳（ear）包括外耳和内耳。

外耳包括耳郭、外耳道和鼓膜。耳郭收集声波，外耳道传导声波到鼓膜。中耳包括鼓室、听小骨和咽鼓管，将声波传到内耳。内耳是位觉和听觉感受器。本节重点展示内耳。内耳位于颞骨岩部，为一些弯曲的管状系统，由于结构复杂，又称迷路。内耳的结构可分为骨迷路和膜迷路两部分。

（1）骨迷路（bony labyrinth） 是颞骨内不规则的腔隙和隧道，腔面覆以骨膜，表面衬以单层扁平上皮。膜迷路位于骨迷路内，两者之间有间隙，其中充满外淋巴。外淋巴间隙借耳蜗导水管与蛛网膜下腔相通连。骨迷路分为骨半规管、前庭和耳蜗。

（2）膜迷路（membranous labyrinth） 位觉感受器和听觉感受器，是一系列的膜性管和囊，悬于骨迷路内。骨半规管内有膜半规管；前庭内有球囊和椭圆囊，二者借Y形小管相连；耳蜗内有膜蜗管。膜半规管、球囊和椭圆囊壁的结构基本相同，都是由上皮和固有层构成。

三、感觉器官图谱

1. 眼 图20-1～图20-28。
2. 耳 图20-29～图20-36。

第二十章 感觉器官 Sensory Organs

图 20-1 马的感觉器官
Fig.20-1 the sensory organ of horse

鼻—nose 眼球—eyeball 外耳—auricle

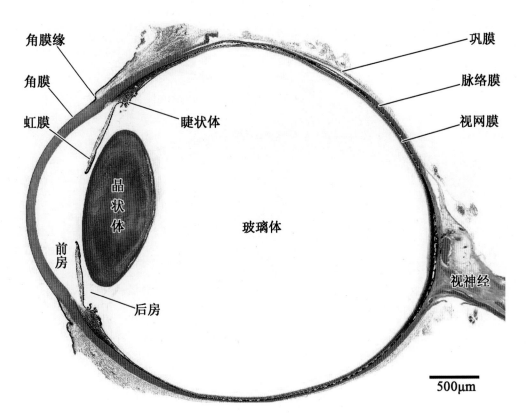

图 20-2 眼球纵切低倍像（HE）
Fig.20-2 low amplification of eyeball longitudinal section（HE）

角膜缘—corneal limbus 角膜—cornea 虹膜—iris 前房—atria 后房—posterior 晶状体—lens 睫状体—ciliary body
玻璃体—vitreum 巩膜—sclera 脉络膜—choroid 视网膜—retina 视神经—optic nerve

739

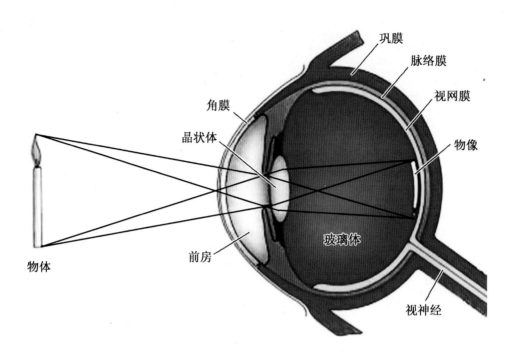

图 20-3　眼球的构造及其成像的原理模式图

Fig.20-3　structure pattern diagram and image-forming principle of eyeball

物体—object　角膜—cornea　晶状体—lens　前房—atria　玻璃体—vitreum　巩膜—sclera
脉络膜—choroid　视网膜—retina　物像— objective image　视神经—optic nerve

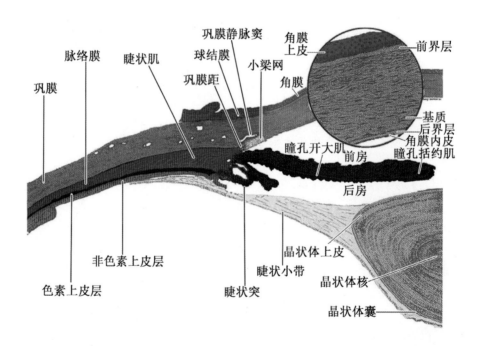

图 20-4　眼球前半部的结构模式图

Fig.20-4　structure pattern diagram of eyeball front part

巩膜— sclera　脉络膜—choroid　睫状肌—ciliaris　巩膜距—scleral spur　球结膜—bulbar conjunctiva
巩膜静脉窦—sinuses circularis iridis　小梁网—trabecular meshwork　角膜—cornea　角膜上皮—corneal epithelium
前界层—anterior limiting lamina　基质—matrix　后界层—posterior limiting lamina　角膜内皮—corneal endothelium
瞳孔括约肌—sphincter pupillae muscle　瞳孔开大肌—dilator pupillae muscle　前房—atria　后房—posterior
睫状突—ciliary process　睫状小带—ciliary zonule　晶状体上皮—epithelium lentis　晶状体核—nuclei lentis
晶状体囊—phacocyst　色素上皮层—pigment epithelial layer　非色素上皮层—non-pigmented epithelium

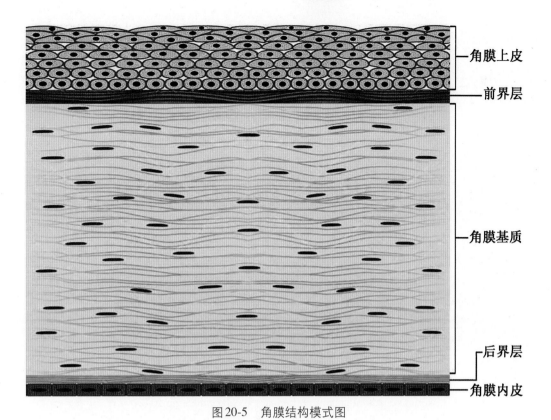

图20-5　角膜结构模式图

Fig.20-5　structure pattern diagram of cornea

角膜上皮—corneal epithelium　前界层—anterior limiting lamina　角膜基质—corneal matrix

后界层—posterior limiting lamina　角膜内皮—corneal endothelium

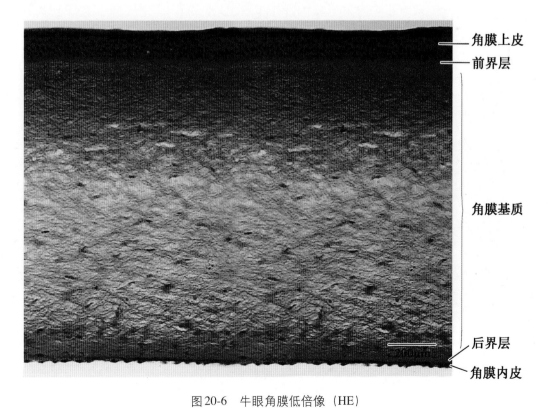

图20-6　牛眼角膜低倍像（HE）

Fig.20-6　low amplification of cow eye cornea（HE）

角膜上皮—corneal epithelium　前界层—anterior limiting lamina　角膜基质—corneal matrix

后界层—posterior limiting lamina　角膜内皮—corneal endothelium

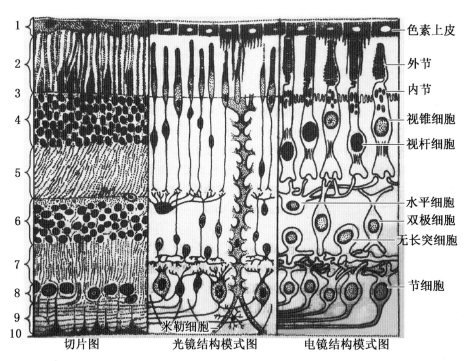

图20-7 视网膜结构模式图（1～10表示视网膜的10层）

Fig.20-7 structure pattern diagram of retina（1～10 indicating 10 layers of retina）

切片图—section at high magnification 光镜结构模式图—light microstructure pattern diagram
电镜结构模式图—ultrastructure pattern diagram 米勒细胞—Müller's cell 色素上皮—pigment epithelium
外节—acromere 内节—inner segment 视锥细胞—cone cell 视杆细胞—rod cell 水平细胞—horizontal cell
双极细胞—bipolar cell 无长突细胞—amacrine cell 节细胞—ganglion cell

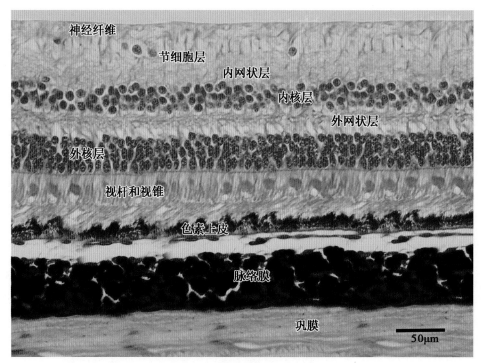

图20-8 牛视网膜高倍像（HE）

Fig.20-8 high magnification of cattle retina（HE）

神经纤维—nerve fiber 节细胞层—ganglion cell layer 内网状层—inner reticular layer 内核层—inner nuclear layer
外网状层—outer reticular layer 外核层—outer nuclear layer 视杆和视锥—rods and cones 色素上皮—pigment epithelium
脉络膜—choroid 巩膜—sclera

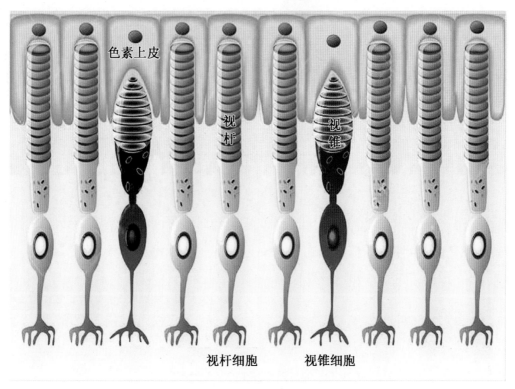

图 20-9　感光细胞模式图

Fig.20-9　pattern diagram of photoreceptor cells

色素上皮—pigment epithelium　视杆—rod　视锥—cone　视杆细胞—rod cell　视锥细胞—cone cell

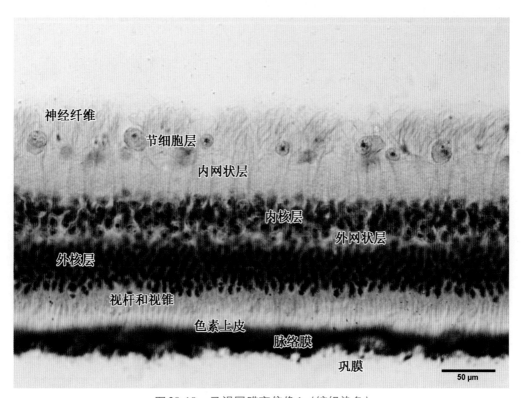

图 20-10　马视网膜高倍像 1（镀银染色）

Fig.20-10　high magnification of horse retina 1（silver stain）

神经纤维—nerve fiber　节细胞层—ganglion cell layer　内网状层—inner reticular layer　内核层—inner nuclear layer
外网状层—outer reticular layer　外核层—outer nuclear layer　视杆和视锥—rods and cones　色素上皮—pigment epithelium
脉络膜—choroid　巩膜—sclera

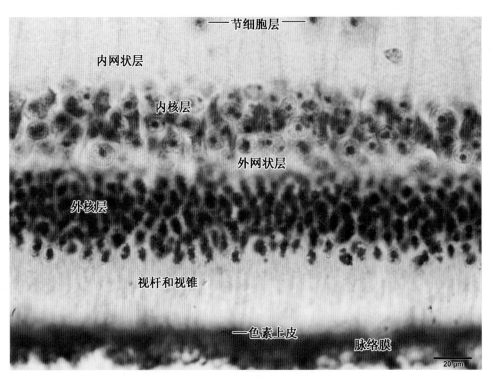

图20-11　马视网膜高倍像2（镀银染色）

Fig.20-11　high magnification of horse retina 2（silver stain）

节细胞层—ganglion cell layer　内网状层—inner reticular layer　内核层—inner nuclear layer
外网状层—outer reticular layer　外核层—outer nuclear layer　视杆和视锥—rods and cones
色素上皮—pigment epithelium　脉络膜—choroid

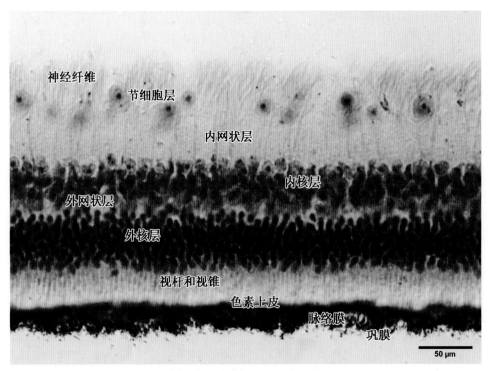

图20-12　猪视网膜高倍像1（镀银染色）

Fig.20-12　high magnification of pig retina 1（silver stain）

神经纤维—nerve fiber　节细胞层—ganglion cell layer　内网状层—inner reticular layer　内核层—inner nuclear layer
外网状层—outer reticular layer　外核层—outer nuclear layer　视杆和视锥—rods and cones　色素上皮—pigment epithelium
脉络膜—choroid　巩膜—sclera

第二十章 感觉器官 Sensory Organs

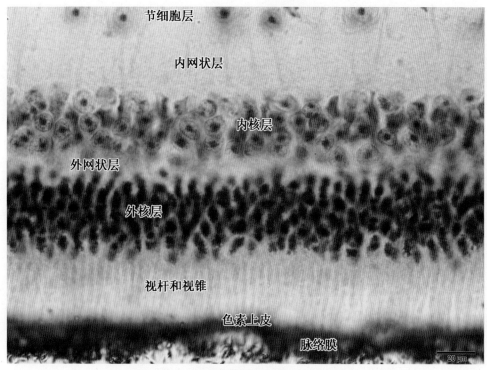

图20-13 猪视网膜高倍像2（镀银染色）

Fig.20-13 high magnification of pig retina 2（silver stain）

节细胞层—ganglion cell layer 内网状层—inner reticular layer 内核层—inner nuclear layer
外网状层—outer reticular layer 外核层—outer nuclear layer 视杆和视锥—rods and cones 色素上皮—pigment epithelium
脉络膜—choroid

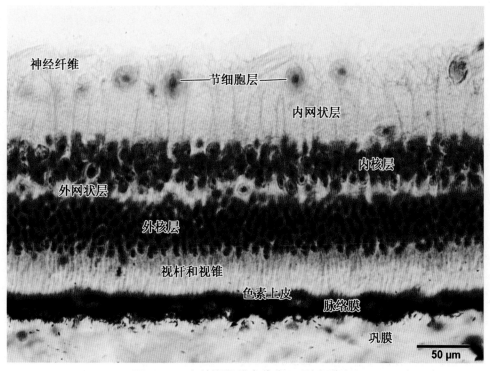

图20-14 山羊视网膜高倍像1（镀银染色）

Fig.20-14 high magnification of goat retina 1（silver stain）

神经纤维—nerve fiber 节细胞层—ganglion cell layer 内网状层—inner reticular layer 内核层—inner nuclear layer
外网状层—outer reticular layer 外核层—outer nuclear layer 视杆和视锥—rods and cones 色素上皮—pigment epithelium
脉络膜—choroid 巩膜—sclera

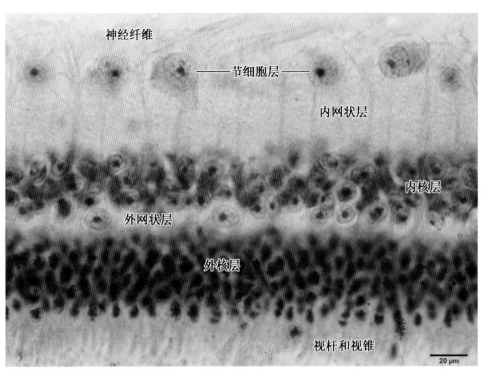

图20-15　山羊视网膜高倍像2（镀银染色）

Fig.20-15　high magnification of goat retina 2（silver stain）

神经纤维—nerve fiber　节细胞层—ganglion cell layer　内网状层—inner reticular layer　内核层—inner nuclear layer
外网状层—outer reticular layer　外核层—outer nuclear layer　视杆和视锥—rods and cones

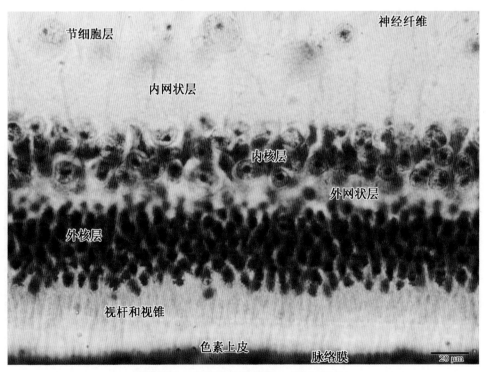

图20-16　犬视网膜高倍像1（镀银染色）

Fig.20-16　high magnification of dog retina 1（silver stain）

神经纤维—nerve fiber　节细胞层—ganglion cell layer　内网状层—inner reticular layer　内核层—inner nuclear layer
外网状层—outer reticular layer　外核层—outer nuclear layer　视杆和视锥—rods and cones　色素上皮—pigment epithelium
脉络膜—choroid

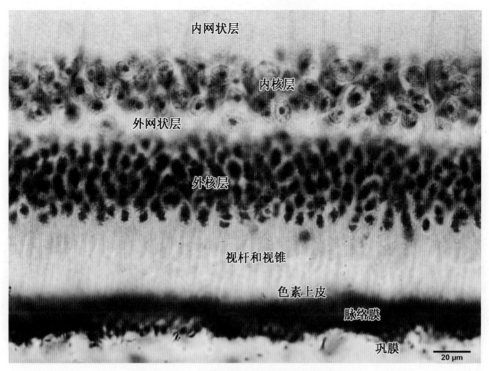

图 20-17　犬视网膜高倍像 2（镀银染色）

Fig.20-17　high magnification of dog retina 2（silver stain）

内网状层—inner reticular layer　内核层—inner nuclear layer　外网状层—outer reticular layer　外核层—outer nuclear layer　视杆和视锥—rods and cones　色素上皮—pigment epithelium　脉络膜—choroid　巩膜— sclera

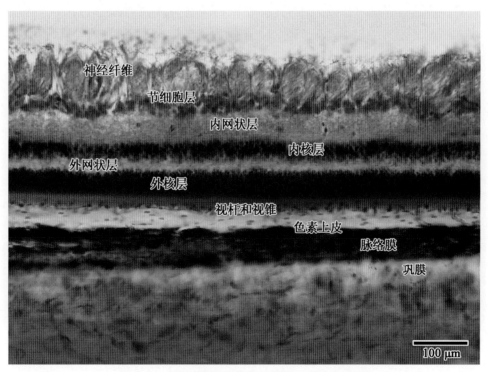

图 20-18　猫视网膜高倍像（镀银+中性红染色）

Fig.20-18　low magnification of cat retina（silver + neutral red stain）

神经纤维—nerve fiber　节细胞层—ganglion cell layer　内网状层—inner reticular layer　内核层—inner nuclear layer　外网状层—outer reticular layer　外核层—outer nuclear layer　视杆和视锥—rods and cones　色素上皮—pigment epithelium　脉络膜—choroid　巩膜— sclera

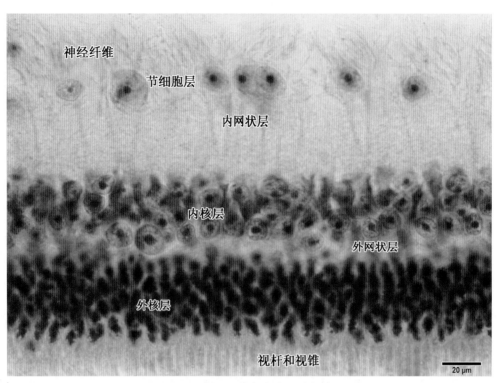

图 20-19 猫视网膜高倍像（镀银染色）

Fig.20-19 high magnification of cat retina（silver stain）

神经纤维—nerve fiber 节细胞层—ganglion cell layer 内网状层—inner reticular layer 内核层—inner nuclear layer 外网状层—outer reticular layer 外核层—outer nuclear layer 视杆和视锥—rods and cones

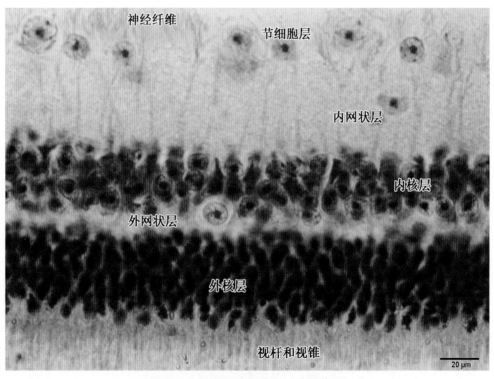

图 20-20 兔视网膜高倍像（镀银染色）

Fig.20-20 high magnification of rabbit retina（silver stain）

神经纤维—nerve fiber 节细胞层—ganglion cell layer 内网状层—inner reticular layer 内核层—inner nuclear layer 外网状层—outer reticular layer 外核层—outer nuclear layer 视杆和视锥—rods and cones

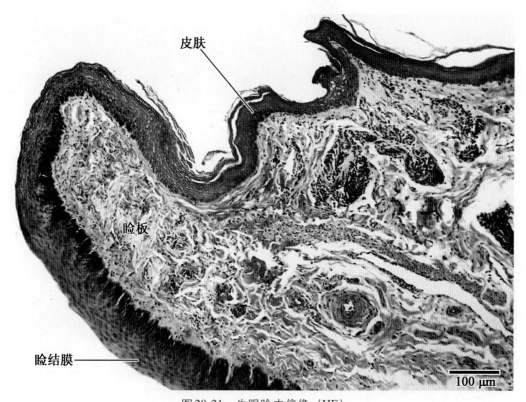

图20-21　牛眼睑中倍像（HE）

Fig.20-21　mid magnification of cattle eyelid（HE）

皮肤—skin　睑板—tarsus　睑结膜—palpebral conjunctiva

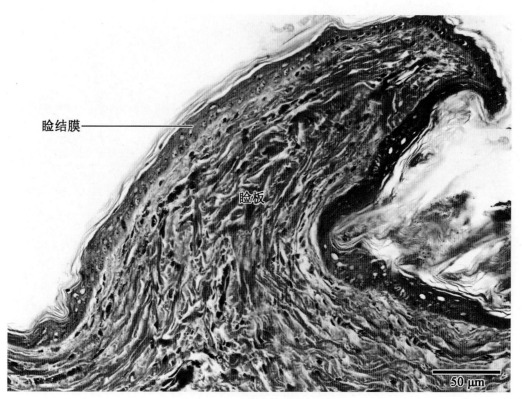

图20-22　马眼睑高倍像（HE）

Fig.20-22　high magnification of horse eyelid（HE）

睑结膜—palpebral conjunctiva　睑板—tarsus

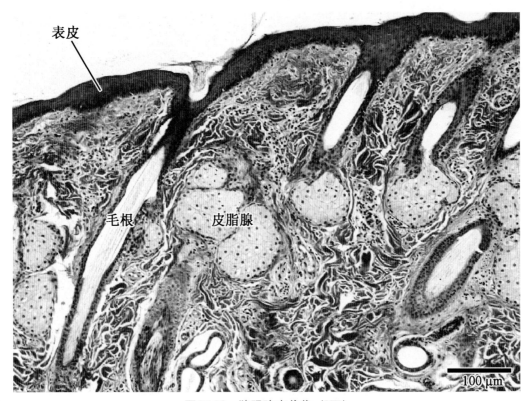

图 20-23　猪眼睑中倍像（HE）

Fig.20-23　mid magnification of pig eyelid（HE）

表皮—epidermis　毛根—hair root　皮脂腺—sebaceous gland

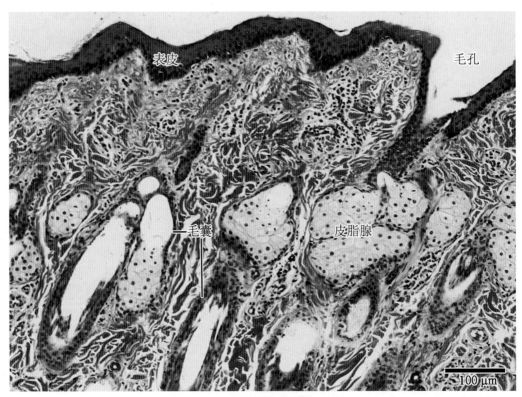

图 20-24　山羊眼睑中倍像（HE）

Fig.20-24　mid magnification of goat eyelid（HE）

表皮—epidermis　毛囊—hair follicle　毛孔—trichopore　皮脂腺—sebaceous gland

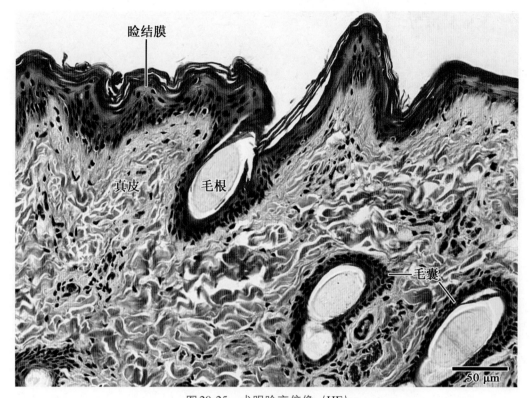

图20-25 犬眼睑高倍像（HE）

Fig.20-25　high magnification of dog eyelid（HE）

睑结膜—palpebral conjunctiva　真皮—dermis　毛根—hair root　毛囊—hair follicle

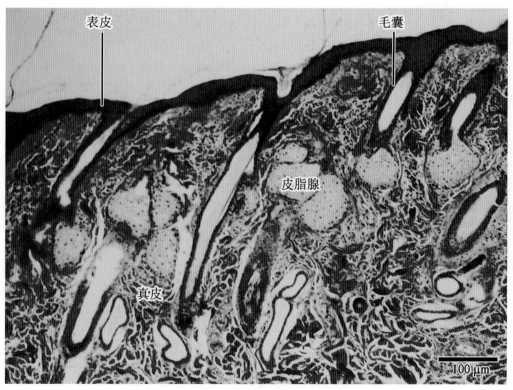

图20-26 猫眼睑低倍像（HE）

Fig.20-26　low magnification of cat eyelid（HE）

表皮—epidermis　真皮—dermis　毛囊—hair follicle　皮脂腺—sebaceous gland

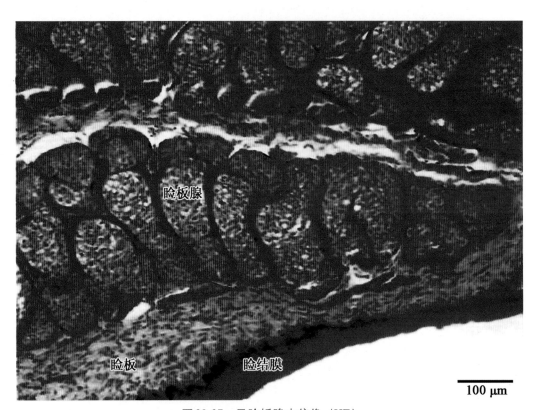

图 20-27　马睑板腺中倍像（HE）

Fig.20-27　mid magnification of horse（HE）

睑板腺—tarsal glands　睑板—tarsus　睑结膜—palpebral conjunctiva

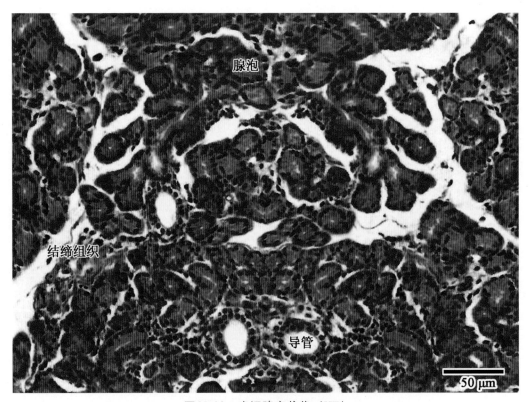

图 20-28　牛泪腺高倍像（HE）

Fig.20-28　high magnification of cattle lacrimal gland（HE）

腺泡—acinus　结缔组织—connective tissue　导管—duct

第二十章 感觉器官　Sensory Organs

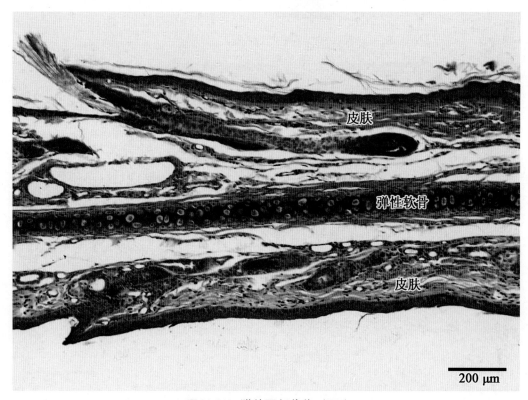

图20-29　猫外耳低倍像（HE）

Fig.20-29　low magnification of cat auricle（HE）

皮肤—skin　弹性软骨—elastic cartilage

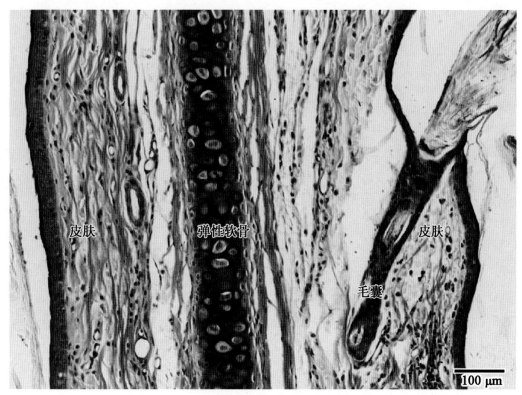

图20-30　猫外耳中倍像（HE）

Fig.20-30　mid magnification of cat auricle（HE）

皮肤—skin　弹性软骨—elastic cartilage　毛囊—hair follicle

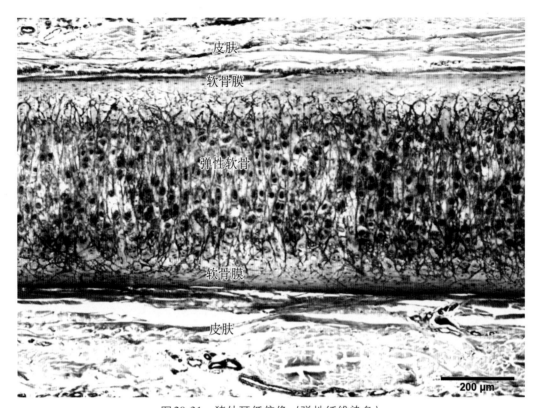

图20-31 猪外耳低倍像（弹性纤维染色）

Fig.20-31 low magnification of pig auricle（elastic fiber stain）

皮肤—skin 软骨膜—perichondrium 弹性软骨—elastic cartilage

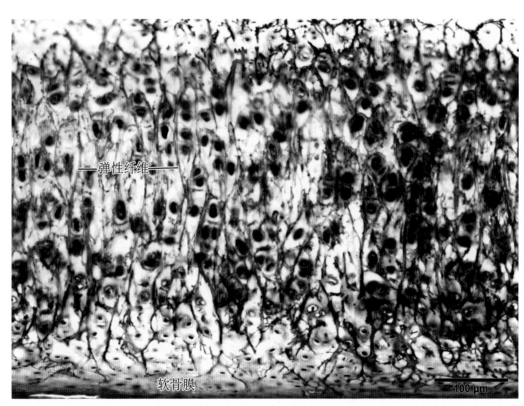

图20-32 猪外耳中倍像（弹性纤维染色）

Fig.20-32 mid magnification of pig auricle（elastic fiber stain）

弹性纤维—elastic fiber 软骨膜—perichondrium

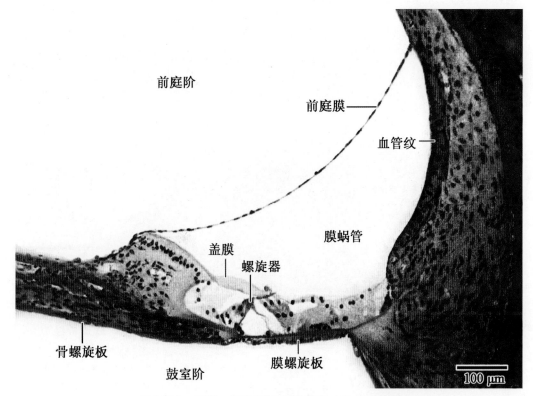

图20-33 耳蜗、膜蜗管和螺旋器的中倍像（HE）

Fig.20-33 mid magnification of cochlea, membranous cochlea and spiral organ（HE）

前庭阶—vestibular canal 前庭膜—vestibular membrane 血管纹—stria vascularis 盖膜—velum 螺旋器（科尔蒂器）—spiral organ (organ of Corti) 膜蜗管—membranous cochlea 骨螺旋板—osseous spiral lamina 鼓室阶—scala tympanum 膜螺旋板—membranous spiral lamina

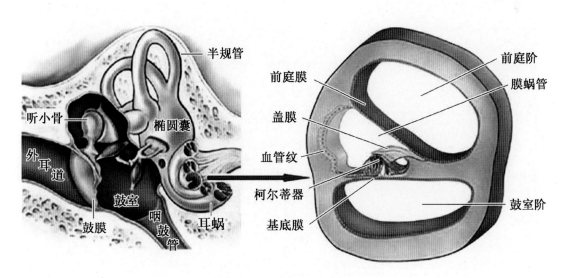

图20-34 耳的结构模式图

Fig.20-34 structure pattern diagram of ear

半规管—semicircular canal 听小骨—auditory ossicle 椭圆囊—alveus communis 外耳道—external auditory canal 鼓膜—eardrum 鼓室—tympanum 咽鼓管—auditory tube 耳蜗—cochlea 前庭膜—vestibular membrane 盖膜—velum 血管纹—stria vascularis 柯尔蒂器—organ of Corti 基底膜—basilar membrane 前庭阶—vestibular canal 膜蜗管—membranous cochlea 鼓室阶—scala tympanum

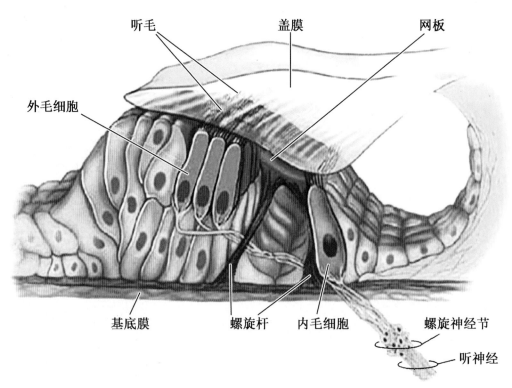

图 20-35　膜蜗管结构模式图

Fig.20-35　structure pattern diagram of membranous cochlea

外毛细胞—outer hair cell　听毛—auditory hair　盖膜—velum　网板—halftone　基底膜—basilar membrane
螺旋杆—spiral rod　内毛细胞—inner hair cell　螺旋神经节—spiral ganglion　听神经—auditory nerve

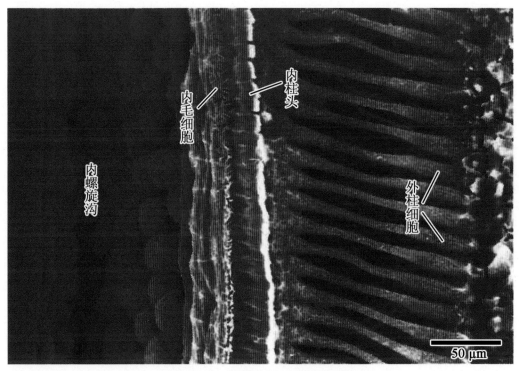

图 20-36　螺旋器局部扫描电镜像

Fig.20-36　scanning electron microphoto of spiral organ

内螺旋沟—inner spiral sulcus　内毛细胞—inner hair cell　内柱头—head of inner pillar　外柱细胞—outer pillar cell

第二十一章
禽类的主要组织结构特征
Main Structural Features of Fowl Tissues

Outline

The microscopic structures of avian tissues are basically similar to that of mammals. This chapter shows the main structural features of poultry.

The erythrocytes of birds are elliptic with oval nuclei. The structural characteristics of avian leukocytes are heterophilic granulocytes, which are equivalent to mammalian neutrophils. Avian thrombocytes are equivalent to mammalian platelets and have a complete cellular structure.

The lymphocytes in the thymus cortex of birds are closely arranged and the germinal center is not obvious. Lymphocytes in the medulla are few, and the distribution is relatively rare. The nuclei of epithelial reticular cells are clearly visible and lightly stained. Thymus corpuscles are small and diffuse.

Bursa of Fabricius is a unique central immune organ of poultry. The bursa of young birds are developed, and the volume of the bursa is the largest when sexually mature. It gradually shrinks and degenerates later. The structure of bursa of Fabricius is similar to that of the digestive tract, with a 4-layer of mucosa, submucosa, muscularis and tunica adventitia. Mucous membrane formed large longitudinal plica to the lumen, and the mucous epithelium was pseudostratified columnar epithelium or simple columnar epithelium. Lamina propria is thick, with many densely arranged bursal nodules. Each nodule is composed of the peripheral cortex and the central medulla. The medulla is composed of epithelial reticular cells, large and medium-sized lymphocytes, and macrophages.

The spleen of poultry is small and the tissue structure is characterized by an indistinct boundary between the red and white pulp. The spleen has no blood storage effect due to its small size.

Avian lymph nodes are found only in waterfowls, and the larger ones include cervical and thoracic lymph nodes and lumbar lymph nodes. Its structure is characterized by no hilum, no cortex and no medulla.

The cervical segment of the poultry esophagus is long, and there are esophageal glands in the wall of the esophagus. The esophagus of chicken and pigeon forms enlarged ingluvies on the ventral side before entering the thorax. Ducks and geese do not form an ingluvies, but form a spindle-shaped bulge in the rear of the esophagus neck. The ingluvies is similar in structure to the esophagus.

The glandular stomach of birds is fusiform, with a thick wall and four layers of structure. There are many round nipples on the mucosal surface with the single columnar epithelium, which can secrete mucus. The lamina propria contains many tubular glands and more immune tissues. There is a developed glandular stomach gland in the submucosa, which is equivalent to the fundus gland of mammals.

The muscular stomach also has four layers. The surface of mucous membrane is covered with a thick and rough keratin-like membrane. It is formed by the combination of secretions from myogastric glands, epithelial secretions and exfoliated epithelial cells in the acidic environment to protect the mucosa. The epithelium is simple columnar and depressions form many infundibular crypts. There are many parallel branching tubular glands in the lamina propria, namely the myogastric glands.

The large intestine of birds has a pair of cecum, no colon, and the end of the rectum expands into a cloaca. Both small intestine and large intestine have intestinal villi, but there is no central chylous duct in the villi. There is no duodenal gland in the submucosa of the duodenum. The entire intestinal wall is rich in diffuse lymphoid tissue or lymphoid nodules.

Poultry liver tissue structure characteristics: the connective tissue between the lobules is not well developed, so the boundary between the lobules is unclear, with the central vein as the center, liver cells are arranged into hepatocyte tubes, arranged in a radial pattern, there are lymphoid tissues in the liver lobules, and the pigeon liver has no gallbladder.

The various bronchi in the lungs of the bird communicate with the air sac. The pulmonary parenchyma is composed of bronchi, pulmonary chambers and pulmonary capillaries. After entering the lung, the bronchus forms a primary bronchus that runs through the entire lung. The diameter of the bronchus gradually narrows, and the end of the bronchus passes through the abdominal airbag. Secondary bronchus of varying thickness are issued along the way. The tertiary bronchi spread throughout the lung, forming loops with each other and communicating with the secondary and primary bronchi. Therefore, the lungs of birds don't form bronchial trees.

The kidney of birds is located between the sacrum and ilium, with anterior, middle and posterior segments. The cortex and medulla are unclearly demarcated, and there is no calyces, renal pelvis and hilum. There is lymphoid tissue in the renal parenchyma.

The testis of birds is located in the abdominal cavity throughout life, without mediastinum or septum, and no lobular structure. The convoluted tubules are slender and curved, branching, and anastomose into a net, so the shape of the cross section is very irregular. The tissue structure is similar to that of mammals.

The female reproductive organs in birds have ovaries and oviducts, only the left side develops normally, and the right side degrades in the embryo period. There are several large follicles and many small follicles on the surface of the ovary during the spawning period. The ovaries retract after spawning stops. Structural characteristics of poultry ovary: there is no follicular cavity or follicular fluid in the follicles, the follicle wall quickly degenerates after ovulation, and no corpus luteum is formed.

Poultry oviduct are long and curved, thickening during the laying period, and shortening and thinning during the rest period. The oviduct is divided into five sections according to structure and function: funnel, dilatation, isthmus, uterus and vagina. The mucosal surface of each segment has plica, the mucosal

第二十一章 禽类的主要组织结构特征 Main Structural Features of Fowl Tissues

epithelium has cilia, and most of the lamina propria has glands and lymph tissues.

The spinal cord of birds extends to the cauda bone in the spinal canal, without forming the "cauda equina". The dorsal part of the lumbar enlargement splits to form a rhomboid sinus with glial cell clusters, also known as glycogen. In the cervical and lumbar enlargements, some neurons in the ventral gray column migrate to the peripheral white matter and form the limbic nucleus.

The brainstem of the bird includes the developed medulla oblongata and midbrain without obvious pons. The dorsolateral part of the midbrain has a well-developed optic lobe that corresponds to the mammalian anterior colliculus. There is a lateral midbrain nucleus behind the optic lobes, which is equivalent to the caudal colliculus of mammals. The diencephalon is shorter and has no mamillary body. Cerebellum is well developed, among them vermis is particularly developed, flanking is fluff ball. The main structure of the brain is the basal ganglia, and the striatum is developed, which is an important motor integration center. The cerebral cortex is thin and smooth without sulci and gyri.

The comb is a special skin fold with a thin, dermal layer rich in capillary plexus. The capillaries of sexually mature roosters and laying hens are highly congested and the crowns are bright red and thick. The deep dermal connective tissue is rich in fibers and mucinous material fills the gaps, keeping the crown upright.

Preen gland is the only skin gland of poultry, located under the skin of the dorsal part of the pygostyle. It is small in chicken and developed in waterfowl. There is a connective tissue capsule on the surface and it penetrates inward to divide the parenchyma into left and right lobes. In the center of each lobe, there is a large glandular cavity, which is full of secretions, containing fat, lecithin, advanced alcohol and glycogen, etc. Coated on the feathers through the beak, which can moisten the feather and avoid flooding. It is very important for waterfowl.

禽类（avian）组织的微细结构与哺乳动物的基本相似。本章展示禽类主要的组织结构特点。

一、禽类的主要组织结构特点概述

（一）血细胞

禽类的红细胞呈椭圆形，有椭圆形的细胞核，胞质内除含有大量血红蛋白外，还有线粒体和高尔基复合体等细胞器。

禽类的白细胞结构的特点是含有异嗜性粒细胞（heterophilic granulocyte），相当于哺乳动物的中性粒细胞。其胞质内有许多呈杆状或纺锤形的嗜酸性颗粒，染成暗红色。

禽类的血栓细胞（thrombocyte）相当于哺乳动物的血小板，并具有完整的细胞结构。胞体呈椭圆形，内有椭圆形的细胞核。细胞质呈弱嗜碱性，其中有少量嗜天青颗粒。

（二）免疫器官

禽类的免疫组织广泛分布于许多器官的壁内和间质内，多为弥散淋巴组织，有的可形成淋巴小结，并有生发中心；在有些部位则形成特殊的免疫组织，如小肠淋巴集结、盲肠扁桃体和哈德腺。禽类免疫组织还形成一些较大的免疫器官。

1. 胸腺 家禽的胸腺呈黄色或灰红色，分布于颈部的两侧，鸡的每侧有5～7叶，鸭的每侧有4～6叶，鹅的每侧有4～5叶。在近胸腔入口处，后部的胸腺常与甲状腺、甲状旁腺及腮后体紧密相连，彼此间无结缔组织分隔。胸腺在性成熟前体积最大，以后逐渐萎缩退化。

胸腺外面均包有一层较薄的结缔组织被膜，其中含较多的胶原纤维和少量弹性纤维。被膜伸入实质，分出许多不完全的小叶，每小叶又分着色较深的皮质和着色较浅的髓质。

皮质的淋巴细胞排列紧密，使上皮网状细胞和其他细胞被掩盖，故生发中心不明显。髓质的淋巴细胞较少，分布较稀。上皮网状细胞的核清晰可见，染色质少，着色较淡。网状细胞退化时，首先胞质出现小空泡，逐步扩大，最后形成有界膜的胸腺小体。因此，家禽胸腺小体小，呈弥散状态。此外，在髓质中还分布着有粒白细胞、浆细胞等。

2. 腔上囊（bursa of Fabricius） 或称法氏囊，是禽类特有的中枢免疫器官，在泄殖腔背侧，球形或长椭圆形，有一短管通肛道。幼禽的腔上囊发达，性成熟时体积最大，以后逐渐萎缩退化。腔上囊起源于泄殖腔，其结构保留与消化管结构相似的特点，由内向外依次分为黏膜、黏膜下层、肌层和外膜4层结构。

（1）黏膜　由上皮和固有层构成，无黏膜肌层。向囊腔突起形成较大的纵行皱襞，大的皱襞之间有许多小皱襞。黏膜上皮为假复层柱状上皮，局部为单层柱状上皮。固有层较厚，有许多密集排列的圆形、卵圆形或不规则形腔上囊小结（bursal nodule）。每个小结由周边的皮质和中央的髓质构成。髓质由上皮网状细胞、大中型淋巴细胞和巨噬细胞组成。上皮网状细胞彼此间借突起相互连接成支架，淋巴细胞位于网眼内并不断进行分裂分化，新形成的部分淋巴细胞被巨噬细胞吞噬。皮质由密集的中小型淋巴细胞、巨噬细胞和上皮网状细胞构成，内有毛细血管分布。在皮质、髓质交界处，有一层连续的上皮细胞和完整的基膜，并与黏膜表面的上皮和基膜相连续。

（2）黏膜下层　较薄，由疏松结缔组织构成，参与形成黏膜皱襞。

（3）肌层　由内纵、外环两层较薄的平滑肌构成。

（4）外膜　为浆膜。

3. 脾　家禽的脾较小，呈红棕色或紫红色，位于腺胃与肌胃交界处的右背侧。鸡脾呈球形，鸭和鹅的脾则为扁卵圆形。家禽脾的组织结构特点是红髓与白髓的分界不明显。脾的血液通过脾动脉、小梁动脉、中央动脉、笔毛动脉，而后汇入动脉毛细血管，此血管直接开口于脾索的网状组织中，因此家禽的脾属开放循环。血液先在网状细胞突起所形成的网眼内流通，再汇入静脉毛细血管。家禽的脾因体积小而无储血作用。

4. 淋巴结　禽类的淋巴结仅见于水禽，体积较大的有颈胸淋巴结和腰淋巴结。其结构特点是无门部，也无皮质、髓质之分，其实质由中央窦、淋巴小结、弥散淋巴组织、淋巴索和周围淋巴窦等构成。中央窦形状不规则，有输入和输出淋巴管与其相连，并有分支与周围淋巴窦相通。

（三）消化器官

1. 食管与嗉囊（ingluvies） 禽类食管的颈段长，食管壁内有食管腺。鸡与鸽的食管入胸腔前，其腹侧形成膨大的薄壁憩室，称嗉囊。鸭、鹅不形成嗉囊，而在食管颈段后部形成一个纺锤形的膨大部。嗉囊的组织结构与食管相似，但固有层富含弹性纤维和免疫组织，鸭、鹅食管膨大部的固有层内有黏液腺分布。鸽的嗉囊是两个对称的囊，黏膜内分布有混合腺。在鸽抱窝后期及育雏早期，雌鸽和雄鸽嗉囊的黏膜上皮迅速增生，浅层细胞大量增生后脱落，与腺体的分泌物共同形成"鸽乳"，通过逆吐，哺育幼鸽。

2. 腺胃（glandular stomach） 腺胃也称前胃，呈纺锤形，壁厚，结构分为4层。

（1）黏膜　表面有许多圆形乳头，乳头中央有深层腺胃腺的开口，孔的周围有呈同心圆排列的皱襞和沟。上皮为单层柱状上皮，胞质弱嗜碱性，可分泌黏液。固有层内含许多管状腺和较多的免疫组织。管状腺较短，由黏膜上皮向固有层内下陷形成单管状腺或分支管状腺，分泌黏液。管壁衬有单层立方上皮或单层柱状上皮。黏膜肌层为薄层纵行平滑肌。

（2）黏膜下层　较厚，有发达的腺胃腺（glandular stomach gland），也称前胃腺，相当于哺乳动物的胃底腺。腺胃腺体积大，数量多，呈圆形或椭圆形，为复管状腺。腺中央为集合窦，窦周围有呈辐射状排列的腺小管。腺小管由单层腺细胞构成，细胞形态呈立方形或矮柱状，胞质嗜酸性，有许多分泌颗粒，胞核呈圆形或卵圆形，位于细胞基部。相邻腺细胞的近游离端彼此间有小间隙，所以腺上皮的游离面呈锯齿状。相邻多个腺小管共同开口于短的三级管，数个三级管开口于集合窦。相邻腺体的集合窦再汇合成大导管，最后开口于黏膜乳头。腺胃腺的细胞可分泌胃蛋白酶原和盐酸，兼有胃底腺内主细胞和壁细胞的双重功能。黏膜和黏膜下层共同形成许多黏膜皱襞。

(3) 肌层　由内纵、中环、外纵3层平滑肌构成。

(4) 外膜　为浆膜。

3. 肌胃（muscular stomach）　肌胃也称砂囊（gizzard），结构分4层。

(1) 黏膜　由上皮和固有层构成，无黏膜肌。黏膜的表面覆盖一层厚而粗糙的类角质膜，即肫皮，中药称鸡内金。它是由肌胃腺的分泌物、上皮的分泌物和脱落的上皮细胞共同在酸性环境下结合硬化而成，可保护黏膜、抵抗蛋白酶和酸性物质侵蚀。上皮为单层柱状上皮。上皮下陷形成许多漏斗状的隐窝。固有层的疏松结缔组织内有许多平行排列的、细而直的分支管状腺，即肌胃腺。腺的顶部开口于隐窝，腺管由单层上皮构成。细胞呈矮柱状或立方形，胞质嗜酸性，内有许多细小颗粒，胞核位于细胞基部。腺腔狭小，充满腺细胞分泌的液态物质。腺细胞的分泌物经隐窝流出，遍布于黏膜上皮表面，位于原已形成的类角质膜的下方。来自腺胃的盐酸可透过类角质膜进入黏膜表面，使液态物质的pH降低而硬化，形成新的类角质膜，以补充表面被磨损的部分。腺底部的细胞有较强的增殖能力，并不断向表面移行以补充脱落的黏膜上皮。

(2) 黏膜下层　很薄，由致密结缔组织构成，内含较多的胶原纤维和弹性纤维。

(3) 肌层　由特别发达的环形平滑肌构成。整个肌层分成2块很厚的侧肌和2块较薄的中间肌，彼此间借腱组织连接，形成肌胃两侧的中央腱膜，称腱镜。由于4块肌肉彼此相连，神经冲动可迅速扩散至整个肌层，可使肌胃发生快速而一致的有力收缩。

(4) 外膜　为浆膜。

4. 肠　禽类肠分小肠和大肠。大肠特殊：盲肠一对，无结肠，直肠末端膨大成泄殖腔。

小肠和大肠的组织结构相似，管壁分4层，均有肠绒毛。十二指肠的绒毛最长，且有分支，向后逐渐变粗变短，分支减少。绒毛内无乳糜管，黏膜上皮吸收的甘油和脂肪酸等被重新合成为乳糜微粒后进入毛细血管。黏膜上皮为单层柱状上皮，由柱状细胞、杯状细胞和内分泌细胞组成。肠腺为单管状腺或分支管状腺，短而直，细胞成分同黏膜上皮。十二指肠的黏膜下层内无十二指肠腺。整个肠壁内富含弥散淋巴组织，局部形成淋巴小结或淋巴集结。黏膜下层很薄，局部缺如。肌层和外膜的组织结构同哺乳动物。

5. 肝　禽类的肝分左右两叶。左叶的肝管直接开口于十二指肠，右叶的脏面有胆囊（鸽无胆囊），肝管先通过胆囊，再由其发出胆管开口于十二指肠。胆囊的结构与哺乳动物相似。

禽类肝的组织有如下结构特点。①小叶间结缔组织不发达，因此小叶分界不清，根据中央静脉和门管区的位置来判定。②以中央静脉为中心，肝细胞排列成肝细胞管呈辐射状排列，且相互吻合。肝细胞管的中央形成胆小管。相邻肝细胞管之间的间隙为肝血窦，窦壁由内皮构成。窦周间隙位于肝细胞管与内皮之间。③在肝小叶内和小叶间结缔组织中有弥散淋巴组织。

(四) 肺

禽类的肺呈鲜红色，紧贴胸壁，其背侧嵌入椎肋间隙。肺内的各级支气管与气囊相通。肺门位于腹侧的前部。肺表面被覆浆膜，内富含弹性纤维。肺实质由各级支气管、肺房及肺毛细管组成。支气管入肺后形成纵贯全肺的初级支气管，其管径逐渐变细，末端与腹气囊相通，沿途发出粗细不等的次级支气管，伸向腹内侧、背内侧、腹外侧和背外侧。次级支气管的末端与颈、胸气囊相通，沿途分出三级支气管。三级支气管遍布全肺，彼此吻合成襻状，并沟通次级和初级支气管。所以，禽肺无哺乳动物支气管树的结构。每条三级支气管与周围许多呈辐射状排列的肺房相通，每一肺房又连着许多肺毛细管。每条三级支气管及其所属分支共同构成一个肺小叶，其横切面呈多边形，中央是一条三级支气管的横切面，周围有许多肺房开口。因相邻肺小叶的部分肺毛细管相互吻合，因此小叶间结缔组织不完整。

三级支气管（tertiary bronchus）或副支气管（parabronchus）相当于哺乳动物的肺泡管，是肺小叶的中心。黏膜表面衬以单层立方或单层扁平上皮，上皮周围有少量结缔组织，内有许多弹性纤维。平滑肌排成较细的螺旋形肌束，在肺房开口处有肌束环绕。肺房（lung atria）为不规则的囊腔，相当于哺乳动物的肺泡囊。肺房内壁为不完整的单层扁平上皮，上皮外有弹性纤维包绕。肺房底壁形成许多小的隐窝，与肺毛细管相通。

（五）肾

禽类的肾呈红褐色，位于综荐骨与髂骨间的肾窝内，质软而脆，长条状，可分前、中、后3段。肾表面无脂肪囊和完整的被膜，无典型的肾锥体和肾叶，皮质和髓质分界不清，没有肾盏、肾盂和肾门。血管、神经和输尿管从不同部位进出肾。

肾表面有极薄的结缔组织被膜。结缔组织伸入实质，形成肾的间质，内有免疫组织和丰富的毛细血管。实质主要由大量肾小叶构成。每个肾小叶有许多肾单位，称皮质。基部狭小，主要由集合小管和髓襻构成，称髓质。部分肾小叶的顶部在肾的表面形成直径1～2mm的圆形隆起。相邻肾小叶被小叶间静脉形成的复杂网络隔开。在肾小叶皮质的中央有一条较大的中央静脉。肾单位有两种类型，即皮质肾单位和髓质肾单位，两者的主要区别是有无髓襻。

（六）睾丸

禽类的睾丸终生位于腹腔，其内侧缘借短的系膜悬于腹腔顶壁，紧贴腹气囊，没有纵隔和小隔，无小叶结构。由于生精小管的管腔较大，内含较多液态物质，故质地较软。

睾丸的生精小管细长而弯曲，有分支，相互吻合成网，因此断面的形状极不规则。性成熟后的生精小管结构与哺乳动物的相似，但各级生精细胞不形成特殊的细胞组合，而是围绕支持细胞构成许多垂直于基膜的细胞柱，其顶端有许多精子的头部插入支持细胞内。每个细胞柱可独立发生精子。生精小管的末端延续为直精小管，其管壁衬以支持细胞。

（七）卵巢和输卵管

禽类的雌性生殖器官有卵巢和输卵管，仅左侧的发育正常，右侧的在胚胎期退化。

1. 卵巢 卵巢位于左肾前端的腹侧，紧邻肾上腺，通过卵巢系膜悬吊于腹腔顶壁。随日龄增长，卵巢的体积增大，其表面呈桑葚状，性成熟时，体积急剧增大。产卵期的卵巢表面常见有多个体积依次递增的大型卵泡和许多小卵泡，最大的直径可达40mm以上，呈黄色或橙色，小卵泡的直径2 mm，呈白色。产卵停止后卵巢回缩，恢复到静止期的形状及大小。卵巢内有数百万个卵母细胞，但只有少数发育成熟排卵。

家禽卵巢的表面被覆单层生殖上皮，细胞形态由扁平到柱状。生殖上皮的下方由致密结缔组织构成白膜。结缔组织伸入卵巢内部形成基质。卵巢实质分周边的皮质和中央的髓质，皮质内含有不同发育阶段的卵泡和闭锁卵泡，髓质内含有丰富的血管、神经和平滑肌纤维。禽类卵巢结构特点：①大的生长卵泡和成熟卵泡不在卵巢基质内，而是突出于卵巢表面，仅借卵泡柄与其相连；②卵泡内无卵泡腔，也无卵泡液；③卵泡膜内富含毛细血管；④排卵后的卵泡壁很快退化，不形成黄体。

2. 输卵管 输卵管长而弯曲，产蛋期增粗变长，管壁肥厚；休产期则缩短变细。依据其结构与功能从前向后分为5段：漏斗部、膨大部、峡部、子宫部和阴道部。各段均由黏膜、肌层和外膜构成。各段的黏膜表面均有皱襞，黏膜上皮有纤毛，大部分固有层内有腺体和免疫组织，无黏膜肌层；肌层由内环、外纵两层平滑肌构成；外膜全为浆膜。

（八）脊髓

禽类的脊髓在椎管内一直延伸到尾综骨，不形成"马尾"，颈膨大部和腰膨大部较粗。腰膨大部的背侧裂开，形成菱形窦（rhomboidal sinus），窦内有胶质细胞团，因细胞内糖原丰富，又称糖原体。在颈膨大部和腰膨大部，其灰质腹侧柱的神经元有一部分迁移到外周的白质中，形成缘核。

（九）脑

禽类的脑干包括发达的延髓和中脑，无明显的脑桥。中脑的背外侧有发达的视叶（视顶盖），相当于哺乳类的前丘。视叶后方有中脑外侧核，相当于哺乳类动物的后丘。间脑较短，无乳头体。小脑发达，其中间的蚓部特别发达，向后上方隆起，两侧为绒球。大脑的主要结构为基底神经节，纹状体发达，突入侧脑室中，是重要的运动整合中枢。大脑皮层薄，表面光滑，不形成沟和回。嗅球很小，胼胝体很薄，以前联合和皮质联合联络两个大脑半球。

（十）冠

冠(comb)为特殊的皮肤褶，结构与皮肤相似。冠的表皮较薄，约有6层细胞。真皮厚，富含毛细血管丛。性成熟公鸡和产卵期母鸡的毛细血管高度充血，因而冠的颜色鲜红且肥厚。真皮深层的结缔组织富含纤维，黏液性物质充填于间隙中，使冠直立。冠的中央层由致密结缔组织构成，内含较大的血管和胶原纤维。冠的大小及颜色受性激素调控，当性激素缺乏时，毛细血管萎缩，黏液性物质消失，体积缩小，颜色变白。

（十一）尾脂腺

尾脂腺(preen gland)是禽类唯一的皮肤腺，卵圆形，在尾综骨背侧皮下，鸡的较小，水禽的发达。尾脂腺表面有结缔组织被膜，并向内深入形成叶间隔，将实质分为左右两叶。每叶中央有一大的初级腺腔，其中充满分泌物。在初级腺腔周围，分布着呈辐射状排列的分支腺小管，腺小管的盲端位于近被膜或叶间隔处，另一端开口于初级腺腔。腺小管分皮脂区和糖原区，皮脂区约占腺管的外2／3，糖原区约占腺管的内1／3。在皮脂区的腺小管横切面上，中央有次级腺腔，内含分泌物。腺小管管壁为复层上皮，近管腔是数层角化的扁平细胞；中间数层多角形细胞较大，胞质内充满脂滴，呈泡沫状；基部几层细胞稍扁平，胞质内含致密颗粒，最底层细胞附着于基膜上。皮脂区腺小管的分泌方式与哺乳动物的皮脂腺相似，属全质分泌型，近腔面的角化细胞不断脱落、解体，由基底层细胞不断增殖分化予以补充。在糖原区腺小管的横切面上，基底层细胞与皮脂区相同；中间层细胞较其增多2～3层，胞质嗜酸性，内含糖原颗粒，形成一厚层的糖原带；近腔管的细胞较少，胞质染色淡。尾脂腺的分泌物含脂肪、卵磷脂、高级醇和糖原等，经喙涂在羽上，可润泽羽毛，以免浸水，对水禽极为重要。

二、禽类的主要组织结构特征图谱

1. **禽类器官形态结构**　图21-1～图21-2。
2. **血细胞**　图21-3～图21-10。
3. **免疫器官**　图21-11～图21-43。
4. **消化器官**　图21-44～图21-97。
5. **肺**　图21-98～图21-108。
6. **肾**　图21-109～图21-117。
7. **睾丸**　图21-118～图21-128。
8. **卵巢**　图21-129～图21-134。
9. **输卵管**　图21-135～图21-152。
10. **脊髓**　图21-153～图21-160。
11. **脑**　图21-161～图21-168。
12. **冠**　图21-169～图21-171。
13. **尾脂腺**　图21-172～图21-174。

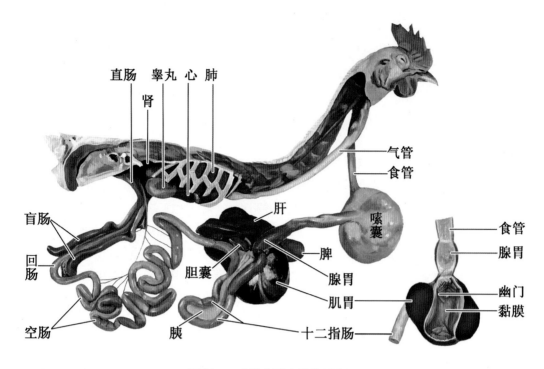

图 21-1 鸡器官形态结构示意图

Fig.21-1 diagram of chicken organs morphology and construction

气管—trachea 食管—esophagus 嗉囊—crop 腺胃—glandular stomach 幽门—pylorus 黏膜—mucosa
肌胃—muscular stomach 十二指肠—duodenum 脾—spleen 肝—liver 胆囊—gallbladder 胰—pancreas
空肠—jejunum 回肠—ileum 盲肠—caecum 直肠—rectum 肾—kidney 睾丸—testis 心—heart 肺—lung

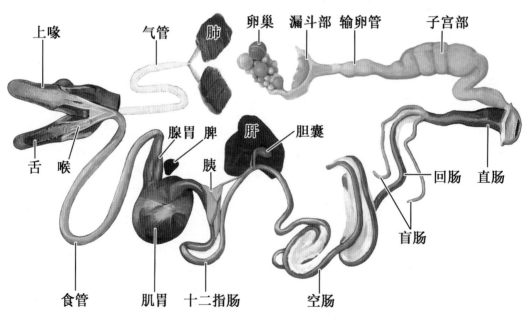

图 21-2 鹅器官形态结构示意图

Fig.21-2 diagram of goose organs morphology and construction

上喙—dorsal beak 气管—trachea 肺—lung 卵巢—ovary 漏斗部—funnel 输卵管—oviduct
子宫部—uterine part 舌—tongue 喉—throat 食管—esophagus 腺胃—glandular stomach
肌胃—muscular stomach 脾—spleen 胰—pancreas 肝—liver 胆囊—gallbladder
十二指肠—duodenum 空肠—jejunum 回肠—ileum 盲肠—caecum 直肠—rectum

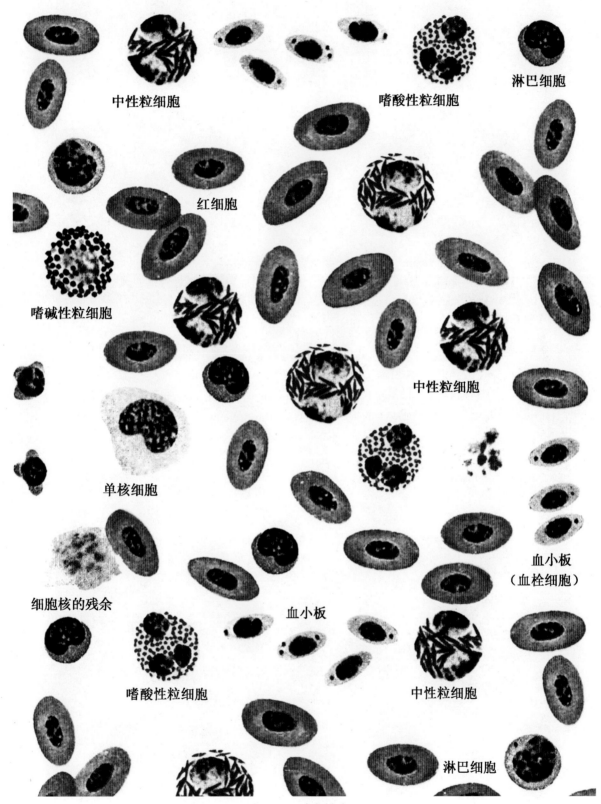

图 21-3 鸡血模式图

Fig.21-3 diagram of chicken blood

中性粒细胞—neutrophil 嗜酸性粒细胞—eosinophil 淋巴细胞—lymphocyte 红细胞—erythrocyte

嗜碱性粒细胞—basophil 单核细胞—monocyte 细胞核的残余—nucleus residue

血小板（血栓细胞）—blood platelet (thrombocyte)

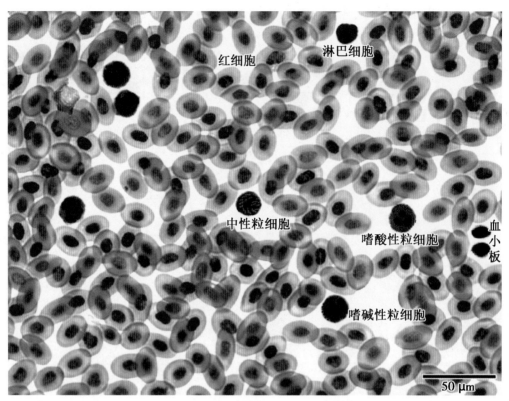

图21-4　鸡血涂片高倍像1（Wright染色）

Fig.21-4　high magnification of chicken blood smear 1 （Wright stain）

红细胞—erythrocyte　淋巴细胞—lymphocyte　中性粒细胞—neutrophil　嗜酸性粒细胞—eosinophil
血小板—blood platelet　嗜碱性粒细胞—basophil

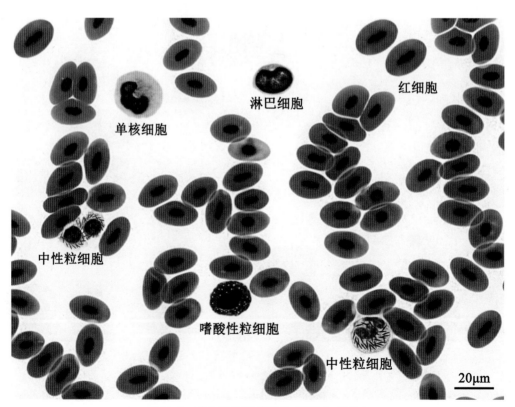

图21-5　鸡血涂片高倍像2（Wright染色）

Fig.21-5　high magnification of chicken blood smear 2 （Wright stain）

单核细胞—monocyte　淋巴细胞—lymphocyte　红细胞—erythrocyte　中性粒细胞—neutrophil　嗜酸性粒细胞—eosinophil

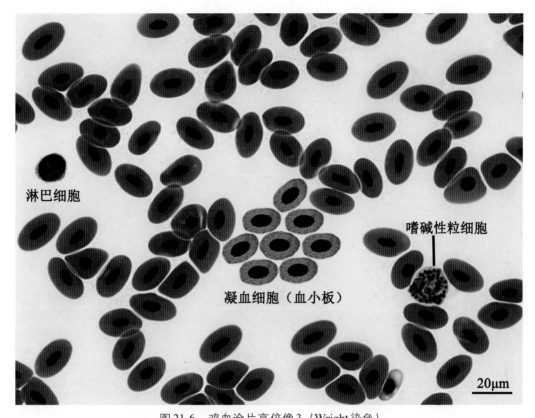

图21-6　鸡血涂片高倍像3（Wright染色）

Fig.21-6　high magnification of chicken blood smear 3（Wright stain）

淋巴细胞—lymphocyte　凝血细胞—thrombocyte　嗜碱性粒细胞—basophil

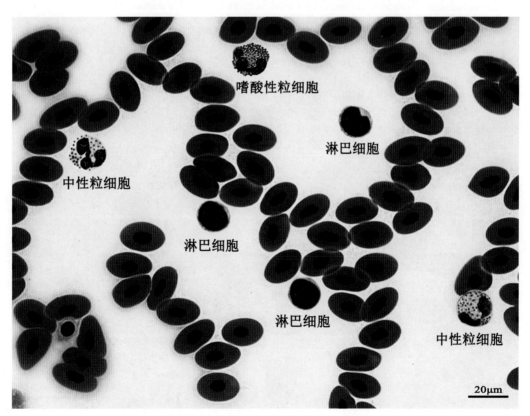

图21-7　鸭血涂片高倍像1（Wright染色）

Fig.21-7　high magnification of duck blood smear 1（Wright stain）

嗜酸性粒细胞—eosinophil　中性粒细胞—neutrophil　淋巴细胞—lymphocyte

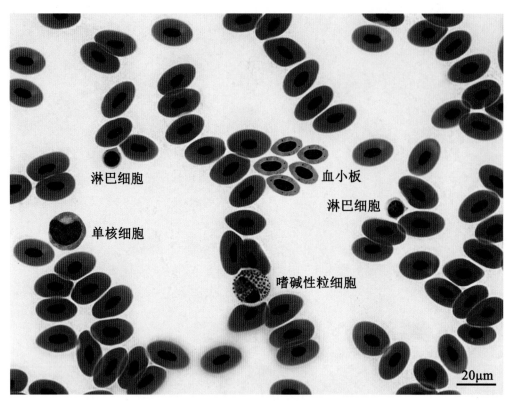

图21-8　鸭血涂片高倍像2（Wright染色）

Fig.21-8　high magnification of duck blood smear 2（Wright stain）

淋巴细胞—lymphocyte　血小板—blood platelet　单核细胞—monocyte　嗜碱性粒细胞—basophil

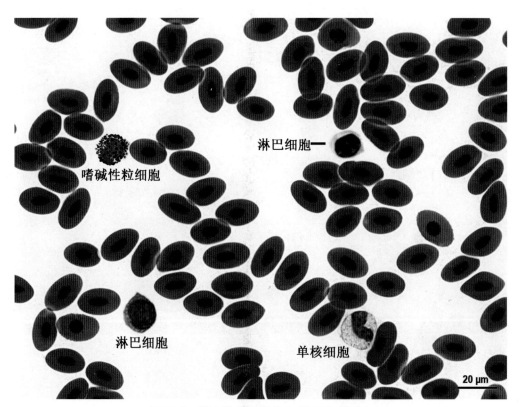

图21-9　鹅血涂片高倍像1（Wright染色）

Fig.21-9　high magnification of goose blood smear 1（Wright stain）

嗜碱性粒细胞—basophil　淋巴细胞—lymphocyte　单核细胞—monocyte

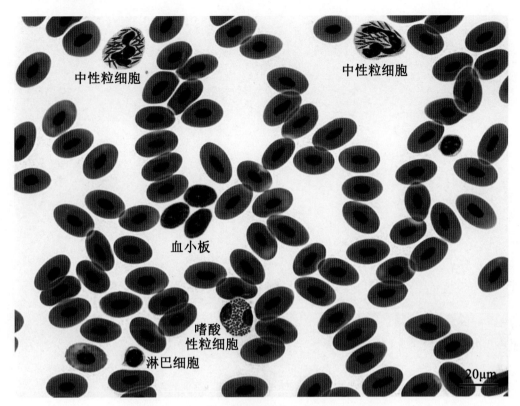

图21-10　鹅血涂片高倍像2（Wright染色）

Fig.21-10　high magnification of goose blood smear 2（Wright stain）

中性粒细胞—neutrophil　血小板—blood platelet　嗜酸性粒细胞—eosinophil　淋巴细胞—lymphocyte

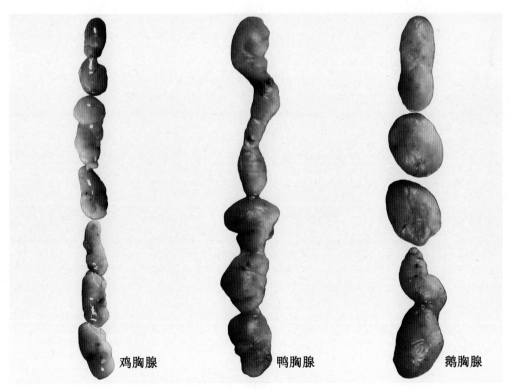

图21-11　家禽的胸腺（原色）

Fig.21-11　avian thymus（original color）

鸡胸腺—chicken thymus　鸭胸腺—duck thymus　鹅胸腺—goose thymus

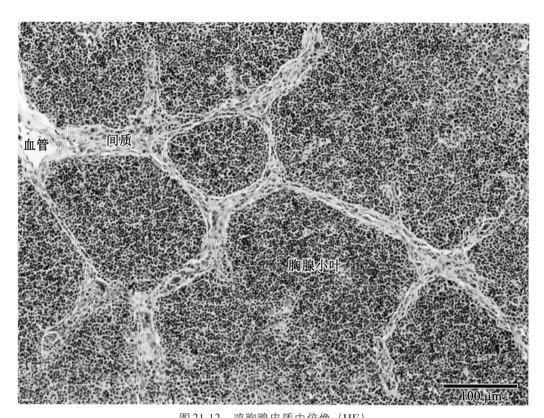

图21-12　鸡胸腺皮质中倍像（HE）

Fig.21-12　mid magnification of chicken thymus cortex（HE）

血管—blood vessel　间质—mesenchyme　胸腺小叶—thymic lobule

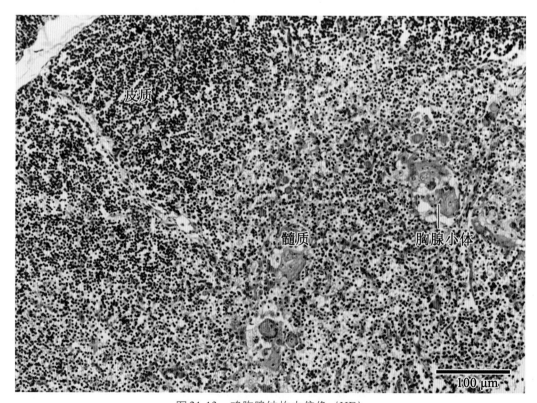

图21-13　鸡胸腺结构中倍像（HE）

Fig.21-13　mid magnification of chicken thymus structure（HE）

皮质—cortex　髓质—medulla　胸腺小体—thymic corpuscle

第二十一章 禽类的主要组织结构特征　Main Structural Features of Fowl Tissues

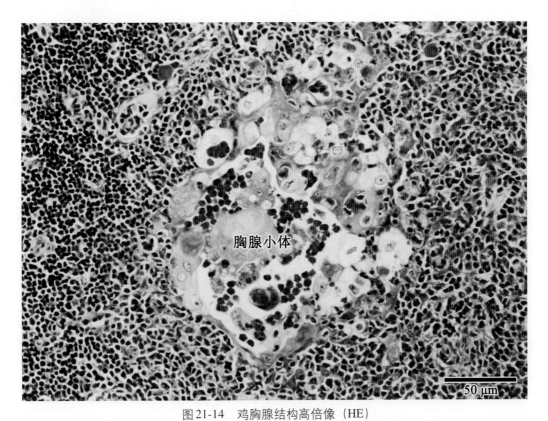

图21-14　鸡胸腺结构高倍像（HE）

Fig.21-14　high magnification of chicken thymus structure（HE）

胸腺小体—thymic corpuscle

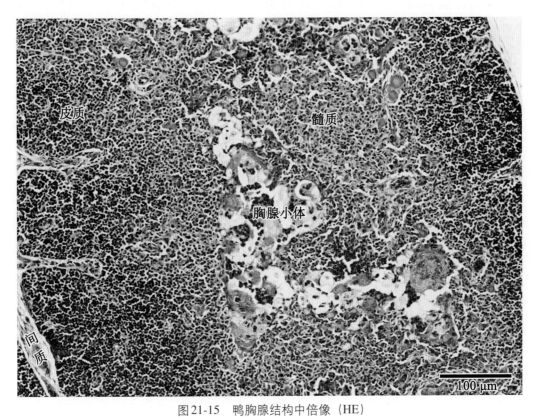

图21-15　鸭胸腺结构中倍像（HE）

Fig.21-15　mid magnification of duck thymus structure（HE）

皮质—cortex　髓质—medulla　胸腺小体—thymic corpuscle　间质—mesenchyme

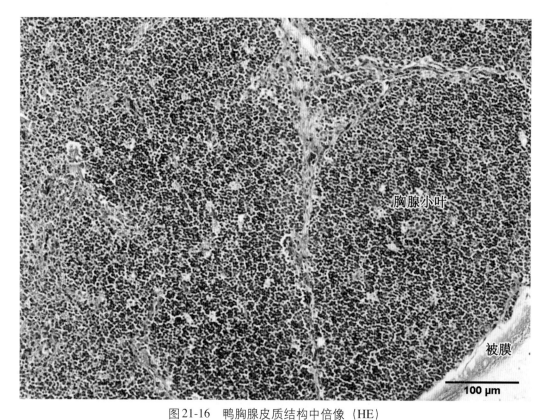

图 21-16　鸭胸腺皮质结构中倍像（HE）

Fig.21-16　mid magnification of duck thymus cortex structure（HE）

胸腺小叶—thymic lobule　被膜—capsule

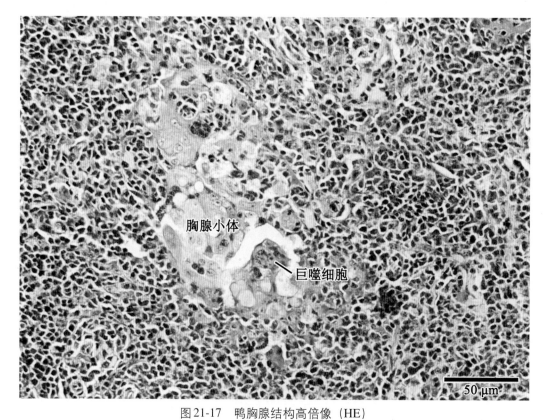

图 21-17　鸭胸腺结构高倍像（HE）

Fig.21-17　high magnification of duck thymus structure（HE）

胸腺小体—thymic corpuscle　巨噬细胞—macrophage

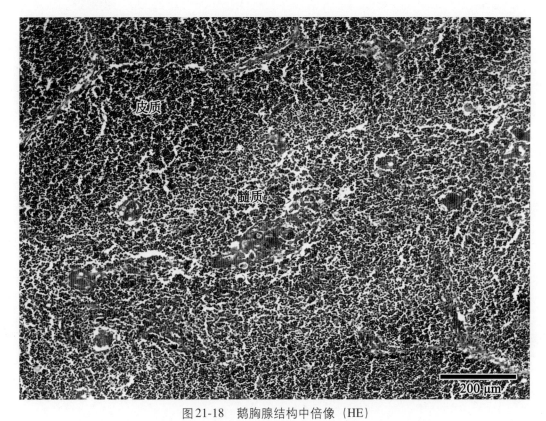

图21-18 鹅胸腺结构中倍像（HE）

Fig.21-18 mid magnification of goose thymus structure（HE）

皮质—cortex 髓质—medulla

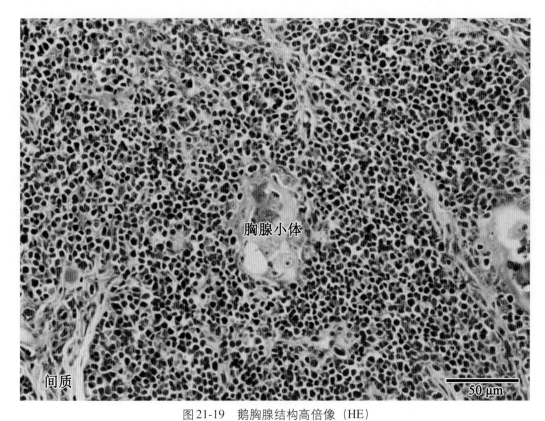

图21-19 鹅胸腺结构高倍像（HE）

Fig.21-19 high magnification of goose thymus structure（HE）

间质—mesenchyme 胸腺小体—thymic corpuscle

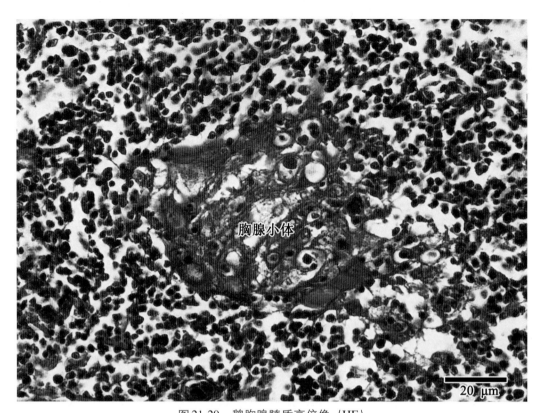

图 21-20　鹅胸腺髓质高倍像（HE）

Fig.21-20　high magnification of goose thymic medulla（HE）

胸腺小体—thymic corpuscle

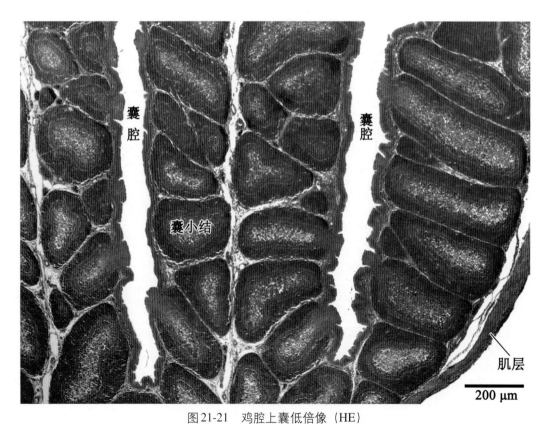

图 21-21　鸡腔上囊低倍像（HE）

Fig.21-21　low magnification of chicken bursa Fabricius（HE）

囊腔—lumen　囊小结—bursa nodule　肌层—muscular layer

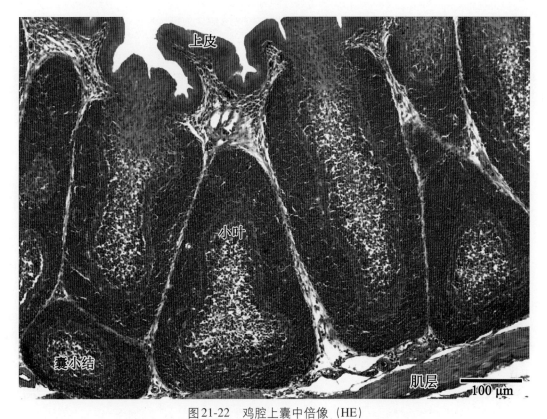

图 21-22　鸡腔上囊中倍像（HE）

Fig.21-22　mid magnification of chicken bursa Fabricius（HE）

上皮—epithelium　小叶—lobule　囊小结—bursa nodule　肌层—muscular layer

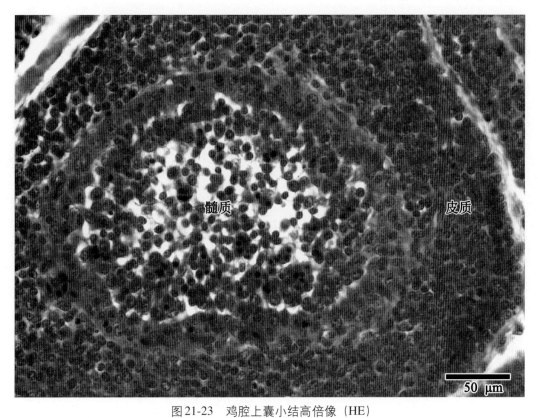

图 21-23　鸡腔上囊小结高倍像（HE）

Fig.21-23　high magnification of chicken bursa nodule（HE）

髓质—medulla　皮质—cortex

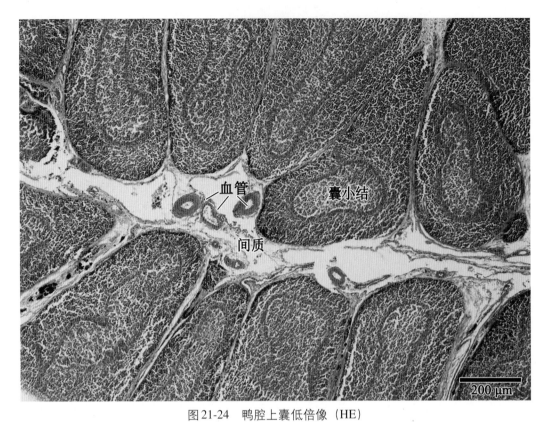

图21-24 鸭腔上囊低倍像（HE）

Fig.21-24 low magnification of duck bursa Fabricius（HE）

血管—blood vessel 间质—mesenchyme 囊小结—bursa nodule

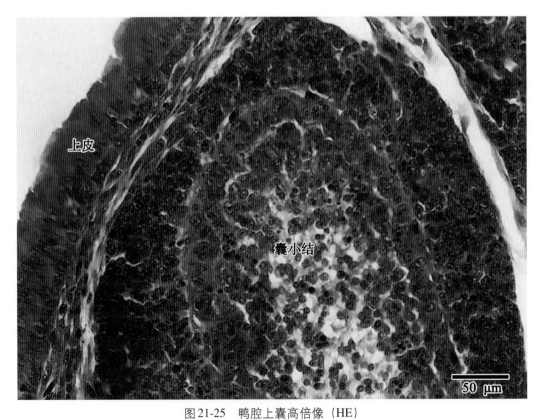

图21-25 鸭腔上囊高倍像（HE）

Fig.21-25 high magnification of duck bursa Fabricius（HE）

上皮—epithelium 囊小结—bursa nodule

第二十一章 禽类的主要组织结构特征　Main Structural Features of Fowl Tissues

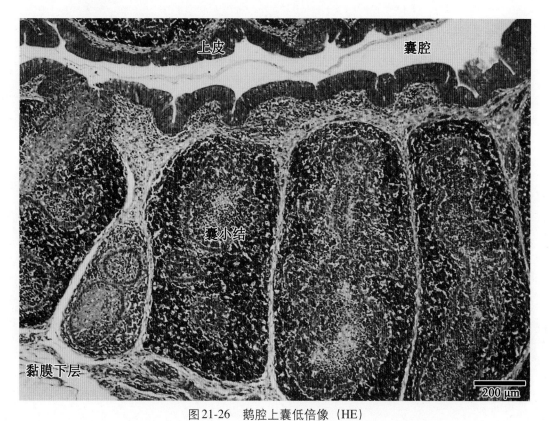

图21-26　鹅腔上囊低倍像（HE）

Fig.21-26　low magnification of goose bursa Fabricius（HE）

上皮—epithelium　囊腔—lumen　囊小结—bursa nodule　黏膜下层—submucosa

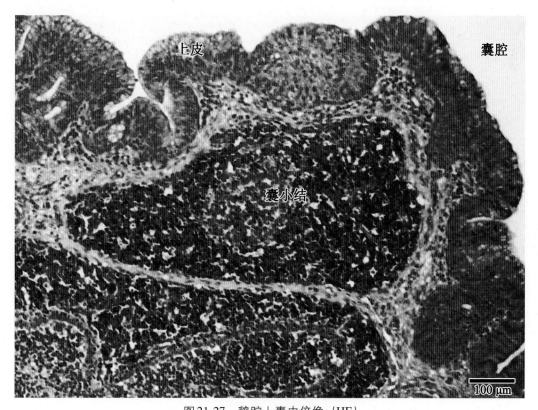

图21-27　鹅腔上囊中倍像（HE）

Fig.21-27　mid magnification of goose bursa Fabricius（HE）

上皮—epithelium　囊腔—lumen　囊小结—bursa nodule

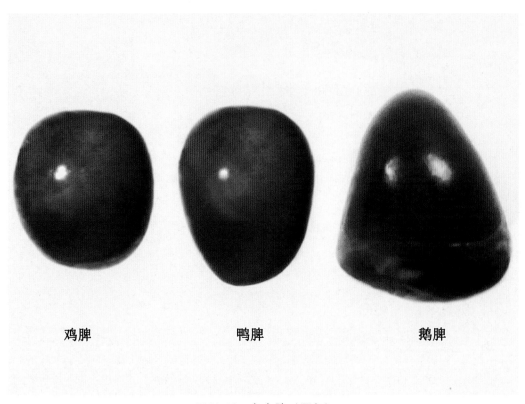

图21-28 家禽脾（原色）

Fig.21-28 avian spleen（original color）

鸡脾—chicken spleen 鸭脾—duck spleen 鹅脾—goose spleen

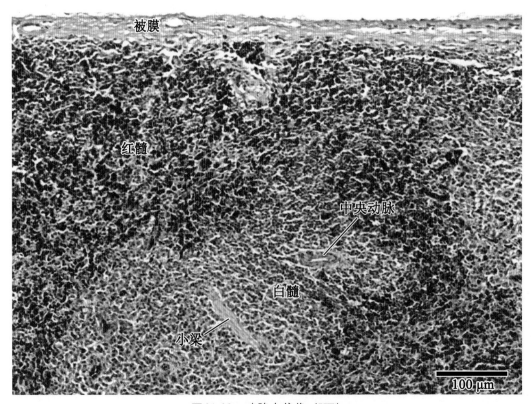

图21-29 鸡脾中倍像（HE）

Fig.21-29 mid magnification of chicken spleen（HE）

被膜—capsule 红髓—red pulp 白髓—white pulp 中央动脉—central artery 小梁—trabecula

第二十一章 禽类的主要组织结构特征 Main Structural Features of Fowl Tissues

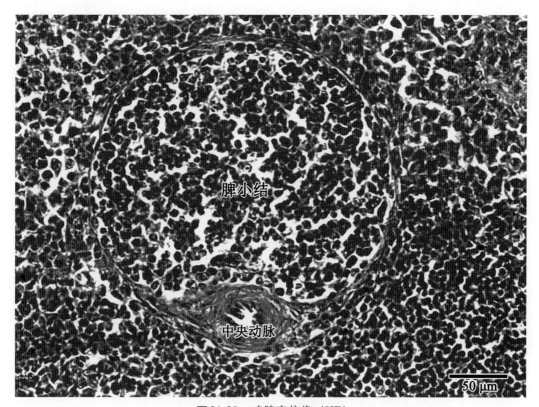

图21-30　鸡脾高倍像（HE）

Fig.21-30　high magnification of chicken spleen（HE）

脾小结—splenic nodule　中央动脉—central artery

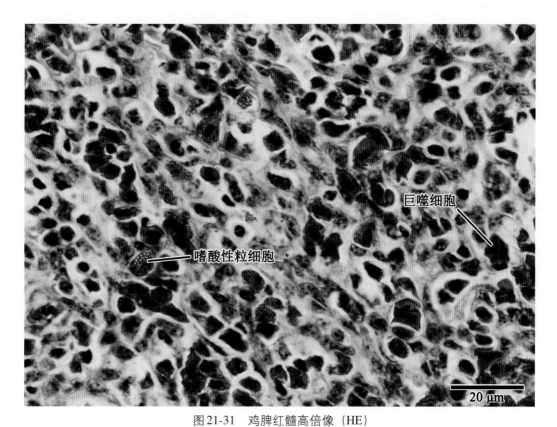

图21-31　鸡脾红髓高倍像（HE）

Fig.21-31　high magnification of red pulp in chicken spleen（HE）

嗜酸性粒细胞—eosinophil　巨噬细胞—macrophage

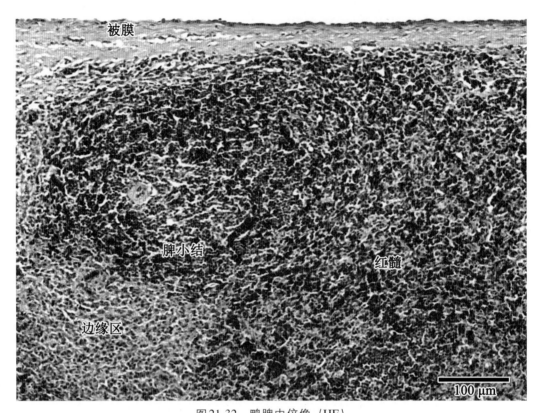

图 21-32　鸭脾中倍像（HE）

Fig.21-32　mid magnification of duck spleen（HE）

被膜—capsule　脾小结—splenic nodule　边缘区—marginal zone　红髓—red pulp

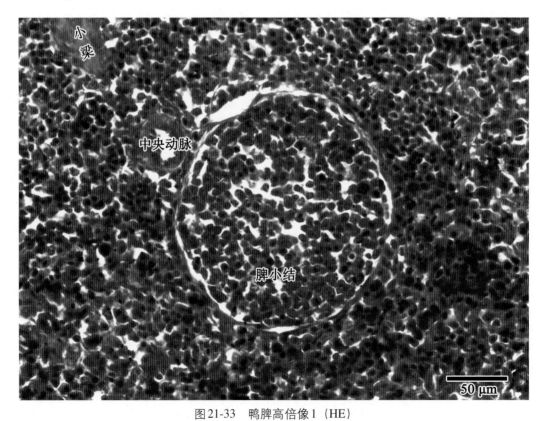

图 21-33　鸭脾高倍像 1（HE）

Fig.21-33　high magnification of duck spleen 1（HE）

小梁—trabecula　中央动脉—central artery　脾小结—splenic nodule

第二十一章 禽类的主要组织结构特征 Main Structural Features of Fowl Tissues

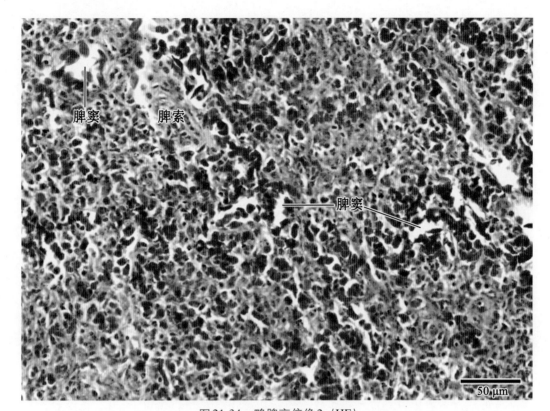

图21-34　鸭脾高倍像2（HE）

Fig.21-34　high magnification of duck spleen 2（HE）

脾窦—splenic sinusoid　脾索—splenic cord

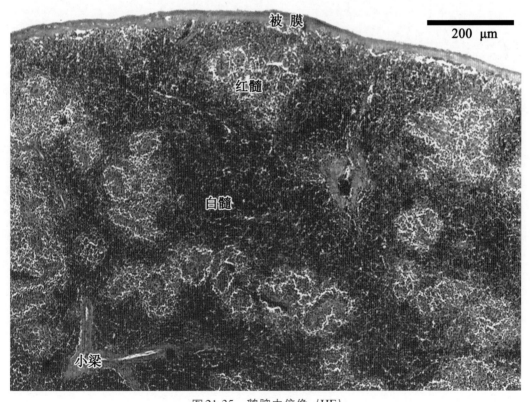

图21-35　鹅脾中倍像（HE）

Fig.21-35　mid magnification of goose spleen（HE）

被膜—capsule　红髓—red pulp　白髓—white pulp　小梁—trabecula

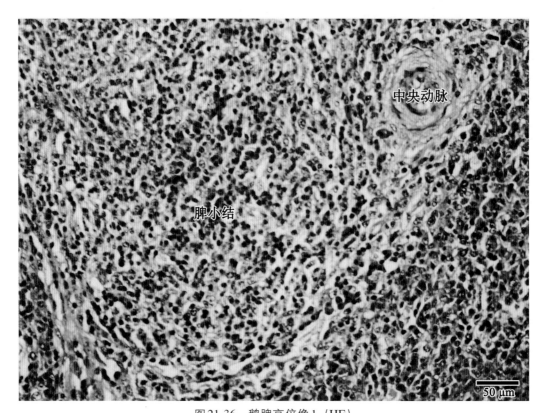

图 21-36 鹅脾高倍像 1（HE）

Fig.21-36 high magnification of goose spleen 1（HE）

脾小结—splenic nodule　中央动脉—central artery

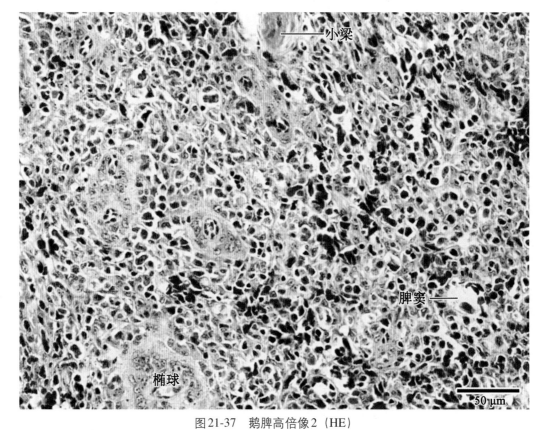

图 21-37 鹅脾高倍像 2（HE）

Fig.21-37 high magnification of goose spleen 2（HE）

小梁—trabecula　脾窦—splenic sinusoid　椭球—ellipsoid

图 21-38 鸭淋巴结低倍像 1（HE）

Fig.21-38 low magnification of duck lymph node 1 (HE)

被膜—capsule 淋巴小结—lymphoid nodule

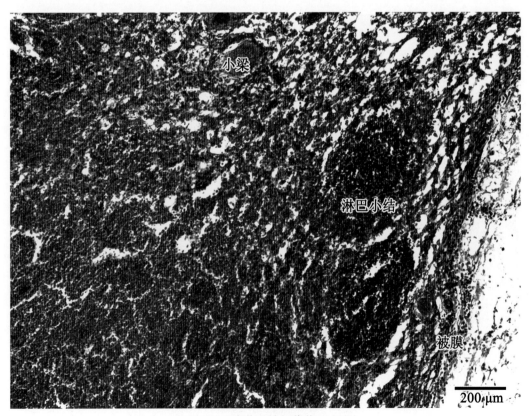

图 21-39 鸭淋巴结低倍像 2（HE）

Fig.21-39 low magnification of duck lymph node 2 (HE)

小梁—trabecula 淋巴小结—lymphoid nodule 被膜—capsule

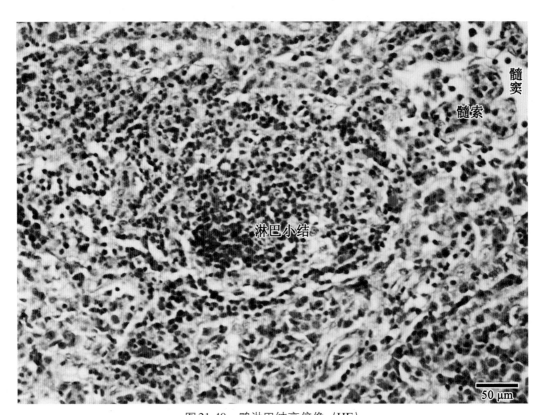

图 21-40 鸭淋巴结高倍像（HE）

Fig.21-40 high magnification of duck lymph node（HE）

淋巴小结—lymphoid nodule 髓索—medullary cord 髓窦—medullary sinus

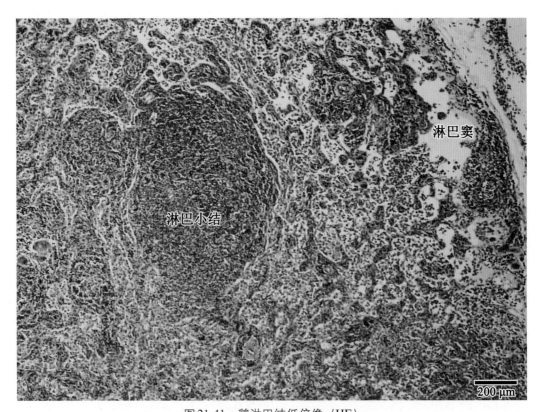

图 21-41 鹅淋巴结低倍像（HE）

Fig.21-41 low magnification of goose lymph node（HE）

淋巴小结—lymphoid nodule 淋巴窦—lymph sinus

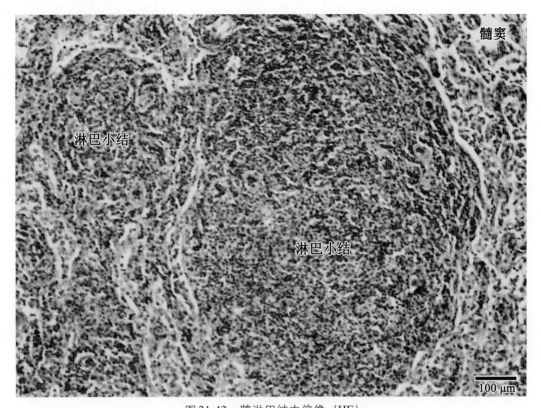

图21-42 鹅淋巴结中倍像（HE）

Fig.21-42 mid magnification of goose lymph node（HE）

淋巴小结—lymphoid nodule　髓窦—medullary sinus

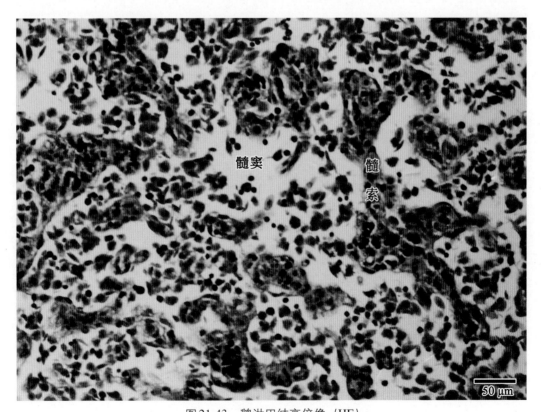

图21-43 鹅淋巴结高倍像（HE）

Fig.21-43 high magnification of goose lymph node（HE）

髓窦—medullary sinus　髓索—medullary cord

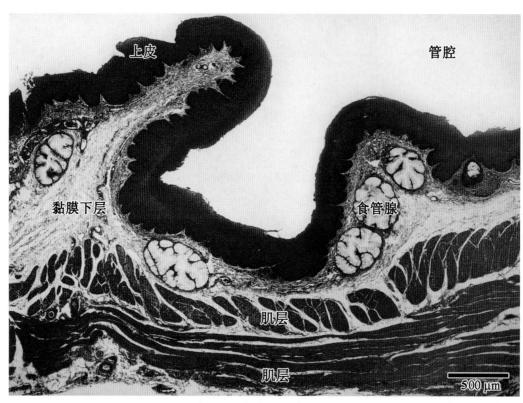

图21-44 鸡食管低倍像（HE）

Fig.21-44 low magnification of chicken esophagus（HE）

上皮—epithelium 管腔—lumen 黏膜下层—submucosa 食管腺—esophageal gland 肌层—muscular layer

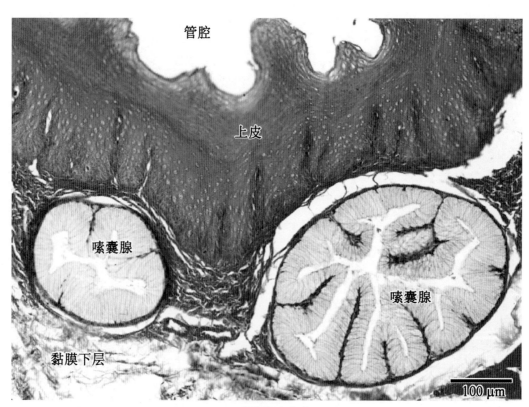

图21-45 鸡嗉囊中倍像（HE）

Fig.21-45 mid magnification of chicken crop（HE）

管腔—lumen 上皮—epithelium 嗉囊腺—crop gland 黏膜下层—submucosa

第二十一章 禽类的主要组织结构特征 Main Structural Features of Fowl Tissues

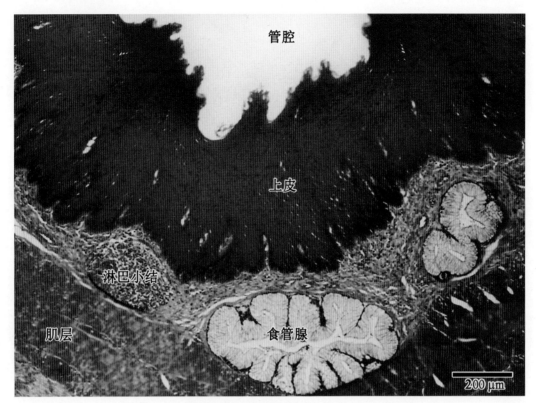

图21-46 鸽食管低倍像（HE）

Fig.21-46 low magnification of pigeon esophagus（HE）

管腔—lumen 上皮—epithelium 淋巴小结—lymphoid nodule 肌层—muscular layer 食管腺—esophageal gland

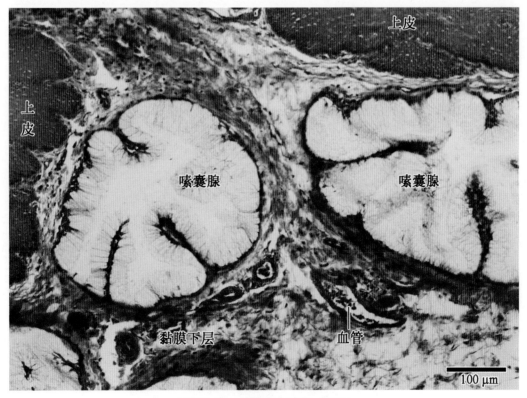

图21-47 鸽嗉囊中倍像（HE）

Fig.21-47 mid magnification of pigeon crop（HE）

上皮—epithelium 嗉囊腺—crop gland 黏膜下层—submucosa 血管—blood vessel

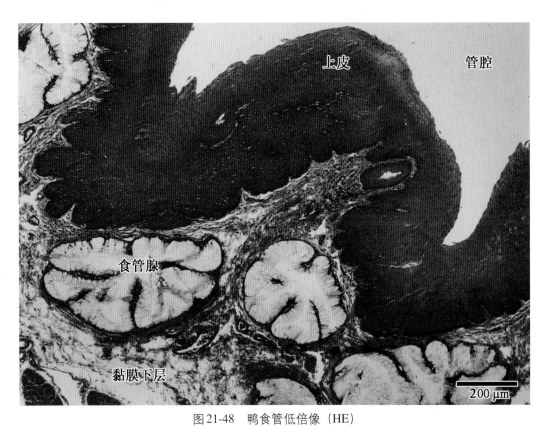

图 21-48　鸭食管低倍像（HE）

Fig.21-48　low magnification of duck esophagus（HE）

上皮—epithelium　管腔—lumen　食管腺—esophageal gland　黏膜下层—submucosa

图 21-49　鸭食管膨大部中倍像（HE）

Fig.21-49　mid magnification of duck esophagus intumescentia（HE）

管腔—lumen　上皮—epithelium　腺管—glandular duct　食管腺—esophageal gland

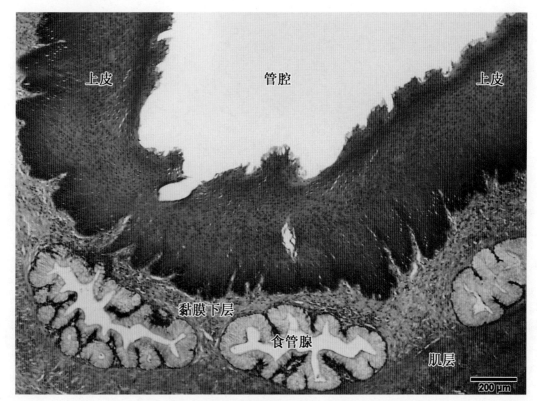

图21-50　鹅食管低倍像（HE）

Fig.21-50　low magnification of goose esophagus（HE）

上皮—epithelium　管腔—lumen　黏膜下层—submucosa　食管腺—esophageal gland　肌层—muscle layer

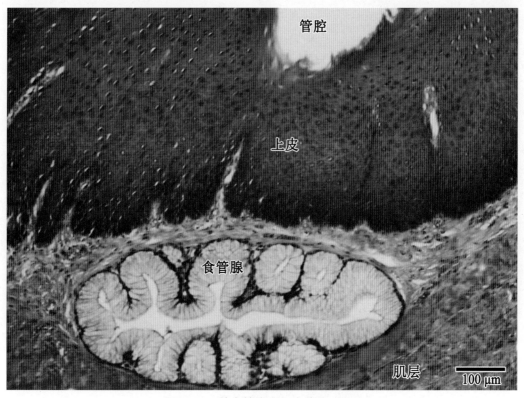

图21-51　鹅食管膨大部中倍像（HE）

Fig.21-51　mid magnification of goose esophagus intumescentia（HE）

管腔—lumen　上皮—epithelium　食管腺—esophageal gland　肌层—muscle layer

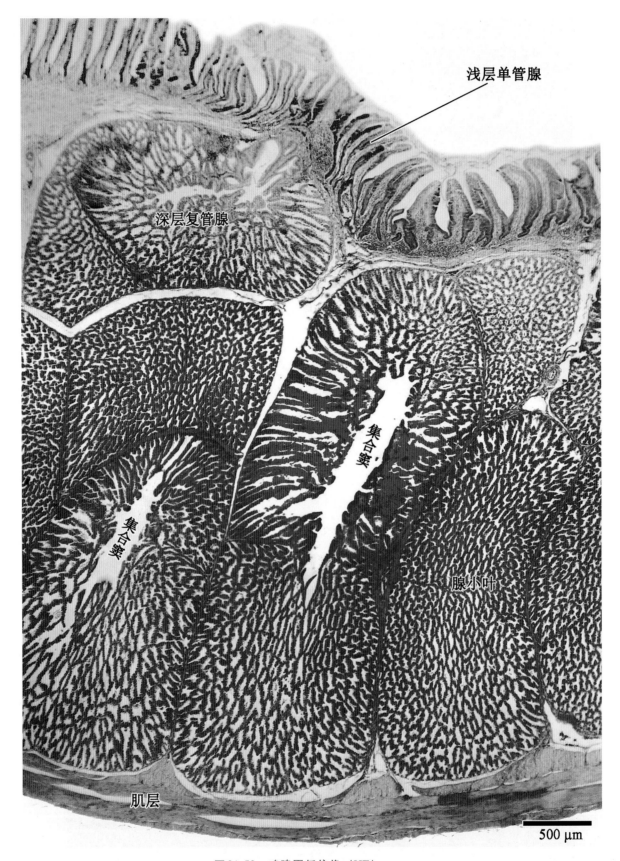

图 21-52　鸡腺胃低倍像（HE）

Fig.21-52　low magnification of chicken glandular stomach（HE）

浅层单管腺—shallow single tubular gland　深层复管腺—deep compound tubular gland

集合窦— collection of sinus　腺小叶—glandular lobule　肌层—muscular layer

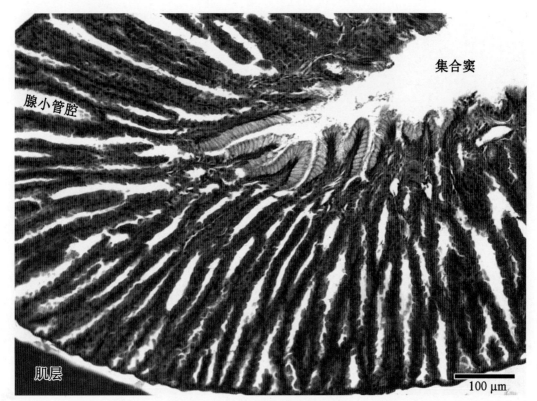

图21-53　鸡腺胃中倍像（HE）

Fig.21-53　mid magnification of chicken glandular stomach（HE）

腺小管腔—glandular tubule　集合窦— collection of sinus　肌层—muscular layer

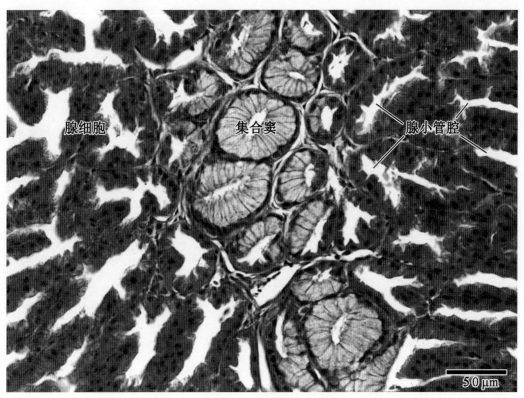

图21-54　鸡腺胃高倍像（HE）

Fig.21-54　high magnification of chicken glandular stomach（HE）

腺细胞—glandular cell　集合窦— collection of sinus　腺小管腔—glandular tubules

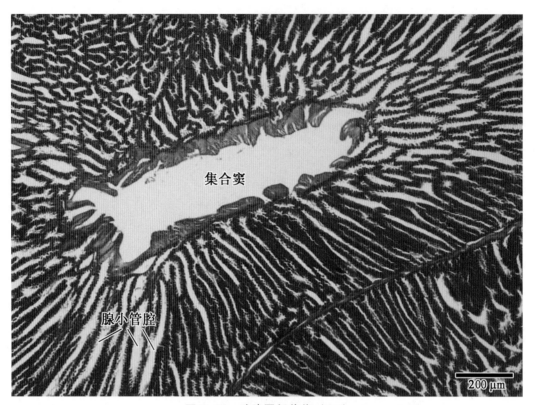

图21-55 鸭腺胃低倍像（HE）

Fig.21-55 low magnification of duck glandular stomach（HE）

集合窦—collection of sinus 腺小管腔—glandular tubules

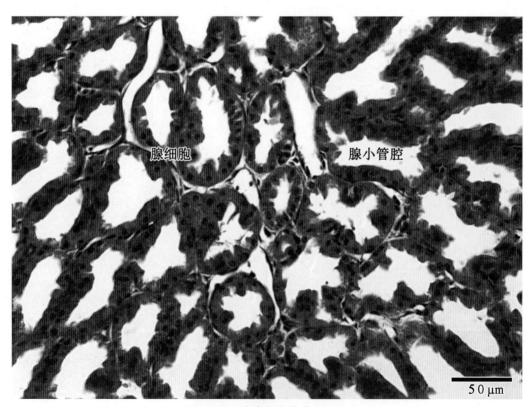

图21-56 鸭腺胃高倍像（HE）

Fig.21-56 high magnification of duck glandular stomach（HE）

腺细胞—glandular cell 腺小管腔—glandular tubule

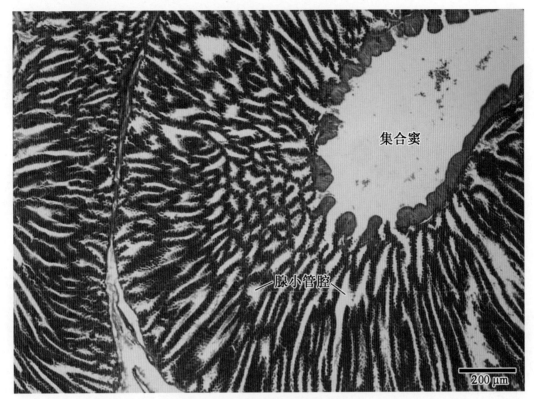

图21-57 鹅腺胃低倍像（HE）

Fig.21-57 low magnification of goose glandular stomach（HE）

集合窦—collection of sinus 腺小管腔—glandular tubules

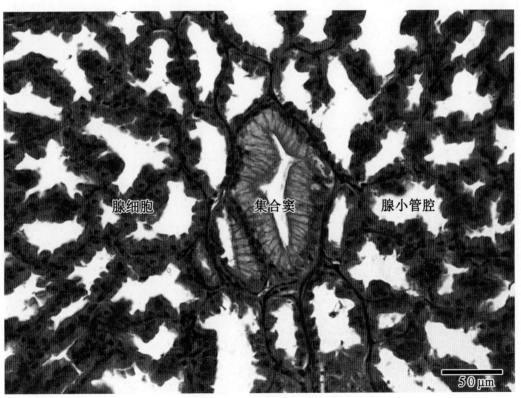

图21-58 鹅腺胃高倍像（HE）

Fig.21-58 high magnification of goose glandular stomach（HE）

腺细胞—glandular cell 集合窦—collection of sinus 腺小管腔—glandular tubule

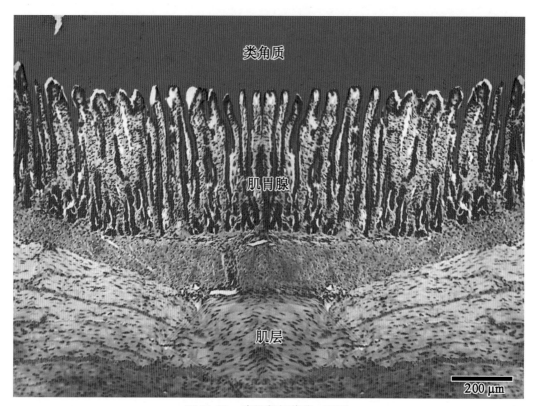

图 21-59　鸡肌胃低倍像（HE）

Fig.21-59　low magnification of chicken muscular stomach（HE）

类角质—keratinoid　肌胃腺—gizzard glands　肌层—muscular layer

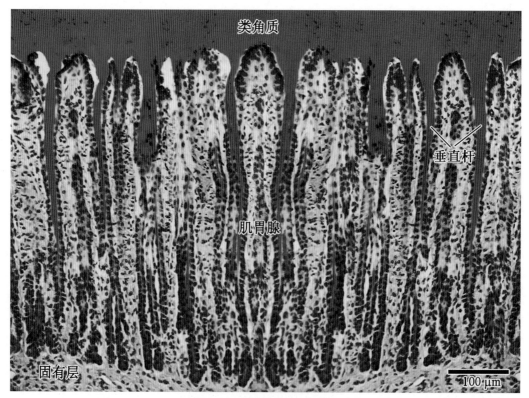

图 21-60　鸡肌胃中倍像（HE）

Fig.21-60　mid magnification of chicken muscular stomach（HE）

类角质—keratinoid　垂直杆—vertical rods　肌胃腺—gizzard glands　固有层—proper layer

第二十一章 禽类的主要组织结构特征　Main Structural Features of Fowl Tissues

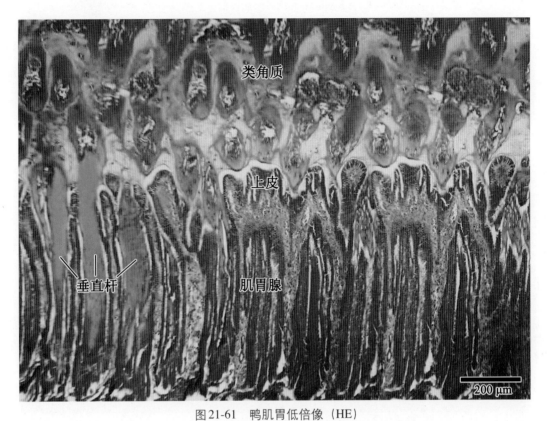

图21-61　鸭肌胃低倍像（HE）

Fig.21-61　low magnification of duck muscular stomach（HE）

类角质—keratinoid　上皮—epithelium　垂直杆—vertical rods　肌胃腺—gizzard glands

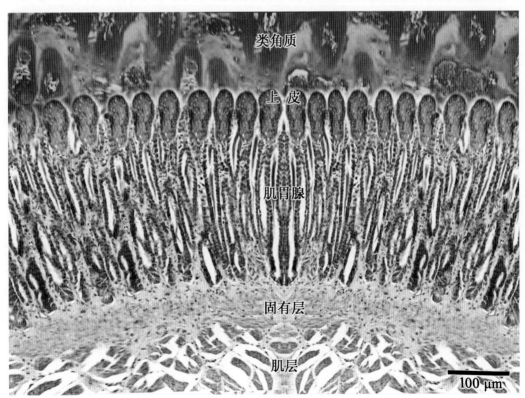

图21-62　鸭肌胃中倍像（HE）

Fig.21-62　mid magnification of duck muscular stomach（HE）

类角质—keratinoid　上皮—epithelium　肌胃腺—gizzard glands　固有层—proper layer　肌层—muscular layer

795

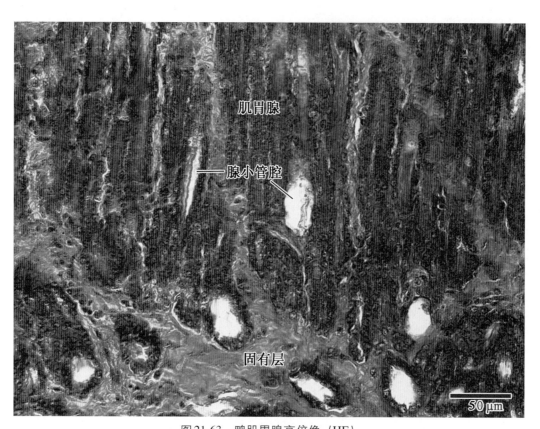

图 21-63　鸭肌胃腺高倍像（HE）

Fig.21-63　high magnification of duck gizzard glands（HE）

肌胃腺—gizzard glands　腺小管腔—glandular tubules　固有层—proper layer

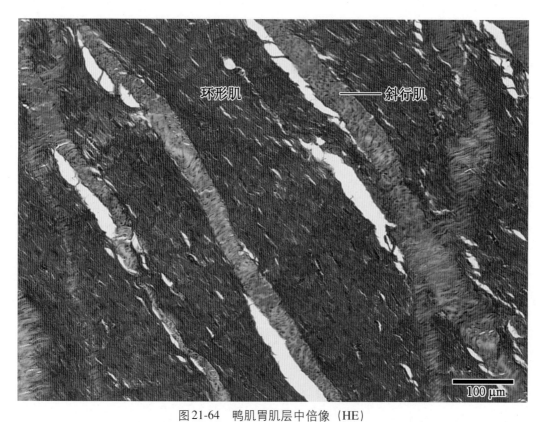

图 21-64　鸭肌胃肌层中倍像（HE）

Fig.21-64　mid magnification of duck muscular stomach's muscular layer（HE）

环形肌—circular muscle　斜行肌—diagonal muscle

第二十一章 禽类的主要组织结构特征　Main Structural Features of Fowl Tissues

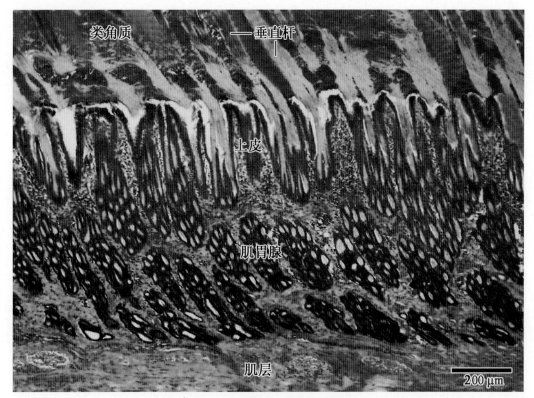

图21-65　鹅肌胃低倍像1（HE）

Fig.21-65　low magnification of goose muscular stomach 1（HE）

类角质—keratinoid　上皮—epithelium　肌胃腺—gizzard glands　肌层—muscular layer　垂直杆—vertical rods

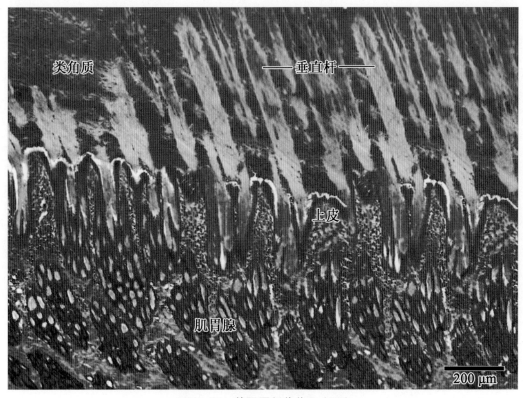

图21-66　鹅肌胃低倍像2（HE）

Fig.21-66　low magnification of goose muscular stomach 2（HE）

类角质—keratinoid　垂直杆—vertical rods　上皮—epithelium　肌胃腺—gizzard glands

797

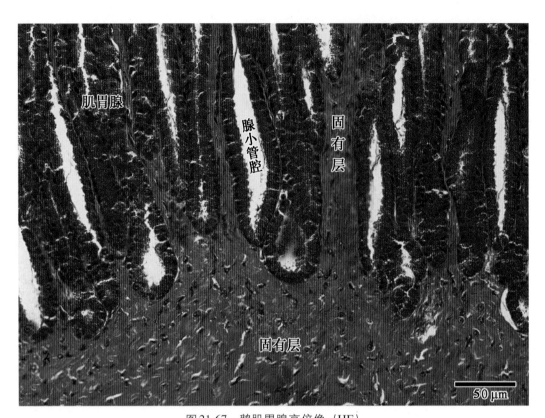

图21-67 鹅肌胃腺高倍像（HE）

Fig.21-67 high magnification of goose gizzard glands（HE）

肌胃腺—gizzard glands　腺小管腔—glandular tubules　固有层—proper layer

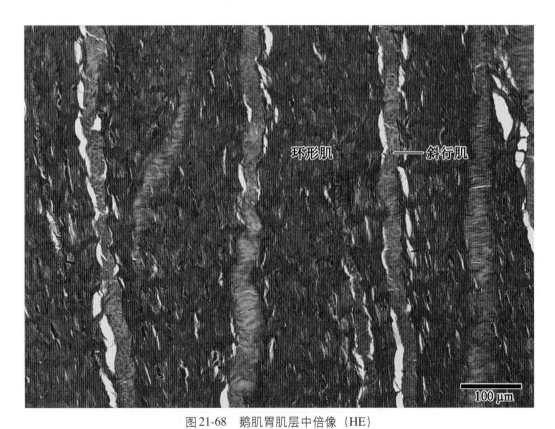

图21-68 鹅肌胃肌层中倍像（HE）

Fig.21-68 mid magnification of goose muscular stomach's muscular layer（HE）

环形肌—circular muscle　斜行肌—diagonal muscle

第二十一章　禽类的主要组织结构特征　Main Structural Features of Fowl Tissues

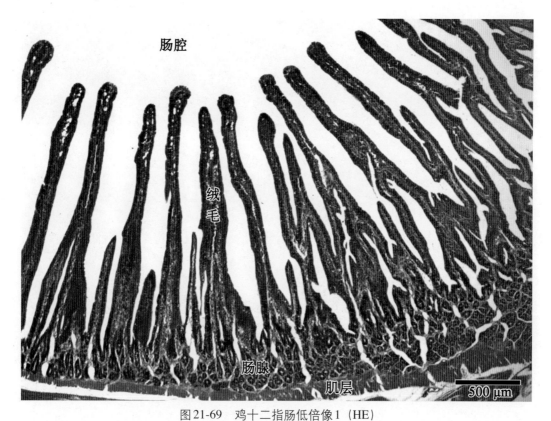

图21-69　鸡十二指肠低倍像1（HE）

Fig.21-69　low magnification of chicken duodenum 1 （HE）

肠腔—intestinal cavity　绒毛—villi　肠腺—intestinal gland　肌层—muscular layer

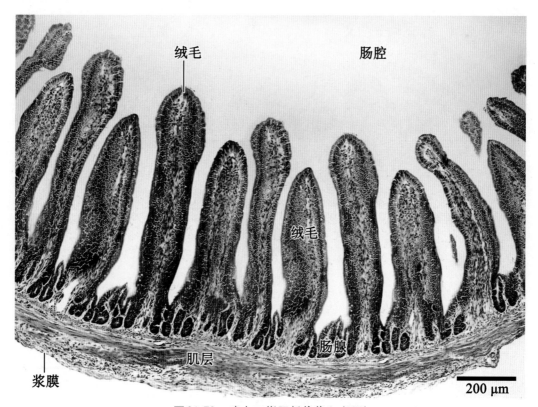

图21-70　鸡十二指肠低倍像2（HE）

Fig.21-70　low magnification of chicken duodenum 2 （HE）

绒毛—villi　肠腔—intestinal cavity　肠腺—intestinal gland　肌层—muscular layer　浆膜—serosa

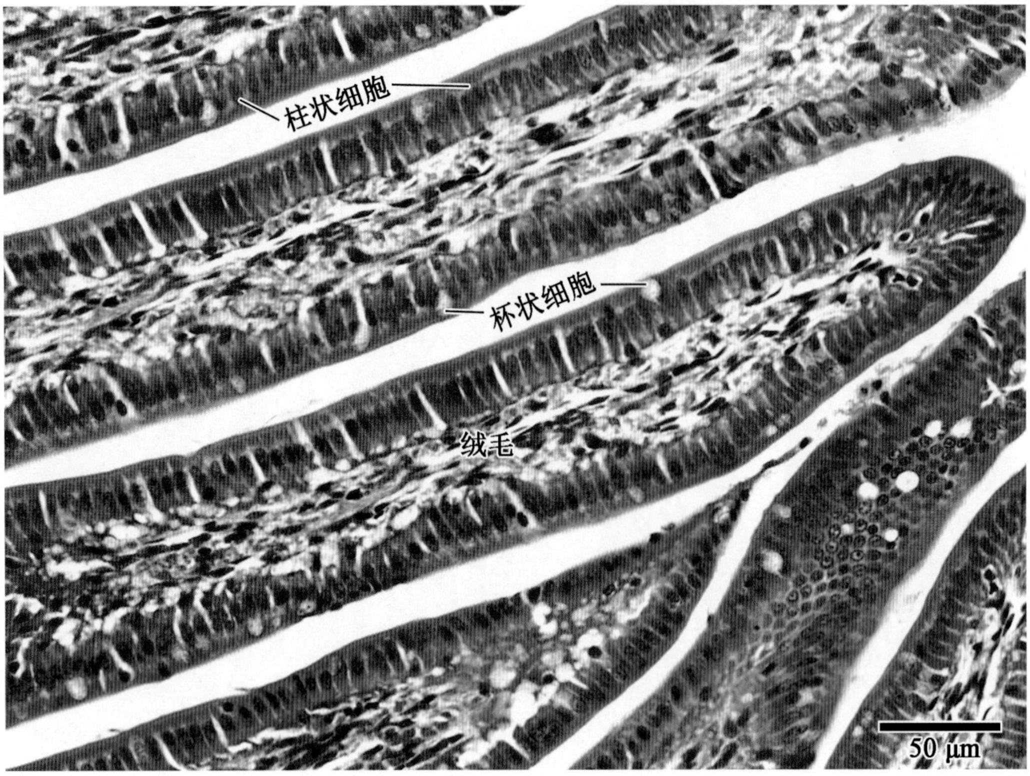

图 21-71 鸡十二指肠高倍像（HE）

Fig.21-71　high magnification of chicken duodenum（HE）

柱状细胞—columnar cell　杯状细胞—goblet cell　绒毛—villi

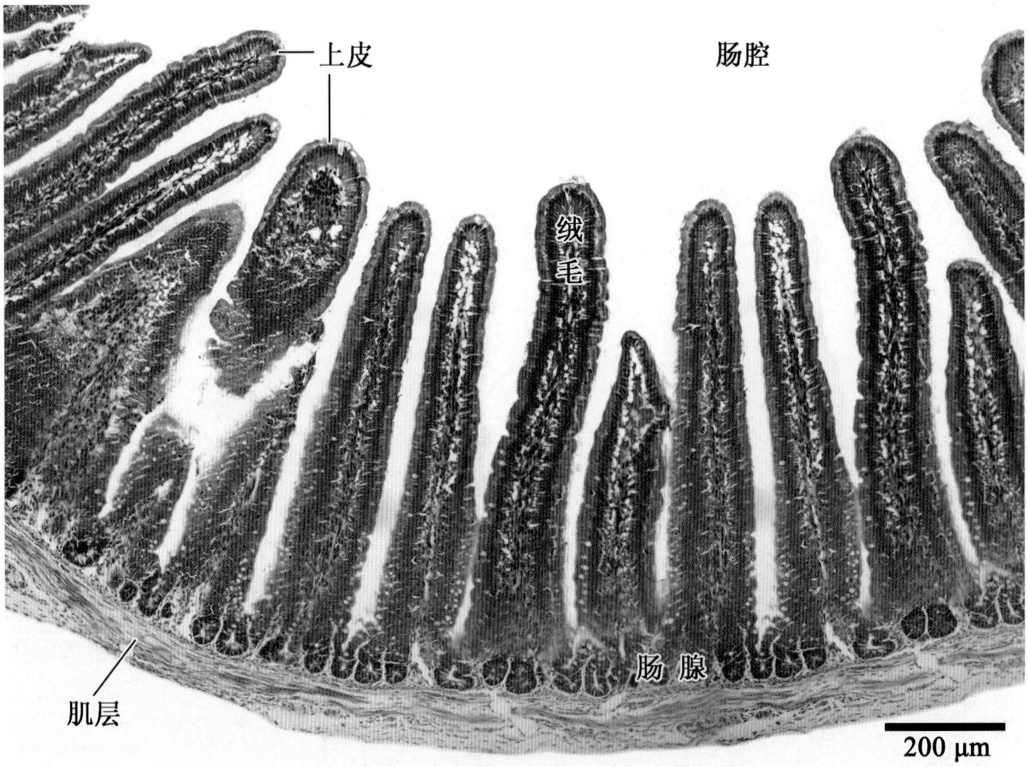

图 21-72 鸭十二指肠低倍像（HE）

Fig.21-72　low magnification of duck duodenum（HE）

上皮—epithelium　肠腔—intestinal cavity　绒毛—villi　肠腺—intestinal gland　肌层—muscular layer

第二十一章 禽类的主要组织结构特征　Main Structural Features of Fowl Tissues

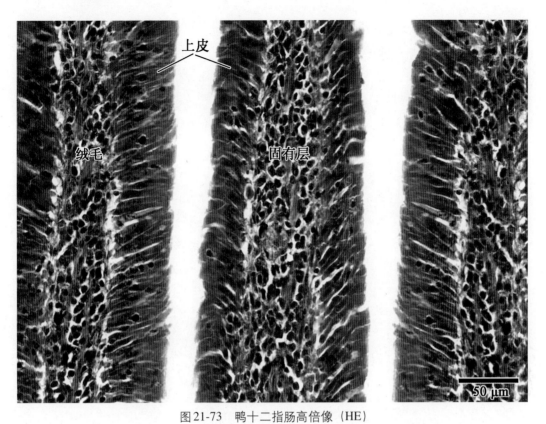

图21-73　鸭十二指肠高倍像（HE）

Fig.21-73　high magnification of duck duodenum（HE）

上皮—epithelium　绒毛—villi　固有层—proper layer

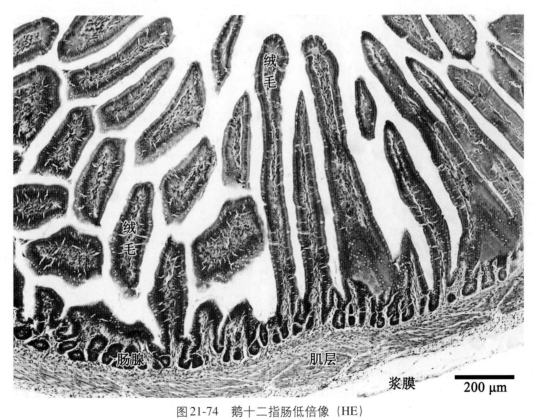

图21-74　鹅十二指肠低倍像（HE）

Fig.21-74　low magnification of goose duodenum（HE）

绒毛—villi　肠腺—intestinal gland　肌层—muscular layer　浆膜—serosa

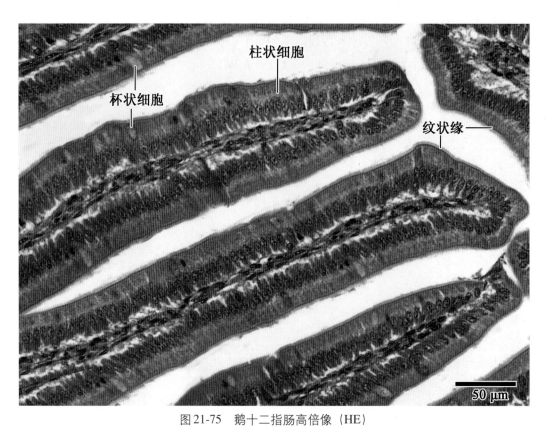

图 21-75 鹅十二指肠高倍像（HE）

Fig.21-75 high magnification of goose duodenum（HE）

杯状细胞—goblet cell 柱状细胞—columnar cell 纹状缘—striated border

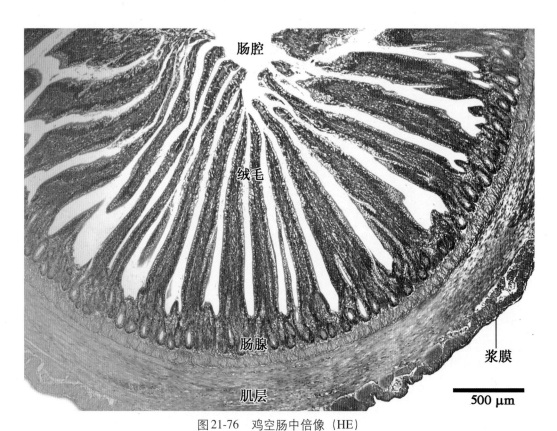

图 21-76 鸡空肠中倍像（HE）

Fig.21-76 mid magnification of chicken jejunum（HE）

肠腔—intestinal cavity 绒毛—villi 肠腺—intestinal gland 肌层—muscular layer 浆膜—serosa

第二十一章 禽类的主要组织结构特征 Main Structural Features of Fowl Tissues

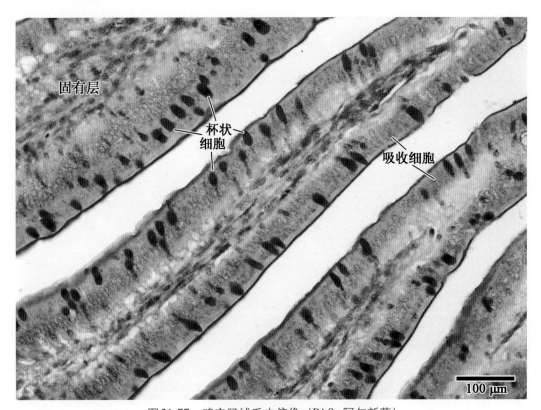

图21-77 鸡空肠绒毛中倍像（PAS+阿尔新蓝）

Fig.21-77 mid magnification of chicken jejunum villi（PAS + Alcian Blue）

固有层—proper layer　杯状细胞—goblet cell　吸收细胞—absorptive cell

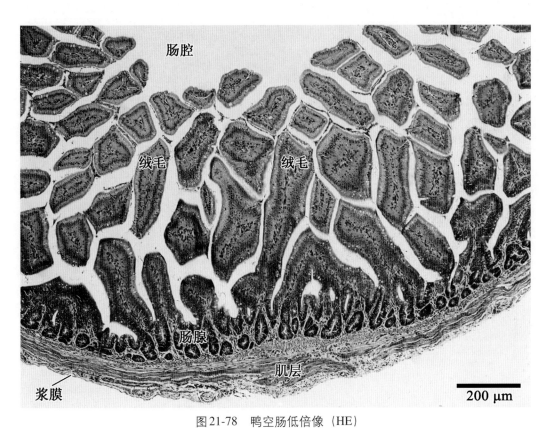

图21-78 鸭空肠低倍像（HE）

Fig.21-78 low magnification of duck jejunum（HE）

肠腔—intestinal cavity　绒毛—villi　肠腺—intestinal gland　肌层—muscular layer　浆膜—serosa

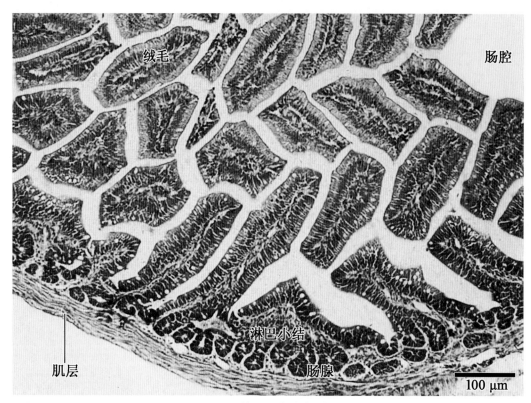

图 21-79　鹅空肠中倍像（HE）

Fig.21-79　mid magnification of goose jejunum（HE）

肠腔—intestinal cavity　绒毛—villi　淋巴小结—lymphoid nodule　肠腺—intestinal gland　肌层—muscular layer

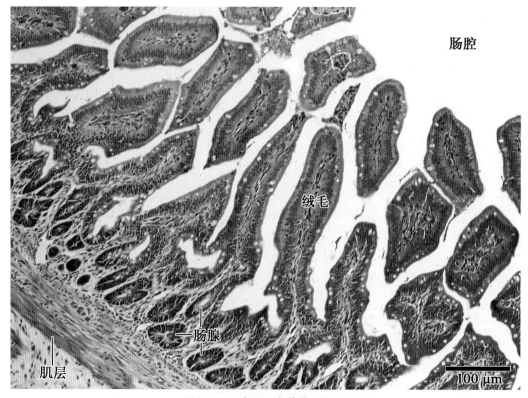

图 21-80　鸡回肠中倍像（HE）

Fig.21-80　mid magnification of chicken ileum（HE）

肠腔—intestinal cavity　绒毛—villi　肠腺—intestinal gland　肌层—muscular layer

第二十一章 禽类的主要组织结构特征　Main Structural Features of Fowl Tissues

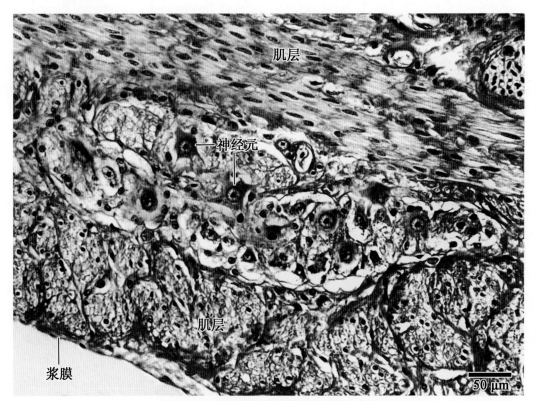

图21-81　鸡回肠肌间神经丛高倍像（HE）

Fig.21-81　high magnification of chicken ileum myenteric nerve plexus（HE）

肌层—muscular layer　神经元—neuron　浆膜—serosa

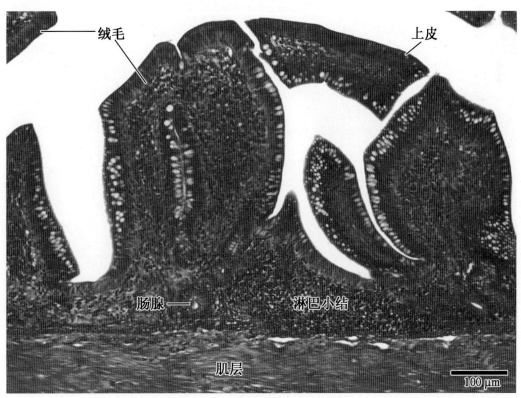

图21-82　鸭回肠中倍像（HE）

Fig.21-82　mid magnification of duck ileum（HE）

绒毛—villi　上皮—epithelium　肠腺—intestinal gland　淋巴小结—lymphoid nodule　肌层—muscle

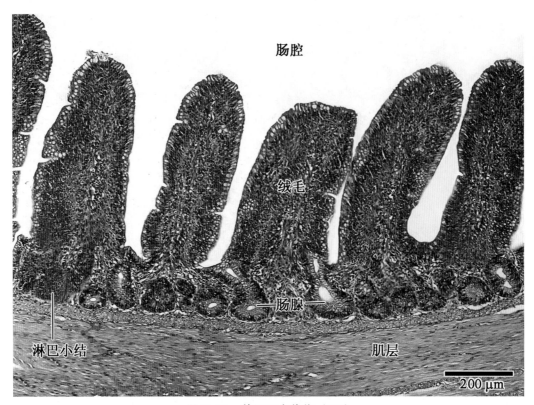

图 21-83　鹅回肠中倍像（HE）

Fig.21-83　mid magnification of goose ileum（HE）

肠腔—intestinal cavity　绒毛—villi　肠腺—intestinal gland　淋巴小结—lymphoid nodule　肌层—muscular layer

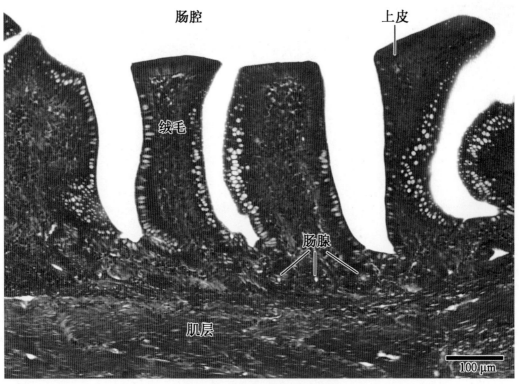

图 21-84　鸡盲肠中倍像（HE）

Fig.21-84　mid magnification of chicken cecum（HE）

肠腔—intestinal cavity　上皮—epithelium　绒毛—villi　肠腺—intestinal gland　肌层—muscular layer

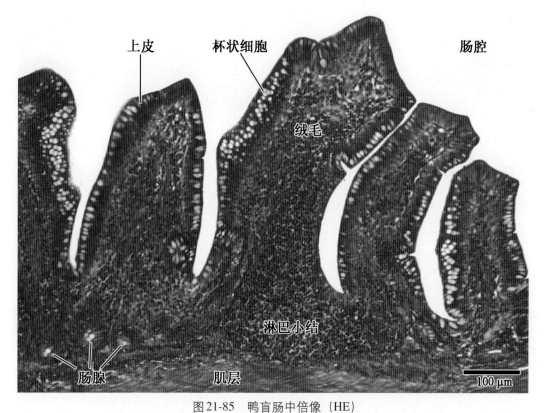

图 21-85　鸭盲肠中倍像（HE）

Fig.21-85　mid magnification of duck cecum（HE）

上皮—epithelium　杯状细胞—goblet cell　肠腔—intestinal cavity　绒毛—villi

淋巴小结—lymphoid nodule　肠腺—intestinal gland　肌层—muscular layer

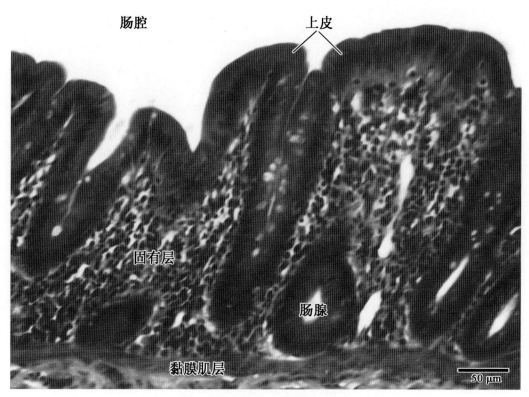

图 21-86　鹅盲肠高倍像（HE）

Fig.21-86　high magnification of goose cecum（HE）

肠腔—lumen　上皮—epithelium　固有层—proper layer　肠腺—intestinal gland　黏膜肌层—muscularis mucosae

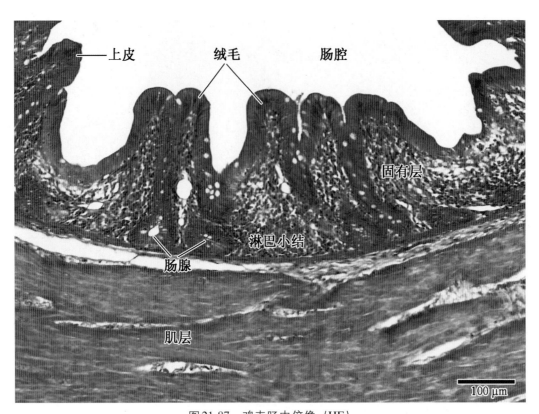

图21-87 鸡直肠中倍像（HE）

Fig.21-87 mid magnification of chicken rectum（HE）

上皮—epithelium 绒毛—villi 肠腔—intestinal cavity 固有层—proper layer
淋巴小结—lymphoid nodule 肠腺—intestinal gland 肌层—muscular layer

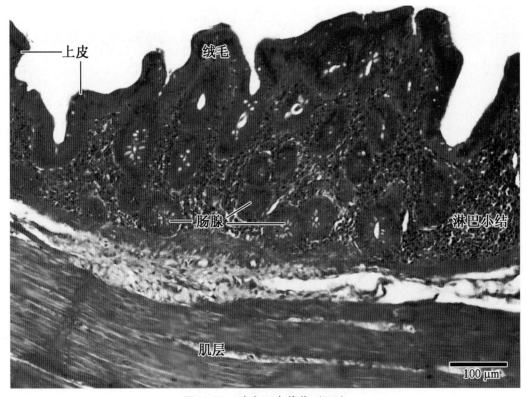

图21-88 鸭直肠中倍像（HE）

Fig.21-88 mid magnification of duck rectum（HE）

上皮—epithelium 绒毛—villi 肠腺—intestinal gland 淋巴小结—lymphoid nodule 肌层—muscular layer

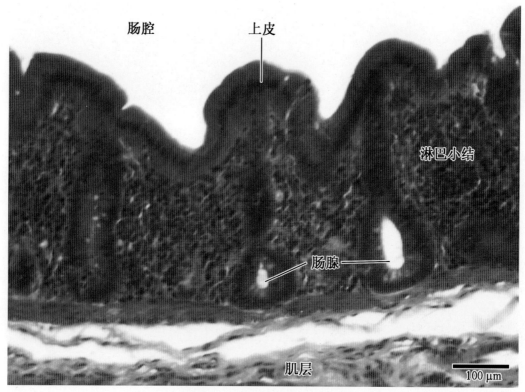

图21-89 鹅直肠中倍像（HE）

Fig.21-89 mid magnification of goose rectum（HE）

肠腔—lumen　上皮—epithelium　淋巴小结—lymphoid nodule　肠腺—intestinal gland　肌层—muscular layer

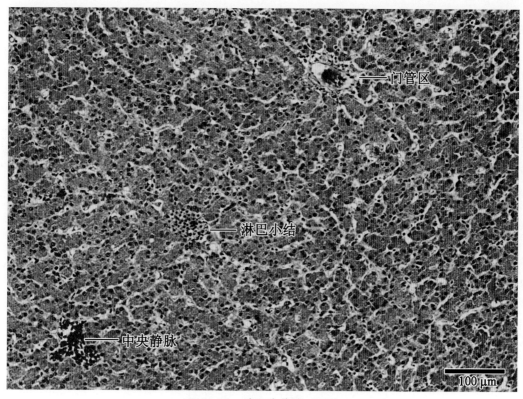

图21-90 鸡肝中倍像（HE）

Fig.21-90 mid magnification of chicken liver（HE）

门管区—portal area　淋巴小结—lymphoid nodule　中央静脉—central vein

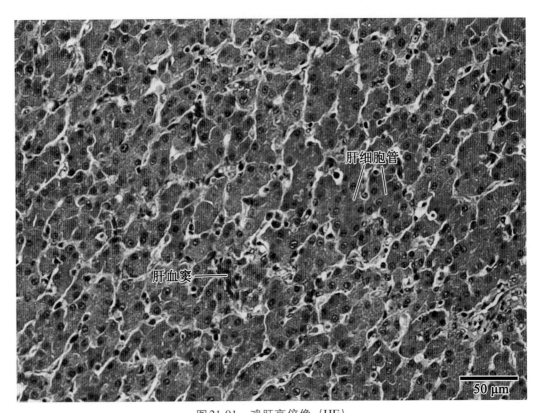

图21-91　鸡肝高倍像（HE）

Fig.21-91　high magnification of chicken liver（HE）

肝细胞管—liver cell tube　肝血窦—hepatic sinusoid

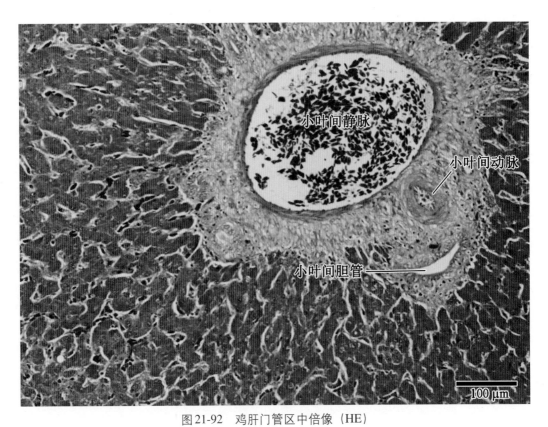

图21-92　鸡肝门管区中倍像（HE）

Fig.21-92　mid magnification of chicken liver portal area（HE）

小叶间静脉—interlobular vein　小叶间动脉—interlobular artery　小叶间胆管—interlobular bile duct

第二十一章 禽类的主要组织结构特征 Main Structural Features of Fowl Tissues

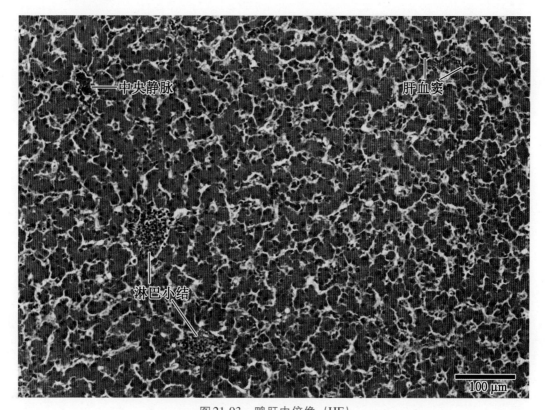

图21-93 鸭肝中倍像（HE）

Fig.21-93 mid magnification of duck liver (HE)

中央静脉—central vein　淋巴小结—lymphoid nodule　肝血窦—hepatic sinusoid

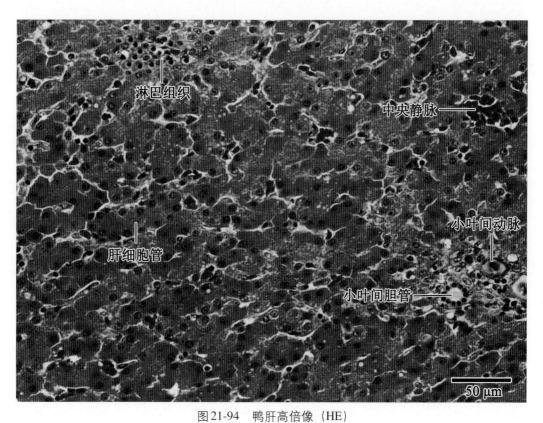

图21-94 鸭肝高倍像（HE）

Fig.21-94 high magnification of duck liver (HE)

淋巴组织—lymph tissue　肝细胞管—liver cell tube　中央静脉—central vein

小叶间动脉—interlobular artery　小叶间胆管—interlobular bile duct

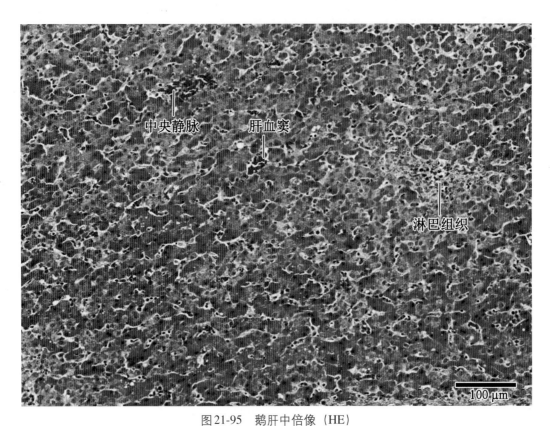

图21-95 鹅肝中倍像（HE）

Fig.21-95 mid magnification of goose liver（HE）

中央静脉—central vein 肝血窦—hepatic sinusoid 淋巴组织—lymph tissue

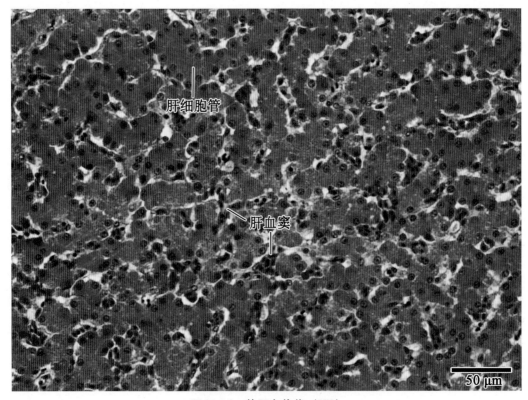

图21-96 鹅肝高倍像（HE）

Fig.21-96 high magnification of goose liver（HE）

肝细胞管—liver cell tube 肝血窦—hepatic sinusoid

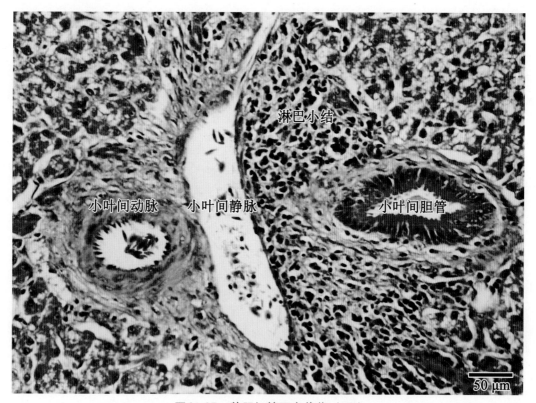

图21-97　鹅肝门管区高倍像（HE）

Fig.21-97　high magnification of goose liver portal area（HE）

淋巴小结—lymphoid nodule　小叶间静脉—interlobular vein　小叶间动脉—interlobular artery

小叶间胆管—interlobular bile duct

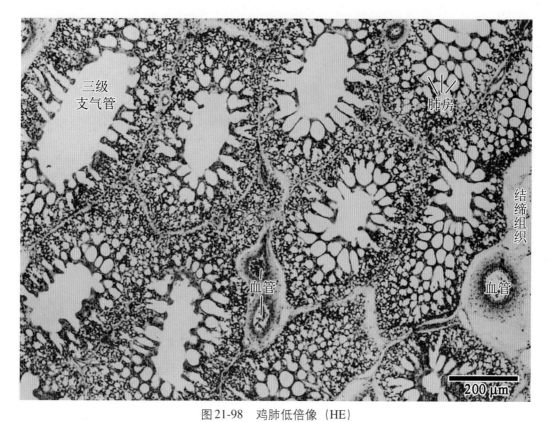

图21-98　鸡肺低倍像（HE）

Fig.21-98　low magnification of chicken lung（HE）

三级支气管—tertiary bronchus　肺房—lung atria　血管—blood vessel　结缔组织—connective tissue

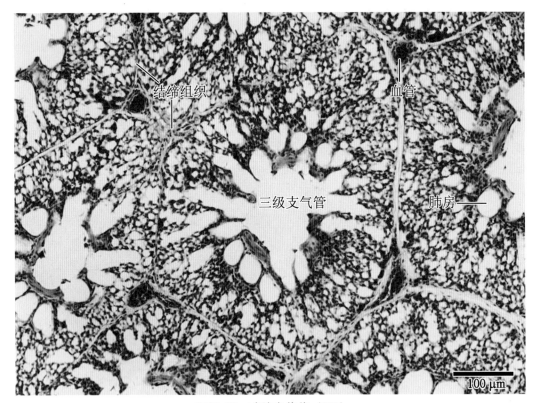

图21-99 鸡肺中倍像（HE）

Fig.21-99 mid magnification of chicken lung（HE）

结缔组织—connective tissue 血管—blood vessel 三级支气管—tertiary bronchus 肺房—lung atria

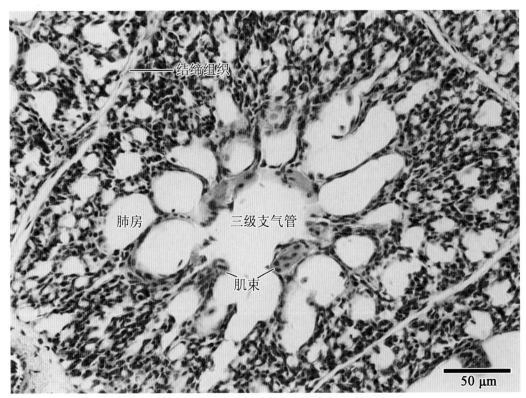

图21-100 鸡肺小叶高倍像（HE）

Fig.21-100 high magnification of chicken lung lobule（HE）

结缔组织—connective tissue 肺房—lung atria 三级支气管—tertiary bronchus 肌束—muscle bundle

第二十一章 禽类的主要组织结构特征 Main Structural Features of Fowl Tissues

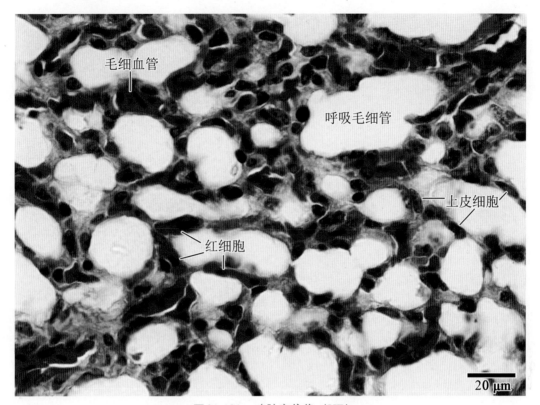

图21-101 鸡肺高倍像（HE）

Fig.21-101 high magnification of chicken lung（HE）

毛细血管—capillary 呼吸毛细管—respiratory capillary 上皮细胞—epithelium 红细胞—red cell

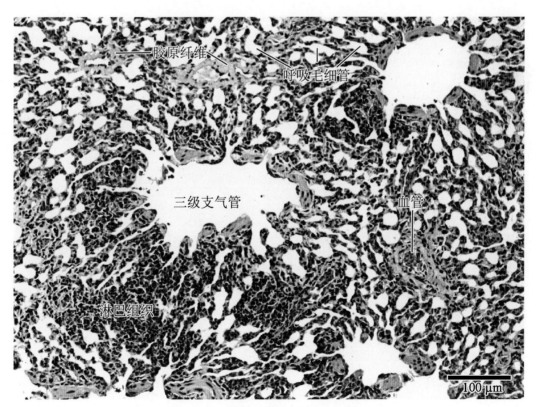

图21-102 鸡肺中倍像（Masson染色）

Fig.21-102 mid magnification of chicken lung（Masson stain）

胶原纤维—collagen fiber 呼吸毛细管—respiratory capillary 三级支气管—tertiary bronchus

血管—blood vessel 淋巴组织—lymph tissue

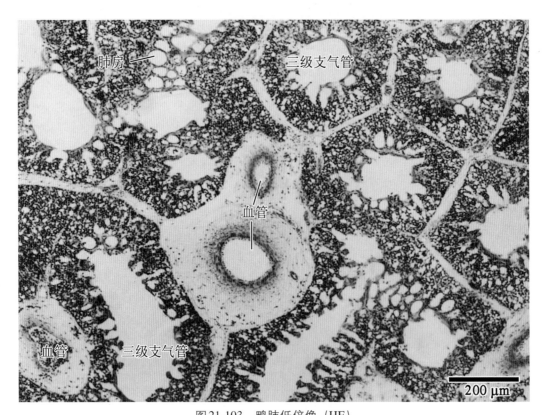

图21-103 鸭肺低倍像（HE）

Fig.21-103 low magnification of duck lung（HE）

肺房—lung atria 三级支气管—tertiary bronchus 血管—blood vessel

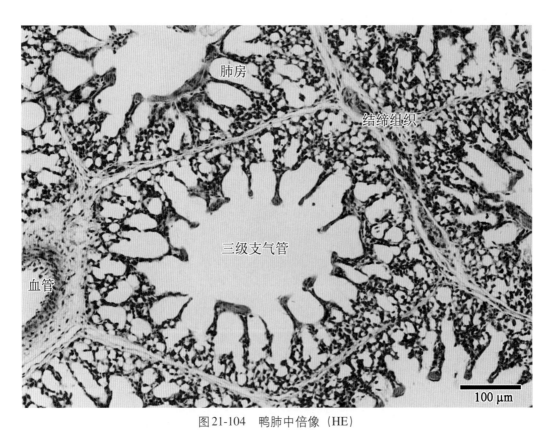

图21-104 鸭肺中倍像（HE）

Fig.21-104 mid magnification of duck lung（HE）

肺房—lung atria 结缔组织—connective tissue 血管—blood vessel 三级支气管—tertiary bronchus

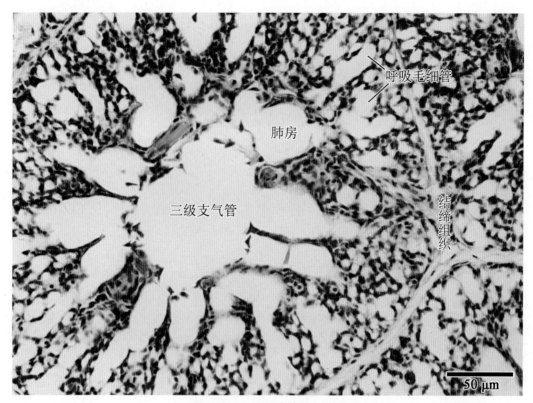

图21-105 鸭肺高倍像（HE）

Fig.21-105 high magnification of duck lung (HE)

呼吸毛细管—respiratory capillary　肺房—lung atria　三级支气管—tertiary bronchus　结缔组织—connective tissue

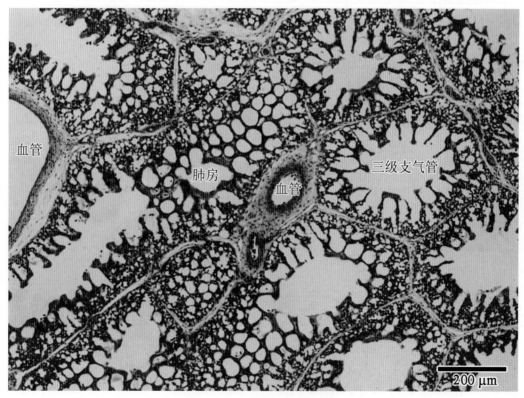

图21-106 鹅肺低倍像（HE）

Fig.21-106 low magnification of goose lung (HE)

血管—blood vessel　肺房—lung atria　三级支气管—tertiary bronchus

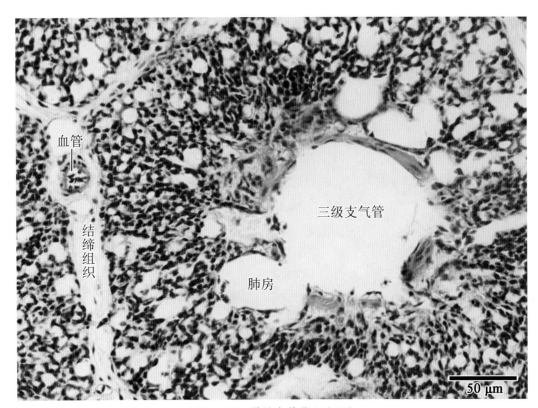

图21-107 鹅肺高倍像1（HE）

Fig.21-107 high magnification of goose lung 1（HE）

血管—blood vessel 结缔组织—connective tissue 肺房—lung atria 三级支气管—tertiary bronchus

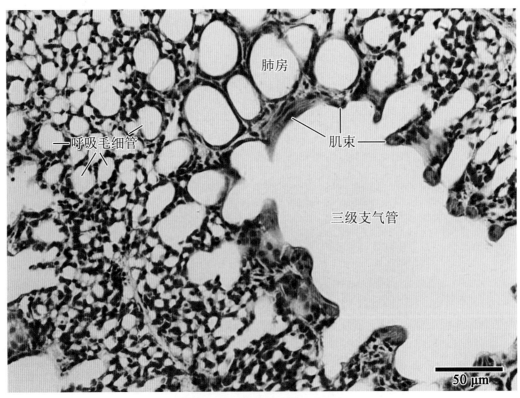

图21-108 鹅肺高倍像2（HE）

Fig.21-108 high magnification of goose lung 2（HE）

呼吸毛细管—respiratory capillary 肺房—lung atria 肌束—muscle bundle 三级支气管—tertiary bronchus

第二十一章 禽类的主要组织结构特征 Main Structural Features of Fowl Tissues

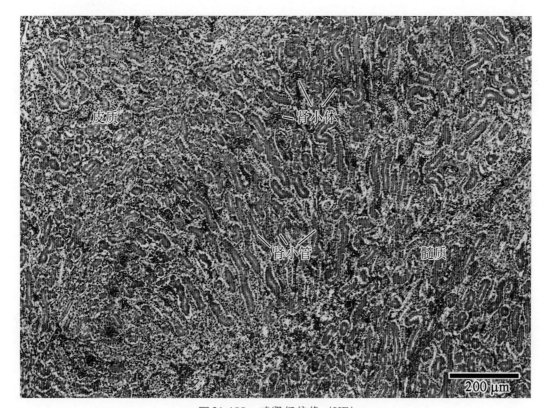

图21-109 鸡肾低倍像（HE）

Fig.21-109 low magnification of chicken kidney（HE）

皮质—cortex　肾小体—renal corpuscle　肾小管—renal tubule　髓质—medulla

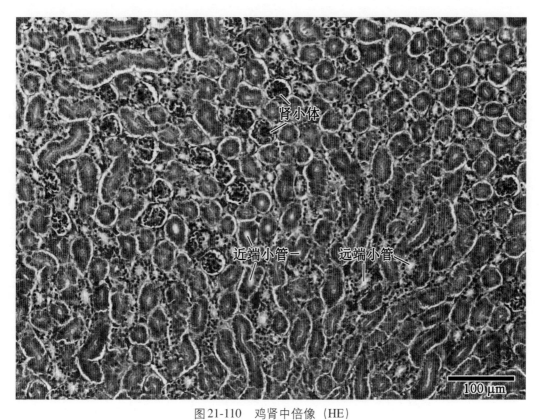

图21-110 鸡肾中倍像（HE）

Fig.21-110 mid magnification of chicken kidney（HE）

肾小体—renal corpuscle　近端小管—proximal tubule　远端小管—distal tubule

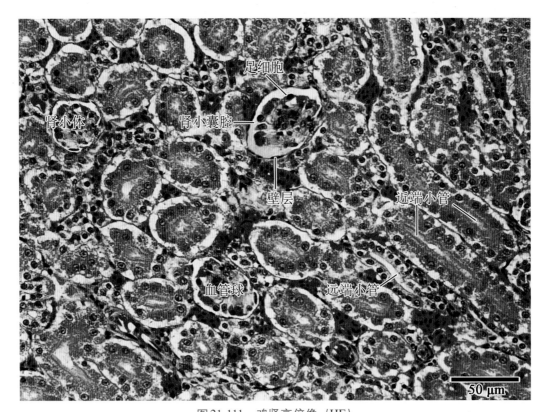

图21-111　鸡肾高倍像（HE）

Fig.21-111　high magnification of chicken kidney（HE）

肾小体—renal corpuscle　肾小囊腔—capsular space　足细胞—podocyte　壁层—parietal layer

血管球—glomus　近端小管—proximal tubule　远端小管—distal tubule

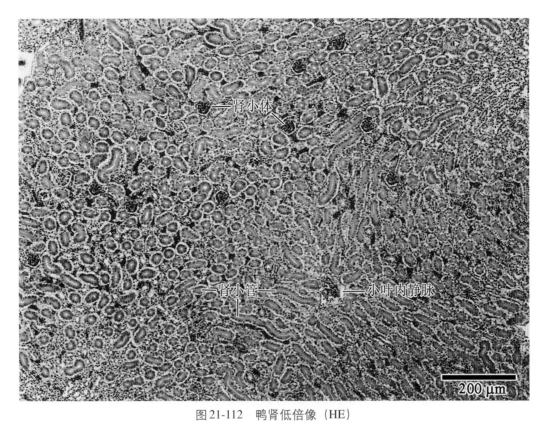

图21-112　鸭肾低倍像（HE）

Fig.21-112　low magnification of duck kidney（HE）

肾小体—renal corpuscle　肾小管—renal tubule　小叶内静脉—intralobular vein

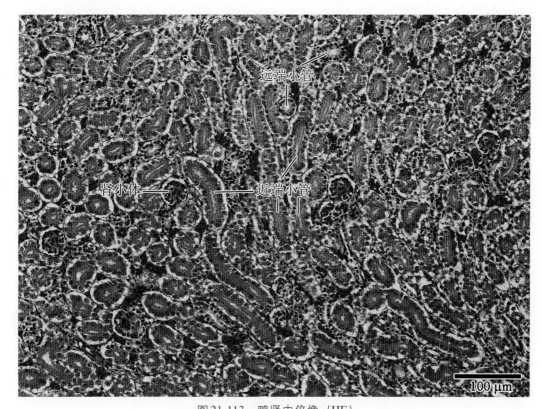

图21-113 鸭肾中倍像（HE）

Fig.21-113 mid magnification of duck kidney（HE）

肾小体—renal corpuscle 远端小管—distal tubule 近端小管—proximal tubule

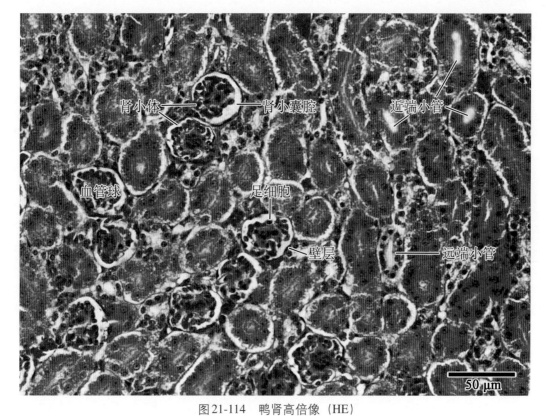

图21-114 鸭肾高倍像（HE）

Fig.21-114 high magnification of duck kidney（HE）

肾小体—renal corpuscle 肾小囊腔—capsular space 血管球—glomus 足细胞—podocyte
壁层—parietal layer 近端小管—proximal tubule 远端小管—distal tubule

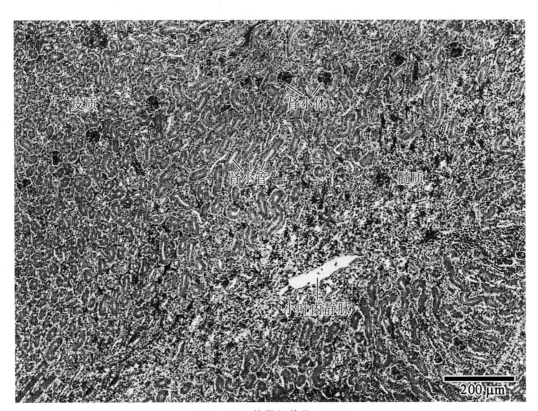

图 21-115 鹅肾低倍像（HE）

Fig.21-115 low magnification of goose kidney （HE）

皮质—cortex　肾小体—renal corpuscle　肾小管—renal tubule　髓质—medulla　小叶内静脉—intralobular vein

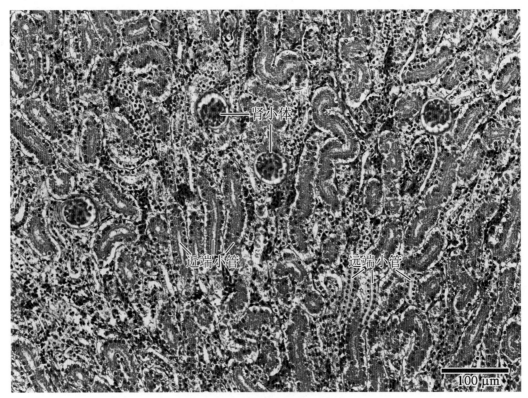

图 21-116 鹅肾中倍像（HE）

Fig.21-116 mid magnification of goose kidney （HE）

肾小体—renal corpuscle　近端小管—proximal tubule　远端小管—distal tubule

第二十一章 禽类的主要组织结构特征 Main Structural Features of Fowl Tissues

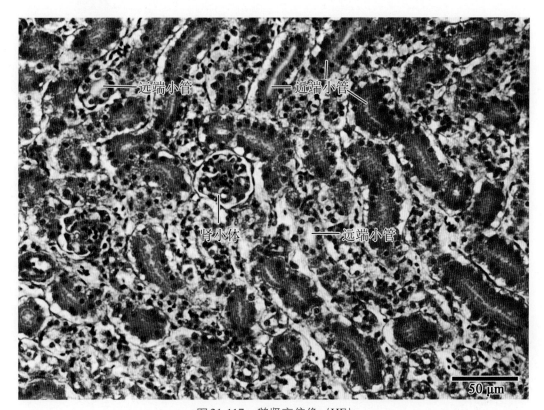

图21-117 鹅肾高倍像（HE）

Fig.21-117 high magnification of goose kidney（HE）

远端小管—distal tubule 近端小管—proximal tubule 肾小体—renal corpuscle

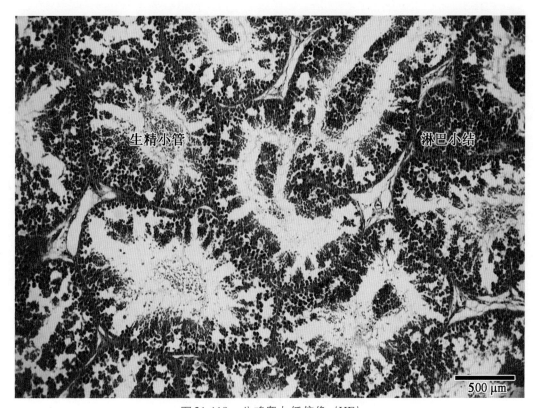

图21-118 公鸡睾丸低倍像（HE）

Fig.21-118 low magnification of cock testis（HE）

生精小管—seminiferous tubule 淋巴小结—lymphoid nodule

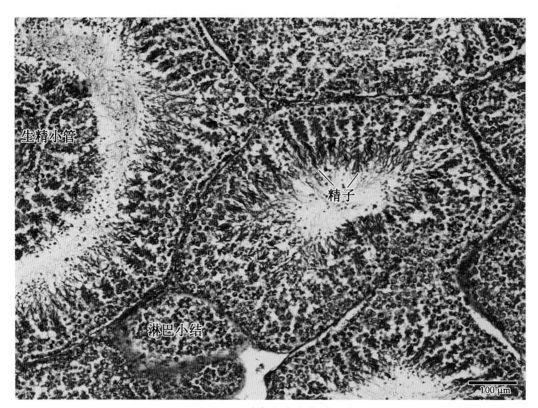

图21-119 鸡睾丸中倍像（HE）

Fig.21-119 mid magnification of cock testis（HE）

生精小管—seminiferous tubule 精子—sperm 淋巴小结—lymphoid nodule

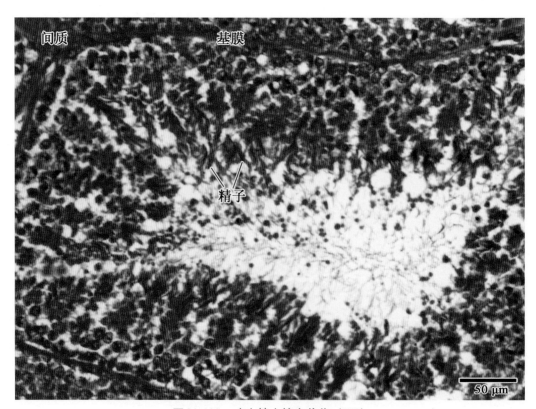

图21-120 鸡生精小管高倍像（HE）

Fig.21-120 high magnification of cock seminiferous tubule（HE）

间质—interstitium 基膜—basal membrane 精子—sperm

第二十一章 禽类的主要组织结构特征　Main Structural Features of Fowl Tissues

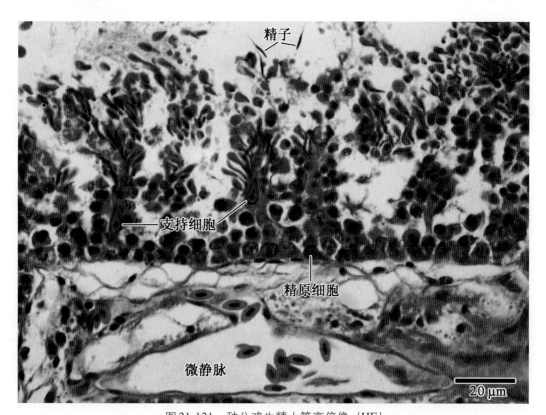

图21-121　种公鸡生精小管高倍像（HE）

Fig.21-121　high magnification of cock seminiferous tubule（HE）

精子—sperm　支持细胞—supporting cell　精原细胞—spermospore　微静脉—venule

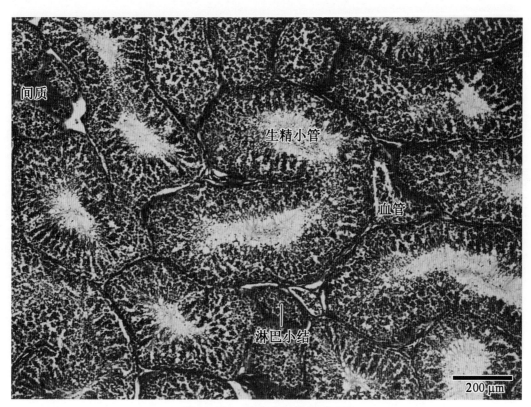

图21-122　鸭睾丸低倍像（HE）

Fig.21-122　low magnification of duck testis（HE）

间质—interstitium　生精小管—seminiferous tubule　血管—blood vessel　淋巴小结—lymphoid nodule

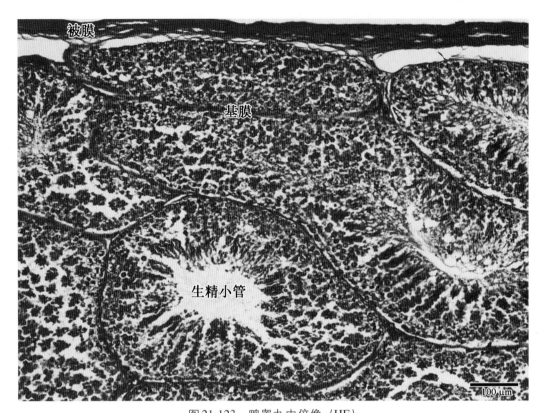

图21-123 鸭睾丸中倍像（HE）

Fig.21-123 mid magnification of duck testis（HE）

被膜—capsule 基膜—basal membrane 生精小管—seminiferous tubule

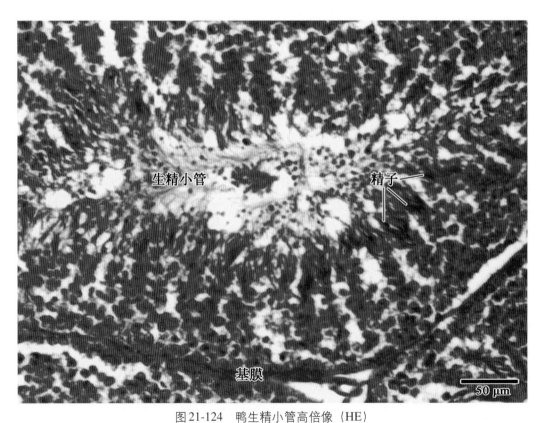

图21-124 鸭生精小管高倍像（HE）

Fig.21-124 high magnification of duck seminiferous tubule（HE）

生精小管—seminiferous tubule 精子—sperm 基膜—basal membrane

第二十一章　禽类的主要组织结构特征　Main Structural Features of Fowl Tissues

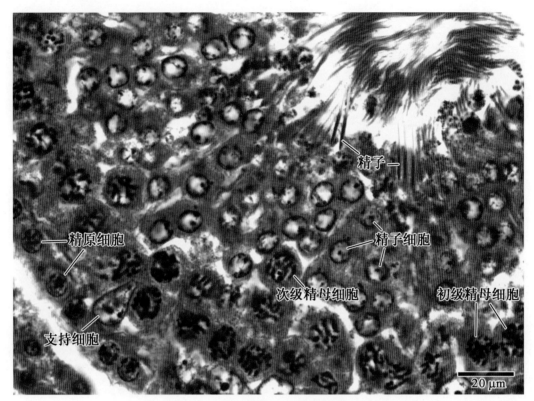

图21-125　种公鸭生精小管高倍像（HE）

Fig.21-125　high magnification of duck seminiferous tubule（HE）

精原细胞—spermospore　支持细胞—supporting cell　精子—sperm　精子细胞—spermatid

次级精母细胞— secondary spermatocyte　初级精母细胞— primary spermatocyte

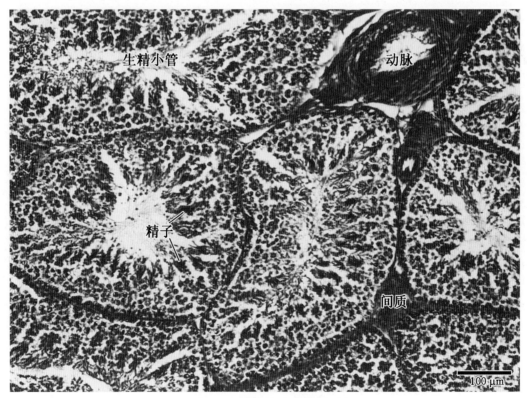

图21-126　鹅睾丸中倍像（HE）

Fig.21-126　mid magnification of goose testis（HE）

生精小管—seminiferous tubule　动脉—artery　精子—sperm　间质—mesenchyme

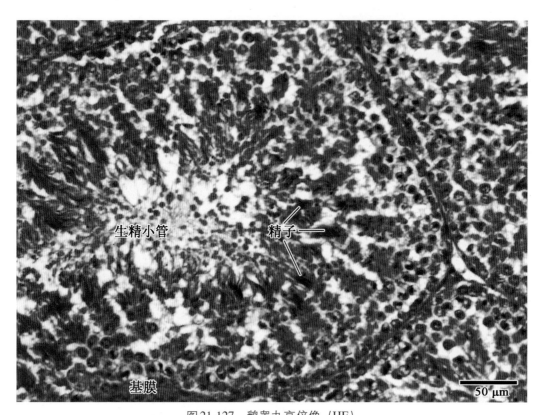

图 21-127　鹅睾丸高倍像（HE）

Fig.21-127　high magnification of goose testis（HE）

生精小管—seminiferous tubule　精子—sperm　基膜—basal membrane

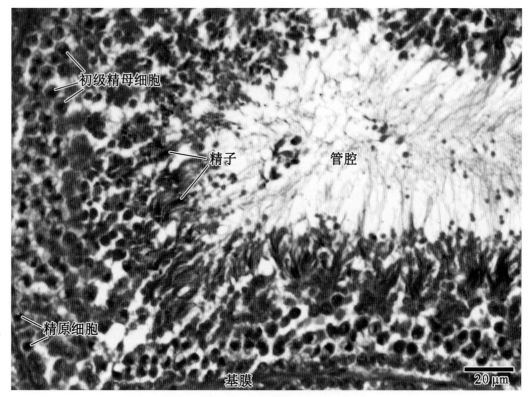

图 21-128　鹅生精小管高倍像（HE）

Fig.21-128　high magnification of goose seminiferous tubule（HE）

初级精母细胞— primary spermatocyte　精子—sperm　管腔—lumen　精原细胞— spermospore　基膜—basal membrane

第二十一章 禽类的主要组织结构特征　Main Structural Features of Fowl Tissues

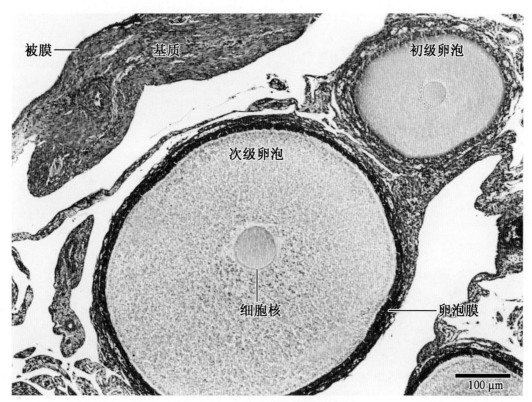

图21-129　鸡卵巢中倍像（HE）

Fig.21-129　mid magnification of hen ovary（HE）

被膜—capsule　基质—stroma　初级卵泡—primary follicle　次级卵泡—secondary follicle

细胞核—nucleus　卵泡膜—follicular theca

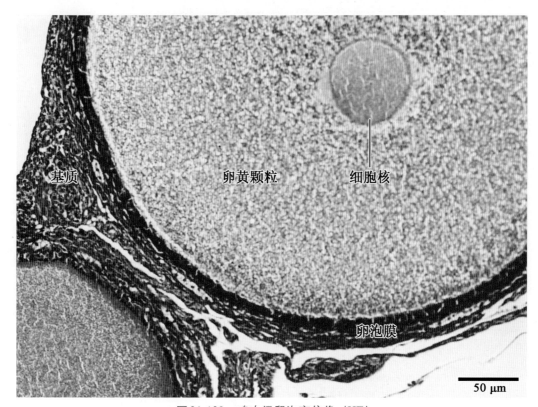

图21-130　鸡次级卵泡高倍像（HE）

Fig.21-130　high magnification of hen secondary follicle（HE）

基质—stroma　卵黄颗粒—yolk granule　细胞核—nucleus　卵泡膜—follicular theca

829

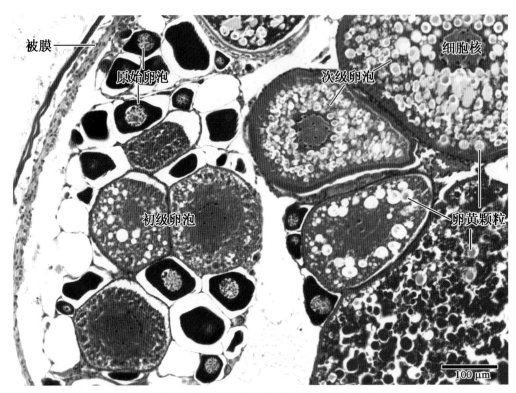

图21-131 鸭卵巢中倍像（三色法染色）

Fig.21-131 mid magnification of duck ovary（trichrome stain）

被膜—capsule 原始卵泡— primordial follicle 初级卵泡— primary follicle 次级卵泡—secondary follicle

细胞核—nucleus 卵黄颗粒—yolk granule

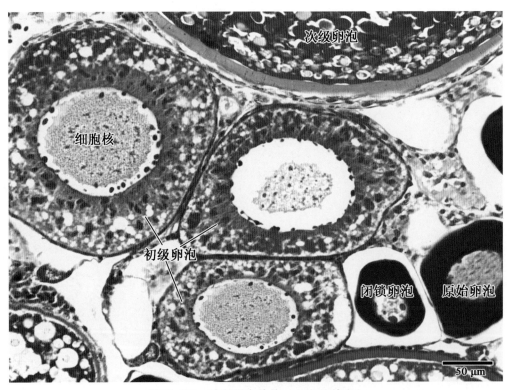

图21-132 鸭卵巢高倍像（三色法染色）

Fig.21-132 high magnification of duck ovary（trichrome stain）

次级卵泡— secondary follicle 细胞核—nucleus 初级卵泡— primary follicle

闭锁卵泡—atretic follicle 原始卵泡— primordial follicle

第二十一章 禽类的主要组织结构特征 Main Structural Features of Fowl Tissues

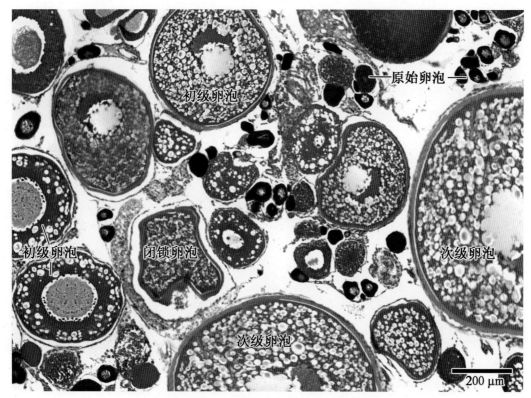

图 21-133　鹅卵巢低倍像（三色法染色）

Fig.21-133　low magnification of goose ovary（trichrome stain）

原始卵泡—primordial follicle　初级卵泡—primary follicle　闭锁卵泡—atretic follicle　次级卵泡—secondary follicle

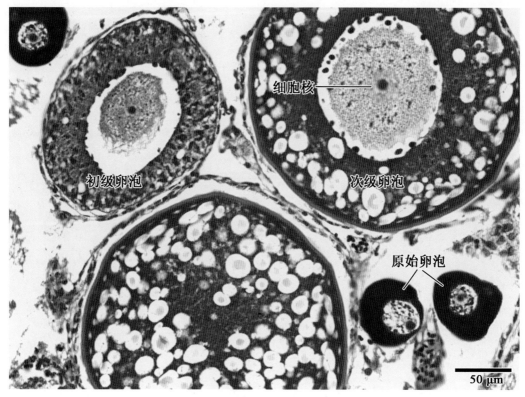

图 21-134　鹅卵巢高倍像（三色法染色）

Fig.21-134　high magnification of goose ovary（trichrome stain）

细胞核—nucleus　初级卵泡—primary follicle　次级卵泡—secondary follicle　原始卵泡—primordial follicle

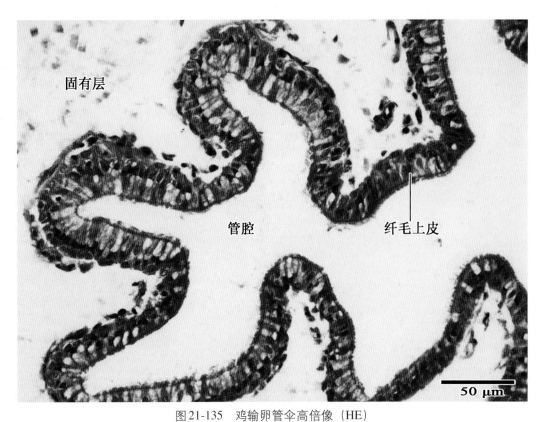

图 21-135　鸡输卵管伞高倍像（HE）

Fig.21-135　high magnification of hen oviduct umbrella（HE）

固有层—proper layer　管腔—lumen　纤毛上皮—ciliated epithelium

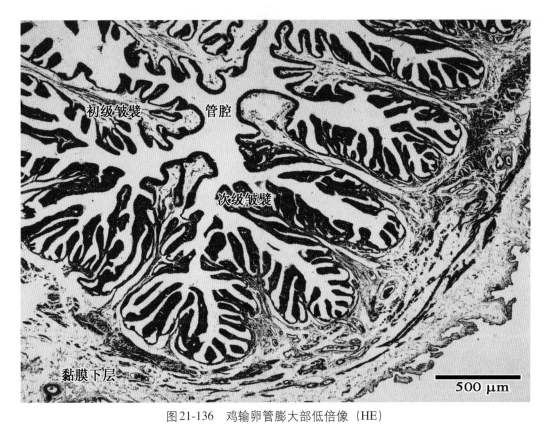

图 21-136　鸡输卵管膨大部低倍像（HE）

Fig.21-136　low magnification of hen oviduct enlargement（HE）

初级皱襞—primary plica　管腔—lumen　次级皱襞—secondary plica　黏膜下层—submucosa

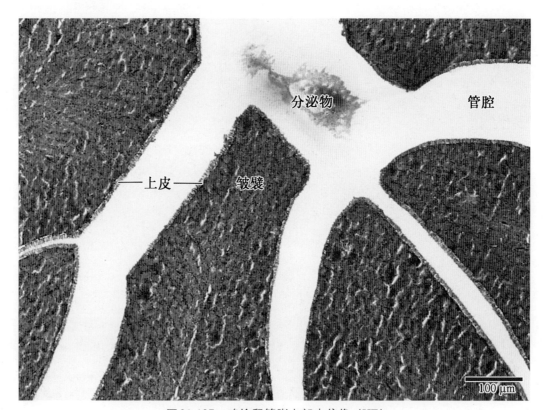

图21-137 鸡输卵管膨大部中倍像（HE）

Fig.21-137 mid magnification of hen oviduct enlargement（HE）

上皮—epithelium 皱襞—plica 分泌物—secretion 管腔—lumen

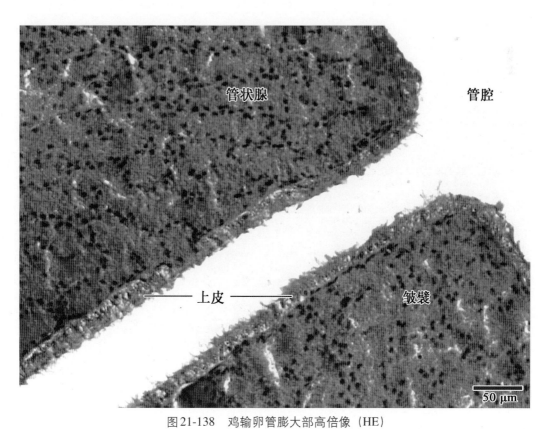

图21-138 鸡输卵管膨大部高倍像（HE）

Fig.21-138 high magnification of hen oviduct enlargement（HE）

管状腺—tubular gland 管腔—lumen 上皮—epithelium 皱襞—plica

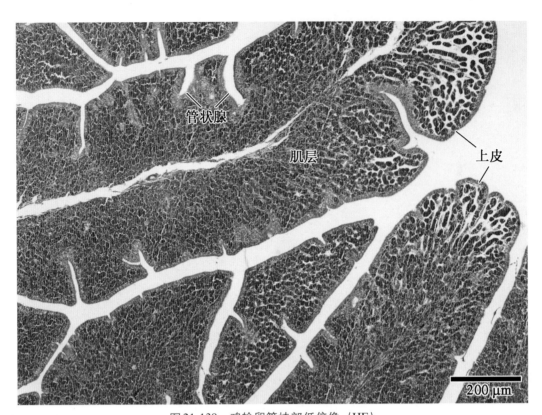

图 21-139 鸡输卵管峡部低倍像（HE）

Fig.21-139 low magnification of hen oviduct isthmus（HE）

管状腺—tubular gland 肌层—muscular layer 上皮—epithelium

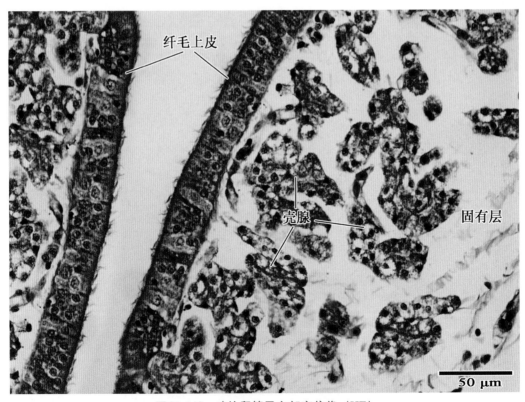

图 21-140 鸡输卵管子宫部高倍像（HE）

Fig.21-140 high magnification of hen oviduct uterus part（HE）

纤毛上皮—ciliated epithelium 壳腺—shell gland 固有层—proper layer

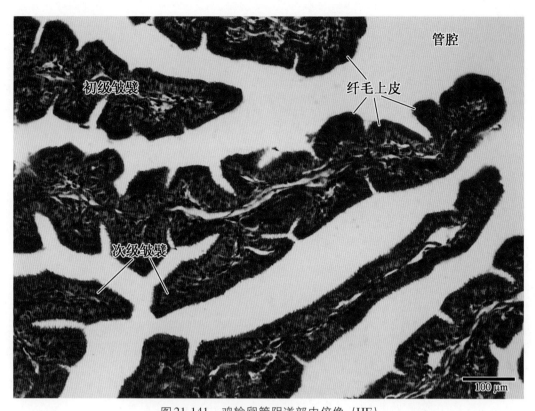

图21-141 鸡输卵管阴道部中倍像（HE）

Fig.21-141 mid magnification of hen oviduct vagina part（HE）

初级皱襞—primary plica 次级皱襞—secondary plica 纤毛上皮—ciliated epithelium 管腔—lumen

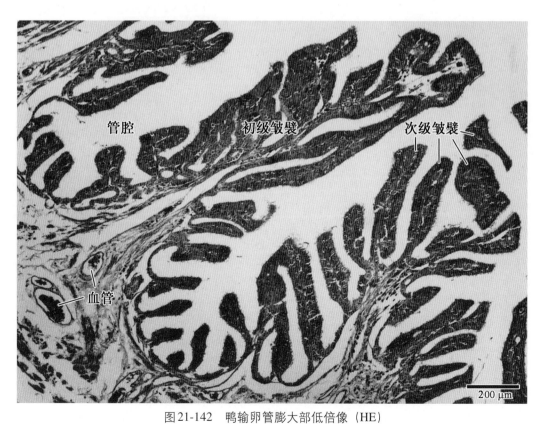

图21-142 鸭输卵管膨大部低倍像（HE）

Fig.21-142 low magnification of duck oviduct enlargement（HE）

管腔—lumen 初级皱襞—primary plica 次级皱襞—secondary plica 血管—blood vessel

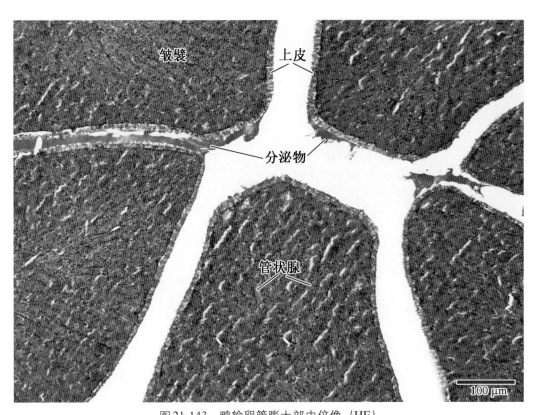

图21-143 鸭输卵管膨大部中倍像（HE）

Fig.21-143 mid magnification of duck oviduct enlargement（HE）

皱襞—plica 上皮—epithelium 分泌物—secretion 管状腺—tubular gland

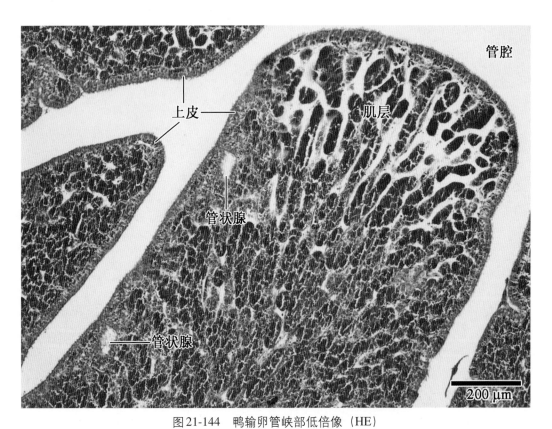

图21-144 鸭输卵管峡部低倍像（HE）

Fig.21-144 low magnification of duck oviduct isthmus（HE）

上皮—epithelium 管状腺—tubular gland 肌层—muscular layer 管腔—lumen

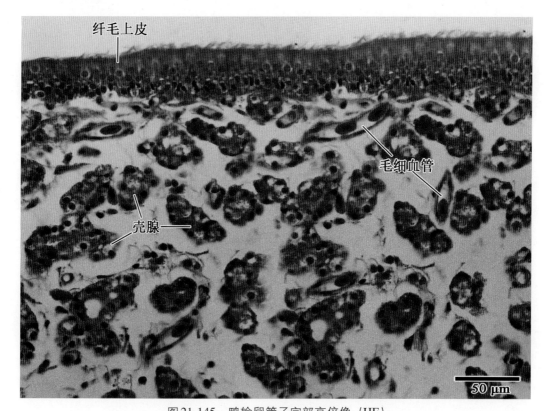

图21-145 鸭输卵管子宫部高倍像（HE）

Fig.21-145 high magnification of duck oviduct uterus part （HE）

纤毛上皮—ciliated epithelium 壳腺—shell gland 毛细血管—capillary

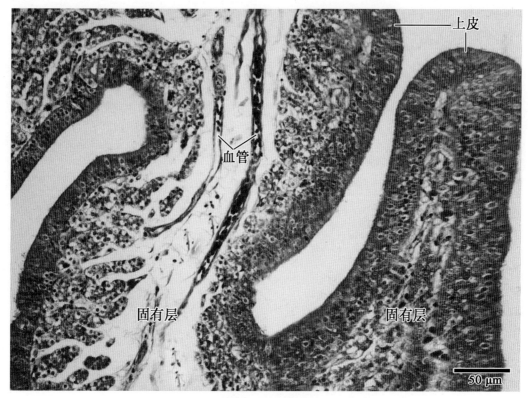

图21-146 鸭输卵管阴道部高倍像（HE）

Fig.21-146 high magnification of duck oviduct vagina part （HE）

血管—blood vessel 固有层—proper layer 上皮—epithelium

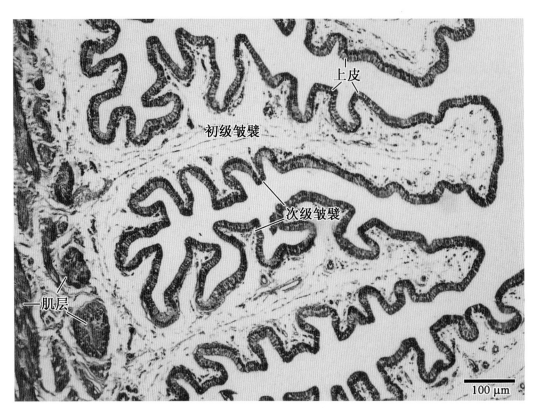

图 21-147　鹅输卵管漏斗中倍像（HE）

Fig.21-147　mid magnification of goose oviduct infundibulum（HE）

上皮—epithelium　初级皱襞—primary plica　次级皱襞—secondary plica　肌层—muscular layer

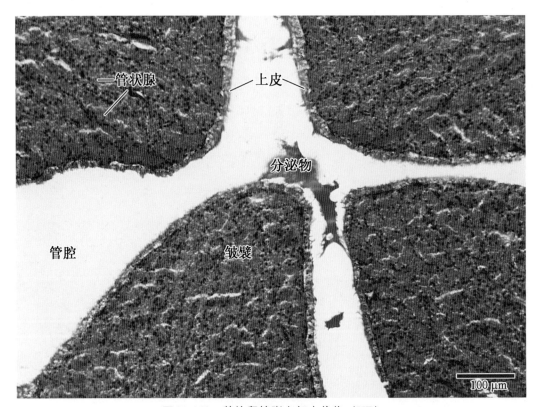

图 21-148　鹅输卵管膨大部中倍像（HE）

Fig.21-148　mid magnification of goose oviduct enlargement（HE）

管状腺—tubular gland　上皮—epithelium　分泌物—secretion　管腔—lumen　皱襞—plica

第二十一章 禽类的主要组织结构特征　Main Structural Features of Fowl Tissues

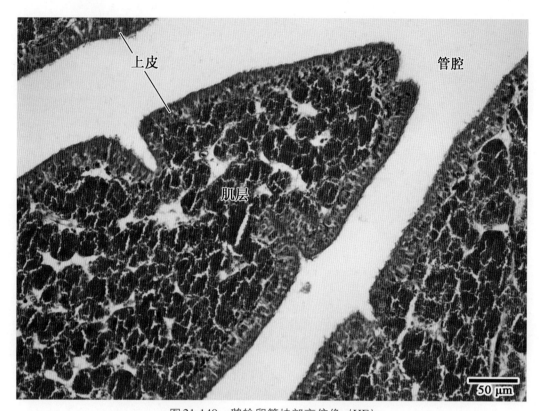

图21-149　鹅输卵管峡部高倍像（HE）

Fig.21-149　high magnification of goose oviduct isthmus（HE）

上皮—epithelium　管腔—lumen　肌层—muscular layer

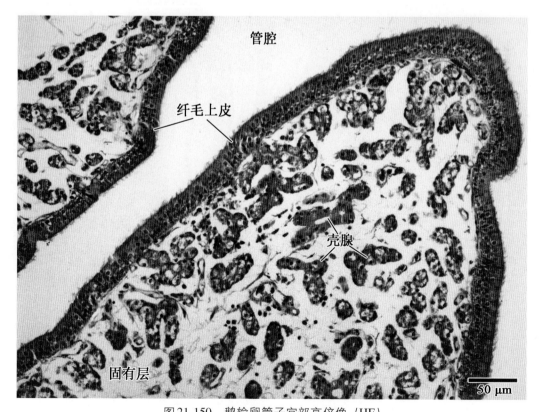

图21-150　鹅输卵管子宫部高倍像（HE）

Fig.21-150　high magnification of goose oviduct uterus part（HE）

管腔—lumen　纤毛上皮—ciliated epithelium　壳腺—shell gland　固有层—proper layer

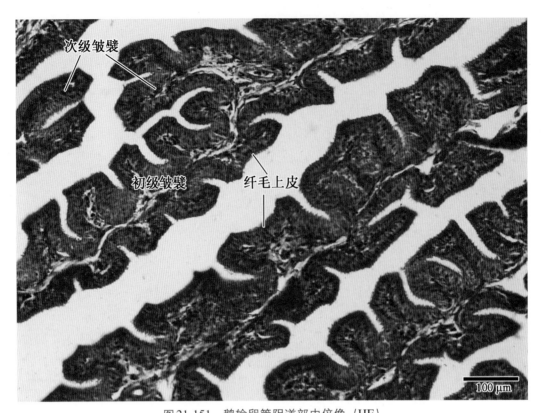

图21-151　鹅输卵管阴道部中倍像（HE）

Fig.21-151　mid magnification of goose oviduct vagina part（HE）

次级皱襞—secondary plica　初级皱襞—primary plica　纤毛上皮—ciliated epithelium

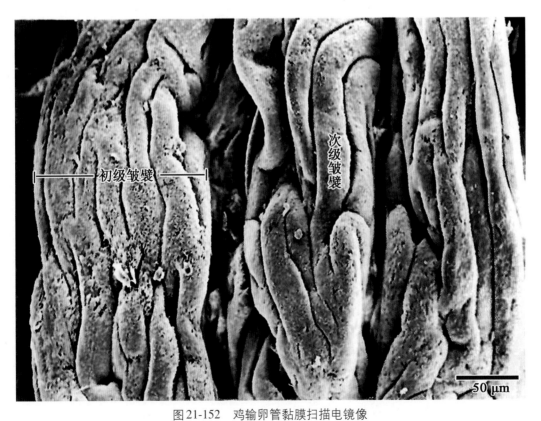

图21-152　鸡输卵管黏膜扫描电镜像

Fig.21-152　scanning electrical image of chicken oviduct tube mucosa

初级皱襞—primary plica　次级皱襞—secondary plica

第二十一章 禽类的主要组织结构特征　Main Structural Features of Fowl Tissues

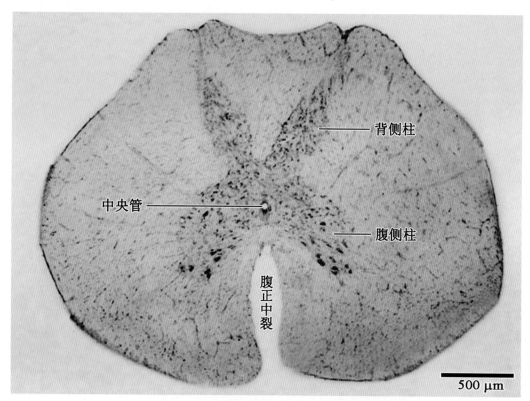

图 21-153　鸡颈段脊髓低倍像（中性红染色）

Fig.21-153　low magnification of chicken cervical spinal cord (neutral red stain)

背侧柱—dorsal column　中央管—central canal　腹侧柱—ventral column　腹正中裂—ventral median fissure

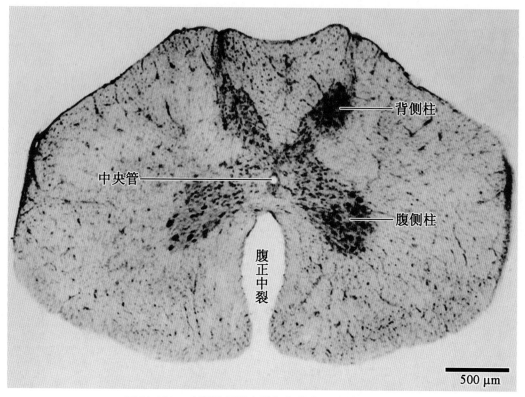

图 21-154　鸡脊髓颈膨大横切低倍像（中性红染色）

Fig.21-154　low magnification of chicken cervical enlargement (neutral red stain)

背侧柱—dorsal column　中央管—central canal　腹侧柱—ventral column　腹正中裂—ventral median fissure

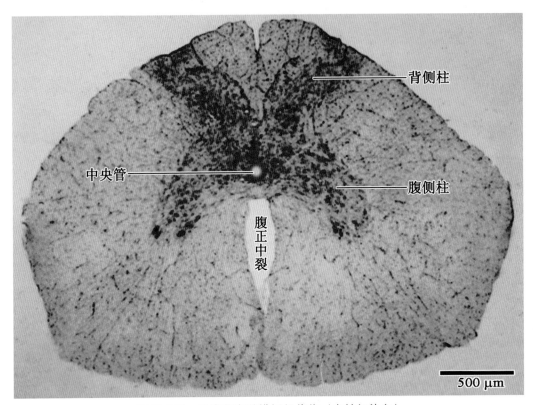

图 21-155 鸡胸段脊髓横切低倍像（中性红染色）

Fig.21-155 low magnification of chicken thoracic spinal cord (neutral red stain)

背侧柱—dorsal column 中央管—central canal 腹侧柱—ventral column 腹正中裂—ventral median fissure

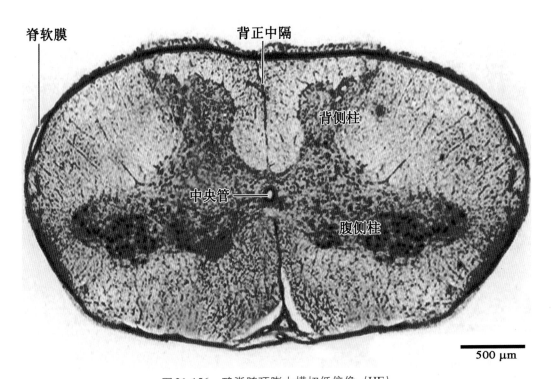

图 21-156 鸭脊髓颈膨大横切低倍像（HE）

Fig.21-156 low magnification of duck cervical enlargement (HE)

脊软膜—spinal pia matter 背正中隔—dorsal septum 背侧柱—dorsal column
中央管—central canal 腹侧柱—ventral column

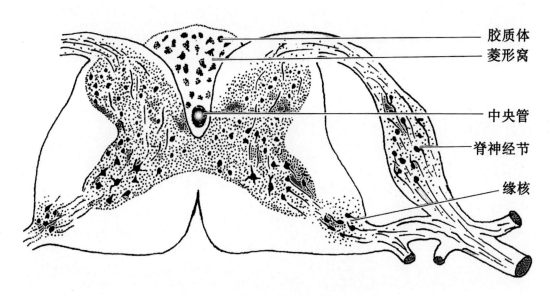

图 21-157　禽类脊髓腰膨大横切模式图

Fig.21-157　transverse cut pattern of lumbar enlargement of avian spinal cord

胶质体—corpus gelatinosum　菱形窝—rhomboid sinus　中央管—central canal

脊神经节—spinal ganglion　缘核—marginal nucleus

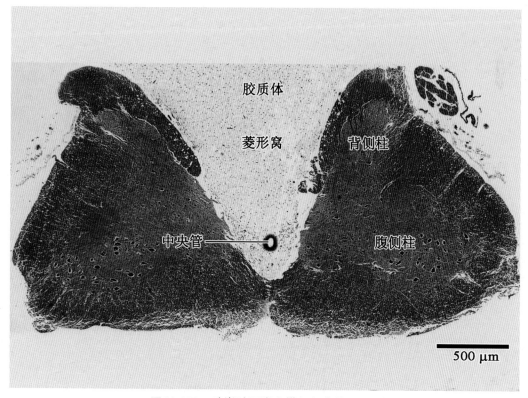

图 21-158　鸭脊髓腰膨大横切低倍像（HE）

Fig.21-158　low magnification of lumbar enlargement of duck spinal cord（HE）

胶质体—corpus gelatinosum　菱形窝—rhomboid sinus　中央管—central canal

背侧柱—dorsal column　腹侧柱—ventral column

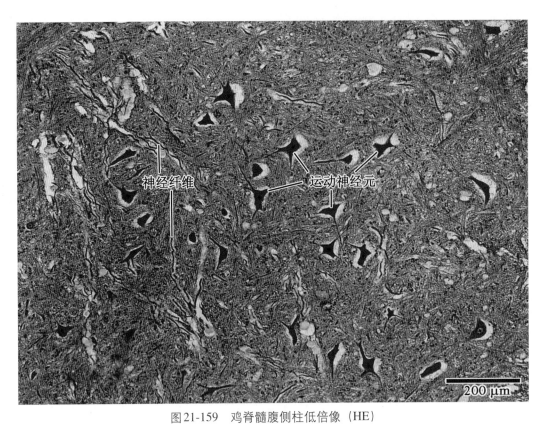

图 21-159　鸡脊髓腹侧柱低倍像（HE）

Fig.21-159　low magnification of ventral column of chicken spinal cord（HE）

神经纤维— nerve fiber　运动神经元—motor neuron

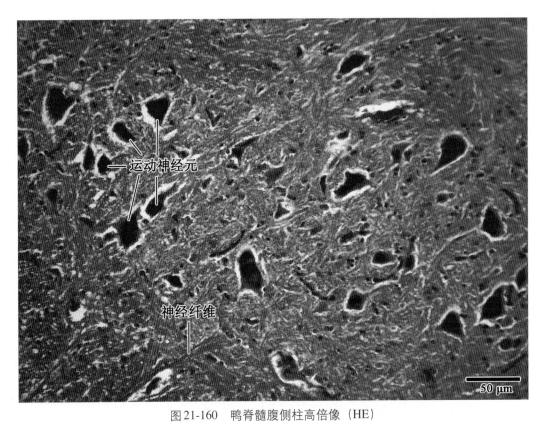

图 21-160　鸭脊髓腹侧柱高倍像（HE）

Fig.21-160　high magnification of ventral column of duck spinal cord（HE）

运动神经元—motor neuron　神经纤维— nerve fiber

第二十一章 禽类的主要组织结构特征 Main Structural Features of Fowl Tissues

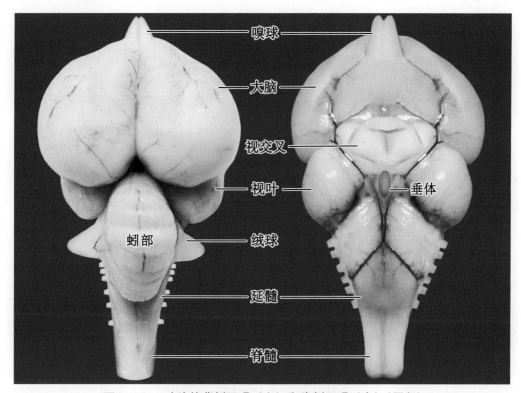

图21-161　鸡脑的背侧面观（左）和腹侧面观（右）（原色）

Fig.21-161　dorsal view (left) and ventral view (right) of chicken brain（primary color）

嗅球—olfactory bulb　大脑—cerebrum　视交叉—optic chiasma　视叶—optic lobe　垂体—hypophysis

蚓部—vermis　绒球—flocculus　延髓—medulla　脊髓—spinal cord

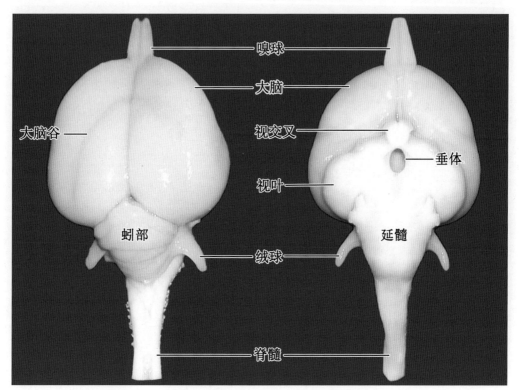

图21-162　鸭脑的背侧面观（左）和腹侧面观（右）（原色）

Fig.21-162　dorsal view（left）and ventral view（right）of duck brain（primary color）

嗅球—olfactory bulb　大脑—cerebrum　大脑谷—cerebrum vally　视交叉—optic chiasma　视叶—optic lobe

蚓部—vermis　绒球—flocculus　延髓—medulla　脊髓—spinal cord　垂体—hypophysis

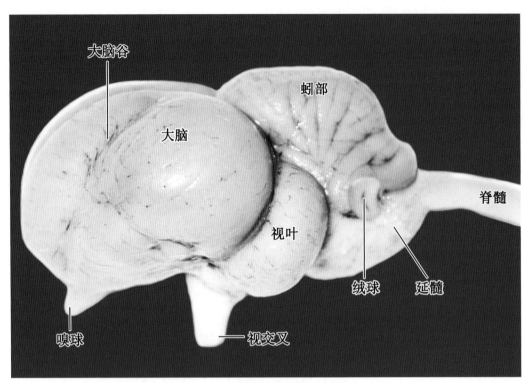

图 21-163 鹅脑的侧面观（原色）

Fig.21-163 lateral view of goose brain（primary color）

大脑谷—cerebrum vally 大脑—cerebrum 蚓部—vermis 嗅球—olfactory bulb

视交叉—optic chiasma 视叶—optic lobe 绒球—flocculus 延髓—medulla 脊髓—spinal cord

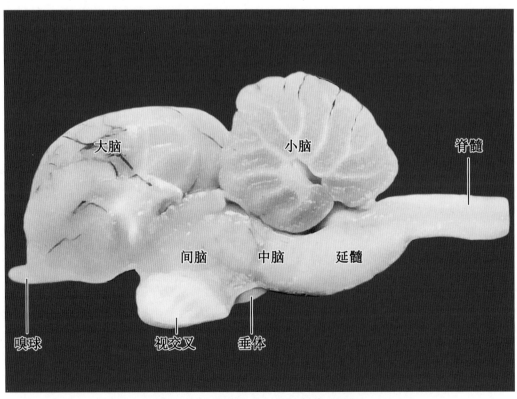

图 21-164 鸡脑正中矢状切面观（原色）

Fig.21-164 median sagittal section view of chicken brain（primary color）

大脑—cerebrum 小脑—cerebellum 脊髓—spinal cord 嗅球—olfactory bulb 间脑—diencephalon

视交叉—optic chiasma 垂体—hypophysis 中脑—midbrain 延髓—medulla oblongata

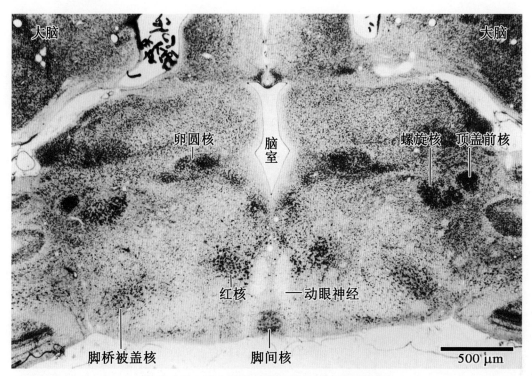

图 21-165 鸭中脑前部横切低倍像（洋红染色）

Fig.21-165 transverse low magnification of duck anterior midbrain (kermes stain)

大脑—cerebrum 卵圆核—oval nucleus 脑室—ventricle 螺旋核—spiral nucleus 顶盖前核—pretectal nucleus 红核—ruber nucleus 动眼神经—oculomotor nerve 脚桥被盖核—pedunculopontine tegmental nucleus 脚间核—interpeduncular nucleus

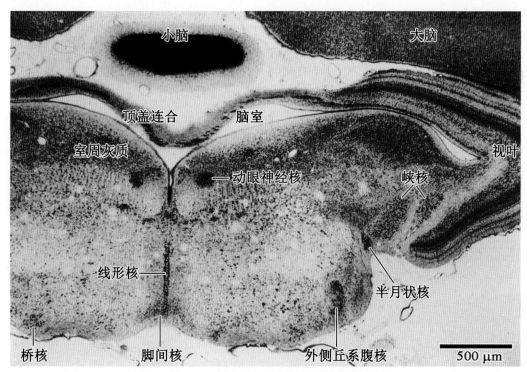

图 21-166 鸭中脑后部横切低倍像（洋红染色）

Fig.21-166 transverse low magnification of duck posterior midbrain (kermes stain)

小脑—cerebellum 大脑—cerebrum 顶盖连合—commissurae tecti 脑室—ventricle 室周灰质—periventricular gray matter 动眼神经核—oculomotor nucleus 峡核—nucleus isthmi 视叶—optic lobe 线形核—linear nucleus 半月状核—semilunar nucleus 桥核—pontine nucleus 脚间核—interpeduncular nucleus 外侧丘系腹核—lateral colliculi ventral nucleus

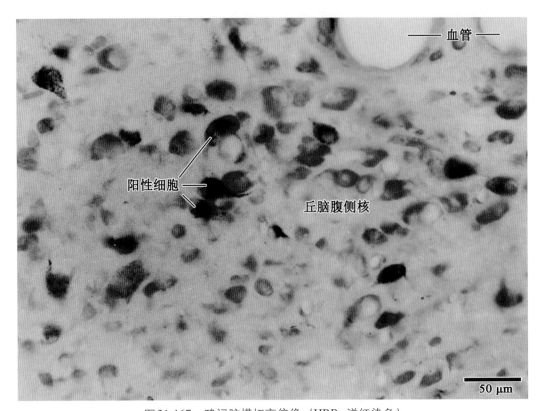

图 21-167　鸭间脑横切高倍像（HRP+洋红染色）
Fig.21-167　transverse high magnification of duck diencephalon(HRP + kermes stain)
血管—blood vessel　阳性细胞—positive cell　丘脑腹侧核—ventral thalamic nucleus

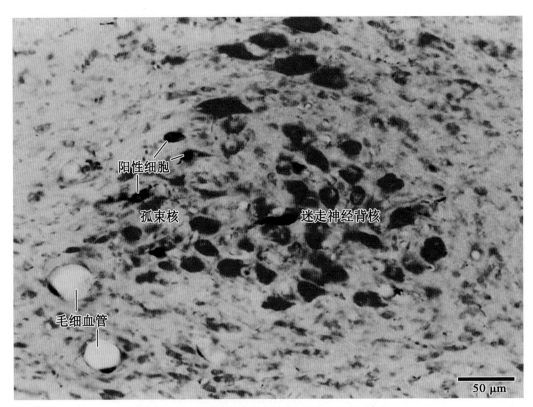

图 21-168　鸭延髓横切高倍像（HRP+洋红染色）
Fig.21-168　transverse high magnification of duck medulla (HRP + kermes stain)
阳性细胞—positive cell　孤束核—nucleus of solitary tract　迷走神经背核—dorsal nucleus of vagus nerve
毛细血管—capillary

第二十一章 禽类的主要组织结构特征　Main Structural Features of Fowl Tissues

图21-169　公鸡的冠（原色）

Fig.21-169　the cock comb（original colour）

冠—comb　喙—beak　肉髯—wattle

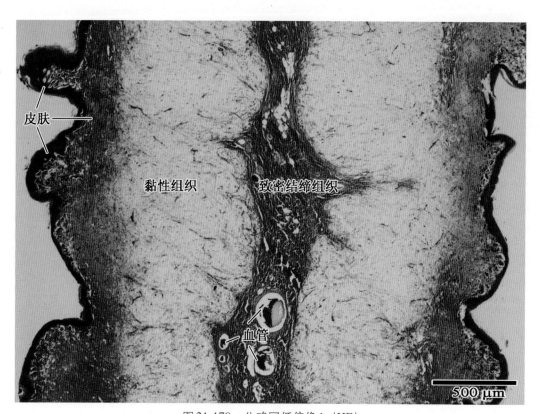

图21-170　公鸡冠低倍像1（HE）

Fig.21-170　low magnification of cock comb 1（HE）

皮肤—skin　黏性组织—mucinous tissue　致密结缔组织—condense connective tissue　血管—blood vessel

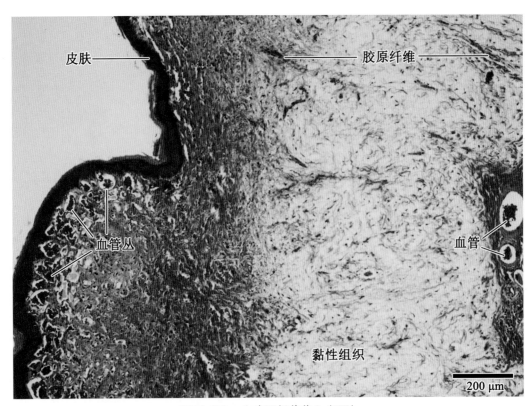

图21-171　公鸡冠低倍像2（HE）

Fig.21-171　low magnification of the cock comb 2（HE）

皮肤—skin　血管丛—vascular plexus　胶原纤维—collagen fiber　黏性组织—mucinous tissue　血管—blood vessel

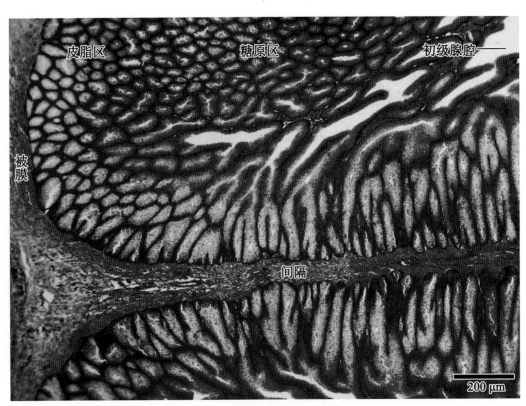

图21-172　鸡尾脂腺低倍像（HE）

Fig.21-172　low magnification of chicken preen gland（HE）

皮脂区—sebum zone　糖原区—glycogen zone　初级腺腔—primary lumen　被膜—capsule　间隔—interval

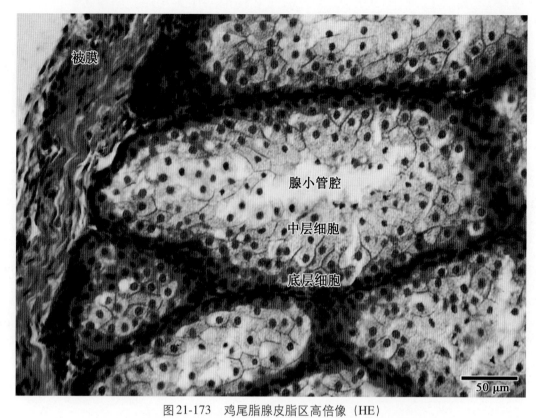

图21-173 鸡尾脂腺皮脂区高倍像（HE）

Fig.21-173 high magnification of chicken preen gland sebum zone（HE）

被膜—capsule 腺小管腔—glandular lumen 中层细胞—mid-layer cell 底层细胞—basal cell

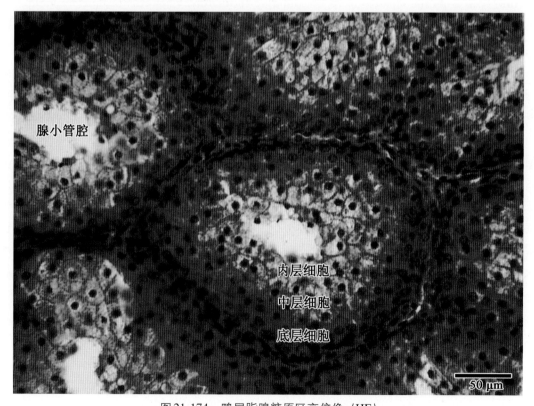

图21-174 鸭尾脂腺糖原区高倍像（HE）

Fig.21-174 high magnification of duck preen gland glycogen zone（HE）

腺小管腔—glandular lumen 内层细胞—inner cell 中层细胞—mid-layer cell 底层细胞—basal cell

第二十二章
畜禽早期胚胎发育
Early Embryonic Development of Livestock and Fowl

> **Outline**
>
> Animal embryology is a science which study the processes and the regulations of the development of the creature fetus from the moment of its inception up to the time when it is born as an infant. Animal embryological development can be divided into three stages which are pre-embryo, embryo and post-embryo.
>
> The pre-embryo stage is the period before embryo genesis including the structure and genesis of gamete. Sperm is the male gamete, which goes through the reproductive period, development period, maturation period and figuration period. At last it changes into the sperm which has flagellum and can moves in testis. Ovum is the female gamete, which comes through the reproductive period, development period and maturation period and finally becomes to the ovum with abundant nutrition.
>
> Embryo stage is the period form impregnation to parturition or hatching. This stage includes impregnation, cleavage, formation of blastula, gastrulae and neurula and formation of organs. Impregnation is the process of sperm and ovum combining into zygote. Through impregnation, cell comebacks to diploid karyotype, and germ plasm recombines. This process not only maintains the stability and continuity of species but also promotes the evolution of species. Cleavage represents the course of successive mitosis of zygote that forms some blastomeres. Egg cleavage can be classified into complete egg cleavage and incomplete egg cleavage. Along with egg cleavage, embryo becomes blastula gradually. Blastula is an embryo of multiple cells which originate from the many times' egg cleavage of zygote. Blastula is separated into five kinds in accordance with the presence or absence of blastocele and the blastocele's location. Fish mostly has two kinds, multilayer blastula and discoblastula. Poultry has discoblastula. Mammalian has blastocyst. Through gastrulation, some cells on the outer surface of blastula make use of all kinds of means such as invagination and ingression in order to get inside and form a gastrulae that has two germ layers:

第二十二章 畜禽早期胚胎发育 Early Embryonic Development of Livestock and Fowl

> endoderm and ectoderm or a gastrulae that has three germ layers: endoderm, mesoderm and ectoderm. Neurula is an important development stage after gastrulae, beginning with the formation of neural board and stopping with the folding of neural tube. At this stage, primordium appears in each germ layer, but tissue is not differentiated. Primordium differentiates further until into corresponding tissues, organs and systems—a process called organogenesis.
>
> Fetal membrane is an accessorial structure which forms in the progress of embryonic development. It is not composed of embryo directly, but it provides protection, nutrition, breath and drainage for the embryo. There are four kinds of fetal membrane: yolk sac, amnion, serosa (birds) or chorion (mammals) and allantois. Placenta of mammalians is allantois chorion together with uterine inner membrane, which can ensure the successful exchanges of matters between embryo and matrix. In light of the different conjoint modes of caul and uterine inside membrane, placenta can be classified into four types: epitheliochorial placenta, syndesmochorial placenta, endotheliochorial placenta and haemochorial placenta.

动物的胚胎（animal embryology）发育包括两性配子的生成、受精、卵裂、囊胚形成、原肠胚形成、胚层分化、器官和系统发生以及胎膜的形成。哺乳类是胎生动物，胚胎发育在母体的子宫内进行，不同的家畜有不同的胎盘类型。鸟类为卵生动物，胚胎发育在卵内进行，在母体外完成。本章展示家畜和家禽的胚胎发育特点两部分内容。

一、家畜的早期胚胎发育

（一）生殖细胞（germ cell）

生殖细胞包括雄配子（精子）和雌配子（卵子）。精子和卵子分别在睾丸和卵巢发生，是高度特化的细胞，其形成和结构完全适应于受精和早期胚胎发育的需要。

1. **精子（sperm）** 产生于睾丸的精子携带有父系基因的单倍体遗传物质，具有定向运动能力和使卵子受精的潜能。家畜的精子形态大同小异，长度在 55～77μm，主要由头部、颈部和尾部组成，尾部从前到后可分为中段、主段和末段。家畜的精子因含遗传物质的不同分为两种，携带X染色体的X精子和携带Y染色体的Y精子。因此，家畜的性别是由精子决定的。家畜的精子在离开睾丸时还不具备前行运动的能力，进入附睾后经过一系列的成熟变化获得运动能力。

2. **卵子（ovum）** 产生于卵巢的卵子携带有母系基因的遗传物质，不具备运动能力，其移动只能靠输卵管黏膜上皮细胞的纤毛摆动和管壁的肌层收缩而实现。卵子呈球形，除了作为遗传物质的载体外，还要为早期胚胎发育准备细胞质条件和营养物质。因此，卵子的体积大而圆。卵子直径依物种不同而异，一般在 120～160μm。大多数家畜从卵巢中排出的卵是处于第二次减数分裂中期的次级卵母细胞，在卵周隙内可见到第一极体。

（二）受精（fertilization）

受精是两性配子互相融合，形成一个新的细胞——合子的过程。它标志着胚胎发育的开始，受精的意义在于双亲遗传物质的融合，恢复二倍体染色体数目。受精的另一个作用是激活卵子，使其继续发育，不受精的卵子就会老化和死亡，在雌性生殖道内家畜卵子保持受精能力的时间一般为 11～24 h。

受精过程中首先是成熟的精子要在雌性生殖道内获得受精能力，简称获能。获能的必然结果是精子发生了顶体反应。顶体反应使精子能穿越透明带进入卵子。精子和卵子相遇融合的过程中，卵细胞被激活并发生了皮层反应。皮层反应使卵细胞和透明带之间出现卵周隙，具有阻止多精入卵的作用。最后，精子核在卵胞质内解凝形成雄原核，卵母细胞完成减数分裂形成雌原核，两原核在卵细胞中央相会后融合，形成合子。受精过程结束，即准备开始第一次卵裂。

（三）卵裂（cleavage）

卵裂是合子最初发生的几次细胞分裂，子细胞称卵裂球（blastomere）。家畜胚胎的卵裂和体细胞的有丝分裂相比有所不同：卵裂间期短，细胞质总量没有增加，卵裂球越来越小；部分细胞质转化成细胞核，卵裂球质和核的比例越来越小。卵裂的结果是细胞的数量增加，最后成为形似桑葚的细胞团，称为桑葚胚（morula）。

（四）家畜的囊胚和胚泡附植（blastula and blastocyst implantation）

随着卵裂的进行，到16细胞期时分裂速度减慢，胚胎内部发生变化，形成囊胚。家畜的囊胚表面是一层大而扁的细胞，称为滋养层细胞，滋养层日后发育成胎膜。囊胚的内部有一细胞团称内细胞团，内细胞团日后发育成胚体。

早期囊胚束缚在透明带中，晚期囊胚（猪受精后第8天）的透明带溶解，囊胚迅速增大发育成胚泡。胚泡呈长带状，如猪的胚泡可长达150cm。

早期囊胚漂浮在子宫腔中，因吸收子宫内膜的分泌物发育成胚泡后，胚泡的滋养层细胞与子宫内膜发生接触并与子宫壁发生了不同的粘贴关系。猪的胚泡与子宫内膜发生贴附关系，日后形成上皮绒毛膜胎盘。反刍动物的胚泡与子宫内膜形成结缔绒毛膜胎盘。犬、猫和灵长类的胚泡滋养层使子宫内膜蜕膜化，胚泡侵入子宫壁中，称植入。犬、猫日后形成内皮绒毛膜胎盘，灵长类则形成血绒毛膜胎盘。

（五）原肠胚和胚层形成（gastrula and blastoderm formation）

囊胚发育到一定阶段形成具有两个胚层的胚胎称为原肠胚，里面新出现的胚层叫内胚层。胚胎进入原肠胚时，细胞核开始起主导作用，合成新的蛋白质，分化成不同的胚层，同时发生细胞迁移，形态也随之变化。内细胞团表面的滋养层细胞脱落后裸露出来称为胚盘，胚盘表面的细胞以集中和卷入的运动方式迁移后形成原条。随着原条的出现，胚胎在内外胚层之间形成中胚层。三胚层形成后胚胎有了两种胚胎性组织：上皮与间充质。外胚层和内胚层基本上为上皮的形态特征，日后大多分化为上皮组织，部分外胚层的细胞在中胚层脊索的诱导下则分化为神经组织。中胚层为间充质的形态特征，以后进一步分化为肌肉、骨骼及结缔组织等。三胚层分化产生四类基本组织，由各类组织相互作用和结合形成各器官。胚层形成以后，胚胎性组织开始分化形成各器官的原基，主要标志是神经胚形成，在猪胚14～16胚龄。

（六）系统发生和器官形成（phylogeny and organogenesis）

胚胎性组织形成器官原基后，胚胎性组织继续分化为四大基本组织的同时，由各类组织互相作用和结合形成各种复杂的器官，称为器官形成。各种器官形成后组成系统，称为系统发生。系统发生过程中胚胎的外部形态和内部结构均发生了巨大的变化，个体有了物种的形态学特征。

（七）胎膜、脐带和胎盘（fetal membrane, umbilical cord and placenta）

家畜囊胚期时分化为两部分：一部分为内细胞团，以后主要分化为胚胎；另一部分为滋养层，以后形成胎膜。胎膜是陆生脊椎动物胚胎发育才有的特殊结构，有四种，卵黄囊、羊膜、绒毛膜和尿囊。家畜胎膜的特点是羊膜、绒毛膜和尿囊较发达，卵黄囊不发达，仅在胚胎的早期明显可见，以后就迅速消失。羊膜能分泌羊水，为胚胎发育提供了水生环境。家畜的胎膜与相应的子宫内膜结合在一起构成胎盘。胎盘的主要功能是通过脐带参与胚胎与母体之间的物质交换。

1. 胎膜（fetal membrane）

（1）绒毛膜（chorion） 是胎膜的最外层，牛、羊的绒毛膜上的绒毛群集成小叶，小叶间的绒毛膜是光滑的。猪和马的绒毛膜上的绒毛均匀分布。兔的绒毛膜上的绒毛则集中于脐部周围。犬和猫的绒毛膜上的绒毛环绕着胎儿腰部分布。

（2）羊膜（amnion） 直接包在胎儿的外周，光滑、薄而透明，羊膜腔内有大量羊水，胎儿浮在羊水中。

（3）尿囊（allantois） 位于羊膜和绒毛膜之间的胚外体腔中，并与绒毛膜相贴，表面有许多血管，囊内有尿囊液。

（4）卵黄囊（yolk sac） 位于胚胎腹侧，胚胎发育早期体积大，随尿囊发育而退化萎缩，形成卵黄蒂。

2. **脐带**（umbilical cord） 脐带呈索状，位于胎儿腹部，长40～50cm。它连接在胎膜和胎儿脐部之间。脐带外包羊膜和胶冻样的黏液结缔组织，内有两条较细的脐动脉和一条粗大的脐静脉。

3. **胎盘**（placenta） 家畜的胎盘多数为尿囊绒毛膜胎盘，也就是说与子宫内膜结合的胎膜是尿囊绒毛膜。不同的家畜有不同的胎盘类型，不同类型的胎盘结构有差异。形态学差异为绒毛膜上绒毛的分布方式不同；显微结构的差异为胎儿的尿囊绒毛膜组织结构虽然变化不大，但母体子宫内膜的组织结构有很大变化。家畜的胎盘可分为四种类型。

(1) 上皮绒毛膜胎盘（epitheliochorial placenta） 又称散布型胎盘。以猪和马的胎盘为例，绒毛膜上的绒毛多而均匀分布，绒毛与子宫内膜的子宫腺等嵌合。母、子胎盘之间关系不密切。

(2) 结缔绒毛膜胎盘（connective chorionic placenta） 又称子叶型胎盘。以牛、羊的胎盘为例，绒毛膜上有许多丛状的绒毛小叶，与对应的子宫内膜上的子宫肉阜相嵌合。嵌合处部分子宫内膜上皮被溶解，使绒毛直接与子宫内膜的结缔组织接触。因此，母、子胎盘关系较上皮绒毛膜胎盘密切。

(3) 内皮绒毛膜胎盘（endotheliochorial placenta） 又称环状胎盘。以犬、猫为例，胎盘呈环状，包绕在胎儿腰部。绒毛膜上的绒毛仅分布于胎儿腰部的绒毛膜上。绒毛膜上的绒毛破坏子宫内膜上皮，与固有层结缔组织血管的内皮接触。母、子胎盘关系密切。

(4) 血绒毛膜胎盘（haemochorial placenta） 又称盘状胎盘。以灵长类和兔为例，胎盘呈圆盘状，绒毛膜上的绒毛集中于该盘状区。绒毛膜上的绒毛破坏了子宫内膜上皮及血管内皮，而浸于血窦中。母、子胎盘关系最密切。

二、家禽的早期胚胎发育

(一) 生殖细胞

禽类精子的结构与家畜的相似，也分头部、颈部和尾部。禽类精子所携带的父系基因的单倍体遗传物质中只有一种Z染色体，而禽类卵细胞所携带的母系基因的单倍体遗传物质中有Z染色体和W染色体两种。因此，禽类的性别是由卵子决定的。禽类的胚胎发育在卵内进行，胚胎发育需要大量的营养物质，所以卵的结构特殊。所谓禽蛋是指禽的卵细胞及其周围的各种卵膜。卵黄是卵细胞，其中的主要物质是卵黄颗粒，细胞核及少量的细胞质被挤向细胞的一端，形成小圆盘称为胚珠。如果卵细胞在输卵管中受精，待禽蛋产出时，大约已发育至原肠早期，此时白色圆盘较大，称为胚盘。卵细胞周围的卵膜是一层厚6～11μm的卵黄被膜。卵黄被膜外是卵白，浓卵白在内，稀卵白在外，部分浓卵白扭曲成索状结构，称卵黄系带，它黏附于卵黄两端，悬吊卵黄于蛋中央。卵白外两层软壳膜，外层紧贴蛋壳，内层与卵白相邻。在蛋的钝端，内、外两层壳膜间形成气室，壳膜有微孔，供空气通过。壳膜外有蛋壳，是表面光滑、内部多孔的坚硬石灰质，除具有保护作用外，还是供给胚胎发育所需要的钙库。蛋壳外有薄的角质层，用以防止蛋内水分蒸发和微生物的入侵。

(二) 受精特点

鸡产蛋后7～74 min排卵，排卵后最快3min即可卷入输卵管漏斗；鸡精子可以在输卵管伞部黏膜皱襞中贮存，存活时间可达7～34 d；鸡卵在输卵管伞部受精，排卵至精子入卵需15 min；多精入卵，单精受精。

(三) 卵裂和囊胚

禽类卵的卵黄多，胞质与核位于卵细胞一端，称为端黄卵。端黄卵的卵裂是不全裂，即卵裂球没有完全分开。禽类胚的卵裂又称为盘状裂，即胚的胚盘表面裂开，腹侧面仍融合，卵裂区呈圆盘形。卵裂的子细胞达到一定数目时，也会出现空腔，称为盘状囊胚。

(四) 原肠胚和胚层形成

禽类的盘状囊胚发育成原肠胚正值蛋刚产出时，经过约16h的孵化后，禽类胚已形成了原条，出现了中胚层。三胚层形成以后，同样由胚层分化出基本组织以及由组织形成器官，逐渐发育成雏禽。

（五）胎膜

禽类的胎膜也有四种：羊膜、浆膜（即绒毛膜）、卵黄囊和尿囊。禽类的卵黄囊极为发达，卵黄囊具有消化和吸收卵黄的作用，尿囊浆膜贴着禽蛋的壳膜，能从外界吸收氧气呼出二氧化碳；尿囊具有贮存水和沉淀代谢废物尿酸盐的作用，并能溶解和吸收蛋壳中的钙。尿囊浆膜还包裹卵白形成了卵白囊，帮助卵白进入卵黄和进入羊水中被胚胎吸收利用。

三、早期胚胎发育图谱

1. **家畜早期胚胎发育**　图22-1～图22-89。
2. **家禽早期胚胎发育**　图22-90～图22-102。

第二十二章 畜禽早期胚胎发育 Early Embryonic Development of Livestock and Fowl

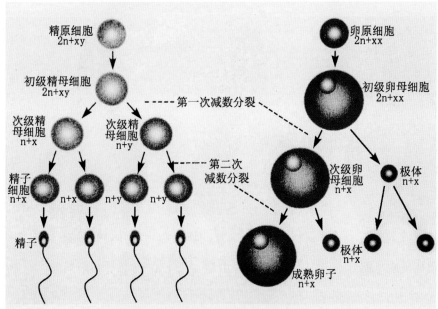

图 22-1 精子与卵子发生示意图
Fig.22-1 generation diagram of sperm and ovum
精原细胞—spermatogonia
初级精母细胞—primary spermatocyte
次级精母细胞—secondary spermatocyte
精子细胞—spermatid
精子—sperm
第一次减数分裂—first meiotic division
第二次减数分裂—second meiotic division
卵原细胞—oogonium
初级卵母细胞—primary oocyte
次级卵母细胞—secondary oocyte
极体—polar body
成熟卵子—mature egg

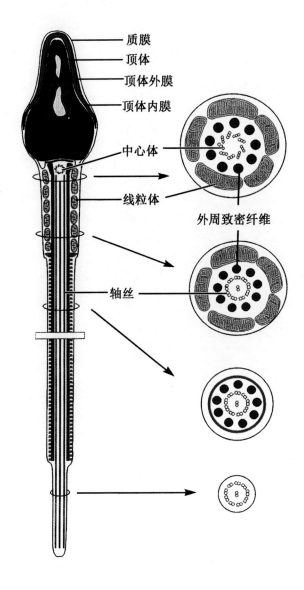

图 22-2 精子电镜结构模式图
Fig.22-2 electron microscope diagram of sperm
质膜—plasma membrane   顶体—perforatorium
顶体外膜—outer acrosomal membrane
顶体内膜—inner acrosomal membrane
中心体—centrosome   线粒体—mitochondria
外周致密纤维—outer dense fiber   轴丝—axoneme

857

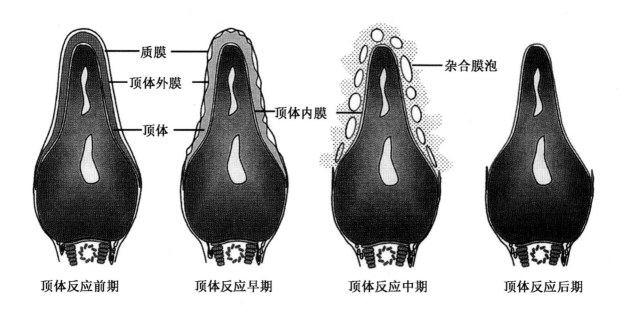

图22-3 精子顶体反应模式图

Fig.22-3 diagram of sperm acrosomal reaction pattern

质膜—plasma membrane　顶体外膜—outer acrosomal membrane　顶体—perforatorium
顶体内膜—inner acrosomal membrane　杂合膜泡—hybrid membrane bubble　顶体反应前期—acrosomal reaction prophase
顶体反应早期—acrosomal reaction early phase　顶体反应中期—acrosomal reaction metaphase
顶体反应后期—acrosomal reaction anaphase

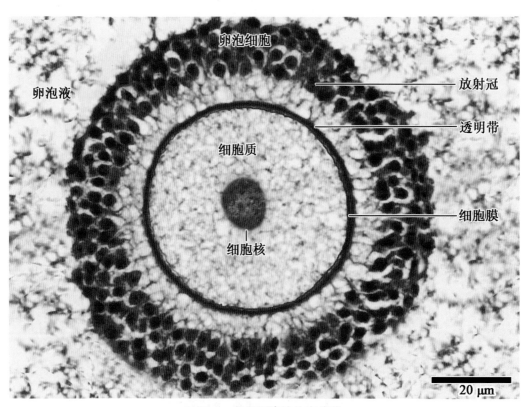

图22-4 卵母细胞结构高倍像

Fig.22-4 high magnification of oocyte structure

卵泡液—follicular fluid　卵泡细胞—follicular cell　细胞质—cytoplasm　细胞核—nucleus
放射冠—corona radiata　透明带—zona pellucida　细胞膜—cell membrane

第二十二章 畜禽早期胚胎发育 Early Embryonic Development of Livestock and Fowl

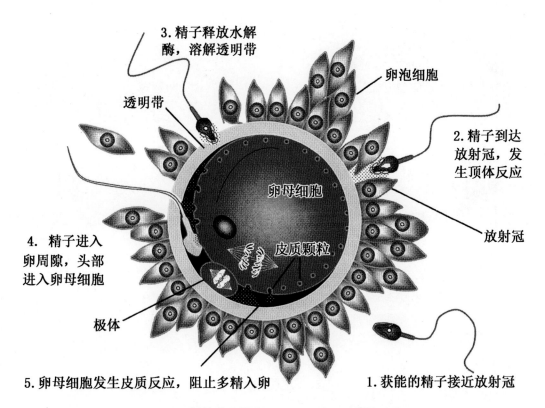

图 22-5 受精过程模式图（1～5表示受精的过程）

Fig.22-5 diagram of fertilization process（1～5 indicating fertilization process）

卵泡细胞—follicular cell 放射冠—corona radiata 透明带—zona pellucida 极体—polar body 卵母细胞—oocyte
皮质颗粒—cortical granule 获能精子接近放射冠—capacitated sperm approaching corona radiata
精子到达放射冠，发生顶体反应—sperms reaching corona radiata and undergoing acrosomal reaction
精子释放水解酶，溶解透明带—sperm releasing hydrolases that dissolve zona pellucida
精子进入卵周隙，头部进入卵母细胞—sperm entering the peri-lacunae and its head entering the oocyte
卵母细胞发生皮质反应，阻止多精入卵—oocytes undergoing cortical response that prevents polyspermy

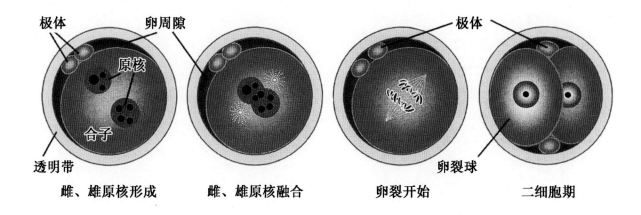

图 22-6 卵裂过程模式图

Fig.22-6 diagram of cleavage process

极体—polar body 卵周隙—perivitelline space 原核—pronucleus 合子—zygote 透明带—zona pellucida
卵裂球—oocyte 雌、雄原核形成—female and male prokaryotic formation
雌、雄原核融合—female and male prokaryotic fusion 卵裂开始—cleavage starting 二细胞期—2 daughter cells period

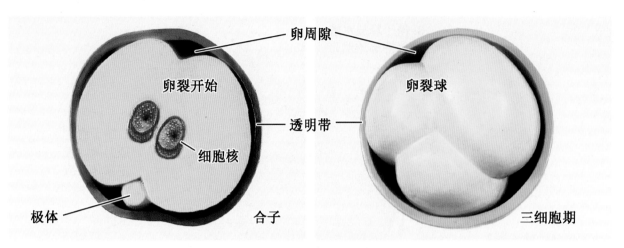

图 22-7 卵裂过程模型像 1

Fig.22-7 model of cleavage process 1

卵周隙—perivitelline space 透明带—zona pellucida 卵裂开始—cleavage starting 细胞核—nuclei 极体—polar body 合子—zygote 卵裂球—oocyte 三细胞期—3 daughter cells period

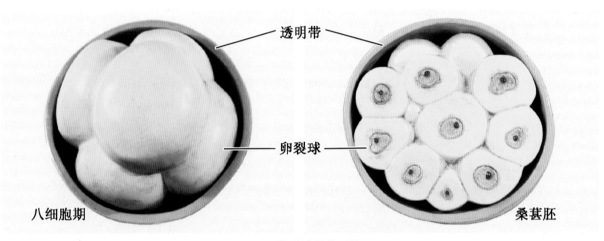

图 22-8 卵裂过程模型像 2

Fig.22-8 model of cleavage process 2

透明带—zona pellucida 卵裂球—oocyte 八细胞期—8 daughter cells period 桑葚胚—morula

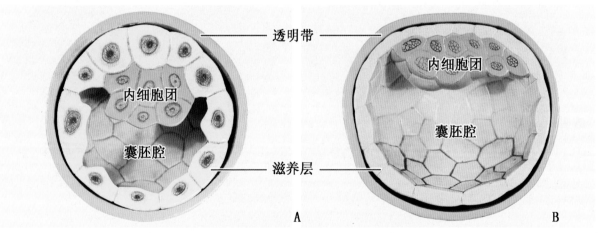

图 22-9 囊胚形成模型像（A、B 代表发育顺序）

Fig.22-9 model of blastocyst formation (A and B indicating developmental process)

透明带—zona pellucida 滋养层—trophoblast 内细胞团—inner cell mass 囊胚腔—blastocele

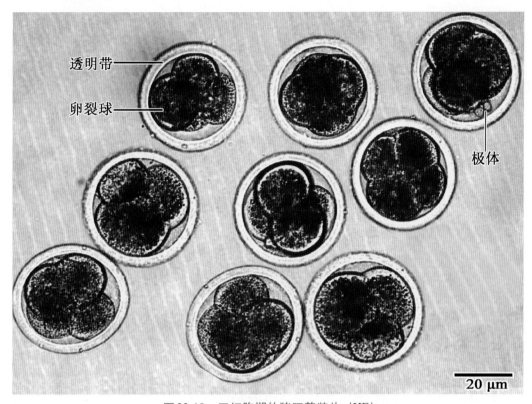

图 22-10 四细胞期的猪胚整装片（HE）

Fig.22-10 preparation of pig embryo at the four-cell stage (HE)

透明带—zona pellucida 卵裂球—oocyte 极体—polar body

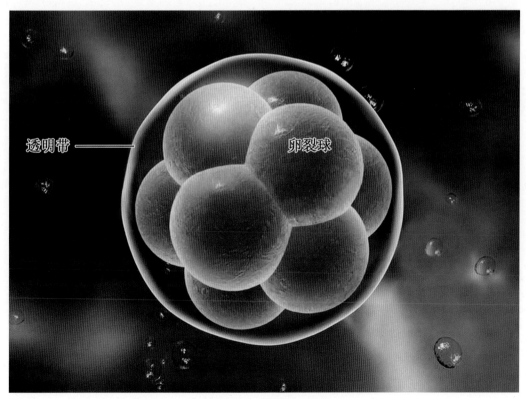

图 22-11 桑葚胚立体图

Fig.22-11 stereogram of morula

透明带—zona pellucida 卵裂球—oocyte

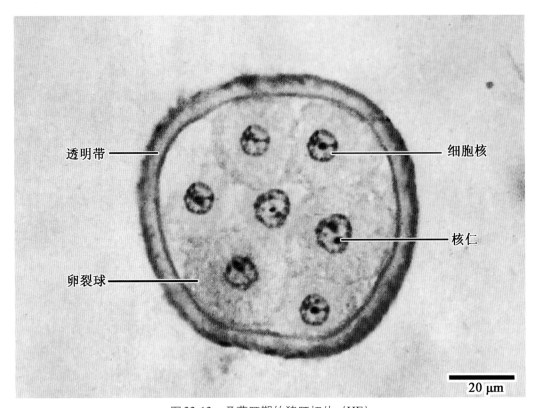

图 22-12 桑葚胚期的猪胚切片（HE）

Fig.22-12　sections of pig embryo at morula stage (HE)

透明带—zona pellucida　卵裂球—oocyte　细胞核—nucleus　核仁—nucleolus

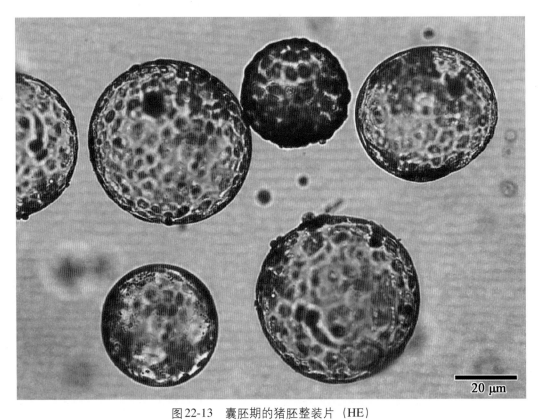

图 22-13 囊胚期的猪胚整装片（HE）

Fig.22-13　preparation of pig embryo at blastocyst stage (HE)

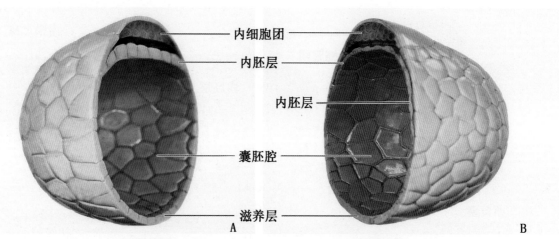

图 22-14 囊胚发育模型（A、B表示发育顺序）

Fig.22-14 model of blastocyst development (A and B indicating developmental process)

内细胞团—inner cell mass 内胚层—endoderm 囊胚腔—blastocele 滋养层—trophoblast

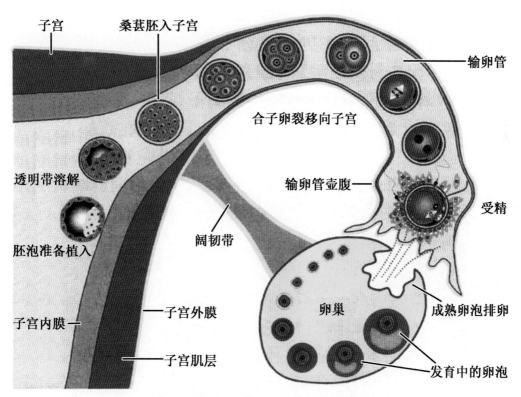

图 22-15 排卵、胚胎发生及其与生殖道的关系

Fig.22-15 ovulation, embryogeny and their relationship with genital tract

子宫—uterus 子宫内膜—endometrium 子宫外膜—perimetrium 子宫肌层—myometrium 阔韧带—broad ligament
输卵管—oviduct 输卵管壶腹—ampullae tubae uterinae 卵巢—ovary 发育中的卵泡—developing follicle
成熟卵泡排卵—mature follicle ovulation 受精—fertilization
合子卵裂移向子宫—zygotic undergoing cleavage and moving to uterus 桑葚胚入子宫—morula entering uterus
透明带溶解—zona pellucida dissolving 胚泡准备植入—blastocyst ready for implantation

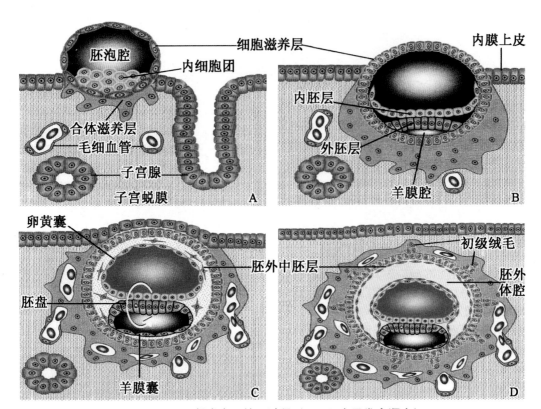

图 22-16　灵长类囊胚植入过程（A～D表示发育顺序）

Fig.22-16　primate blastocyst implantation process (A～D indicating developmental process)

子宫蜕膜—uterus decidua　子宫腺—uterine gland　毛细血管—capillary　合体滋养层—syncytiotrophoblast
内细胞团—inner cell mass　胚泡腔—blastocyst cavity　细胞滋养层—cytotrophoblast　内膜上皮—endometrial epithelium
内胚层—entoderm　外胚层—ectoderm　羊膜腔—amnion cavity　卵黄囊—yolk sac　胚盘—germinal disc
羊膜囊—amniotic sac　胚外中胚层—extraembryonic mesoderm　初级绒毛—primary villi　胚外体腔—exocoelom

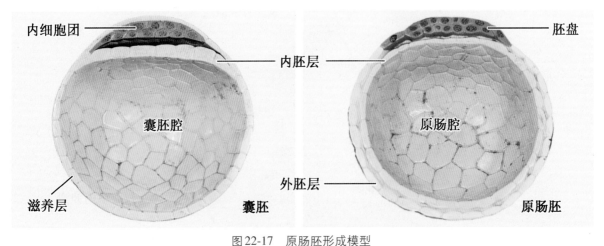

图 22-17　原肠胚形成模型

Fig.22-17　model of gastrula formation

内细胞团—inner cell mass　囊胚腔—blastocele　滋养层—trophoblast　囊胚—blastula　内胚层—endoderm
外胚层—ectoderm　胚盘—embryonic disc　原肠腔—gastrocoele　原肠胚—gastrula

第二十二章 畜禽早期胚胎发育 Early Embryonic Development of Livestock and Fowl

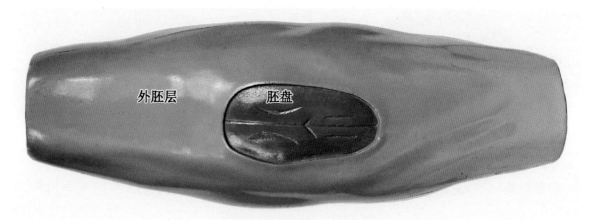

图 22-18　原肠胚背侧面观 1

Fig.22-18　dorsal view of gastrula model 1

外胚层—ectoderm　胚盘—embryonic disc

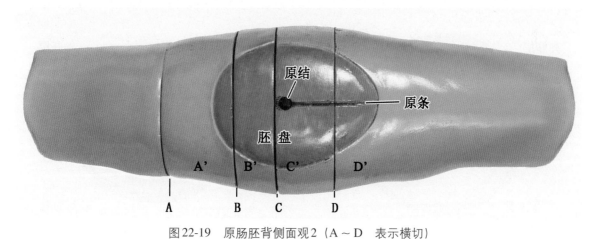

图 22-19　原肠胚背侧面观 2（A～D 表示横切）

Fig.22-19　dorsal view of gastrula model 2 (A～D indicating transection)

原结—primitive knot　原条—primitive streak　胚盘—embryonic disc

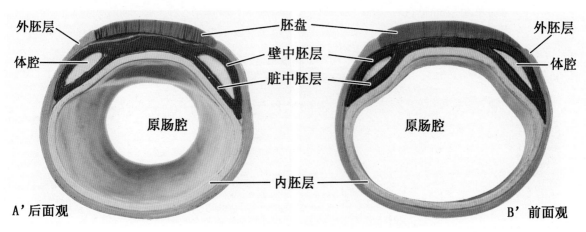

图 22-20　原肠胚横切面观 1

Fig.22-20　transection view of gastrula model 1

A'后面观—A' posterior view　B'前面观—B' anterior view

外胚层—ectoderm　体腔—coelom　原肠腔—gastrocoele　胚盘—embryonic disc　壁中胚层—somatic mesoderm

脏中胚层—splanchnic mesoderm　内胚层—endoderm

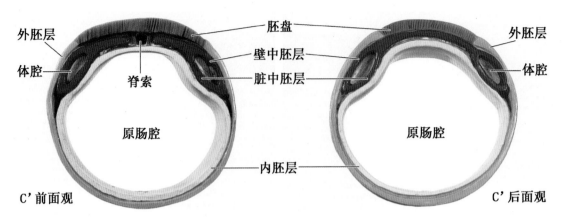

图 22-21 原肠胚横切面观 2

Fig.22-21 transection view of gastrula model 2

C'前面观—C' anterior view　C'后面观—C' posterior view

外胚层—ectoderm　体腔—coelom　脊索—notochord　原肠腔—gastrocoele　胚盘—embryonic disc

壁中胚层—somatic mesoderm　脏中胚层—splanchnic mesoderm　内胚层—endoderm

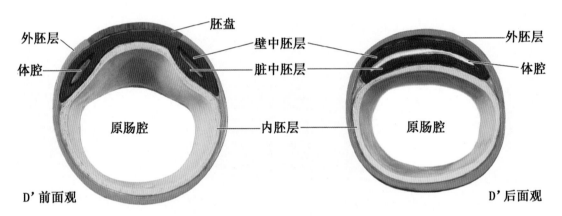

图 22-22 原肠胚横切面观 3

Fig.22-22 transection view of gastrula model 3

D'前面观—D' anterior view　D'后面观—D' posterior view

外胚层—ectoderm　体腔—coelom　原肠腔—gastrocoele　胚盘—embryonic disc

壁中胚层—somatic mesoderm　脏中胚层—splanchnic mesoderm　内胚层—endoderm

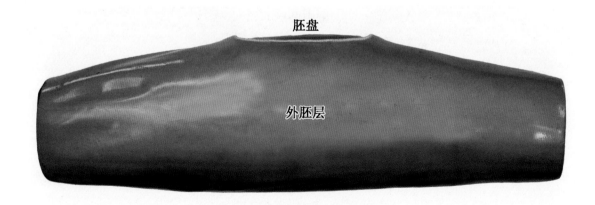

图 22-23 原肠胚侧面观

Fig.22-23 lateral view of gastrula model

胚盘—embryonic disc　外胚层—ectoderm

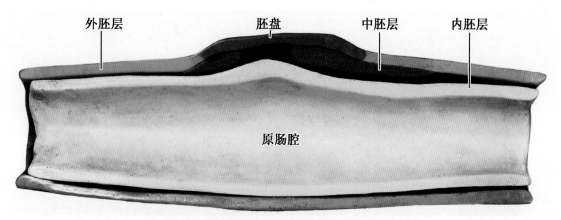

图 22-24 原肠胚纵切 1
Fig.22-24 longitudinal section of gastrula model 1
外胚层—ectoderm　胚盘—embryonic disc　中胚层—mesoderm　内胚层—endoderm　原肠腔—gastrocoele

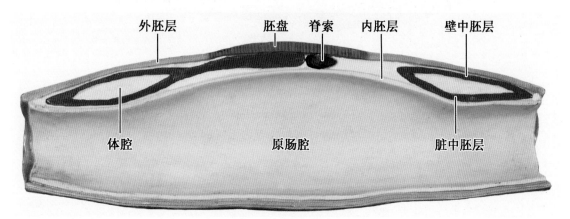

图 22-25 原肠胚纵切 2
Fig.22-25 longitudinal section of gastrula model 2
外胚层—ectoderm　胚盘—embryonic disc　脊索—notochord　内胚层—endoderm　壁中胚层—somatic mesoderm　体腔—coelom　原肠腔—gastrocoele　脏中胚层—splanchnic mesoderm

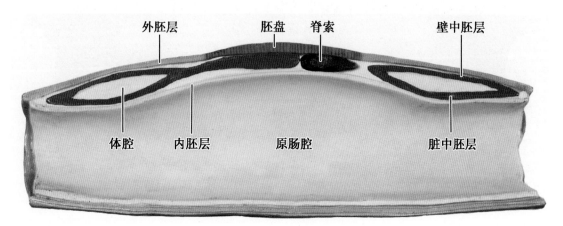

图 22-26 原肠胚纵切 3
Fig.22-26 longitudinal section of gastrula model 3
外胚层—ectoderm　胚盘—embryonic disc　脊索—notochord　壁中胚层—somatic mesoderm　体腔—coelom　内胚层—endoderm　原肠腔—gastrocoele　脏中胚层—splanchnic mesoderm

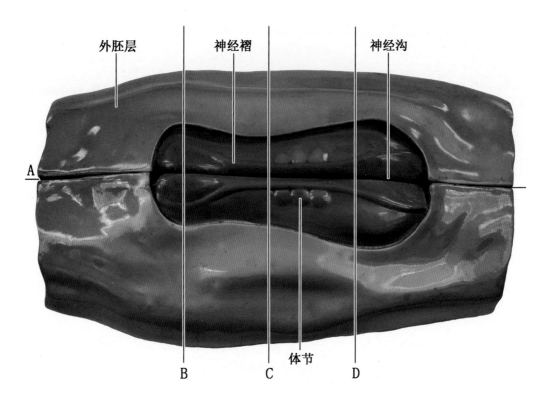

图22-27　原肠胚的发育（A表示纵切，B～D表示横切）

Fig.22-27　development of gastrula (A indicating longitudinal section, B～D indicating transection)

外胚层—ectoderm　神经褶—neural fold　神经沟—neural groove　体节—somite

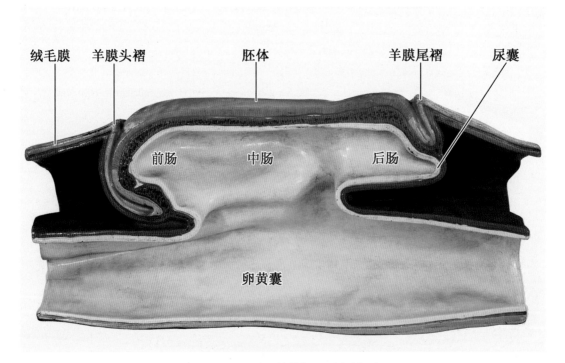

图22-28　原肠胚的纵切面（右侧面）

Fig.22-28　longitudinal section of gastrula model（right side）

绒毛膜—chorion　羊膜头褶—amniotic head fold　胚体—embryoid　羊膜尾褶—amniotic tail fold
前肠—foregut　中肠—midgut　后肠—hindgut　尿囊—allantois　卵黄囊—yolk sac

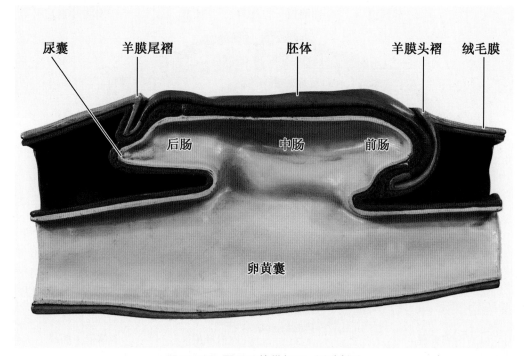

图22-29 原肠胚的纵切面（左侧面）

Fig.22-29 longitudinal section of gastrula model (left side)

尿囊—allantois 羊膜尾褶—amniotic tail fold 胚体—embryoid 羊膜头褶—amniotic head fold
绒毛膜—chorion 后肠—hindgut 中肠—midgut 前肠—foregut 卵黄囊—yolk sac

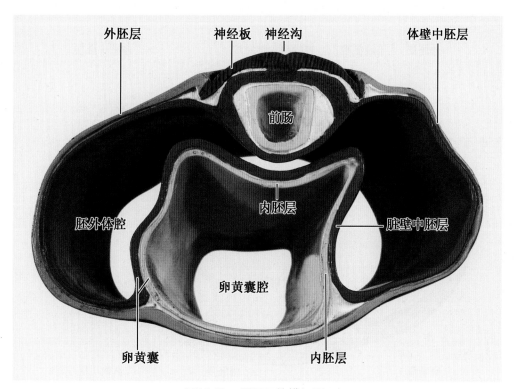

图22-30 原肠胚的横切面1

Fig.22-30 transection of gastrula model 1

外胚层—ectoderm 神经板—neural plate 神经沟—neural groove 体壁中胚层—somatoderm 前肠—foregut
胚外体腔—exocoelom 内胚层—endoderm 脏壁中胚层—splanchnic mesoderm 卵黄囊—yolk sac
卵黄囊腔—yolk sac cavity

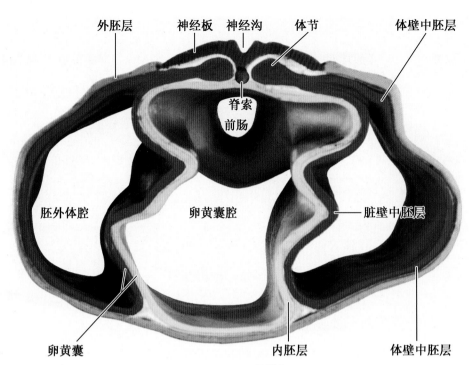

图 22-31　原肠胚的横切面 2

Fig.22-31　transection of gastrula model 2

外胚层—ectoderm　神经板—neural plate　神经沟—neural groove　体节—somite　体壁中胚层—somatoderm
脊索—notochord　前肠—foregut　胚外体腔—exocoelom　卵黄囊腔—yolk sac cavity　脏壁中胚层—splanchnic mesoderm
卵黄囊—yolk sac　内胚层—endoderm

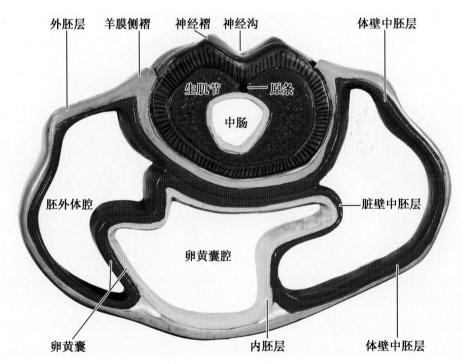

图 22-32　原肠胚的横切面 3

Fig.22-32　transection of gastrula model 3

外胚层—ectoderm　羊膜侧褶—amnion side fold　神经褶—neural fold　神经沟—neural groove　体壁中胚层—somatoderm
生肌节—myotome　原条—primitive streak　中肠—midgut　胚外体腔—exocoelom　卵黄囊腔—yolk sac cavity
脏壁中胚层—splanchnic mesoderm　卵黄囊—yolk sac　内胚层—endoderm

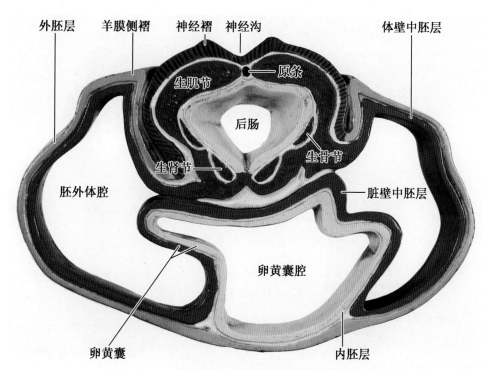

图 22-33 原肠胚的横切面 4

Fig.22-33 transection of gastrula model 4

外胚层—ectoderm 羊膜侧褶—amnion side fold 神经褶—neural fold 神经沟—neural groove 体壁中胚层—somatoderm 生肌节—myotome 原条—primitive streak 后肠—hindgut 生肾节—nephrotome 生骨节—sclerotome 胚外体腔—exocoelom 脏壁中胚层—splanchnic mesoderm 卵黄囊—yolk sac 卵黄囊腔—yolk sac cavity 内胚层—endoderm

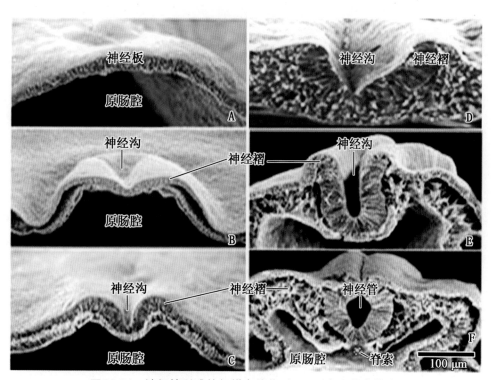

图 22-34 神经管形成的扫描电镜像（A～F 表示发育顺序）

Fig.22-34 scanning electrical image of neural tube formation (A～F indicating developmental process)

神经板—neural plate 原肠腔—gastrocoele 神经沟—neural groove 神经褶—neural fold 神经管—neural tube 脊索—notochord

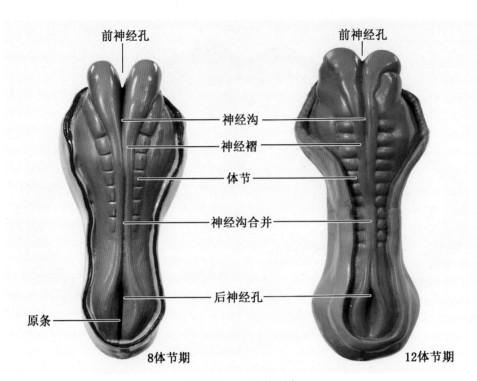

图 22-35　胚体的发育

Fig.22-35　development of embryoid

前神经孔—anterior neuropore　神经沟—neural groove　神经褶—neural fold　体节—somite
神经沟合并—neural groove fusion　后神经孔—posterior neuropore　原条—primitive streak
8体节期—8 somite stage　12体节期—12 somite stage

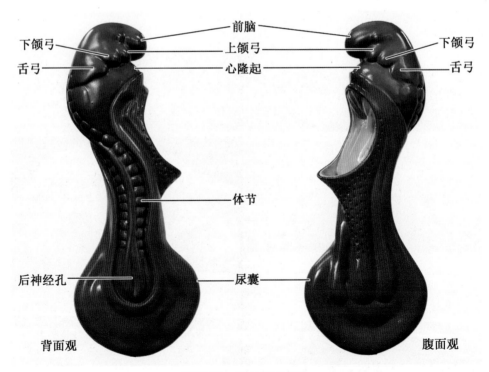

图 22-36　18体节的胚体

Fig.22-36　embryoid of 18 somites stage

前脑—forebrain　下颌弓—mandibular arch　上颌弓—maxillary arch　舌弓—hyoid arch　心隆起—heart prominence
体节—somite　后神经孔—posterior neuropore　尿囊—allantois　背面观—dorsal view　腹面观—ventral view

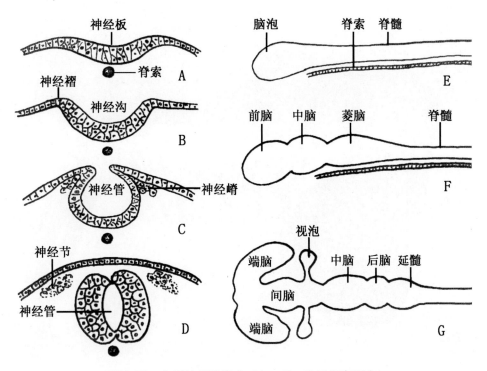

图 22-37 中枢神经的发生（A～G 表示发育顺序）

Fig.22-37　generation of central nervous system (A～G indicating developmental process)

神经板—neural plate　脊索—notochord　神经褶—neural fold　神经沟—neural groove　神经管—neural tube　神经嵴—neural crest　神经节—ganglion　脑泡—brain vesicles　脊髓—spinal cord　前脑—forebrain　中脑—midbrain　菱脑—rhombencephalon　端脑—telencephalon　间脑—diencephalon　视泡—optic vesicle　后脑—hindbrain　延髓—medulla oblongata

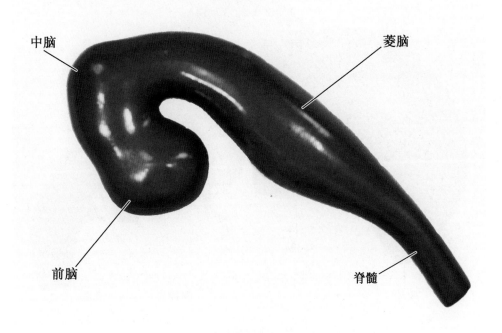

图 22-38　中枢神经的发育 1

Fig.22-38　development of central nervous system 1

前脑—forebrain　中脑—midbrain　菱脑—rhombencephalon　脊髓—spinal cord

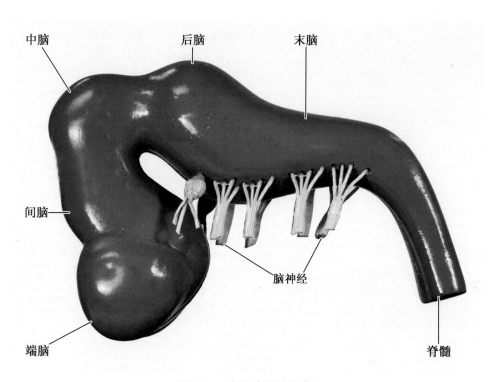

图 22-39　中枢神经的发育 2

Fig.22-39　development of central nervous system 2

端脑—telencephalon　间脑—diencephalon　中脑—midbrain　后脑—metencephalon

末脑—myelencephalon　脑神经—cerebral nerves　脊髓—spinal cord

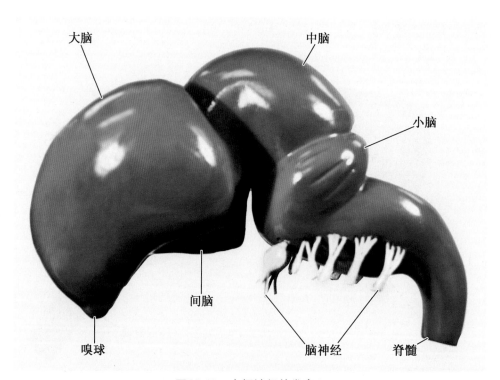

图 22-40　中枢神经的发育 3

Fig.22-40　development of central nervous system 3

嗅球—olfactory bulb　大脑—cerebrum　间脑—diencephalon　中脑—midbrain

小脑—cerebrum　脑神经—cerebral nerves　脊髓—spinal cord

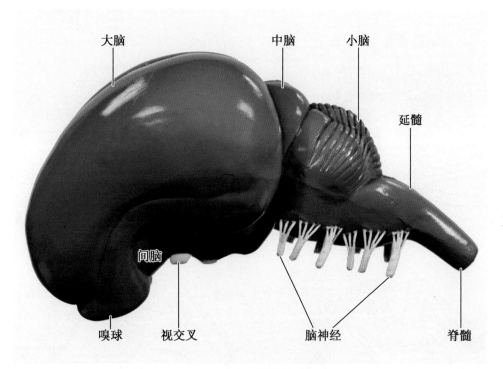

图22-41 中枢神经的发育4

Fig.22-41 development of central nervous system 4

大脑—cerebrum 中脑—midbrain 小脑—cerebrum 延髓—medulla oblongata 嗅球—olfactory bulb
间脑—diencephalon 视交叉—optic chiasma 脑神经—cerebral nerves 脊髓—spinal cord

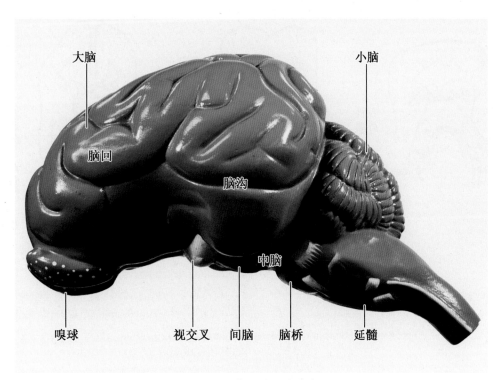

图22-42 中枢神经的发育5

Fig.22-42 development of central nervous system 5

大脑—cerebrum 脑回—gyrus 脑沟—sulcus 小脑—cerebrum 嗅球—olfactory bulb 视交叉—optic chiasma
间脑—diencephalon 中脑—midbrain 脑桥—pons 延髓—medulla oblongata

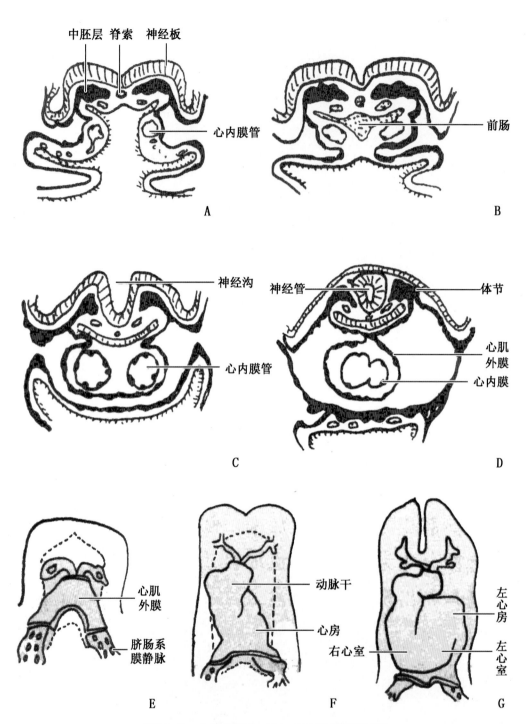

图 22-43　心脏的发生（A～G 表示发育顺序）

Fig.22-43　generation of heart (A～G indicating developmental process)

中胚层—mesoderm　脊索—notochord　神经板—neural plate　心内膜管—endocardial tube　前肠—endocardial tube
神经沟—neural groove　神经管—neural tube　体节—somite　心肌外膜—epimyocardium　心内膜—endocardium
脐肠系膜静脉—omphalomesenteric vein　动脉干—arterial trunk　心房—atrium　右心室—right ventricle
左心房—left atrium　左心室—left ventricle

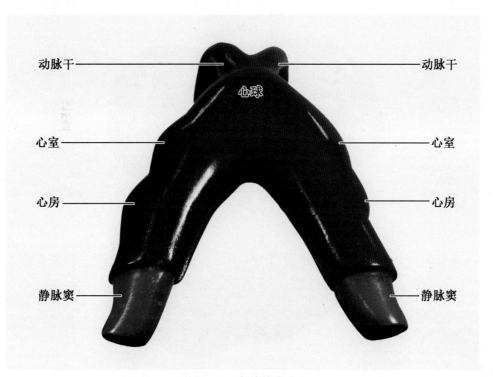

图22-44 心脏的发生2

Fig.22-44 generation of heart 2

动脉干—arterial trunk 心球—bulbus cordis 心室—ventricle 心房—atrium 静脉窦—venous sinus

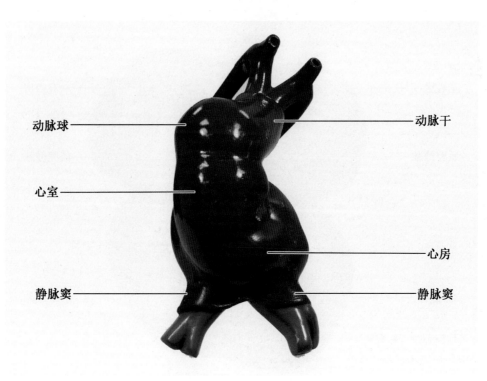

图22-45 心脏的发育1

Fig.22-45 heart development 1

动脉球—arterial bulbus 动脉干—arterial trunk 心室—ventricle 心房—atrium
静脉窦—venous sinus

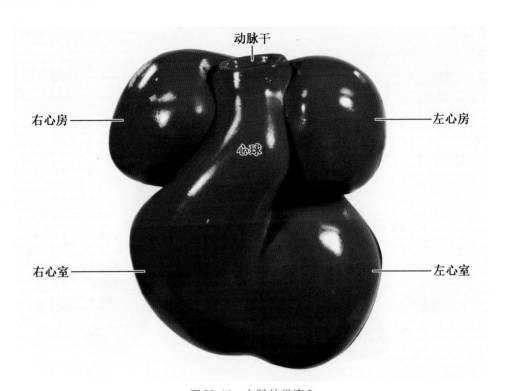

图22-46　心脏的发育2

Fig.22-46　heart development 2

动脉干—arterial trunk　心球—bulbus cordis　右心房—right atrium　右心室—right ventricle

左心房—left atrium　左心室—left ventricle

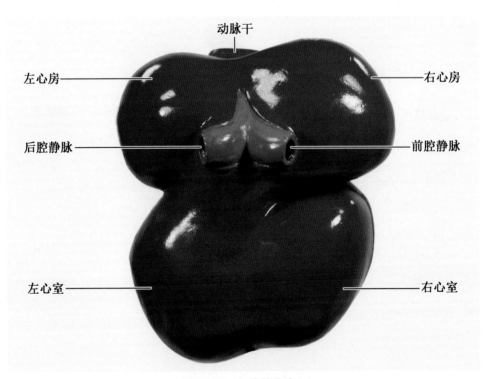

图22-47　心脏的发育3

Fig.22-47　heart development 3

动脉干—arterial trunk　左心房—left atrium　后腔静脉—postcaval vein　左心室—left ventricle

右心房—right atrium　前腔静脉—precaval vein　右心室—right ventricle

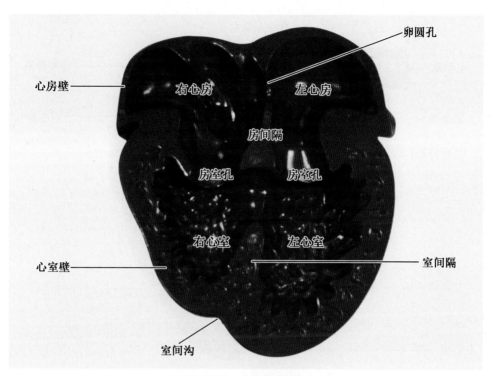

图 22-48 心脏的发育 4

Fig.22-48 heart development 4

心房壁—atrial wall　右心房—right atrium　左心房—left atrium　卵圆孔—oval foramen　房间隔—atrial septum
房室孔—atrioventricular orifice　心室壁—ventricular wall　右心室—right ventricle
左心室—left ventricle　室间隔—interventricular septum　室间沟—interventricular groove

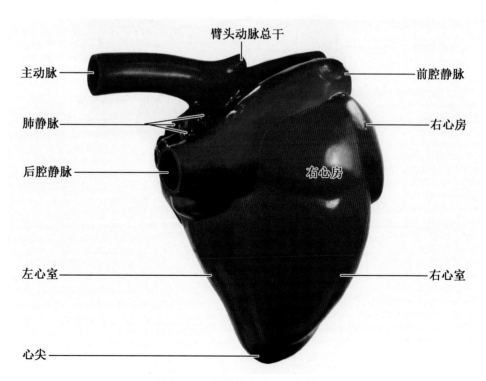

图 22-49 心脏的发育 1（右侧观）

Fig.22-49 heart development 1 (right side view)

臂头动脉总干—brachiocephalic artery trunk　主动脉—aorta　肺静脉—pulmonary vein　前腔静脉—precaval vein
后腔静脉—postcaval vein　右心房—right atrium　左心室—left ventricle　右心室—right ventricle　心尖—cardiac apex

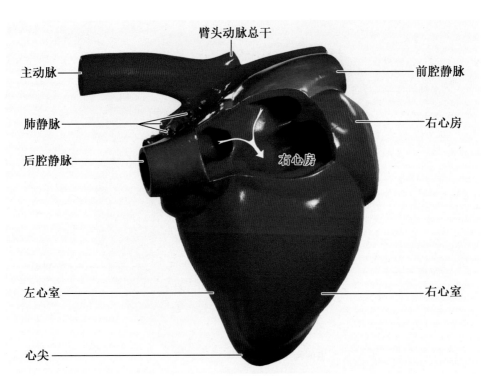

图 22-50 心脏的发育 2（右侧观）

Fig.22-50 heart development 2 (right side view)

臂头动脉总干—brachiocephalic artery trunk　主动脉—aorta　肺静脉—pulmonary vein　前腔静脉—precaval vein　后腔静脉—postcaval vein　右心房—right atrium　左心室—left ventricle　右心室—right ventricle　心尖—cardiac apex

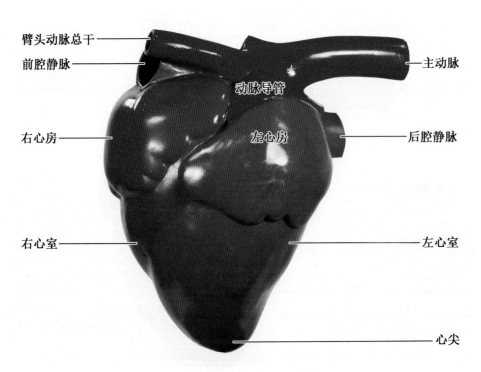

图 22-51 心脏的发育 1（左侧观）

Fig.22-51 heart development 1 (left side view)

臂头动脉总干—brachiocephalic artery trunk　前腔静脉—precaval vein　主动脉—aorta　动脉导管—arterial duct　右心房—right atrium　左心房—left atrium　后腔静脉—postcaval vein　右心室—right ventricle　左心室—left ventricle　心尖—cardiac apex

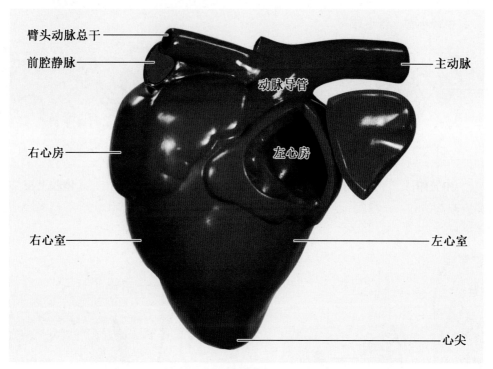

图 22-52　心脏的发育 2（左侧观）

Fig.22-52　heart development 2（left side view）

臂头动脉总干—brachiocephalic artery trunk　前腔静脉—precaval vein　主动脉—aorta　动脉导管—arterial duct
右心房—right atrium　左心房—left atrium　右心室—right ventricle　左心室—left ventricle　心尖—cardiac apex

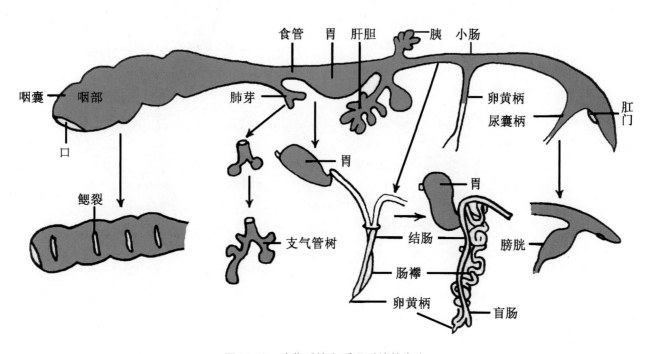

图 22-53　消化系统和呼吸系统的发生

Fig.22-53　development of digestive and respiratory system

咽囊—pharyngeal pouch　咽部—pharynx　口—mouth　鳃裂—branchial cleft　肺芽—lung bud　支气管树—bronchial tree
食管—esophagus　胃—stomach　肝胆—liver and gall　胰—pancreas　小肠—small intestine　卵黄柄—yolk stalk
尿囊柄—allantois stalk　肛门—anus　结肠—colon　肠襻—intestinal loop　膀胱—bladder　盲肠—cecum

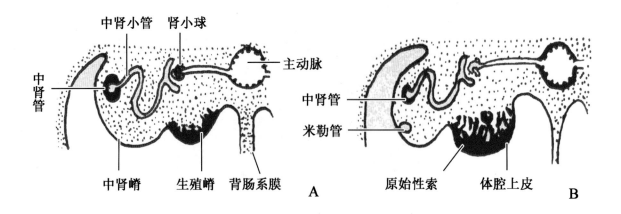
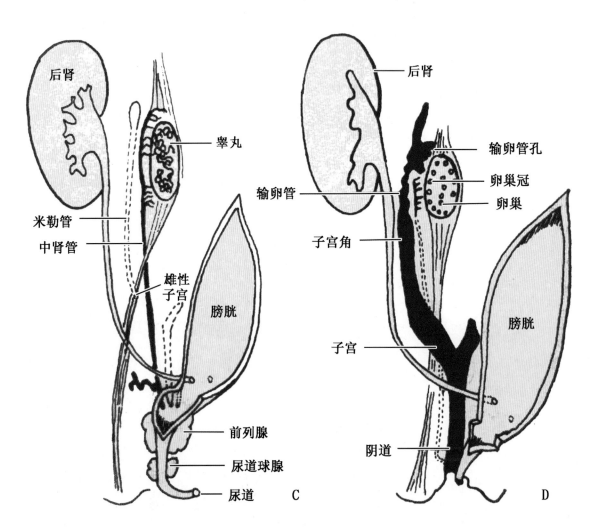

图 22-54 泌尿和生殖器官的发生 1（A 和 B 表示发育顺序；C 为雄性，D 为雌性）

Fig.22-54 development of urinary and reproductive organs 1 (A and B indicating developmental process. C indicating male, D indicating female)

中肾小管—mid-renal tubules　肾小球—renal glomerulus　中肾管—mid-renal tube　主动脉—aorta　中肾嵴—mid-renal crest
生殖嵴—reproductive ridge　背肠系膜—dorsal mesentery　米勒管—Müllerian duct　原始性索—primary germ cord
体腔上皮—coelomic epithelium　后肾—metanephros　睾丸—testis　雄性子宫—male uterus　膀胱—urinary bladder
前列腺—prostate　尿道球腺—bulbourethral gland　尿道—urethra　输卵管孔—fallopian tube hole　输卵管—fallopian tube
卵巢冠—parovarium　卵巢—ovary　子宫角—uterine horn　子宫—uterus　阴道—vagina

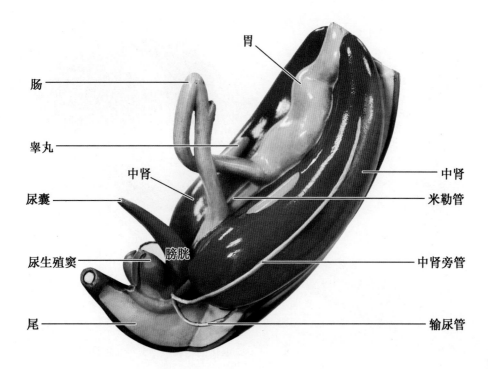

图 22-55　泌尿和生殖器官的发生 2

Fig.22-55　development of urinary and reproductive organs 2

胃—stomach　肠—intestine　睾丸—testis　中肾—mid kidney　尿囊—allantois　尿生殖窦—urogenital sinuses
膀胱—urinary bladder　尾—tail　米勒管—Müllerian duct　中肾旁管—paramesonephric duct　输尿管—ureter

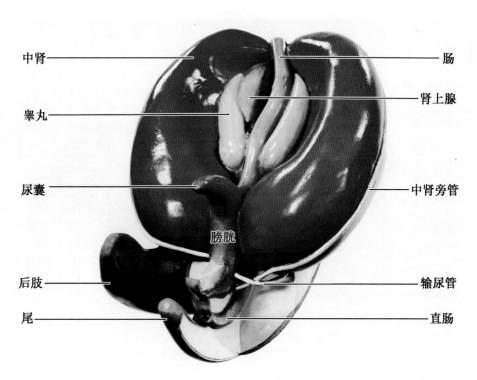

图 22-56　泌尿和生殖器官的发生 3

Fig.22-56　development of urinary and reproductive organs 3

中肾—mid kidney　肠—intestine　睾丸—testis　肾上腺—adrenal gland　尿囊—allantois　中肾旁管—paramesonephric duct
膀胱—urinary bladder　后肢—hind leg　尾—tail　输尿管—ureter　直肠—rectum

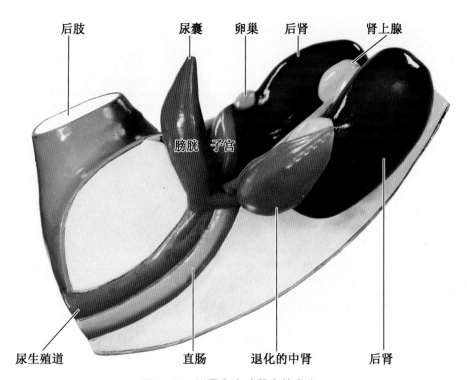

图 22-57　泌尿和生殖器官的发生 4

Fig.22-57　development of urinary and reproductive organs 4

后肢—hind leg　尿囊—allantois　卵巢—ovary　后肾—hind kidney　肾上腺—adrenal gland　膀胱—urinary bladder　子宫—uterus　尿生殖道—urogenital tract　直肠—rectum　退化的中肾—vestigial mid kidney

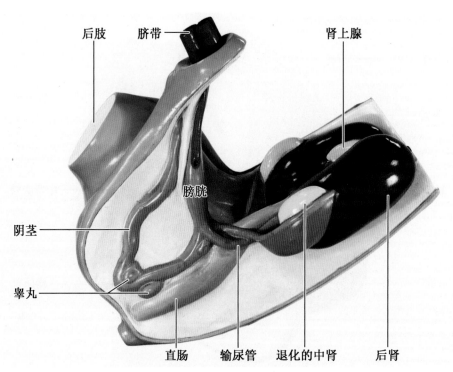

图 22-58　泌尿和生殖器官的发生 5

Fig.22-58　development of urinary and reproductive organs 5

后肢—hind leg　脐带—umbilical cord　肾上腺—adrenal gland　膀胱—urinary bladder　阴茎—penis　睾丸—testis　直肠—rectum　输尿管—ureter　退化的中肾—vestigial mid kidney　后肾—hind kidney

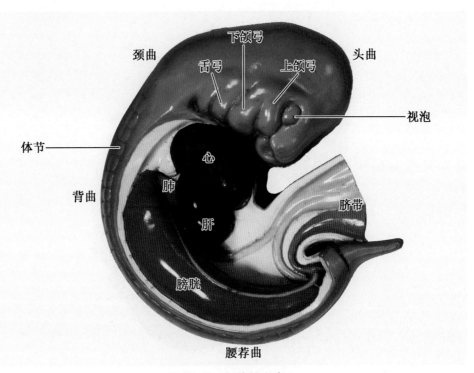

图22-59　胚体的发育1

Fig.22-59　embryo development 1

头曲—cranial flexure　颈曲—cervical flexure　背曲—dorsal flexure　腰荐曲—lumbo-sacral flexure　视泡—optic vesicle
上颌弓—maxillary arch　下颌弓—mandibular arch　舌弓—hyoid arch　体节—somite　心—heart　肺—lung　肝—liver
膀胱—urinary bladder　脐带—umbilical cord

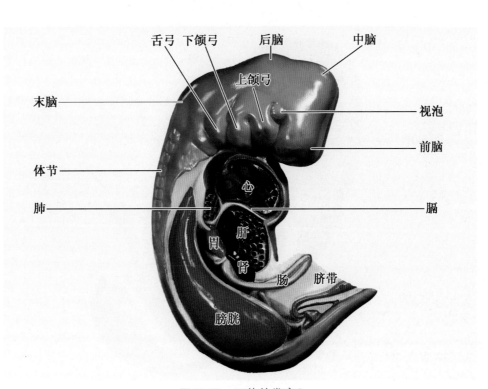

图22-60　胚体的发育2

Fig.22-60　embryo development 2

前脑—forebrain　中脑—midbrain　后脑—hindbrain　末脑—marrowbrain　视泡—optic vesicle　上颌弓—maxillary arch
下颌弓—mandibular arch　舌弓—hyoid arch　体节—somite　心—heart　肺—lung　膈—diaphragm　肝—liver
胃—stomach　肾—kidney　肠—intestine　膀胱—urinary bladder　脐带—umbilical cord

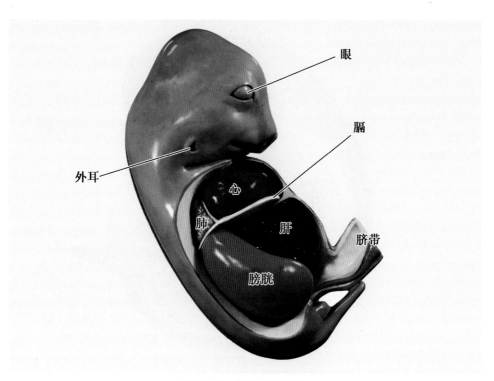

图 22-61　胚体的发育 3

Fig.22-61　embryo development 3

眼—eye　外耳—external ear　心—heart　肺—lung　膈—diaphragm　肝—liver　膀胱—urinary bladder　脐带—umbilical cord

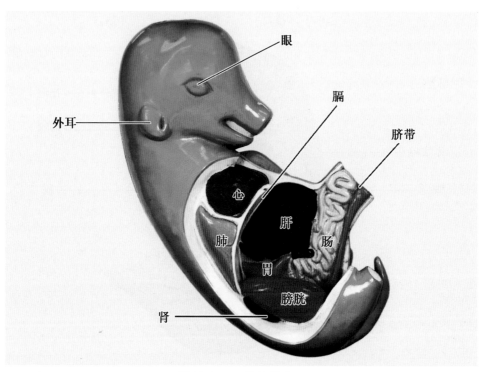

图 22-62　胚体的发育 4

Fig.22-62　embryo development 4

眼—eye　外耳—external ear　心—heart　肺—lung　膈—diaphragm　肝—liver　胃—stomach
肠—intestine　肾—kidney　膀胱—urinary bladder　脐带—umbilical cord

第二十二章 畜禽早期胚胎发育　Early Embryonic Development of Livestock and Fowl

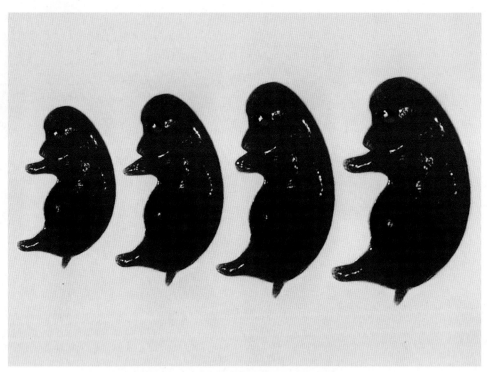

图22-63　发育中的猪胚胎（伊红染色）
Fig.22-63　developing pig embryos（eosin stain）

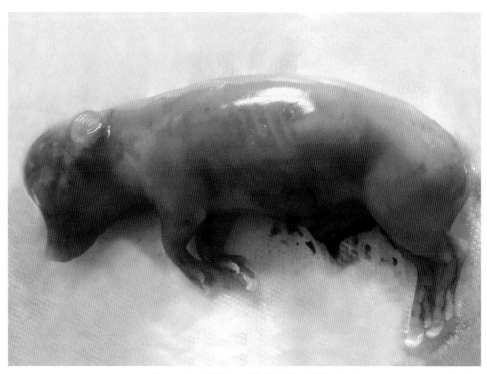

图22-64　妊娠35d的猪胎儿（原色）
Fig.22-64　pig embryo at 35 days gestation (original color)

图22-65　出生后的小香猪
Fig.22-65　miniature pig after birth

图22-66　仔兔透明标本
Fig.22-66　transparent specimen of rabbit

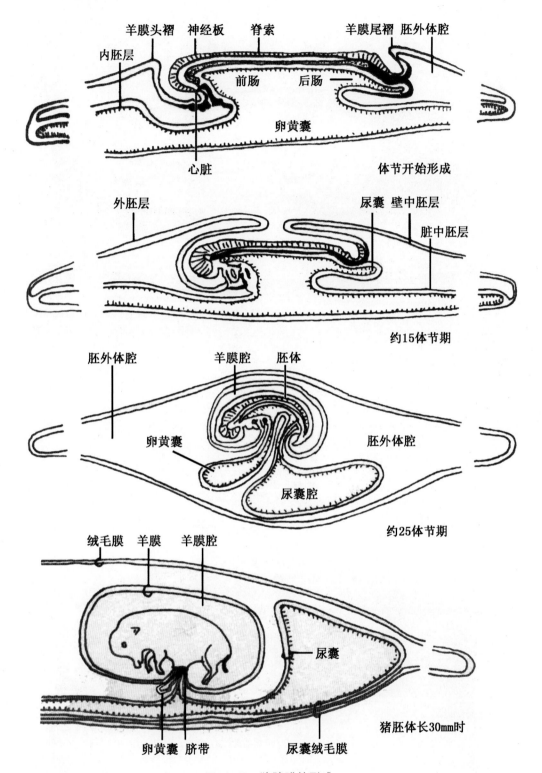

图 22-67 猪胎膜的形成

Fig.22-67 formation of pig foetal membranes

体节开始形成—somite beginning to form 羊膜头褶—amniotic head fold 神经板—neural plate 脊索—notochord
羊膜尾褶—amniotic tail fold 胚外体腔—exocoelom 内胚层—endoderm 前肠—foregut 后肠—hindgut
卵黄囊—yolk sac 心脏—heart 约15体节期—about 15 somites stage 外胚层—ectoderm 尿囊—allantois
壁中胚层—parietal mesoderm 脏中胚层—visceral mesoderm 约25体节期—about 25 somites stage
胚外体腔—exocoelom 羊膜腔—amnion cavity 胚体—embryoid 卵黄囊—yolk sac 尿囊腔—allantoic cavity
猪胚体长30mm时—pig embryo length is 30mm 绒毛膜—chorion 羊膜—amnion 羊膜腔—amnion cavity
卵黄囊—yolk sac 脐带—umbilical cord 尿囊绒毛膜—chorio-allantois

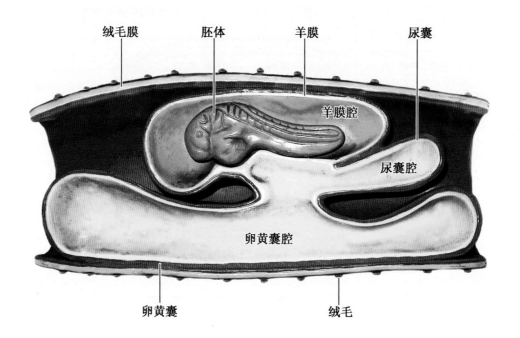

图 22-68　胎膜的形成 1

Fig.22-68　formation of foetal membranes 1

绒毛膜—chorion　胚体—embryoid　羊膜—amnion　羊膜腔—amnion cavity　尿囊—allantois
尿囊腔—allantoic cavity　卵黄囊—yolk sac　卵黄囊腔—yolk sac cavity　绒毛—villus

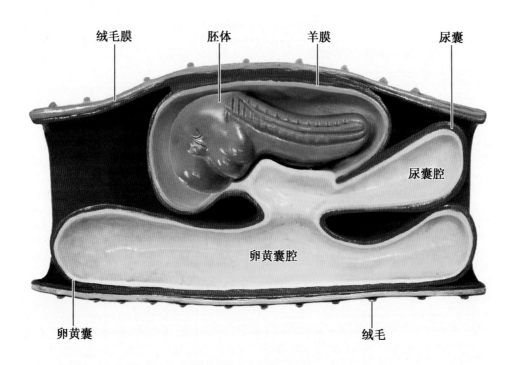

图 22-69　胎膜的形成 2

Fig.22-69　formation of foetal membranes 2

绒毛膜—chorion　胚体—embryoid　羊膜—amnion　尿囊—allantois
尿囊腔—allantoic cavity　卵黄囊—yolk sac　卵黄囊腔—yolk sac cavity　绒毛—villus

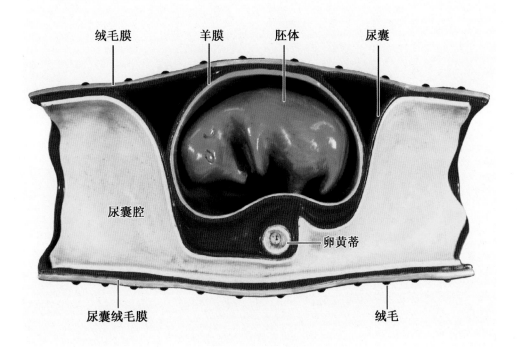

图 22-70 胎膜的形成 3

Fig.22-70 formation of foetal membranes 3

绒毛膜—chorion 羊膜—amnion 胚体—embryoid 尿囊—allantois 尿囊腔—allantoic cavity 卵黄蒂—yolk stalk
尿囊绒毛膜—chorio-allantois 绒毛—villus

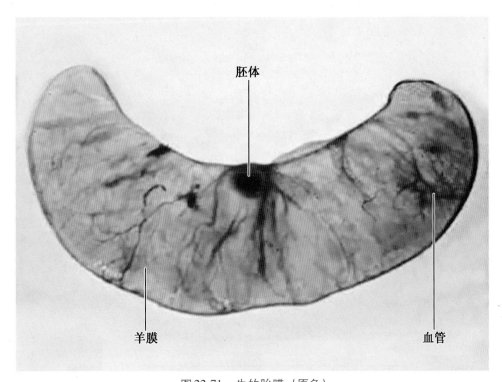

图 22-71 牛的胎膜（原色）

Fig.22-71 foetal membranes of cow（original color）

胚体—embryoid 羊膜—amnion 血管—blood vessel

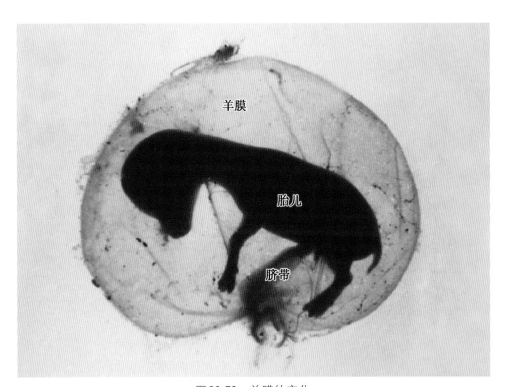

图 22-72　羊膜的变化

Fig.22-72　amnion development

羊膜—amnion　胎儿—foetus　脐带—umbilical cord

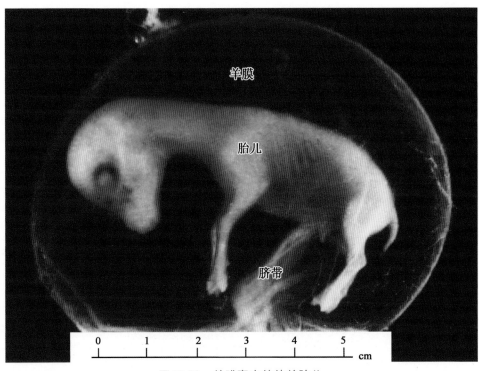

图 22-73　羊膜囊内的绵羊胎儿

Fig.22-73　sheep foetus in amnion sac

羊膜—amnion　胎儿—foetus　脐带—umbilical cord

第二十二章 畜禽早期胚胎发育 Early Embryonic Development of Livestock and Fowl

图22-74 哺乳动物的脐带（原色）
Fig.22-74 umbilical cord of mammals (original color)
胎盘—placenta 脐带—umbilical cord 血管—blood vessel

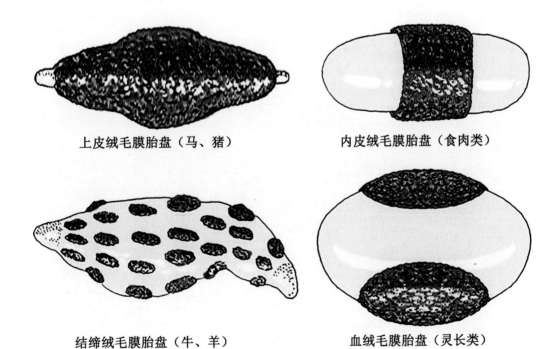

图22-75 胎盘分类模式图（外面观）
Fig.22-75 diagram of placenta classification (outside view)
上皮绒毛膜胎盘（马、猪）—epithelial chorionic placenta (horse, pig) 结缔绒毛膜胎盘（牛、羊）—connective chorionic placenta (cow, sheep) 内皮绒毛膜胎盘（食肉类）—endothelial chorionic placenta (carnivore)
血绒毛膜胎盘（灵长类）—blood chorionic placenta (primate)

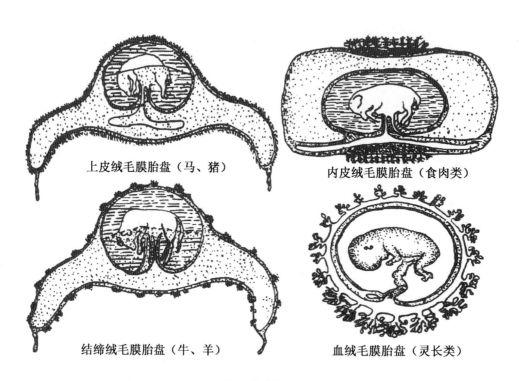

图 22-76 胎盘分类模式图（剖面）

Fig.22-76 diagram of placenta classification（cutaway view）

上皮绒毛膜胎盘（马、猪）—epithelial chorionic placenta（horse, pig）

结缔绒毛膜胎盘（牛、羊）—connective chorionic placenta（cow, sheep）

内皮绒毛膜胎盘（食肉类）—endothelial chorionic placenta (carnivore)

血绒毛膜胎盘（灵长类）—blood chorionic placenta (primate)

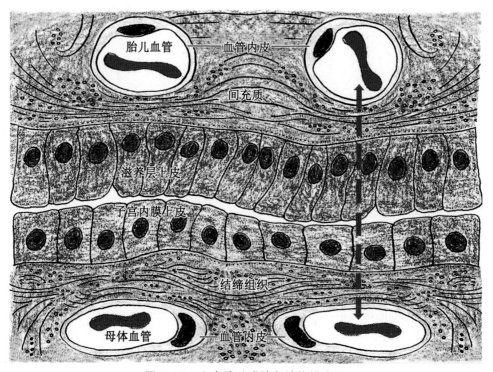

图 22-77 上皮绒毛膜胎盘结构模式图

Fig.22-77 structural diagram of epithelial chorionic placenta

胎儿血管—fetal blood vessel　血管内皮—vessel endothelium　间充质—mesenchyme　滋养层上皮—trophoblast epithelium

子宫内膜上皮—endometrial epithelium　结缔组织— connective tissue　母体血管—mother blood vessel

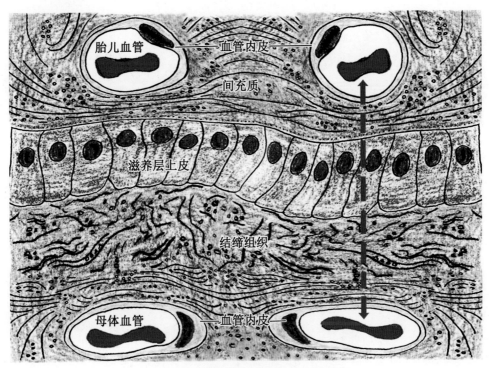

图 22-78 结缔绒毛膜胎盘结构模式图

Fig.22-78 structural diagram of connective chorionic placenta

胎儿血管—fetal blood vessel 血管内皮—vessel endothelium 间充质—mesenchyme

滋养层上皮—trophoblast epithelium 结缔组织—connective tissue 母体血管—mother blood vessel

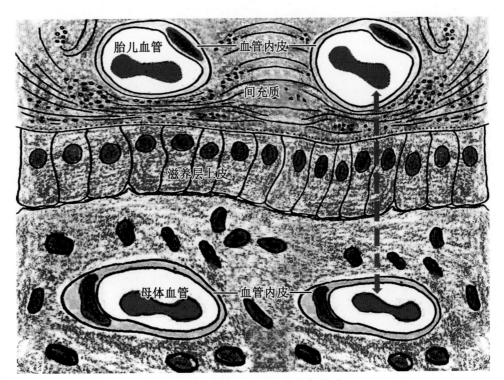

图 22-79 内皮绒毛膜胎盘结构模式图

Fig.22-79 structural diagram of endothelial chorionic placenta

胎儿血管—fetal blood vessel 血管内皮—vessel endothelium 间充质—mesenchyme

滋养层上皮—trophoblast epithelium 母体血管—mother blood vessel

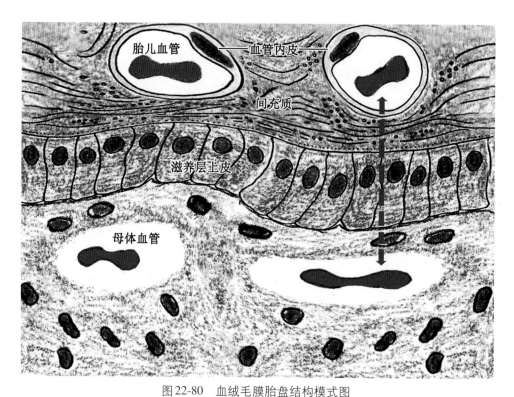

图 22-80　血绒毛膜胎盘结构模式图

Fig.22-80　structural diagram of blood chorionic placenta

胎儿血管—fetal blood vessel　血管内皮—vessel endothelium　间充质—mesenchyme

滋养层上皮—trophoblast epithelium　母体血管—mother blood vessel

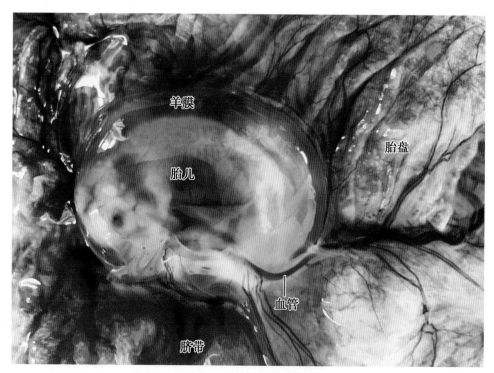

图 22-81　猪胎儿及上皮绒毛膜胎盘（原色）

Fig.22-81　pig fetus and epithelial chorionic placenta（original color）

羊膜—amnion　胎儿—foetus　胎盘—placenta　血管—blood vessel　脐带—umbilical cord

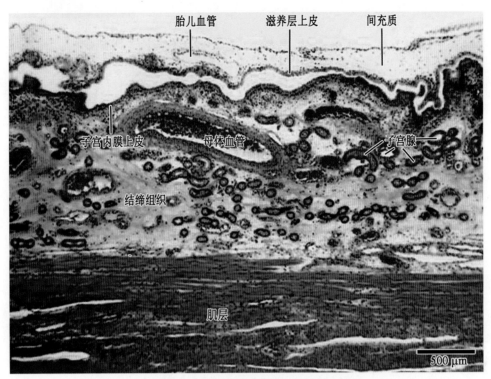

图22-82 猪上皮绒毛膜胎盘低倍像（HE）

Fig.22-82 low magnification of pig epithelial chorionic placenta（HE）

胎儿血管—fetal blood vessel 滋养层上皮—trophoblast epithelium 间充质—mesenchyme
子宫内膜上皮—endometrial epithelium 母体血管—mother blood vessel 子宫腺—uterine gland
结缔组织—connective tissue 肌层—muscular layer

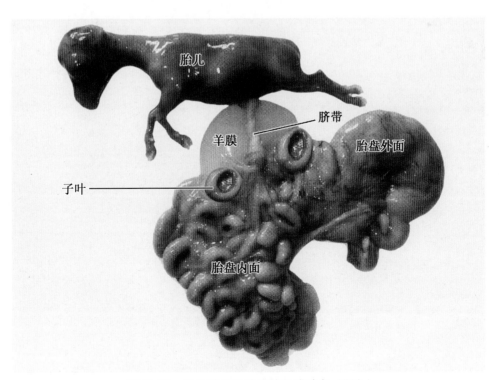

图22-83 绵羊胎儿及结缔绒毛膜胎盘（原色）

Fig.22-83 sheep fetus and connective chorionic placenta（original color）

胎儿—foetus 羊膜—amnion 脐带—umbilical cord 子叶—cotyledon
胎盘外面—outside of placenta 胎盘内面—inside of placenta

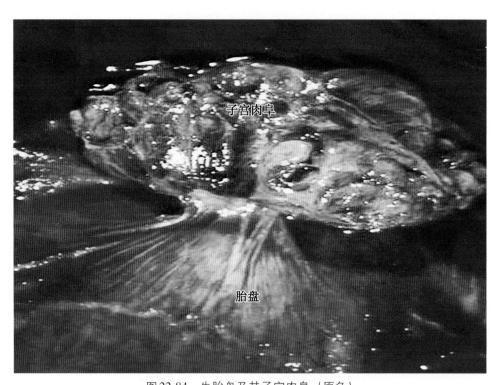

图22-84 牛胎盘及其子宫肉阜（原色）

Fig.22-84 cattle placenta and uterine caruncle (original color)

子宫肉阜—uterine caruncle　胎盘—placenta

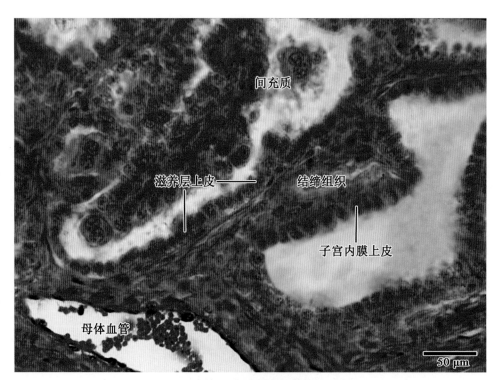

图22-85 牛结缔绒毛膜胎盘高倍像（HE）

Fig.22-85 high magnification of cattle connective chorionic placenta (HE)

间充质—mesenchyme　滋养层上皮—trophoblast epithelium　结缔组织—connective tissue

子宫内膜上皮—endometrial epithelium　母体血管—mother blood vessel

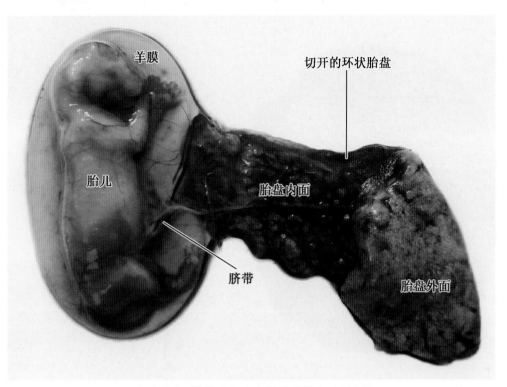

图22-86 猫胎儿及内皮绒毛膜胎盘（原色）
Fig.22-86 cat fetus and endothelial chorionic placenta（original color）
羊膜—amnion 胎儿—foetus 脐带—umbilical cord 切开的环状胎盘—opened zonary placenta
胎盘内面—inside of placenta 胎盘外面—outside of placenta

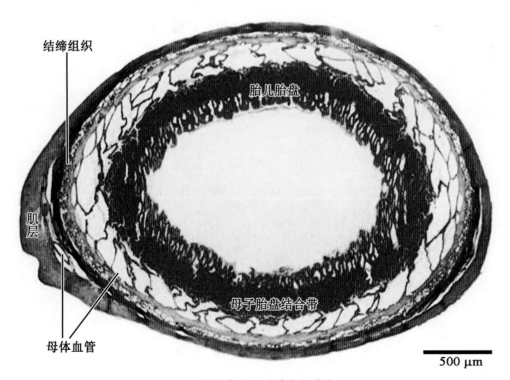

图22-87 猫内皮绒毛膜胎盘低倍像（HE）
Fig.22-87 low magnification of cat endothelial chorionic placenta（HE）
肌层—muscular layer 结缔组织—connective tissue 母体血管—mother blood vessel
胎儿胎盘—fetal placenta 母子胎盘结合带—junctional zone of mother and fetal placenta

图 22-88　血绒毛膜胎盘和脐带（原色）
Fig.22-88　blood chorionic placenta and umbilical cord（original color）
胎盘—placenta　脐带—umbilical cord

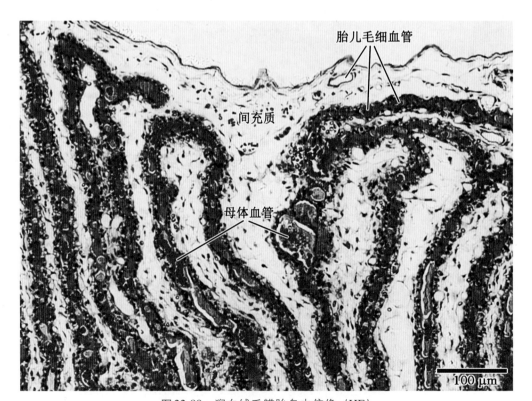

图 22-89　猴血绒毛膜胎盘中倍像（HE）
Fig.22-89　mid magnification of monkey hemochorial placenta（HE）
胎儿毛细血管—fetal capillary　间充质—mesenchyme　母体血管—mother blood vessel

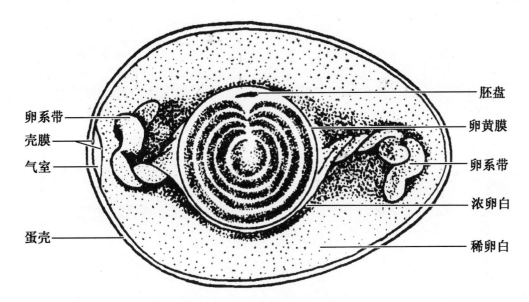

图 22-90　鸡蛋结构模式图

Fig.22-90　structure pattern of egg

卵系带—chalaza　壳膜—shell theca　气室—air chamber　蛋壳—eggshell　胚盘—embryonic disc
卵黄膜—vitelline membrane　浓卵白—thick albumen　稀卵白—lean albumen

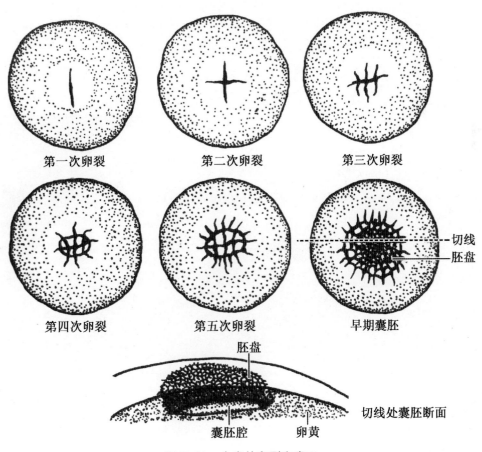

图 22-91　禽类的卵裂和囊胚

Fig.22-91　cleavage and blastocyst of bird

第一次卵裂—first cleavage　第二次卵裂—second cleavage　第三次卵裂—third cleavage
第四次卵裂—fourth cleavage　第五次卵裂—fifth cleavage　早期囊胚—early blastula　切线—cut line
胚盘—embryonic disc　囊胚腔—blastocele　卵黄—yolk　切线处囊胚断面—blastocyst section

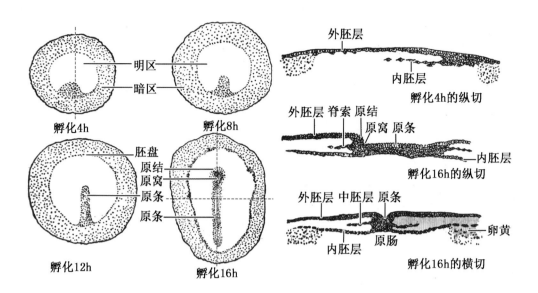

图 22-92 鸡胚的原条形成和原肠形成

Fig.22-92 formation of primary streak and gut of chicken embryo

孵化 4h—incubation 4 hours 孵化 8h—incubation 8 hours 孵化 12h—incubation 12 hours

孵化 16h—incubation 16 hours 孵化 4h 的纵切—slitting of incubation 4 hours

孵化 16h 的纵切—slitting of incubation 16 hours 孵化 16h 的横切—cross section of incubation 16 hours

明区—light area 暗区—dark area 胚盘—blastoderm 原结—primitive knot 原窝—primitive pit 原条—primitive streak

外胚层—ectoderm 内胚层—entoderm 脊索—notochord 中胚层—mesoderm 原肠—primitive gut

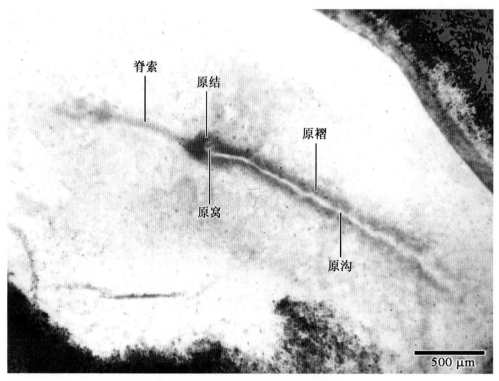

图 22-93 原条期鸡胚整装片（HE）

Fig.22-93 mounting of chicken embryo at primitive streak stage（HE）

脊索—notochord 原结—primitive knot 原窝—primitive pit 原褶—primitive fold 原沟—primitive groove

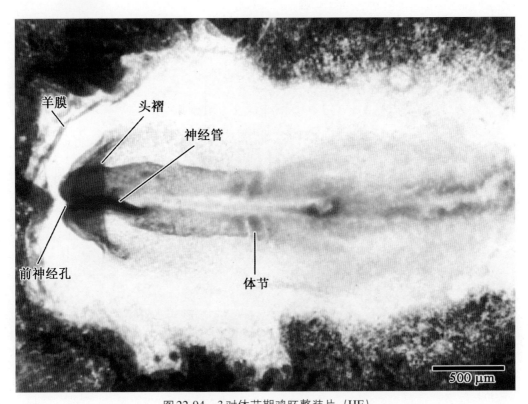

图 22-94　3 对体节期鸡胚整装片（HE）

Fig.22-94　mounting of chicken embryo at 3-somite stage（HE）

羊膜—amnion　头褶—head fold　前神经孔—anterior neuropore　神经管—neural tube　体节—somite

图 22-95　6 对体节期鸡胚整装片（HE）

Fig.22-95　mounting of chicken embryo at 6-somite stage（HE）

羊膜头褶—amniotic head fold　前神经孔—anterior neuropore　神经管—neural tube　体节—somite

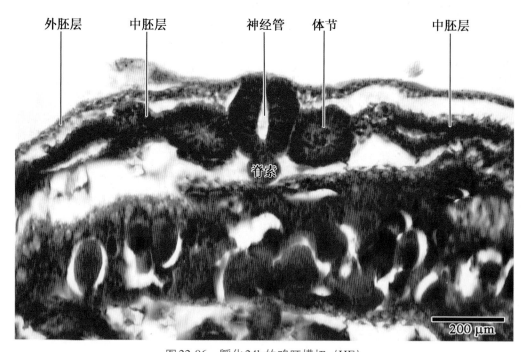

图 22-96　孵化 24h 的鸡胚横切（HE）

Fig.22-96　chicken embryo transection after incubation 24hours（HE）

外胚层—ectoderm　中胚层—mesoderm　神经管—neural tube　体节—somite　脊索—notochord

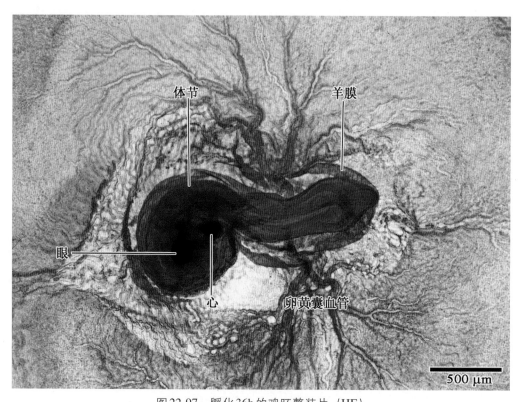

图 22-97　孵化 36h 的鸡胚整装片（HE）

Fig.22-97　mounting of chicken embryo after incubation 36hours（HE）

眼—eye　体节—somite　羊膜—amnion　心—heart　卵黄囊血管—yolk sac blood vessel

第二十二章 畜禽早期胚胎发育 Early Embryonic Development of Livestock and Fowl

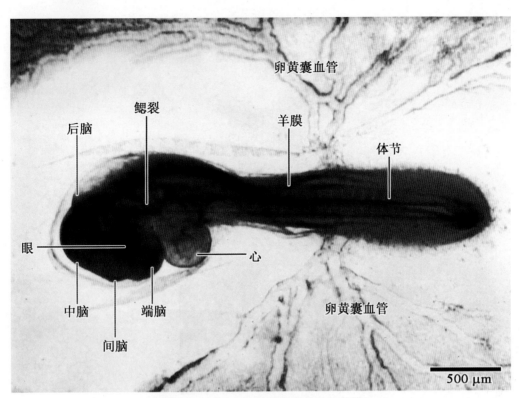

图22-98 孵化48h的鸡胚整装片1（HE）

Fig.22-98 mounting of chicken embryo after incubation 48hours 1（HE）

眼—eye 后脑—hindbrain 中脑—midbrain 间脑—diencephalon 鳃裂—gill cleft 端脑—telencephalon
心—heart 羊膜—amnion 体节—somite 卵黄囊血管—yolk sac blood vessel

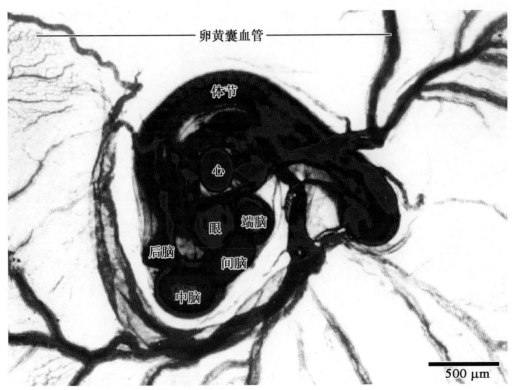

图22-99 孵化48h的鸡胚整装片2（HE）

Fig.22-99 mounting of chicken embryo after incubation 48hours 2（HE）

卵黄囊血管—yolk sac blood vessel 体节—somite 心—heart 眼—eye 端脑—telencephalon
后脑—hindbrain 间脑—diencephalon 中脑—midbrain

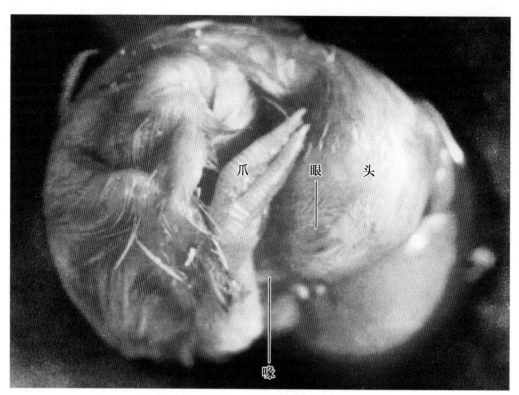

图22-100 孵化20d的鸡胚全貌（原色）
Fig.22-100 whole chicken embryo after incubation 20days（original color）
爪—claw 眼—eye 头—head 喙—beak

图22-101 鸡胚发育成熟出壳
Fig.22-101 mature chicken embryos hatch out

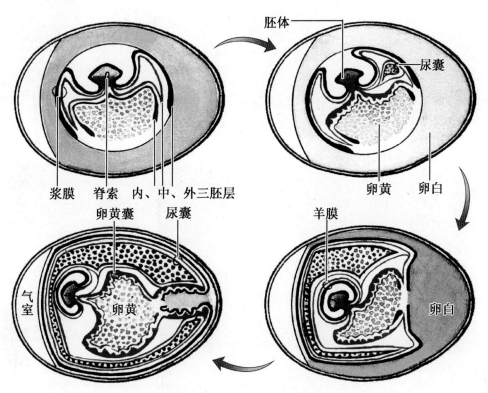

图22-102 鸡胚胎膜形成模式图

Fig.22-102 chicken embryo membrane formation pattern

浆膜—chorion 脊索—notochord 内、中、外三胚层—endoderm, mesoderm and ectoderm 胚体—embryoid 尿囊—allantois 卵黄—yolk 卵白—albumen 羊膜—amnion 卵黄囊—yolk sac 气室—air chamber

主要参考文献

Bacha J. William, Jr., Bacha M. Linda, 2007. 兽医组织学彩色图谱[M]. 陈耀星, 主译. 北京: 中国农业大学出版社.

陈秋生, 2019. 动物组织学与胚胎学[M]. 北京: 科学出版社.

陈耀星, 崔燕, 2018. 动物解剖学及组织胚胎学: 全彩版 [M]. 北京: 中国农业出版社.

成令忠, 冯京生, 冯子强, 2000. 组织学彩色图鉴[M]. 北京: 人民卫生出版社.

成令忠, 王一飞, 钟翠平, 2003. 组织胚胎学: 人体发育和功能组织学[M]. 上海: 上海科学技术文献出版社.

成令忠, 钟翠平, 蔡文琴, 2003. 现代组织学[M]. 上海: 上海科学技术文献出版社.

高英茂, 2005. 组织学与胚胎学: 双语版 [M]. 北京: 科学出版社.

李德雪, 栾维民, 岳占碰, 2003. 动物组织学与胚胎学[M]. 长春: 吉林人民出版社.

李德雪, 尹昕, 1995. 动物组织学彩色图谱[M]. 长春: 吉林科学技术出版社.

李继承, 2003. 组织学与胚胎学[M]. 杭州: 浙江大学出版社.

李子义, 岳占碰, 张学明, 2014. 动物组织学与胚胎学[M]. 北京: 科学出版社.

刘斌, 2005. 组织学与胚胎学[M]. 北京: 北京大学医学出版社.

彭克美, 2005. 畜禽解剖学[M]. 北京: 高等教育出版社.

彭克美, 2009. 畜禽解剖学[M]. 2版. 北京: 高等教育出版社.

彭克美, 2009. 动物组织学与胚胎学: 彩色版 [M]. 北京: 高等教育出版社.

彭克美, 2016. 畜禽解剖学[M]. 3版. 北京: 高等教育出版社.

彭克美, 2016. 动物组织学与胚胎学[M]. 2版. 北京: 高等教育出版社.

彭克美, 王政富, 2016. 动物组织学及胚胎学实验: 彩色版 [M]. 北京: 高等教育出版社.

秦鹏春, 2001. 哺乳动物胚胎学[M]. 北京: 科学出版社.

沈霞芬, 卿素珠, 2015. 家畜组织学与胚胎学[M]. 5版. 北京: 中国农业出版社.

唐军民, 李英, 卫兰, 2012. 组织学与胚胎学彩色图谱[M]. 北京: 北京大学医学出版社.

杨倩, 2008. 动物组织学与胚胎学[M]. 北京: 中国农业大学出版社.

中国人民解放军兽医大学, 1979. 马体解剖图谱[M]. 长春: 吉林人民出版社.

Bacha J. William, Jr., Bacha M. Linda, 2000. Color Atlas of Veterinary Histology [M]. USA: Lippincott Williams and Wilkins.

Banks J. William, 2001. Applied Veterinary Histology [M]. Baltimore, USA: William and Wilkins.

Gartner P. Leslie, Hiatt L. James, 2006. Color Atlas of Histology [M]. 4th ed. USA: Lippincott Williams and Wilkins.

Gartner P. Leslie, Hiatt L. James, 2017. Textbook of Histology [M]. 4th ed. Philadelphia PA, USA: Elsevier.